The Oil and Gas Habitats of the South Atlantic

Cover illustration. GEOSAT satellite free-air gravity image for the southern Atlantic region. Illumination is from the north. Total gravity colour range is –100 to +40 mgals (see gravity legend of Fig. 1, Karner & Driscoll). Gravity contours less than –100 mgals are coloured black whilst contours greater than +40 mgals are white. The free-air gravity anomaly, because it is dominated by near-field density contrasts, is particularly useful for defining the general bathymetry and structure of both thinned continental and oceanic crust. For example, the present-day mid-Atlantic ridge spreading centre, ridge jumps, fracture zone trends, continental margin rift structures and their continuity, the ocean/continent boundary, the 'edge-effect' anomaly that characterizes the shelf break, hot-spot traces, and possible reactivation structures are clearly delineated. Further, the early opening trends of magnetic quiet zone oceanic crust are well-imaged, allowing for improved plate tectonic reconstructions for this early part of the post-rift history to be mapped. Four hotspots have likely influenced significantly oceanic crust chemistry and thickness during the late Cretaceous period. From south to north, these hotspots are Tristan da Cunha, Saint Helena, Ascension, and Fernando de Noronha. Note also that the large river deltas (e.g. Congo, Niger, Ogooué and São Francisco) are all associated with extreme positive gravity anomalies, reflecting the large flexural strength of the underlying lithosphere at the time of loading (see Karner & Driscoll, pages 11–40, for details).

It is recommended that reference to all or part of this book should be made in one of the following ways:

CAMERON, N. R., BATE, R. H. & CLURE, V. S. (eds) 1999. *The Oil and Gas Habitats of the South Atlantic*. Geological Society, London, Special Publications, **153**.

CLEMSON, J., CARTWRIGHT, J. & SWART, R. 1999. The Namib Rift: a rift system of possible Karoo age, offshore Namibia. *In:* CAMERON, N. R., BATE, R. H. & CLURE, V. S. (eds) *The Oil and Gas Habitats of the South Atlantic*. Geological Society, London, Special Publications, **153**, 381–402.

GEOLOGICAL SOCIETY SPECIAL PUBLICATION NO. 153

The Oil and Gas Habitats of the South Atlantic

EDITED BY

N. R. CAMERON
Imperial College of Science, Technology and Medicine
London
UK

R. H. BATE
Lacustrine Basin Research Ltd
Ansty
UK

V. S. CLURE
Technical Outsourcing
Wantage
UK

SAMC
IGCP 381
IUGS - UNESCO

IUGS
UNESCO
IGCP

1999

Published by

The Geological Society

London

THE GEOLOGICAL SOCIETY

The Society was founded in 1807 as The Geological Society of London and is the oldest geological society in the world. It received its Royal Charter in 1825 for the purpose of 'investigating the mineral structure of the Earth'. The Society is Britain's national society for geology with a membership of around 8500. It has countrywide coverage and approximately 1500 members reside overseas. The Society is responsible for all aspects of the geological sciences including professional matters. The Society has its own publishing house, which produces the Society's international journals, books and maps, and which acts as the European distributor for publications of the American Association of Petroleum Geologists, SEPM and the Geological Society of America.

Fellowship is open to those holding a recognized honours degree in geology or cognate subject and who have at least two years' relevant postgraduate experience, or who have not less than six years' relevant experience in geology or a cognate subject. A Fellow who has not less than five years' relevant postgraduate experience in the practice of geology may apply for validation and, subject to approval, may be able to use the designatory letters C Geol (Chartered Geologist).

Further information about the Society is available from the Membership Manager, The Geological Society, Burlington House, Piccadilly, London W1V 0JU, UK. The Society is a Registered Charity, No. 210161.

Published by The Geological Society from:
The Geological Society Publishing House
Unit 7, Brassmill Enterprise Centre
Brassmill Lane
Bath BA1 3JN
UK
(*Orders*: Tel. 01225 445046
Fax 01225 442836)

First published 1999

British Library Cataloguing in Publication Data
A catalogue record for this book is available from the British Library.

ISBN 1-86239-030-4
ISSN 0305-8719

Typeset by Cambridge University Press, Cambridge, UK
Printed by Cambridge University Press, Cambridge, UK

Distributors
USA
AAPG Bookstore
PO Box 979
Tulsa
OK 74101-0979
USA
(*Orders*: Tel. (918) 584-2555
Fax (918) 560-2652)

Australia
Australian Mineral Foundation
63 Conyngham Street
Glenside
South Australia 5065
Australia
(*Orders*: Tel. (08) 379-0444
Fax (08) 379-4634)

India
Affiliated East-West Press PVT Ltd
G-1/16 Ansari Road
New Delhi 110 002
India
(*Orders*: Tel. (11) 327-9113
Fax (11) 326-0538)

Japan
Kanda Book Trading Co.
Cityhouse Tama 204
Tsurumaki 1-3-10
Tama-Shi
Tokyo 0206-0034
Japan
(*Orders*: Tel. (0423) 57-7650
Fax (0423) 57-7651)

Contents

Oil and gas habitats of the South Atlantic: Introduction

NICK CAMERON[1], RAY BATE[2], VAL CLURE[3] & JEREMY BENTON[4]

[1]*T. H. Huxley School of Environment, Earth Science and Engineering, Imperial College of Science, Technology and Medicine, Royal School of Mines, Prince Consort Road, London SW7 2BP, UK*

[2]*Lacustrine Basin Research Ltd, Little Lower Ease, Cuckfield Road, Ansty, West Sussex RH17 5AL, UK*

[3]*Technical Outsourcing, Court Hill House, Letcombe Regis, Wantage, Oxfordshire OX12 9JG, UK*

[4]*IHS Energy, Enterprise House, Cirencester Road, Ilsem, Tetbury GL8 8RX, UK*

The South Atlantic hydrocarbon province defined

For the purposes of this volume, the South Atlantic comprises the Atlantic Ocean south of a direct line from Tangier (Morocco) to the Guyana–Venezuela border and north of a direct line from Cape Town (South Africa) and Cape Horn (Argentina). At the end of 1997 this region contained 4.3% of the world's total discovered oil reserves and 2.4% of the world's total discovered gas reserves (Petroconsultants 1998).

On the basis of the 1997 figure, six countries – Brazil, Argentina, Nigeria, Gabon, Congo (Brazzaville) and Angola – host 96.3% of the region's oil reserves and 92.1% of the region's gas reserves. Figure 1 illustrates the geographical distribution of the reserve splits, Table 1 gives the details.

One hundred years of discoveries

Voluminous basin margin asphalts in West Africa (Côte d'Ivoire to Angola), Argentina, Brazil and Surinam attracted the serious attention of entrepreneurs from the late 1800s. As discussed in more detail by **Dolan** shallow wells were drilled soon afterwards on seep locations, though with little commercial success. More systematic explorations began on both margins in the 1930s and in 1941 the first large discovery, the Candeias Field in the Recôncavo Basin of Brazil, was made: the reserves of this field are now placed at 120 MMbo. Success did not materialize in West Africa until 1955 when the small Benfica Field was discovered in the Kwanza Basin of Angola. Thereafter, following the first giant field discovery in the Niger Delta in 1958, West African reserves rose rapidly. Their subsequent growth history is depicted graphically in Figs 2–4. Key discoveries and events are indicated.

The South Atlantic proved, as is demonstrated in this book, to contain a wide range of source rocks, reservoirs and trap types, and 40 years on both substantial new plays and fairway extensions continue to be encountered. Thus, as one play matured and discovery sizes fell, equally good, and in many cases superior, replacements emerged. These discovery pulses are most evident on Fig. 4 (Brazil) and Fig. 3 (Gabon to Angola), but even in Fig. 2, which is dominated by the scale of the initial Niger Delta discoveries, subsequent peaks are evident. The current sharp rise in reserve additions followed the technological advances that permitted the exploration and the subsequent development of deepwater objectives. The results also reflect the continuing favourable geology of the South Atlantic, notably the presence of younger Tertiary aged fans and channel complexes.

Deepwater exploration commenced in the Campos Basin of Brazil in 1984, but it was not until ten years later that the first deepwater well was drilled in West Africa. The location was the Lower Congo Basin of Angola. A year afterwards deepwater exploration began in Nigeria. Current accounts of the resulting successes may be found in Dolivo (1997) and Bentley & Kübler (1998). By May 1998 an estimated 17 000 MMbo of reserves had been discovered in Brazil, Angola, Congo (Brazzaville) and Nigeria. The intense excitement generated by these deepwater results is most simply presented by noting the jump in size of the largest fields. In Brazil the largest deepwater field is Marlim with reserves in excess of 2400 MMbo; the previous largest field was Marimba (reserves 400 MMbo), itself sited just beyond the shelf edge. In West Africa, the biggest pre-deepwater fields had reserves in the 1000–1200 MMbo range, but this figure was doubled with the Angolan deepwater Dalia discovery with reserves conservatively placed at 2000 MMbo. As is the case in the Campos Basin, there are adjacent giant accumulations. Significant discoveries are expected to continue, as is evidenced

Cameron, N. R, Bate, R. H., Clure, V S. & Benton, J. 1999. Oil and Gas Habitats of the South Atlantic: Introduction. *In:* Cameron, N. R., Bate, R. H. & Clure, V. S. (eds) *The Oil and Gas Habitats of the South Atlantic.* Geological Society, London, Special Publications, **153**, 1–9.

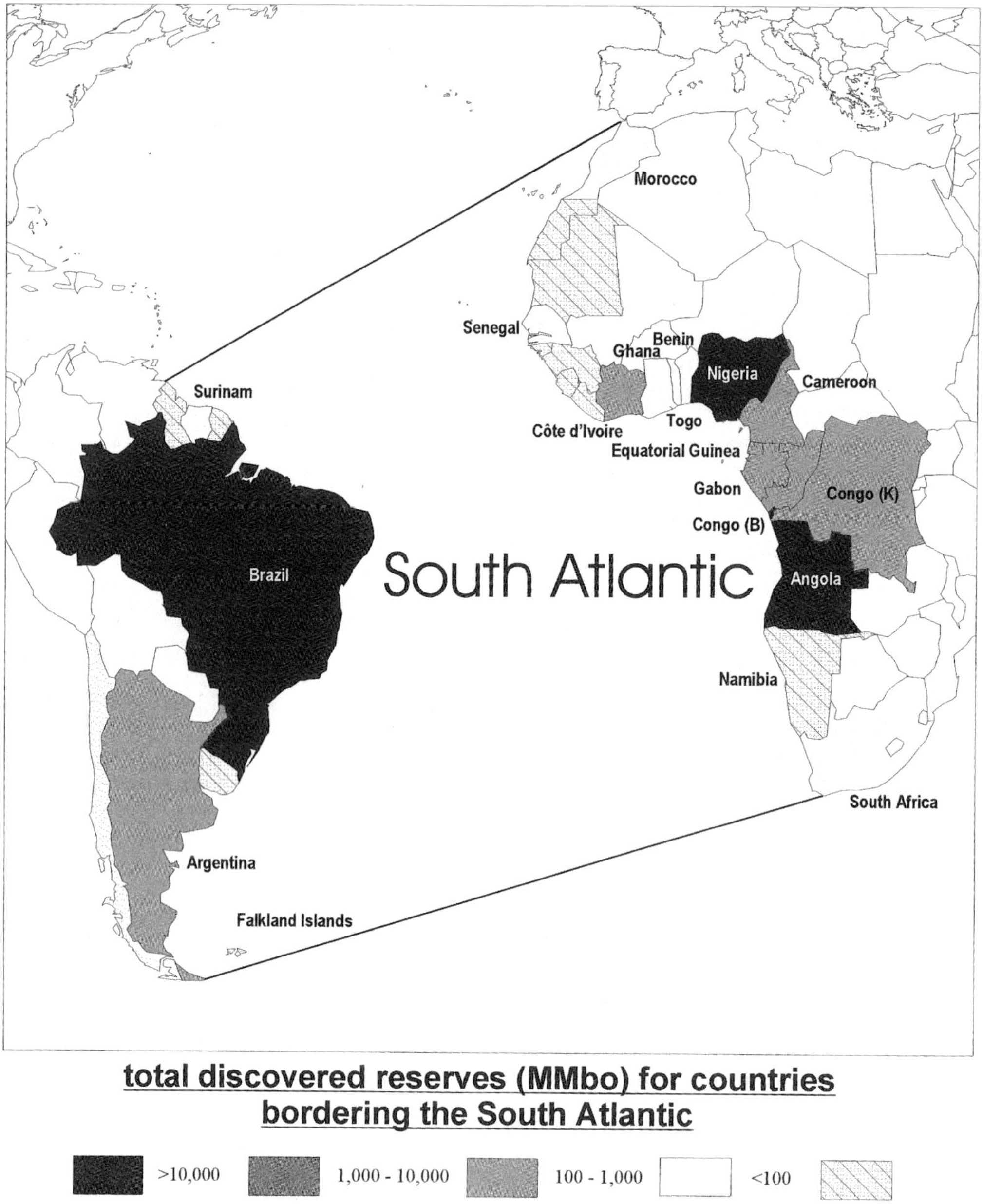

Fig. 1. The South Atlantic as defined for this volume. The country oil reserves correspond to those presented in Table 1.

by the high prices being paid for prime acreage. Statistical methods also suggest further successes are probable (Kronman *et al.* 1995), while a preliminary evaluation of the ultimate reserves is provided by Campbell (1998).

Whilst most attention has been focused on the deepwater, significant discoveries continue in more conventional settings. These discoveries, often the result of new seismic, logging and drilling techniques, have seen the renaissance of the Côte d'Ivoire as an exploration focus (200 MMbo of new reserves plus gas) and the discovery in 1995 of oil in Equatorial Guinea: Zafiro Complex reserves are placed at some 350 MMbo.

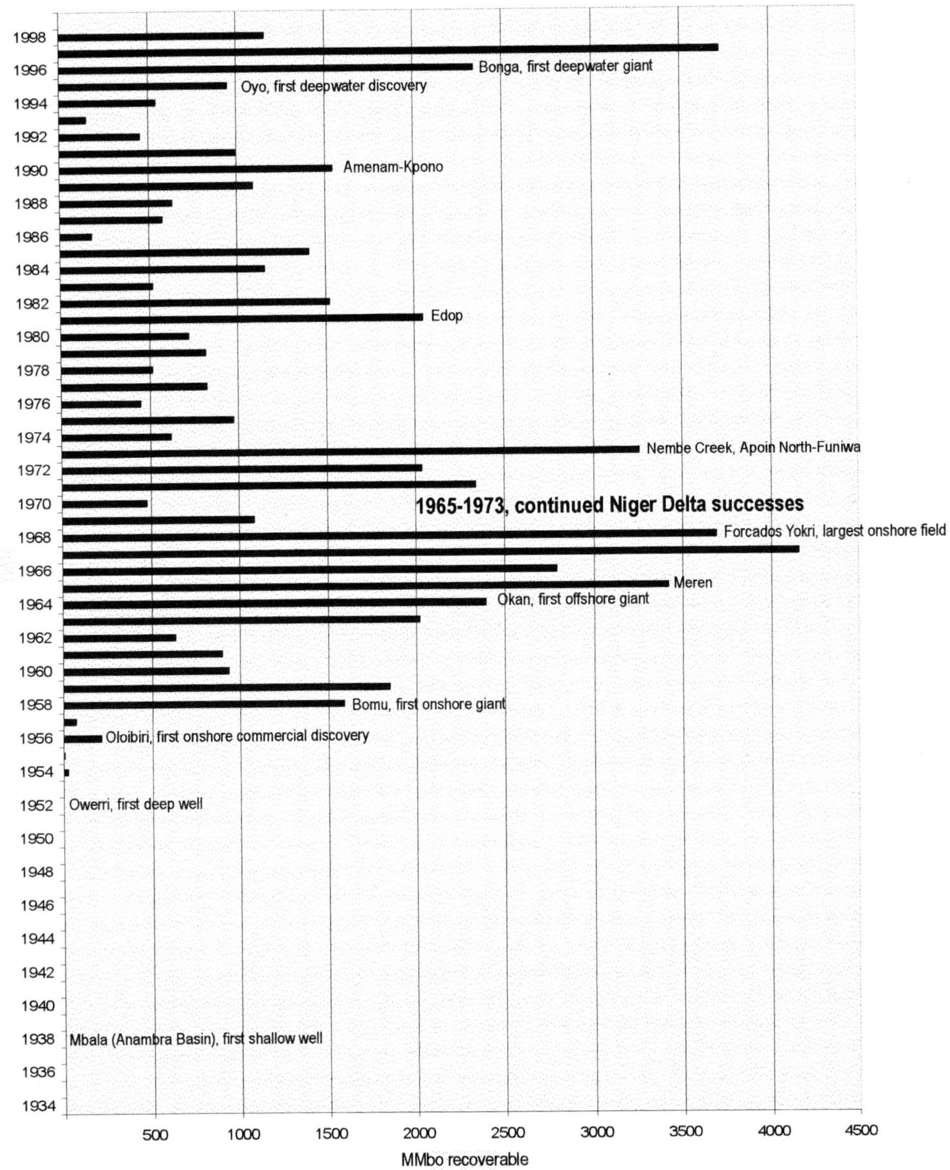

Fig. 2. Oil discovery by year plot for West Africa. The chart, which includes all countries from the Côte d'Ivoire to South Africa, was prepared using Petroconsultants' database. The database attributes reserve revisions to the discovery year. This compilation includes reserve estimates for natural gas liquids. Due to the dominance of the Niger Delta discoveries, the itemized discoveries/events in this figure relate only to Nigeria. Note that the discovery peaks in 1996 and 1997 are primarily the result of the Angola deepwater discoveries, these are detailed in Fig. 3.

Discoveries continue in the Congo Republic (Moho Field 1995, reserves 400 MMbo plus), northern Angola and to a lesser extent in Brazil. At the same time the first of the West African offshelf fields, N'Kossa in Congo (Brazzaville) (reserves in excess of 550 MMbo), has begun production as have the western Pinda discoveries in Cabinda (Angola) (total reserves in excess of 870 MMbo) and the Zafiro Field in Equatorial Guinea. The Kudu Gas Field reserves in Namibia are estimated to be 3.2 Tscf following a successful well in 1996, while the upside is thought to be considerably higher. The first well in the Falklands was spudded in May 1998 and the results are considered encouraging. Licensing rounds are either in progress or about to

Table 1. *Total discovered oil and gas reserves to the end of 1997. The country portion of the table was compiled from Petroconsultants' database. The world reserves tie to Petroconsultants (1998). The figures for Argentina and Brazil exclude the interior basins. The Tethyan basins of Morocco are also excluded*

Country	Oil (MMbo)	Regional total (%)	Gas (Bscfg)	Regional total (%)
Angola	10 549	13.5	8544	4.6
Benin	30	0.0	60	0.0
Cameroon	1319	1.7	4348	2.3
Congo (Kinshasa)	340	0.4	113	0.1
Congo (Brazzaville)	2643	3.4	2623	1.4
Equatorial Guinea	491	0.6	1212	0.7
Gabon	4251	5.4	3080	1.7
Ghana	50	0.1	177	0.1
Côte d'Ivoire	232	0.3	1896	1.0
Morocco	11	0.0	83	0.0
Namibia		0.0	3200	1.7
Nigeria	38 559	49.3	123 378	66.6
Senegal	10	0.0	82	0.0
South Africa	200	0.3	3624	2.0
Togo		0.0		0.0
Argentina	4723	6.0	18 251	9.9
Brazil	14 647	18.7	14 557	7.9
Surinam	166	0.0		0.0
Regional total	78 224		185 228	
Global total	1 829 883		7 577 357	
Global total (%)		4.3		2.4

The chart was prepared using Petroconsultants' database. The database attributes reserve revisions to the discovery year. This compilation includes reserve estimates for natural gas liquids.

open in almost every country bordering the South Atlantic.

The objectives

It is against this background of continuing success and increasing expectations that this volume has been brought together. Indeed, between its inception in mid-1995 and the time of writing this introduction (May 1998), some 10 000 MMbo of new reserves have been added.

The key objective of the volume is to supply, for an area having only a fragmentary English language literature base, an overview of the geology that will permit companies to obtain the maximum return from the new technologies available to them. An equally important objective is to provide a platform from which service companies and universities can demonstrate that their products can more than satisfactorily replace the diminishing output of industry research and regional study groups. A further reason this volume is to progress the work of IGCP Project 381, 'South Atlantic Mesozic Correlation', with which two of the editors (Nick Cameron and Ray Bate) are closely involved. IGCP 381 was instigated in 1995 with a mandate to provide products that would assist the petroleum industry. Since we have progressed this mandate, it is appropriate to dedicate the entire volume to IGCP 381.

The contributions

Twenty-six papers form the rest of this volume. Commercial sensitivity has precluded some of the topics we would have liked to include. Despite their loss, and our consequent inability to deliver a balance of papers that adequately reflects the discovery pattern, we do not consider that we have failed in our objectives. For example, we have assembled, using contributions based both on new and traditional tools, the material required to permit the South Atlantic to be explored as a single entity. The provision of multiple papers on a single subject, or an area, demonstrates, exactly as we intended, the insight deliverable to industry by service companies and universities.

The book is laid out in five thematic sections: The geological and geophysical framework; Oil and gas habitats; Organic geochemistry; Biostratigraphic applications; Exploration studies and issues.

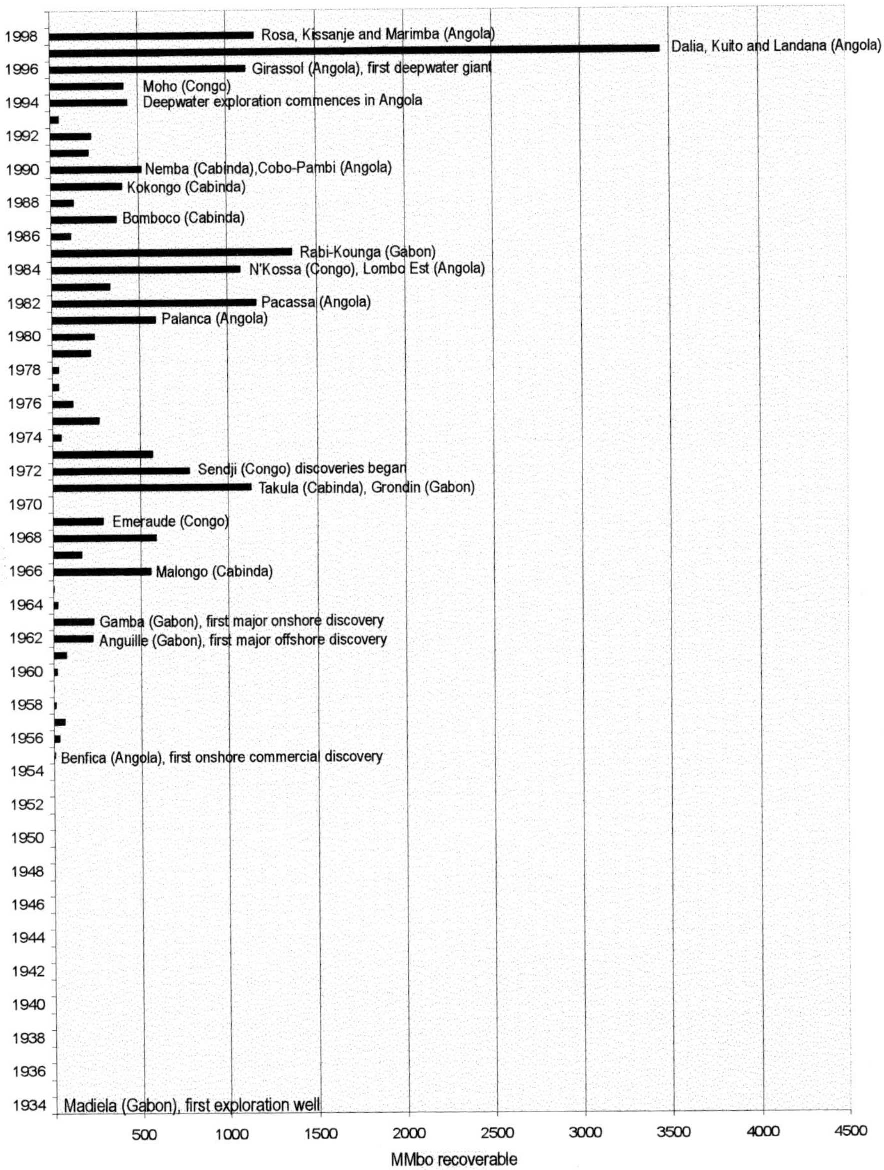

Fig. 3. Oil discovery by year plot for Gabon, Congo (Brazzaville), Congo (Kinshasa) and Angola.

The geological and geophysical framework

Karner & Driscoll demonstrate, using the evolution of hinge lines through time, the reasons why common features characterize plays in West Africa and the eastern seaboard of Brazil. Major strike variations in plays formed during the drift phase were found to be related to switches in the entry points of the largest rivers following interior capture. This compilation sets the scene for understanding and further progressing the deepwater exploration programme. For Namibia and south-east Brazil, **Gallagher & Brown** find, from apatite track fission track studies, that denudation increased during the Tertiary from a low in the late Cretaceous. This contribution is equally relevant to deepwater exploration as the results can be used to understand and exploit changes through time in the delivery rate of sediments offshore. **De Matos** was invited to prepare an update

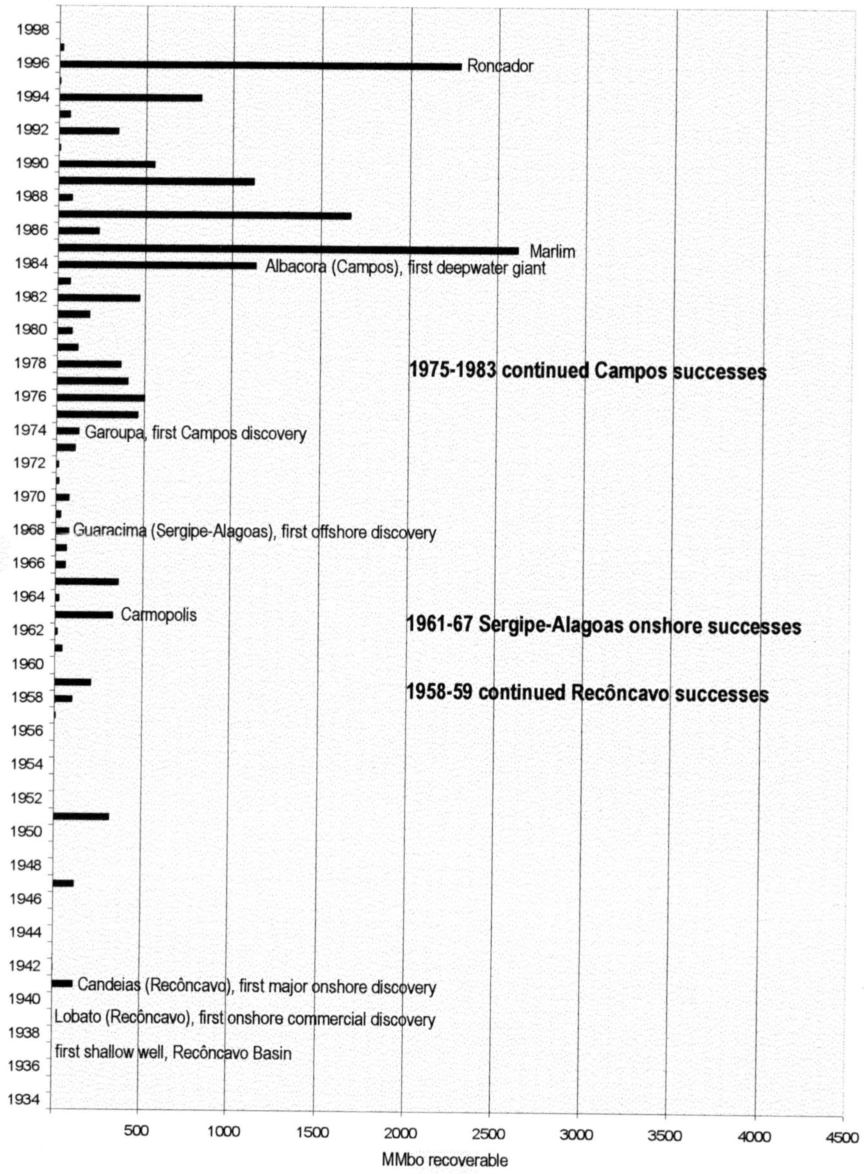

Fig. 4. Oil discovery by year plot for Brazil. Amazon Basin discoveries are excluded. The chart was prepared using Petroconsultants' database. The database attributes reserve revisions to the discovery year. This compilation includes reserve estimates for natural gas liquids.

of his 1992 paper on northeast Brazil (published in *Tectonics*, **11**, 766–791) in view of the importance of his results to companies building their understanding of this region. **Rosendahl & Groschel-Becker's** paper is an entirely new compilation of their ongoing rework of the Project Probe II seismic shot some ten years ago in the Gulf of Guinea. Here we feel the results will aid interpreters working deepwater opportunities in this poorly understood region. **Crossley & Cripps** use their field expertise to demonstrate how the present-day geography and geology of Africa and the Red Sea can be applied to reduce risk in otherwise poorly controlled objectives.

Oil and gas habitats

Dolan summarizes the progress of exploration in western Africa since the arrival of the first entrepreneurs 100 years ago; he also examines where future opportunities could be. **Coward *et al.*** examine the distribution of oil and gas reserves by regional and structural settings. This paper provides previously unavailable information on the geographical extent and the size of plays. As such, it is an excellent example of the aims of the book and we predict that it will see wide use. **Davison** provides a comprehensive and timely account of the Brazilian margin plays. His contribution includes a first for a Geological Society Special Publication, a CD-ROM map supplement. This holds the tectonic features map of the South Atlantic that accompanies his paper, together with play related diagrams for the major offshore basins of Brazil and the West African offshore basins between Nigeria and Angola. **Jungslager** uses colour illustrations to provide an equally useful account of the petroleum geology and opportunities of South Africa's Atlantic margin basins.

Organic geochemistry studies

Schiefelbein *et al.* determine, in a short contribution, the nature and distribution of oil families in the Atlantic margin basins of Brazil and Africa, from the Côte d'Ivoire to Angola. **Burwood** focuses, in a highly topical presentation, on the oil families of Angola.

Applications

The first three contributions in this section utilize palaeontology, sedimentology and organic geochemistry to examine the environmental factors determining the distribution of Middle–Upper Cretaceous marine source rocks. **Holbourn *et al.*** review the entire African margin from Morocco to southern Africa. **El Albani *et al.*** examine the rich, but immature, source horizons exposed in the onshore Tarfaya Basin of southwest Morocco. Their paper has important consequences for the companies holding the recently awarded offshore permits. **Wagner & Pletsch** use ODP Leg 159 results to review Cretaceous black shale habitats and organic richness in the Equatorial Atlantic. **Preece *et al.*** employ similar techniques to offshore Cabinda (Angola) in their examination of a Miocene aged, regional oxygen minimum zone. Their paper is an excellent example of the objectives of the book in that it shows how skilfully applied palaeontology can advance exploration frontiers. **Bate** uses ostracods as a novel means of determining megasequences within the lacustrine Pre-Salt succession. **Dingle** also utilizes ostracods, this time to explore the occurrence of source rocks through time and location in the South Atlantic. Both papers bring with them important exploration consequences.

Exploration studies and issues

The southern Rio Muni region of Equatorial Guinea has yet to enjoy the success of the north of the country. The paper by **Turner** sets the scene as to how this might be achieved. Four papers are included on Namibia in the anticipation that they will prove timely in assisting the progress of exploration and development. **Bagguley & Prosser** provide an account of the seismic stratigraphy of the Tertiary specifically requested to catalogue stratigraphic opportunities. **Stanistreet & Stollhofen**'s contribution provides onshore templates for assessing Kudu-style traps. The paper also carries with it important regional implications. **Jerram *et al.*** amplify the template detail. **Clemson *et al.*** offer suggestions as to the nature and age of the underlying and undrilled, up to 6 s two-way time travel thick, Namib Rift. They conclude that the succession most probably defines an aborted Karoo rift event. **Barker** looks at the problems associated with models for the opening history of the southernmost Atlantic. This short paper has considerable implications for the three final contributions, all of which are on the Falkland Islands. **Lawrence *et al.*** use a variety of techniques to prepare an exploration model for the North Falkland region. **Bransden *et al.*** concentrate on seismic stratigraphy to derive their play model for the same region. Finally, **Meadows** provides a new account of the stratigraphy of the onshore Palaeozoic succession based on an extensive programme of fieldwork. The locality descriptions, which have been deposited with the Geological Society's librarian, are confidential. Information on access may be obtained from the author. In due course it is expected that this restriction will be removed.

The legacy

We anticipate that the pace of exploration in the South Atlantic will necessitate a follow-on regional volume within the next five years. We also anticipate that the opportunities for both service company work and applied university research will grow significantly. IGCP projects, because of their international basis and strong links with government bodies through UNESCO, offer an excellent vehicle for facilitating this work.

Pursuit of three topics, all of which are referred to in this book, will be especially rewarding to exploration as they will provide the insight required to understand the factors localizing the quality and extent of plays.

- Improved understanding of the separation history of the South Atlantic. At present there is a wide disparity in rift models, notably in the Gabon–northeast Brazil and the Falkland Islands–South Africa region. Indeed, contrasting yet equally plausible viewpoints on these areas may be found in this volume. Just as important is further understanding as to the causes and exploration consequences of the post-separation uplift events and volcanicity that effect both margins of the South Atlantic. Until the variance in these issues is considerably reduced, reliable palaeogeographies will not be possible. In turn meaningful basin modelling will not be achievable for many time intervals. As an example, it is still not possible to depict with any degree of certainty the detailed palaeogeography of the late Aptian salts despite the crucial nature of this interval to almost every aspect of the play geology of West Africa and eastern Brazil.
- Improved understanding of the environmental and facies changes associated with the rift-induced, continental margin hinge lines. Once the topographic evolution of these features through time is understood, it will be possible to construct, using the geological, geochemical and palaeontological techniques described in this book, transects showing how source rock, seal and reservoir units developed in the dip direction through time. As an example, Fig. 5, constructed from Henry *et al.* (1995), illustrates, as a cross-section, the oceanwards expansion of the rift and drift successions across the Atlantic Hinge in central and northern offshore Angola. Particularly relevant is the incoming, to the west of the hinge, of an oil-prone source horizon within the expanded sag phase. The extent and quality of these source rocks will be crucial to deepwater exploration in Angola where the post-salt source section is immature and the previously known pre-salt section is overmature.
- Improved understanding of the timing, volumes, nature and places of delivery of sediments offshore. Classic geomorphology will have a major role to play here, as will the pursuit of the processes controlling the previously noted, young continental margin processes. This information is required to provide control on the strike behaviour of plays.

Once reliable dip and strike control is in place, it will be possible to tackle the 'why' questions posed by so many of our reviewers; for example, 'Why is the central region of the South Atlantic so productive?' We anticipate that the answers to many of the posed questions will exercise a prominent place in any follow-on publication.

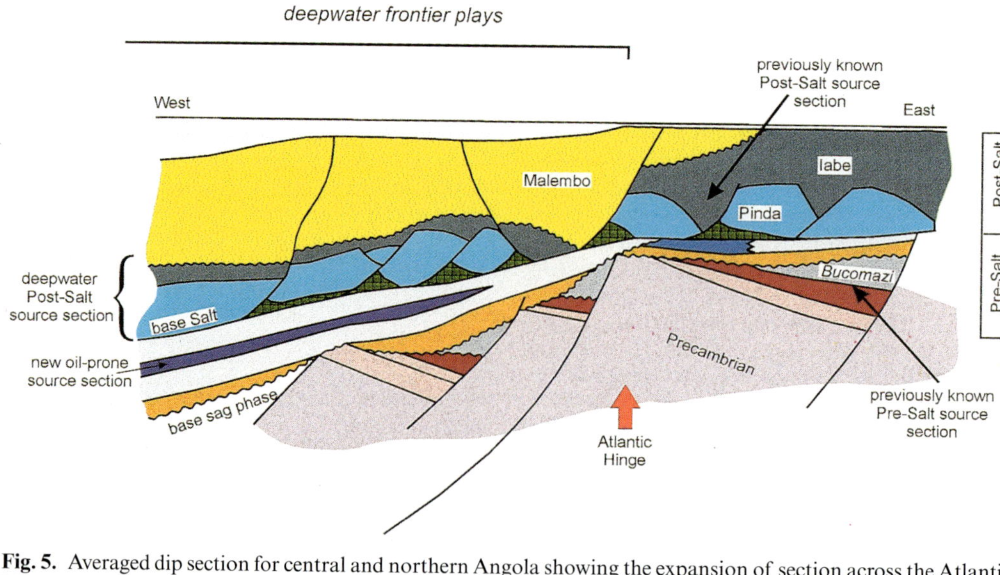

Fig. 5. Averaged dip section for central and northern Angola showing the expansion of section across the Atlantic Hinge and the incoming of a new source rock within the sag phase of the Pre-Salt section [constructed from Henry *et al.* (1995)]. The vertical axis is in two-way travel time; the horizontal scale is schematic.

This volume arises from a meeting held at the Geological Society in February 1997. Neither the meeting nor this volume would have been possible without the ready assistance provided by Amanda Harris (Geological Society), and David Ogden (Geological Society Publishing House), who contributed their able assistance to the production on this volume. The success of the meeting owes much to the financial and administrative support provided by BP Exploration, and to further sponsorship provided by Geolab UK, GeoMark Research Inc., Lacustrine Basin Research, Lasmo and Petroconsultants. Dave Bamford, Joe Pape and Dave Roberts of BP were, together with Nick Cameron and Ray Bate, the meeting convenors.

Generous sponsorship for the book was supplied by Amerada Hess, Chevron, Lasmo, Norsk Hydro, Phillips, Saga and the Petroleum Group of the Geological Society.

We wish to thank both the authors and reviewers for making this volume possible, in particular Bob Stoneley and Phil Nelson who so willingly handled much of the work required to ensure technical standards. Finally, we wish to acknowledge the numerous delegates who managed to (rightly) persuade us that a publication was essential. Val Clure joined the editors at this crucial and decisive time. Jeremy Benton teamed up with us to prepare the foundation material for this chapter.

References

BENTLEY, J. & KÜBLER, M. 1998. Sub-Saharan Africa's oil and gas activity. *In*: BLAKEY G. (ed.) *African Business Focus Yearbook for 1998*. British African Business Association, Reddy, London, 57–60.

CAMPBELL, C. 1998. The Future of oil. *Special Lecture to the PESGB Membership. Petroleum Exploration Society of Great Britain, May 1998.*

DOLAN, P. 1999. Western Africa: an unfinished story of oil and gas exploration. *This volume.*

DOLIVO, E. 1997. What's been drilled, found in world's deepwater areas. *Oil and Gas Journal*, **95**, 104–110, 112.

HENRY, S. G., BRUMBAUGH, W. D. & CAMERON, N. R. 1995. Pre-salt source rock development on Brazil's conjugate margin: West African examples (extended abstract). *1st Latin American Geophysical Conference, Rio de Janeiro, August*.

KRONMAN, G. E., RUSHWORTH, S. W., JAGIELLO, K. & ALEMAN, A. 1995. Oil and gas discoveries and basin resource predictions in Latin America. *In*: TANKARD A. J., SUÁREZ, R. & WELSINK, H. J. (eds) *Petroleum Basins of South America*. AAPG Memoir **62**, 53–61.

PETROCONSULTANTS. 1998. World Petroleum Trends. *In*: *Petroleum Economics & Policy Solutions*. Petroconsultants (UK) Ltd, London, Second Quarter 1998 (CD-ROM).

Tectonic and stratigraphic development of the West African and eastern Brazilian Margins: insights from quantitative basin modelling

GARRY D. KARNER[1] & NEAL W. DRISCOLL[2]

[1]*Lamont-Doherty Earth Observatory, Palisades, NY 10964, USA*
[2]*Woods Hole Oceanographic Institution, Woods Hole, MA 02543, USA*

Abstract: By applying an integrated basin modelling approach that combines quantitative kinematic and isostatic basin modelling with the principles of sequence stratigraphy the tectonic significance of the major syn- and post-rift stratigraphic packages that comprise the West African and eastern Brazilian Margins have been determined. Rifting occurred in three phases: Berriasian–Hauterivian, Hauterivian–late Barremian and late Barremian–early Aptian, with each phase being recorded by an onlap surface followed by an overall regressive package representing the subsequent infilling of the basin. Each rift phase resulted in the formation of deep, anoxic, lacustrine systems. The spatial partitioning of extension was responsible for the development of two major tectonic hinge zones, an inner–onshore and outer–offshore hinge zone subparallel to the margin. Seaward of the offshore hinge, individual rift basins tend to show an en-echelon arrangement. Deposition in these basins resulted in the production of thick sediment wedges. Using the modelling results from the West African Margin, together with palaeontological and stratigraphic correlations, the tectonic history of the rift basins comprising the eastern Brazilian Margin (i.e. Jequitinhonha to Sergipe–Alagoas) can be inferred. For example, the unconformities bounding the rift packages within the Camamu–Almada Basins (late Berriasian, Valanginian–Hauterivian and late Barremian–early Aptian) are identical to those on the West African Margin. Further, the rift architecture of the eastern Brazilian Margin is controlled by an onshore and an offshore hinge zone, similar to that observed on the West African Margin.

The late Barremian–early Aptian depositional packages of the Congo margin are bounded by a number of regional unconformities. The unconformity separating the Marnes Noires and the Djeno formations registers the onset of the third phase of the rifting. The top of the third rift phase sediment package is bounded by two unconformities; a sequence boundary termed the pre-Chela unconformity associated with an early Aptian sea-level fall, and the slightly younger Chela unconformity produced during the ensuing flooding of the margin. The resulting Gamba and Chela formations are transgressive sequences comprising fluvial sandstones at the base grading up into lagoonal facies, and imply sub-aerial exposure followed by marine transgression. The lagoonal facies of the upper Gamba and Chela Formations pass conformably into the evaporites of the regionally extensive Ezanga and Loeme formations. Significant post-rift accommodation occurs in the same region characterized by only minor brittle, syn-rift deformation of the upper crust and negligible erosional truncation of earlier syn-rift units. This subsidence form implies that extension was partitioned as a function of depth, with plastic deformation thinning the lower crust with brittle deformation of the upper crust occurring to the west near the ocean-continent boundary. Lower crustal extension beneath an intracrustal detachment maintains the region near sea level during the final stages of the rifting process (i.e. as break-up is approached). These shallow water regions are extremely sensitive to small-scale eustatic fluctuations. Repeated marine incursion and desiccation cycles could feasibly explain the thick salts of the Ezanga and Loeme formations. The evaporites are thus late syn-rift deposits, placing the timing of continental break-up as early Aptian. The existence of a mid-crustal detachment beneath the west African margin, active during the late stages of continental rifting, has profound implications on the hydrocarbon maturation history of the margin, both between the hinges and in the deep water rift basins seaward of the Atlantic hinge.

Although the post-rift development of the West African and eastern Brazilian Margins was dominated by carbonate deposition, major rivers delivered clastics at select points along the margins. Given that local tectonics controls sediment delivery to the evolving rift basins, it is the development and interaction of continental drainage systems that ultimately determines the history, volume and distribution of clastic delivery to the post-rift development of the margin. In particular, the process of river capture in which the headwaters of one drainage system are captured from another is undoubtedly a common, but poorly documented, process controlling landscape evolution and margin development. We suggest that: (1) the antecedent Congo River System, which once flowed in the valley now occupied by the Ogooué River, migrated southward as a function of time, as evidenced by the age of the major unconformities and turbidite deposits developed along the margin. This southward migration from Gabon in the Senonian to Congo in the Oligocene Period predicts that reservoir formation along the

KARNER, G. D & DRISCOLL, N. W. 1999. Tectonic and stratigraphic development of the West African and eastern Brazilian Margins: insights from quantitative basin modelling. *In:* CAMERON, N. R., BATE, R. H. & CLURE, V. S. (eds) *The Oil and Gas Habitats of the South Atlantic.* Geological Society, London, Special Publications, **153**, 11–40.

margin is linked to the migration of the Congo River; (2) the Parnaíba River of eastern Brazil, being the major source of clastics for the northeast Brazilian Margin (Mundaú Basin), was captured by the São Francisco River in the mid-Eocene Period, abruptly reducing clastic delivery and the frequency of turbidites to the Mundaú Basin margin while accentuating clastic delivery to the Sergipe Basin Margin; (3) the southwest headward retreat of the Paraíba River, a subsequent northeast flowing river of southeast Brazil, progressively captured the consequent south and southeasterly flowing streams delivering clastics to the Santos Basin, which decreased the sediment supply to the Santos Basin while increasing clastic delivery to the Campos Basin during the Tertiary Period.

The culmination of continental rifting often results in the beaching of extended lithosphere and the onset of seafloor spreading. It is during this rifting stage that the structural fabric and segmentation of rift margins are developed. The distribution and geometry of rift basins comprising passive margins tend to be asymmetric in terms of both their location along and across the margin. In addition, rift basin development is also typically discontinuous with time, with many extensional basins showing a complex assemblage of rift 'pulses' separated by periods of tectonic quiescence (e.g. Driscoll *et al.* 1995). Each rift pulse is characterized by divergent reflectors that onlap onto a single surface. Since the divergence and rotation of seismic reflectors is indicative of differential subsidence and block rotation, each rift pulse is recorded by a regressive stratigraphic package bounded by unconformities.

Where a margin proceeds to break-up, the last of these unconformities is often termed the break-up unconformity, being correlative to the cessation of extension across the region and the onset of seafloor spreading (e.g. Flavey 1976). For the West African Margin (Gabon–northern Angola), a series of rift pulses with varying duration culminated in the initiation of seafloor spreading between Brazil and Africa. Efforts to reconstruct the timing and spatial distribution of these rift pulses immediately prior to the break-up of Africa and Brazil, and thus the age of the break-up unconformity, have been hampered by: (1) the lack of seismic and well data in the deep-water offshore regions of West Africa and Brazil; (2) the existence of a magnet quiet zone that limits the use of correlatable marine magnetic anomalies to define either the timing or location of the ocean–continent boundary; (3) thick wedges of clastic sediments, often disrupted by subsequent salt diapirism, that tend to obscure the syn-rift structures of the margin. While the crustal Bouguer gravity anomaly has proved particularly useful in delineating the major structural features of the West African and Brazilian Margins, such as the location and along-strike continuity of hinge zones, rift basins, major accommodation zones, transform faults and the ocean–continent boundary (e.g. Karner *et al.* 1997; Davison 1999), the timing of the rift-drift transition remains a problem.

The preserved sedimentary record can be used to constrain and map the deformation of the lithosphere responsible for the development of a basin system and thus help define the age of break-up. In particular, the sedimentary record is punctuated by unconformities, the timing, spatial distribution and formation of which is critical in evaluating the importance of both the syn-rift and post-rift tectonics of the basin (e.g. Driscoll *et al.* 1995; Karner *et al.* 1997). Further, since the spatial and temporal distribution of these rift pulses can vary across and along the margin, then the resulting interplay between the earlier and later rift-induced subsidences and uplifts can result in complex stratigraphic stacking patterns in response to changing sediment source regions (i.e. basement, pre-rift or earlier syn-rift sediments), physiography, drainage modification and the exploitation of accommodation zones, and relative sea level (e.g. Driscoll & Hogg 1994). By iterating between the modelled time-line basin stratigraphy and the observed stratal patterns, sedimentary facies and estimated palaeoenvironments allows us to establish a quantitative relationship between the tectonic deformation of the lithosphere and the formation of stratigraphic sequences and their bounding unconformities (termed 'Quantitative Basin Analysis'; Karner *et al.* 1997; Driscoll & Karner 1998). From a hydrocarbon exploration perspective, Quantitative Basin Analysis also predicts the basal heat flow, source rock maturation and regional migration pathways consistent with the reconstructed tectonic history.

It will be shown that, during the final and protracted phase of extension in the South Atlantic that led to break-up, earlier rift-induced accommodation and post-rift accommodation induced by the third rift phase were infilled by a series of primarily regressive sequences with the first major marine incursion into the basin leading to the deposition of evaporites in areas that were suitably restrictive. After deposition of the evaporite sequence, carbonate platforms developed on both the West African and Brazilian Margins. Clastic input was limited to the sites of major river discharges on the margin. Major river systems appear to have had a fundamental control on the post-rift depositional history of many margins (e.g. Mississippi–Gulf of Mexico,

Orinoco–offshore Venezuela, Zambesi–East African Margin and the Niger–Benue Margin of West Africa). It is the clastic input of large river systems, together with changes in base level that result in the potential incisement and reworking of the margin, and a basinward shift of clastic delivery to deeper water parts of the basin. Often these lowstand deposits form major hydrocarbon reservoirs, thus it is particularly important to understand the development of drainage systems and how they change with time. For example, Driscoll & Karner (1994) showed that the uplift associated with the peripheral bulge engendered by the Amazon fan sediment load provided a tectonic feedback mechanism that affected both the coastal and fluvial depositional processes. In particular, the development of the peripheral bulge appears to have deflected the sediment-laden waters of the Amazon River progressively southward through time over a lateral distance of *c.* 600 km. This feedback between sediment loading and the peripheral bulge uplift modifies the spatial and temporal history of sediment delivery to the margin and provides a viable way to change base level without having to invoke glacial-eustasy. In this case, a relationship exists between river input of clastics and the formation and timing of major erosional unconformities and reservoir formation along the margin. By inference, this hypothesis implies a correlation between the timing of reservoir formation and the lateral migration of river systems.

The objective of this paper is to highlight the first-order processes responsible for margin formation and development and how these processes can be quantified, modelled and integrated to determine the timing and distribution of accommodation creation and destruction and thus the style and magnitude of lithospheric deformation. The particular aims are: (1) to determine the distribution, magnitude and timing of the various tectonic events across the West African Margin and, by inference, the rift history of the conjugate Brazilian Margin; (2) to use this information to constrain the age of the break-up unconformity on the Gabon–Angolan Margin; (3) to map the major syn-rift structural features of the offshore basins of West Africa and eastern Brazil by analysing the crustal Bouguer anomaly of the margins; (4) using recently released digital elevation data of Africa and Brazil generated from the Defense Mapping Agency's (DMA) 1:1 000 000 scale maps, point elevation and hydrology data, and general chronstratigraphic charts of select offshore basins, identify times and regions where stream capture and major drainage readjustment have occurred.

Structural setting of the West African and eastern Brazilian Margins

The major tectonic and structural features of the Gabon–Angola Margin superposed onto the crustal Bouguer gravity anomaly, are illustrated in Fig. 1. The gravity map was constructed by determining the Bouguer gravity anomaly from the 2 min × 2 min global free-air gravity grid of the world's oceans (GEOSAT) generated by Sandwell & Smith (1992) and the ETOP05 topographic grid using a sediment–water density contrast of (2200–1030 kg m^{-3}). From the Bouguer gravity anomaly a least-squares bicubic trend surface was subtracted to produce the crustal Bouguer gravity map shown in Fig. 1. This rendition of the gravity can be used to help delineate a series of small sub-basins separated by transfer faults that occur along the length of the West African Margin (e.g. Kambla and Takula Sub-basins, McHargue 1990; Vera, Dentale and Dianongo Troughs, Teisserenc & Villemin 1990). The gravity anomalies indicate the approximate along-strike continuity of the various sub-basins (Fig. 1). Two tectonic hinge zones, the Eastern and Atlantic, trend subparallel to the Cabinda, Congo and southern Gabon Margins (Fig. 1). However, the Atlantic Hinge Zone steps onshore in southern Gabon at *c.* 2° latitude to connect with the Lambarene Horst. It is therefore suggested that the stratigraphic development of the Ogooué Delta region i.e. west of the Atlantic Hinge Zone) differs significantly from the geology of the sub-basins existing between the Eastern and Atlantic Hinge Zones along the southern Gabon, Congo and Cabinda Margins.

The onshore Eastern Hinge Zone (thin red line, Fig. 1) demarcates the eastern limit of extension and separates syn-rift sediments from Precambrian basement. While minor faulting may exist to the east of the hinge zone, the hinge zone nevertheless separates a zone of negligible extension from a zone of significantly increased extension (Watts 1981). Syn-rift sediments have not been observed or recognized to the east of the Eastern Hinge (N. Cameron, pers. comm.). The Atlantic Hinge Zone is located further basinward (bold red line, Fig. 1) and occurs beneath the outer shelf–upper slope transition. Analysis of the gravity data supplemented with local seismic grids and well data, indicates that the Atlantic Hinge Zone is a series of short segment, echelon high-standing blocks (e.g. Kaba High, Fig. 1). Basement depths are a maximum offshore southern Gabon (*c.* 10 km) seaward of the Atlantic Hinge Zone (Fig. 1). Landward of the Atlantic Hinge Zone there are a series of local depocentres with basement ranging from 2 to 4 km depth located on the southern

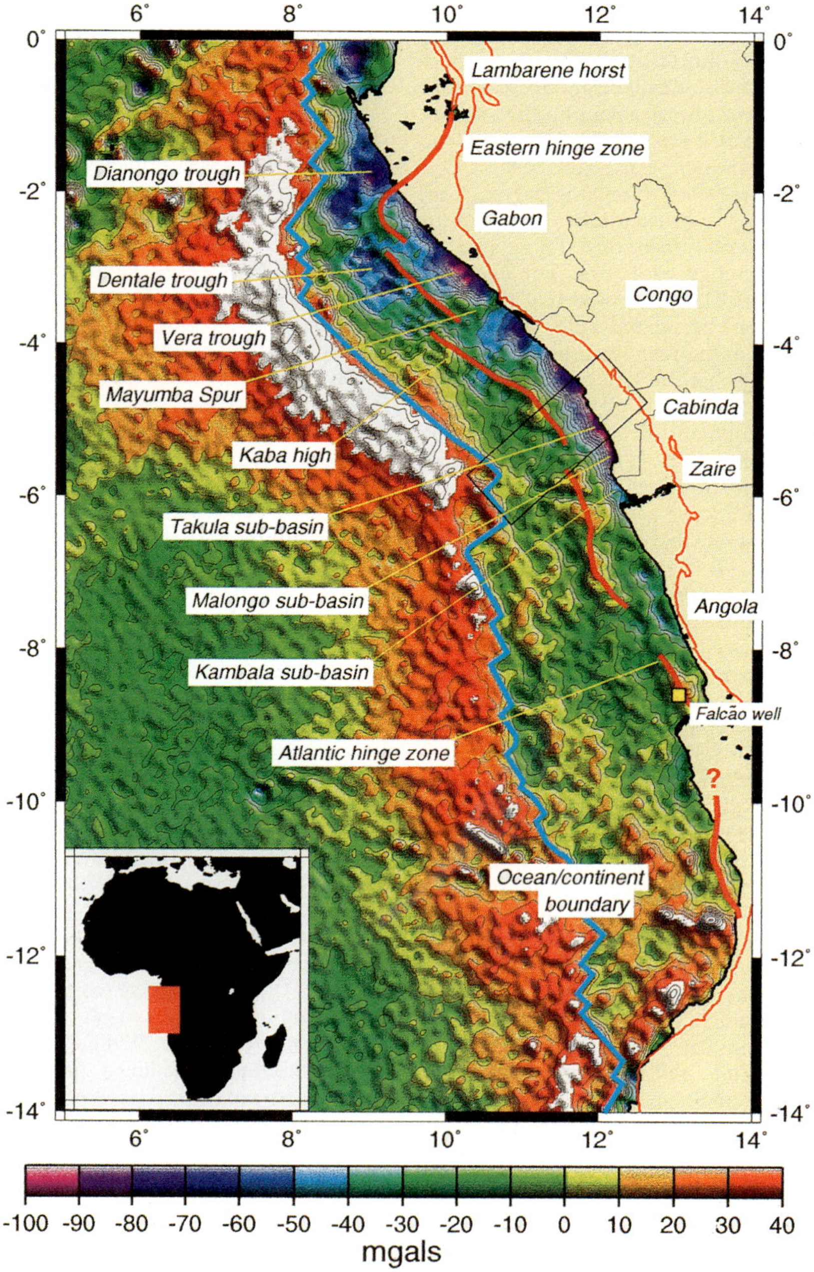

Fig. 1. Location map showing the major tectonic and structural features along the Gabon–Cabinda continental margin. The base map is an offshore crustal Bouguer gravity anomaly of the region. Scale is in mgals with values > +40 being shown in white. The modelled section is located in the outlined box. Two tectonic hinge zones, the Eastern (thin red line) and Atlantic (bold red line), trend subparallel to the margin. The onshore Eastern Hinge Zone demarcates the eastern limit of Neocomian extension and separates continental margin sediments from Precambrian basement. The Atlantic Hinge Zone occurs beneath the outer shelf–upper slope transition as a series of en-echelon high-standing blocks. A series of small but, in general, unconnected sub-basins, identified as localized negative gravity anomalies, exist on either side of the Atlantic Hinge Zone. The ocean–continent boundary, approximately delineated by the strong positive–negative gradient in the gravity anomaly, is shown as a bold aqua line.

Gabon, Congo and Cabinda Margins. These portions of the margin are associated with large Bouguer gravity lows of < 100 mgal (Fig. 1). The ocean–continent boundary (bold aqua line, Fig. 1) is interpreted as the strong positive–negative gradient in the gravity anomaly. The gravity amplitude across the boundary ranges from 20 to 30 mgals (Fig. 1). The source of the gravity gradient is in part a consequence of the juxtaposition of relatively thick, old oceanic crust adjacent to thinned continental crust. The thickened oceanic crust is a function of the decompressive melting of relatively hot mantle, most likely induced by the presence of the Tristan da Cunha, Saint Helena and Ascension Hotspots (e.g. Gladczenko *et al.* 1997).

Seaward of the Atlantic Hinge, individual rift basins tend to show an en-echelon arrangement but with limited along-strike connectivity. This implies that the along-strike basin, and hence rift-lake communication, water chemistry and possibly hydrocarbon source rock quality and preservation, is spatially limited. As with many segmented basin-bounding faults (e.g. Leeder & Gawthorpe 1987; Frostick & Reid 1989; Driscoll & Hogg 1995; Lambiase & Bosworth 1995; Driscoll & Karner 1998), both hinge zones controlled the spatial distribution and development of river drainage systems while their segmentation influenced river access to the newly formed rift basins. In turn, sites of clastic input and/or starvation will impact both hydrocarbon source and reservoir rock quality.

Brazilian basins conjugate to the Gabon–Angola Margin (from south to north) are the Cumuruxatiba, Jequitinhonha, Almada, Camamu, Jacuípe and Sergipe–Alagoas Basins (Santos *et al.* 1994; Netto *et al.* 1994; Feijó 1994; Fig. 2). The major tectonic and structural features of the eastern Brazilian Margin are superposed onto the offshore crustal Bouguer gravity anomaly (Fig. 2). The gravity anomalies indicate the approximate along-strike continuity of various basins and associated hinge zones that comprise the margin (Fig. 2). Two tectonic hinge zones, the western/onshore and an offshore hinge (locally termed the Alagoas Hinge, adjacent to the Sergipe–Alagoas Basin; Lana 1990), trend subparallel to the eastern Brazilian Margin (Fig. 2). A major exception is the rift 'excursion' that formed the onshore Tucano and Jatobá Basins. The onshore/western hinge zone (thin red line, Fig. 2) demarcates the western limit of Neocomian extension and separates syn-rift sediments from Precambrian basement. While minor faulting exists within the adjacent basement to the west of the Tucano Basin (e.g. Roguel 1990), the timing of this faulting is not known primarily because of the lack of any preserved sediment. As with the conjugate West African Margin, the hinge zone represents a boundary between a region of negligible deformation from a region characterized by significant normal faulting and accommodation generation. The offshore hinge zone is located further basinward (bold red line, Fig. 2) and consists of a series of high-standing blocks that separate onshore– nearshore basins from deeper water offshore equivalents (Fig. 2). The intersection of the Maceió Fracture Zone with the rift system has led to a major offset of the Alagoas hinge zone (at the junction of the Sergipe and Alagoas Basins; Fig. 2).

In marked contrast to the Atlantic Hinge Zone, the Alagoas Hinge Zone appears to be more continuous rather than en-echelon. For example, the offshore hinge zone of the Alagoas and Sergipe Basins (Fig. 3) separates onshore/nearshore sub-basins from deeper offshore sub-basins. While the hinge zone is continuous, the sub-basins and their associated border faults are en-echelon. This en-echelon fault arrangement suggests that this segment of the margin was produced by wrenching with the transport direction sub-parallel to the oceanic crust fracture zones (yellow arrow, Fig. 3).

The width of the offshore rift zone, defined by the distance between the western hinge zone and the ocean–continent boundary (bold aqua line, Fig. 2), changes dramatically along the Brazilian Margin. For example, it is particularly narrow between the Recôncavo and southern Sergipe Basins, abruptly widening across the Maceió Fracture Zone. Despite the correlation of rift width and fracture zones, the width of the rift zone might also be controlled by plumes (e.g. Bahia Seamounts; Macedo 1990). Anomalously high asthenosphere temperatures, consistent with plume activity during rifting, is evidenced by the existence of seaward-dipping reflectors along the offshore Jacuípe, Sergipe and Alagoas Basins (Mohriak *et al.* 1998). The large amplitude gravity anomalies across the ocean–continent boundary may, in part, be associated with the extrusives of the seaward-dipping reflectors. Plume-introduced heat possibly accentuated the plastic deformation of the lithosphere during extension, the Maceió Fracture Zone decoupling highly extended crust within the Alagoas Basin from regions of smaller amounts of extension associated with the Sergipe and Jacuípe Basins to the south (Fig. 2). It is interesting to note that the reciprocal relationship exists on the conjugate West African Margin; i.e. areas of accentuated rift zone width along the West African Margin are conjugate to areas with reduced rift zone width along the Brazilian Margin.

accentuated
Reduce of
Rift
Zone

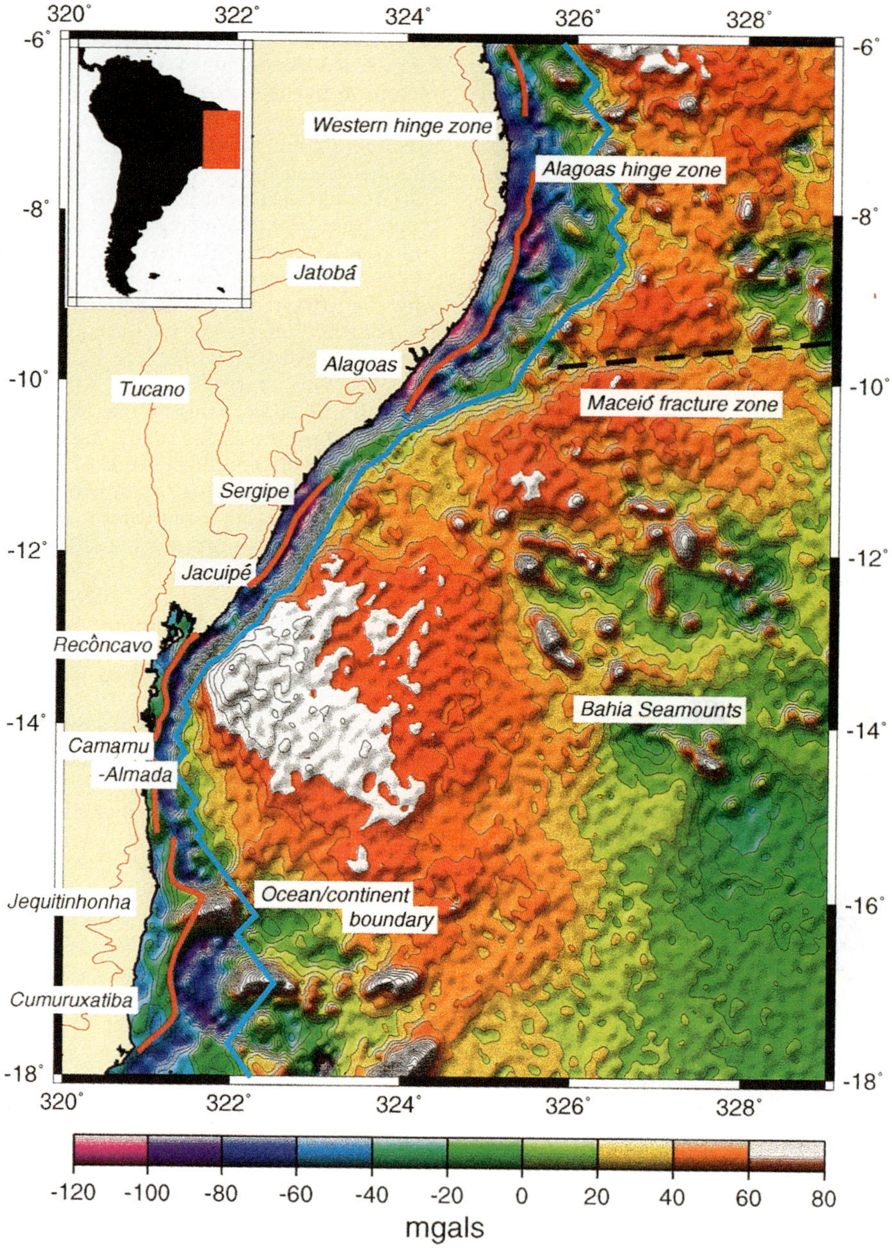

Fig. 2. Location map showing the major tectonic and structural features along the eastern Brazilian continental margin in addition to the main basins comprising the margin. The base map is an offshore crustal Bouguer gravity anomaly of the region. Scale is in mgals with values > +60 being shown in white. The onshore western Hinge Zone, identified by a thin red line, demarcates the western limit of Neocomian extension and separates continental margin sediments from Precambrian basement. The position of the ocean–continent boundary, approximately delineated by the strong positive–negative gradient in the gravity anomaly, is shown as a bold aqua line.

Lithosphere and crustal deformation kinematics

A kinematic model is used for mapping the deformation of the lithosphere in which extension within the crust is accommodated by a combination of brittle (upper crustal) and plastic (lower crustal) processes. The brittle collapse of the

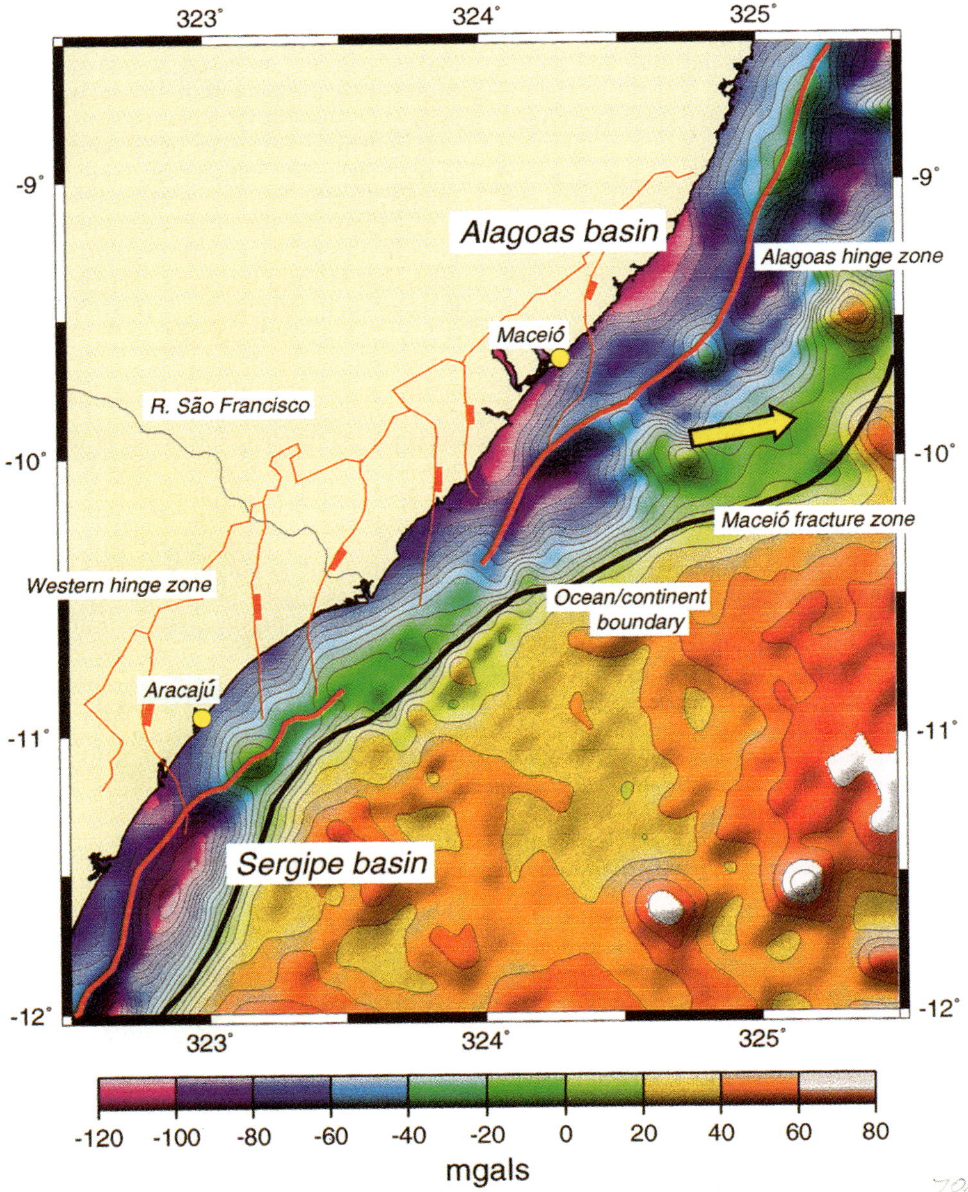

Fig. 3. Enlargement of the major tectonic and structural features along the Sergipe–Alagoas Margin superposed into the crustal Bouguer gravity. Two tectonic hinge zones, the Western (thin red line) and Alagoas (bold red line), are also shown. The Alagoas Hinge Zone occurs beneath the outer self–upper slope transition as a high-standing ridge. The Maceió Fracture Zone, a major oceanic transform fault, segments the Alagoas Hinge Zone while facilitating differential extension between the Sergipe and Alagoas Basins. A series of small but, in general, unconnected en-echelon basins, identified as localized negative gravity anomalies, exist on either side of the Alagoas Hinge. The ocean–continent boundary, approximately delineated by the strong positive–negative gradient in the gravity anomaly, is shown as a bold black line. The onshore basin border faults are clearly en-echelon, the extension transport direction being sub-parallel to the Maceió Fracture Zone (yellow arrow; transport direction).

upper crust in response to extension results in thinning of the crust that produces spatially discrete rift basins. Within the lithospheric mantle, extension is accommodated by plastic deformation and induces a passive rise of the lithosphere–asthenosphere boundary and, in so doing, introduces heat to the base of the thinned lithosphere. Ultimately, it is the spatial distribution of lithosopheric mantle thinning that governs the location and magnitude of post-rift or thermal subsidence. During rifting, the partial removal of the hanging wall block unloads the lithosphere inducing a flexural rebound (e.g. Weissel & Karner 1989). The form of this rebound is controlled by the flexural strength of the lithosphere during rifting. The flexural strength of the lithosphere or, equivalently, its effective elastic thickness is important in controlling the regional architecture of the rift and post-rift basins. The topography engendered by lithospheric deformation influences the drainage and, therefore, the sedimentation patterns in and around the evolving basin. In addition to the flexural rebound induced by crustal unloading, thinning of the lithosphere adds heat to the base of the lithosphere causing density variations that represent a buoyant load. With increasing time after rifting, the perturbed lithospheric temperature structure re-equilibrates allowing the lithosphere to cool, contract and subside. Thus, the rapid initial subsidence associated with the brittle deformation of the rift phase is replaced by a slower subsidence associated with the conductive cooling of the rifted lithosphere (McKenzie 1978).

Referring to Fig. 4, the response of the lithosphere to extension can be described by considering the distribution and magnitude of the various kinematically produced loads. Starting with a pre-rift lithosphere of thickness a and crustal thickness t_c, slip along a basin-bounding fault produces a surifical hole by collapse of the hanging wall block. The heave across the fault or, equivalently, the amount of extension, is e_0. The ratio of the pre-rift to the post-rift crustal thickness is termed $\delta(x)$. The delta function determines the distribution of the rift-phase subsidence and the total sediment thickness that can accumulate within a basin. Similarly, the ratio of the pre-rift to the immediate post-rift lithospheric mantle thickness is termed $\beta(x)$. The beta function principally controls the distribution of the post-rift subsidence. Both $\delta(x)$ and $\beta(x)$ determine the amplitude of the post-rift subsidence. The relative distribution and magnitude of the estimated $\delta(x)$ and $\beta(x)$ functions, obtained by matching the observed distribution and thicknesses of syn- and post-rift sequences, will ultimately determine if some form of detachment was operative during the extension process rather than an a priori modelling assumption, exhibiting itself as either an offset between the relative distributions of $\delta(x)$ and $\beta(x)$ (e.g. the 'offset' shown in Fig. 4) or numerically different values of the stretching functions for a given location.

The computer code used to investigate margin evolution is a two-dimensional kinematic forward modelling program which tracks the thermal and mechanical response of the lithosphere both during and after rifting, and thus the history of rift and post-rift basin subsidence. Features of this model include:

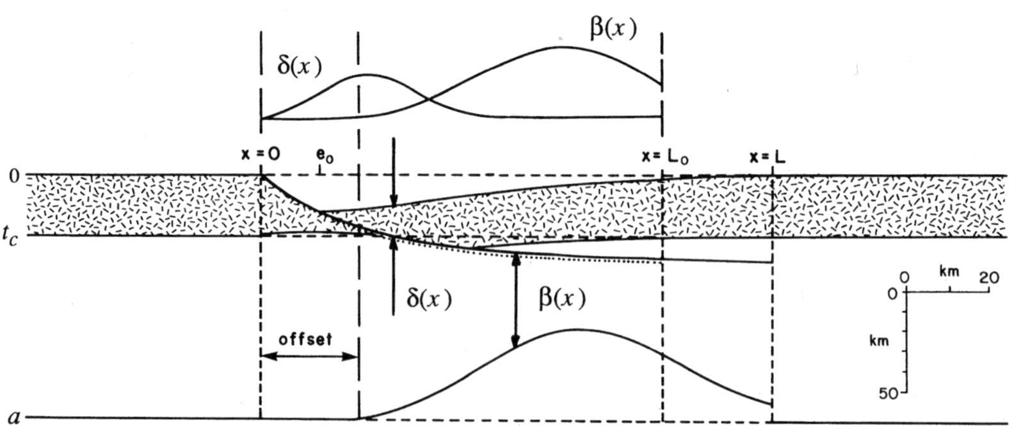

Fig. 4. Kinematic description of instantaneous extension of the lithosphere involving simple slip with heave, e_0, along a detachment (equal to the amount of extension L-L_0) combined with pure shear $\delta(x)$ in the upper plate or hanging wall and pure shear $\beta(x)$ in the lower plate or footwall. The deformed configuration of the kinematic model is found by using the extension parameters δ and β shown above the model as mapping functions. The initial shape of the detachment is denoted by the dotted curve while its shape in the deformed configuration is the solid curve. The initial crustal and lithosphere thickness are t_c and a, respectively. The lithosphere thickness is defined as the depth where temperature reaches T_m, the asthenosphere temperature.

- basin subsidence and rift flank uplift are calculated as a function of time;
- a kinematic description of the amount and distribution of crustal extension and lithospheric mantle thinning (termed delta and beta functions, respectively);
- crustal extension in hanging wall blocks can be incorporated;
- the extension computer code incorporates finite rifting and finite re-rifting capabilities;
- basin fill with compacting sediments is included. The compaction characteristics are spatially variable;
- erosion of emergent and submarine topography (producing peneplains and unconformities);
- eustatic variations can be included;
- flexural isostasy is employed throughout the calculations, during both the rift and post-rift development; i.e. the effective elastic thickness of the lithosphere changes as a function of space and time;
- profiles of time-line stratigraphy and the basin geometry at any time during its evolution (i.e. both during and after rifting).

Many rift systems are characterized by multiple fault systems with the dominant faults changing their importance as a function of space and time during extension (e.g. Central and Viking Graben of the North Sea and sub-basins of the Australian northwest shelf). The kinematic and flexural interaction of the hanging wall and footwall blocks helps to explain the complexity of rift basin architecture and the pattern of clastic delivery to the evolving rifts. The incorporation of several finite rifting phases in this modelling approach is an important aspect in simulating the architecture and geological development of the West African Margin. To illustrate the flexural interaction of multiple border fault systems and the effect of rift-induced topography on fluvial network development, the resultant basin and rift flank geometry for a range of fault configurations have been calculated (Fig. 5).

It is assumed that a rift phase comprises a series of pulses (and, in turn, the tectonic event termed break-up comprises a series of phases). Extension occurs by multiple slip across an eastward-dipping border fault (Fig. 5a). The amplitude and wavelength of the footwall uplift is a function of the heave across the fault ($e_0 = 5$ km), the dip of the fault (30°) and the effective elastic thickness of the lithosphere (T_e, assumed a constant value of 20 km in this example) during rifting. A characteristic structural and stratigraphic response is generated; rifting induces both the creation and destruction of accommodation (Christie-Blick & Driscoll 1995). The collapse and rotation of the hanging wall block results in the generation of a series of onlap surfaces and a thickening syn-rift sediment wedge towards the border fault. In contrast, the flexural uplift of the rift flank represents a relative sea-level fall allowing erosional reworking of pre-rift sediment and/or basement. During the first rift phase, fluvial drainage networks will tend to flow down the back side of the footwall away from the rift basin. In general, these drainage networks can only gain access to the rift basin in structurally low-lying regions between fault segments (i.e. accommodation zones). Because of both the lag in clastic production by the erosion of rift flank topography and the preferential delivery of clastics to the accommodation zones, the facies within the rift basin proper will tend to consist of regressive packages built on top of a condensed section. Notable exceptions exist when the antecedent rivers are particularly youthful, such as the Orange River that continued to deliver clastics to the evolving Orange Basin on the South African Margin despite the uplift and rotation of the coastal ranges (Gerrard & Smith 1982).

To investigate the structural and stratigraphic effects of synthetic and antithetic border fault systems, it is assumed that the first rift phase is associated with a eastward-dipping border fault. A second rift phase is accommodated by extension across a westward dipping fault located to the east of the previous basin-bounding fault (Fig. 5b). The hanging wall to the initial deformation now becomes the footwall to this phase of deformation. The uplifted footwall is capped by deposits associated with the first rifting phase with the basin depocentre migrating from west to east. Alternatively, if the second rifting phase is accommodated by extension across a westward-dipping fault located significantly to the west of the first border fault (Fig. 5c), then the footwall to the initial deformation is also the footwall to this second phase of deformation, leading to the formation of a horst block. The sediments associated with the first rifting phase are rotated and uplifted above base level as part of the footwall to the second phase of deformation. Finally, the case of a second rifting phase in which extension is accommodated across an eastward-dipping fault, located to the east of the original basin-bounding fault (Fig. 5d), is explored. The hanging wall to the initial deformation is now part of the footwall to this phase of deformation and is consequently uplifted and deformed.

This alternating footwall–hanging wall role of border fault systems is an important factor determining the architecture and facies development of the various syn-rift basins that comprise the

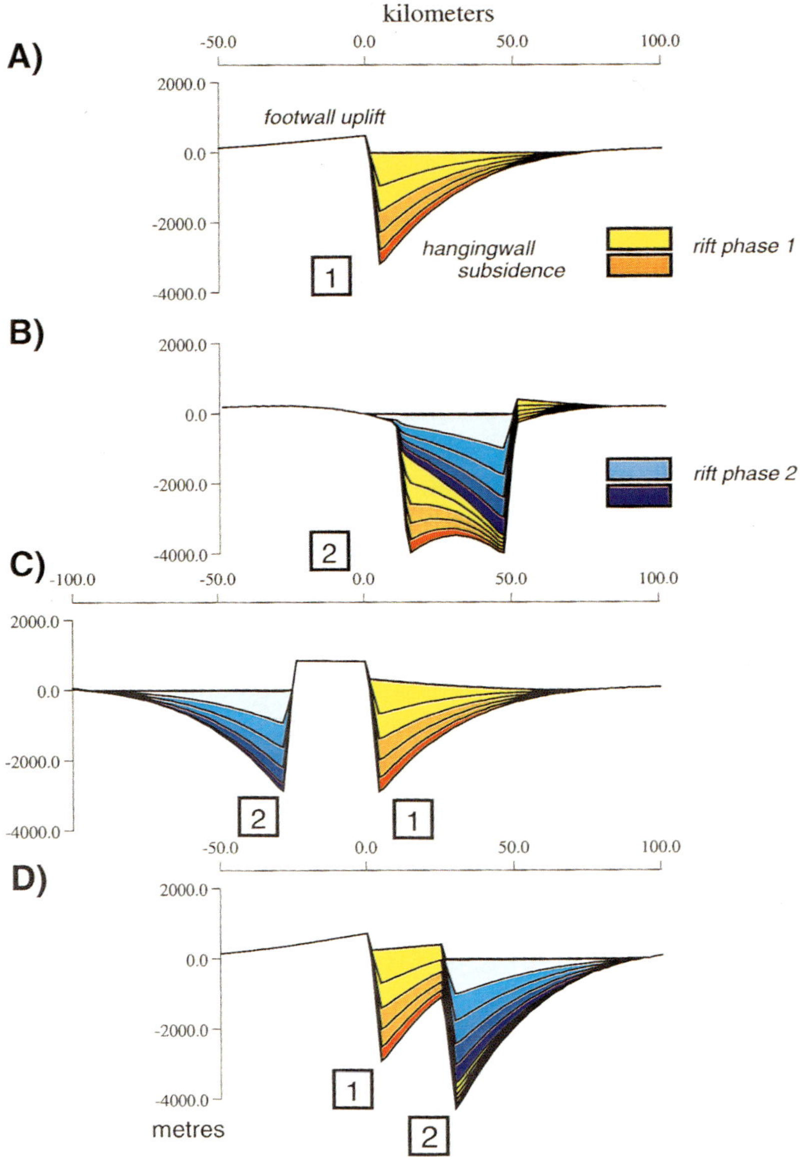

Fig. 5. Characteristic responses to multiple faulting associated with synthetic and antithetic border fault systems. (**a**) Multiple slip across a single fault system. Total fault heave (e_0) is 5 km, fault dip 30° and effective elastic thickness of the lithosphere, T_e, assumed a constant value of 20 km in this example during rifting. Hanging wall collapse and rotation results in the generation of a series of onlap surfaces and a thickening syn-rift sediment wedge towards the border fault. Note that during rifting accommodation is both created (collapsed hanging wall) and destroyed (uplifted footwall). Fluvial drainage networks will tend to flow down the back side of the footwall away from the rift basin proper, only gaining access to the basin in structurally low-lying regions between fault segments (i.e. accommodation zones). (**b**) Structural interactions induced by antithetic faulting. The hanging wall of the initial deformation now becomes the footwall to this second phase of deformation. The uplifted footwall is capped by deposits associated with the first rifting phase with the basin depocentre migrating from west to east. (**c**) Structural interactions induced by antithetic faulting. The footwall to the initial deformation is also the footwall to this second phase of deformation leading to the formation of a horst block. The sediments associated with the first rifting phase are rotated and uplifted above base level as part of the footwall to the second phase of deformation. (**d**) Structural interactions induced by synthetic faulting. The hanging wall to the initial deformation is now part of the footwall to this phase of deformation and is consequently uplifted and deformed.

West African and eastern Brazilian Margins. Understanding this interplay between rift-induced topography and fluvial networks allows predictions to be made concerning source and reservoir quality along and across the evolving rift basins. As stated earlier, during the initial rift phase, fluvial drainage networks flowed down the back side of the footwall away from the rift basin proper. Renewed extension, depending on the synthetic antithetic nature of the second border fault system, will either promote major capture of the footwall river systems or uplift, structuring and reworking of the earlier rift sediments.

Multi-phase syn-rift basin development

Using the kinematic model outlined above, and constrained by seismic reflection and well data, the present-day sediment thicknesses, stratal geometries and palaeowater depths across the Congo Margin have been forward modelled. In

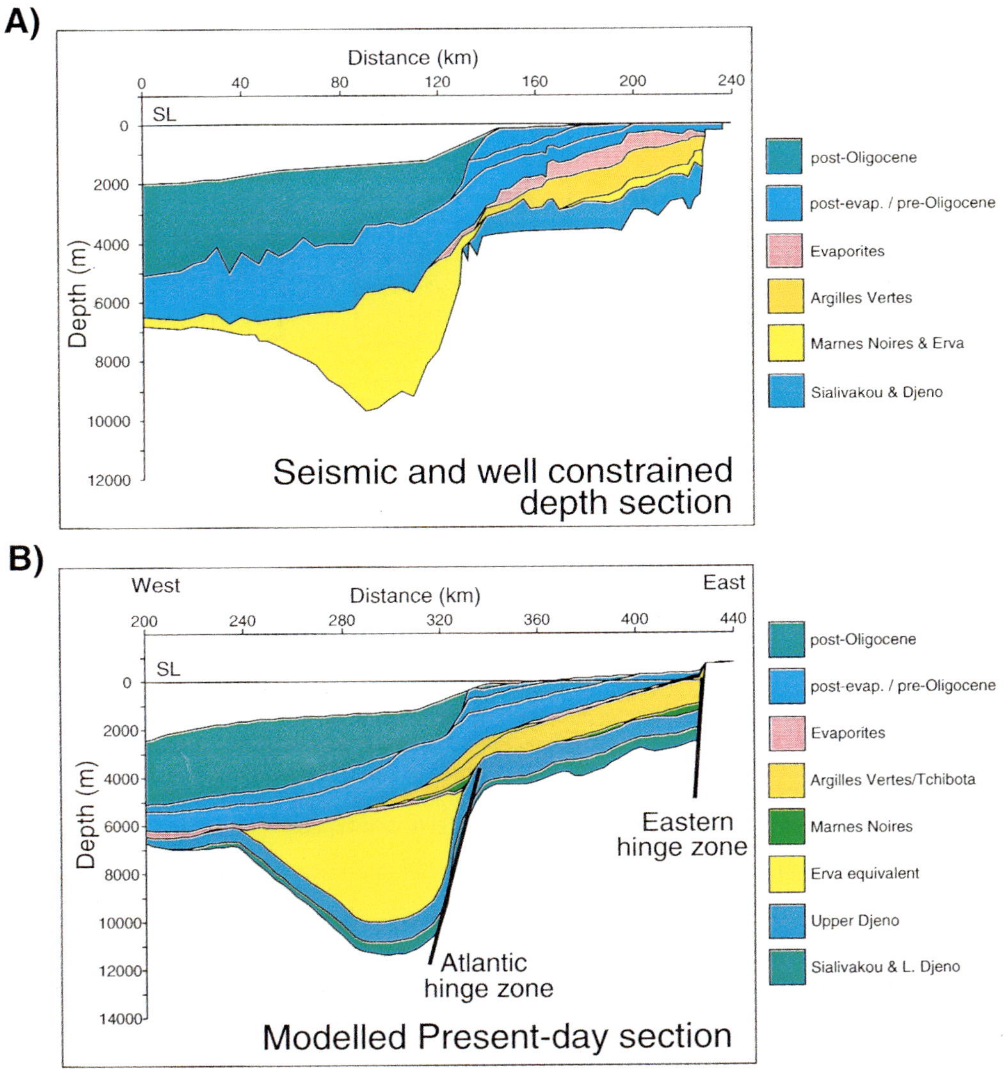

Fig. 6. Representative depth section across the Congo continental margin constructed from seismic reflection and exploratory well data (**a**) showing the major syn- and post-rift stratigraphic sequences that comprise the margin. This depth section was used to constrain the kinematic and isostatic basin modelling of the Congo Margin (**b**) which shows the modelled basin geometry and time-line stratigraphy for the entire rift and post-rift development of the margin. SL, present-day sea level.

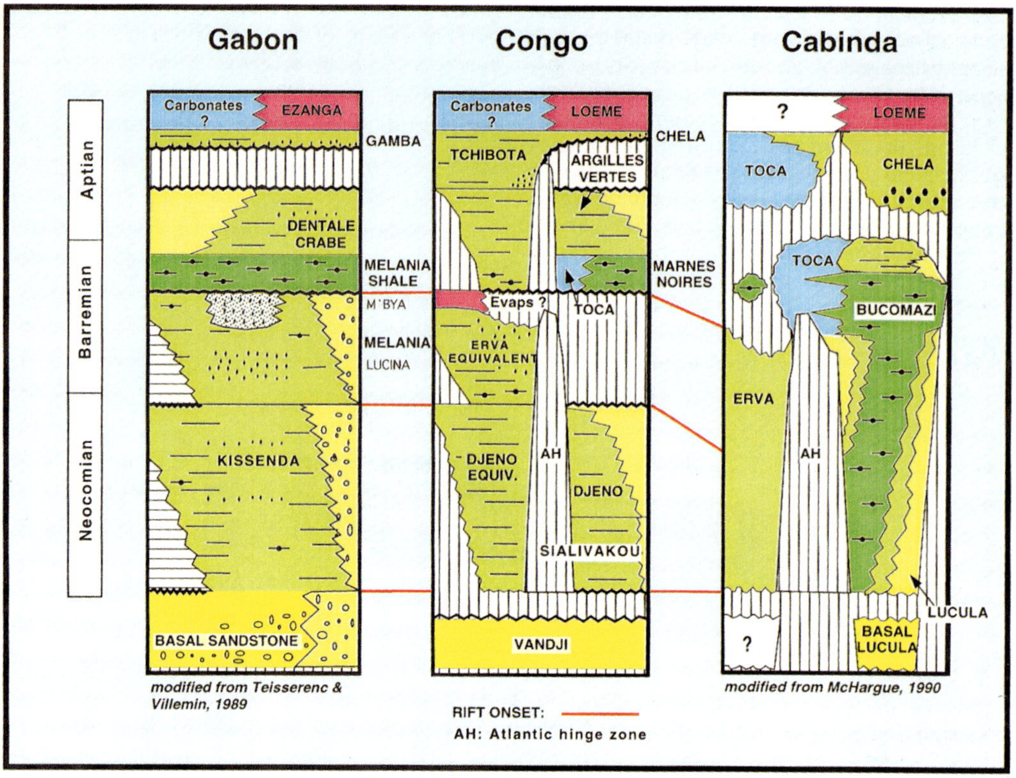

Fig. 7. Generalized chronostratigraphy for the Gabon, Congo and Cabinda Margins illustrating the link between extensional phases and the stratigraphic development of the margin. Bold red lines mark the onset of rifting for the three main episodes of tectonic deformation. Sandstones are shown in yellow, shales in dark green, light green represents a composite of sandstones and shales, evaporites are shown in pink and limestones are shown in blue. Stratigraphic variability between the sand-prone Congo and shale-prone Cabinda Margins is consistent with the Congo and northern Gabon Margins being part of a low-relief accommodation zone while the Cabinda and southern Gabon Margins are within a major border-fault segment.

particular, a depth section (Fig. 1, outlined box) from across the Congo Margin was generated from a series of seismic sections, well log and formation velocity data (Fig. 6a). Figure 6b represents the simulated development of the Congo Margin through various rifting events up to the Present. It is the close similarities between the observed and modelled stratigraphy (cf. Figs 6a and b) that gives us confidence in reconstructing the margin during Neocomian rifting. The results from this modelling, coupled with the generalized chronostratigraphy (Fig. 7) of the rift section for the Congo, Gabon and Cabinda Margins, allow reconstruction of the stratigraphy and basin geometry at critical times during the development of the margin; namely, immediately preceding each rift pulse up to and immediately following break-up. Three rifting phases comprise the break-up history between Africa and South America. As demonstrated by Karner *et al.* (1997) for the Congo Margin, the rift event itself is relatively rapid, leading to the formation of a single onlap surface – the rift-onset unconformity. The remainder of the rift phase relates to the partial or complete sediment infilling of the rift-generated space.

Berriasian–Hauterivian rift phase

Broadly distributed continental rifting (*c.* 200–500 km) between the Gabon–Cabinda Margin and its conjugate Brazilian Margin (see Fig. 2) commenced in the early Cretaceous with the rapid creation of a series of deep anoxic lakes (Fig. 8a). The fast rate of accommodation production is consistent with the fact that divergent reflectors indicative of block rotation onlap onto a single surface and by the formation of an underfilled basin. The initial stage of extension was characterized by large fault-controlled half-graben accompanied by the local development of a series of horst and graben structures (Fig. 8a; Karner *et al.* 1997). Basin infill consists of deep-water shales (Sialivakou Formation; Kissenda: Gabon–lower Bucomazi:Cabinda) over which prograded sandstones and conglomerates of the Djeno Formation. The upper Djeno Formation represents a progradation clastic wedge. The relief of these clinoforms defines a minimum lake water depth of 500 m. In Fig. 7, the Berriasian–Hauterian syn-rift package (Sialivakou and lower Djeno Formations) is bounded by a lower erosional unconformity, the rift-onset unconformity, that separates a lower from an upper Vandji sandstone, and an upper unconformity characterized by minor erosion and subaerial exposure as evidenced by the existence of palaeosol horizons with iron-stained sands. The lower sands of the Vandji and Lucula Formations are pre-rift units, laterally equivalent to the Brazilian upper Jurassic Sergi Formation of the Recôncavo and Tucano Basins (*et al.* 1994).

The syn-rift sections of the Cumuruxatiba Basin consists of high-energy fan–deltas overlain by dark grey–black lacustrine shales of the Porto Seguro Formation deposited during the late Berriasian (Santos *et al.* 1994; i.e. coeval with the Gabon–Cabinda margin). A similar situation exists for the Camamu and Almada Basins. This suggests that a series of broad lakes existed throughout much of the offshore syn-rift stage of the South America/African region. Upper Berriasian accommodation was subsequently infilled by shoaling upward sedimentary packages; the upper Djeno Formation on the West Africa Margin and the Cricaré and Morro do Barro Formations on the Brazilian Margin. In general, the rifting pulses that characterize the break-up of Africa from South America are particularly abrupt, separated by relatively long periods of tectonic quiescence. Given that the rift phase spans 10–15 Ma implies that the rift event is substantially faster. For example, Driscoll *et al.* (1995) determined that, in the Jenne d'Arc Basin, the rift event was typically 2–4 Ma. As a result, the rate of generating accommodation is greater than the rate of clastic production, thus explaining why extension tends to produce large palaeowater depths that are subsequently infilled by progradational (regressive) systems (Driscoll & Hogg 1994). These regressive phases are a direct response to a decrease in the rate of relative sea-level rise and/or increasing sediment supply. The resulting fluvio-deltaic systems form the potential hydrocarbon reservoirs with the next rift phase supplying the necessary seal and possible source rocks (i.e. pro-delta shales).

The Eastern Hinge Zone demarcates the eastern limit of Beriasian Extension (Fig. 8a). As much as 1.5–2 km of uplift is predicted across the hinge zone at this time, being a consequence of the mechanical unloading and flexural rebound of the lithosphere in response to extension. The predicted rift flank topographies, asymmetric with their steepest face to the west (Fig. 8a), helped to create a time lag between the formation of the topography, the development of fluvial drainage networks and the input of clastics into the evolving rift basin. This resulted in the formation of condensed sections in the basins away from the accommodation zones. Crustal and lithospheric mantle extension associated with this rifting phase increases from the Eastern Hinge Zone into the central portion of the basin, reaching maximum values of 1.2–1.3 (Karner *et al.* 1997). The estimated distribution and magnitude of $\delta(x)$ and $\beta(x)$ are similar, implying that the extension in the lithosphere was depth independent.

Hauterivian–late Barremian rift phase

During the second phase of rifting on the African Margin the main axis of extension stepped westward by *c.* 100 km relative to the earlier Berriasian rift phase and resulted in the formation of the Atlantic Hinge Zone and a major depocentre termed the Outer Basin Sediment Wedge (Figs 5d, 6 and 8b). This wedge, where drilled, consists of mudstone and matrix-supported, coarse-grained argillaceous sandstones and conglomerates, and poorly-sorted, micaceous and feldspathic lithic wacke or arenites (McHargue 1990). The timing of formation of the Atlantic Hinge, assumed to be the consequence of rift-induced uplift, is defined by the first appearance and landward progradation of the Lucula Sandstones across the shales of the lower Bucomazi Formation (Fig. 7). The timing of the Lucula Sandstones is defined by McHargue (1990) and Bracken (1994). Extension was localized with maximum extension values calculated to be of the order of 1.2–1.3. Whereas between the hinge zones the stratigraphy is characterized by a succession of condensed sections followed by regressive packages, the stratigraphy west of the Atlantic Hinge Zone is dominated by rapidly deposited shales equivalent to the Erva Formation of offshore Cabinda (e.g. McHargue 1990). Limited drilling data indicate that, by the end of the second rift phase, the outer basin had basically filled to become a shallow lake, whereas between the hinge zones, the Djeno Formation was subaerially exposed. Although an Atlantic Hinge Zone equivalent along the conjugate

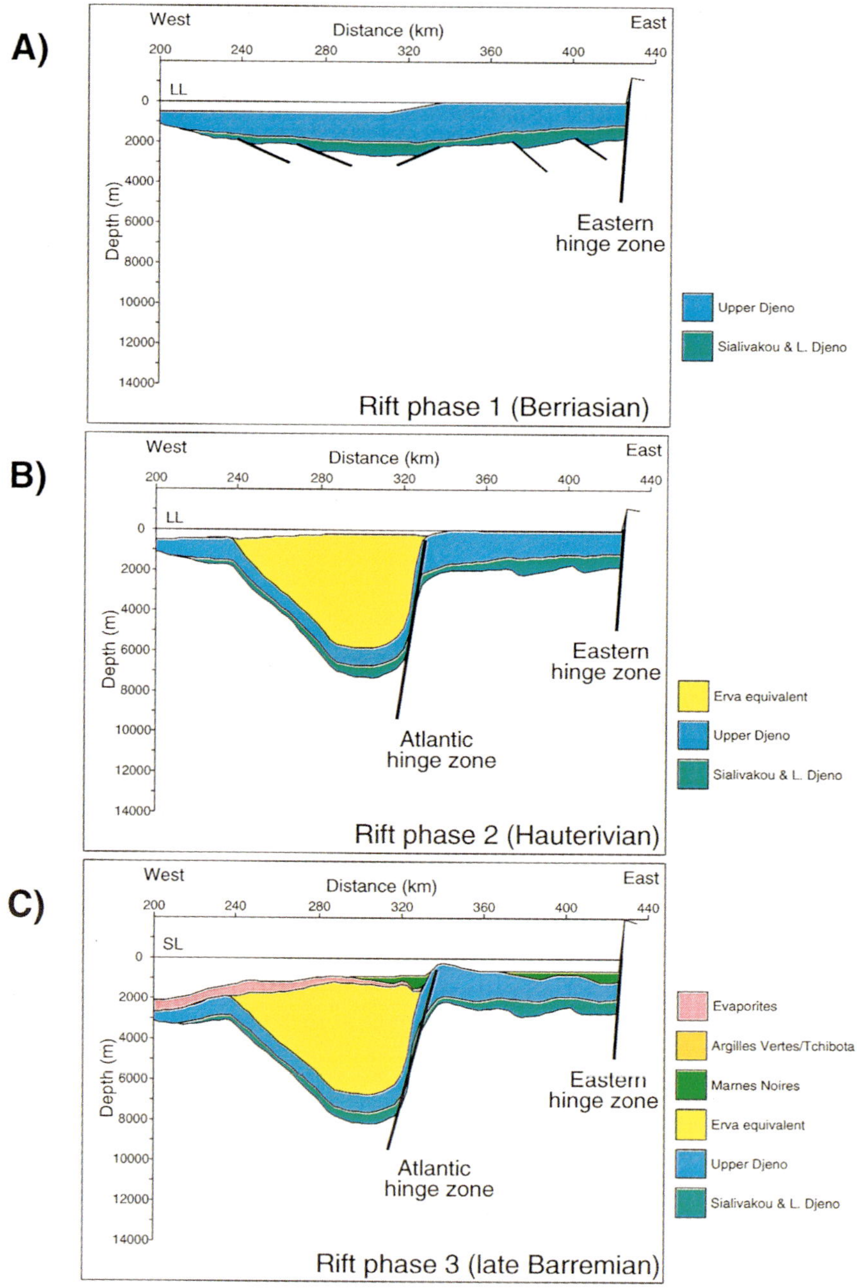

Fig. 8. Modelled basin geometry and time-line stratigraphy for the three rifting events responsible for the development of the Gabon–Cabinda Margin (**a**) Berriasian; (**b**) Hauterivian–late Barremian; (**c**) late Barremian–early Aptian. Initial broadly distributed continental rifting became more focused with time as the main axis of extension migrated to the west. Abrupt jumps in the spatial location of extension as a function of time was responsible for the development of a double hinge zone system (the Eastern and Atlantic Hinge Zones). The Eastern Hinge Zone demarcates the eastern limit of the broadly distributed Berriasian extension. The Atlantic Hinge, located *c*. 90 km west of the Eastern Hinge, marks the eastern limit of Hauterivian extension. A third phase of extension reactivated both the Eastern and Atlantic Hinge Zones and was responsible for creating the accommodation space for the Marnes Noires Formation (and equivalent) source rock deposition across the Gabon–Cabinda Margin. LL is lake level; SL is sea level.

Brazilian Margin is recognized in the Bouguer gravity (Figs 2 and 3), and a major depocentre seaward of this hinge zone exists (e.g. Lana 1990), no drilling data is presently available from this region to define either the sedimentary facies or age. Lana (1990) termed this wedge as an offshore equivalent to the onshore Coquiero Seco Formation.

The topography developed across the Atlantic Hinge Zone competed with the subsidence induced by crustal extension and sediment loading, resulting in commensurately smaller relief (100–200 m) compared with the Eastern Hinge Zone. Varying amounts of erosional truncation along the Atlantic Hinge Zone are a function of the distribution and magnitude of lithospheric extension. For example, offshore Gabon, the Melania Formation is well developed between the Atlantic and Eastern Hinge Zones, and records the erosion of the Atlantic Hinge and the existence of palaeowater space. Conversely, the absence of the Melania Formation on the Congo Margin suggests that, in this region, only minor flank uplift occurred. The lack of uplift is consistent with this segment of the Congo Margin being an accommodation zone. The Eastern Hinge Zone underwent minor reactivation during this rift phase. As with the first rift phase, the distribution and magnitude of $\delta(x)$ and $\beta(x)$ are similar.

Late Barremian–early Aptian rift phase

A third and final phase of rifting reactivated both the Eastern and Atlantic Hinge Zones and created the accommodation necessary for deposition of the deepwater (*c.* 400 m) lacustrine Marnes Noires Formation on the Congo Margin and a second offshore clastic wedge as part of the Dentale Formation on the Gabon Margin (Figs 7 and 8c). The rapid transition from palaeosol horizons of the upper Djeno Formation to the thick (200–600 m) anoxic black shales of the Marnes Noires Formation (Melania Black Shale:Gabon/middle Bucomazi:Cabinda) in the late Barremian documents the timing and distribution of this new accommodation space. Following Marnes Noires deposition, prograding clastics of the Argilles Vertes and Tchibota Formations (Dentale Formation:Gabon/TOCA B & C and upper Bucomazi:Cabinda) register the infilling of accommodation between the hinges (Figs 7 and 9). Apart from minor intrasedimentary faulting associated with differential compaction, these formations appear to be relatively unfaulted. This is a critical piece of evidence. Since the post-Marnes Noires and equivalent formations are unfaulted implies that the brittle extension has ceased on this margin. As demonstrated by Driscoll & Karner (1998), however, the cessation of brittle deformation along proximal portions a margin does not necessarily require that the extension process is complete. In fact, the large post-rift accommodation that characterizes many continental margins but is also associated with little syn-rift crustal deformation implies that lower crustal and lithospheric mantle extension is occurring and is separated from the upper crust by an intracrustal detachment. The balancing brittle deformation is focussed in a narrow region across the nascent ocean-continent boundary.

A number of regional unconformities bound the depositional packages existing between the onset of the third rift phase and the evaporites blanketing the margin. The unconformity between the Marnes Noires and the Djeno formations registers the onset of the third phase of rifting on the Congo margin. The sediments infilling the newly created accommodation consists of organic rich shales (Red Cuvo, Falcão, Marnes Noires, and Melania Shale source rocks) and regressive packages of the Argilles Vertes, Dentale and Gray Cuvo formations deposited in a brackish to marginal marine environment. While this rift infill package purportedly has sediment thicknesses of 4000–6000 m in the Kwanza basin (Henry *et al.* 1995), the equivalent sediments on the Congo margin are considerably thinner (600–800 m). At the top of this syn-rift package are two unconformities; a sequence boundary termed the pre-Chela unconformity associated with an early Aptian sea-level fall, and the slightly younger Chela unconformity produced during the ensuing flooding of the margin. The resulting Gamba and Chela formations are transgressive sequences comprising fluvial sandstones at the base grading up into lagoonal facies, suggesting sub-aerial exposure followed by marine transgression. The lagoonal facies of the upper Gamba and Chela Formations pass conformably into the evaporites of the regionally extensive Ezanga and Loeme formations. The observation of a regional shallow water environment at the end of rifting is critical in constraining the development of the margin and, in particular, the distribution of lithospheric extension with depth.

The modelled distribution of lithospheric extension is significantly asymmetric, with extension values ranging from 1.08 to 1.2 between the hinges and rapidly increasing from 1.2 to >4 towards the site of the ocean–continent boundary (Fig. 1; Karner *et al.* 1997). Compared with the location of the earlier rift events, the region of

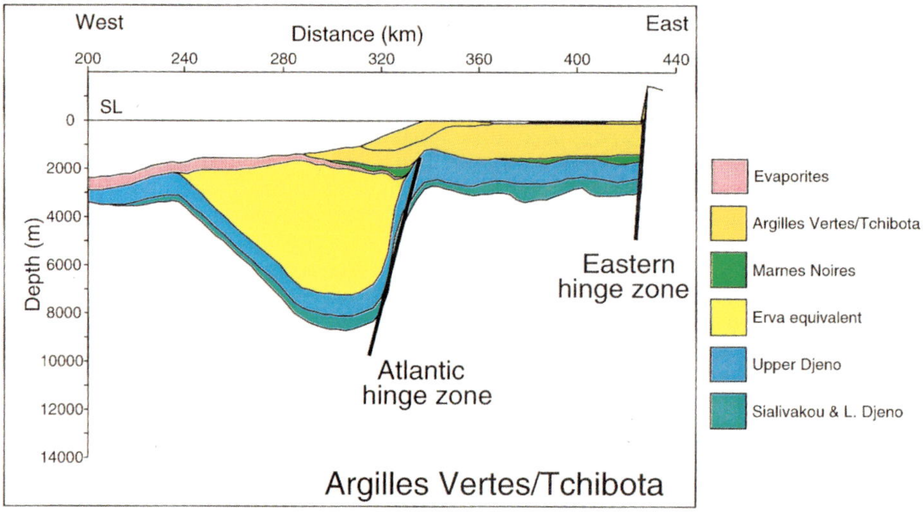

Fig. 9. Modelled basin geometry and time-line stratigraphy for the late syn-rift phase. The Argilles Vertes and Tchibota Formations consist of interacting deltaic deposits that partially infilled the region between the hinge zones following deposition of the Marnes Noires. Water depths continue to increase to the west of the Atlantic Hinge. Two unconformities exist within the post-Marnes Noires–pre-Loeme sediment package: (1) a sequence boundary, termed the pre-Chela unconformity; and (2) a 'ravinement' unconformity termed the Chela unconformity. The pre-Chela unconformity separates the Argilles Vertes from the Tchibota Formation and records non-deposition and sediment bypass of the Argilles Vertes Formation and subsequent onlap of the Tchibota onto the Argilles Vertes. Canyon cutting into the Argille Vertes and earlier formations is observed on seismic reflection sections indicating a relative sea-level fall during the development of this regional unconformity.

maximum extension and eventual oceanic crust emplacement appears to have stepped significantly westward (Fig. 1). Rift-induced uplift across the Atlantic Hinge Zone created shallow water conditions and the development of restricted carbonates (i.e. the TOCA Formation; Figs 7 and 8c; Lomando 1996; McHargue 1990). The magnitude of uplift of the Atlantic Hinge Zone during the third rift phase influenced the palaeo-environments developed and lithologies deposited across the Atlantic Hinge and between the hinge zones. For example, when the flank uplift is minor, deep and anoxic lakes tended to form between the hinge zones. Increasing the uplift across the Atlantic Hinge promoted either fringe reef development and/or subaerial exposure, resulting in denudation and clastic production. Subaerial exposure was also conducive to the development of secondary porosity, permeability and dolomitization of the TOCA carbonates (e.g. Lomando 1996). East of the Atlantic Hinge Zone, deposition was dominated by reworked siliciclastics deposited in a lacustrine environment, while west of the hinge zone, carbonates, organic-rich shales and interbedded sandstones accumulated as part of a westward-thickening wedge.

Along the Congo section of the Eastern Hinge Zone, Marnes Noires sediments (presumably more sand prone) prograded into lakes, downlapping onto the underlying Djeno Formation. The increasing complexity of ostracod form and shape observed within the Marnes Noires Formation reflects increasing lake salinity as a function of time (R. Bate, pers. comm.). This change in lake salinity suggests that either the lakes were drying out or that they were being contaminated by marine waters. Interestingly, unambiguous marine faunas do not appear until the development of the Chela unconformity and deposition of the Chela and Gamba Formations (Teisserenc & Villemin 1990). Although open marine conditions across the entire region were only established during the late Aptian–early Albian (Dale *et al.* 1995), it is possible that the lake systems between the Eastern and Atlantic Hinge Zones, being protected from open marine conditions by Atlantic Hinge topography and archipelagoes, can be influenced by marine waters without the accompanying biota (R. Bate & A. Lomando; pers. comm.).

In contrast to the restricted depositional environments between the hinge zones, significant accommodation was created seaward of the Atlantic Hinge Zone resulting in the deposition of 500–6000 m of late Barremian–early Aptian aged sediment, depending on the locality. For example, in southern Gabon, the easternmost unit of the Dentale Formation (the Crabe Facies) is a deltaic lacustrine shale having a maximum thickness of *c.* 600 m. The Dentale Formation records northwest to southeast sediment progradation along the Dianongo Trough (Fig. 1), the axis of a major pre-salt basin. In contrast, in regions away from the influence of accommodation zones, upwelling accentuated by the archipelagos of the Atlantic Hinge Zone allowed the development and deposition of offshore organic-rich shales, such as the Falcão source rock recognized in the Kwanza Basin (Henry *et al.* 1995), that are equivalent to the Marnes Noires Formation between the Atlantic and Eastern Hinge Zones and the Lagoa Feia source rock of the Brazilian Campos Basin. According to Henry *et al.* (1995), the depositional environment responsible for the development of the Falcão source rock is brackish to marginal marine.

Following the cessation of rifting and deposition of the Marnes Noires, residual accommodation was infilled by a series of regressive sequences with the first major marine incursion into the basin leading to the deposition of evaporites in areas that were suitably restrictive. The most important of the progradational packages are the Argilles Vertes and Dentale Formations (Figs 7 and 9). The pre-Chela unconformity separates the Argilles Vertes Formation from the Tchibota Formation. This sequence boundary records non-deposition and sediment bypass of the Argilles Vertes Formation and subsequent onlap of the lowstand and transgressive sequences of the Tchibota Formation onto the Argilles Vertes Formation (Fig. 7). Canyon cutting into the Argilles Vertes Formation, and earlier formations, is observed on seismic reflection sections indicating a relative sea-level fall during the development of this regional unconformity. The regional destruction of accommodation across the margin may be a combination of eustasy, correlating with a major late Barremian–early Aptian sea-level fall seen on the Haq *et al.* (1988) eustatic curve, and tectonic uplift associated with lower crustal and lithospheric mantle extension during this rift phase. Alternatively, this lowstand may be a function of decreasing fresh water input into the lake system, as a consequence of an increasingly drier climate at this time, consistent with increasing lake salinity. This hypothesis can be rejected because: (1) as noted by Siesser (1978), this time period in West Africa is temperate and dry, based on the abundance of plant remains and the low diversity of pollen species, i.e. there were no major fluctuations in climate at this time; (2) the large clastic wedges of the Argilles Vertes and Dentale Formations indicate that fluvial systems remained active at this time; (3) the Chela and Tchibota Formations contain marine fauna. These observations, the hydrothermal dolomitization of the TOCA carbonates and the marine affinity of the Marnes Noires equivalent Falcão source rocks of the Kwanza Basin, are interpreted as circumstantial evidence for the general establishment of marine conditions prior to oceanic crust emplacement to the west of the Atlantic Hinge Zone.

The ensuing flooding of the lowstand palaeorelief resulted in the formation of a minor erosional unconformity (the Chela 'ravinement' unconformity) between the hinge zones and, in some cases, out across the Atlantic Hinge Zone (Figs 7 and 9). The amount of erosion associated with either the pre-Chela or Chela unconformities might be overestimated in regions where the unconformities coalesce. As a consequence of this transgression, locally unfilled accommodation space created between the interacting deltaic systems became the depocentre for the lagoonal and evaporitic deposits of the upper Chela and Loeme formations. As would be expected, the Chela Formation (Gamba:Gabon/Chela: Cabinda and Congo) consists of basal conglomerates and sandstones overlain by lagoonal sediments (Fig. 7), i.e. a restricted marine environment.

Several observations lead to the conclusion that a mid-crustal detachment was operative during this third rift phase: (1) the development of the regional Chela unconformity and overlying fluvial sandstones at the base of the Gamba and Chela formations requiring an environment at or close to base level; (2) the transition from the deep water restricted lake environment of Melania Shale deposition to the regionally extensive, shallow water, restricted marine conditions that facilitated the development of evaporites of the Ezanga and Loeme formations, and (3) the development of significant post-rift accommodation (deposition of the late Cretaceous, Paleogene and Neogene formations) in the same region characterized by only minor brittle, syn-rift deformation of the upper crust and negligible erosional truncation of earlier syn-rift units. In order to collectively explain the above observations and borrowing from the results of Driscoll & Karner (1998) for the development of the northwestern Australian margin, we suggest that during this third rift phase, extension became partitioned across a crustal detachent, characterised by plastic thinning of the lower crust with brittle deformation of the upper crust occurring farther to the west near the incipient ocean–continent boundary. It is because of this extension

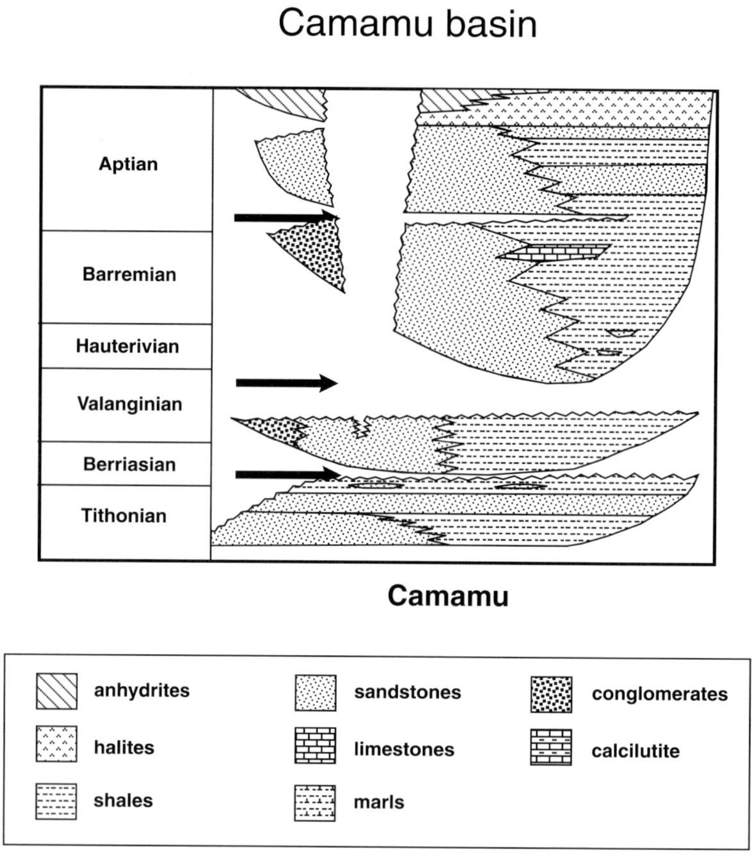

Fig. 10. Chronostratigraphy of the Camamu Basin of the eastern Brazilian Margin. Note that the unconformities bounding the syn-rift packages are late Berriasian, Valanginian–Hauterivian and late Barremian–early Aptian (bold arrows). Modified from Netto *et al.* (1994).

partitioning with depth that allows the rift process to continue in the absence of brittle deformation between, and immediately to the west, of the Eastern and Atlantic hinge zones. The detacment acts to maintian the region at sea level, making the margin extremely sensitive to small-scale eustatic fluctuations, thus enabling repeated marine incursion and desiccation cycles to build up thick salts similar to those of the Ezanga and Loeme formations. In effect, the third rift phase subsidence consists of two components, an initial phase associated with the localized, brittle deformation of the crust being replaced by a rapid but regional subsidence phase induced by the preferential plastic deformation of the lower crust with little involvement of the upper crust. We suggest that the smoothly varying form of the predicted crustal extension obtained by Karner *et al.* (1997) for the Congo margin, given the absence of observed crustal faulting, is a proxy for plastic deformation of the lower crust.

In summary, the formation of the West African Margin was the consequence of rift propagation both along and across the region as a function of time, resulting in the generation of two hinge zones, the Eastern and Atlantic Hinges. Rifting commenced in the Berriasian (143 Ma) and ceased in the early Aptian (118 Ma) with the emplacement of oceanic crust (although the oldest recognized magnetic anomaly is 34 (Santonian; 84 Ma). Rifting between Africa and Brazil occurred in three phases: (1) Berriasian–Hauterivian (143–134 Ma); (2) Hauterivian–late Barremian (134–126 Ma); (3) late Barremian–early Aptian (126–118 Ma). Within the resolution of inter-Atlantic palaeontological correlations, the eastern Brazilian rift basins (Jequitinhonha to Sergipe–Alagoas) show the same general structuring (e.g. the double hinge zone of the Sergipe–Alagoas Basin) and timing of rift events, although the relative importance of each event differs along the margin. In particular, the Camamu and Almada Basins are characterized by three rift packages bounded by major unconformities, the age of which are late Berriasian, Valanginian–Hauterivian, and late Barremian–early Aptian (Fig. 10; Netto *et al.* 1994).

Post-rift margin development: the establishment of drainage networks and the relationship between large fluvial systems and canyon development

Although the post-rift development of the West African and eastern Brazilian Margins was dominated by carbonates, major rivers delivered clastics at select points along the margin. The volume and distribution of clastic delivery to the evolving margin was basically independent of the processes responsible for margin subsidence. The stratal geometry and facies architecture preserved along continental margins record the complex interplay between tectonics, eustatic variations, sediment supply, physiography and climate (Vail *et al.* 1977; Posamentier *et al.* 1988). However, these processes that shape continental margins are not independent of one another and many feedbacks occur (Christie-Blick 1991). Accommodation is a critical factor governing facies and stratal architecture in non-marine environments and sediment delivery to the margin. Non-marine accommodation is created or destroyed when the top of the fluvial–coastal plain system subsides below or is uplifted above the 'theoretical equilibrium profile' (e.g. Posamentier *et al.* 1988). The theoretical equilibrium profile is defined as a graded or dynamic surface upon which there is no net sediment accumulation or erosion (Mackin 1948), and this concept is the foundation for many sequence stratigraphic models that predict how relative sea-level fluctuations affect fluvial processes (e.g. Posamentier *et al.* 1988). Coastal plain tilting induced by regional or local tectonics (e.g. Harris & Laws 1997), in-plane force variations (Karner *et al.* 1992) and sediment loading/unloading (e.g. Whiting *et al.* 1994; McGinnis *et al.* 1993) can potentially cause a base-level change across the coastal plain, thus modifying the equilibrium profile. This change in accommodation affects the lower reaches of the fluvial network which, in turn, modifies the volume and distribution of sediment input to a margin.

Continental-scale river migration: the antecedent Congo River System

The antecedent Congo River System appears to have had a fundamental control on the post-rift history of the West African Margin. In particular, the present-day drainage system and divides of West Africa (Fig. 11) suggest that the antecedent Congo River System migrated southward as a function of time, as evidenced by the age of the major unconformities developed across the margin; Gabon–Senonian and Congo–Oligocene (Teisserenc & Villemin 1990). If there is a feedback between sediment loading and flexural deformation of the basement (e.g. Driscoll & Karner 1994), then a relationship exists between the input of clastics and the formation and timing of major canyon downcutting and erosional unconformities along the margin, especially when accentuated during lowstands. Perhaps more importantly, this hypothesis implies a correlation between the timing of reservoir formation and the migration of the antecedent Congo River southward along the margin. Consequently, there may be a tectonic control on the time-transgressive nature of deep-water turbidite sands along margins in general. The behaviour of the Kwanza River System and its antecedents may play a similar role in the development of unconformities along the northern Angola Margin.

The present-day Congo Canyon System is associated with the Congo (Zaire) River network, which drains a major portion of the African continent and has the largest water discharge in Africa (Fig. 11). At present, the headwaters of the river system originate near the East African Rift System and the high topography of southern Africa. The onset of deformation in the East African Rift System and the inferred development of the Congo Cone are approximately coeval, occurring sometime in middle–late Oligocene time (e.g. Ebinger *et al.* 1991). This apparent synchroneity led many researchers to conclude that the formation of the Congo River was directly linked to the rift-flank uplift associated with extensional deformation in East Africa. However, the present-day topography indicates the possible location of antecedent lower tributaries of the Congo River (Fig. 11, purple boxes). The boxes enclose a number of abandoned river channels linking the Congo River to the Ogooué, Kouilou and present-interior Congo drainage systems. An antecedent Congo River System appears to have drained large portions of the African continent, flowing north of its present location and discharging near Port Gentil, northern Gabon, during the late Cretaeous and early Tertiary. The magnitude of late Cretaceous erosion and downcutting along the outer Gabon shelf and slope, and the immense size of the late Cretaceous–early Tertiary delta, cannot be explained by the present-day coastal river sys-

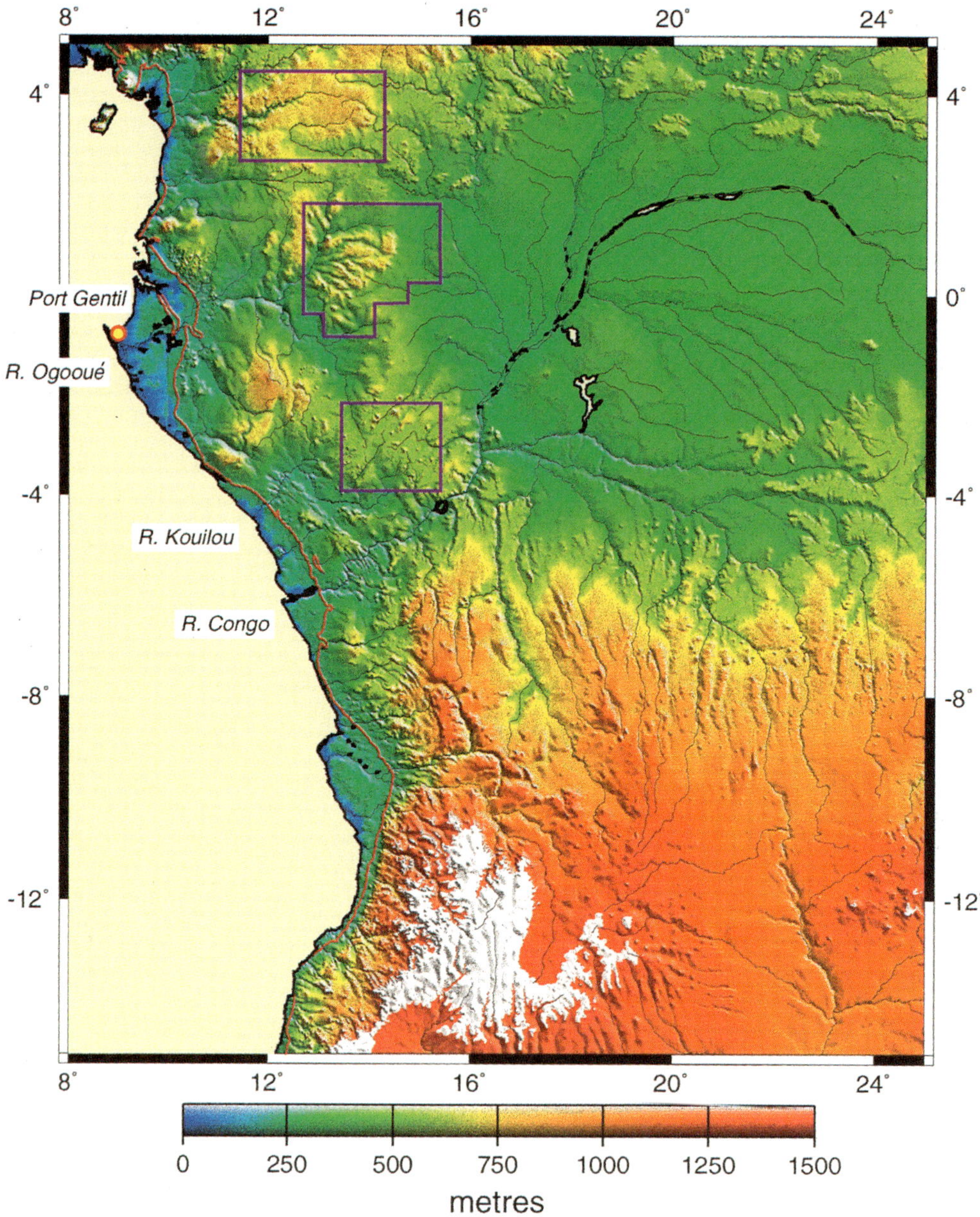

Fig. 11. Present-day topography of the West African region constructed from a 30 × 30 arc-second DEM of Africa. This DEM was generated by the Defense Mapping Agency using 1:1 000 000 Operational Navigational Charts, point elevation and hydrology data. Shown are the three main rivers that presently empty along the West African Margin; Ogooué, Kouilou and Congo. Shown are a series of potential abandoned channels that imply a connectivity between these contemporary rivers and the interior Congo River System (purple boxes). The present-day drainage system and divides suggest that these channels were once part of the antecedent Congo River System that migrated southward as a function of time as evidenced by the age of the major unconformities developed across the margin; Gabon–Senonian and Congo–Oligocene. The magnitude of the late Cretaceous erosion and downcutting along the outer Gabon shelf and slope and the size of the large late Cretaceous–early Tertiary delta cannot be explained by the present-day coastal river systems alone, suggesting that the Ogooué River System is a misfit stream. River migration along the Gabon–Congo Margin is consistent with the age of the major unconformities and turbidite deposits developed along the margin – Gabon–Senonian and Congo–Oligocene – implying a correlation between the timing of reservoir formation and the migration of the Congo drainage system.

tems alone. Examination of remote sensing (LANDSAT, DTM and SAR images) and geological maps suggests that the abandoned river channels once used by the antecedent Congo River are now occupied by the present Ogooué and Kouilou River Systems (Fig. 11; centre and lower boxes), implying that the present rivers are misfit streams within the valley they now occupy.

Possible explanations for why the antecedent Congo River abandoned this path for southerly routes to the sea are not clear, although sediment loading and flexural uplift of the hinterland associated with the huge Ogooué Delta of central Gabon may have been sufficient to redirect the course of the river. Preliminary evidence for our hypothesis is found in seismic reflection data across the West African continental slope and shelf, which shows that certain regional unconformities systematically change their character along the margin through time. For example, the Senonian unconformity is characterized by erosional downcutting and incision along northern Gabon in the vicinity of the present-day Ogooué River (Teisserenc & Villemin 1990). In contrast, the Congo region is characterized by either non-deposition and bypass, or carbonate platform construction at this time. The opposite pattern is observed for the Oligocene. In the north, a minor Oligocene unconformity registers non-deposition and bypass and, towards the south along the Congo Margin, the unconformity is characterized by extreme erosional downcutting and incision. The erosional downcutting is interpreted to be indicative of where large river systems were located along the margin during falls of relative sea level, suggesting that the antecedent Congo River mouth was located in northern Gabon in Senonian time and migrated to its present location during the late Palaeogene. Furthermore, the development along the northern Gabon Margin (Port Gentil) of a large late Cretaceous–early Cretaceous delta, which is absent along the Congo Margin, is also consistent with this hypothesis.

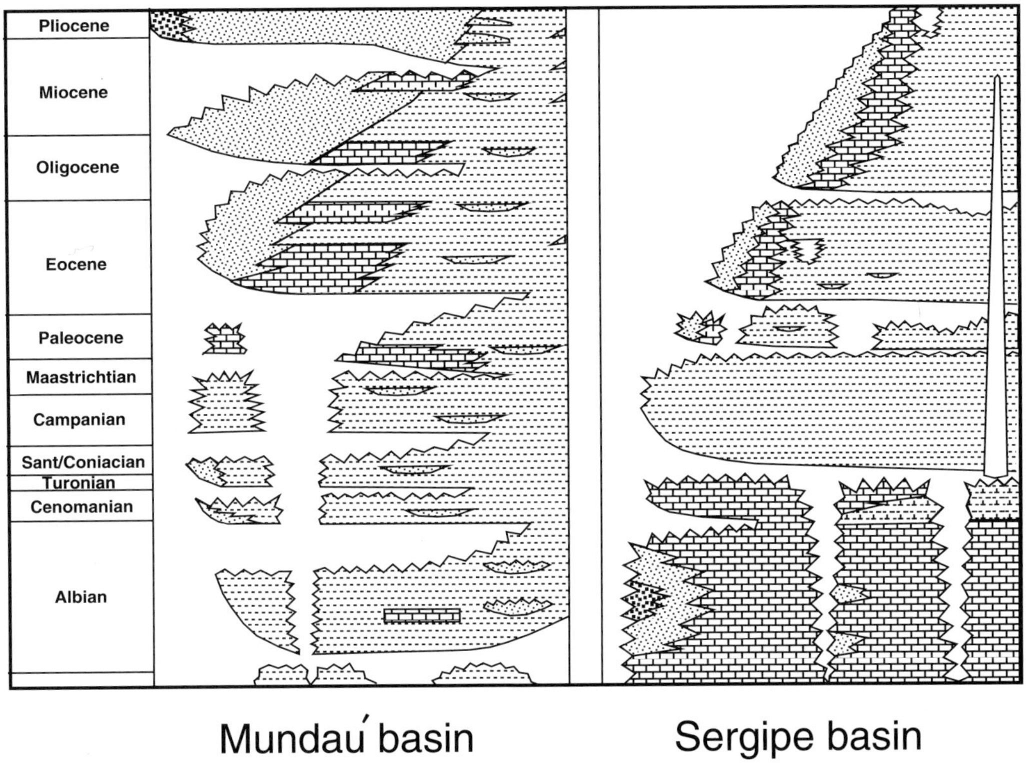

Fig. 12. Chronostratigraphy of the Mundaú and Sergipe Basins of the northeastern and eastern Brazilian Margins, respectively. It is assumed that the abrupt mid-Eocene disappearance of canyon systems on the Mundaú Margin indicates a major decrease in the importance of fluvial systems on this margin. In contrast, continued shelf incision and turbidite production on the Sergipe Margin suggests that fluvial systems always played a dominant role in the development of the margin. Modified from Beltrami *et al.* (1994) and Feijó (1994).

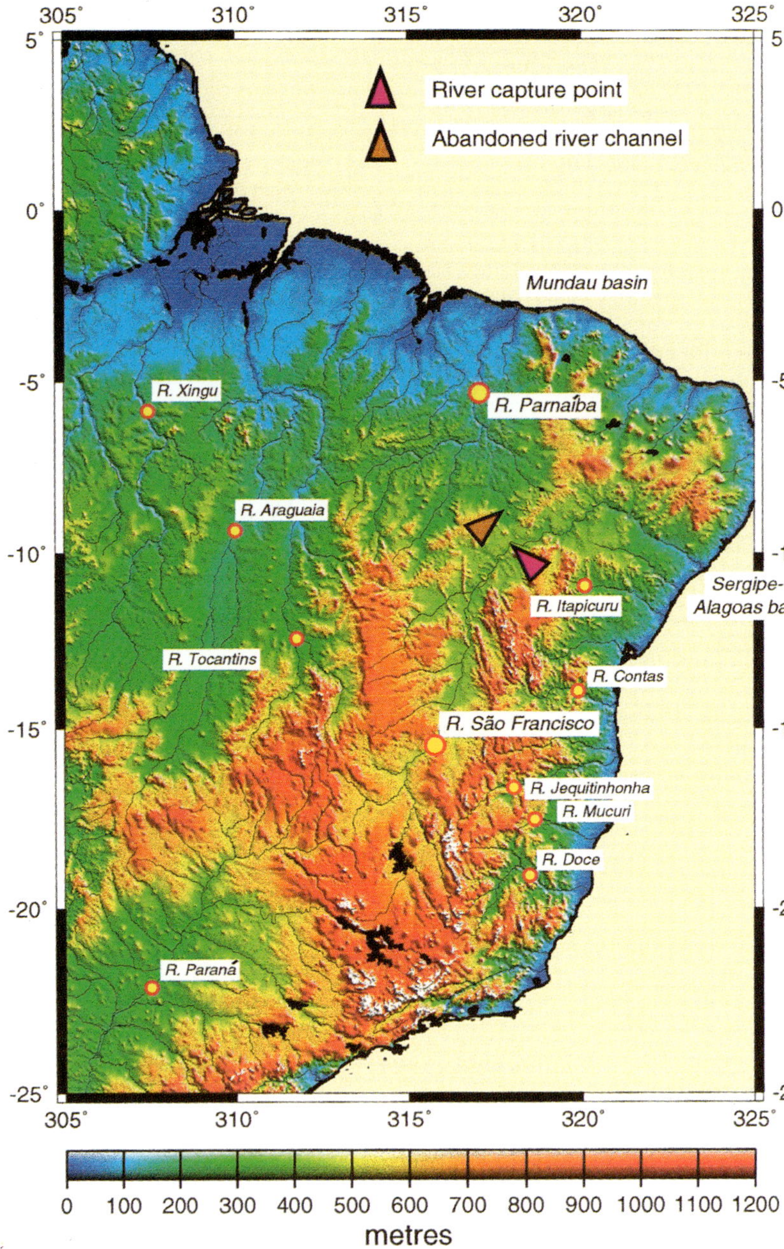

Fig. 13. Present-day topography of the eastern Brazil region constructed from a 30 × 30 inch arc-second DEM of South America. This DEM was generated by the Defense Mapping Agency using 1:1 000 000 Operational Navigational Charts, point elevation and hydrology data. Shown are the main north–south continental (Araguaia, Tocantins, Parnaíba and Paraná) and east–west Atlantic (São Francisco, Itapicuru, Contas, Jequitinhonha, Mucuri and Doce) rivers that presently empty along the margin. Shown also are abandoned channels that show a connectivity between the Parnaíba and São Francisco Rivers (brown arrows). Capturing of the Parnaíba River by the São Francisco River in the mid-Eocene abruptly reduced clastic delivery and the frequency of turbidites to the Mundaú Basin. Continental drainage modification may directly influence both clastic delivery and turbidite frequency to the margin over scales of thousands of kilometres.

Intra-continental drainage development and modification: the dualling Parnaíba and São Francisco River Systems

Continental landscapes are constantly in a state of flux as drainage systems evolve and interact. This evolution leads to the transport and delivery of clastics by fluvial systems to the margins. Rivers therefore represent point sources of clastics along margins. Once deposited, marine processes, such as longshore drift, rework and redistribute the sediments along the margin. The three-dimensional headward advance and lateral migration of continental drainage systems implies that the distribution of rivers are likewise in a state of flux. In particular, the process of river capture in which the headwaters of one drainage system are captured or pirated from another, leading to stream rejuvenation at the expense of the other, is undoubtedly a common but poorly documented process controlling both landscape evolution and the history of clastic delivery to margins. The impact of continental scale river capture in controlling the volume, timing and location of clastic delivery to the basins comprising the eastern Brazilian Margin using the Mundaú (northeastern margin) and Sergipe (eastern margin) Basins as examples (Figs 12 and 13) will now be explored.

A proxy is required that allows recognition of the existence of major rivers from the chronostratigraphy of a margin. It is noted that the erosional and depositional interaction between rivers and the shoreline are accentuated by changes in relative sea level. In particular, during a sea-level fall, margins characterized by relatively large fluvial systems will tend to be associated with significant river incision of the exposed shelf and well-developed canyon development and associated fan deposits downslope of the shelf–slope break. Given suitable conditions, this setting will be conducive to the generation of turbidities. In contrast, margins characterized by minor rivers will tend to result in non-deposition and bypass of the exposed shelf with little to no downslope canyon development. Further, canyons that do develop will tend to be relatively small and show little relationship to the location of the onshore streams, being more a function of *in situ* failure conditions rather than erosive instabilities caused by fluvial incision. Thus, it is assumed that margins showing major canyon development and erosion are associated with large fluvial systems. For example, the chronostratigraphic charts of the Mundaú (northeastern Brazilian; Beltrami *et al.* 1994) and Sergipe (eastern Brazilian; Feijó 1994) Margins suggest that their development has been dominated by fluvial activity (Fig. 12). In particular, the Parnaíba River empties directly onto the Mundaú Margin while the large São Francisco River disgorges onto the Sergipe Margin (Fig. 13). Surprisingly, while fluvial incision had modified the stratigraphy of the Sergipe Basin throughout its development, fluvial influence dramatically ceased in the Mundaú Basin during the Eocene (Fig. 12).

From the drainage distribution of eastern Brazil it is noticed that there are two continental drainage systems (Fig. 13). Firstly, an approximately east–west drainage system along the eastern Brazilian Margin, of which the lower reaches of the São Francisco River are part. Other major rivers with this trend are the Itapicuru, Contas, Mucuri, Doce and Jequitinhonha. In marked contrast, further west, away from the localized eastern drainage, the continent is drained by north–south flowing rivers, such as the Tocantins, Araguaia, Xingu and Paraná. It is interesting to note that the upper reaches of the São Francisco River parallel this trend whereas the lower reaches parallel the 'eastern Atlantic' trend. Further, compared with the length of other north–south continental Brazilian rivers, the Parnaíba River appears to be truncated (Fig. 13). It is postulated that the Parnaíba and São Francisco Rivers are both long-lived (late Cretaceous–Present) river systems, and that the São Francisco River captured the Parnaíba River midlength, dramatically accentuating waters to the São Francisco system while significantly reducing the flow of the Parnaíba River. The point of capture is identified in Fig. 13, where both an inflexion in the course of the São Francisco River and an abandoned river channel immediately north of the postulated capture point can be observed. This abandoned channel, it is suggested, is the old course of the Parnaíba River. From the chronostratigraphy of the Mundaú and Sergipe Basins (Fig. 12), the timing of capture and modification of the Parnaíba drainage system was probably during the middle Eocene. As a direct consequence of this stream reconfiguration, clastic delivery and margin incision of the Mundaú Margin was abruptly curtailed, reducing significantly the frequency of turbidite input to the outer slope during lowstands. In contrast, turbidite frequency increased markedly in the Sergipe Basin at the same time (e.g. Lana 1990). If this hypothesis is correct, then the timing of capture and modification of the drainage systems directly influences both clastic delivery and turbidite frequency to the margin.

Coastal v. inland river interactions: asymmetric clastic delivery to the margin

The orthogonal interaction of the São Francisco

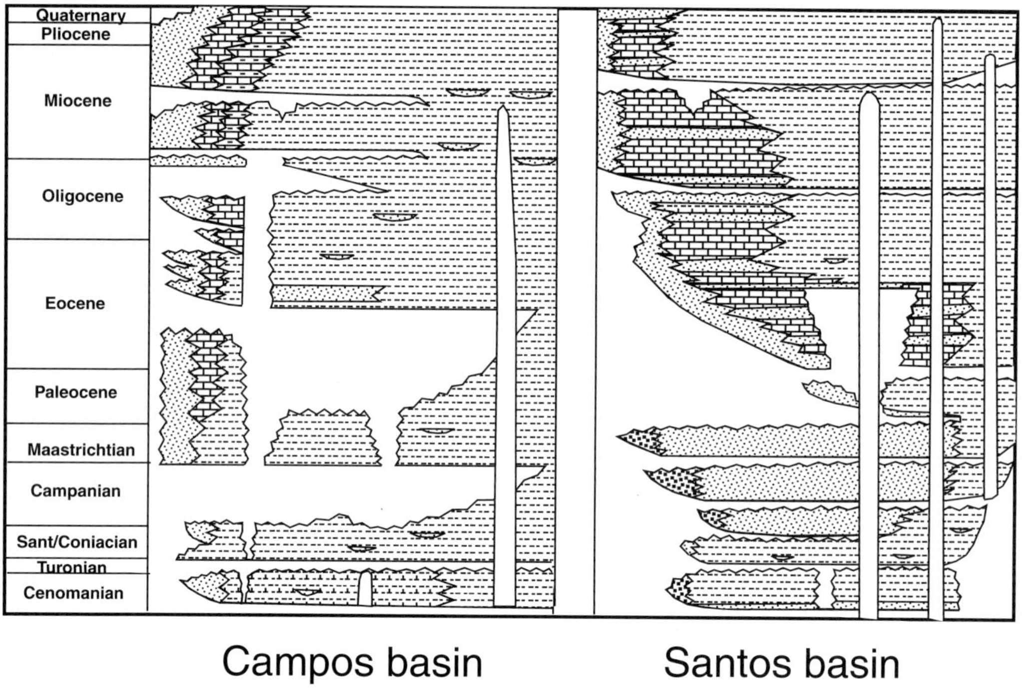

Fig. 14. Chronostratigraphy of the Campos and Santos Basins of the southeastern Brazilian margin. Note the asymmetry in the mega sequences of each basin; while conglomerates and sands were being deposited across the Santos Basin (i.e. Santos Formation), the Campos Basin was undergoing bypass and erosion of the shelf (lower Campos Formation). From the Tertiary–Present, the depositional systems slowly but progressively changed as the Campos Basin shoreline and associated facies stepped basinward while the southerly, but adjacent, Santos Basin shoreline and associated facies stepped landward. This depositional pattern is most simply explained in terms of variations in sediment supply to the respective margins. Modified from Rangel *et al.* (1994) and Pereira & Feijó (1994).

and Parnaíba Rivers involved the capture and reconfiguration of drainage systems over continental scales (i.e. 500–1000 km). A smaller scale, but no less dramatic, example of river capture deals with the erosional history of the Serra do Mar and Serra da Mantiqucira Mountains of southeast Brazil, and the development of the adjacent offshore Campos and Santos Basins (Figs 14 and 15). The southeastern Brazilian highlands consist of two components (Fig. 15): (1) a broad plateau with a width of 1500–2000 km and a relief ranging from 500 to 1500 m relative to sea level; (2) a region of coastal topography represented by a combined Serra do Mar and Serra da Mantiqueira with a width of 100–200 km and a local relief of 500–2800 m. This marginal topography is most simply explained as part of the rift flank produced in response to the extension that formed the Campos and Santos Basins (Weissel & Karner 1989; Gallagher *et al.* 1994). The present Serra do Mar is the erosional coastal remnant of a scarp retreat process initiated during Neocomian rifting. According to Karner *et al.* (1993), the southeastern highlands are a consequence of either (1) magmatic underplating associated with the emplacement of the Serra Geral basalts (at 133 ± 5 Ma); and/or (2) residual lithospheric heat introduced by the same mantle plume responsible for the tholeiitic basalts.

The Campos and Santos Basins were formed in the early Cretaceous during the break-up of Africa and South America. The tripartite rift history that characterizes the northern rift basins of the West African and eastern Brazilian Margins is not observed, possibly because of complications introduced by the intrusive and extrusive tholeiites of the Serra Geral Formation. Because the rift sequences of the Santos and Campos Basins are built directly on an early Cretaceous volcanic basement, and that the syn-rift sequence is represented solely by the Barremian–lower Aptian Lagoa Feia Formation, it is possible that only the last rift phase responsible for the structuring of the margin is being observed. The implication is that the sub-basins produced by

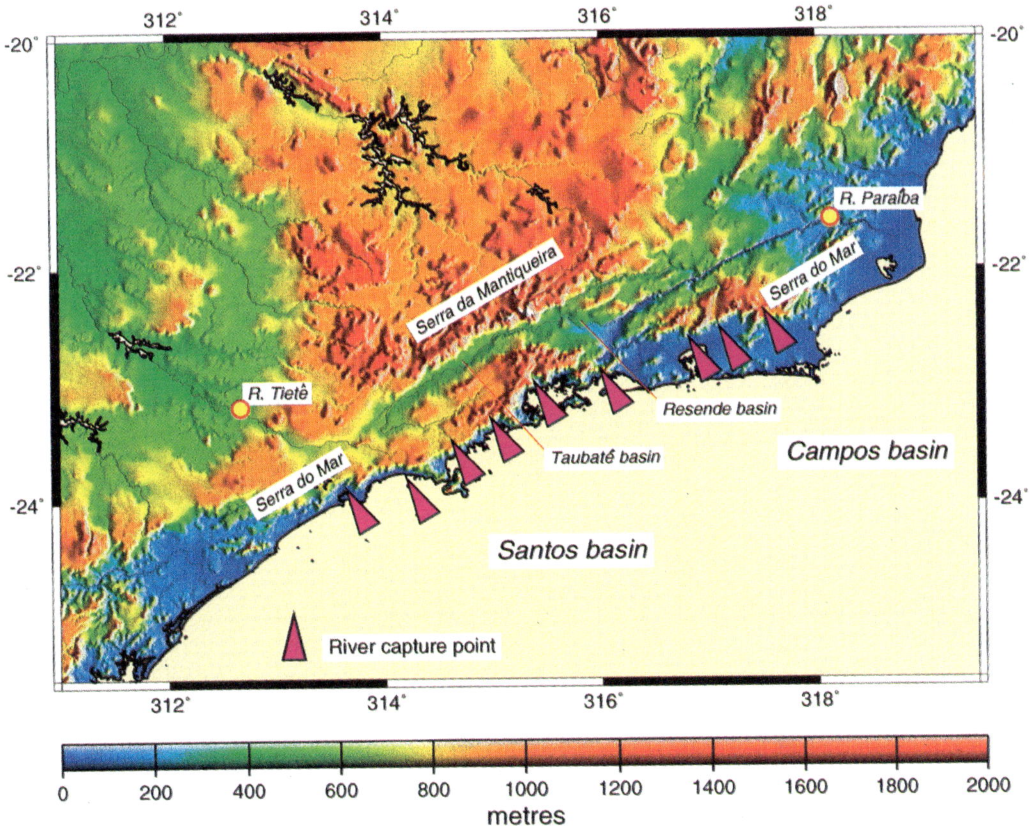

Fig. 15. Present-day topography of the southeastern Brazil region constructed from a 30 × 30 arc-second DEM of South America. This DEM was generated by the Defense Mapping Agency using 1:1 000 000 Operational Navigational Charts, point elevation and hydrology data. Shown also are the abandoned or misfit consequent streams (flowing down the rift flank of the Serra do Mar into the Santos Basin; cyan arrows) and the subsequent Paraíba River flowing parallel to the Serra da Mantiqueira. The change in sediment supply to the Santos and Campos Basins was a result of southwest headward propagation of the Paraíba River and progressive capture of the consequent Serra do Mar streams (capture points identified by the cyan arrows). As such, headwaters and sediments were progressively robbed from the Santos Basin while concomitantly augmenting clastic delivery to the Campos Basin.

earlier rift phases, if they exist, have been infilled primarily by sub-aerial(?) basalts with intercalated sediments. Irrespective of this, attention is drawn to the regional Tertiary post-rift stratigraphic development of the Campos (Rangel *et al.* 1994) and Santos (Pereira & Feijó 1994) Basins and, in particular, the deposition of the Campos, Santos and Iguape Formations (Fig. 14). The stratigraphic development of the basins show an interesting asymmetry; while conglomerates and sands were being deposited in the Santos Basin (i.e. Santos Formation), the Campos Basin was undergoing bypass and erosion of the shelf (lower Campos Formation). Starting in the Tertiary, the depositional system slowly, but progressively, changed as the Campos Basin shoreline and associated facies stepped basinward, while the southerly but adjacent Santos Basin shoreline and associated facies stepped landward. This synchronous reciprocal relationship of the Campos and Santos Basins cannot be explained by eustasy alone and most likely reflects variations in sediment supply to the respective basins.

This progressive change in the sediment supply to the Campos and Santos Basins, following from the work of Azevedo (1991), is a function of the efficiency of the consequent streams (i.e. those streams following the initial slope of the topography and, in this case, flowing east) to become established along the coastal or Serra do Mar Ranges v. the efficiency of the subsequent streams (i.e. streams developing along zones of weakness) to incise and become established inland but parallel to the Serra do Mar. Production of the rift flank topography allowed small consequent streams to develop down the escarpment, directly delivering coarse clastics to both the Santos and Campos Basins. However, the Proterozoic granitic gneisses that comprise the Serra do Mar topography proved difficult to erode, such that the consequent streams failed to establish a well-connected drainage system. During this time, subsequent streams began to develop along a series of small high-altitude rift basins inland of the Serra do Mar (Tabuaté and Resende Basins; Fig. 15), exploiting the regional fabric of the exposed Proterozic terranes (e.g. Mohriak *et al.* 1990). These streams, as they coalesced by headward advance, widened their valleys and, in doing so: (1) initiated a second westward-migrating slope that formed the Serra da Mantiqueira Escarpment; and (2) progressively captured the headwaters of the consequent streams feeding the Santos Basin. The resulting subsequent stream became the Paraíba River and, though the progressive capturing of consequent stream watersheds, increased clastic delivery to the Campos Basin. The progressive decrease in sediment supply to the Santos Basin and the concomitant increase of clastic delivery to the Campos Basin during the Tertiary is therefore interpreted to be a consequence of this fluvial interaction. Figure 15 identifies the sites of river capture along the drainage divide of the Serra do Mar, the arrows marking the position of abandoned river channels and/or misfit streams. Prior to the development of the Paraíba River these, now abandoned, river channels were relatively major consequent streams feeding clastics to the Santos Basin. Their decapitation by the southward-migrating Paraíba River robbed headwaters and clastics from the Santos Basin and redirected them to the Campos Basin.

Conclusions

By applying an integrated basin modelling approach, that combines quantitative kinematic and isostatic basin modelling, with the principles of seismic sequence stratigraphy, the tectonic significance of the major stratigraphic packages on the West African and eastern Brazilian Margins has been determined. In addition, the rift history of the respective margins has been refined. From this work the following conclusions are drawn.

- The formation of the Congo Margin was the consequence of rift propagation across the region as a function of space and time. Rifting commenced in the Berriasian and terminated in the late Barremian–early Aptian with the emplacement of oceanic crust. Rifting between Africa and Brazil occurred in three phases: Berriasian, Hauterivian and late Barremian–early Aptian. Initial broadly-distributed continental rifting became more focused with time as extension migrated to the west. The abrupt jumps in the spatial location of extension as a function of time were responsible for the development of a double hinge zone system (the Eastern and Atlantic Hinge Zones). The Eastern Hinge Zone demarcates the eastern limit of Berriasian extension whereas the Atlantic Hinge Zone marks the eastern limit of Hauterivian extension. The hinge zones represent rift flanks associated with crustal unloading during the formation of the onshore and outer rift basins. The eastern Brazilian margin shows the same sequence of events, both in space and time. Along-strike variations in the distribution and amplitude of continental extension are responsible for both the segmentation of the Atlantic Hinge and offsets in the ocean–continent boundary.
- The segmented character of basin-bounding faults, the interplay between along-strike footwall–hanging wall alternation and the rift flank topography engendered by crustal unloading (e.g. hinge zones) controls the spatial distribution of river drainage systems and their access to the newly formed rift basins. These fluvial systems control the delivery of clastics to the evolving basin and hence influence both source and reservoir rock quality. The segmentation of the Eastern and Atlantic Hinge Zones controlled where clastic sediment gained access to the onshore and offshore basins. The Atlantic Hinge Zone is a series of short-segment, en-echelon high-standing blocks. Landward of the hinge zone there are a series of local depocentres, while seaward of the hinge individual rift basins tend to show an en-echelon arrangement but with limited along-strike continuity. Similarly, the ocean–continent boundary is highly segmented, being offset by numerous fracture zones. The Brazilian equivalents, the western/onshore and offshore Hinge Zones, also controlled clastic input to the basins. In contrast to the Atlantic Hinge Zone, the Brazilian offshore hinge zone appears to be co-linear and significantly less segmented. While the hinge zone is continuous, the Brazilian subbasins and their associated border faults are en-echelon. A major accommodation zone, the width of the Sergipe Basin, occurs at the intersection of the Alagoas Hinge Zone and

the Maceió Fracture Zone. This fracture zone also appears to have facilitated differential extension between the Sergipe and Alagoas Basins, the extension perhaps being augmented by syn-rift hotspot activity.

- Each rift event is characterized by seismic reflectors that exhibit rotation and divergence, indicative of differential subsidence and block rotation. The divergent reflectors generally onlap onto a single surface, the rift-onset unconformity, suggesting that the rifting process was rapid. The stratigraphic response to these rifting events was a number of syn-rift packages comprising a condensed section followed by regressive units representing the subsequent infilling of the rift-induced accommodation. Each rift phase resulted in the development of deep, anoxic, lacustrine systems. The eastern Brazilian rift basins (Jequitinhonha to Sergipe–Alagoas) show the same general structuring and timing of rift events as seen on the West African Margin, although the relative importance of each event differs along the margin.
- Based on seismic reflection data from the West African margin, the third and final phase of normal faulting and block rotation occurred during the late Barremian–early Aptian. It was this third rifting event that created the necessary restrictive environments and space for the various source rocks to be deposited along the Gabon–Cabinda Margin, both between the Eastern and Atlantic Hinge Zones (e.g. Marnes Noires Formation) and seaward of the Atlantic Hinge (e.g. Falcão source rock equivalent). Post-Marnes Noires and equivalent formations are unfaulted implying that the brittle extension had ceased on this margin at this time. However, the subsequent depositional packages and the nature of their bounding unconformities suggests that extension had not ceased. For example, the transition from the deep water restricted lake environment of the Marnes Noires and Melania Shale formations to the regionally extensive, shallow water, restricted marine conditions that facilitated the development of evaporites of the Ezanga and Loeme formations was replaced by slow subsidence and sediment starvation along the entire west African margin. This sequence of events is not diagnostic of post-rift subsidence. Further, two unconformities occur within the post-Marnes Noires/pre-Loeme sediment package; a sequence boundary termed the pre-Chela unconformity associated with an early Aptian relative sea-level fall, and the younger Chela unconformity produced during the ensuing flooding of the margin. The development of the regional Chela unconformity and the overlying fluvial sandstones at the base of the Gamba and Chela formations require a depositional environment at or close to lake level. These observations suggest that during the third rift phase, extension became progressively partitioned with depth, with plastic deformation thinning the lower crust while the brittle deformation of the upper crust was confined to the incipient ocean-continent boundary located farther west. It is because of this extension partitioning with depth that allows the rifting process to continue in the absence of brittle deformation between, and immediately to the west of, the Eastern and Atlantic hinge zones. This style of extension maintains the region near sea level, making the margin extremely sensitive to small-scale eustatic fluctuations. Repeated marine incursion and desiccation cycles can potentially form the thick salts represented by the Ezanga and Loeme formations.
- The process of river capture in which the headwaters of one drainage system are captured from another appears to have played a major role in controlling landscape evolution and development of both the West African and Brazilian Margins. Examination of remote sensing (LANDSAT, DTM and SAR images) and geological maps suggests that the abandoned river channels once used by the antecedent Congo River are now occupied by the present Ogooué River (Fig. 11, centre box), implying that the present Ogooué River is a misfit stream within the river valley it now occupies. Abandoned river channels linking the Congo River to the Ogooué, Kouilou and present Congo drainage systems are also evident. In particular, an antecedent Congo River System appears to have drained large portions of the African continent, flowing north of its present location and discharging near Port Gentil, northern Gabon, during the late Cretaceous and early Tertiary. The timing is inferred from the distribution, intensity and timing of erosion responsible for the major unconformities existing in the preserved sedimentary section. If these abandoned river channels were active in the late Cretaceous–early Tertiary, then the outflow of the antecedent river system was spatially coincident with the regions where offshore Senonian erosional downcutting and incision are observed. Slope modification of the hinterland region (i.e. the site of the observed abandoned channels), perhaps in response to flexural uplift engendered by the sediment load of the

delta, forced the antecedent Congo River System to migrate southwards as a function of time. This migration along the Gabon–Congo Margin is consistent with the age of the major unconformities and turbidite deposits developed along the margin– Gabon–Senonian and Congo–Oligocene, implying a correlation between the timing of reservoir formation and the migration of the Congo drainage system.

- By assuming that, during a sea-level fall, margins characterized by relatively large fluvial systems will tend to be associated with shelf incision and well-developed canyon development, the chronostratigraphy of Mundaú and Sergipe Basins has been used to understand the interaction and river capture between the Parnaíba and the São Francisco Rivers. The Parnaíba River of eastern Brazil, being the major source of clastics for the northeast Brazilian Margin (Mundaú Basin), was captured by the São Francisco River mid-length in the mid-Eocene, abruptly reducing clastic delivery and the frequency of turbidites to the Mundaú Basin. Local shelf incision and turbidite delivery to the offshore regions of the margin stopped abruptly. In contrast, this time period saw a major increase in clastic delivery to the Sergipe Basin.
- Late Cretaceous–Tertiary megasequences within the Santos and Campos Basins of southeast Brazil show a reciprocal relationship; while conglomerates and sands were being deposited in the Santos Basin (i.e. Santos Formation), the Campos Basin was undergoing bypass and erosion of the shelf (lower Campos Formation). Starting in the Tertiary, the depositional system slowly, but progressively, changed as the Campos Basin shoreline and associated facies stepped basinward while the southerly, but adjacent, Santos Basin shoreline and associated facies stepped landward. This synchronous reciprocal relationship of the Campos and Santos Basins cannot be explained by eustasy alone and most likely reflects variations in sediment supply to the respective basins. The change in sediment supply, it is suggested is a result of southwest headward propagation of the Paraíba River, a subsequent river that emptied into the Campos Basin and flowed parallel to the Serra da Mantiqueira, progressively captured the consequent easterly and southeasterly flowing streams that flowed down the rift flank topography (i.e. the Serra do Mar) of the Santos and Campos Basins. This river capture process progressively robbed clastics from the Santos Basin while concomitantly augmenting clastic delivery to the Campos Basin.

The authors acknowledge stimulating discussions and financial support from members of the West African Continental Margin (WACOMA) consortium: ARCO, Amoco, British Gas, CONOCO, Elf-Aquitaine, Geco-Prakla, Mobil, Occidental, TGS-Calibre and Texaco. I. Davison, J. F. Dewey and N. Cameron are thanked for their critical review of the manuscript. The authors would also like to express their thanks for the support and patience of the editors of this volume; N. Cameron, R. Bate and V. Clure. The Wessel and Smith GMT software was used extensively in the construction of the figures shown in this paper. Lamont-Doherty Earth Observatory publication #5804. Woods Hole Oceanographic Institution publication #9748.

References

AZEVEDO, M. F. 1991. *Evolução cenozóica das bacias de Campos e Santos suas implicacões tectônicas e geomorfológicas.* MSc Thesis, Universidade Federal de Ouro Preto, Brazil.

BELTRANI, C. V., ALVES, L. E. M. & FEIJÓ, F. J. 1994. Bacia de Ceará. *Boletim de Geosciências Petrobrás*, **8**, 117–125.

BRACKEN, B. R. 1994. Syn-rift lacustrine beach and deltaic sandstone reservoirs – pre-salt (Lower Cretaceous) of Cabinda, Angola, west Africa. *In:* LOMANDO, A. J. SCHREIBER, B. C. & HARRIS, P. M. (eds) *Lacustrine Reservoirs and Depositional Systems.* SEPM Core Workshop, No. 19, 173–200.

BRADLEY, C. A. & FERNANDEZ, M. N. 1992. Early Cretaceous paleogeography of Gabon/northeastern Brazil: a tectono-stratigraphic model based on propagating rifts. *Géologie Africaine: Coll. Géol. Libreville*, recueil des Communic., 17–30.

CAIXETA, J. M., BUENO, G. V., MAGNAVITA, L. P. & FEIJÓ, F. J. 1994. Bacias do Recôncavo, Tucano e Jatobá. *Boletim de Geosciências Petrobrás*, **8**, 163–172.

CANDE, S. C., LABRECQUE, J. L. & HAXBY, W. F. 1988. Plate kinematics of the South Atlantic: Chron 34 to present. *Journal of Geophysical Research*, **93**, 13 479–13 492.

CHRISTIE-BLICK, N. 1991. Onlap, offlap, and the origin of unconformity bounded depositional sequences. *Marine Geology*, **97**, 35–56.

—— & DRISCOLL, N. W. 1995. Sequence stratigraphy. *Annual Review of Earth and Planetary Sciences*, **23**, 451–478.

DALE, C. T., LOPES, J. R. & ABOLIO, S. 1992. Takula oil field and the greater Takula area, Cabinda, Angola. *In:* HALBOUTY, M. T. (ed). *Giant Oil and Gas Fields of the Decade 1978–1988*, American Association of Petroleum Geologists Memoir, **54**, 197–215.

DRISCOLL, N. W. & HOGG, J. R. 1995. Statigraphic response to basin formation: Jeanne d'Arc Basin, offshore Newfoundland. *In*: LAMBIASE J. J. (ed.) *Hydrocarbon Habitat in Rift Basins*. Geological Society, London, Special Publications, **80**, 145–163.

—— & KARNER, G. D. 1994. Flexural effects of

Amazon fan loading: a potential tectonic feedback mechanism on depositional processes. *Geology*, **22**, 1015–1018.

—— & —— 1998. Lower crustal extension across the northern Carnarvon Basin, Australia: evidence for an eastward dipping detachment. *Journal of Geophysical Research*, **103**, 4975–4992.

—— HOGG, J. R., KARNER, G. D. & CHRISTIE-BLICK, N. 1995. Extensional tectonics in the Jeanne d'Arc Basin: implications for the timing of break-up between Grand Banks and Iberia. *In*: SCRUTTON, R. A., STOKER, M. S., SHIMMIELD, G. B. & TUDHOPE A. W. (eds) *The Tectonics, Sedimentation and Palaeoceanography of the North Atlantic Region*. Geological Society, London, Special Publications, **90**, 1–28.

EBINGER, C. J., KARNER, G. D. & WEISSEL, J. K. 1991. Mechanical strength of extended continental lithosphere: Constraints from the Western rift system. *East Africa, Tectonics*, **10**, 1239–1256.

FALVEY, D. A. 1974. The development of continental margins in plate tectonic theory. *Australian Petroleum Exploration Association Journal*, **14**, 95–106.

FEIJÓ, F. J. 1994. Bacias de Sergipe e Alagoas. *Boletim de Geosciências Petrobrás*, **8**, 149–162.

FROSTICK, L. E. & REID, I. 1989. Is structure the main control of river drainage and sedimentation in rifts? *Journal of African Earth Science*, **8**, 165–182.

GLADCZENKO, T. P., HINZ, K., ELDHOLM, O., MEYER, H., NEBEN, S. & SKOGSEID, J. 1997. South Atlantic volcanic margins. *Journal of the Geological Society, London*, **154**, 465–470.

GALLAGHER, K., HAWKESWORTH, C. J. & MANTOVANI, M. S. 1994. The denudation history of the onshore continental margin of SE Brazil inferred from apatite fission track data. *Journal of Geophysical Research*, **99**, 18 117–18 146.

GERRARD, I. & SMITH G. C. 1982. Post-Paleozoic succession and structure of the southwestern African continental margin. *In*: (eds) WATKINS, J. S. & DRAKE, C. L. *Studies in Continental Margin Geology. AAPG Memoir*, **34**, 49–76.

HAQ, B. U., HARDENBOL, J. & VAIL, P. R. 1988. Mesozoic and Cenozoic chronostratigraphy and eustatic cycles. In: WILGUS, C. K., HASTINGS, B. S., KENDALL, C. G. STC., POSAMENTIER, H. W., ROSS, C. A. & VAN WAGONER, J. C. (eds) *Sea-Level Changes: An Integrated Approach*. SEPM Special Publication, **42**, 71–108.

HARRIS, W. B. & LAWS, R. A. 1997. Paleogene-stratigraphy and sea-level history of the North Carolina Plain: global coastal onlap. *Sedimentary Geology*, **108**, 91–120.

HENRY, S. G., BRUMBAUGH, W. & CAMERON, N. 1995. Pre-salt source rock development on Brazil's conjugate margin: West African examples. *1st Latin American Geophysical conference, Rio de Janeiro*, extended abstracts, p3.

KARNER, G. D., DRISCOLL, N. W. & WEISSEL, J. K. 1992. Response of the lithosphere to in-plane force variations. *Earth and Planetary Science Letter*, **114**, 397–416.

——, USSAMI, N. & ALKMIM, F. F. 1993. Tectonic significance of the topography of southeastern Brazil: Speculations and implications (abst). *10th Anniversary conference UFOP/Petrobrás*, Universidade Federal de Ouro Preto, Ouro Petro, Brazil, 2–3 Dec.

——, DRISCOLL, N. W., MCGINNIS, J. P., BRUMBAUGH W. D. & CAMERON, N. 1997. Tectonic significance of syn-rift sedimentary packages across the Gabon–Cabinda continental margin. *Marine and Petroleum Geology*, **14**, 973–1000.

LAMBIASE, J. J. & BOSWORTH, W. 1995. Structural controls on sedimentation in continental rifts. *In*: LAMBIASE, J. J. (ed.) *Hydrocarbon Habitat in Rift Basins*. Geological Society, London, Special Publications, **80**, 117–144.

LANA, M. C. 1990. Bacia de Sergipe–Alagoas: Uma hipótese de evolução tectono-sedimentar. *In*: DE RAJA GABAGLIA, G. P. & MILANI, E. J. (eds) *Origem e evolução de bacias sedimentares*, Petrobrás.

LEEDER, M. R. & GAWTHORPE, R. L. 1987. Sedimentary models for extensional tilt-block/half-graben basins. *In*: COWARD, M. P., DEWEY, J. F. & HANCOCK P. L. (eds) *Continental Extensional Tectonics*. Geological Society, London, Special Publications, **28**, 139–152.

LOMANDO, A. J. 1996. Exploration for lucustrine carbonate reservoirs: Insights from west Africa. AAPG Bulletin, **80**, 1308–1309.

MACEDO, J. M. 1990. Evolução tectônica da bacia de Santos e áreas continentais adjacentres. *In*: DE RAJA GABAGLIA, G. P. & MILANI E. J. (eds). *Origem e evolução de bacias sedimentares*. Petrobrás.

MACKIN, J. H. 1948. Concept of the graded river. *Geological Society of America*, **59**, 463–512.

MCGINNIS, J. P., DRISCOLL, N. W., KARNER, G. D. BRUMBAUGH, W. D. & CAMERON, N. 1993. The flexural response of passive margins to deep-sea erosion and slope retreat: Implications for relative sea-level change. *Geology*, **21**, 893–896.

MCHARGUE, T. R. 1990. Stratigraphic development of proto-South Atlantic Rifting in Cabinda, Angola – a petroliferous lake basin. *In*: KATZ, B. J. (ed.) *Lacustrine Basin Exploration. Case Studies and Modern Analogs*. AAPG Memoir, **50**, 307–326.

MCKENZIE, D. P. 1978. Some remarks on the development of sedimentary basins. *Earth and Planetary Science Letters*, **40**, 25–32.

MOHRIAK, W. U., BASSETTO, M. & VIEIRA, I. S. 1998. Crustal architecture and tectonic evolution of the Sergipe–Alagoas and Jacuípe basins, offshore northeastern Brazil. *Tectonophysics*, **288**, 199–220.

——, HOBBS, R. & DEWEY, J. F. 1990. Basin-forming processes and the deep structure of the Campos basin, offshore Brazil. *Marine and Petroleum Geology*, **7**, 94–122.

NETTO, A. S. T., FILHO, J. R. W. & FEIJÓ, F. J. 1994. Bacias de Jacuípe, Camamu e Almada. *Boletim de Geosciências Petrobrás*, **8**, 173–184.

PERIERA, M. J. & FEIJÓ, F. J. 1994. Bacia de Santos. *Boletim Geosciências Petrobrás*, **8**, 219–234.

POSAMENTIER, H. W., JERVEY, M. T. & VAIL, P. R. 1988. Eustatic controls on clastic deposition, I. *In*: WILGUS, C. K. *ET AL.* (eds) *Sea Level Changes: An Integrated Approach*. Society of Economic Paleontologists and Mineralogists, Special Publication, **42**, 109–124.

RANGEL, H. D., MARTINS, F. A. L., ESTEVES, F. R. & FEIJO, F. J. 1994. Bacia de Campos. *Boletim Geosciências Petrobrás*, **8**, 203–217.
ROGUEL, N. C. 1990. *Analise estructural das falhas ocurrentes nas sub-bacias do Tucano sul e central Bahia.* Masters Thesis, Universidade Federal de Ouro Preto.
SANDWELL, D. T. & SMITH, W. H. F. 1992. Global marine gravity from ERS-1, Geosat and Seasat reveals new tectonic fabric. *EOS trans. AGU*, **73**, 133.
SANTOS, C. F., GONTIJO, R. C. ARAÚJO, & FEIJÓ F. J. 1994. Bacias de Cumuruxatiba e Jequitinhonha. *Boletin Geoscience Petrobrás*, **8**, 185–190.
SCHUMM, S. A. 1993. River response to baselevel change: Implications for sequence stratigraphy. *Journal of Geology*, **101**, 279–294.
SIESSER, W. G 1978. Leg 40 results in relation to continental shelf and onshore geology. *In:* BOLLI, H. M., RYAN, W. B. F. *et al.* Reports of the Deep Sea Drilling Project, Washington, D.C., (US Govt. Print. Office), 965–979.
TEISSERENC, P. & VILLEMIN, J. 1990. Sedimentary basin of Gabon – geology and oil systems. *In*: EDWARDS, J. D. & SANTOGROSSI, P. A. (eds) *Divergent/Passive margin Basins.* AAPG Memoir, **48**, 117–199.
VAIL, P. R. 1987. Seismic stratigraphy interpretation, utilizing sequence stratigraphy. *In*: BALLY, A. W. (ed.) *Atlas of Seismic Stratigraphy, Volume 1.* AAPG Studies in Geology, **27**.
——, MITCHUM, R. M. JR., TODD, R. G. 1977. Seismic stratigraphy and global changes of sea level. *In*: CLAYTON, C. E. (ed.) *Seismic Stratigraphy – Application to Hydrocarbon Exploration.* AAPG Memoir, **26**, 49–212.
WATTS, A. B. 1981. The U.S. Atlantic continental margin: Subsidence history, crustal structure and thermal evolution. *In*: *Geological of Passive Continental Margins: History, Structure and Sedimentologic Record (with special emphasis on the Atlantic margin).* AAPG, Education course note series, **19**.
WEISSEL, J. K. & KARNER, G. D. 1989. Flexural uplift of rift flanks due to mechanical unloading of the lithosphere during extension. *Journal of Geophysical Research*, **94**, 13 919–13 950.
WEISSEL, P. & SMITH, W. H. F. 1995. New version of the generic mapping tools released. *EOS. Trans. AGU*, **76**, 329.
WHITING, B. W., KARNER, G. D. & DRISCOLL, N. W. 1994. Flexural and stratigraphic development of the west Indian continental margin. *Journal of Geophysical Research*, **99**, 13 791–13 811.

The Mesozoic denudation history of the Atlantic margins of southern Africa and southeast Brazil and the relationship to offshore sedimentation

KERRY GALLAGHER[1] & RODERICK BROWN[2]

[1]*T. H. Huxley School of Environment, Earth Science and Engineering, Imperial College of Science, Technology and Medicine, South Kensington, London SW7 2AS, UK*
[2]*Department of Earth Sciences, La Trobe University, Bundoora, Victoria, 3083, Australia*

Abstract: A new interpretation of regional apatite fission track data suites from western southern Africa (South Africa, Namibia) and southeast Brazil, in terms of the post-break-up denudation chronologies, is presented. The data reveal that although significant denudation occurs parallel to the present-day margin (associated with the inland escarpment) there are regional variations along-strike and with time. Moreover, the data reveal a significant amount of post-break-up denudation from within the continental interior of southern Africa. The data indicate relatively little post-rift denudation but the rates increase dramatically in both regions in the late Cretaceous–Early Tertiary, more than 50 Ma after initial rifting. The inferred denudation chronologies are broadly consistent with the sedimentary record in the offshore basins, which can show increases in sedimentation rate, contemporaneous unconformities and major changes in depositional character.

Some of the earliest geodynamic models for passive continental margin evolution explicitly incorporated denudation as a means of thinning the crust (e.g. Hsü 1965; Sleep 1971). Since then, of course, the role of tectonic thinning has been widely accepted as the primary means for driving subsidence in the offshore basins (e.g. McKenzie 1978). However, denudation is still a key issue for a comprehensive passive margin evolution model in that it represents a primary means of providing sediment to the basins. Furthermore, the onshore structural history, often manifested as discrete periods of locally enhanced denudation, is likely to be reflected to a variable degree in the offshore basin (e.g. Clemson *et al.* 1997), and so the denudation chronology has direct relevance for offshore exploration strategies.

There are various geodynamic models which have considered the formation of long-term (> 50–100 Ma) passive margin topography (Braun & Beaumont 1989; Weissel & Karner 1989; Gilchrist & Summerfield 1990). In practice, however, it is not possible to determine how surface elevation changes over time. The process of denudation removes the evidence of palaeoelevation (e.g. marine sediments). Consequently, unlike the stratigraphy in the offshore basin, there is no direct record of vertical motion onshore. To some extent, the offshore basins provide some useful information about volumes, rates and lithological nature of sediment supply over time (Poag & Sevon 1989; Rust & Summerfield 1990; Pazzaglia & Brandon 1996; White & Lovell 1997). In terms of understanding long-term denudation, the major limitation is that these data do not provide direct evidence of the sediment source region, i.e. where was being eroded at a given time. Therefore, it is necessary to consider the onshore region explicitly.

However, there are a variety of geochronological methods which can provide key information relevant to the denudation history. Cosmogenic surface exposure dating (e.g. Cerling & Craig 1994; Gillespie & Bierman 1995; Brook *et al.* 1995; Nishiizumi *et al.* 1993; Burbank *et al.* 1996) is one method which yields an estimate of the longevity of a given surface, or the rate of erosion. This method is limited to timescales of *c.* 5–6 Ma. Wolf *et al.* (1997) have recently revived (U–Th)/He dating of apatite as a thermochronological tool, previously considered by Zeitler *et al.* (1987). This system, based on the thermally activated diffusion of He, has great promise in denudation studies, particularly because of its apparently low-temperature sensitivity (40–75°C), although there is still considerable work to be done in terms of calibrating the system kinetics. Perhaps the most well known and widely used thermochronological technique applied to denudation studies is fission track analysis. A detailed overview of this method is beyond the scope of this paper, but it has been recently reviewed in the context of denudation by Brown *et al.* (1994*a*), Gallagher & Brown (1997), Gallagher *et al.* (1998) and Dumitru (1998).

In this paper, a summary of apatite fission track data from the Atlantic margins of southern Africa and southeast Brazil are presented. These data are used to determine the post-break-up thermal history for temperatures < 130°C. From this a denudation chronology for each margin is constructed, which is subsequently related to the offshore sedimentary record.

GALLAGHER, K. & BROWN, R. 1999. The Mesozoic denudation history of the Atlantic margins of southern Africa and southeast Brazil and the relationship to offshore sedimentation. *In:* CAMERON, N. R., BATE, R. H. & CLURE, V. S. (eds) *The Oil and Gas Habitats of the South Atlantic.* Geological Society, London, Special Publications, **153**, 41–53.

Fission track analysis

Fission tracks form by the spontaneous fission of trace amounts (1–1000 ppm) of ^{238}U present in minerals such as apatite and zircon. The tracks are discrete, linear zones of damage within the crystal lattice. These form at an essentially constant rate, determined by the spontaneous fission decay constant of ^{238}U. The number of tracks present in a given apatite crystal depends on both the uranium concentration and the duration over which tracks have been accumulating. Once the uranium concentration is determined, the number of tracks intersecting an internal surface of a crystal can be used to determine a fission track age. The major difference between fission track dating and a conventional isotopic dating system is that the daughter product is physical damage to the crystal lattice (i.e. fission tracks) rather than a daughter isotope.

However, fission tracks shorten, or anneal, as the temperature increases so there will be a reduction in the fission track age, as fewer tracks will intersect the surface of observation. The length of a given track depends on the maximum temperature to which it was exposed since it was formed. As each track forms at a different time, each experiences a different portion of the total thermal history of the host rock. Therefore, track length data provide information on the temperature variations experienced by the host rock, while the fission track age provides information about the duration of these variations. At a high level of annealing (e.g. temperatures of 110–130°C acting over timescales of 10^{6-7} years) all tracks are completely annealed, and their etchable lengths and the fission track age are effectively reduced to zero. The dependence of the fission track age on the underlying track length distribution means that a meaningful interpretation of a measured fission track age, in terms of timing an event, such as continental break-up or a discrete cooling event, requires track length data.

To appreciate what fission track data can tell us about passive margin evolution, what controls the thermal evolution of a margin needs to be considered. In the onshore region, where the fission track samples are collected, the thermal gradients in the shallow crust are largely unaffected by the rifting process (e.g. Buck *et al.* 1988; Gallagher *et al.* 1994; Brown *et al.* 1994*b*). Consequently, fission track ages recording a thermal perturbation from rifting-enhanced heat flow are not expected. The low-temperature thermal history of rocks in the onshore region is primarily controlled by their displacement relative to the Earth's surface, i.e. denudation. The process of denudation, both erosional and tectonic, is therefore the major factor controlling the rate of cooling (Gallagher *et al.* 1994, 1995; Brown *et al.* 1994*a*, 1998). Therefore, it should be possible to extract a chronology of denudation from a regional suite of fission track data.

Fission track data for southern Africa and southeast Brazil

Fission track age and mean track length data collected from outcrop samples on the South Atlantic continental margins have been discussed in detail by Brown (1992), Gallagher *et al.* (1994, 1995) and Brown *et al.* (1998). The locations of the samples are shown with the present-day topography in Fig. 1 and the fission track data are presented in map form in Fig. 2. Although the data coverage is sparse in places, it is clear from Fig. 2 that the fission track ages proximal to the present-day coast are not only relatively young (60–70 Ma), but also considerably younger than the inferred age of break-up (*c.* 130 Ma). Furthermore, these younger ages tend to be associated with longer mean track lengths (13–14 μm) and unimodal, but slightly negatively, skewed track length distributions. Fission track ages older than the break-up age occur only in the interior parts of the sampled region and, as a consequence of the topographic morphology of these margins (Fig. 1), tend to be associated with higher elevations. Significantly, all the fission track ages are younger than the stratigraphic age of the host rock, indicating that these have all been exposed to elevated temperatures in the past, and many samples have experienced temperatures in excess of 110–120°C.

Although there is a general trend of age younging towards the coast in passive continental margins (Gallagher & Brown 1997), it is apparent from the southern Africa data that relatively young ages can occur well inland. In central Namibia there is a distinct corridor of young ages extending in a northeasterly direction inland. Again, these ages are much younger than the break-up age and are associated with longer mean track lengths. There are less data from southeast Brazil, but even here it is apparent there are some young fission track ages well inland in the northern area. In this case, these samples were collected from a late Cretaceous intrusion [Poços da Caldas, Gallagher *et al.* (1994), dated with K–Ar as 63–80 Ma, Amaral *et al.* (1967)] and so a post-break-up age is expected. In contrast, the data from southern Africa are from rocks ranging in age from Precambrian to Triassic and it is clear that the young ages reflect extensive post-break-up denudation in the continental interior (Brown *et al.* 1998).

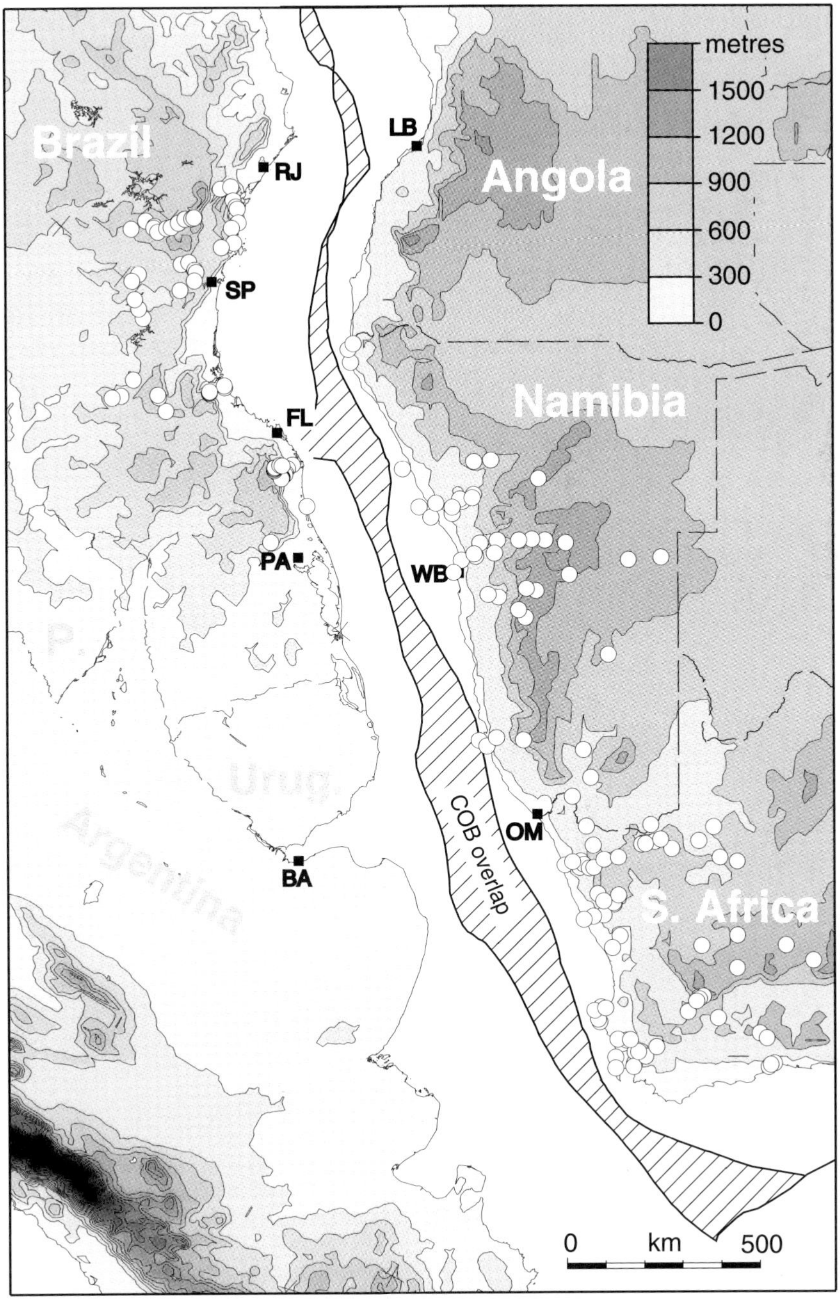

Fig. 1. Fission track samples (open circles) and present-day elevation in southern Africa and southern Brazil (with the continents in a schematic pre-break-up configuration). RJ, Rio de Janeiro; SP, Sao Paulo; Fl, Florinopolis; BA, Buenos Aries; LB, Lobito; WB, Walvis Bay; OM, Oranjemund; COB, continent–ocean boundary.

The relationship between the fission track and the mean track length is typically referred to as a boomerang plot, and such plots are given in Fig. 3 for the two data suites. Green (1986) demonstrated that if a region experiences a discrete phase of accelerated denudation then the apatite fission track ages, and the mean track lengths, define a distinctive pattern, provided that the

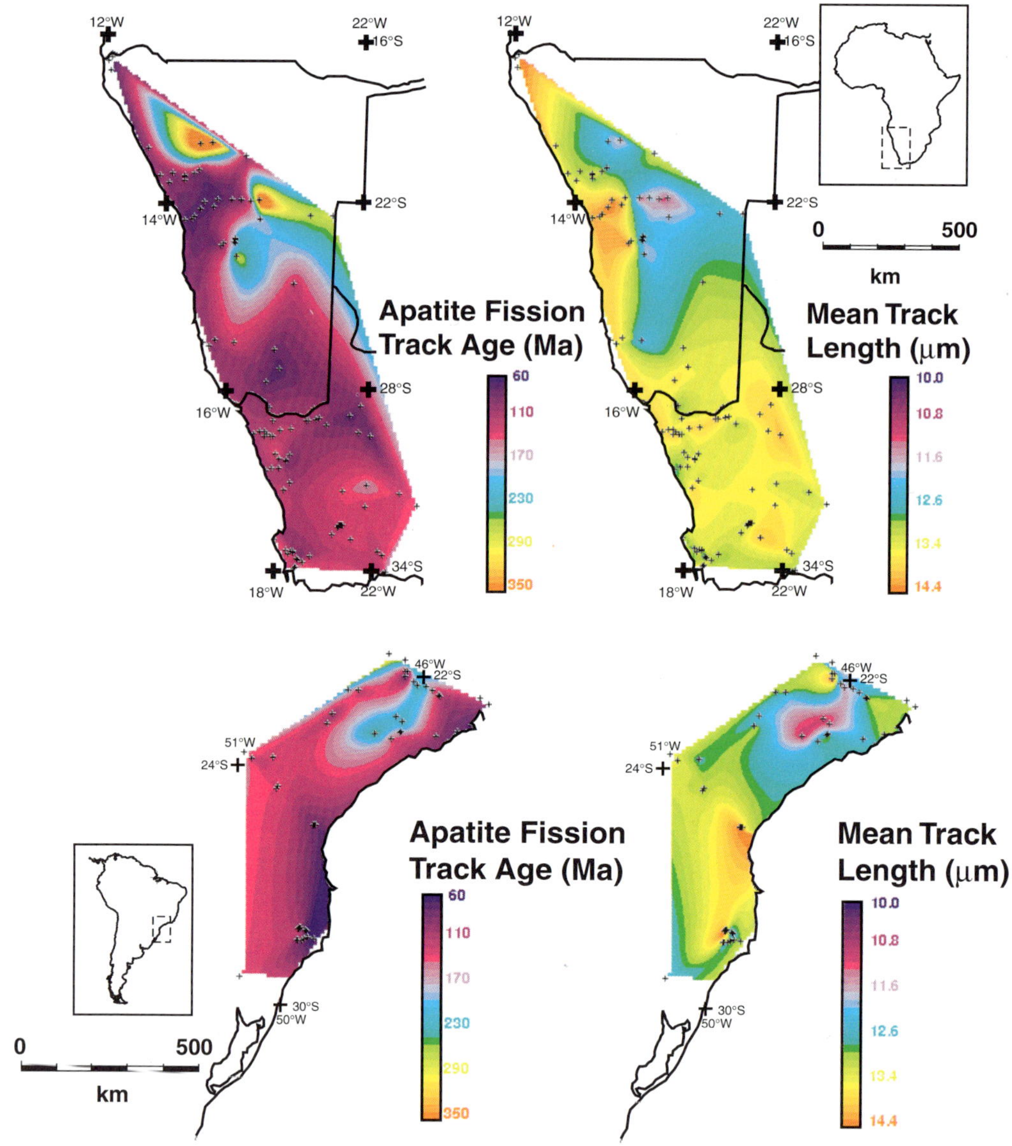

Fig. 2. Contour maps of apatite fission track age and mean track length for western southern Africa and southeast Brazil. The sample locations are marked by +. The fission track ages near the present-day coast are all considerably younger than the inferred age of break-up (*c*. 130 Ma) and similarly young ages well inland in southern Africa.

samples have experienced different amounts of cooling, i.e. differential denudation. This pattern is characterized by an initial decrease in the mean track length with decreasing age to a minimum mean length for intermediate ages, followed by a progressive increase in mean length for successively younger ages. This concave-up pattern defines the boomerang shape and the interpretation of this plot in terms of palaeotemperature has been discussed in detail by Brown *et al.* (1994*a*) and Gallagher & Brown (1997).

In essence, the youngest ages with long track lengths (e.g. ≥ 14 µm) indicate that the sample cooled rapidly from temperatures in excess of *c*. 120 ± 10°C (i.e. the total annealing temperature) at the time indicated by the fission track age. In this situation, effectively all of the tracks formed after the rapid cooling event. The intermediate

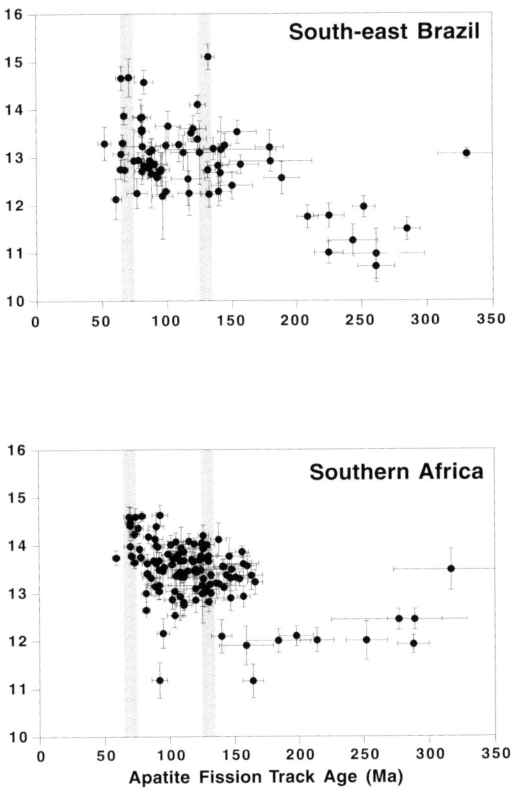

Fig. 3. Relationship between the apatite fission track age and the mean track length (typically referred to as a boomerang plot). The two shaded bands indicate the approximate time of break-up (*c.* 130 Ma) and the most recent significant discrete cooling event inferred from these data (*c.* 70 Ma).

ages are taken to reflect the same cooling episode, but from pre-cooling temperatures < 120 ± 10°C. In this case, the track length distribution reflects a combination of tracks formed either before or after the rapid cooling event. The later tracks will all be relatively long provided that the sample has remained at temperatures *c.* < 60°C since the cooling episode. The tracks which formed earlier will have lengths which reflect the pre-cooling ambient palaeotemperature. For apatite with a uniform chemical composition, the shorter the track, the higher the equivalent palaeotemperature. The oldest ages with long track lengths do not record the cooling event and resided at temperatures < 60°C for a duration approximately equal to the measured fission track age.

In the context of passive continental margin evolution, it might be expected that apatite fission track data will record the time of break-up as a discrete event. The fact that the data with fission track ages *c.* 130–140 Ma do not have the longest mean track lengths shows this is not the case for southern Africa and southeast Brazil. Two samples from Brazil with long mean track lengths (15.1± 0.3 and 14.1 ± 0.19 μm) and fission track ages (132 ± 5 and 127 ± 6 Ma) close to the inferred time of break-up are both from the Paraná continental flood basalt. This basalt has been dated by $^{40}Ar/^{39}Ar$ at *c.* 132 ± 1 Ma (Turner *et al.* 1994) in the vicinity of the fission track samples. Although the formation of the flood basalt is clearly intimately linked to the break-up process, the fission track age reflects rapid magmatic cooling rather than break-up related heat flow. These two samples imply maximum post-magmatic cooling temperatures of *c.* 30–40°C, or *c.* 800–1000 m of denudation (Gallagher *et al.* 1994, 1995; Brown *et al.* 1998).

Returning to the boomerang plot (Fig. 3), it can be seen that the longest track lengths from southern Africa are associated with fission track ages of *c.* 70 Ma. Geographically, these occur in a corridor of young ages running from the coast northeast through central Namibia and in the interior of southern Namibia/northern South Africa. A similar age peak is present in Brazilian data, but these samples are from the Poços da Caldas intrusion and the fission track ages reflect the timing of intrusion, as mentioned earlier. The young ages in southern Africa tend to be associated with major Pan-African structures, such as the northern central zone of the Damara metamorphic belt in central Namibia, and the fault-bounded Karasberg Mountains in southern Namibia (Brown *et al.* 1998). The track length distributions have long means (*c.* 14 μm) and are unimodal, with a restricted spread. This form indicates rapid cooling at a time similar to the fission track age (*c.* 60–80 Ma).

The youngest ages in the southern region of Brazilian localities were from samples not directly associated with intrusions but collected from a Jurassic sandstone directly below the Paraná basalt, within < 2 m of the contact. As discussed by Gallagher *et al.* (1994, 1995), these samples were totally reset (in terms of their fission track age) as a consequence of the thermal influence of the hot basalt at *c.* 132 Ma. However, this would have been effectively instantaneous and so it might be expected that fission track ages similar to those of the basalt would be measured. However, they are 40–50 Ma less than the basalt age and similar to those from basement samples close to the present-day margin further north. The track length distributions of these data do not have particularly long means (< 14 μm), although they are negatively skewed with a pronounced tail of shorter tracks. Such track length distributions can be qualitatively interpreted in terms of protracted cooling, as a consequence of not particularly rapid denudation occurring well after break-up.

To summarize, the distribution of young ages along the margins, and within the interiors, indicate widespread post-break-up denudation, although variably distributed. As well as a broad geographical trend of young ages close the margin, there are clear associations with pre-existing regional scale structures. The fission track data imply that the rate of denudation (and cooling) was variable both spatially and in time, with discrete periods of enhanced denudation occurring well after the time of continental break-up.

Quantifying the denudation chronology

As demonstrated by Brown *et al.* (1998), regional fission track data sets can be used to construct quantitative denudation maps. This requires modelling the fission track data in terms of a temperature history and then converting this temperature history to an equivalent depth as a function of time. To solve the first part of this problem the stochastic search methodology is used, described by Gallagher (1995), incorporating the fission track annealing algorithm of Laslett *et al.* (1987) to model the data. This process involves a large search of possible thermal histories (here a maximum of 80 000 models have been considered). The outcome of this modelling exercise is an optimal thermal history for each sample, with 95% confidence intervals about the best solution. In terms of fitting the observed data, contour maps of the predicted fission track ages and mean track lengths are effectively identical to the observed data maps (Fig. 2), and so are not reproduced here.

Initially, the results of this modelling exercise are considered in terms of palaeotemperature maps for selected times. In practice, computer animations of these maps can be made to provide a visual monitor of the spatial evolution of the palaeotemperature field. Here, three time-slice maps are shown (Fig. 4), chosen to be represent various stages of the post-break-up thermal history. The striking feature of these maps is the inference that the samples close to the present-day margin were nearly all at temperatures >130°C at 120 Ma, and many of these had cooled to >80°C by 60–80 Ma. The results for southern Africa show that by 80 Ma some of the margin samples cooled to lower temperatures than some of the interior samples. This is consistent with the qualitative interpretation of significant late Cretaceous–Early Tertiary denudation in the interior of western south Africa discussed earlier.

In order to convert the palaeotemperature maps to equivalent depths, or denudation, some informed assumptions must be made regarding temperature gradients. The temperature gradient is simply the ratio of heat flow (Q) and thermal conductivity (k), and the equivalent depth (z) is given as

$$z = k(T - T_s)/Q$$

where T is the temperature (estimated from modelling the fission track data) and T_s is the assumed surface temperature.

The present-day heat flow data (from the data base of Pollack *et al.* (1993)) were used for southern Africa and Brazil, incorporating the heat flow data from the Paraná Basin of Hurter & Pollack (1996). It was assumed that the heat flow was constant over time (but varied spatially as shown in Fig. 5) and we used a constant thermal conductivity of 2.5 $Wm^{-1}K^{-1}$. Finally, a constant surface temperature of 20°C was used. It is straightforward to incorporate time varying heat flow and surface temperature, and variations in thermal conductivities, both spatial and temporal. However, it was decided to illustrate the modelling results with these poorly-constrained extra layers of complexity. As mentioned earlier, modelling studies imply that the thermal effects of rifting are unlikely to be significant within the shallow crust of the onshore regions of the margin. Furthermore, the global heat flow data set indicates that the spread of temperature gradients in continental interiors is relatively restricted with a mean between 20 and 30°C km^{-1} (Pollack *et al.* 1993). Uncertainties arising from anomalous transient thermal gradients (e.g. associated with rapid erosion) are unlikely, even in extreme cases, to be greater than a factor of two (Brown & Summerfield 1997).

In Fig. 6 the inferred cumulative denudation between three discrete time intervals is shown. The maximum total denudation is *c.* 4500 m and, in general, the largest amount of post-rift denudation is predicted along the coastal region with decreasing amounts further inland. However, geographically distinct anomalies are evident. For example, over 4 km of post-rift denudation is predicted for a relatively restricted region over 400 km inland of the margin near 22°E, 29°S. According to the model thermal histories for the samples in this region, the majority of this denudation occurs between 100 and 60 Ma, which post-dates the time of rifting by *c.* 50–60 Ma (Brown *et al.* 1998). The location of the anomaly broadly coincides with a complex,

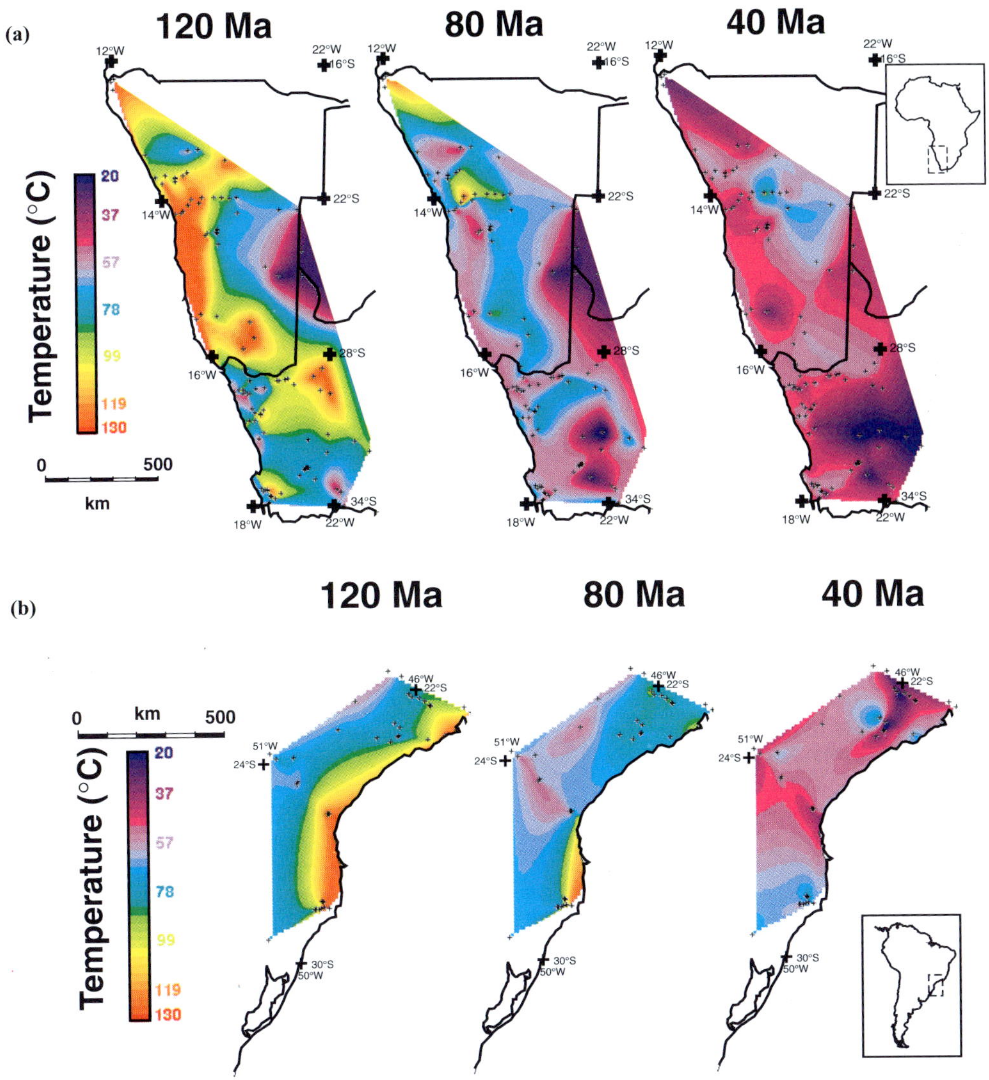

Fig. 4. (**a**) Palaeotemperature (°C) estimates from the southern Africa samples for three post-break-up times. The strongest red colour code indicates a minimum temperature of 130°C (sufficient to anneal fission tracks in apatite totally for heating durations of 1–10 Ma). (**b**) As for (a) but for southeastern Brazil.

northwest trending structural zone marking the western boundary of the Kaapvaal Craton. The location, restricted areal distribution of the anomaly and the timing of the predicted denudation suggests that it is unrelated to the initial formation of the margin, but may rather be associated with later tectonic reactivation of the craton boundary during the Late Cretaceous.

In Fig. 7 the spatially integrated denudation chronology is shown, which was obtained by taking the difference between the calculated depth at successive time intervals (i.e. time derivative) and summing these over the total area. The inferred 95% confidence intervals on the temperature histories were mapped into the denudation chronology using the lower and upper bounds on these temperature histories. Although these bounds give upper and lower limits on the cumulative denudation curve, the fact that the time derivative of the cumulative curves is taken means that denudation chronology bounds can cross-over and overlap. The denudation chronology is also shown, calculated with the optimal thermal history models and a constant heat flow (equal to the average value of the maps in Fig. 5, *c*. 64 m Wm^{-2} for Brazil and *c*. 53 m Wm^{-2} for southern Africa).

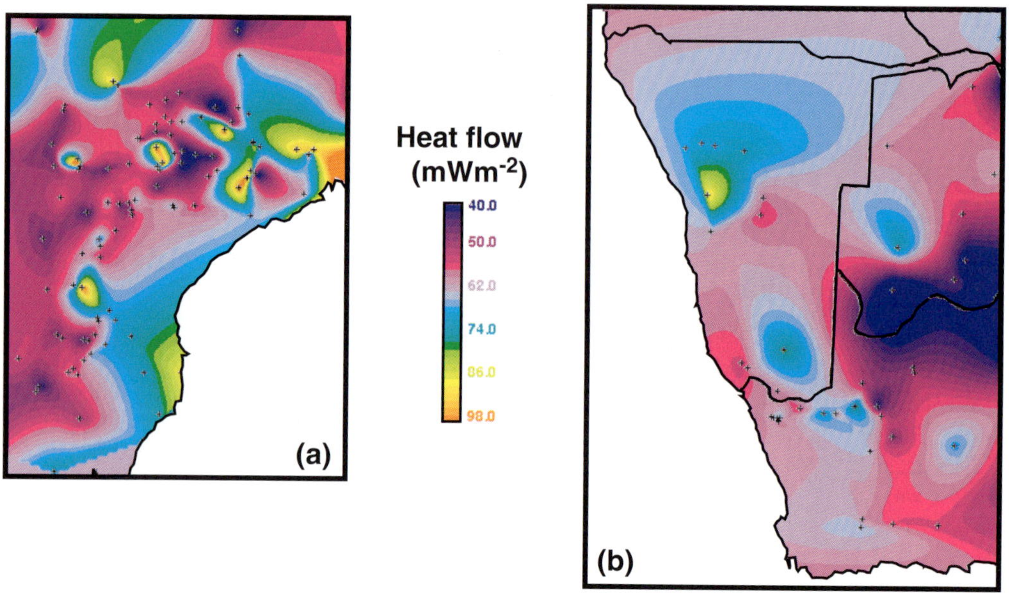

Fig. 5. Present-day heat flow maps for southeast Brazil and southern Africa [based on the data of Pollack *et al.* (1993) and Hurter & Pollack (1996)].

The first point to make concerning these curves is that the general shape is relatively insensitive to the spatial variation in heat flow. As the variation is relatively restricted (Fig. 5) this is not surprising. The other characteristic is that the timing of periods of enhanced denudation seems to be robust to the uncertainties in modelling the data (i.e. all curves show essentially the same features). The potential bias introduced by individual points having an undue influence in the interpolation procedure (e.g. where data are sparse) was also considered. To interpolate the denudation at a given point, the method described by Sambridge *et al.* (1995) was used. This calculates a weighted average of the neighbouring true data points, where the weighting depends on the spatial distribution of the points. Dummy data points inserted between the real data points were experimented with. These dummy points were assigned a constant temperature history, so that they do not contribute to the denudation chronology. The outcome of this was that the absolute magnitude of the denudation was reduced but the relative variations over time were essentially the same as shown in Fig. 7.

In terms of the geological interpretation, the obvious point is that these periods of enhanced denudation occur well after the time of break-up. Overall, south east Brazil appears to have experienced enhanced denudation *c.* 100, 75 and 40 Ma, while southern Africa has fairly high denudation rate between 120 and 80 Ma, and a minor peak *c.* 60 Ma. Both regions appear to show an increasing rate of denudation to the present day, particularly from *c.* 20 Ma. This is likely to be partly an artifact, reflecting the fact that the fission track annealing algorithms tend to underpredict the amount of annealing at temperatures < 50°C over geological timescales (Gallagher *et al.* 1998). The effect of this is that the modelling results commonly show late-stage cooling from *c.* 50 to 60°C (otherwise the predicted track lengths would be too long).

Gallagher *et al.* (1998) and Brown *et al.* (1998) have discussed this fission track in the context of geomorphic models for long-term landscape evolution in passive margins. They conclude that the data are broadly consistent with a model where the majority of denudation occurs in what is now the low-elevation coastal plain, seaward of the topographic escarpment (e.g. Gilchrist & Summerfield 1990; Kooi & Beaumont 1994), with regional anomalies in the interior and along-strike of the margin related to post-break-up reactivation of regional structures.

Denudation and sedimentation

As mentioned earlier, offshore sediment volumes have been used in the past to try and constrain the onshore denudation chronology. In Fig. 8 the

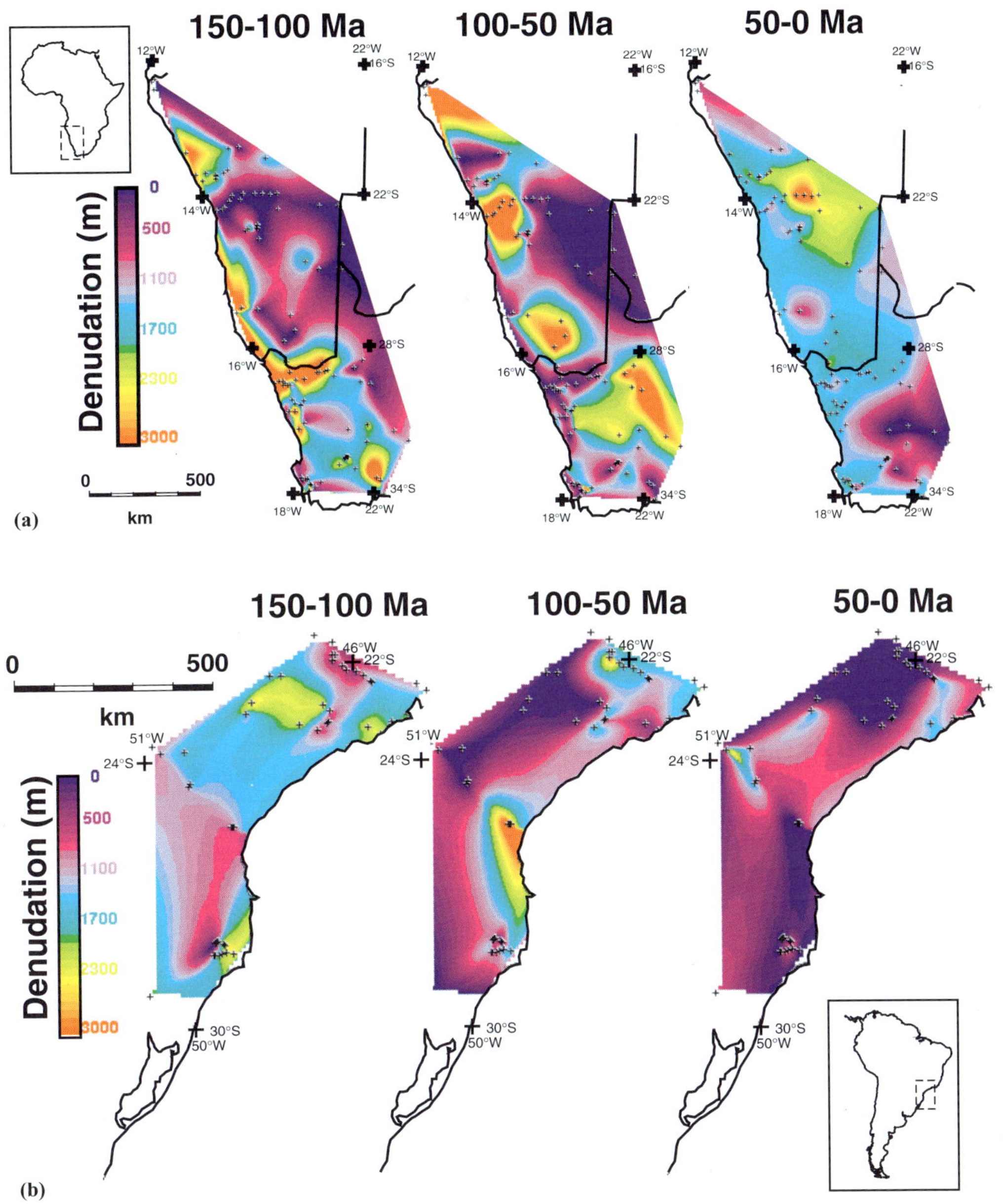

Fig. 6. (**a**) Estimated denudation (metres) for 150–100, 100–50 and 50–0 Ma for southern Africa. (**b**) As for (a) but for southeastern Brazil.

inferred denudation flux (i.e. rate of removal of material) is compared with estimated sediment fluxes for the parts of the respective margins. The sediment volume data for southern Africa is based entirely on isopachs constructed from seismic data from offshore Namibia (the Namibe, Walvis, Luderitz and northern Orange Basins – see Maslanyj *et al.* 1992; Light *et al.* 1992, 1993; Clemson *et al.* 1997). Consequently, these data are representative of only about half of the margin considered in terms of denudation. The two curves of Brazil are based on average sedimentation rates in the Santos Basin (Williams & Hubbard 1984) and decompacted well subsidence analyses for the Campos Basin (Chang *et al.* 1992). In the absence of more detailed data, a

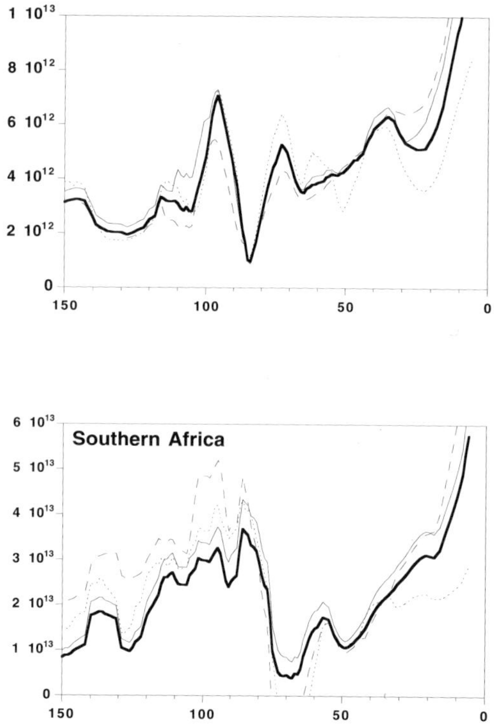

Fig. 7. Denudation chronologies for western southern Africa and southeast Brazil. The heavy line is the calculated denudation flux using the maximum likelihood temperature history estimates, and the dashed and dotted lines are the denudation fluxes calculated using the 95% confidence limits (upper and lower, respectively) about the maximum likelihood temperature estimate. The thin solid line is the denudation flux calculated assuming a constant heat flow (see text for details).

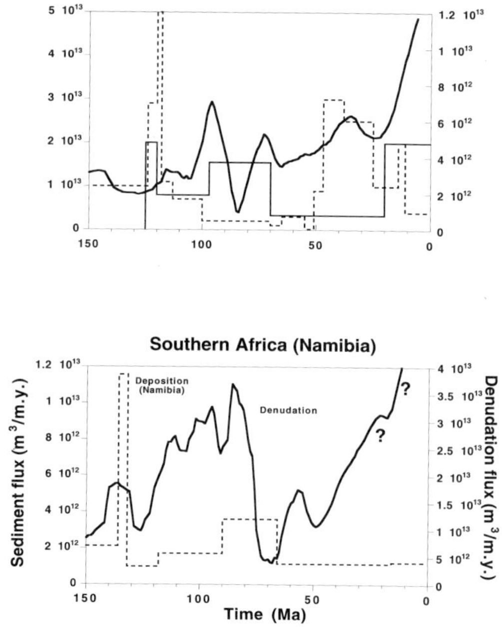

Fig. 8. The maximum likelihood denudation chronologies (from Fig. 7) as heavy solid lines together with estimates of the sediment flux for adjacent offshore basins. Note: the denudation and sediment fluxes have different scales (right- and left-hand sides, respectively). The sediment flux for southern Africa represents only the basins offshore of Namibia, whilst the sediment flux for southeast Brazil is presented for the Campos (heavy dotted line) and Santos Basins (thin solid line).

constant depositional area of 100 000 km^2 is assumed for each basin (e.g. Williams & Hubbard 1984; Mohriak *et al.* 1989) to convert sedimentation rates to volumes. This is clearly a first-order approximation and it is reiterated here that the assumptions made to construct the denudation chronologies, i.e. constant heat flow over time (but spatially variable), and constant surface temperature and thermal conductivity (of the eroded material). These assumptions also limit the accuracy of the absolute denudation rates. Given these restrictions, the sedimentation and denudation curves are plotted on different axes and the correlation is discussed in time rather than the absolute values.

The sedimentation curve for Namibia is dominated by the early syn-rift sequence (132–136 Ma), although the height of this peak is sensitive to the duration of this interval. Although this sedimentation peak does correlate with a denudation peak *c.* 135 Ma, the more interesting feature is the increase in sedimentation from *c.* 120 to *c.* 70 Ma. During this period the total depocentres doubled in size, when the basin evolved from the rift-transitional phase to the thermal sag phase (Light *et al.* 1993). The denudation rate also increases significantly over this time interval and peaks in the Late Cretaceous. This was previously inferred by Brown *et al.* (1990), based on earlier fission track analyses. Much of the Late Cretaceous cooling/denudation occurred where there are now fission track ages of a similar age (i.e. northern South Africa, central Namibia). The depositional history in the late Cretaceous–Early Tertiary is characterized by progradational build-outs, with extensive deltas and channel cutting (Light *et al.* 1993), consistent with a rapid influx of detrital sediment and either regional uplift and/or sea-level fall. The Early Tertiary peak in the denudation chronology does not appear to be reflected in the sediment isopach data, although the relative amplitudes between the sediment and denudation rates at *c.* 80 and *c*. 60 Ma are similar. The apparent increase in denudation from *c*. 50 Ma to the present day is also not obviously reflected in the overall sediment flux data. However, part of the denudation contribution in the Early Tertiary is attributable to enhanced erosion in northern Namibia, where the fission track data are somewhat scarce. The Namibe Basin, north of the Walvis Ridge, has a thick Tertiary sequence relative to the Luderitz, Walvis and Orange Basins south of the ridge (Light *et al.* 1993). Thus, the denudation chronology appears to be consistent in terms of providing a relatively local source for these sediments, although more fission track analyses are required to confirm this.

The two sets of sedimentation rate data for Brazil are from adjacent basins, although these are separated by a long-lived structural boundary and have undergone somewhat different evolutions (e.g. Williams & Hubbard 1984; Mohriak *et al.* 1989, 1995; Chang *et al.* 1992). The early history of these basins is characterized initially by evaporites and then carbonates up to the end of the Albian (*c*. 95 Ma). During this time, however, clastic sediments were deposited in the coastal region, and there is one fairly widespread period of rapid clastic sedimentation in the early Albian identified in both basins (Williams & Hubbard 1984; Change *et al.* 1992). Spatially, the majority of denudation occurs in what is now the low-elevation coastal pain, with relatively minor contributions from the inland samples (and those currently at high elevation in the southern part of the study region). The denudation chronology implies that the first major increase in denudation rate occurred around the end of the Albian, with a second peak *c*. 70 Ma. This time interval of *c*. 25–30 Ma corresponds to a period of regressive clastic sedimentation in the Santos Basin (Williams & Hubbard 1984), while a Late Cretaceous–Early Tertiary regional unconformity is present in the Campos Basin (Mohriak *et al.* 1989). Overall, the sedimentary record is consistent with the inferred two phases of enhanced denudation in the Late Cretaceous given that it appears that the Campos Basin also probably experienced erosion/non-deposition at this time. Much of the rapid denudation in these two phases occurred in the Ponta Grossa Arch. This is a geomorphic and structural feature that runs northwest from the coast near São Paulo and is currently a region of relatively subdued elevation creating a distinct embayment in the outcrop of the Parana flood basalt (Fig. 1; Gallagher & Hawkesworth 1994).

The Campos Basin shows enhanced turbiditic sediment flux and rapidly decreasing palaeowater depths in the Eocene–Oligocene (50–25 Ma). This timing correlates with a peak in the denudation chronology, as does the recent (< 20 Ma) increase in sediment flux in the Santos Basin, and these peaks reflect denudation in different regions of the margin. This implies that the basins were sourced from different regions at different times, and/or the regional drainage changed in the Miocene. It is considered that the progressively increasing denudation rate, from 20 Ma, is partly attributable to problems with the annealing model adopted, as mentioned earlier. Until these annealing models are improved it is not possible to determine how much of the recent increase in denudation is real.

Conclusions

In this paper the denudation chronology of the South Atlantic margins of southern Africa and southeast Brazil have been addressed. This has been quantified using apatite fission track data and conventional optimization methods to construct thermal history models for individual sample locations. These thermal histories were combined with the presence-day heat flow maps to produce maps of cumulative denudation over time. The maps reveal that although most denudation occurs adjacent to the present-day margin, there are regions inland which also underwent significant post-break-up denudation as a result of structural reactivation. By integrating the denudation maps spatially, a generalized denudation chronology was constructed for each margin. These show periods of enhanced denudation well after break-up although the most recent (< 20 Ma) chronology is suspect as a consequence of uncertainties in the extrapolation of fission track annealing models to low temperature (< 50–60°C), long timescale (10–100 Ma) scenarios.

Given the many assumptions made in the modelling of the fission track data, it is believed that there is reasonable agreement between the margin denudation chronology with the inferred offshore sediment flux. The real problem associated with such analysis is how much confidence we have in our understanding of the relationship between the onshore sediment source region and the offshore depocentres. Ultimately, this is largely controlled by the regional drainage, which is clearly linked to topography, climate and regional elevation, as well the depositional architecture and structural evolution of the basins. These factors themselves will also be influenced by the large-scale geodynamic evolution of the basin–margin system.

We would like to thank the Namibian Oil Corporation (NAMCOR) and Geoquest for access to seismic and isopach data from Namibia, and the Namibia Geological Survey, Anglo American Corporation, SOPEMI (Brazil), Mike de Wit, Chris Hawkesworth and Marta Mantovani for help in collecting the fission track samples. This work has also been supported by NERC, The Nuffield Foundation, The Royal Society and the University of London Central Research Fund.

References

AMARAL, G., BUSHEE, J., CORDANI, U. G., KAWASHITA, K. & REYNOLDS, J. H. 1967. Potassium–argon ages of alkaline rocks from southern Brazil. *Geochimical et Cosmochimica Acta*, **31**, 117–142.

BRAUN, J. & BEAUMONT, C. 1989. A physical explanation of the relationship between flank uplifts and the break-up unconformity at rifted continental margins. *Geology*, **17**, 760–764.

BROOK, E. J. & BROWN, E. T., KURZ, M. D., ACKERT, R. P., RAISBACK, G. M. & YIUO, F., 1995. Constraints on age, erosion, and uplift of Neogene glacial deposits in the Transantarctic Mountains determined from in situ cosmogenic ^{10}Be and ^{26}Al. *Geology*, **23**, 1063–1066.

BROWN, R. W. 1992. *A fission track study of the subaerial continental margins of southern Africa*. PhD Thesis, University of Melbourne.

—— & SUMMERFIELD, M. A. 1997. Some uncertainties in the derivation of rates of denudation from thermochronologic data. *Earth Surface Processes and Landforms*, **22**, 239–248.

——, RUST, D. J., SUMMERFIELD, M. A., GLEADOW, A. J. W. & DE WIT, M. C. J. 1990. An early Cretaceous phase of accelerated erosion on the southwestern margin of Africa: Evidence from apatite fission track analysis and the offshore sedimentary record. *Nuclear Tracks and Radiation Measurements*, **17**, 339–351.

——, SUMMERFIELD, M. A. & GLEADOW, A. J. W. 1994*b* Apatite fission track analysis: Its potential for the estimation of denudation rates and implications of long-term landscape development. *In*: KIRKBY, M. I. (ed.) *Process Models and Theoretical Geomorphology*. Wiley, New York, 23–53.

——, GALLAGHER, K. & DUANE, M. 1994*a*. A quantitative assessment of the effects of magmatism on the thermal history of the Karoo sedimentary sequence. *Journal of African Earth Sciences*, **18**, 227–243.

——, GALLAGHER, K., GLEADOW, A. J. W. & SUMMERFIELD, M. A. 1998. Morphotectonic evolution of the South Atlantic margins of Africa and South America. *In*: SUMMERFIELD, M. A. (ed.) *Geomorphology and Global Tectonics*. Wiley, Chichester.

BUCK, W. R., MARTINEZ, F., STECKLER, M. S. & COCHRAN, J. R. 1988. Thermal consequences of lithospheric extension: pure and simple. *Tectonic*, **7**, 213–234.

BURBANK, D. W., LELAND, J., FIELDING, E., ANDERSON, R. S., BROZOVIC, N., REID, M. R. & DUNCAN, C. 1996. Bedrock incision, rock uplift and threshold hillslopes in the northwestern Himalayas. *Nature*, **379**, 505–510.

CERLING, T. E. & CRAIG, H. 1994. Geomorphology and *in-situ* cosmogenic isotopes. *Annual Review of Earth and Planetary Sciences*, **22**, 273–317.

CHANG, H. K., KOWSMANN, R. O., FIGUERIREDO, A. M. F. & BENDER, A. A. 1992. Tectonics and stratigraphy of the East Brazil Rift system: an overview. *Tectonophysics*, **213**, 97–138.

CLEMSON, J., CARTWRIGHT, J. & BOOTH, J. 1998. Structural segmentation and the influence of basement structure on the Namibian passive margin. *Journal of the Geological Society, London*, **154**, 477–482.

DUMITRU, T. A. 1998. Fission-track geochronology. *In*: NOLLER, J. S., SOWERS, J. M. & LETTIS, W. R. (eds) *Quaternary Geochronology: Applications to Quaternary Geology and Paleoseismology*. Geological Society of America, Special Publication, in press.

GALLAGHER, K. 1995. Evolving temperature histories from apatite fission-track data. *Earth and Planetary Science Letters*, **136**, 421–435.

—— & BROWN, R. W. 1997. The onshore record of passive margin evolution. *Journal of the Geological Society, London*, **154**, 451–457.

—— & HAWKESWORTH, C. 1994. Mantle plumes, continental magmatism and asymmetry in the South Atlantic. *Earth and Planetary Science Letters*, **123**, 105–117.

——, BROWN, R. W. & JOHNSON, C. 1998. Fission track analysis and its application to geological problems. *Annual Review of Earth and Planetary Sciences*, **26**, 519–572.

——, HAWKESWORTH, C. & MANTOVANI, M. 1994. The denudation history of the onshore continental margin of S.E. Brazil inferred from fission track data. *Journal of Geophysical Research*, **99**, 18 117–18 145.

——, —— & —— 1995. Denudation, fission track analysis and the long-term evolution of passive margin topography: application to the S.E. Brazilian margin. *Journal of South American Earth Sciences*, **8**, 65–77.

GILCHRIST, A. R. & SUMMERFIELD, M. A. 1990. Differential denudation and flexural isostasy in formation of rifted-margin upwarps. *Nature*, **346**, 739–742.

——, KOOI, H. & BEAUMONT, C. 1994. Post-Gondwana geomorphic evolution of southwestern Africa: Implications for the controls on landscape development from observations and numerical experiments. *Journal of Geophysical Research*, **99**, 12 211–12 228.

GILLESPIE, A. R. & BIERMAN, P. R. 1995. Precision of terrestrial exposure ages and erosion rates estimated from analysis of cosmogenic isotopes produced *in-situ*. *Journal of Geophysical Research*, **100**, 24 637–24 649.

GREEN, P. F. 1986. On the thermo-tectonic evolution of Northern England: evidence from fission track analysis. *Geological Magazine*, **153**, 493–506.

Hsü, K. J. 1965. Isostasy crustal thinning, mantle changes and the disappearance of ancient land masses, *American Journal of Science*, **263**, 97–109.

Hurter, S. J. & Pollack, H. N. 1996. Terrestrial heat flow in the Paraná basin, southern Brazil. *Journal of Geophysical Research*, **101**, 8659–8671.

Kooi, H. & Beaumont, C. 1994. Escarpment evolution on high-elevation rifted margins: Insights derived from a surface processes model that combines diffusion, advection, and reaction. *Journal of Geophysical Research*, **99**, 12191–12209.

Laslett, G. M., Green, P. F., Duddy, I. R & Gleadow, A. J. W. 1987. Thermal annealing of fission tracks in apatite 2. A quantitative analysis. *Chemical Geology (Isotope Geoscience)*, **65**, 1–13.

Light, M. P. R., Maslanyj, M. P. & Banks, N. L. 1992. New geophysical evidence for extensional tectonics on the divergent margin offshore Namibia. *In*: Storey, B. C., Alabaster, T & Pankhurst, R. J. (eds) *Magmatism and the Causes of Continental Break-up*. Geological Society, London, Special Publications, **68**, 257–270.

——, ——, Greenwood, R. J. & Banks, N. L. 1993. Seismic sequence stratigraphy and tectonics offshore Namibia. *In*: Williams, G. D. & Dobb, A. (eds) *Tectonics and Seismic Sequence Stratigraphy*. Geological Society, London, Special Publications, **71**, 163–191.

Maslanyj, M. P., Light, M. P. R., Greenwood, R. J. & Banks, N. L. 1992. Extension tectonics offshore Namibia and evidence for passive rifting in the South Atlantic. *Marine and Petroleum Geology*, **9**, 590–601.

McKenzie, D. 1978. Some remarks on the development of sedimentary basins. *Earth and Planetary Science Letters*, **40**, 25–32.

Mohriak, W. U., Mello, M. R., Karner, G. D., Dewey, J. F. & Maxwell, J. R. 1989. Structural and stratigraphic evolution of the Campos Basin, offshore Brazil. *In*: Tankard, A. J. & Balkwill, H. R (eds) *Extensional Tectonics and Stratigraphy of the North Atlantic Margins*. AAPG Memoir **46**, 577–598.

—— et al. 1995. Salt tectonics and structural styles in the deep-water province of the Cabo Frio region, Rio de Janeiro, Brazil. *In*: Jackson, M. P. A., Roberts, D. G. & Snelson, S. (eds) *Salt Tectonics: A Global Perspective*. AAPG Memoir, **65**, 273–304.

Nishiizumi, K., Kohl, C. P., Arnold, J. R., et al. 1993. Role of *in-situ* cosmogenic nuclides ^{10}Be and ^{26}Al in the study of diverse geomorphic processes. *Earth Surfaces, Processes and Landforms*, **18**, 407–425.

Pazzaglia, F. J. & Brandon, M. T. 1996. Macrogeomorphic evolution of the post-Triassic Appalachian mountains determined by deconvolution of the offshore basin sedimentary record. *Basin Research*, **8**, 255–278.

Poag, C. W. & Sevon, W. D. 1989. A record of Appalachian denudation in postrift Mesozoic and Cenozoic sedimentary deposits of the U.S. middle Atlantic continental margin. *Geomorphology*, **2**, 119–157.

Pollack, H. N., Hurter, S. J. & Johnson, J. R. 1993. Heat flow from the Earth's Interior; Analysis of the Global Data Set. *Reviews of Geophysics*, **31**, 267–280.

Rust, D. J. & Summerfield, M. A. 1990. Isopach and borehole data as indicators of rifted margin evolution in southwestern Africa. *Marine and Petroleum Geology*, **7**, 277–287.

Sambridge, M., Braun, J. & McQueen, H. 1995. Geophysical parameterization and interpolation of irregular data using natural neighbours. *Geophysical Journal International*, **122**, 837–857.

Sleep, N. H. 1971. Thermal effects of formation of Atlantic continental margins by continental breakup. *Geophysical Journal of the Royal Astronomical Society*, **24**, 325–351.

Steckler, M. S. & Watts, A. B. 1978. Subsidence of the Atlantic-type continental margin off New York. *Earth and Planetary Science Letters*, **42**, 1–13.

Turners, S., Regelous, M., Kelley, S., Hawkesworth, C. & Mantovani, M. 1994. Magmatism and continental break-up in the South Atlantic: High precision ^{40}Ar/^{39}Ar geochronology. *Earth and Planetary Science Letters*, **121**, 333–348.

van der Beek, P., Andriessen, P. A. M. & Cloetingh, S. A. P. L. 1995. Morphotectonic evolution of rifted continental margins: inferences from a coupled tectonic-surface processes model and fission track thermochronology. *Tectonics*, **14**, 406–421.

Weissel, J. K. & Karner, G. D. 1989. Flexural uplift of rift flanks due to mechanical unloading of the lithosphere during extension. *Journal of Geophysical Research*, **94**, 13919–13950.

White, N. & Lovell, B. 1997. Measuring the pulse of a plume with sedimentary record. *Nature*, **387**, 888–891.

Williams, B. G. & Hubbard, R. J. 1984. Seismic stratigraphic framework and depositional sequences in the Santos Basin, Brazil. *Marine and Petroleum Geology*, **1**, 91–104.

Wolf, R. A., Farley, K. A. & Silver, L. T. 1987. Assessment of (U–Th)/He thermochronometry: The low-temperature history of the San Jacinto mountains, California. *Geology*, **25**, 65–68.

Zeitler, P. K., Herczeg, A. L., McDougall, I. & Honda, M. 1987. U–Th–He dating of apatite; A potential thermochronometer. *Geochimical et Cosmochimica Acta*, **51**, 2865–2868.

History of the northeast Brazilian rift system: kinematic implications for the break-up between Brazil and West Africa

RENATO M. DARROS DE MATOS

PETROBRAS, E&P–RNCE, Av. Int. Mario Camara 2783-Bl. F/07, Natal-RN, Brazil 59074-600

Abstract: The record of diachronic rift stages and widespread crustal deformation within the Brazilian–West African rift basins indicates changes in plate tectonics regimes and associated stress fields during the opening of the South Atlantic. Similarly to the presently active East African Rift System, the intracontinental rift basins of northeast Brazil provide a good example of an ancient intracontinental rift zone. They illustrate the three-dimensional geometries of rift systems and the control exercised by the basement structural framework, as well as important chronostratigraphic data that constrain the break-up of Gondwanaland between Brazil and West Africa.

The stepwise opening of the South Atlantic led to considerable deformation, recorded in northeast Brazil as magmatic and sedimentary events, and by at least three tectonic stages, Syn-rift I, II and III. The tectonic stages are defined in northeast Brazil by discrete episodes of half-graben formation distributed over three main rift axes: (1) the Gabon–Sergipe–Alagoas trend; (2) the Recôncavo–Tucano–Jatobá trend; (3) the Cariri–Potiguar trend.

Extensional deformation commenced in the Late Jurassic–early Berriasian and culminated in the late Aptian, when oceanic crust began to be emplaced between northeast Brazil and West Africa. Rift propagation is a matter of debate in the Equatorial Atlantic. It is suggested that the break-up progressed from south to north in the southern South Atlantic (Southern Branch) and concurrently from east and west in the Equatorial Branch. The rifting has been dated as Early Cretaceous (Neocomian–Barremian time) in the Southern Branch, and younger (Aptian–Albian) in the Equatorial Branch Palaeontological data indicate that these two branches intersected between northeast Brazil and Cameroon, Africa.

Numerous reconstructions of the opening history of the South Atlantic have been proposed. However, most of these models are based on partial observations or an incomplete understanding of the sedimentary and tectonic record, and have resulted in derivation of improbable regional stress fields. This paper provides a regional view of a series of intracontinental rift basins in northeast Brazil, emphasizing their differences and similarities, as well as the diachronism within the equatorial and the southern basins. These basins yield substantial chronostratigraphic data that constrain the temporal and spatial interaction of the rift phases, especially during the Neocomian. Palaeostress directions for the northeast Brazilian rift basins are, even with the best available data sets, difficult to determine because different tectonic regimes can result in similarly oriented rift basins, especially in the Borborema Province, northeast Brazil, where large Proterozoic shear zones determine the pre-rift tectonic framework. Examination of this region demonstrates the results of changes in the regional kinematic framework through time and the control exercised by the extensively deformed Proterozoic continental crust of rift geometries.

The easternmost region of the Equatorial Atlantic encompass a series of initially intracontinental rift basins associated with the intersection of the Southern and Equatorial Branches of the Atlantic Rift System (Fig. 1). Besides constraining the timing of rifting, the northeast Brazilian rift basins show that large continental masses did not behave rigidly, as they are commonly treated in plate reconstructions. The rift basins of northeast Brazil demonstrate diffuse Early Cretaceous extension, with deformation distributed over a 600 km wide region. As such, these rifts provide invaluable information concerning the kinematic evolution of complex rift system during continental break-up.

Conceição *et al.* (1988) have indicated that the early stages of deformation associated with the South Atlantic opening began during Jurassic–Triassic time (Early Sul Atlantiano Event, 225–160 Ma), considerably earlier than the major Cretaceous rifting event (Sul Atlantiano Event, 160–115 Ma). While the first event was characterized by magmatic activity and sedimentation in southern and northern Brazil, and neighbouring regions, the Sul Atlantiano Event marked the opening of the South Atlantic. It was accompanied by the development of major magmatic and tectono-sedimentary phenomena, including intraplate deformations in the intracontinental northeast Brazilian rift basins, and some of the earliest tectonic activity in the Benue Trough. However, most Early Cretaceous

Darros de Matos, R. M. 1999. History of the northeast Brazilian rift system: kinematic implications for the break-up between Brazil and West Africa. *In:* Cameron, N. R., Bate, R. H. & Clure, V. S. (eds) *The Oil and Gas Habitats of the South Atlantic.* Geological Society, London, Special Publications, **153**, 55–73.

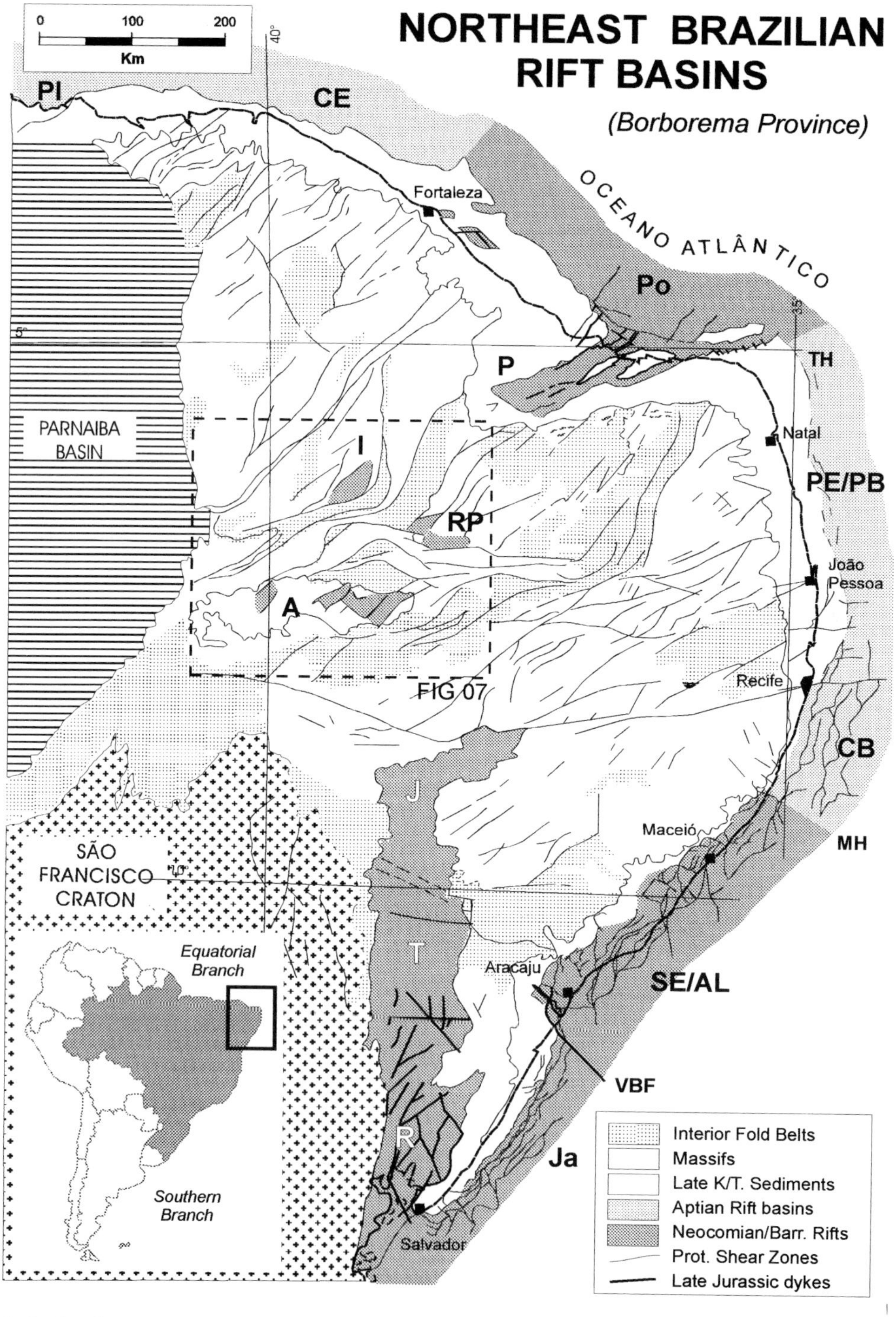

Fig. 1. Regional tectonic map of northeast Brazil illustrating the distribution of Early Cretaceous rift basins (two groups: Aptian and Neocomian) and the main geotectonic characteristics of the Precambrian Borborema Province. Basins: R. Recôncavo; T, Tucano; J, Jatobá; Ja, Jacuípe, SE/AL, Sergipe–Alagoas; CB, Cabo; A, Araripe; I, Iguatu; RP, Rio do Peixe; P, onshore Potiguar; Po, offshore Potiguar, PE/PB, Pernambuco–Paraíba; C, Ceará; PI, Piauí; between I and RP, Malhada Vermelha, Lima Campos and Icó.

intraplate deformation in Africa is post-Albian and occurred in the central and northern regions (Unternehr *et al.* 1988). Chronologic distribution of stratigraphic sequences within Brazil and West Africa (Fig. 3), indicates genetic links between the onshore northeast Brazilian rift basins and the Southern Branch of the Atlantic Rift System, while the Benue Trough had a rift phase synchronous with that of the Equatorial Branch. Paleogeographic reconstructions from Neocomian through Barremian and Aptian time are discussed based on detailed tectonic maps of the Brazilian and African basins, and regional potential field data from Petrobras and Geosat data bases (Fig. 2).

The pre-rift framework

Upper crustal extensional deformations were strongly influenced by the complex Proterozoic heritage as illustrated on the pre-drift reconstructions of northeast Brazil and western Africa (Fig. 4*a*). Dominant are the east–west trending megashear zones of Pernambuco (Brazil) and Ngaoundere (Africa) and the northeast–southwest trending shear zones between the Borborema (Almeida *et al.* 1977) and the Nigerian–Cameroon shields. These megashears are the product of the complex evolution of the Brasiliano–Pan–African orogenic chain (0.45–0.7 Ga). This orogeny resulted in a complex network of northeast trending fold belts/shear zones and regional east–west trending shear zones (Jardin de Sá 1994). Crustal heterogeneities are controlled by this network of shear zones and by two kinds of terrains (Fig. 1): (1) rigid gneiss migmatite terrains/granulite (massifs); (2) weak metasedimentary/volcanic fold belts. Both are bordered by the São Francisco, Congo, West African and São Luis Cratons.

Initial extension events

By the Late Jurassic, as deformational extension was spreading northwards in the Southern Atlantic Branch, the entire region may have undergone a general northwest–southeast extension. This is suggested by the consistency of the northeast oriented rift basins (Fig. 1) and by the regional north-northeast axis of the proto-South Atlantic. Structural style, rift dynamics and stratigraphic variations through time indicate at least three syn-rift stages: Syn-rift I, II and III (Chang *et al.* 1988; Matos 1987, 1992*a*). The Syn-rift I (Late Jurassic) sequence records the commencement of extensional deformation by deposition of clastics in a widespread and shallow depression (Fig. 3). The Syn-rift II (Neocomian) stage is characterized by the occurrence of taphrogenic strata deposited within linked intracontinental rift basins. Rifting reached its climax in the Southern Branch/northeast Brazil during the Syn-rift III stage (Barremian) when extension began to focus along the future continental margin.

Syn-rift phase I

The Syn-rift I stage comprises remnants of a regional intracontinental basin, the 'African–Brazilian Depression' of Ponte (1971). It is preserved in the Gabon, Congo and Cabinda Basins of Africa, and in the northeast Brazilian basins of Recôncavo, Tucano–Jatobá, Sergipe–Alagoas and Araripe (Netto & Oliveira 1985). The succession comprises alluvial fans and fluvial deposits with subordinate evaporites and aeolian sediments. Even though this unit is commonly grouped with the pre-rift section (Viana *et al.* 1971; Netto & Oliveira 1985; Magnavita 1992; Karner *et al.* 1992), it is assigned as a rift unit, recording the initial stages of rift-related extension, as proposed by Chang *et al.* (1988), and corroborated by the works of Figueiredo (1981), Conceição *et al.* (1988), Teisserence & Villemin (1989), Chang *et al.* (1992) and Matos (1992*a*).

Assigning a pre- or syn-rift significance to this Upper Jurassic–lower Berriasian section is a matter of recognizing, or not, whether extensional deposits are only characterized by the presence of fault-related conglomerates and/or by a set of divergent wedge-like seismic reflectors with subordinate onlap patterns. Even though this pattern does indicate a creation of space faster than the rate of sediment supply, it may not indicate the onset of extensional deformation. Spaces created earlier may be the product of a mild upper-crustal response to a synchronous extension. The onset of rifting has been dated as coeval with the occurrence of the oldest fanglomerates of the Salvador Formation in the Recôncavo, Tucano and Jatobá Basins, which are considered to be of late Berriasian age (Magnavita 1996). On the other hand, Silva (1996) relocated the onset of rifting within the lower Berriasian strata of the Recôncavo–Tucano–Jatobá rift basins, based primarily on palaeogeographical studies, including grain-size distribution, external geometry and erosional surfaces in the Agua Grande Formation, which would indicate a time-lag between the pre- and syn-rift sediments.

However, lithospheric weakening started earlier, during the Late Jurassic. The African–Brazilian

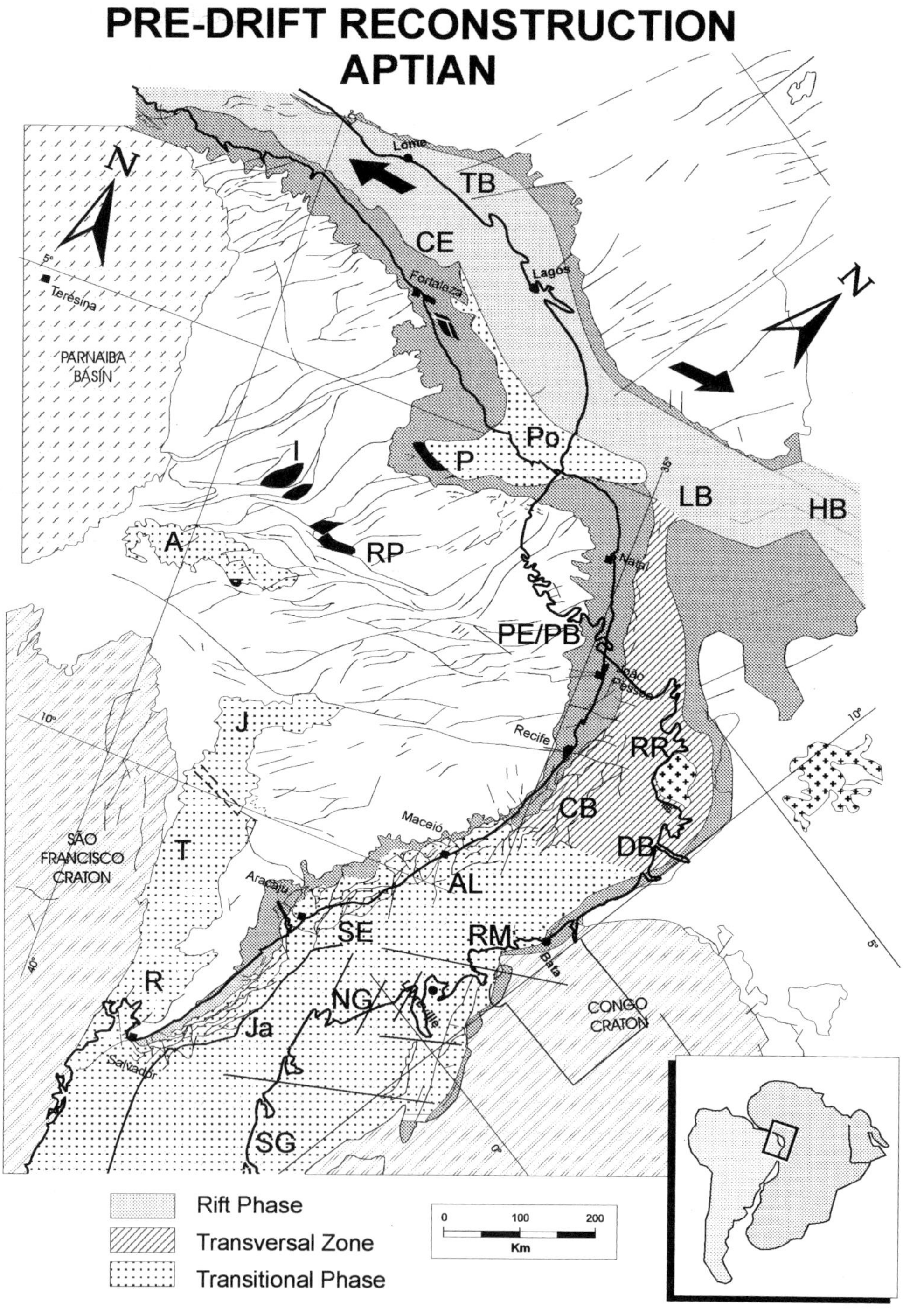

Fig. 2. The Aptian pre-drift Scenario: palaeogeography and the spatial distribution of rift basins immediately prior to the break-up of the South Atlantic. The figure illustrates the very different stages of development of the branches of the South Atlantic Rift System. The large arrows indicate the direction of tectonic transport in the Equatorial Branch (after Matos 1992*b*). African basins: TB, Togo–Benin; LB, Lower Benue; HB, High Benue; RR, Rio del Rey; DB, Douala; RM, Rio Muni; NG, North Gabon; SG, South Gabon.

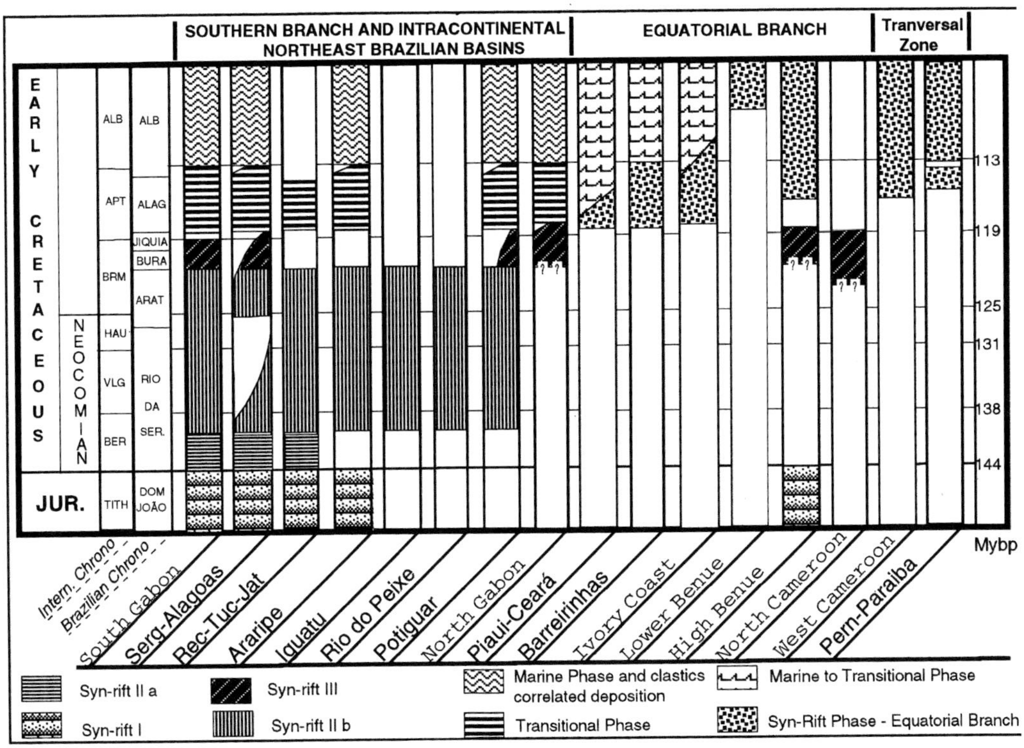

Fig. 3. Chart demonstrating the spatial and temporal distribution of sediments in northeast Brazil and West Africa during the Late Jurassic–Cretaceous (references in Matos 1992*a*). The relationship between the international chronostratigraphic scale (Harland *et al.* 1982) and the Brazilian chronostratigraphy is on the left (Regali & Viana 1989).

depression is elongated in a north–south direction at a few degrees to the later north-northeast–south-southwest orientated Neocomian rift basins (Fig. 4*a*). Also, deposition was synchronous with Late Jurassic extensional-related magmatism recorded in the proto-equatorial and southern South Atlantic (Conceição *et al.* 1988). Some conglomerates and coarse- to medium-grained sandstones of the basal N'Dombo Formation of the Gabon Basin may be related to syntectonic faulting (Teisserence & Villemin 1989).

The absence of major syn-depositional faults may indicate an elastic response of the upper crust to ductile subcrustal stretching and thinning; while the lack of the Late Jurassic magmatism in the Recôncavo–Tucano–Jatobá and Gabon rift systems, and the lack of a thermal-domed palaeogeography, may indicate a passive or cold non-volcanic rift disassociated with an asthenospheric plume. However, magmatic activity occurred north of the Patos shear zone during the Syn-rift I stage. In Brazil, it is represented by an arcuate array of tholeiitic dykes, locally en-echelon, trending northeast–southwest to east–west, that intrude the Precambrian basement at the southern edge of the onshore Potiguar Basin (Fig. 4*a*). This alkaline magmatism (Rio Ceará Mirim Magmatism; Gomes *et al.* 1981), yielded K–Ar ages ranging between 150 and 120 Ma (Neocomian–Late Jurassic), with two peaks of activity at 145 and 130 Ma (Oliveira 1992): the remnant magnetism best fits a Jurassic palaeomagnetic pole position (Bucker *et al.* 1986). The arcuate trajectory of these dykes led Oliveira (1992) to suggest a change in the least-principal stress axis from northwest–southeast to north–south around the Potiguar Basin. This change in the stress field corroborates the suggestion of a clockwise rotation of the least-principal compressive stress north of the Patos Shear Zone made by Matos (1992*a*). Magmatism in Benue Trough are represented by the Burashika volcanic complex (147+7 Ma, K–Ar; Popoff *et al.* 1988) and earlier granitic intrusions of the Jos Plateau (Nigeria), with Triassic–Jurassic ages (213–141 Ma, Rb–Sr; Vail 1989).

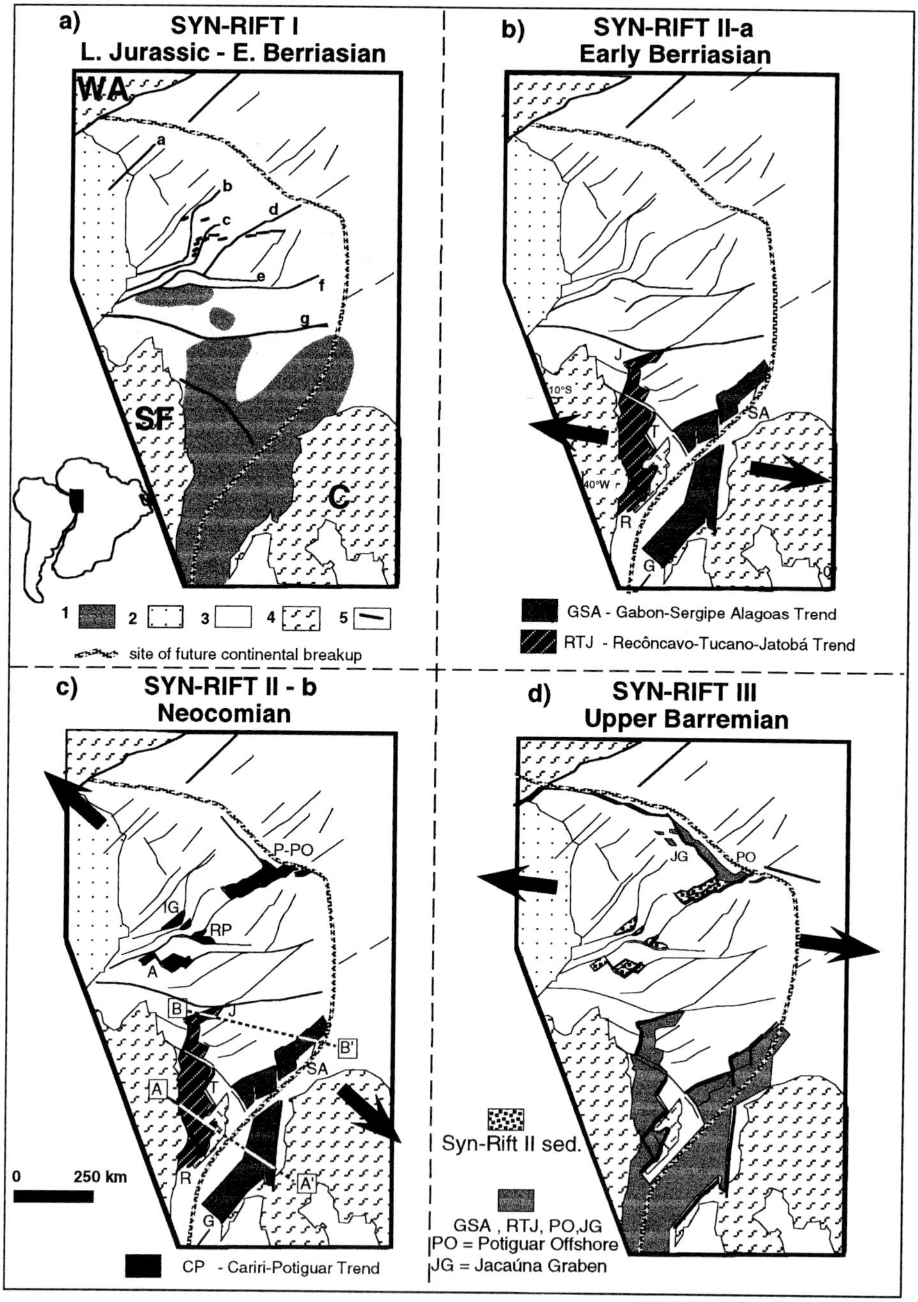

Fig. 4. Pre-drift reconstructions of northeast Brazil and West Africa during four tectonic stages (modified after Matos 1987, 1992*a*). The large solid arrows indicate the main direction of tectonic transport (**a**) Syn-rift I (Late Jurassic) and the development of the Afro–Brazilian Depression. Thick lines with letters are locations of the main Proterozoic shear zones: a, Sobral Pedro II; b, Orós; c, Jaguaribe; d, Porta Alegre; e, Malta; f, Patos; g, Pernambuco; h, Macureré; i, Ngaundere; j, Dahomeyan; l, Late Jurassic sediments; 2, Parnaíba Basin; 3, Proterozoic terrains; 4, Precambrian cratons WA, West African–São Luis Craton; SF, São Francisco; C, Congo; 5, Late Jurassic dykes. (**b**) Syn-rift IIa (early Berriasian). Crustal fracturing was initiated at this time in the Recôncavo–Tucano–Jatobá and Gabon–Sergipe–Alagoas trend, but extension in the Cariri–Potiguar trend further north had yet to begin. Rift kinematics in the Recôncavo–Tucano–Jatobá trend as discussed by Magnavita (1992)

Syn-rift phase II

During the main rift stage (Neocomian–early Barremian) diffuse continental rifting was responsible for stretching and thinning the crust over a 600 km wide region in three major rift segments: Gabon–Sergipe–Alagoas, Recôncavo–Tucano–Jatobá and Cariri–Potiguar. These intracratonic rift valleys (Fig. 1) consist of a set of northeast trending, asymmetric-tilted half-grabens separated by basement highs, transfer faults and/or accommodation zones. These features are controlled by the distribution of Proterozoic lithotectonic units, major faults and transverse megashear zones. The basins share the same tectonic signature: they are typically a few tens of kilometres wide and a couple of hundreds kilometres long. The northeast–southwest trending border faults lie almost perpendicular to an inferred northwest–southeast extension. The azimuth of the planar fabrics (schistosity, gneissosity and shear zones) controlled the location and nucleation of the basin-bounding normal faults, as well as along-axis changes in rift polarities (Matos 1992*b*).

Recently, based on detailed biostratigraphic research at PETROBRAS, a subtle, but significant, diachronism for the onset of rifting between the southern basins (Recôncavo–Tucano–Jatobá and Gabon–Sergipe–Alagoas trends and the Cariri–Potiguar trend has demonstrated the northwards propagation of rifting through time. This led to the division of the Syn-rift II stage into units, IIa (Fig. 4b) and IIb (Fig. 4c). During the early Berriasian (Syn-rift IIa), crustal fracturing was restricted to the south of the Pernambuco lineament, with the deposition of the N'Dombo, lower Barra de Itiuba and 'Itaparica–Agua Grande–Lower Tauá' Formations in the Gabon, Sergipe–Alagoas and Recôncavo–Tucano–Jatobá Basins, respectively. These narrow rift valleys represent the onset of the Syn-rift II stretching phase. Based on fault kinematics indicators observed in the Recôncavo and Jatobá Basins, Magnavita (1992) proposed an east–west extension during this stage.

During the Syn-rift IIb (late Berriasian–early Barremian) stage, extensional deformation jumped westward (Fig. 4c) from the easternmost basins (Gabon–Sergipe–Alagoas and Recôncavo–Tucano–Jatobá trends) to the northeast trending intracratonic basins making up the Cariri–Potiguar trend. The trend comprises the Araripe, Rio de Peixe, Iguatu, Malhada Vermelha, Lima Campos and Potiguar Basins. Each basin forms an asymmetric half-graben separated from the next basin by basement highs, transfer faults and/or accommodation zones. A general north-northwest–south-southeast tectonic transport direction is consistent with the development of these northeast–southwest half- grabens throughout the Borborema Province. At this time, the megashear zones of Pernambuco (Brazil) and Ngaundere (Africa) apparently behaved like a huge accommodation zone balancing extensional deformation along the Recôncavo–Tucano–Jatobá/ Gabon–Sergipe–Alagoas trends with simultaneous extensions along the Cariri– Potiguar trend.

The Recôncavo–Tucano–Jatobá and Gabon–Sergipe–Alagoas trends

The Recôncavo–Tucano–Jatobá northeast–southwest trending asymmetric grabens separated by basement highs and northwest–southeast trending transfer faults (Fig. 5). Total syn-rift thickness reaches > 10 000 m in some of the depocentres of the Recôncavo and South Tucano Basins. The Vaza Barris Arch accommodated motion between opposite-facing half-grabens (the central Tucano and north Tucano Grabens, Fig. 5) which developed as disconnected sub-basins with (probably) distinctive rift sedimentation during the Syn-rift II phase (Magnavita & Cupertino 1988). Figure 5 shows five geologic cross-sections through the Tucano–Jatobá rift valley and one dip-section in the offshore Sergipe Basin. The section shows that the maximum syn-rift sediment thicknesses are positioned on the west side in the North Tucano and Jatobá Grabens, while in the central–south Tucano and Recôncavo Grabens they are located on the east side of the rifts. Even though the sedimentary record ranges from Late Jurassic to late Aptian, only minor thermal subsidence, post-rift sediments are present in this rift system. The absence of a normal post-rift, thermal sag-phase has been used to support a simple shear crustal detachment model for this rift system (Ussami *et al.*

and after Matos (1994). (**c**) Syn-rift IIb (late Berriasian–early Barremian) period when widespread fracturing developed in the Borborema Province. The primary results were the consolidation of the Recôncavo–Tucano–Jatobá and Gabon–Sergipe–Alagoas rift trends, and the formation of the Cariri–Potiguar rift valley. Brazilian and African basins as in Figs 1 and 2. A–A′ and B–B′ mark the location of the schematic cross-sections illustrated in Fig. 6. (**d**) Syn-rift III (upper Barremian). Major chances in rift kinematics occurred throughout the region at this time with diffuse deformation in the Borborema Province and Nigerian Shield.

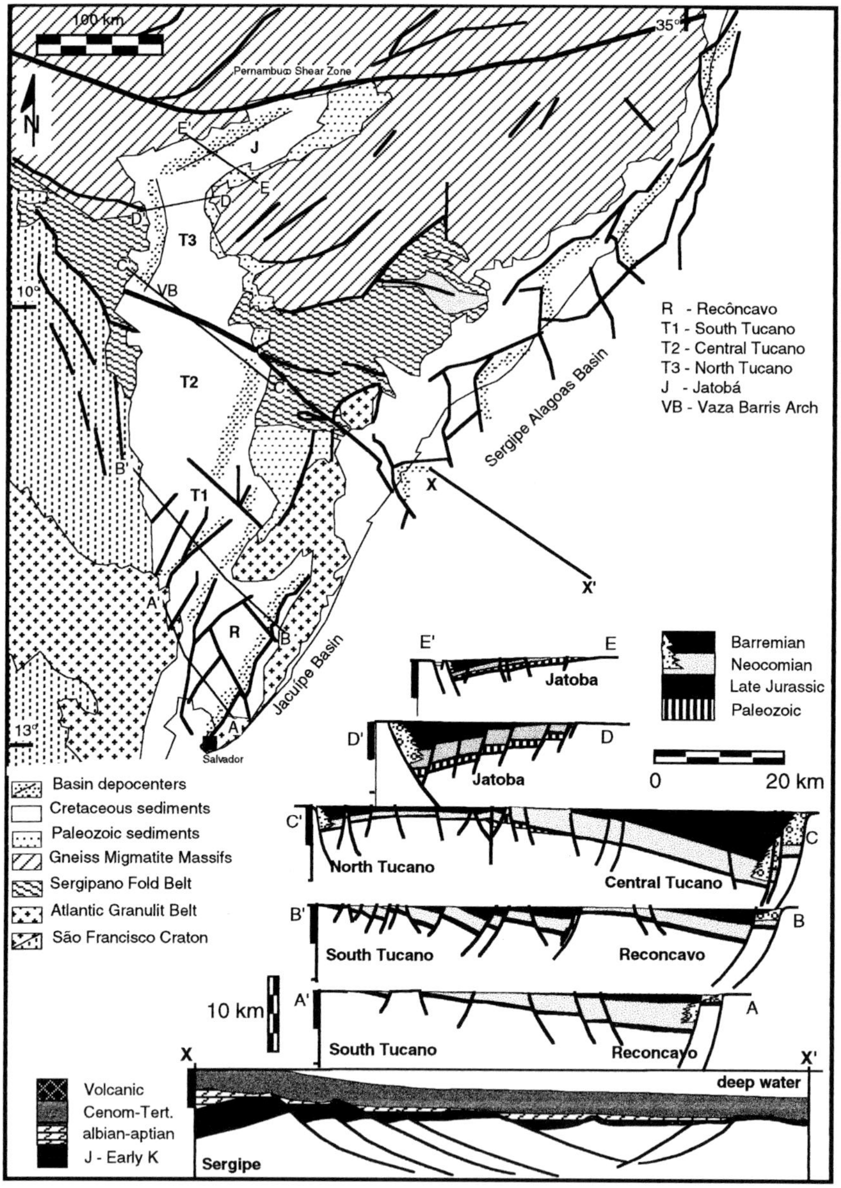

Fig. 5. Regional tectonic framework of the Recôncavo–Tucano–Jatobá/Sergipe–Alagoas Basins. Northeast trending border faults dominate the rift architecture which is characterized by half-grabens with common azimuths displaced by northwest–southeast lineaments. These comprise a mixture of sharp transfer faults, diffuse, faults and extensional folds. The upper five cross-sections illustrate the Recôncavo–Tucano–Jatobá trend and the lower section the offshore Sergipe–Alagoas Basin [after Magnavita & Cupertino (1988), Milani & Davison (1988), Matos (1992*a*), Mohriak *et al.* (1995)].

1986; Kusznir *et al.* 1987; Castro Jr 1987), or to advocate magmatic underplating and uplift/erosion of the faulted rift flanks (Magnavita *et al.* 1994).

The onshore/offshore Sergipe–Alagoas Basin is the most complex of the northeastern Brazilian basins (Fig. 5). Virtually all the rift and post-rift sedimentary record is present. There are strong links with the southern branch of the Atlantic Rift System. The rift architecture of the Sergipe–Alagoas Basin is composed of a series of north–northeast trending half-grabens, controlled by an array of north–south, northwest and northeast trending bounding faults (Bacellar & Costa 1993), and bounded by the Maragogi High and Vaza Barris Fault Systems (MH and VBF in Fig. 1). Sedimentary thickness may reach up to 10 000 m in local depocentres (Cainelli 1992; Mohriak *et al.* 1995). Pre-rift, syn-rift and post-rift sediments outcrop onshore along a narrow coast-parallel strip.

The north Gabon Basin is the African counterpart of the Sergipe–Alagoas Basin. It illustrates features typical of a rifted continental margin (Teisserence & Villemin 1989). The south sub-basin of Gabon contains a complete rift section, with ages ranging from Neocomian to Barremian. This succession is subdivided by Teisserence & Villemin (1989) into three stages: early extension (Syn-rift I), followed by a late Neocomian to early Barremian rift phase (Syn-rift II), and a late Barremian phase of rapid sedimentation and shift westward of the main depocentres of the south sub-basin (Syn-rift III). As with the Brazilian Margin, a transitional Aptian section (with salt) and an Albian–Eocene drift sequence is developed. The N'Komi Fracture Zone divides the south and north basins. This fracture separates the predominantly Neocomian aged north basin from the near-complete rift and drift section occupying the south basins (Teisserence & Villemin 1989).

Ussami *et al.* (1986) provide a regional view of the development of rift depocentre, in the Recôncaro–Tucano–Jatobá and Gabon–Sergipe–Alagoas rift trends by proposing an asymmetrical, crustal-scale detachment linking the Tucano Basin and the Gabon Basin in Africa. Castro Jr (1987) proposed an elegant double rift system model with multiple crustal detachments accommodating the approximately northwest–southeast extension direction. Davison (1988) pointed out that detailed gravity data suggested localized crustal thinning below the Recôncavo–Tucano–Jatobá rift. Matos (1992*a*, *b*) proposed an alternative model to Castro Jr which involves mid-crustal detachment surfaces (Fig. 6). The fact that the basin-bounding faults of the north

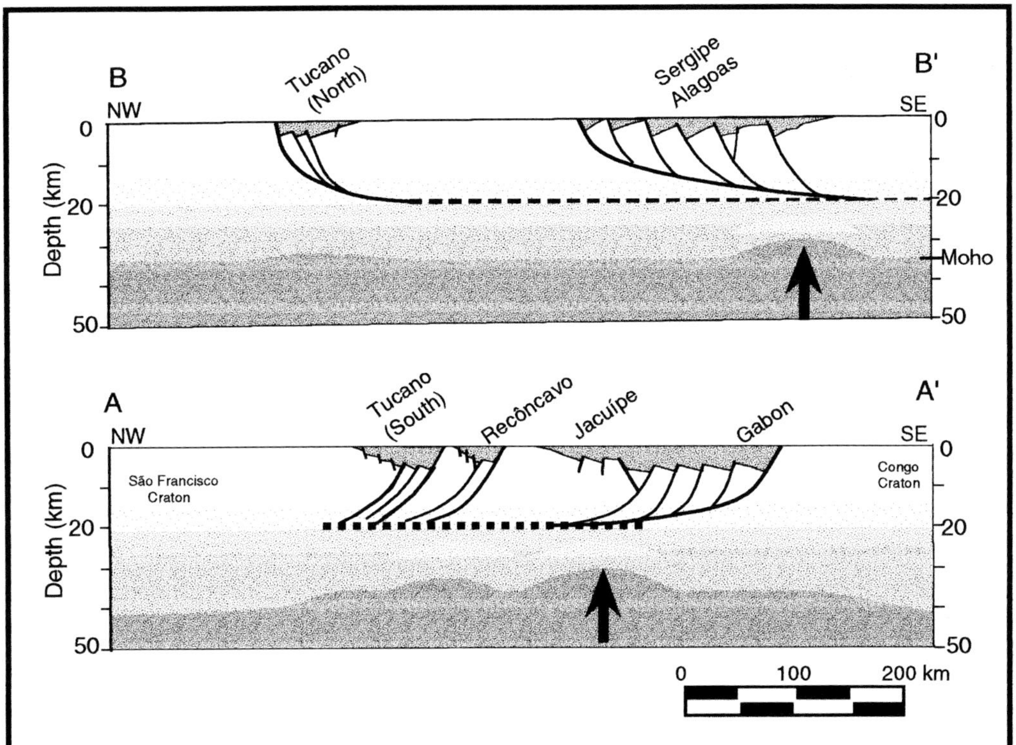

Fig. 6. Crustal scale schematic structural sections across the Recôncavo–Tucano–Jatobá and Gabon–Sergipe–Alagoas rift trends [after Castro Jr (1987) and modified from Matos (1992*a*)] whose location is shown in Fig. 4c. The dashes show the intracrustal horizon along which the major, graben-bounding faults would be generated. Note the opposite azimuths between the northern and the southern rifts. The arrows represent the future site of continental break-up.

Tucano Basin are not parallel to the Precambrian fold belt structures does not necessarily indicate that the detachment model is not applicable, as suggested by Davison (1988).

Although many authors have described and proposed different kinematic models for the Recôncavo–Tucano–Jatobá and Gabon–Sergipe–Alagoas trends (Milani 1985; Milani & Davison 1988; Szatmari *et al.* 1985; Ussami *et al.* 1986; Magnavita & Cupertino 1988; Matos 1992*a*; Magnativa 1992), a common characteristic of all these models is an acknowledgement of the importance of the Precambrian heritage, which controlled the simultaneous birth of both of the trends through a rift bifurcation which developed close to the site of the city of Salvador in Brazil (Fig. 1). This rift bifurcation was controlled by the rigidity of the São Francisco–Congo Cratons granulite belt and the presence of the Pernambuco–Ngaoundere shear zone further north, which prevented the subsequent northward propagation of both of the trends. Since rift fault orientations are strongly controlled by the anisotropy of the basement section, differences in fault trends need not be the product of regional kinematic variations. Moreover, since rifting was taking place from south to north, changes in deformation rates and space creation through time will result in polyphasic rift development. As the majority of border faults are northeast–southwest trending, it is conceivable that the direction of tectonic transport during rifting was roughly perpendicular to that direction, at least during the main stretching phase. Magnavita (1992) proposed a double-stage evolution model for Recôncavo–Tucano–Jatobá trend based on fault kinematic indicators in the Recôncavo and Jatobá Basins. This model envisages east–west extension during the early rifting stages (Berriasian) followed by a northwest–southeast extension during the late Barremian–early Aptian).

The Cariri Potiguar trend

During the Syn-rift IIb phase, rifting migrated northwestwards as the zone of rifting widened to form a series of northeast trending intracratonic basins: Araripe, Rio do Peixe, Iguatu, Malhada Vermelha, Lima Campos and Potiguar Basins (Figs 1 and 4c). The megashear zones of Pernambuco (Brazil) and Ngaoundere (Africa) Basins apparently acted as a large accommodation zone, balancing extensional deformation along the Recôncavo–Tucano–Jatobá and Gabon–Sergipe–Alagoas trends with simultaneous extension along the Cariri Potiguar trend. The shear tracks, the framework of supracrustal sequences (interior fold belts) within the Borborema province (Matos 1987), and upper crustal heterogeneity controlled the nucleation of the resulting extensional faults.

The Cariri Valley rift basins (Fig. 7) encompass the Araripe Basin and a series of small basins in central northeast Brazil characterized by a synchronous Neocomian aged rift phase. Only portions of these basins are preserved following regional Late Cretaceous–Early Tertiary uplift and the formation of the Araripe Plateau. The remnants form a series of 5–20 km wide half-grabens named the Rio do Peixe, Iguatu, Malhada Vermelha, Icó and Lima Campos Basins. These are filled with sandstones, conglomerates, siltstones, shales and rhythmites. Neither Late Jurassic or late Barremian–Aptian aged beds are known.

The concave shape of the master normal faults bordering the Iguatu and Lima Campos Basins is the result of the reactivation of the Limoeiro and Orós shear zones, respectively, suggesting a roughly orthogonal, northwest–southeast extension. The Rio do Peixe Basin is located where the east–west-trending Patos megashear zone fans out into a series of northwest to northeast trending shear zones that resemble, in map form, large-scale sigmoids (Fig. 7). The western sub-basin (northeast–southwest trending Brejo das Freiras) shows typical half-graben geometries, ranging in width from 5 to 15 km (Françolin & Szatmari 1987). The east–west trending Souza sub-basin is generally controlled by the east–west trending segment of the Malta Fault and shows structural dips ranging from southeast (western part), to south (central) and to northwest (eastern part). The main structural characteristic is the northeast–southwest trending faults truncated by the east–west or northwest–southeast trending fault sets.

The rift sediments of the Araripe Basin lie above basement or Devonian aged remnants of the neighbouring Palaeozoic Parnaíba Basin. The Jurassic sequence of the Araripe Basin represents the northernmost occurrence of sediments associated with the Syn-rift I phase. Gravity, magnetic and regional seismic surveys indicate the presence of two sub-basins, the Feira Nova sub-basin in the east and the Crato sub-basin in the west. Both are bounded by northeast–southwest trending normal faults (Fig. 7), and possibly also northwest–southeast trending transfer faults and/or accommodation zones. Up to 1800 m of Neocomian aged fluvial deltaic sediments are present (Arai *et al.* 1989).

The basement of the Cariri Valley is composed of a complex Proterozoic sequence of folded

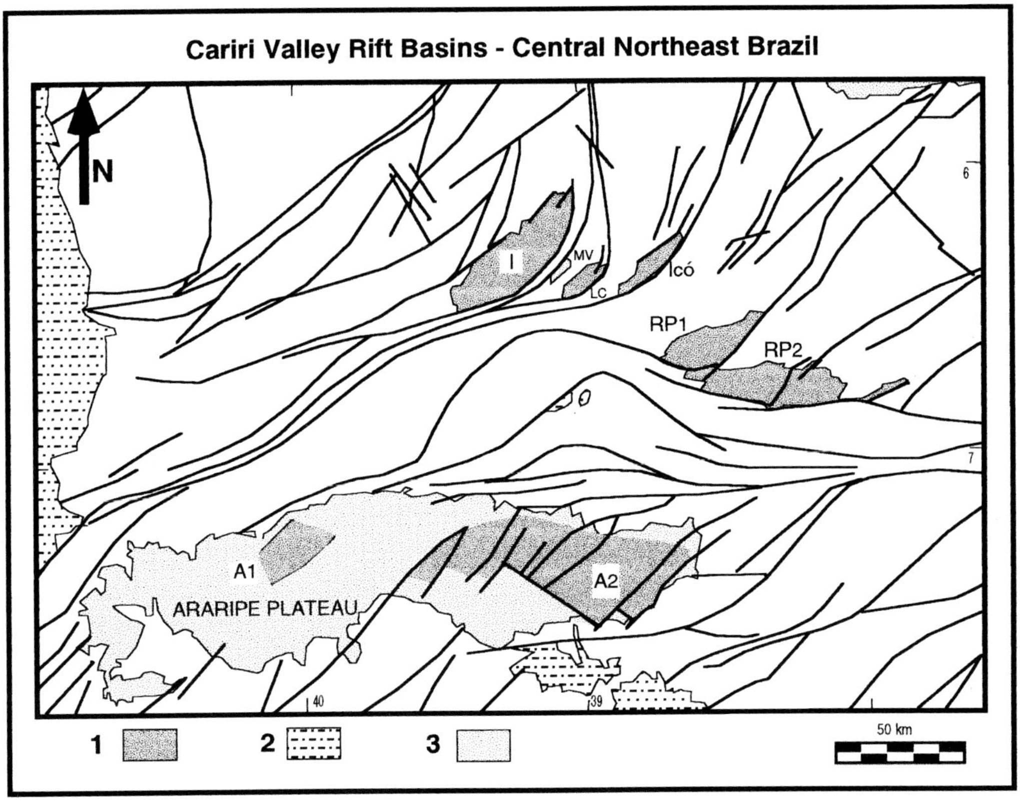

Fig. 7. Structural framework and syn-rift distribution of the Cariri Valley rift basins in central northeast Brazil (see Fig. 1 for location), after Matos (1987). Thick lines with letters are locations of the main Proterozoic shear zones. 1, Distribution of Neocomian rift sediments deposited during the Syn-rift IIb stage; 2, Paleozoic sediments – Parnaíba Basin; 3, Post rift sediments; A, Araripe Basin; A1, Feira Nova sub-basin; A2, Crato sub-basin; I, Iguatu; MV, Malhada Vermelha; LC, Lima Campos; Icó, Icó Basin; RP, Rio do Peixe Basin; RP1, Brejo das Freiras sub-basin; RP2, Souza sub-basin.

belts and massifs cut by a network of sigmoidal shear zones related to the Pernambuco and Patos megashears (Fig. 1 and 8). During the Neocomian–early Barremian, northwest–southeast extension inverted the former northeast–southwest transpressional faults as northeast trending normal faults, and reactivated a small segment of the east–west trending Patos shear zone under transtensional conditions to form the Souza sub-basin (Matos 1987; Fig. 8).

The northernmost expression of Neocomian rifting occurs in the Potiguar Basin, the easternmost equatorial marginal basin of Brazil (Figs 1 and 9). The Neocomian–Barremian rift sequence is capped by an Aptian–Campanian aged fluvial to marine transgressive sequence, and a regional progradational sequence of Tertiary clastics and carbonates (Souza 1982). Structural style, rift architecture, facies distribution and rift v. post-rift thickness are the most significant differences between the onshore and offshore Potiguar Basin (Matos *et al.* 1987). During the stretching phase, the onshore rift was filled with fluvial, deltaic, lacustrine and deltaic fan sediments ranging in age from early Neocomian to early Barremian. The offshore rift remained active, as described in next section, until the late Barremian–early Aptian.

The onshore rift is characterized by a rhombohedral map form geometry made up of northeast–southwest trending half-grabens and intervening highs controlled by a double system of intracrustal listric normal faults. These are considered to overlie favourably aligned northeast trending basement faults formed during the Brasiliano–Pan–African Orogeny (Matos 1987, 1992*a*). Maximum onshore syn-rift thickness are 5–6 km along the Carnaubais fault system. The

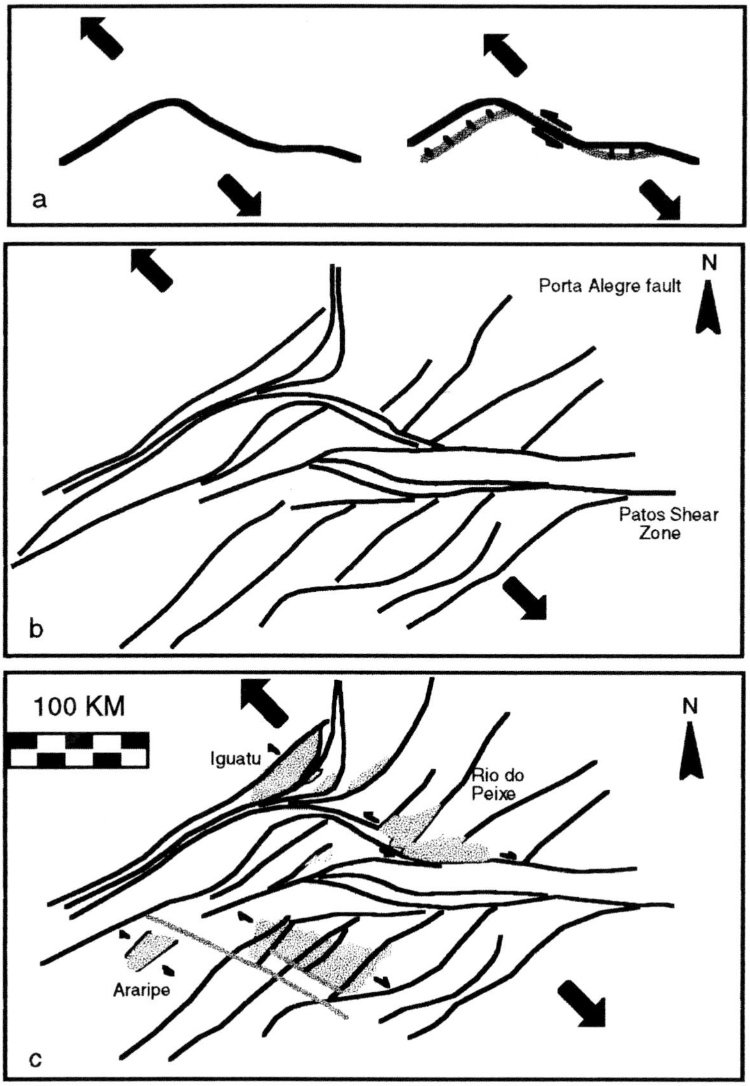

Fig. 8. Schematic illustration of the tectonic evolution of the Cariri Valley rift basins [after Matos (1992*a*)]. Northwest–southeast extension would reactivate the most favourable segments of pre-existing sigmoidal-like shear zones [(**a**) and (**b**)] to generate northeast–southwest trending half-grabens and the transtensional Souza sub-basin of the Rio do Peixe Basin (**c**).

associated gravity anomaly indicates a major northeast–southwest axis of high-density material (Matos 1992*b*) beneath the low-density sediments of the onshore rift; evidence which corroborates the inference of a roughly north-northwest–south-southeast extensional stress field.

Just offshore, the Macau High separates the onshore pattern of northeast–southwest half-grabens from a Syn-rift III aged set of northwest–southeast trending half-grabens. Neither Syn-rift I nor Syn-rift II aged sediments are known westward of the Macau Low in the offshore Potiguar Basin (Fig. 9; Cremonini *et al.* 1996). Highly oblique, shear conditions in the western offshore Potiguar Basin may have prevented the deposition and/or preservation of Syn-rift II sediments, while in the eastern part (north-northeast of the Macau High) a deep and wide basin developed (Fig. 9 and 10a). The Syn-rift II stage was responsible for weakening the lithosphere in the Potiguar Basin and probably

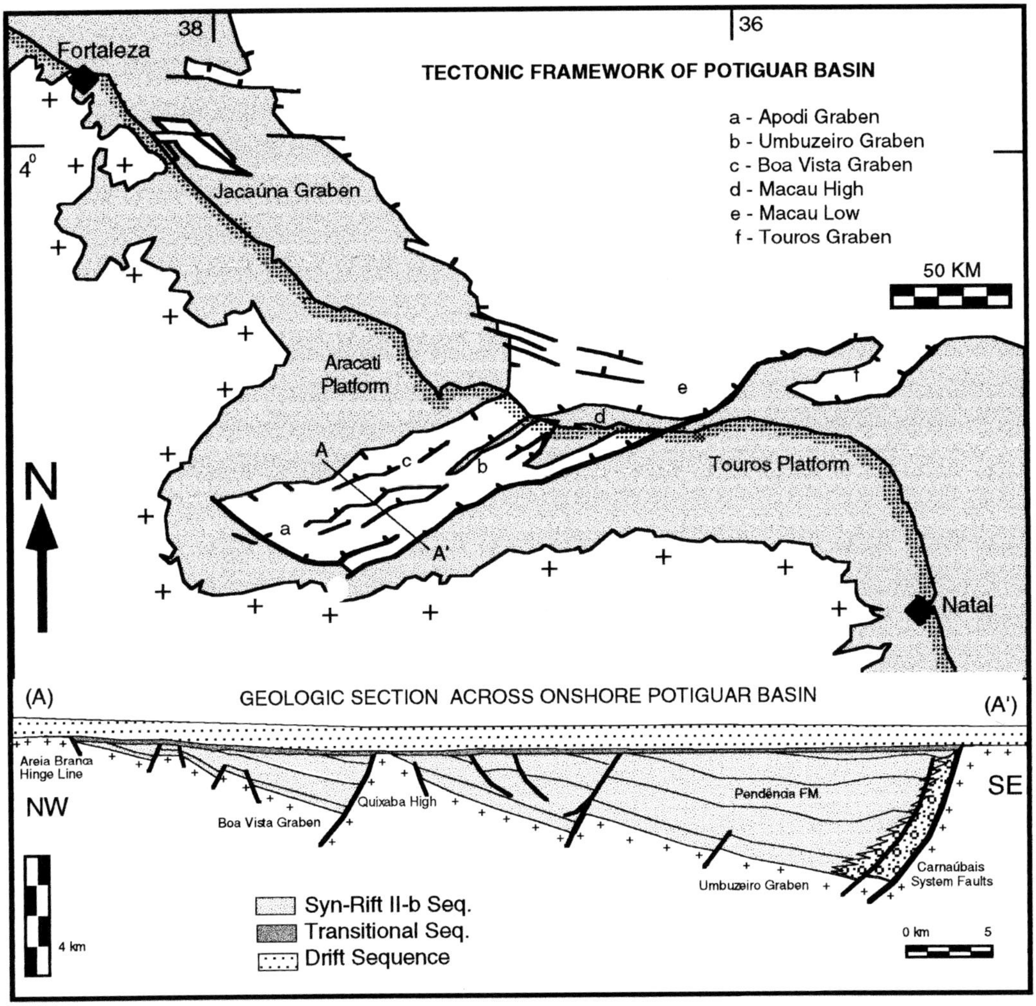

Fig. 9. The Potiguar Basin – tectonic framework and an onshore geologic cross-section.

helped initiate the subsequent northwest–southeast trending offshore grabens by reactivating the former northwest–southeast transfer faults/accommodation zones.

Syn-rift phase III

The Syn-rift III phase (late Barremian) records a major change in the style of deformation related to the initial linkage between the Southern Branch of Atlantic rifting and the early stages of rifting in the eastern equatorial domain (Fig. 4d). The highest strain/depositional rates are recorded in the Sergipe–Alagoas and Gabon Basins, where 2000–3000 m of rift sediments of this age are present (Chang *et al.* 1988; Matos 1992*a*; Teisserence & Villemin 1989). During this period, the final rift architecture of the Gabon–Sergipe–Alagoas trend was delineated, the rift sedimentation of the Cariri Valley and onshore Potiguar Basin was aborted, and major deformation was initiated in the Equatorial Branch. In the Recôncavo–Tucano–Jatobá trend, the subsidence rate decreased and the grabens were fully filled by a fluvio-deltaic system. In the north Gabon sub-basin, late Barremian rift sediments are represented by the N'Toum Formation (Teisserence & Villemin 1989). This major change in rifting kinematics was contemporaneous with early stages of intracontinental shear and rifting in the Benue Trough in Africa. The Pan-African dextral shear zones (Benkhelil 1988) and/or the Late Jurassic fracture zones (Popoff 1988) transmitted crustal deformation toward the interior of the African continent. Barremian basal conglomerates are developed in the North Cameroon Basin (Popoff 1988), as are possible Barremian clastics (Benkhelil 1988; Popoff 1988).

As the Syn-rift III phase progressed in the Benue Trough in Africa, continuous rift sedimentation became possible in the offshore Potiguar Basin. Deep northwest–southeast trending tilted half-grabens formed related to a northwest–southeast trending stretching event that overprinted the previously described northeast–southwest rift trend and propagated towards the northwest. Rift sediments deposited on the Fortaleza High (Fig. 4c, d, 9, 10b) where a series of small rhomb-type grabens developed and were filled with late Barremian sediments: this event provides direct evidence of an east–west extension synchronous with early stages of rifting in the Benue Trough. The Jacaúna Graben (Fig. 9) is a typical transtensional graben characterized by two opposing half-grabens, pull-apart geometry and bounded by east–west transtensional border faults that became northwest–southeast trending normal faults at the releasing bend. The slight diachronism and the significant change in strain orientation during the evolution of the onshore Potiguar rifts strongly affected their evolution. Even though the onshore Potiguar rift was still subject to an extensional stress field, late Barremian sediments were not deposited in the basin. This is because large volumes of rock were removed by erosion as the onshore basin had become, in the new tectonic regime, the rift shoulder for the offshore basin.

These changes suggest that the main direction of tectonic transport switched from north-northwest (Syn-rift IIb) to a roughly east–west orientation (Syn-rift III; Fig. 10). However, based on extensive field work, Magnavita (1992) deduced, for the inland Recôncavo–Tucano–Jatobá trend, the presence at this time of a northwest–southeast extension. However, local stress field variations may explain this discrepancy. At this time (late Barremian), lithospheric rupture was taking place to the south and generating sufficient momentum for a divergent movement. A triangular crustal block between the Recôncavo–Tucano–Jatobá and Gabon–Sergipe–Alagoas trends, termed the 'East Brazilian Microplate' by Milani (1985), here considered to be a microcontinent named 'Northeast Brazilian Microcontinent', would have rotated counterclockwise as an independent entity to accommodate the differential distribution of forces generated by the clockwise and counter-clockwise movement of Africa and South America, respectively. The result was minor left lateral motion in the Sergipe–Alagoas Basin and a northwest–southeast oriented tectonic transport in the Recôncavo–Tucano–Jatobá trend (Fig. 11). The main difference to Milani's model is in the timing of these movements. While Milani proposed microplate rotation during the early stages of rifting, in this model this could only happen during late stages of rifting, when the lithosphere would be stretched enough to allow relative movements of individual crustal blocks. High strain rates observed in the Southern Branch during the Syn-rift III stage had to be accommodated southward of the Pernambuco– Ngaoundere shear zone because of the presence of this major, lithospheric-anchored transversal structure. Anomalous local stress fields were developed, changing through space and time, which led to along-axis variations in the direction of tectonic transport in the Recôncavo–Tucano– Jatobá rift valley.

During the Aptian (Fig. 2), contrasting depositional environments are observed: the Southern Branch developed a 'Transitional Evaporitic Megasequence' (Chang *et al.* 1988) during the time that the Equatorial Branch experienced its main rift stage, with simultaneous rift propagation throughout the west-northwest trending equatorial rupture zone. While the Equatorial Branch and Benue Trough experienced the main rift phase, the Recôncavo–Tucano–Jatobá trend was aborted and the Gabon–Sergipe–Alagoas trend developed a transitional phase between the rift and drift stage, and the Southern Branch resumed its northward propagation, when a set of west-dipping half grabens were developed in the Cabo Basin (south of the Pernambuco Paraíba Basin in Brazil).

With the cessation of extension, marine-influenced evaporites accumulated in the Gabon–Sergipe–Alagoas trend discordantly above the former rift sequences (Chang *et al.* 1988): this period marks the transition between the rift and drift phases. Even though subsequent regional uplift may have removed part of the post-rift thermal subsidence of the Recôncavo–Tucano–Jatobá Basins, their Aptian-post-rift sediments consist solely of fluvial conglomerates and alluvial fans. Among the Cariri Valley rift basins, only the Araripe Basin contains Aptian sediments (Arai *et al.* 1989). These comprise lacustrine evaporites interbedded with layers of limestones, shales and sandstones (Silva 1986). In the onshore and shallow-water Potiguar Basin, the transitional Aptian sediments are shales, sandstones and calcilutites, which recorded fluvial to deltaic/lacustrine deposits at the basal sequence, covered by marine deltaic units at the top. However, lows in the offshore Potiguar Basin

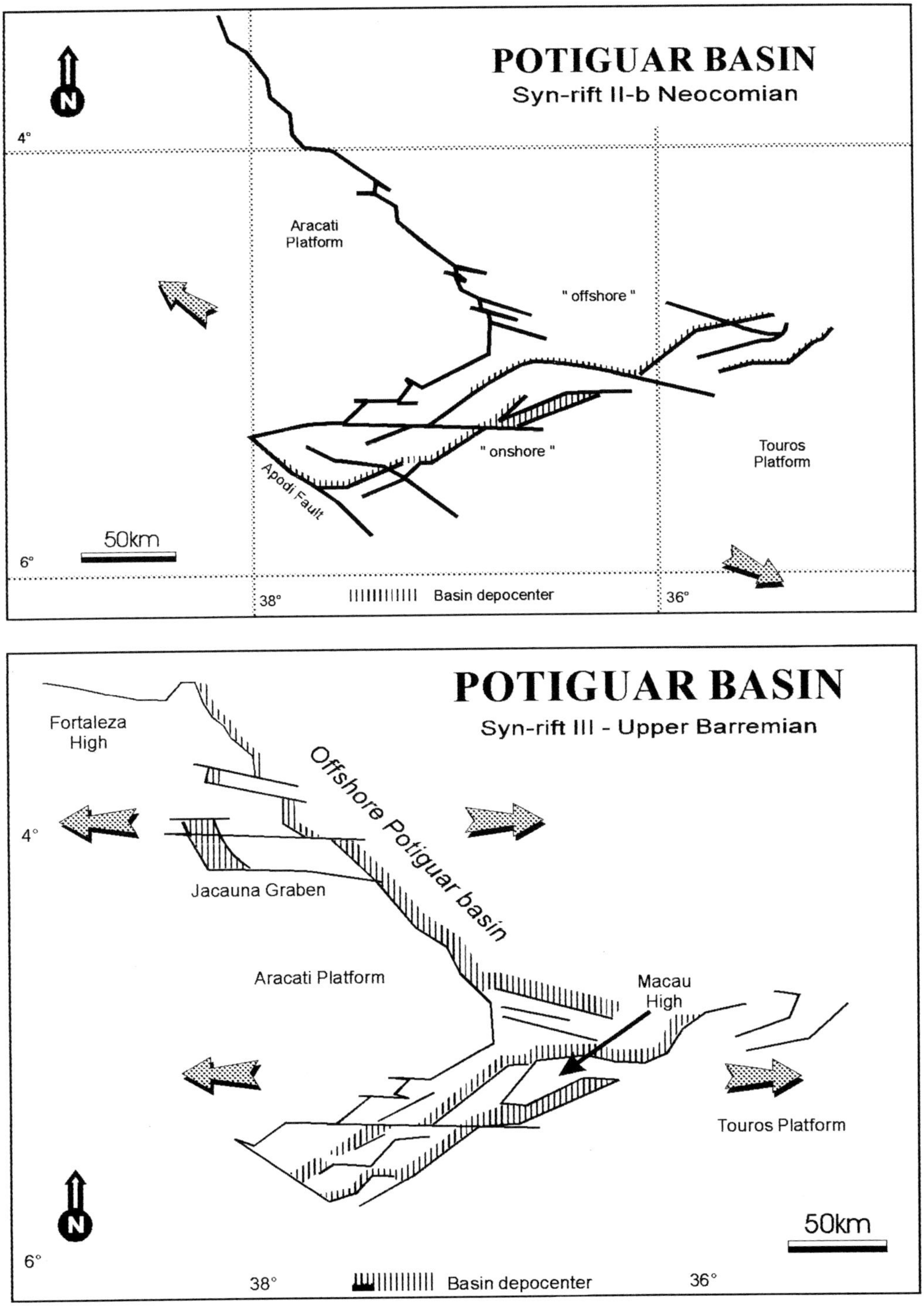

Fig. 10. Diagrams illustrating the changes in tectonic setting times of the Potiguar Basin between Syn-rift IIb and (**a**) and III (**b**) times. Note the growth of the offshore basin in Syn-rift III times and the formation of the Macau High. The arrows indicate the direction of tectonic transport [after Matos (1987*a*)].

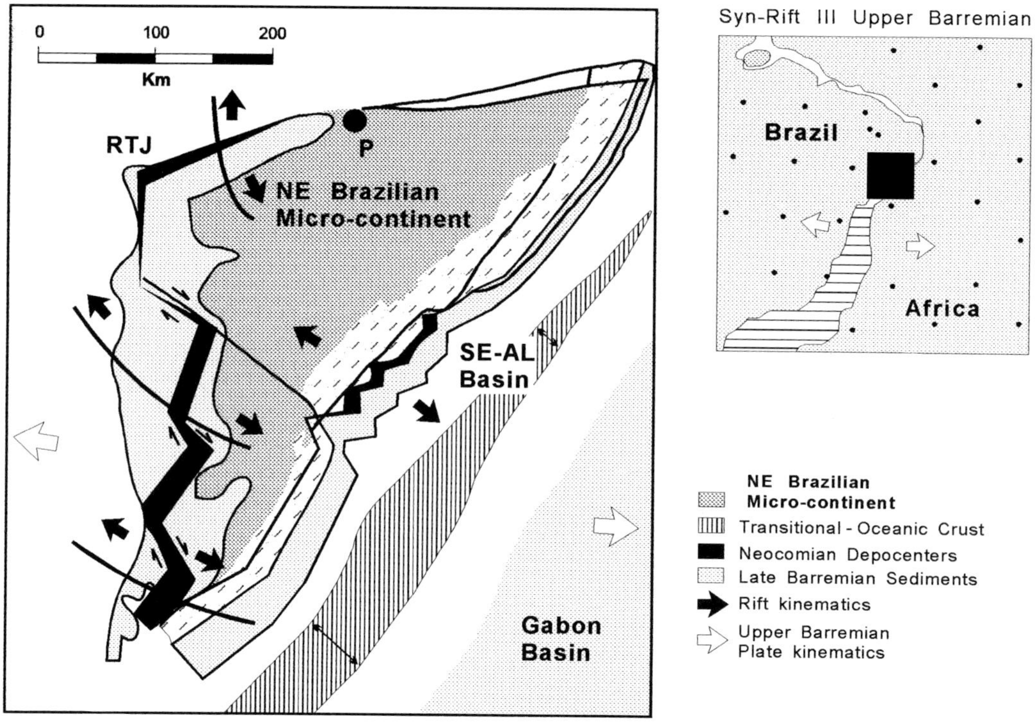

Fig. 11. Tectonic transport directions within the northeastern Brazilian Microcontinent during the late stages of rifting. High strain rates in the Southern Branch during Syn-rift III led to the counter-clockwise rotation, relative to Brazil, of the triangular crustal block defined by the bifurcation between the Recôncavo–Tucano–Jatobá and Gabon–Sergipe–Alagoas trends. The white arrows show the regional separation direction at this time between Brazil and Africa while the black arrows indicate local tectonic transport directions induced by the rotation.

north of the Macau High are occupied by early Aptian rift sediments (Teixeira 1990) and to the north, closer to the main shear zone, by even younger syn-rift units. The widespread Aptian extension triggered the equatorial syn-rift phase in the Upper Benue Trough (Aptian shales of the Bima Formation; Popoff 1988).

The rigidity of the West African–São Luiz Craton and the Borborema–Nigerian Shield (Fig. 2) prevented the onset of rifting in the far east of Brazil and Equatorial Guinea–Cameroon (Africa) until the early Aptian. Stress in the equatorial domain began to dissipate as the entire region started to break apart. Transtensional conditions created a series of northwest–southeast trending en-echelon basins throughout the Equatorial domain. This entire region was affected by two major tectonic phases: a rift stage (Aptian–early Albian) and a shear-dominated phase (early Albian–Cenomanian), as discussed by Zalán & Warme (1985), Azevedo (1986, 1991) and Szatmari *et al.* (1987). Even though shearing signatures and pull-apart features are easily recognized throughout the Brazilian Equatorial Margin, their magnitude and rift architecture varies significantly accordingly to the distance from the main transform faults, generating significant differences in the thermal evolution, basin subsidence and geometry, facies distribution and uplift history. Moreover, local tectonics and magmatism had a key hole in the post-break-up subsidence.

Despite some controversies, it is conceivable that rifts propagated concurrently from east and west to complete the crustal rupture process across the Equatorial Branch. Rifting onset in the Equatorial domain and the Benue Trough is simultaneous with the opening of the Cabo Basin in northeast Brazil (Matos 1987, 1992*a*). The Cabo Basin is the southernmost basin of the Transversal Zone (Mabesoone & Alheiros 1988; Conceição *et al.* 1988), a roughly orthogonal structure between the Pernambuco shear zone and the Benue Trough (Touros High in Brazil), encompassing at least five basins: the Cabo (Brazil), the Rio del Rey (Africa), the

Pernambuco–Paraíba, the Benue Trough–Niger Delta and the easternmost portion of the Potiguar (Touros High). A very fast and focused stretching phase was responsible for overcoming the last link between Africa and Brazil, generating a roughly linear hinge zone in the Transversal Zone, cutting at high angle the Braziliano–Pan-African structures. This unique and short-lived (5 Ma) stress regime resulted in a transient thermal anomaly and minor localized topographic uplift. This Transversal Zone acted as a giant accommodation zone, balancing extension between the Southern Branch and the Equatorial Branch–Benue Trough (Matos 1987).

The Albian stage encompasses the 'Shallow Carbonate Platform Megasequence' (Chang *et al.* 1988) in the Southern Branch and a typical transtensional rift sequence and/or shallow carbonate platform throughout the Equatorial Branch–Benue Trough (Szatmari *et al.* 1987; Benkhelil 1989). The Gabon–Sergipe–Alagoas trend, Pernambuco–Paraíba/Cameroon and offshore Potiguar/western Benue Trough evolved into a marginal open basin defining the corner of a new-born South Atlantic margin. During the early middle Albian, seawaters from the Equatorial and southern Atlantic Oceans met to complete dating the final link between the two branches (Dias Brito 1987).

Conclusions

The history of the northeast Brazilian Rift System (NBRS) yields critical data for evaluating how the process of continental fragmentation can vary in space and time. In this example the process develops as diffuse magmatic and sedimentary events because the grain of the highly deformed and heterogeneous basement lies almost perpendicular to the northwards propagation direction of the South Atlantic Rift System. The distribution of interior fold belts (low strength schists to more rigid granite–gneisses) and major shear zones controlled the pattern of opening and linkage of the rift zones.

Most of the extensional events are marked by the widespread occurrence of tilted half-grabens distributed within three main rift axes: (1) the Gabon–Sergipe–Alagoas trend; (2) the Recôncavo–Tucano–Jatobá trend, (3) the Cariri–Potiguar trend. Mainly on the basis of changes in structural style, facies association, deformation rates and duration of subsidence, three syn-rift phases are recognized, namely Syn-rift I–III.

The Syn-rift I sequence, often referred to as a pre-rift section, encompasses Upper Jurassic–lower Berriasian clastics deposits that filled the broad, shallow 'African–Brazilian Depression'. The main rift stage (Syn-rift II, Neocomian–early Barremian) resulted in widespread block faulting over the Gabon–Sergipe–Alagoas, Recôncavo–Tucano–Jatobá and Cariri–Potiguar trends defined by sets of asymmetric half-grabens separated by basement highs, transfer faults and/or accommodation zones. However, fine biostratigraphic control indicates that during the early Berriasian (Syn-rift IIa), rifting was confined with the Recôncavo–Tucano–Jatobá and Gabon–Sergipe–Alagoas trends. During the Syn-rift IIb (late Berriasian–early Barremian), extensional deformation jumped west from these easternmost basins to the Cariri–Potiguar trend. The Pernambuco–Ngaoundere lineament acted as a regional accommodation zone balancing extensional deformation along the Recôncavo–Tucano–Jatobá/Gabon–Sergipe–Alagoas trends with simultaneous extension along the Cariri–Potiguar trend. A general north-northwest–south-southeast tectonic transport direction is consistent with the development of the northeast–southwest half-graben sets throughout the Borborema Province.

Late Barremian sediments (Syn-rift III) document a major change in rifting kinematics related to the synchronous development of two deformational trends: the initiation of the equatorial, shear-dominated northern margin and the further development of the northwards-propagating southern Atlantic Rift System. Major changes in the style and rate of deformation occurred. Extensional deformation became more focused, and the direction of tectonic transport switched from north-northwest (Syn-rift IIb) to a roughly east–west direction. The Cariri–Potiguar trend was aborted, rifting was initiated at the Equatorial Branch–Benue Trough and the Southern Branch entered its main stretching phase with the accumulation of thick syn-rift sediments in the Gabon and Sergipe–Alagoas Basins. High strain rates in the Gabon–Sergipe–Alagoas trend were accommodated in the neighbouring basins of the Recôncavo–Tucano–Jatobá rift valley. The clockwise rotation of a triangular crustal block between these trends, the East Brazilian Microplate, resulted in left lateral motion in the Sergipe–Alagoas Basin, and a northwest–southeast oriented tectonic transport in the Recôncavo–Tucano–Jatobá trend.

During the Aptian, while the Recôncavo–Tucano–Jatobá and Gabon–Sergipe–Alagoas trends entered the transitional post-rift thermal phase, sea-floor spreading in the south led to a northward propagation of rifting, giving birth to the Cabo Basin, the southernmost basin of the

Transversal Zone, cutting at high angle the Braziliano–Pan–African structures. This Transversal Zone acted as a giant accommodation zone, balancing extension between the Southern Branch and the Equatorial Branch–Benue Trough (Matos 1987). The final rupture of the West African–São Luiz Craton in the Equatorial Branch was completed to trigger the main episode of rifting throughout the Equatorial Atlantic. While the Recôncavo–Tucano–Jatobá trend was aborted, the Gabon–Sergipe–Alagoas trend and the offshore Potiguar Basin subsequently evolved as a drift.

The author is grateful to PETROBRAS for permission to publish this paper. Comments and corrections by Nick Cameron improved earlier versions of the manuscript.

References

ALMEIDA, F. F. M. DE, HASUI, Y. BRITO NEVES, B. B. DE & FUCK, R. A. 1977. Provincias estruturais brasileiras. *Atas VIII Simp. Geol. Nord.*, Campina Grande, 363–391.

ARAI, M., COIMBRA, J. C. & DA S. TELLES, A. C. JR 1989. Estudo bioestratigráfico das bacias interiores Mesozóicas do Nordeste Brasileiro. Relataório de progresso, Petrobrós–Cenpes–Sebipe. Unpublished Report.

AZEVEDO, R. P. DE 1986. Interpretação geodinãmica da evolução mesovóica da Bacia de Barreirinhas. *Proceedings of the XXXIV Congres. Bras. de Geologia*, SBG, Goiãnia, GO, **3**, 1115–1130.

—— 1991. *Tectonic evolution of Brazilian equatorial continental margin basins*. PhD Thesis, Imperial College, London.

BACELLAR, P. B. & COSTA, M. N. C. 1993. Expressão Sísmica da Discordancia Pré-Neo-Alagoas-porção nordeste da Sub-bacia Terrestre de Alagoas. *Congresso Brasileiro de Geofísica, 3*. Rio de Janeiro, Bol. Res. Exp. SBGf, **1**, 163–167.

BENKHELIL, J. 1988. Structure and geodynamic evolution of the intracontinental Benue Trough (Nigeria). *Bulletin des Centres de Recherches Exploration–Production Elf Aquitaine*, **12**, 29–125.

—— 1989. The origin and evolution of the Cretaceous Benue Trough (Nigeria). *Journal of African Earth Sciences*, **8**, 251–282.

BUCKER, C., SCHULT, A., BLOCH, W. & GUERREIRO, S. D. C. 1986. Rock magnetism and paleomagnetism of an Early Cretaceous/Late Jurassic dike swarm in Rio Grande do Norte, Brazil, *Journal of Geophysics*, **60**, 129–135.

CAINELLI, C. 1992. *Sequence stratigraphy, canyons, and gravity mass-flow deposits in the Piaçabuçu Formation, Sergipe–Alagoas Basin, Brazil.* PhD Thesis, University of Texas at Austin.

CASTRO, A. C. M. JR 1987. The Northeastern Brazil and Gabon Basins: A double rifting system associated with multiple crustal detachment surfaces. *Tectonics*, **6**, 727–738.

CHANG, H. K. KOSWMANN, R. O. & FIGUEIREDO, A. M. F. 1988. New concepts on the development of East Brazilian Marginal Basins. *Episodes* **11**, 194–202.

——, ——, —— & BENDER, A. A. 1992. Tectonics and stratigraphy of the East Brazil Rift system: an overview. *Tectonophysics*, **213**, 97–138.

CONCEIÇÃO, J. C. J., ZALAN, P. V. & WOLFF, S. 1988. Mecanismo, evolução e cronologia do Rift Sul-Atlantico. *Boletim de Geociências da Petrobrás*, **2**, 255–265.

CREMONINI, O. A., GOULART, J. P. DE M. & SOARES, U. M. 1996. O Rifte Potiguar: Novos dados e implicações tectonicas. *Boletim do 4o. Simp. sobre o Cretáceo do Brasil*, Unesp, 89–93.

DAVISON, I. 1988. Comment on 'The Northeastern Brazil and Gabon basins: A double rifting system associated with multiple crustal detachment surfaces' by Castro, A. C. M. Jr. *Tectonics*, **7**, 1385–1391.

DIAS BRITO, D. 1987. A bacia de Campos no MesoCretáceo: Uma contribuição à paleoceanografia do Atlântico Sul primitivo. *Revista Brasileira de Geociências*, **17**, 162–167.

FIGUEIREDO, A. M. F. 1981. *Depositional systems in the Lower Cretaceous Morro dos Chaves and Coqueiro Seco formations and their relationship to petroleum accumulations: middle rift sequence, Sergipe–Alagoas basin, Brazil.* PhD Thesis, University of Texas at Austin.

FRANÇOLIN, J. B. L. & SZATMARI, P. 1987. Mecanismo de rifteamento da porção oriental da margem norte brasileira. *Revista Brasileira de Geociências*, **17**, 196–207.

GOMES, J. R. C. & GATTO, C. M. P. P. *ET AL.* 1981. Projeto Radambrasil, Folhas SB 24–25, Jaguaribe e Natal, Geologia, geomorfologia, pedologia, vegetação e uso potencial da terra. Rio de Janeiro, MME/Sec. Geral.

HARLAND, W. B., COX, A. V. *ET AL.* 1982. *A Geologic Time Scale.* Cambridge University Press.

JARDIN DE SÁ, E. F. 1994. *A Faixa Seridó (Província Borborema, NE do Brasil) e seu significado geodinâmico na cadeia Brasiliana/Panafricana.* PhD Thesis, Universidade de Brasilia, DF.

KARNER, G. D., EGAN, S. S. & WEISSEL, J. K. 1992. Modeling the tectonic development of the Tucano and Sergipe–Alagoas rift basins, Brazil. *Tectonophysics*, **215**, 133–160.

KUSZNIR, N., KARNER, G. D. & EGAN, S. S. 1987. Geometric, thermal and isostatic consequences of detachment in continental lithosphere extension and basin formation. *In*: BEAUMONT, C. & TANKARD, A. J. (eds) *Sedimentary Basins and Basin-forming Mechanisms*. Memoirs of the Canadian Society of Petroleum Geologists, **12**, 185–203.

MABESOONE, J. M. & ALHEIROS, M. M. 1988. Origem da bacia sedimentar costeira Pernambuco–Paraíba. *Revista Brasileira de Geociências, SBG*, **18**, 476–482.

MAGNAVITA, L. P. 1992. *Geometry and kinematics of the Recôncavo–Tucano–Jatobá Rift, NE Brazil.* PhD Thesis, University of Oxford.

—— 1996. Sobre a implantação da fase sin-rifte em riftes continentais. *Proceedings of the XXXIX Congresso Brasileiro de Geologia*, Salvador, BA, **5**, 335–338.

—— & Cupertino, J. A. 1988. A new approach to the geological configuration of the lower Cretaceous Tucano and Jatobá basins, Northeastern Brazil. *Revista Brasileira de Geociências*, **18**, 222–230.

——, Davison, I. & Kusznir, N. J. 1994. Rifting, erosion, and uplift history of the Recôncavo–Tucano–Jatobá Rift, northeast Brazil. *Tectonics*, **13**, 367–388.

Matos, R. M. D. de 1987. Sistema de rifts Cretáceos do Nordeste Brasileiro. *Proceedings of 'Tectos', Petrobrás-Depex*, Rio de Janeiro, 126–159.

—— 1992*a*. The Northeast Brazilian Rift System. *Tectonics*, **11**, 766–791.

—— 1992*b*. *Deep seismic profiling, basin geometry and tectonic evolution of intracontinental rift basins in Brazil*. PhD Thesis, Cornell University, Ithaca, NY.

—— 1994. South Atlantic opening and the Northeast Brazilian Rift System. *The 14th International Sedimentological Congress*, Abstracts, Recife, Brazil, S9-9.

——, Lima Neto, F. F. de, Alves, A. C. & Waick, R. N. 1987. O rift Potiguar, gênese, preenchimento e acumulações de hidrocarbonetos. *Proceedings of Rifts intracontinentais, Petrobrás-Depex*, Rio de Janeiro, 160–177.

Milani, E. J. 1985. Tectônica cisalhante na evolução do rift do Recôncavo–Tucano–Jatobá. *Revista Brasileira de Geociências*, **15**, 287–292.

—— & Davison, I. 1988. Basement control and transfer tectonics in the Recôncavo–Tucano–Jatobá rift, Northeast Brazil. *Tectonophysics*, **154**, 41–70.

Mohriak, W. U., Rabelo, J. H. L., Matos, R. M. D. de & Barros, M. C. de 1995. Deep Seismic reflection profiling of sedimentary basins offshore Brazil: Geological objectives and preliminary results in the Sergipe basin. *Journal of Geodynamics*, **20**, 515–539.

Netto, A. S. T. & Oliveira, J. J. 1985. O preenchimento do rift-valley na bacia do Recôncavo. *Revista Brasileira de Geociências, SBG*, **15**, 97–102.

Oliveira, D. C. 1992. *O papel do enxame de diques Rio Ceará Mirim na evolução tectônica do Nordeste Oriental (Brasil): Implicações na formação do rifte Potiguar*. MSc Dissertation, UPOP, Ouro Preto.

Ponte, F. C. 1971. Evolução paleogeológica do Brasil oriental e Africa ocidental. *Petrobras, SETUP-Salvador-BA, CEPG-IV*, ,.

Popoff, M. 1988. Du Gondwana à l'Atlantique sud: les connexions du fossé de la Bénoué avec les bassins du Nort-Est brésilien jusqu'a l'ouverture du golfe de Guinée au Crétacé inféreur. *In*: Sough, J. & Rodgers, J. (eds) *The West African Connection*. Journal of African Earth Sciences, Special Publication, Oxford, 409–431.

Regali, M. S. P. & Viana, C. F. 1989. Late Jurassic–Early Cretaceous in Brazilian Sedimentary Basins: Correlation with the International Standard Scale. *Petróleo Brasileiro S. A.–Petrobrás*, Rio de Janeiro, Brazil, Editora Gavea.

Silva, M. A. M. da 1986. Lower Cretaceous sedimentary sequences in the basin, Northeastern Brazil, a revision. *Revista Brasileira de Geociências, SBG*, **16**, 311–319.

Silva, H. T. F. da 1996. Caracterização do inicio da fase sin-rifte na Bacia do Recôncavo, Estado da Bahia: discussão sobre a formação Agua Grande e o reconhecimento da erosão no inicio do rifteamento. *Proceedings of the XXXIX Congresso Brasileiro de Geologia*, Salvador, BA, **5**, 325–328.

Souza, S. M. 1982. Atualização da litoestratigrafia da Bacia Potiguar. *Proceedings of the Anais do Congress Brasileiralde Geologia, 32*, Salvador, SBG, **5**, 2392–2406.

Szatmari, P., Françolin, J. B. L., Zanotto, O. & Wolff, S. 1987. Evolução Tectônica da margem equatorial Brasileira. *Revista Brasileira de Geociências, SBG*, **17**, 180–88.

——, Milani, E., Lana, M., Conceição, J. & Lobo, A. 1985. How South Atlantic rifting affects Brazilian oil reserves distribution. *Oil and Gas Journal*, **83**, 107–113.

Teisserence, P. & Villemin, J. 1989. Sedimentary basin of Gabon – geology and oil systems. *In*: Edwards, J. D. & Santagrossi, P. A. (eds) *Divergent/Passive Margin Basins*. AAPG Memoir, **48**, 117–199.

Teixeira, L. B. 1990. Reinterpretação sismica do 3D de Pescada e adjacencias. *Petrobras*, Rio de Janeiro, Brasil, Unpublished Report.

Unternehr, P., Curie, D., Olivet, J. L., Goslin, J. & Beuzart, P. 1988. South Atlantic fits and intraplate boundaries in Africa and South America. *Tectonophysics*, **155**, 169–179.

Ussami, N., Karner, G. D. & Bott, M. H. P. 1986. Crustal detachment during South Atlantic rifting and formation of Tucano–Gabon basin system. *Nature*, **322**, 629–632.

Vail, J. R. 1989. Ring complexes and related rocks in Africa. *Journal of African Earth Sciences*, **8**, 19–40.

Viana, C. F., Carozzi, A. V., Araujo, M. B., Cesero, P. de, Fonseca, J. R. & Silva, V. J. L. da 1971. Revisão estratigráfica das bacias do Recôncavo e Tucano. *Boletin Téc. Pet.*, **14**, 157–192.

Zalan, P. V. & Warme, J. E. 1985. Tectonics and sedimentation of the Piauí–Camocim sub-basin, offshore Northeastern Brazil. *Petrobrás–Cenpes, Ciência Técnica Petróleo, seção exploração de petróleo no. 17*.

Deep seismic structure of the continental margin in the Gulf of Guinea: a summary report

BRUCE R. ROSENDAHL & HENRIKE GROSCHEL-BECKER

Rosenstiel School of Marine and Atmospheric Science, 4600 Rickenbacker Causeway, Miami, FL 33149, USA

Abstract: Post-stack reprocessing of the PROBE seismic data set has resulted in a refined tectonic model of the continental margin between Cameroon and southern Gabon. The margin is divided into the North Douala and Gabon Basins by the Kribi Fracture Zone, a 75 km wide transform fault zone that trends northeast–southwest and intersects the coastline between 2 and 3°N. North of the Kribi Fracture Zone, oceanic crust extends essentially to the coastline and floors the narrow margin off Cameroon. Little or no offshore rift margin exists in the northern Douala Basin. Reflection Moho beneath oceanic crust is relatively weak but continuous throughout the northern Douala Basin. Oceanic crustal thickness is extremely uniform in the North Douala Basin, averaging *c.* 1.75 s in two-way travel time. A distinct northeast–southwest trending structural grain at both basement and reflection Moho levels occurs within the Kribi Fracture Zone. Much of the margin off Equatorial Guinea is influenced or controlled by fracture zone tectonics. South of the Kribi Fracture Zone, oceanic crust is offset *c.* 350 km to the southwest, resulting in a broad rift margin off Gabon that is unaffected by major-offset fracture zones. This margin seems to have evolved by a form of imperfect pure shear, involving brittle, upper continental crustal sheets overlying a series of ductile lower sheets that generally dip easterly to southeasterly. The ductile sheets may unroof at outer highs, which mark the boundary between oceanic and continental lithosphere. Dominant reflection events are often associated with the ductile shear zones that define the lower sheets, but a regionally coherent event representing reflection Moho cannot be delineated. The concept of reflection Moho beneath rifted continental crust may be invalid. The tectonic and rheological models presented here can be used to frame hydrocarbon exploration tactics and strategies along this passive margin. In a broader context, elucidation of margin architecture may be a cost-effective way to approach frontier passive margins in the next century.

Under the aegis of the PROBE Study, a grid of nearly 8000 line-km of multichannel seismic (MCS) data were acquired from the region 4° N to 5° S in the Gulf of Guinea. Line coverage spanned the margin from northern Cameroon to southern Gabon and traversed the transition zone from continental to oceanic lithosphere (Fig. 1). The data were acquired in a manner to maximize imaging of the sub-basement architecture of the area. Most of the data were shot with a 7600 cu in airgun array and 7100 m long cable and recorded to 20 s of two-way travel time (TWTT).

Since early 1996 the PROBE seismic data have been undergoing systematic reprocessing to resolve some of the ambiguities raised in earlier interpretations (Groschel-Becker 1996; Meyers 1995; Meyers *et al.* 1996*a*, *b*; Rosendahl *et al.* 1991, 1992). Some of the major ambiguities concern the identification and positioning of fracture zones and crustal boundaries, the significance and meaning of reflection Moho, and the nature of the coupling of the crust with upper mantle. The quality and quantity of the reprocessed PROBE data permit these interpretational problems to be addressed in a more critical fashion than was previously possible. The technical side of the reprocessing will be presented elsewhere.

Results

The locations of key margin-crossing seismic lines are shown in Fig. 1, superimposed on the margin architecture inferred from this study. All of the lines indicated, representing *c.* 4500 km of data, have been reprocessed and converted into line-drawing interpretations. Five of these interpretations are shown in Figs 2 and 3.

In plan view, the overall architecture is deceivingly simple. The main elements are depicted in Fig. 1 and discussed below. In reality there is a large amount of information contained in the reprocessed data (Figs 2 and 3), indicating that the margin structure is far more complex than implied by Fig. 1. Many of these structural details cannot be mapped using the PROBE lines because the wavelength of change is much shorter than the spacing between lines. It is estimated that a grid spacing of the order of 10 km would be necessary to map the deep structure properly. Accordingly, much of the detail contained in the line-drawings is not useful in a mapping context, but the details still have conceptual value.

The dominant transverse structure in the Gulf of Guinea is the Kribi Fracture Zone, a 350 km transform offset that divides the study area into

ROSENDAHL, B. R. & GROSCHEL-BECKER, H. 1999. Deep seismic structure of the continental margin in the Gulf of Guinea: a summary report. *In:* CAMERON, N. R., BATE, R. H. & CLURE, V. S. (eds) *The Oil and Gas Habitats of the South Atlantic.* Geological Society, London, Special Publications, **153**, 75–83.

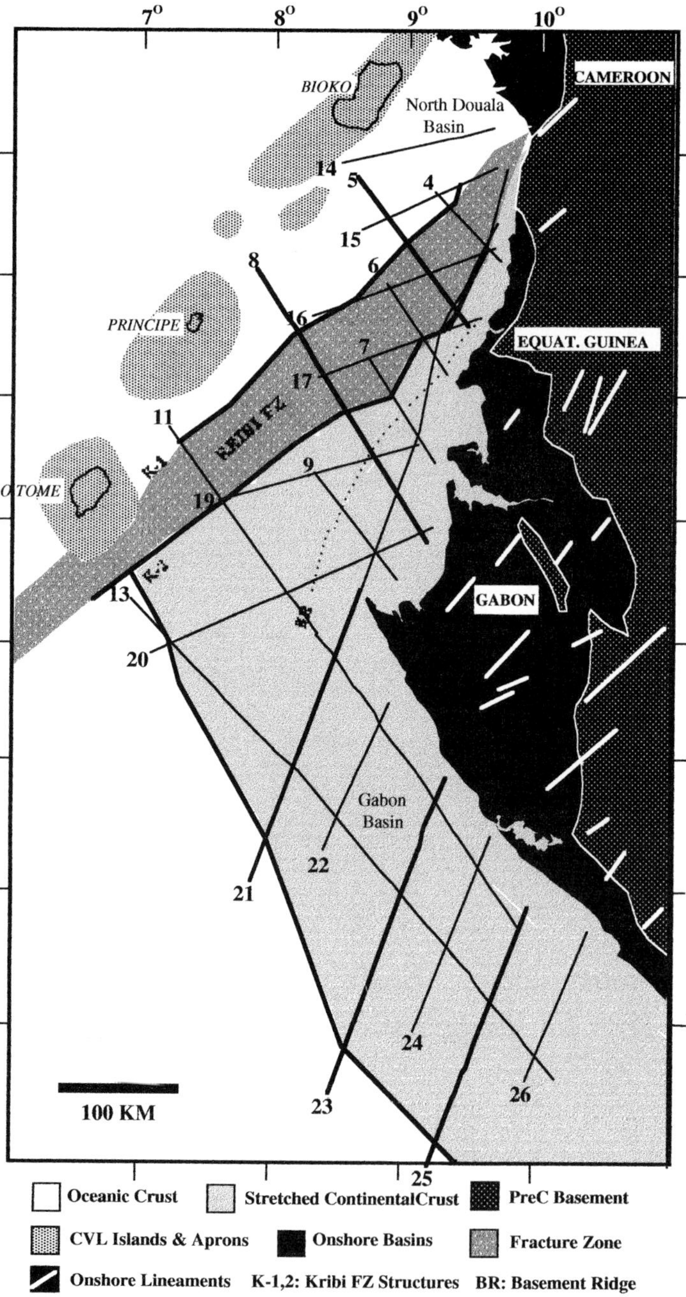

Fig. 1. Locations of reprocessed PROBE seismic lines superimposed on the tectonic framework of the West African margin between 4° N and 5° S. Heavy line segments show locations of line-drawing interpretations in Figs 2 and 3. Onshore lineaments are from Groschel-Becker (1996) and Meyers (1995).

the North Douala and Gabon Basins. In the area north of the Kribi Fracture Zone (the North Douala Basin), oceanic crust extends essentially to the shoreline (Fig. 1). In the area south of the Kribi Fracture Zone (the Gabon Basin), a broad rift margin exists. It does not appear to be dissected or offset by any major-offset fracture zones. This means that PROBE lines south of the Kribi

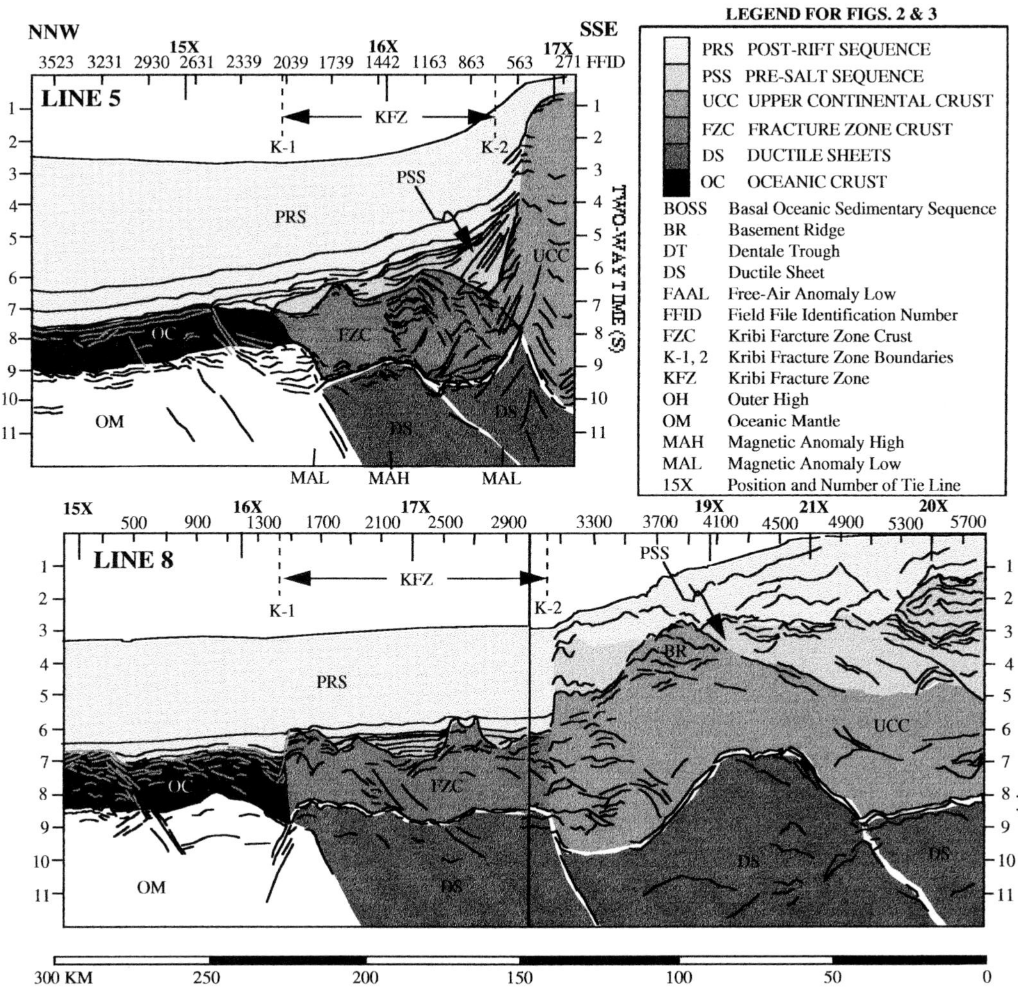

Fig. 2. Line-drawing interpretation of Lines 5 and 8. The legend applies to Figs 2 and 3. See Fig. 1 for the line locations and the correlations with the mapped structures.

Fracture Zone image the rift margin where the plates have separated between major-offset fracture zones. This is the 'clean' part of the rift margin in the PROBE study area, where the data offer the highest likelihood of delineating the processes that produced this margin. This sector is referred to in subsequent discussions as the 'rift margin'.

Oceanic Crust

Both the basement and reflection Moho are relatively clearly imaged in the North Douala Basin. The post-rift succession rests directly on the basaltic basement, equivalent to the break-up unconformity in the continental sector. Reflection Moho at the base of oceanic crust is moderately weak but extremely continuous, except under the Cameroon volcanic aprons where reflection Moho is largely absent (Meyers & Rosendahl 1991). Oceanic reflection Moho is generally a simple reflection couplet that closely parallels the basement surface. (Reflection Moho is the reflection couplet flooring the solid black unit in Figs 2 and 3.) The correlation between basement and reflection Moho is so close that the thickness of oceanic crust (layer 2+layer 3) in TWTT does not vary by more than *c.* 5% throughout the North Douala Basin, except in

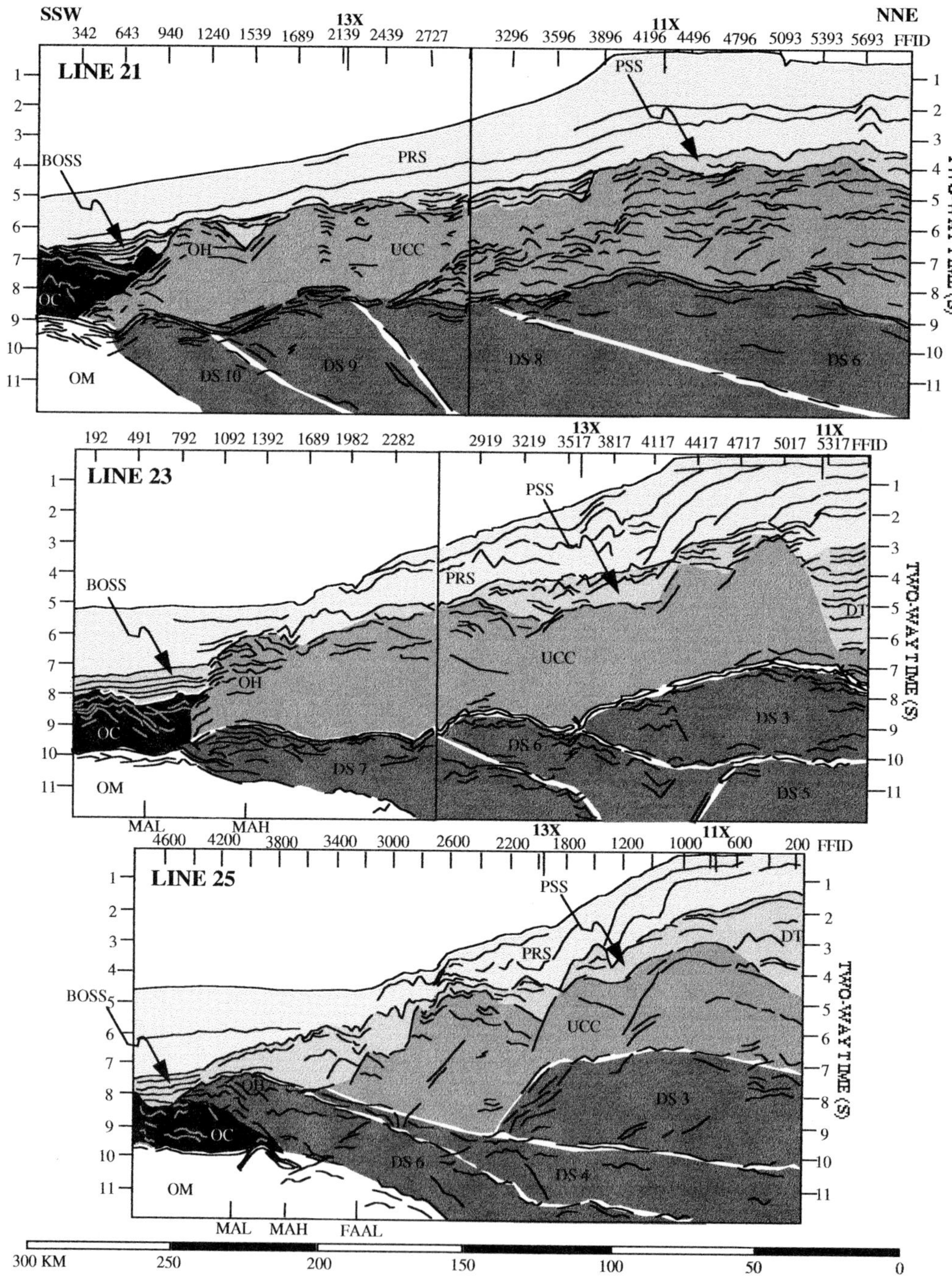

Fig. 3. Line-drawing interpretation of Lines 21, 23 and 25. See Fig. 1 for the line location and the correlations with the mapped structures. See Fig. 2 for the legend.

the vicinity of the Kribi Fracture Zone. The mean thickness in TWTT is 1.75 s. In nominal depth, oceanic crust is 5.63 ± 0.231 km thick. (Nominal depth is defined as the depth computed using an assumed velocity structure of 3.0 km s^{-1} for the sedimentary succession, 6.5 km s^{-1} for material designated as crust and 8.1 km s^{-1} for the mantle.) The thickness values in time and nominal depth are somewhat less than the Atlantic mean, but well within published values. Considering the variability in reported thicknesses of Mid-Atlantic Ridge generated oceanic crust, the uniformity in the North Douala Basin is surprising. This leads to the conclusion that either oceanic crustal genesis in the Gulf of Guinea was extremely uniform, or reflection Moho is coupled to basement in some dynamic way that maintains an apparent constant oceanic crustal thickness. Either conclusion is possible and neither is mutually exclusive. It may be significant that there is a clear relationship between loading and total depth to reflection Moho. The oceanic reflection Moho is depressed or elevated in direct relation to the mass of the overburden; Fig. 2, Line 5 shows this relationship best. A plot of lithostatic pressure v. total depth to reflection Moho (or total depth to basement) for the oceanic sector of Line 5 reveals a tight, linear trend. Hence, there is a long wavelength component of reflection Moho topography beneath oceanic crust related to isostatic loading. In this sense, oceanic crust and reflection Moho are dynamically coupled.

Although oceanic crust appears to 'climb' toward the east in time sections of Lines 14 and 15 (Fig. 1), it actually maintains a relatively uniform total depth of 14–16 km in nominal depth. The oceanic crust along Line 14 is mapped to the landward end of the profile. Here, the landward edge of oceanic crust may actually occur on land. Hence, the narrow continental shelf in the Douala Basin north of 3° N is constructed entirely on oceanic crust and lithosphere.

Oceanic crust in the Gabon Basin is imaged along the ends of Lines 13, 20, 21, 23, and 25, but the areal coverage is limited to *c.* 50 km or less for any given line. Although the average thickness of oceanic crust here remains *c.* 1.75 s TWTT, there is a somewhat larger range in apparent thickness than in the North Douala Basin. The line-drawings in Fig. 3 suggest that oceanic crust is 'lumpier' here. This may relate to the occurrence of outer highs farther eastwards.

There are fewer than a dozen instances in which oceanic basement offsets can be confidently connected to reflection Moho, or sub-reflection Moho events, with actual imaged faults. An example of through-going faulting in oceanic crust occurs in the vicinity of FFID 500 on Line 8 (Fig. 2); this also may be a minor fracture zone (Groschel-Becker 1996). It is concluded that oceanic crust can be ruptured throughout its entire extent, although it seems to occur rather sparingly and it may happen mainly within fracture zones.

Kribi Fracture Zone

The Kribi Fracture Zone is crossed in whole or part by 11 different PROBE lines in three different trackline orientations. Two orthogonal crossings are shown in Fig. 2. The boundary between normal oceanic crust and the northwestern edge of the Kribi Fracture Zone is denoted by K-1 in both Figs 1 and 2. The southeastern edge of the Kribi Fracture Zone is placed at element K-2, which is structurally sensible on most crossings. However, it is conceivable that the southeastern edge extends inboard to the feature labelled BR in Figs 1 and 2. (This matter is returned to in a later discussion.) Crust within the Kribi Fracture Zone is thicker than oceanic crust immediately to the northwest and it is fractured into blocks that apparently tilt towards the southeast. The major block-bounding faults seem to involve the entire crust. Features interpreted as volcanic or intrusive highs are common within the fracture zone and help define its existence. The reflectivity of reflection Moho in the Kribi Fracture Zone is markedly greater than beneath normal oceanic crust, especially in the region of line crossings 5 and 16 (Fig. 1). It is suspected that fluid penetration has enhanced reflection Moho reflectivity here by the hydration of lower crustal or upper mantle rocks.

The internal structure of the Kribi Fracture Zone varies considerably along its length. Along Line 4 the fracture zone essentially juxtaposes oceanic and continental crust across a narrow zone in a rather spectacular, textbook, fashion. The Kribi Fracture Zone is broader and more complex on Line 5, but the expression is quite consistent with that on Line 16 (and on Line 6 to a lesser degree). Between Lines 6 and 8 the Kribi Fracture Zone becomes relatively subdued at the level of reflection Moho. The Kribi Fracture Zone is very subdued where it intersects Line 11. Between Lines 5 and 8 (Fig. 2) the Kribi Fracture Zone has a mapped width of *c.* 75 km, which is consistent with a fracture zone showing a 350 km offset.

The changes in the expression of the Kribi Fracture Zone along-strike make it difficult to map an exact trend. Based upon three iterations of mapping various Kribi Fracture Zone structures the authors believe the overall trend shown in Fig. 1 is good to within *c.* 5° as far west as Line 8.

It is likely that the Kribi Fracture Zone is the eastern end of the Ascension Fracture Zone, as mapped almost three decades ago by Burke (1969) on the basis of bathymetry. This corresponds to the 30° and 31° synthetic flow-lines of Francheteau & Le Pichon (1972). The Kribi Fracture Zone mapped here includes the Kribi,

Bata and Ascension Fracture Zones of both Meyers (1995) and Groschel-Becker (1996). The Kribi Fracture Zone equivalent at the conjugate Brazilian Margin may be the Pernambuco–Paraíba Fracture Zone of Mohriak *et al.* (1995), meaning that the Pernambuco–Paraíba Basin is the conjugate equivalent of the Douala Basin north of the Kribi Fracture Zone. It also is possible that the Madeio Fracture Zone and Alagoas Basin are the conjugate equivalents.

The Rift Margin

The most stunning aspect of the original PROBE data was the widespread occurrence of strong, laterally-coherent reflection packages 7–11 s TWTT within the sector denoted as 'stretched continental rust' in Fig. 1. Previous interpretations of the PROBE data correlated these packages to reflection Moho. Similar events at similar depths at similar positions have been observed at other passive margins and they were generally assumed to represent reflection Moho. The reprocessing of the PROBE data improved the resolution and coherency of these events and initially reinforced the notion that they represented reflection Moho.

Three separate attempts were made to identify and map a regionally coherent reflection Moho event throughout the reprocessed PROBE data set, extending from oceanic crust to the shoreward ends of lines. This proved to be a simple task within the oceanic crustal sector, even within the Kribi Fracture Zone, which left little doubt in the authors' view that reflection Moho beneath oceanic crust is a close proxy of true Moho. The task proved futile beneath the rift margin proper. In the final analysis, a consistent and reproducible set of interpretations could not be generated in the stretched continental crustal sector using the premise that this type of crust was floored by reflection Moho. Although it was true that strong, deep reflection packages could be identified and followed for significant distances on many lines, it was concluded that these packages could not represent the crust–mantle boundary in a regional context.

Attention on the reflection Moho concept was defocused and a more holistic approach adopted to define and tie deep reflection packages; this resulted in the line-drawing interpretations in Figs 2 and 3. The overall picture consists of one or two upper continental crustal sheets overlying a series of lower ductile sheets arranged in an imbricate fashion. Extension in the upper continental sheets is generally accommodated by brittle deformation, with the caveats provided below. Both landward- and seaward-dipping faults occur in the upper crustal sheets. The former create the Dentale Trough landward of FFID 5000 on Line 23 (Fig. 3). There are many instances in which faults offset apparent basement and then disappear at deeper depths in the upper sheets. Often, the projections of the inferred faults coincide with relief on the top surfaces of the ductile sheets. However, there is not a single instance in which fault plane continuity can be unequivocally demonstrated between the top surface of the upper crustal sheets and the ductile lower sheets, in spite of the high quality and quantity of our deep-imaging data. Thus, a mechanical linkage between the uppermost crust and the lower ductile sheets cannot be demonstrated at this passive margin. Either the upper crustal sheets and ductile lower sheets are not mechanically linked, or such linkages are not imaged very well. The upper crustal sheets also tend to be non-reflective at depth (for example as on Line 23), except in the area northeast of FFID 2450 on Line 21, where the upper crustal sheet is distinctively laminated. These observations seem to suggest that the rheology of the upper crustal sheets is not uniformaly brittle in either a vertical or lateral sense.

The rheological behaviour of this rift margin is dictated by the foundation, in this case the imbricate, lower ductile sheets. Extension in the lower sheets is accommodated by ductile shear within and between the sheets. Most of the strong, deep reflection events are created at the boundaries of major ductile sheets. In the southern part of the region (Fig. 3), about ten major ductile sheets have been identified and tied with reasonable confidence. These ductile sheets apparently dip easterly-southeasterly. Sheet numbering reflects apparent stratigraphic order; the higher the number, the lower the apparent stratigraphic level. The actual number of ductile sheets is arbitrary in the sense that perhaps twice as many discrete sheets could have been identified (but not tied with any degree of confidence). Not every ductile sheet occurs along each line. These sheets are markedly three-dimensional and quite variable from line to line, which means that compilation of an apparent stratigraphic order requires good line ties. Near the Kribi Fracture Zone sheet identities cannot be specified because the imaging quality of the critical tie-lines (Lines 19 and 20) is relatively poor.

A basement ridge, or high (labelled BR in Figs 1 and 2), occurs just south of the Kribi Fracture Zone (Fig. 2, Line 8). The feature is loosely situated above a large nose, or hump, in the lower ductile sheets. On Line 20 (Fig. 1) the basement ridge corresponds to the Loiret Volcanic Plug

(Reyre 1984; Teisserence & Villemin 1989), suggesting an igneous origin for the basement ridge. Overall, the basement ridge has a trend of *c.* N30° E, which is also the trend of the Cameroon Volcanic Line. Teisserence & Villemin (1989) and Meyers (1995) relate the Loiret Volcanic Plug to the Cameroon Volcanic Line and assign an early Miocene age to the intrusion, which implies that the basement ridge is a parallel, coeval structure to it. It is equally plausible that the basement ridge is related to the transform margin, especially east of Line 7 where it becomes part of the continental 'wall' of the Kribi Fracture Zone.

The lines shown in Fig. 3 span the entire Gabon Basin and definitively reach oceanic crust without crossing any major-offset fracture zones. In all three cases, the transition from stretched continental lithosphere to oceanic crust is marked by an apparent unroofing of the lower ductile sheets. In the case of Line 25, the seaward-most ductile sheet is modelled as impinging on oceanic crust; in the other two cases an upper crustal sheet bounds the oceanic crust. These differences may be entirely interpretational. All three lines show outer highs at the juncture with oceanic crust. A modest magnetic anomaly is associated with the oceanic crust–outer high transition, which presumably records a change in remanent magnetization or magnetic susceptibility at this transition. A 70 mgal free air gravity low also occurs at the inside edge of the outer high on Line 25.

Comparable outer highs have been imaged at other passive margins (Austin & Uchupi 1982; Beslier *et al.* 1993; Keen & de Voogd 1988; Lehner & De Ruiter 1977; Mohriak *et al.* 1995; Montadert *et al.* 1979; Mutter *et al.* 1982; Schuepach & Vail 1980) and variously interpreted as uplifted blocks of continental crust, volcanic structures of oceanic crustal origin, mafic volcanic blocks, shear zones, fracture zones, serpentinite and serpentinized peridotite. Meyers (1995) interpreted these same highs as faulted, mafic volcanic blocks intermixed with slivers of continental crust. Recent drilling along the Iberian margin (Whitmarsh *et al.* 1996) suggests an ultramafic origin, specifically ridges of peridotite, serpentinite and lower continental rocks. The apparent unroofing of the imbricate lower ductile sheets toward the oceanic crust boundary is quite consistent with the Iberian drilling results. If ductile lower crustal or upper mantle rocks are emplaced at upper crustal levels at the outer highs, the consequent loss in shear strength may explain how seafloor spreading commences.

The rift margin interpretations offered above are rheological in nature. They are not based upon a mapped reflection Moho and they contain few assumptions about what is crust and upper mantle, beyond the suggestion that the upper crustal sheets represent the brittle portion of the stretched continental crust. This brings us to the question of what the boundary between the upper crustal sheets and lower ductile sheets signifies in a geophysical sense. The interface is often highly reflective and laterally coherent over many tens of kilometres, but not universally so. The dominant reflection package eventually switches back and forth between different ductile shear zones, or disappears entirely in some areas. This is why the notion that reflection Moho could be mapped beneath the rift margin was abandoned. It might be significant that the reflectivity of the brittle–ductile boundary tends to be uniformly greater toward the higher numbered ductile sheets, which is sensible if the higher numbered sheets represent lower ductile sheet levels (i.e. larger acoustic impedance contrasts). Could the interface between the upper crustal sheets and lower ductile sheets represent petrologic Moho, replete with the rheological changes that should be manifested across this boundary? This question cannot yet be answered.

Conclusions

In the context of this volume, it is useful to consider the explorational value of these analyses. Figure 1 clearly shows why almost all of the major offshore drilling successes on this margin have occurred in the Gabon Basin. If the information contained in Fig. 1 had been available at the outset of exploration activities on this margin, the exploration history may have been significantly more cost-effective. For example, it now seems futile to seek out offshore syn-rift plays in the North Douala Basin (north of the Kribi Fracture Zone) because almost all the rift geology must be located onshore in Cameroon. Carrying this line of thinking one step further, it might be surmized that most of the rift geology is probably at the conjugate position off Brazil, with only a narrow band of syn-rift section in onshore Cameroon. This is not to say that the offshore area north of the Kribi Fracture Zone is unprospective, but it means that potential sources and targets lie entirely within the post-rift section.

Figure 1 implies that the Niger Delta is essentially constructed on oceanic crust. Although not shown herein, PROBE lines north of Line 14 (Fig. 1) verify this implication. This means that Niger Delta diapiric structures cannot be caused by Aptian salt. It also means that the subsidence history of the delta may follow the 'oceanic' relationship noted earlier for differential loading along Line 5. The reprocessed PROBE results strongly support our earlier contentions (e.g. Meyers & Rosendahl 1991) that the Cameroon Volcanic Line is built entirely on oceanic crust. Hence, the oil seeps on São Tome–Principe must originate in a truly oceanic crustal setting even though the sources of the oils (windows of Douala Basin sediments incorporated into the volcanic edifices) may be similar to those observed on the West African shelf.

Now consider the case of the eastern terminus of the Kribi Fracture Zone, where it abuts the continental shelf at depths potentially within range of the drill bit, say *c*. 2° N. The setting is a continent-bounded transform fault. The floor is presumably fractured and intruded oceanic crust, with local fragments of continental basement. The depositional arenas are narrow troughs flanked by volcanic–plutonic ridges. The fill is predominately continental clastic sediments. Salt may have been locally present. The initial thermal regime was high and variable, and hydrothermal activity would have been prevalent early on. This is the exploration scenario faced by those working offshore Equatorial Guinea. It is not an environment where standard exploration strategies could be employed with much hope for success. These sorts of unconventional settings may play a larger exploration role in the coming century if they can be properly exploited. This requires a proper understanding of tectonic provenance.

Exactly how this rheological model aids the explorationist is not so clear, but the following thoughts are offered. There is usually a crude correlation between the relief on the lower ductile sheets and major basement topography. Along many parts of this margin, the ductile sheets are much better imaged than the basement–syn-rift or syn-rift–post-rift boundaries. It is conceivable that the existence of major basins in frontier areas could be deduced by first mapping the topography of the ductile sheets, rather than trying to resolve the shallower basinal details from the outset. Acquisition programs then could be tailored to specifically extract the basinal details. At the very least, it is suggested that the deep structure should be an imaging target of acquisitional programs in frontier areas. As more is learned about the relationships between basinal development and the underlying margin architecture, this type of information could represent the 'exploration edge' in the next century.

About 4500 line-km of data interpretations are represented by this study. Although this is one of the most extensive deep-imaging seismic collections available for any passive margin, the collection is arguably an order of magnitude too small to unravel the detailed architecture of this margin. Over US$1 million has been spent acquiring, processing and interpreting this collection; US$10 million would be necessary to do the job thoroughly. These are sobering sums to academic planners of future deep-imaging seismic programs on passive margins, but they are not large figures in the petroleum industry. The cost to thoroughly understand the deep structure of this margin is equivalent to a few dry holes.

Much more needs to be done with the PROBE data. Ongoing work includes: (1) conversion of time into nominal depth in a uniform, line-tied fashion; (2) computation and removal of the effects of loading to produce corrected nominal depth sections; (3) construction of complete Bouguer gravity models back-stripped to the brittle–ductile boundary; (4) integration of these results into a conjugate margin analysis incorporating data owned by Petrobras. It is expected that the ongoing work will go a long way towards resolving many of the uncertainties expressed herein, and undoubtedly raise a host of new questions to guide future research on the architecture and evolution of this South Atlantic passive margin.

We thank the governments of Cameroon, Equatorial Guinea, Gabon and São Tome–Principe for permitting this field work to be done. We thank Agip, Amoco, Ark Geophysical, Geco-Prakla and Jebco for support and assistance. This manuscript has benefitted from the laboratory efforts of Zan-Dong Ding and Jason Meyers. Reviews or comments have been kindly provided by Steve Henry, Phil Nelson and Nick Cameron.

References

AUSTIN, J. A. & UCHUPI, E. 1982. Continental–oceanic crustal transition off Southwest Africa. *AAPG Bulletin*, **66**, 1328–1347.

BESLIER, M. O., ASK, M. & BOILLOT, G. 1993. Ocean–continent boundary in the Iberian Abyssal Plain from multichannel seismic data. *Tectonophysics*, **218**, 383–393.

BURKE, K. 1969. Seismic areas of the Guinea coast where Atlantic fracture zone reach Africa. *Nature*, **222**, 655–657.

FRANCHETEAU, J. & LE PICHON, X. 1972. Marginal fracture zones as structural framework of continental margins in South Atlantic Ocean. *AAPG Bulletin*, **56**, 991–1007.

GROSCHEL-BECKER, H. M. 1996. *Formational processes of oceanic crust at sedimented spreading centers: perspectives from the West African continental margin and Middle Valley, Juan de Fuca Ridge.* PhD Thesis, University of Miami, FL.

KEEN, C. E. & DE VOOGD, B. 1988. The continent–ocean boundary at the rifted margin off eastern Canada: New results from deep seismic reflection studies. *Tectonics*, **7**, 101–124.

LEHNER, P. & DE RUITER, P. A. C. 1977. Structural history of the Atlantic margin of Africa. *AAPG Bulletin*, **61**, 961–981.

MASCLE, J., LOHMANN, G. P. & CLIFT, P. 1996. *Proceedings ODP, Initial Reports.* College Station, Texas (Ocean Drilling Program), **159**.

MEYERS, J. B. 1995. *Rifted continental margin architecture off West Africa, as revealed by deep-penetrating multi-channel seismic reflection and potential field data.* PhD Thesis, University of Miami, FL.

—— & ROSENDAHL, B. R. 1991. Seismic reflection character of the Cameroon volcanic line: evidence for uplifted oceanic crust. *Geology*, **19**, 1072–1076.

——, —— & AUSTIN, J. A. 1996*a*. Deep-penetrating MCS images of the South Gabon Basin: Implications for rift tectonics and post-breakup salt remobilization. *Basin Research*, **8**, 65–84.

——, ——, GROSCHEL-BECKER, H., AUSTIN, J. A. & RONA, P. A. 1996*b*. Deep-penetrating MCS imaging of the rift-to-drift transition, offshore Douala and North Gabon Basins, West Africa. *Marine and Petroleum Geology*, **13**, 791–835.

MOHRIAK, W. U., RABELO, J. H. L., DE MATOS, R. D. & DE BARROS, M. C. 1995. Deep seismic reflection profiling of sedimentary basins offshore Brazil: geological objectives and preliminary results in the Sergipe Basin. *Journal of Geodynamics*, **20**, 515–539.

MONTADERT, L., ROBERTS, D. G., DE CHARPAL, D. & GUENNOC, P. 1979. Rifting and subsidence of the northern continental margin of the Bay of Biscay. *Initial Report of the Deep Sea Drilling Project*, **48**, 1025–1060.

MUTTER, J. C., TALWANI, M. & STOFFA, P. L. 1982. Origin of seaward-dipping reflectors in oceanic crust off the Norwegian margin by 'subaerial seafloor spreading'. *Geology*, **10**, 353–357.

REYRE, D. 1984. Caracteres petroles et evolution geologique d'une marge passive. Le cas du bassin bas Congo-Gabon. *Bulletin des Centres de Recherches Exploration–Production Elf Aquitaine*, **8**, 303–332.

ROSENDAHL, B. R. 1997. The seismic roots of the West African passive margin: northern Cameroon to southern Gabon (abstract). *Fifth International Congress of the Brazilian Geophysical Society, Volume of Abstracts*, Rio de Janeiro.

——, GROSCHEL-BECKER, H., MEYERS, J. & KACZMARICK, K. 1991. Deep seismic reflection study of a passive margin, southeastern Gulf of Guinea. *Geology*, **19**, 291–295.

——, MEYERS, J., GROSCHEL-BECKER, H. & SCOTT, D. 1992. Nature of the transition from continental to oceanic crust and the meaning of reflection Moho. *Geology*, **40**, 721–724.

SCHUEPBACH, M. A. & VAIL, P. R. 1980. Evolution of outer highs on divergent continental margins. *In*: *Studies in Geophysics; Continental Tectonics.* National Academy Sciences, Washington, DC, 50–61.

TEISSERENCE, P. & VILLEMIN, J. 1989. Sedimentary basin of Gabon – geology and oil systems. *In*: EDWARDS, J. D. & SANTOGROSSI, P. A. (eds) *Divergent/Passive Margin Basins. AAPG Memoir*, **48**, 117–199.

WHITMARSH, R. B., SAWYER, D. S., KLAUS, A. & MASSON, D. G. (eds) 1996. *Proceedings ODP, Science Results.* College Station, Texas (Ocean Drilling Program), **149**.

Templates from mainland Africa and the Red Sea for interpreting the early evolution of the South Atlantic

R. CROSSLEY & D. CRIPPS

Robertson Research International Ltd, Llandudno, LL30 1SA, North Wales, UK

Abstract: Findings drawn from the Tertiary pre-rift, rift and post-rift sequences of Africa and the Red Sea are applied to the South Atlantic margins with a view of aiding hydrocarbon exploration. In relation to gravity and magnetic prediction of top basement, template rifts tend to follow pre-existing lines of weakness and the upper crust underlying these rifts may have geophysical characteristics which are different from those of the rift flanks. Template rifts may or may not be associated with volcanicity. In sparsely volcanic rifts, syn-rift volcanism is located on rift axis structural highs. Extensive volcanism is accompanied by regional doming and erosional loss of the pre-rift section is likely in rift sectors preceded by volcanism. Consequently, it is doubtful whether pre-rift sediments are preserved beneath the major volcanics offshore from the Parana–Etendeka areas in the South Atlantic. The Kalahari Group of southern Africa is identified as a good analogue for pre-rift reservoir objectives in the South Atlantic, and facies and thickness criteria are discussed that help discriminate pre-rift sequences from rift sequences. Where sedimentation has kept pace with rift subsidence, so that fault scarp facies have not developed, then the Okavango Delta provides a useful analogue. Parallels are drawn between major sand dykes interpreted from seismic in the deep-water deposits of Lake Tanganyika and similar units within similar facies in the Recôncavo Basin of Brazil. Direct analogue links between the Red Sea and the South Atlantic are less strong, mainly because marine incursions, and salt deposition, occurred at a later structural stage in the South Atlantic. The onset of salt deposition in the Gulf of Suez–Red Sea occurred whilst rift block faulting was still in progress, and the post-rift sag phase, though partly salt filled, rapidly turned to fully-marine deposition. The earliest part of the post-rift sag phase in the South Atlantic is represented by laterally extensive lacustrine source rocks. Salt deposition in the South Atlantic occurred only after full development of the sag phase.

In many South Atlantic margin areas, seismic imaging at pre-rift to early drift reservoir horizons is poor and/or the number of well penetrations is low. Exploration risk reduction in this context, therefore, depends upon recognizing play possibilities from a very limited data set. Understanding key sequence controls in analogue systems can help this process.

This paper aims to highlight a few issues that the authors feel have not yet received as much attention as they deserve. This paper is, therefore, neither, comprehensive nor full of supporting detail. The significance of the issues addressed can be summarized as a series of questions: (1) what basement rock properties should be used in gravity/magnetic modelling of depth to basement? This question is particularly important for interpretation of early evolution of the basins because this part of the section is often poorly imaged on seismic records; (2) could significant rift-basin sediments underlie the thick volcanic sequences of southern Brazil, southern Angola and Namibia? Thousands of kilometres of these continental margins are virtually unexplored because of the problems of seismic imaging beneath the volcanics, consequently, alternative sources of information on the risks associated with this play are of particular significance; (3) how can pre-rift sediments be separated from early syn-rift sediments? Differentiation can be extremely difficult if sediment flux into the rifts keeps pace with subsidence, hence preventing development of classic rift fault scarps and their readily recognizable facies associations. These two sedimentary sequences will have very different distributions in relation to early structures in the South Atlantic rifts and so will require different exploration and development strategies (documentation of recent pre-rift template sequences will help start to counterbalance the emphasis on syn-rift sedimentary models that exist in current South Atlantic geological literature); (4) how can near-homogenous sandstone bodies with a 1200 m vertical thickness, and with a wall-like geometry, develop within syn-rift mudstone sequences? At least one such unit has been documented in the Recôncavo Basin of Brazil and understanding its origin will aid exploration in similar sequences comprising oil-prone source rocks with minor volumes of reservoir sandstones; (5) does the presence of salt basins on a continental margin mean that a specific stage of structural development has been reached? If it does, then systems such as the Gulf of Suez–Red Sea can be used as exploration templates for the salt basins of the South Atlantic succession.

CROSSLEY, R. & CRIPPS, D. 1999. Templates from mainland Africa and the Red Sea for interpreting the early evolution of the South Atlantic. *In:* CAMERON, N. R., BATE, R. H. & CLURE, V. S. (eds) *The Oil and Gas Habitats of the South Atlantic.* Geological Society, London, Special Publications, **153**, 85–96.

In addressing these issues, templates from mainland Africa and the Red Sea are drawn upon. The area from southern to east Africa encompasses a range of arid to moist climates similar to those expected during rifting and drifting in the South Atlantic. Parts of southern and central Africa provide insights into pre-rift and rift structural settings, the analogue potential of the rift system of East Africa is well-known, whilst the Red Sea and the Gulf of Aden are classic examples of rift to early drift settings.

Terminology

A variety of terms, and different usages of the same term, are encountered in publications on rift sequences. This section explains the terminology used in this paper. Some of the main features of reservoirs developed in basement, pre-rift and syn-rift sequences are summarized in Table 1.

Basement – in the South Atlantic margin context this may include crystalline to non-deformed Precambrian basement, and sediments of Carboniferous and Permian age. Only those components of potential basement reservoirs that are a fundamental characteristic of the rock unit, such as matrix porosity in Permian sandstones or orogenic fracture porosity in metamorphic rocks, should be ascribed to the basement.

Pre-rift sequences – these are relatively thin, but areally extensive, sequences of sediments, and contemporary erosion and weathering surfaces, developed on basement prior to the onset of rift subsidence. This definition includes changes imposed on basement units by erosional unloading and weathering. Sedimentary sections older than the rift sediment package under investigation are considered part of the basement.

Rift sequences – these include all sediments, and contemporary erosion and weathering surfaces, developed within the rifts during fault-controlled rift basin subsidence. The scarps, fault blocks and monoclines, whether emerged, submerged or buried, are important controls on sediment thickness and facies. There may be several stacked rift sequences if rift tectonic history is complex.

Post-rift sequence – sediments, and contemporary erosion and weathering surfaces, developed as tectonics become dominated by the stretching and general subsidence, here termed sag phase, which accompanies the shift towards basaltic ocean crust creation and the onset of drifting.

Basement

In hydrocarbon exploration, combinations of gravity, magnetic and seismic methods are used to estimate basin-fill and crustal thickness. If these parameters are estimated correctly, then modelling of basin thermal and subsidence histories can be of predictive value. Individually, these methods do not provide unique solutions and, even when used in combination, the estimated thicknesses may have a high level of uncertainty. In the absence of direct information from the drill bit, it is important that the densities, magnetic properties and sonic velocities used for thickness estimation are appropriate.

The African rift system tends to follow suitably oriented structural grain in the metamorphic basement or to reactivate favourably oriented Karoo Basin faults (which often, in turn, follow structural grain in the metamorphic basement). Consequently, the basement rocks underlying rift basins are often significantly different from those uplifted on the rift margins. In addition, metamorphic changes are likely to occur within the crust in areas of greatest subsidence (Flavey 1974).

Implications for the South Atlantic

The importance of these observations for exploration of the Atlantic (or any other) ocean margin is that geophysical modelling based on the character of rift margin basement could be erroneous. It may be wiser to consider using rock property data from along-strike within the rift, even if the nearest data source is many tens of kilometres distant, rather than use data from a margin adjacent to the rift.

Volcanic activity and preservation of pre-rift sequences.

Substantial areas of volcanics occur along both margins of the South Atlantic south of the Kwanza and Campos Basins (Fig. 1). Seismic imaging of pre-volcanic sections is, with current technology, extremely difficult or impossible. The problem for the explorationist is to assess whether it is worthwhile to drill 'blind' through the volcanics in the hope of encountering sediments bearing hydrocarbons beneath. In this context, it is important to determine whether there is any systematic, and hence predictable, general relationship between volcanism, rifting and rift sedimentation.

Most rift systems are, for much of their length, devoid of volcanic rocks. However, volcanic-rich sectors of rifts provide a wealth of geological data thanks to their radiometrically dateable, often well-exposed, sequences. The consequent geological bias has created a possibly misleading

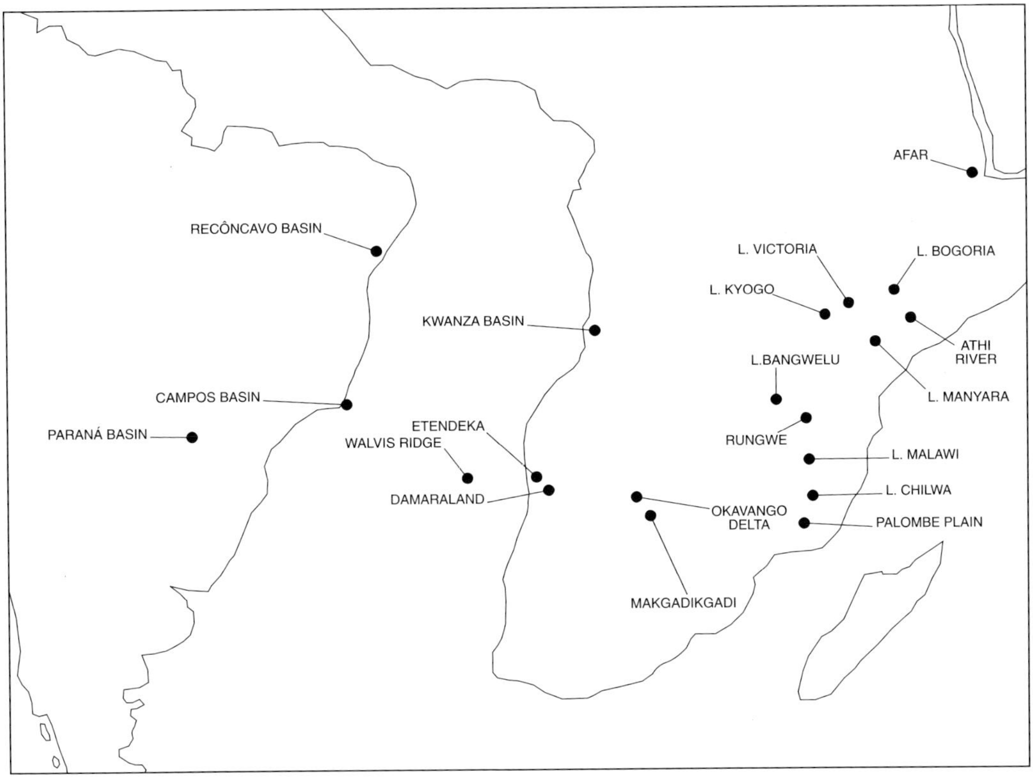

Fig. 1. Locality map.

association between volcanics and rifting. Some resulting models see a direct link between the two processes, either (1) through lithospheric extension during rifting causing volcanism or (2) through updoming over mantle plumes causing rifting (Sengõr & Burke 1978).

The authors believe that the Tertiary of Africa argues for a third, less direct, model (3), in which substantial rift volcanism occurs when an unrelated rift system cuts across a mantle hot welt or hotspot. There is a significant difference between the models in terms of predicting whether pre-rift or rift sediments will occur beneath a major rifted volcanic sequence: for (1) sediments are likely; for (2) sediments are unlikely; for (3) sediments are unlikely, but the possibility that the mantle anomaly post-dated rifting must be tested.

In Africa, major Tertiary volcanic areas are underlain by unusually low-density mantles and are associated with lithospheric doming (Hoggar, Tibesti, Kenya, Ethiopia) or occupy structural highs within the rift system (Rungwe, Kivu).

The Rungwe volcanic complex occurs where the western rift system cuts across a major belt of low-density mantle which extends from the northern end of the Walvis Ridge, across southern Angola, through Kenya and Ethiopia, connecting to the oceanic ridge system at Afar in Eritrea (Fairhead 1986). Formation of the mantle anomaly is inferred to pre-date rifting since the oldest associated undersaturated volcanics are substantially older than the onset of Tertiary rifting. The anomaly axis also correlates with the location of the pre-rift African watershed, suggesting that buoyant support of the lithosphere pre-dates rifting. Significant thicknesses of pre-rift sediment are therefore unlikely to be preserved beneath the Kenyan rift-fill since erosion was likely in the uplifted area prior to rifting.

In the Kenyan and Ethiopian rifts, faults have throws of hundreds to thousands of metres and copious volcanics dominate the rift-fill. The location of the rifts over the mantle anomaly are viewed as essentially accidental and eruption of the large volumes of volcanics as the outcome of geographical coincidence.

Implications for the South Atlantic

The Parana–Etendeka flood basalts of Brazil and Namibia, and the Damaraland igneous complexes

Table 1. *South Atlantic 'pre-salt' reservoir summary*

Reservoir unit	Structure	Weathering, erosion, sedimentation	Generalized comments
Basement			
Basement fractures	Tectonic fractures and foliation, cooling fractures in igneous and metamorphic rocks, and overburden unloading fractures	Weathering usually enhances porosity and permeability of fractures, gravity slippage increases porosity and permeability of fractures	Enhanced reservoir potential due to gravity slippage is best developed around the margins of those pre-rift palaeohighs formed by differential erosion, along the upthrown edges of rift-fault scarps, and along the upthrown subsurface margins of faults of rift or post-rift age. Clastic sediments may act as natural propants in pre-rift and rift-fault scarp fractures
Basement matrix	Microfractures can improve matrix permeability significantly in brittle rocks such as quartzites	Weathering may enhance porosity and permeability of matrix	Rocks with a ductile original matrix, such as mica schist, and rocks with a ductile microporous matrix due to a high proportion of weathering products, such as some horizons in syenite and granite, do not benefit from enhanced permeability
Clastic sediments			
Clastic pre-rift sand-rich sediments		High proportion of polygenetic sand, aeolian and fluvial processes dominant, some major beach systems	High mineralogical maturity, high textural maturity, with high depositional connectivity and potentially large lateral sandbody dimensions
	Crest of fault block	Pre-rift sediments present	Good reservoir potential
	Crest of fault block	Pre-rift sediments eroded off	Reservoir potential dependent upon quality of basement
Clastic syn-rift sand-rich sediments			
		Reworked pre-rift clastic sediments	Potential for high mineralogical and textural maturity is inherited from the pre-rift sediments
		Sediments derived from pre-rift basement	Typically low mineralogical maturity Rarely, high mineralogical maturity may be inherited from quartzitic basement. Textural maturity can be moderate to high depending upon depositional processes
		Sediments derived from rift volcanics	Usually, rift volcanics occur at structural highs along the axes of rift systems; volcanic-rich sandstones are commonest in axial fluvial systems. Mineralogical maturity is low, leading to poor reservoir quality through diagenetic damage of pore systems
	Fault block dip slope		Relatively large volumes of reworked pre-rift sediments at a high structural position, with good depositional

Table 1. *Contd.*

			connectivity. High inherited maturity retained due to limited contamination by basement clasts
	Fault block scarp slope		Relatively small volumes of reworked pre-rift sediment, occurring at low structural positions, showing moderate to low depositional connectivity. Low textural and mineralogical maturity due to rapid deposition and to contamination by basement clasts
	Rift floor	Axial fluvial systems	Large volumes of relatively well connected sandstones. In non-volcanic sectors, early axial fluvial sandstones may contain a high proportion of reworked pre-rift sediment, resulting in good reservoir potential
		Lateral fluvial systems	Small volumes of relatively poorly connected sandstones, except where a major pre-determined drainage system is intersected by the rift basin
		Aeolian systems	In arid areas, large volumes may accumulate with a high proportion of the sand being sourced from pre-rift deposits outside the rifts. Significant volumes can also develop upwind from beach systems even in moist climates
		Upper shoreface lacustrine systems	In rift basins that fill to spill, at least periodically, major beach complexes develop. Where the hydrologic budget does not allow the rift basin to fill to spill, beach complexes are weakly developed
		Middle and lower shoreface lacustrine systems	Extensive sheet sands can develop due to repeated passage of the surf zone across the rift floor in those lakes which show lake level fluctuations. The volumes of sand sheets are determined by the amount of fluvial and aeolian sand supplied to the lake margin
Carbonate sediments			
Carbonate syn-rift sediments	Crest of fault block	Moderate to high probability of exposure-related dissolution creating secondary porosity	Isolated, shallow water, intra-lake highs, starved of siliciclastic sediment, are ideal sites for the localized accumulation of autochthonous freshwater carbonates. The best reservoir potential occurs in winnowed bivalve or gastropod sands. However, algal carbonates can also develop reservoir potential, depending on diagenesis
	Fault block scarp slope	Lower probability of karstic enhancement of reservoir potential	Resedimented carbonate wedges, reworked from fault block crest platforms, can have moderate reservoir potential but generally low predictability

of Namibia, have yielded radiometric dates in the range 150–110 Ma (Milner *et al.* 1995). By analogy with the African Tertiary volcanic provinces it is inferred that this area was underlain by an anomalous mantle, and hence was an area of regional uplift prior to onset of this major phase of igneous activity.

Whether the older dates in the range 150–110 Ma represent a spread in the age of onset of igneous activity or anomalous results from rocks whose true age spans 137–124 Ma, as Milner *et al.* (1995) prefer, the dates give only a minimum age for uplift. The fact that the Entendeka lavas rest directly on Karoo sedimentary rocks and on metamorphic basement, and that only thin deposits of mainly aeolian sediments are interbedded within the lowermost flows (Milner *et al.* 1995), supports the inference that this area was a regional high in the early Cretaceous. The regional uplift may have extended at least as far north as the Campos Basin of Brazil and the Kwanza Basin of Angola, since basaltic volcanics dated at *c.* 120 Ma underlie rift sediments in these areas (Ponte & Asmus 1978; Crossley *et al.* 1993).

The onset of subsidence on Atlantic margin-related rift faults is dated at *c.* 145–140 Ma in basins further north, such as the Recôncavo Basin in Brazil (Magnavita & Silva 1995) and Gabon (Bradley & Fernandez 1993). The relative ages of earliest rifting and earliest volcanism (and, by implication, doming), are therefore not unambiguously established by the data reviewed here. However, if it is assumed that thermal regional uplift pre-dated earliest volcanism, it is, on balance, more likely that rift subsidence occurred too late to preserve significant thicknesses of pre-rift sediments along the Paraná/Campos–Namibia/Kwanza section of the Atlantic margin.

By the early Cretaceous, at the latest, latitudes equivalent to the Campos–Kwanza Basins may have been receiving volcanic-rich sediments from the higher ground to the south. Conversely, the largely non-volcanic sectors of the rift system north of the Campos–Kwanza Basins would have been relative topographic lows from the earliest stages of rifting, offering good prospects for preservation of pre-rift sequences.

Pre-rift sequences

Seismic imaging of deeply buried pre-rift sequences is generally poor along both margins of the South Atlantic. Templates for these sequences, particularly in terms of thickness, lateral continuity and reservoir potential, is important if understanding of the exploration possibilities is to be maximized from the relatively limited number of wells penetrating such deposits. Documentation of sedimentological characteristics of pre-rift sequences is also required to improve discrimination between pre-rift and syn-rift successions in the wellbore. The two issues of predicted lateral continuity and sedimentological characteristics might assist recognition of situations where the crestal area of a prospect, or field, has lost pre-rift reservoirs through erosion over a syn-rift high. This, in turn, may lead to successful drilling for reservoirs way from the crest of such 'bald' structures.

Pre-rift sequences comprise basement with erosional unloading joints, weathered basement (*in situ*, preserving relict basement textures), soil (material not *in situ*, strongly modified by weathering and bioturbation), sediment and, occasionally, volcanic rocks.

The Kalahari Group sediments of south central Africa are a useful analogue for pre-rift sequences in arid to seasonally moist climatic regimes similar to those envisaged in the South Atlantic. Thomas & Shaw (1990) provide an overview of this group, of which the main points of relevance to this discussion are:

- the sediment sequence covers an area of more than 1 million square kilometres and is typically 10–100 m thick outside rift basins; accumulation occurred mainly within regional downwarps, with local thickness changes reflecting burial of erosional topography on the basement;
- the basal sediments of The Kalahari Group, and the weathered top of underlying basement, are characterized by chalcedonic silica cementation;
- the group shows rapid lithological variations, both vertically and horizontally, with the sequence comprising a mix of fluvial conglomerates (mainly, but not exclusively, in the basal part of the sequence). Variably ferruginous fluvial and aeolian sandstones, duricrusts and occasional lacustrine/pan deposits, such as silts, sandstones, marls, limestones and silicified limestones, also occur.

Analogues for most of the deposits in the Kalahari Group can be found amongst the present-day sediments of the region. Consequently, some understanding of controls on the modern/recent sedimentary systems in central and southern Africa is relevant to predicting sediment distributions within equivalent ancient sequences buried deeply beneath the South Atlantic area:

- regional warping, which impedes flow in the drainage net, can promote deposition of alluvial plain–fan deposits (Palombe Plain, Malawi). Regional warping may also result in alluvial plain–swamp systems (Vwaza Marsh, Malawi) or in lacustrine basins (Lakes Victoria and Kyoga, Uganda), depending upon hydrological and sediment budgets. The lacustrine systems may be shallow saline (Lake Makgadikgadi, Botswana) to freshwater (Lake Bangwelu, Zambia) depending upon the hydrological budget;

- since lacustrine hydrological budgets are partly controlled by the area of drainage net, there may be an irreversible change in the lacustrine system through addition or removal of the drainage net as a consequence of warping or erosional capture. It is likely, for example, that the present strongly saline Lake Makgadikgadi (Botswana) was much less saline prior to capture of Okavango and upper Zambezi drainage by the present Zambezi system (Cooke 1980);
- in normally closed lake basins, transgression during wet climatic phases drives the surf zone across alluvial deposits to create beach systems well beyond the usual margins of the lake. If, as in the cases of Lakes Chilwa (Malawi) and Makgadikgadi (Botswana), maximum levels are controlled by overflow, then an amalgamated cluster of beach sands and back beach aeolian dunes accumulates around the basin at overflow level;
- the largest beach complexes develop along the palaeolake margins with maximum fetch in the prevailing winds. The Gidikwe Ridge, along the western side of the Makgadikgadi Basin, is 180 km long (Cooke 1980), and the Chilwa sand bar on the north side of Lake Chilwa (Malawi) is 25 km long, up to 2 km wide and typically 10–15 m thick (Lancaster 1981);
- these sandbodies have moderately high textural and high mineralogical maturities and offer excellent reservoir potential. Feldspar contents in the Chilwa sand bar drop rapidly from *c.* 20% at the western end of the bar (near-source) to 2–3% along the rest of the bar, and heavy minerals (mainly magnetite) are relatively abundant but decrease eastwards from *c.* 5 to 2% (Lancaster 1981);
- in normally open lake basins, beach systems also develop, but amalgamation is less common since the shorelines are prograding due to the flux of alluvial sediment to the lake margin. In the case of Lake Victoria, beach amalgamation has, in addition, been lessened because reduction of maximum lake levels due to lowering of the overflow has left early beach ridges abandoned around the lake margin. A series of beach ridges separated by lagoonal deposits therefore tends to develop (Temple 1964). The beach deposits surveyed by Temple (1964) are relatively small, being typically a few kilometres long and a few metres thick;
- the modern Lakes Victoria and Chilwa are formed outside the current rifts by warping of the rift margins. They are not, therefore, strictly pre-rift lakes, although, in facies terms, they represent useful analogues for ancient pre-rift lake deposits;
- long-lived open lacustrine basins are more likely in arid areas, since a high proportion of the water reaching the basin centre arrives via groundwater flow and thus carries little clastic sediment. Fine clastics and chemical sediments deposited in the lakes are repeatedly deflated out of the basin during dry phases. In moist climate areas, long-lived, non-rift lacustrine basins are only likely if both the area of the drainage net and hinterland erosion rates are small, thus contributing little clastic sediment;
- warping sympathetic with non-rift drainage directions results in accelerated erosion, exposing bedrock over wide areas (parts of the Athi River headwaters, Kenya). In this circumstance there may be little, or no, sediment preserved on the basement surface.

Implications for the South Atlantic

Pre-rift sediments can represent good exploration targets in terms of thickness, depositional lateral continuity and mineralogical maturity. They can also provide sources of mature sand that may later be reworked into otherwise mineralogically immature rift sequences.

In the Recôncavo Basin of Brazil, the conglomerates, sandstones and minor red shales of the Boipeba Member form a sequence typically 100 m thick, which is interpreted as the base of the pre-rift sequence (Caixeta *et al.* 1994). Underlying red shales and evaporites of the Afligidos Formation, once thought part of the pre-rift sequence (Figueiredo *et al.* 1994), have been reassigned to the Permian (Caixeta *et al.* 1994). Dating is also a recurrent problem in predominantly terrestrial pre-rift sediments on the eastern margin of the South Atlantic (e.g. the Lucula Formation of Angola and the Grès de Bas of Gabon; Bradley & Fernandez 1993). Using the Kalahari Group as an analogue, it is likely that the various pre-rift sequences represent a mix of sediments of various ages, as warping, drainage net erosion and climatic changes cause shifts in sediment flux and reworking of deposits across the basement surface.

Structurally, pre-rift sandstones offer good reservoir prospects since they will occur on all structural highs, whether the highs developed during rifting or much later, provided that these highs did not become unroofed and erosionally stripped (e.g. Magnavita & da Silva 1995, Fig. 3).

Rift basin sequences

Rift development in the South Atlantic area resulted in a mix of clastic and freshwater carbonate reservoirs, of freshwater to saline lacustrine source rocks, and of tectonic and stratigraphic trapping mechanisms. This produced a large number of hydrocarbon plays but geometrical complexity makes their prediction difficult. As seismic imaging of prospects is often poor, templates from the African Rift System may assist interpretation.

Early rift-fill

The characteristics of the sedimentary fill can vary greatly, particularly at the earliest stages of rifting, depending upon hydrologic and sediment fluxes. For example, a spectacular rift fault scarp, *c.* 150 m high, bounds the western margin of Lake Manyara (Tanzania). Inflow and sediment flux to the basin is minimal so topography mimics structure and the arrangement of facies in relation to rift structure is obvious, since the clastic sediments are largely locally derived. In contrast, the rift fault bounding the southern margin of the Okavango Swamp (Botswana) is at least of comparable throw but has minimal topographic expression because the large hydrologic and sediment flux from the Okavango River System has kept the graben infilled with peat, silt and sand. Most of the sand is polycyclic and sourced from outside the Okavango Rift Basin.

Implications for the South Atlantic

Discrimination of the basal part of the rift sequence from the pre-rift sequence is a recurrent problem in the South Atlantic basins. In the Recôncavo Basin, for example, the base of the rift sequence is sometimes taken at the onset of deposition of coarse conglomerates sourced from the Salvador Fault, yet, as Magnavita (1996) notes, such coarse facies simply record the development of fault scarps within the rift.

Using the Kalahari Group as an analogue, the so-called pre-rift sequences *c.* > 100 m thick deserve further attention. In the Recôncavo Basin, the pre-rift Boipeba Member, *c.* 100 m thick (see above), is overlain by perhaps 80 m of laterally persistent red lacustrine mudstone, the Capianga Member (Figueiredo *et al.* 1994). The scale of the mudstone implies significant subsidence, interpreted here as the start of the rift sequence. This accords with the proposal by de Matos (1992, 1989) that much of the supposed pre-rift section along the Brazilian east coast is, in fact, infilling rift basins.

The Capianga Member is buried basin-wide by the Sergi Formation, a sequence of predominantly fine-grained fluvial and locally aeolian sandstones, which range from *c.* 100 to 350 m thick. In part, the thickness changes are controlled by the Mata–Catu Fault, which plays an important role at several stages in later rift development (Figueiredo *et al.* 1994); there seems little doubt that substantial rift subsidence was occurring. The volume of Sergi Formation sand implies input from one or more major drainage systems, and the pre-dominantly fine-grained, quartz-rich character of the sand indicates that it is probably polycyclic and sourced from outside the rift basin. The modern Okavango System would appear to be a good analogue for the Sergi Formation.

Horst, or block-fault, rotation may expose previously buried pre-rift and early rift sequences, with consequent stripping of these sediments from the crest of some structural highs (Brice *et al.* 1982). Salah & Alsharhan (1996) illustrate analogous structural relationships in the Gulf of Suez, Egypt. This mechanism represents a risk if pre-rift/early rift reservoirs are the exploration target but, in rift settings, the eroded sediments are often not transported far, so reworked high maturity sands with good reservoir potential may be found around the margins of the structural palaeohighs.

Clastic facies in large rift lakes

Major African rift lakes, such as Lakes Malawi and Tanganyika, have long been recognized as analogues for some South Atlantic syn-rift source rocks and lacustrine turbidites, such as in the Lower Bucomazi (Angola) and Candeias Formation (Brazil).

The Malawi and Tanganyika Rift Valleys show tectonic throws of hundreds to thousands of metres. Lake Tanganyika overflows via a rift flank and thus has a higher 'strandline', relative to rift structures, than does Lake Malawi, which overflows along the rift axis. Drainage capture gives the additional hydrologic inflow to maintain these large lakes but sediment supply cannot keep pace with subsidence, so large topographic and bathymetric relief develops. There is now a substantial literature on clastic facies in the larger African rift lakes. A non-exhaustive list of basic references includes Baltzer (1991), Bouroullec *et al.* (1991), Cohen *et al.* (1986), Crossley (1984), Owen & Crossley (1989) and Scott *et al.* (1991).

Implications for the South Atlantic

Once rift basins have reached the stage of developing substantial topographic/bathymetric relief, then a number of diagnostic clastic facies associations develop. Consequently, many explorationists feel relatively comfortable about applying rift models to their basin-fill sequences and the issues will not be dwelt upon in this paper. However, there is one post-depositional curiosity worthy of comment.

The Candeias Formation in the thickest, axial, part of the Recôncavo Basin contains the enigmatic Pitanga Member that comprises laterally discontinuous units of massive, highly argillaceous sandstone. Some individual units show uniform wireline log responses through 1200 m of vertical wellbore and have abrupt vertical contacts against deformed mudstones of the Maracangalha Member (Figueiredo *et al.* 1994). At outcrop, the sandstones are mainly structureless, or show soft sediment deformation and near-vertical water escape features.

These sandstone–mudstone relationships have been tentatively interpreted as representing localized sand accumulations that supplied gravitational loading, which in turn triggered mud diapirism. However, geometrically similar units are seen on seismic sections from Lakes Malawi and Tanganyika, and some of the units in Lake Tanganyika are interpreted as large-scale sand-rich dykes (Bouroullec *et al.* 1991). The dykes are believed to represent relatively coarse-grained, initially permeable, strata that became overpressured and thus intrude through the overlying muds. This mechanism would account for the features observed in the Pitanga Member. The significance for exploration in the South Atlantic margin is that undisturbed, possibly more permeable sandstones, which sourced equivalents of the Pitanga Member, should be sought in the deeper, rather than shallower, parts of the stratigraphy.

Carbonate facies in large rift lakes

The open waters of Lake Tanganyika are, in practical terms, fresh, but are more saline than those of Lake Malawi, and have an unusually high Mg/Ca ratio. Carbonate facies, all exhibiting high Mg/Ca ratios, comprising *Chara* beds in sheltered shallows, ooid shoals, shelly bioclastic debris sheets and algal bioherms in the deeper waters, occur in Lake Tanganyika (Cohen & Thouin 1987). The authors are not aware of any significant development of carbonate facies in Lake Malawi. The reasons for this difference are not clear, although evidence for direct links between water chemistry and controls on biota are hard to find even in the simplest systems (Beadle 1974).

Patchy accumulations of gastropod shells do occur in the shallow waters of Lake Malawi, but the authors have not observed volumes that would give a build-up of any geological significance. Indeed, gastropod shells in the open waters of Lake Malawi appear to suffer rapid abrasion by grazing organisms and ingestion by fish, and are likely to be poorly represented (except in coprolites) in the future fossil record.

More than 200 soft sediment cores have been collected from the open waters of Lake Malawi (Owen & Crossley 1989), but the authors have only noticed ostracods in a few cores collected from the sediment plume of the Linthipe River, whose hinterland includes carbonate-rich anorthosites.

In contrast, marls and shelly layers are common in sediments flooring lagoons around the lake margin. Plio-Pleistocene sediments, the Chiwondo Beds, on the northwestern margin of Lake Malawi (Crossley 1982), include well-developed algal and bioclastic limestones, interbedded with sandstones and marls. Their depositional setting ranges from shallow lacustrine through swamp to lakeshore plain (Kaufulu *et al.* 1981). It is, therefore, not clear whether the Chiwondo deposits indicate that the northern part of the rift basin contained a major lake more saline than today or whether they represent development of local salinity in a lagoonal system analogous to the present day.

Implications for the South Atlantic

Freshwater carbonate bioclastic reservoirs are known from the rift sequences of the Campos (Bertani & Carozzi 1985*a*, *b*) and Cabinda Basins (Brice & Pardoe 1980). The shallow saline lacustrine–mudflat–sand flat–alluvial fan depositional settings envisaged by Bertani & Carozzi (1985*a*, *b*) do not find an obvious parallel in any African rift lake setting known to the authors. A weak analogue combination might involve bioclast beaches on volcanics at Lake Bogoria (Renaut &

Owen 1991), the Chiwondo Beds and Lake Tanganyika. Whether comparisons could be improved with more information from the African rift lakes or the Campos Basin is not clear. The possibility remains that, in the same setting, carbonate facies developed from Mesozoic biota may be significantly different to those of the present day.

Rift, sag, salt and early drift basins

The depositional distribution of Aptian salt in the South Atlantic area played a key role in trapping hydrocarbons sourced from the syn-rift successions. Early drift, marine carbonate build-ups, such as in the Pinda Formation of Angola, can represent important reservoir objectives.

The Red Sea and Gulf of Aden areas provide a classic template for rift to drift tectonics and sedimentation:

- early rift phase: characterized by tectonic throws of thousands of metres and exhibiting abundant contemporary volcanic debris only where the system crosses the mantle anomaly at Afar. Thus, the early rift phase was similar to the Malawi and Tanganyika rifts throughout much of the Red Sea and Gulf of Aden. Continued rifting caused marine incursions, deposition of restricted marine muds and limited formation of carbonate build-ups on those bathymetric highs that were starved of siliciclastic sediment;
- late rift to early sag phase: some 10 Ma after the onset of rifting, deposition of evaporites commenced in the northern Red Sea basins and, after *c.* 5 Ma, rapid extension and sag was accompanied by evaporite deposition throughout the Red Sea. Open marine conditions persisted south of the Afar structural/volcanic high in the Gulf of Aden;
- late sag and halt: some 20 Ma after the onset of rifting, open marine conditions, with limited carbonate build-ups, returned throughout the Red Sea basins, but extension halted briefly whilst Arabia locked with Asia, giving compressional features in the Gulf of Suez. Open marine conditions continued in the Gulf of Aden, where parting of continental crust was occurring;
- early drift: about 25 Ma after the onset of rifting, continental crust parted within the Red Sea and fully marine conditions have continued, with the development of carbonate build-ups around faulted bathymetric highs in the north and across extensive offshore bathymetric shallows in the south. Movement of salt, driven by tectonics and sediment loading, has become an important local control on offshore bathymetry and facies. Siliciclastic sediment is supplied by predetermined sources (pre-rift drainage and local rift structures).

Implications for the South Atlantic

Geometrical separation in the Red Sea and the Gulf of Aden areas was greater in the south than the north but, at Afar, the general structural high and copious volcanics limited marine incursion from the south. Consequently, a periodically closed evaporite basin formed in the Red Sea whilst, to the south of the Afar barrier, the Gulf of Aden was fully marine (Crossley *et al.* 1992). The general structural high and volcanic accumulations of the Paraná–Etendeka area (early Walvis Ridge) acted as a similar barrier in the South Atlantic system. However, the authors do not believe that the salt sequences of the Red Sea can be used as a direct analogue for the Aptian salt of the South Atlantic (e.g. the Ezanga and Loeme Formations), since it appears that marine salt deposition started at an earlier structural stage (at least in the northern part of the Red Sea) than in the South Atlantic area.

The evaporites of the South Gharib and Zeit Formations of Egypt, and their stratigraphic equivalents throughout the Red Sea, were deposited prior to the sag phase (Crossley *et al.* 1992). Consequently, depositional thickness and facies variations in Red Sea salt formations are partly controlled by contemporary movements on rift fault blocks (Salah & Alsharhan 1996). The sag phase in the northern part of the Red Sea is represented by the Wardan Formation and Recent deposits and represents a transgressive transition from evaporite to the fully marine conditions of the present day.

In contrast, most authors would agree that the Aptian salt formations of the South Atlantic were deposited after onset of the sag phase, so depositional salt thicknesses varied only gradually (e.g. Brink 1974). The late rift and early sag phase off West Africa would be expected, in the absence of marine incursions, to develop major thermally or chemically stratified lakes offering excellent source rock potential. This is compatible with the observation that major lacustrine sediments of Barremian age (e.g. the Melania Shale of Gabon and the Organic Bucomazi of Angola) appear to be correlatable over many hundreds of kilometres along the west coast of Africa (Bradley & Fernandez 1993). Such correlation would not be expected if these source rocks were deposited in isolated active rift basins (Bate 1999).

The supply of clastic sediment during the rift phase was largely controlled by a limited number of major, long-lasting river systems (e.g. the Palaeo-Ogooué and Palaeo-Congo). Halokinetic movement of the thick salt succession off West Africa began relatively early and was controlled, in part, by minor movements of the underlying fault blocks. Shallow marine carbonates accumulated extensively on both margins of the South Atlantic during the early drift phase (e.g. the Pinda Group, and the Sendji, Madiela and Macae Formations). Carbonate deposition will have been concentrated on shallow shelves away from major siliciclastic sediment sources and on tectonically and/or diapirically controlled palaeohighs bypassed by siliciclastic sediment.

The best carbonate reservoir potential will tend to occur in the highest energy deposits forming over the crests of the highs (although Cretaceous shoal deposits will differ from the coral/algal reefal deposits that characterized the Red Sea analogue). Low relief coastal areas, with limited clastic input, will have been characterized by deposition of supratidal evaporites, forming potential intraformational seals.

Basinfloor topography caused by salt movement also influenced the distribution of subsequent drift phase clastic sediments (e.g. the turbiditic Batanga sandstones of the Ogooué Delta in Gabon), as well as forming structural and stratigraphic traps.

Controls on clastic sediment supply, on carbonate build-ups and on bathymetric changes due to salt movement in the South Atlantic, all have useful analogues in the Red Sea. However, the explorationist typically has much better seismic control in these post-salt sequences so, in these sections, the role of rift-drift analogues becomes less significant.

References

BALTZER, F. 1991. Late Pleistocene and Recent detrital sedimentation in the deep parts of northern Lake Tanganyika (East African rift). *Lacustrine Facies Analysis, Special Publication of the International Association of Sedimentologists*, **13**, 147–173.

BATE, R. H. 1999. Non-marine ostracod assemblages of the pre-salt rift basins of West Africa and their role in sequence stratigraphy. *This volume*.

BEADLE, L. C. 1974. *Inland Waters of Tropical Africa*. Longman, London.

BERTANI, R. T. & CAROZZI, A. V. 1985*a*. Lagoa Feia Formation (Lower Cretaceous) Campos Basin offshore Brazil: rift valley stage carbonate reservoirs I. *Journal of Petroleum Geology*, **8**, 37–58.

—— & —— 1985*b*. Lagoa Feia Formation (Lower Cretaceous) Campos Basin offshore Brazil: rift valley stage carbonate reservoirs II. *Journal of Petroleum Geology*, **8**, 199–220.

BOUROULLEC, J.-L., REHAULT, J.-P., ROLET, J., TIERCELIN, J.-J. & MONDEGUER, A. 1991. Quaternary sedimentary processes and dynamics in the northern part of the Lake Tanganyika trough, East African Rift System. Evidence of lacustrine eustatism? *Bulletin Centres Recherches Exploration–Production, Elf-Aquitaine*, **15**, 343–368.

BRADLEY, C. A. & FERNANDEZ, M. N. 1993. Early Cretaceous paleogeography of Gabon/Northeastern Brazil. A tectono-stratigraphic model based upon propagating rifts. *Géologie Africaine: Collection Géologie Libreville, recueil des Communications*, **6–8 May, 1991**, 17–30.

BRICE, S. E. & PARDO, G. 1980. Hydrocarbon occurrences in nonmarine, pre-salt sequences of Cabinda, Angola. *AAPG Bulletin*, **64**, 681.

——, COCHRANE, M. D., PARDO, G. & EDWARDS, A. D. 1982. Tectonics and sedimentation of the South Atlantic Rift Sequence: Cabinda, Angola. *Memoir of the American Association of Petroleum Geologists*, **34**, 5–18.

BRINK, A. H. 1974. Petroleum geology of Gabon basin. *AAPG Bulletin*, **58**, 216–235.

CAIXETA, J. M., BUENO, G. V., MAGNAVITA, L. P. & FEIJO, F. J. 1994. Bacias do Reconcavo, Tucano e Jatoba. *Boletin Geociencias Petrobras*, **8**, 163–172.

COHEN, A. S. 1989. Facies relationships and sedimentation in large rift lakes and implications for hydrocarbon exploration: examples from Lakes Turkana and Tanganyika. *In*: TALBOT, M. R. & KELTS, K. (eds) *The Phanerozoic Record of Lacustrine Basins and Their Environmental Signals*. Palaeography, Palaeoclimatology Palaeoecology, **70**, 65–80.

—— & THOUIN, C. 1987. Nearshore carbonate deposits in Lake Tanganyika. *Geology*, **15**, 414–418.

——, FERGUSON, D. S., GRAM, P. M., HUBLER, S. L. & SIMS, K. W. 1986. The distribution of coarse-grained sediments in modern Lake Turkana, Kenya: implications for clastic sedimentation models of rift lakes. *In*: FROSTRICK, L. E., RENAUT, R. E., REID, I. & TIERCELIN, J.-J. (eds) *Sedimentation in the African Rifts*. Geological Society, London, Special Publications, **25**, 127–139.

COOKE, H. J. 1980. Landform evolution in the context of climatic change and neo-tectonism in the Middle Kalahari of north central Botswana. *Transactions Institute of British Geographers, New Series*, **5**, 80–99.

CROSSLEY, R. 1982. Late Cenozoic stratigraphy of the Karonga area in the Malawi rift. *Palaeoecology of Africa*, **15**, 139–144.

—— 1984. Controls of sedimentation in the Malawi Rift Valley, Central Africa. *In*: JANSA, L. F., BUROLLET, P. F. & GRANT, A. C. (eds) *Basin Analysis: Principles and Applications*. Sedimentary Geology, **40**, 33–50.

——, CAMERON, N. & RAFALSKA, J. K. 1993. Hydrothermal fluids, organic geochemical maturation and hydrocarbon migration in pre-salt sequences of Angola. *Proceedings of the Geofluids '93 Conference*, Torquay, UK.

——, WATKINS, C., RAVEN, M., CRIPPS, D., CARNELL, A. & WILLIAMS, D. 1992. The sedimentary evolution of the Red Sea and Gulf of Aden. *Journal of Petroleum Geology*, **15**, 157–172.

FAIRHEAD, J. D. 1986. Geophysical controls on sedimentation within the African Rift Systems. *In*: FROSTICK, L. E., RENAUT, R. W., REID, I. (eds) *Sedimentation in the African Rifts*, Geological Society, London, Special Publications, **25**, 19–27.

FALVEY, D. H. 1974. The development of continental margins in plate tectonic theory. *Australian Petroleum Exploration Association Journal*, **14**, 95–106.

FIGUEIREDO, A. M. F., BRAGA, J. A. E., ZABALAGA, J. C. *ET AL*. 1994. Reconcavo Basin: a prolific intracontinental rift basin. *In*: LANDON, S. M. (ed.) *Interior Rift Basins*. American Association of Petroleum Geologists Memoir, **59**, 157–203.

KAUFULU, Z., VRBA, E. S. & WHITE, T. D. 1981. Age of the Chiwondo Beds, northern Malawi. *Annals of the Transvaal Museum*, **33**, 1–8.

LANCASTER, N. 1981. Formation of the Holocene Lake Chilwa sand bar. *Catena*, **8**, 369–382.

MAGNAVITA, L. P. 1996. Sobre a implantacao da fase sin-rifte em riftes continentais. *XXXIX Congresso Brasileiro de Geologia*, Salvador, 335–338.

—— & DA SILVA, H. T. F. 1995. Rift border system: the interplay between tectonics and sedimentation in the Reconcavo Basin, northeastern Brazil. *AAPG*, **79**, 1590–1607.

MATOS, R. M. D. DE 1992. The northeast Brazilian rift system. *Tectonics*, **11**, 766–791.

—— 1999. History of the northeast Brazilian rift system: kinematic implications for the break-up between Brazil and West Africa. *This volume*.

MILNER, S. C., LE ROEX, A. P. & O'CONNOR, J. M. 1995. Age of Mesozoic igneous rocks in northwestern Namibia, and their relationship to continental breakup. *Journal of the Geological Society, London*, **152**, 97–104.

OWEN, R. B. & CROSSLEY, R. 1989. Rift structures and facies distributions in Lake Malawi. *Journal of African Earth Sciences*, **8**, 415–427.

PONTE, F. C. & ASMUS, H. E. 1978. Geological framework of the Brazilian Continental Margin. *Geologische Rundschau*, **68**, 210–235.

RENAUT, R. W. & OWEN, R. B. 1991. Shore-zone sedimentation and facies in a closed rift lake: the Holocene beach deposits of Lake Bogoria, Kenya. *In*: *Lacustrine Facies Analysis*. Special Publication of the International Association of Sedimentologists, **13**, 175–195.

SALAH, M. G. & ALSHARHAN, A. S. 1996. Structural influence on hydrocarbon entrapment in the northwestern Red Sea, Egypt. *AAPG Bulletin*, **80**, 101–118.

SCOTT, D. L., NG'ANG'A, P., JOHNSON, T. C. & ROSENDAHL, B. R. 1991. High-resolution acoustic character of Lake Malawi (Nyasa), East Africa and its relationship to sedimentary processes. *In*: *Lacustrine Facies Analysis*. Special Publication of the International Association of Sedimentologists, **13**, 175–195.

SENGÖR, A. H. C. & BURKE, K. 1978. Relative timing of rifting and volcanism on earth and its tectonic implications. *Geophysical Research Letters*, **5**, 419–421.

TEMPLE, P. H. 1964. Raised features along the southern shoreline of Lake Victoria. *Proceedings of the East African Academy*. First Symposium, Makere University College, June 1963, 13–22 (Longmans, London).

THOMAS, D. S. G. & SHAW, P. A. 1990. *The Kalahari Environment*. Cambridge University Press.

Western Africa: an unfinished story of oil and gas exploration

P. DOLAN

IKODA Limited, 5 Old Lodge Place, Twickenham, Middlesex TW1 1RQ, UK

Abstract: Understanding the future oil and gas exploration potential of western Africa's South Atlantic margin depends not only on access to modern, high-quality geological and geophysical data, but also on correctly interpreting the motivations for earlier exploration campaigns and the basic geological characteristics which can be used to sub-divide the margin of western Africa.

For the purposes of summarizing exploration history, and anticipating the future, western Africa's margin has been simplistically divided by Dolan (1998) into five sectors, with broadly discrete geological characteristics (Fig. 1):

- **Area 1** (Mauritania to Guinea Bissau). Triassic salt to Tertiary clastics and carbonates in Atlantic-type pull-apart basins.
- **Area 2** (Guinea to Benin). Lower Cretaceous to Tertiary mostly clastics in transcurrent basins resting unconformably on Palaeozoic, often Devonian, clastics.
- **Area 3** (Nigeria to Equatorial Guinea). Niger Delta influenced Tertiary plays clastics.
- **Area 4** (Gabon to Angola). Lower Cretaceous clastics below Aptian salt with Upper Cretaceous and Tertiary carbonates and clastics above salt.
- **Area 5** (Namibia). Permo-Triassic clastics through Upper Cretaceous clastics and carbonates to Tertiary clastics in Atlantic-type pull-apart basins with Orange River Delta locally superimposed.

In addition to these five areas, a further Area 6 now needs to be defined, comprising deepwater areas (typically 200–3000 or 4000 m of water) along the entire western Africa margin.

- **Area 6** ('outboard' of Areas 1–5). Tertiary turbidite clastics.

Pre-World War II

Probably the earliest drilling in search of commercial oil production was in Ghana (then the Gold Coast) when four onshore wells were drilled by British interests between 1886 and 1897, the deepest to 1402 ft, with one well achieving production of 5 bopd – not much, but a start! Soon after, Societe Française de Petrole (S.F.P.) drilled seven wells in Côte d'Ivoire near seeps in the Tano Basin, with further modest recoveries being achieved. At about the same time, 15 onshore wells were drilled further to the east in the Niger Delta from 1907 to 1913; many oil and gas shows were encountered, but none of commercial interest.

After World War I, there was little activity until the mid and late 1920s, when exploration occurred in four countries. First was a Sinclair (later to become Arco) subsidiary, which is reported to have been evaluating the coastal regions of Angola, where there are extensive onshore basin margin asphaltic oil seeps, from 1922 to 1938. About the same time, from 1923 to 1925, two further wells were drilled in Ghana (neither successful) and there was a sustained program of nine core holes and two deep wells drilled in Gabon to investigate good oil seeps. Finally, in 1928, further oil seeps were reported in Nigeria.

Very little exploration activity has been reported for western Africa during the 1930s apart from that of Sinclair, mentioned above, and a Shell/D'Arcy (BP) combine in Nigeria from 1937 to 1941.

Post-World War II

World War II brought virtually all activity to a standstill but then, in contrary fashion, stimulated an explosion of interest in oil exploration during the 1950s. It is interesting to note that onshore surface evidence of oil had by this time been found in geological Areas 1–5. Of 20 littoral countries, from Mauritania to Namibia, 17 attracted some form of activity during the 1950s, ten had exploration wells drilled leading to six oil and/or gas discoveries. These, in turn, lead to the first production in Angola, Gabon and Nigeria. This post-World War II activity was, in large part, driven by the European colonial powers wishing to secure access to strategic, non-Middle East, supplies of crude oil. The successes of the 1950s were nicely balanced between the dominant colonial powers, Angola (Portugal), Gabon (France) and Nigeria (Great Britain). Spanish activity in Equatorial Guinea [previously Spanish Guinea and incorporating Bioko (previously Fernando Po)] was limited during the 1950s to field work. Somewhat anomalously, Belgium did not establish a significant foothold in its natural

Dolan, P. 1999. Western Africa: an unfinished story of oil and gas exploration. *In:* Cameron, N. R., Bate, R. H. & Clure, V. S. (eds) *The Oil and Gas Habitats of the South Atlantic.* Geological Society, London, Special Publications, **153**, 97–99.

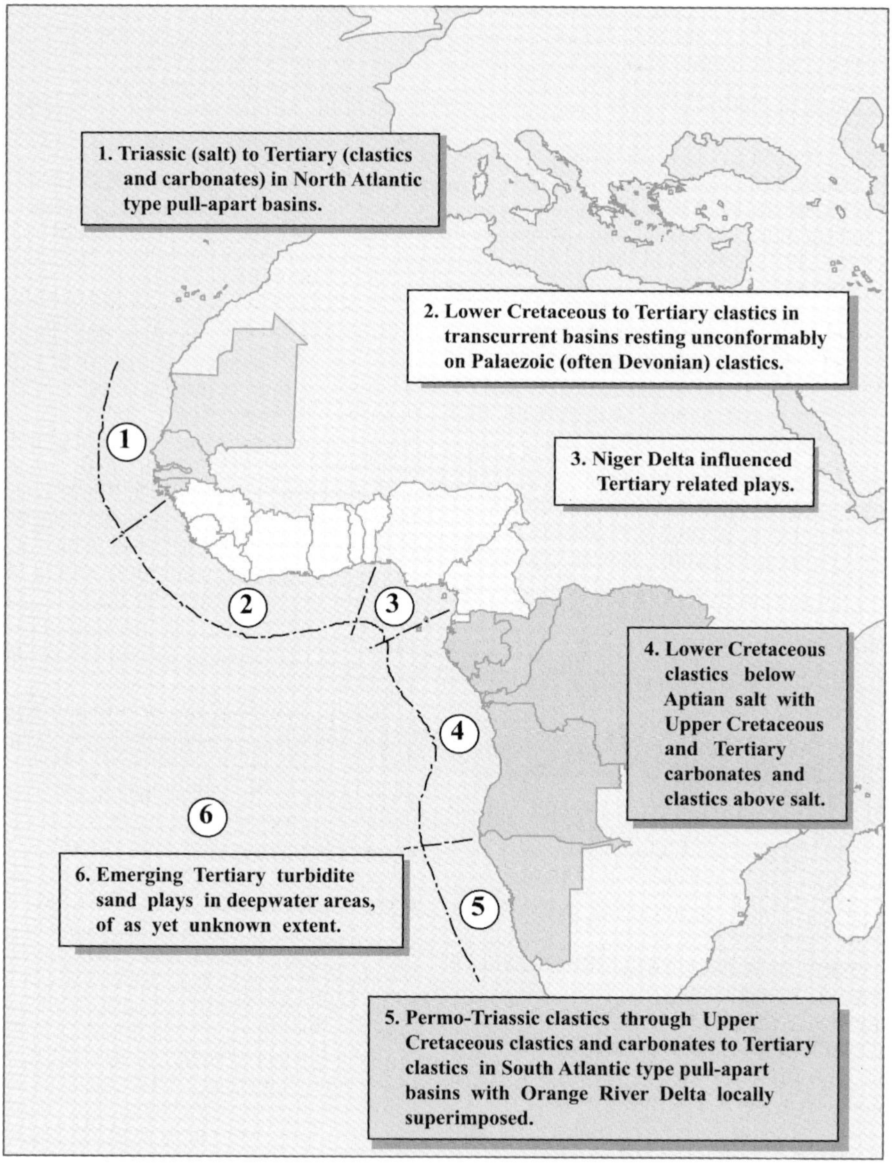

Fig. 1. Principal petroleum provinces of Western Africa.

domain of the Democratic Republic of Congo (then the Belgian Congo) but rather onshore Angola and Cabinda, thereby sharing the benefits with Portugal.

The various forms of 'companies' by which the colonial powers achieved these results were partially government-owned, or at least officially inspired, e.g. numerous forerunners of Elf. However, they were not alone.

During the 1950s, Cities Services (in Mauritania and Liberia), Conoco (in Mauritania), Gulf (now Chevron) (in Ghana and Angola), Esso (in Guinea Bissau) and Mobil (in Nigeria) are a few examples of US companies establishing a presence. From these early American activities to the 1950s, some sustained campaigns developed. Notably, Esso concentrated on geological Areas 1 and 2, Mobil on Areas 1, 3 and 4, and Gulf on Areas 2, 3 and 4.

The success of the 1950s may have contributed to the political changes which swept Africa in the 1960s; ten of the 20 littoral countries obtained independence in 1960 alone. Independence from the colonial powers made exploration desirable as a potential means of economic development.

The 1960s consequently witnessed continuing oil industry interest but with attention switching to the offshore as:

- offshore development technology became available;
- plate tectonic concepts supported ideas about the prospectivity of continental shelves;
- the Geneva Convention, which helped to define territorial waters, was ratified.

Consequently, 19 out of 20 countries had E&P activity in the 1960s, ten experienced their first offshore wildcats and seven had commercial offshore discoveries, namely Angola, Benin, Cameroon, Congo (Brazzaville), Gabon, Nigeria and Senegal.

During the 1970s, western Africa confirmed its position as a world-class petroleum province, with activity ranging from rapidly increasing production in Gabon and Nigeria to two shallow wells in Sao Tome and Principe, where some enigmatic oil seeps had been recorded 100 years previously.

With the likes of Agip, BP, Elf, Gulf and Shell then established as producing operators in western Africa, many other large companies started to become involved. One notable example was the Seagap Group [a consortium of Agip, Getty (now Texaco), Hispanoil (now Repsol) and Phillips], which acquired reconnaissance seismic data on continental shelves around the world. It was this data which, in part, led Phillips into Côte d'Ivoire and Ghana, and Hispanoil into Equatorial Guinea. In Phillips' case it led to discoveries in both Côte d'Ivoire and Ghana which, due to their modest size, political unrest and other difficulties, were relinquished.

As it became clear that, in the medium term, the bulk of discoveries and production would occur in geological Areas 3 and 4, i.e. Nigeria to Angola, it is not surprising that this is where the major oil companies focused their interest throughout the late 1970s and 1980s.

This shift of emphasis left geological Areas 1 and 2 ripe for a second generation of smaller companies and latecomers to establish a presence. Names of companies that appear regularly in activity reports include (but are not limited to) Amoco, Aracca, Frontier (Reading and Bates), Oxoco, Signal, Union Carbide, Unocal, etc.

The 1990s started with relatively low levels of exploration activity due to generally depressed world oil prices. However, four factors have combined to make western Africa a favoured place for E&P investment at the end of the millennium:

- the availability of deepwater E&P technology and resultant high success rates in finding giant oil accumulations, so far notably off Angola, Congo and Nigeria; this is geological Area 6;
- a desire by many reserve-constrained North American independents to participate in international exploration;
- an awareness that gas flaring should be reduced, although finding viable markets is often difficult;
- a nascent recognition by host governments of the desirability of coordinating infrastructure.

Examples of these trends are Exxon's acquisition of equity positions in about 20 deepwater licences between 1990 and 1998, the mooted gas transmission line from Nigeria through Benin and Togo to Ghana and the rash of smaller companies that have totally licenced the continental shelf off Ghana in just two years.

Post-2000

The next few years should see deepwater Area 6 become a world-class producing area, particularly off Angola, Congo and Nigeria, and maybe Equatorial Guinea and Gabon. More speculatively, significant discoveries may be made in deepwaters 'outboard' of Areas 1 and 2. These areas will be dominated by the major oil companies with fresh impetus being given to the continental shelf and onshore by a third generation of mid-sized to small companies (Bentley & Kübler 1998). A good example of this is the rejuvenation of the continental shelf of Côte d'Ivoire by independents who have, through changed circumstances, found it economic to develop the unproduced reserves relinquished by Phillips and others in the 1980s.

References

BENTLEY, J. & KÜBLER, M. 1998. Sub-Saharan Africa's oil and gas activity. *In*: BLAKEY, G (ed.) *African Business Focus Yearbook for 1998*. British African Business Association, Reddy, London, 57–60.

DOLAN, P. 1998. *Western Africa (Mauritania to Namibia), 100 years of oil and gas E&P activity*. IKODA Limited, London, in press.

—— 1947–1989. Foreign Developments (1947–1979) and World Energy Developments (1980–1989) annual accounts of international E&P activity. *AAPG Bulletin*, various issues.

The distribution of petroleum reserves in basins of the South Atlantic margins

M. P. COWARD[1], E. G. PURDY[2], A. C. RIES[3] & D. G. SMITH[4]

[1] *Ries–Coward Associates Ltd, Caversham, Reading RG4 5ES & T. H. Huxley School of the Environment, Earth Sciences and Engineering, Imperial College, London SW7 2BP, UK*
[2] *PetroQuest International Inc., Weybridge, Surrey KT13 8YA, UK*
[3] *Ries–Coward Associates Ltd, Caversham, Reading RG4 5ES, UK*
[4] *Petroconsultants (UK) Ltd, London SW15 6TQ, UK*

Abstract: The opening of the South Atlantic created a series of passive margin basins on both sides of the new ocean. This paper reviews the distribution of petroleum reserves in these basins in terms of their tectono-stratigraphic position within the framework of the rift–drift succession. Seven megasequences are recognized. Three are in the drift succession, three are in the rift succession and one comprises the pre-rift succession. Ninety-three per cent of the presently discovered recoverable hydrocarbons are reservoired in the drift succession, 6% are located in the rift succession and 1% is associated with the pre-rift units. The basins group geographically into seven sectors within which both margins share some common features. Sixty-five per cent of the reserves are contained in Sector V which is dominated by the Niger Delta, and 28% are positioned in Sector III, which includes the Campos and Lower Congo Basins. The new deepwater giant discoveries of Angola, which are located in Sector III, are the Africa counterparts of the earlier deepwater Campos Basin discoveries. In general sectors III–V, located between the Walvis Ridge and the Equatorial Atlantic transforms, are most favourable as they are characterized by the most prolific source rocks at all horizons.

As part of Petroconsultants' multiclient Basin Monitor project, the authors have recently completed a series of studies on the petroleum geology of the principal basins along both margins of the South Atlantic Ocean. This study is possibly the first to take a completely systematic approach to all such basins, drawing on a comprehensive and independent database of well, field and reserves data. The present paper draws on the results of this project, to present an overview of the distribution of oil and gas reserves in the principal basins of the South Atlantic, looking particularly at the distribution of the reserves in terms of the large-scale tectono-stratigraphic units that can be recognized in all of the basins.

For the purposes of this paper, the South Atlantic margins extend from the Cape of Good Hope in South Africa to Cape Palmas at the border between Liberia and Côte d'Ivoire, and from the Falkland Islands to the Amazon Cone on the South American side. It includes only true continental margin basins on each side, and thus excludes basins such as the San Jorge Basin of Argentina and the Recôncavo–Tucano Basins of Brazil which are intracratonic in origin.

From these basin studies, a summary account of the petroleum geology of the basins is presented, grouping them into seven 'sectors' in order to examine: (1) the similarities across the South Atlantic; (2) the differences between adjacent sectors on the same margin. From this analysis, the reserves statistics have been summarized in terms of the broad-scale megasequence stratigraphy of the basins. The result shows some huge contrasts between the seven sectors. In some sectors the basins 'work' on both sides of the ocean (although sometimes at different stratigraphic levels); other sectors are characterized by far less rewarding exploration results.

Rift and drift evolutionary phases are present in all the basins (Fig. 1). However, the age of the break-up unconformity is not constant and there is much variation between basins. Some of this variation is related to diachronous events, in other cases it is nomenclatural. To enable a review of where oil and gas reserves are positioned within the evolutionary history of the basins, the rift–drift package has been separated into seven evolutionary megasequences, based on the location of major unconformities within the successions. From the base upwards they comprise the Pre-Rift, the Early Rift, the Middle Rift, the Late Rift, the Early Drift, the Middle Drift and the Late Drift Megasequences.

The study area includes two major deltas of Tertiary age, of which the Niger Delta is by far the richest of all the basins in this study; in contrast, results from the Amazon Cone have been disappointing.

COWARD, M. P., PURDY, E. G., RIES, A. C. & SMITH, D. G. 1999. The distribution of petroleum reserves in basins of the South Atlantic margins. *In:* CAMERON, N. R., BATE, R. H. & CLURE, V. S. (eds) *The Oil and Gas Habitats of the South Atlantic.* Geological Society, London, Special Publications, **153**, 101–131.

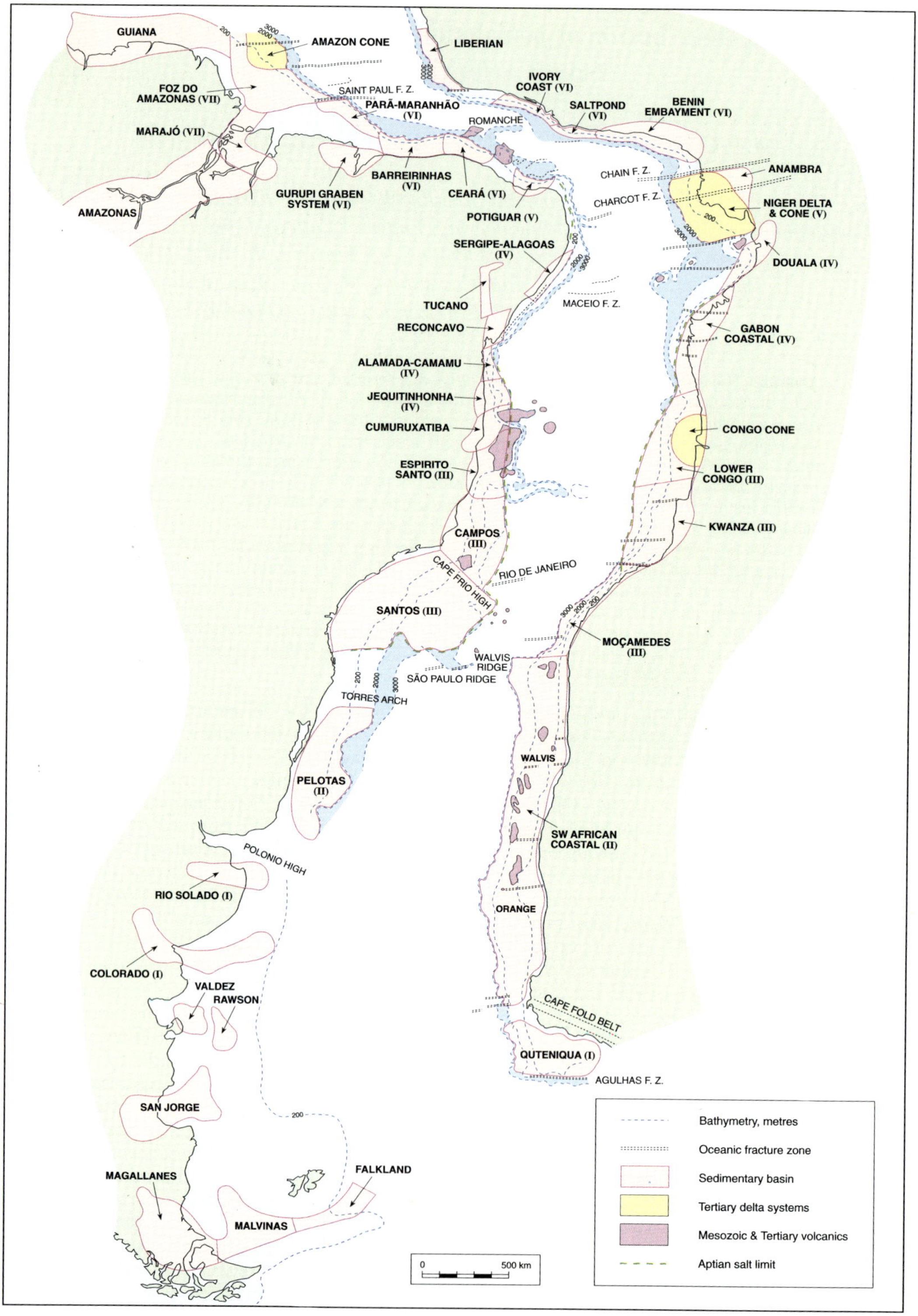

Fig. 1. Palaeoseparation of South America and Africa showing the distribution of Atlantic margin basins and grouping of the basins (Sectors I–VII) referred to in the text. The reconstruction is modified from the map presented by Davison (1999) and does not correspond to a specific geographic position at a specific time. Seaward limit of the Aptian salt is used as the corresponding seaward limit of the pertinent South American and African basins. North–south basin boundaries are geologically well-defined in South America but less so in Africa. F. Z.: Fracture Zone.

Basin evolution

Events controlling the basin development and evolution of eastern Brazil and western Africa are summarized below. The principal basement structures are summarized in Fig. 2.

I. Upper Proterozoic–Cambro–Ordovician (Brazilian–Pan-African Cycle)

The basement is composed of folded metamorphosed and sheared Precambrian rocks and igneous intrusives of Archean and Proterozoic age. They formed during periods of plate accretion and plate collision, and generally represent the deeper parts of mountain belts. The youngest belt of metamorphic rocks formed during what has been termed the Brazilian (or Pan-African) Cycle during the late Proterozoic and Palaeozoic. Structures of this age form the basement fabric to the Mesozoic–Cenozoic basins of the South Atlantic. Foreland basins, formed by flexural loading of the thickened upper Precambrian–lower Palaeozoic lithosphere, occur along the margins of the fold belts. Preserved foreland basin sediments occur in central Brazil and West Africa along the western edge of mobile belt, in

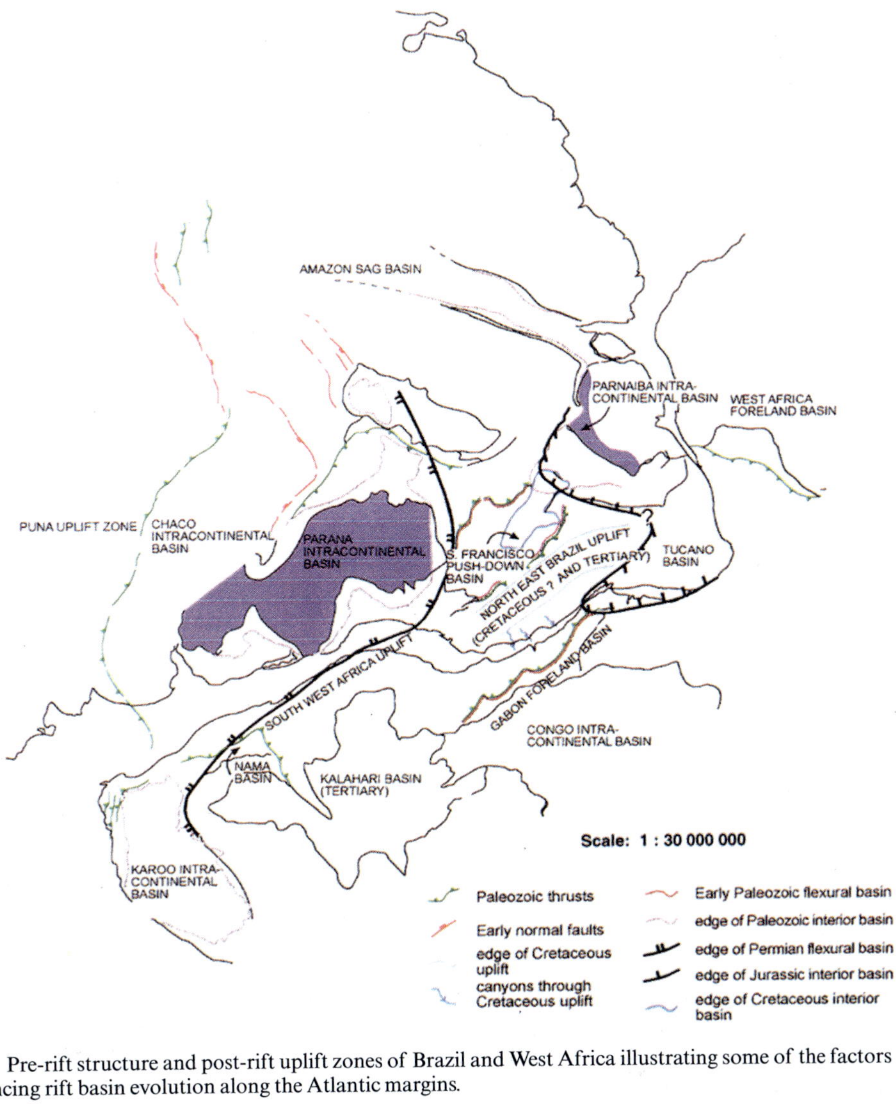

Fig. 2. Pre-rift structure and post-rift uplift zones of Brazil and West Africa illustrating some of the factors influencing rift basin evolution along the Atlantic margins.

foreland basins along the eastern edge of the Pan-African Belt in Gabon and Namibia–South Africa, and in the São Francisco Basin which formed as an intra-mobile belt pushdown basin in eastern Brazil.

II. Middle–Upper Palaeozoic basin development

Thick blankets of coarse sandstones and basal conglomerates were deposited over the Precambrian–Ordovician erosional surface of Brazil, with an irregular angular unconformity. The sediments were deposited in the Solimoes, Amazonas, Parana and Parnaiba Basins. These four basins are characterized by three Palaeozoic, almost complete, transgressive–regressive marine cycles, Ordovician–Silurian, Devonian–early Carboniferous and late Carboniferous–Permian in age. Each basin has at least one glacial horizon. The southern basins, including the Parana and Karoo, were probably influenced by flexural subsidence at the edge of the upper Palaeozoic Andean orogeny and its continuation into the Cape Fold Belt.

III. Mesozoic volcanics and rifting

Triassic and Jurassic rifts in the South Atlantic trended northwest–southeast, parallel to the trend of the underlying Permo-Carboniferous structures. The basins onshore and offshore Argentina, and the south coast of South Africa, were formed at this stage. In the Parana Basin, and its eastward continuation into northern Namibia, the subaerial sedimentation, which began in the Triassic, ceased with the outpouring of the Serra Geral basaltic flows in the early Cretaceous. These basalts, which covered an area of $> 800\,000\ km^2$ and reached thicknesses of > 2000 m, are the world's largest flood basalt complex (Milani *et al.* 1994). Basaltic rocks of similar age underlie the Santos, Campos and Espirito Santo Basins, where they are also associated with the initiation of South Atlantic rifting. The widespread volcanism suggests the presence of a mantle hotspot, causing localized uplift and erosion of the underlying sediments. The hotspot caused a lithospheric weakness and deflected the rifting in Gondwana away from the upper Palaeozoic belt of northern Argentina and along the Brazilian Atlantic margin, locally reworking the upper Precambrian orogenic structures. The Rio Grande–Walvis Rise formed at this time, probably associated with crustal underplating by basaltic magma.

Rifting associated with opening of the South Atlantic began during the early Cretaceous with a pole of rotation in the vicinity of eastern Nigeria. Maps of the rift system are given in Davison (1999). Continental separation, and the formation of ocean floor volcanics, only occurred in the southern part of this system, south of the Walvis Ridge. Further north, margin evaporitic sequences accumulated in a restricted basin which was bounded to the south by the Walvis Ridge and extended from the Santos and Kwanza Basins in the south to the Sergipe–Alagoas and Douala Basins in the north (Ponte *et al.* 1980).

Along the coast of northeast Brazil and West Africa, an en-echelon set of basins formed during the Aptian in an east–west trending zone *c.* 400 km wide between the western continuations of the Saint Paul and Romanche Fracture Zones. These basins linked northwards into the Central Atlantic. The en-echelon nature of the basins suggests that they formed under right lateral shear conditions between the precursors of these two oceanic fracture zones. The shear systems continue into Central Africa and link the northeast Brazilian rifts with rift systems of similar age in Niger and Sudan. The tip of the Central Atlantic Rift System lies in East Africa, although failed tips occur south of the basins, inland from the northeast coast of Brazil.

IV. Continental separation and thermal subsidence

During the late Aptian–early Albian the southeast propagating rift zone from the Central Atlantic interfered with the South Atlantic Rift Zone and allowed lithospheric separation between the African and South American plates. The marine sediments, which accumulated in the marginal basins from Albian to Recent times, comprise Albian–Santonian mainly carbonate sequences, while the Campanian to Recent sequences are dominantly clastic (Ponte *et al.* 1980).

Thick clastic sequences developed at the mouths of the major rivers during the Tertiary. In basins of eastern Brazil the marine sequence is overlain by Eocene–Pliocene coarse terrigenous clastics deposited in a shallow marine setting. This renewed clastic sedimentation resulted from a phase of uplift which affected the eastern part of Brazil during the later Palaeogene (Fig. 2). The origin of the uplift is uncertain; it may be due to Andean tectonic processes on the western margin of South America, or alternatively it may be due to slight changes in oceanic spreading direction causing compression and uplift close to old transform structures. This uplift caused exhumation of the Cretaceous–Tertiary sequence. A similar uplift formed along the African coast, north of Gabon. It is the positions of these uplifts and the resulting Tertiary major river systems that influenced much of the hydrocarbon distribution in the marginal basins.

Zonation of South Atlantic margin basins

The South Atlantic margin basins of Africa and Brazil, shown on Fig. 1, group into seven sectors, numbered from south to north. Figure 2 depicts these sectors in terms of the pre-separation geography. Figures 3–10 illustrate the tectono-stratigraphy and hydrocarbon reserves of the main producing regions.

Sector I

South America. This sector comprises the basins of the South American margin to the south of Brazil. Extensional deformation commenced during the Permo-Triassic, immediately following completion of the Cape Fold Belt Orogeny. There was some basin subsidence during the early and middle Jurassic (Urien *et al.* 1995), but it was not until the late Jurassic that the main phase of rifting commenced (Urien *et al.* 1976; Tankard *et al.* 1995). From this time, volcanism and dyke emplacement became important along the future South Atlantic margin, and the associated rifts of northern Argentina and Uruguay. The largest volcanic outpourings occurred during the early Cretaceous in the Chaco–Paraná Basin to the west (Milani *et al.* 1994) to form the Serra Geral Formation. In the latest Jurassic–early Cretaceous, extension and faulting intensified and was concentrated along a sector that extended from the Malvinas Plateau to the Rio Grande/Walvis Bay Rise (Kelley & Light 1993). The rifting created grabens, such as the Salado and Colorado Basins, orientated sub-perpendicular to the direction of Atlantic opening, but parallel to the trend of late Palaeozoic structures. These rifts accumulated sediments from Neocomian times, continental at first, but with an increasing marine component associated with a transgression from the south and east along the South Atlantic Fracture System. Exploration in these basins has not yet yielded any hydrocarbon reserves.

Africa: Outeniqua Basin. The Outeniqua Basin lies off the south coast of South Africa and comprises a number of en-echelon northwest–southeast trending sub-basins. Rifting began in the middle–late Jurassic and continued into the early Cretaceous. The rift sequence comprises continental clastic sediments, including red beds, passing up into lagoonal and shallow marine sequences. An upper Valanginian unconformity marks the onset of drifting (Tankard *et al.* 1995). In the western sub-basins the drift phase was accompanied by thermal subsidence forming a shelf slope–deep marine basin infilled with turbidite fans. However, during this early stage of drifting there was renewed extension in the eastern sub-basins, close to the Agulhas–Falkland Transform Fault. This was followed by basin inversion, particularly in the southeast. Deep canyons were cut into the uplifted syn-rift sequence and were infilled with clastic sediments. Following continental separation during the Albian, thermal subsidence affected the South African Coastal Basin and shelf sediments gradually prograded southwards over the basin. During the early Tertiary, uplift of South Africa caused erosion along the northern edge of the basins, synchronous with subsidence farther south, causing tilting of the basins.

Sector II

South America: Pelotas Basin. This sector encompasses the Pelotas Basin (Fig. 4a) in southern Brazil, which has been summarized by Milliman (1978), Goncalves *et al.* (1979), Pees & Delaney (1985), Dias *et al.* (1994). Continental stretching probably began in the middle–late Jurassic. In this sector terrestrial sedimentation was followed by volcanism. The area was uplifted and remained above sea level during much of the rifting. Aeolian sandstones are overlain by seaward-dipping basaltic lavas with K–Ar dates of 124 ± 8.6 Ma (Dias *et al.* 1994). North of the Florianopolis High, at the western end of the Walvis Ridge–Rio Grande Rise, evaporites accumulated following rifting. However, in the Pelotas Basin to the south there are age-equivalent bioclastic, oolitic calcarenites and calcilutites which pass laterally and shoreward into a shallow water sequence of fine-grained clastics. Throughout the late Cretaceous and Tertiary the Atlantic margin, east of the present-day shelf, was buried under an eastward-prograding wedge of sediments, inferred from seismic velocities to be 8–9 km thick seaward of the shelf. The Rio Grande Cone was formed at this time.

Africa: Southwest African Coastal Basin (*Fig. 4b*). The Southwest African Coastal Basin includes what is usually called the Walvis and Orange River Basins (Gerrard & Smith 1982; Muntingh 1993; Light *et al.* 1993; Letullier 1995). There is no fundamental geological distinction between the two, although the southern part of the Orange River Basin overlies Upper Palaeozoic structures and can be considered as part of Sector I. The Walvis Basin, like the Pelotas Basin, overlies Precambrian basement. Upper Jurassic–lower Cretaceous syn-rift sediments were deposited in north–south-trending grabens under initially non-marine conditions; these basins are parallel to the grain of the Palaeozoic basement. The extension direction was at a high angle to that in Sector I, suggesting the development of a triple junction whose position was partly controlled by Precambrian basement structures. The extension was accompanied by volcanicity and late rift-fill accumulations were terminated by a period of plateau-forming volcanic activity. Volcanicity was most intense west of the graben of the Southwest African Coastal Basin and was coincident with the onset of early Cretaceous continental spreading and the opening of the South Atlantic Ocean. Individual subbasins lost their identities after break-up and the thermal subsidence sequences have a well-defined palaeogeographic subdivision into marine shelf, slope and basinal environmental settings. Numerous unconformities are related to sea-level changes superimposed on the thermal subsidence. These unconformities have been correlated along the length of the basin. During the late Cretaceous and Tertiary, large-scale, gravity driven, extensional fault structures, several hundred kilometres across, disturbed the drift sediments. Additional information on this rift system may be found in Jungslager (1999).

Sector III

South America: Santos, Campos and Espirito Santo Basins. This structural sector encompasses the Santos (Fig. 5a), Campos (Fig. 6a) and Espírito Santo (Fig. 7a) Basins (Pereira & Macedo 1990; Guardado *et al.* 1990; Mohriak *et al.* 1990; Vieira *et al.* 1994). In this sector, volcanism pre-dated and accompanied rifting, leading to the development of thick volcaniclastic deposits. Pre-rift uplift, associated with the volcanism, led to the erosion of any pre-rift sediments on the basement. The rifting was essentially Valanginian in age, although minor pulses of rifting continued into the early Aptian. The volcanics are overlain by lacustrine shales, lacustrine carbonates alternating with black shales followed by evaporites, reflecting the gradual infilling of the rift topography. The Atlantic opening and continent separation during the Albian was recorded by a major marine incursion which led to the build-up of a carbonate platform in the west, deepening upward into calcareous mudstones. Thermal subsidence, accompanied by uplift of the hinterland, allowed the carbonate bank to submerge and become covered with a thick prograding sequence of clastic sediments. Gravity gliding occurred on the salt, developing major growth faults with associated sediment thickness changes.

Africa: Moçamedes, Kwanza and Lower Congo Basins. The Moçamedes (Fig. 5b), Kwanza (Fig. 6b) and Lower Congo (Fig. 7b) Basins are part of the series of basins that extended southward from the Cameroon Volcanic Ridge in the north to the Walvis Ridge in the south. All were characterized by similar basin development (Duval *et al.* 1992; McHargue 1990; Lundin *et al.* 1991; Lundin 1992), including the occurrence of Aptian salt. Figure 4b exemplifies the stratigraphy and basin evolution. The common history of these basins appears to have begun in late Jurassic time, when they were initiated as a series of northwest–southeast trending intracratonic basins. These formed in response to tensional forces resulting from the onset of South Atlantic separation of South America from Africa. The initial depocentres were filled with coarse continental clastics that occasionally contain interbedded volcanics, radiometrically dated as being erupted at the Jurassic–Cretaceous boundary.

Further pull-apart resulted in the formation of horsts and grabens that are generally coast-parallel. Erosion of the horsts provided thick sequences of fluvio-clastic sediments that were deposited in the adjoining grabens. Locally, source rocks are developed, as are reservoirs of secondary regional importance.

Toward the end of Neocomian time, a second major phase of rifting occurred, producing a distinct unconformity above which one or more deep graben lakes formed. Within these lakes a bituminous marl was deposited which constitutes the major regional source rock. Carbonates were deposited around some of the lacustrine shoals, and where known are exceptionally good reservoirs. By late Barremian time, the graben lakes had been filled by a succession of siliciclastics that includes some reservoir-quality sands. This major phase of non-marine rift sedimentation was followed by widespread upper Barremian–lower Aptian peneplanation.

The upper Barremian–lower Aptian surface is overlain by transgressive sands of late Aptian age. These have excellent reservoir properties, but are of relatively minor production importance in the Lower Congo Basin mainly due to

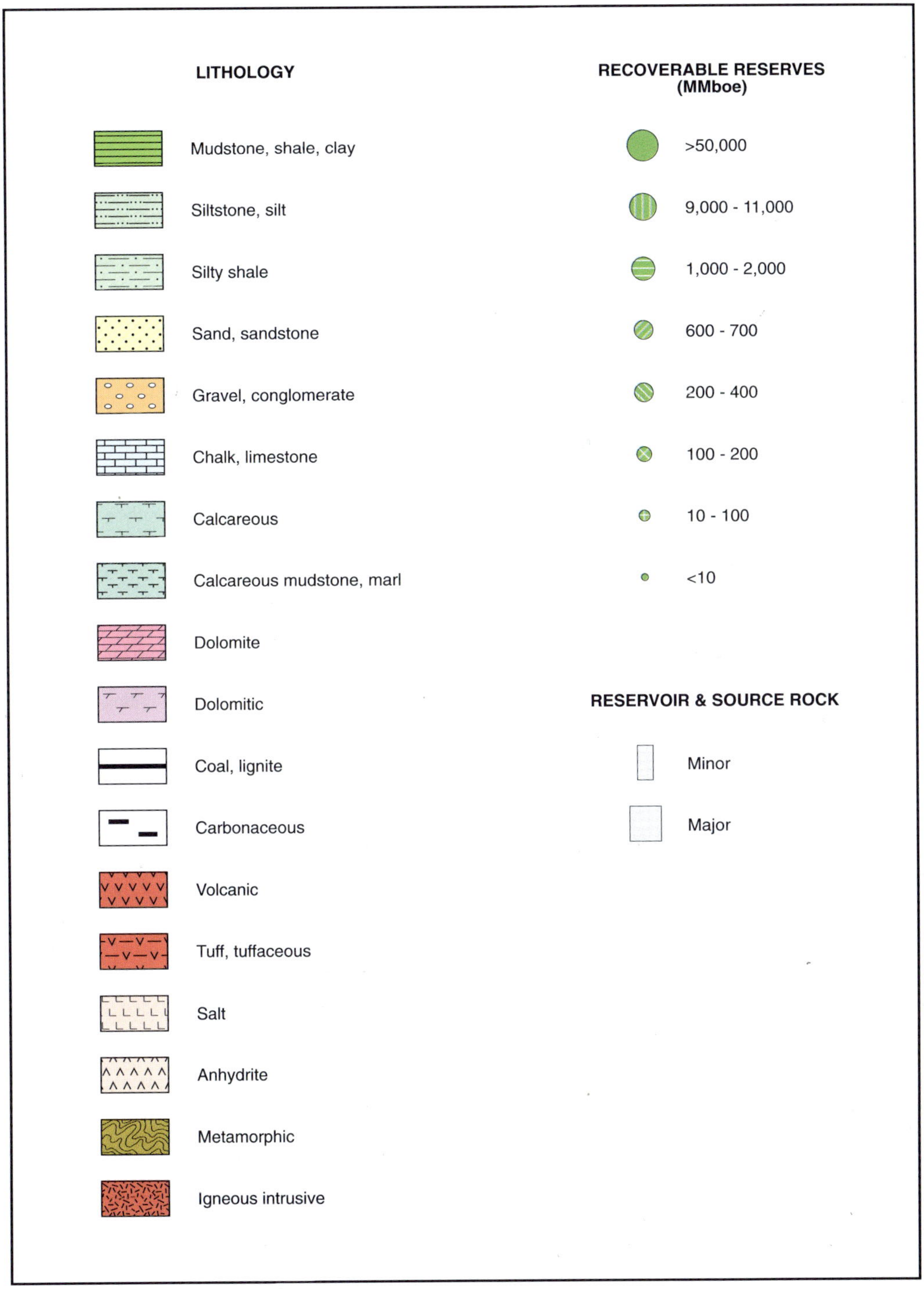

Fig. 3. Legend for the stratigraphy, depositional environment and hydrocarbon attributes and reserves shown in Figs 4–10.

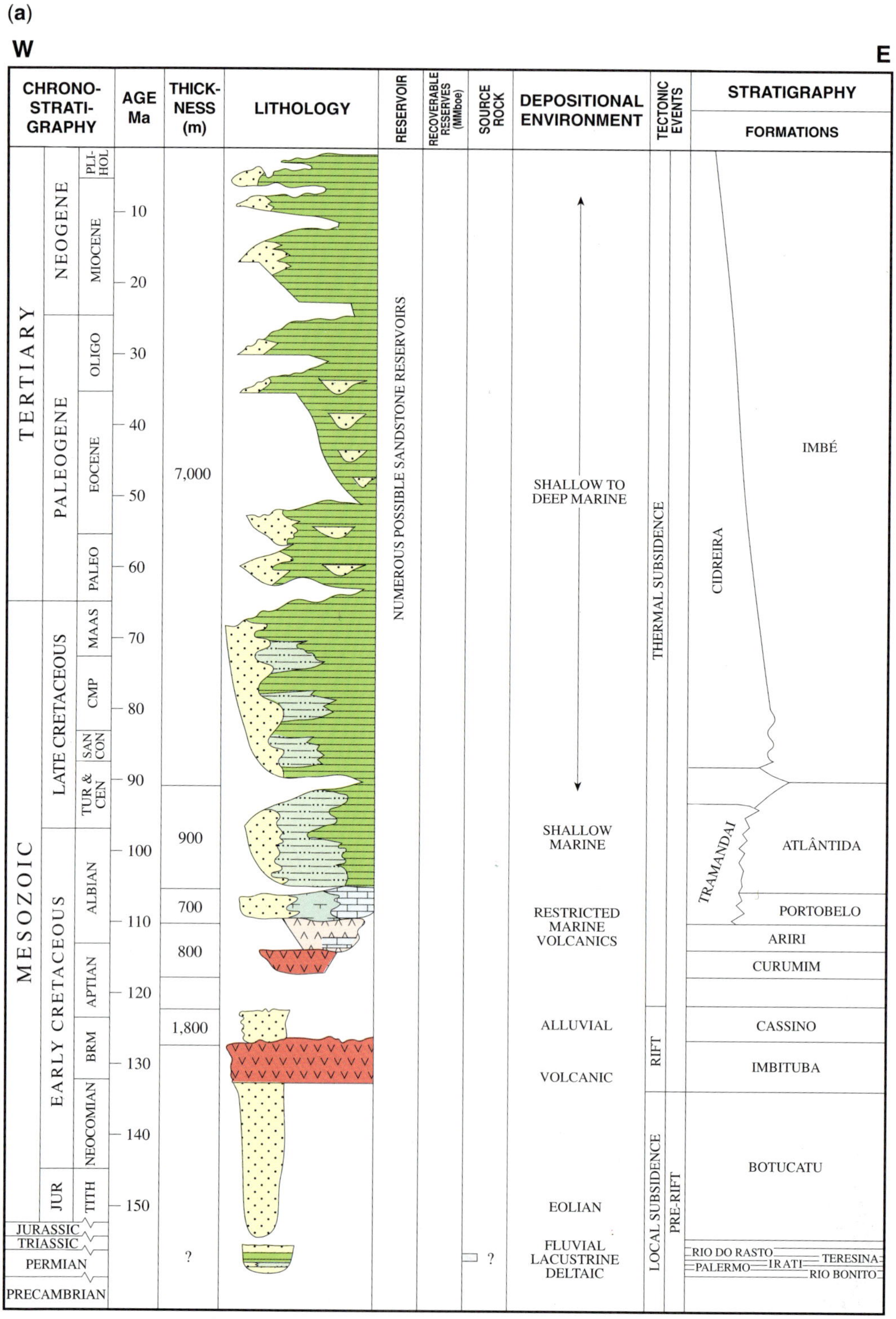

Fig. 4. Stratigraphy, depositional environment and hydrocarbon reserves of: (**a**) the Pelotas Basin; (**b**) the Southwest African Coastal Basin.

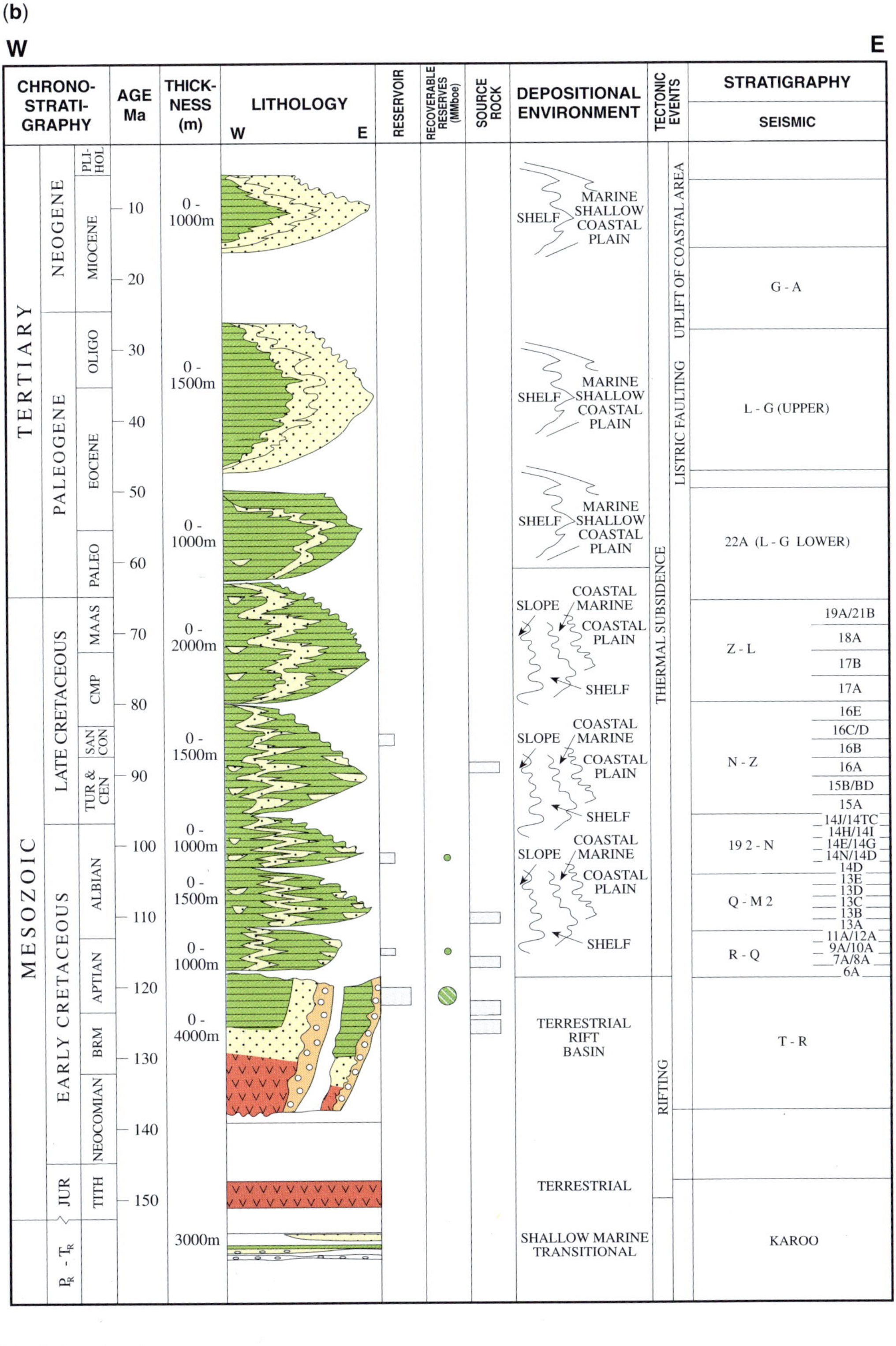

Fig. 4. (*continued*)

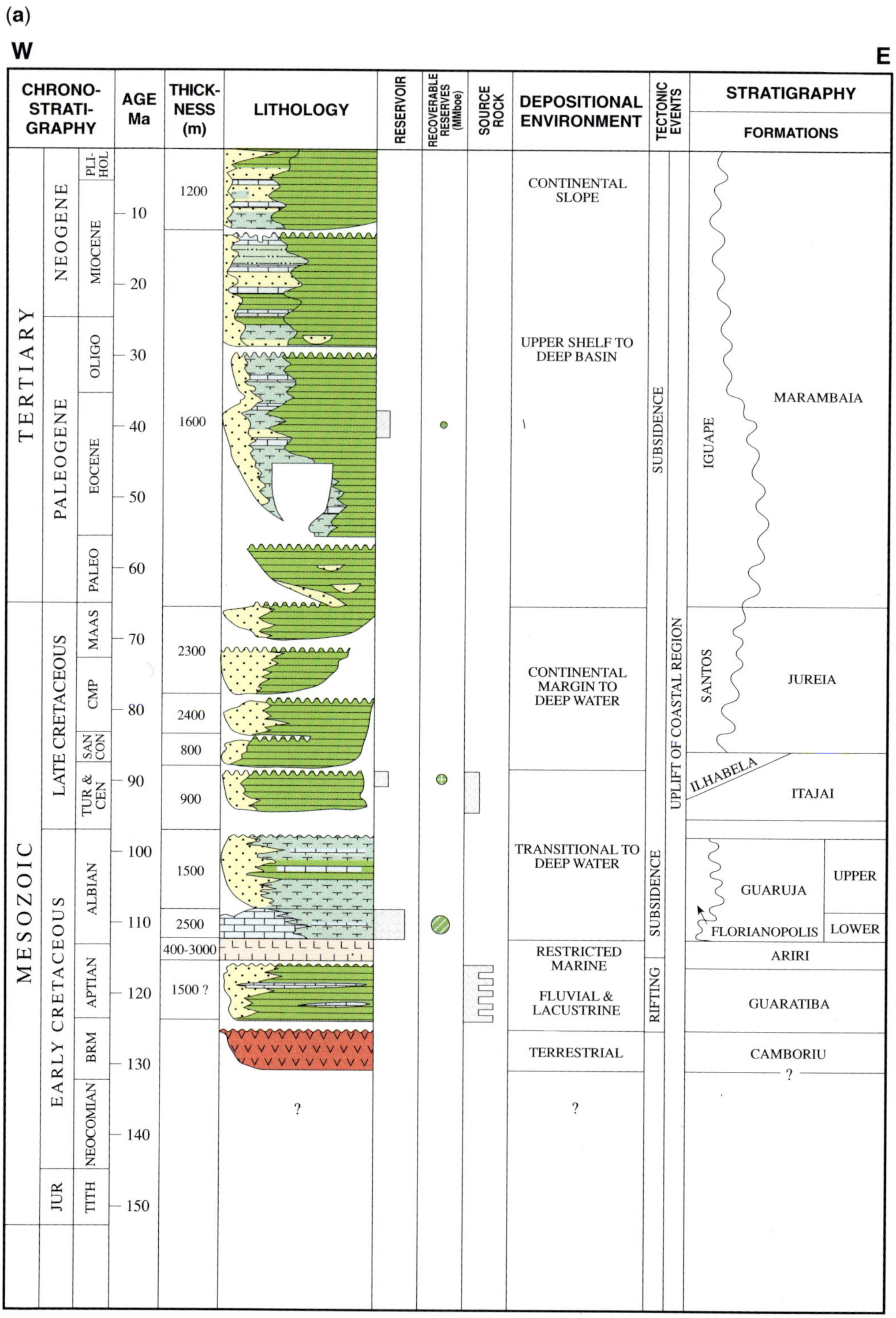

Fig. 5. Stratigraphy, depositional environment and hydrocarbon reserves of: (**a**) the Santos Basin; (**b**) the Moçamedes Basin.

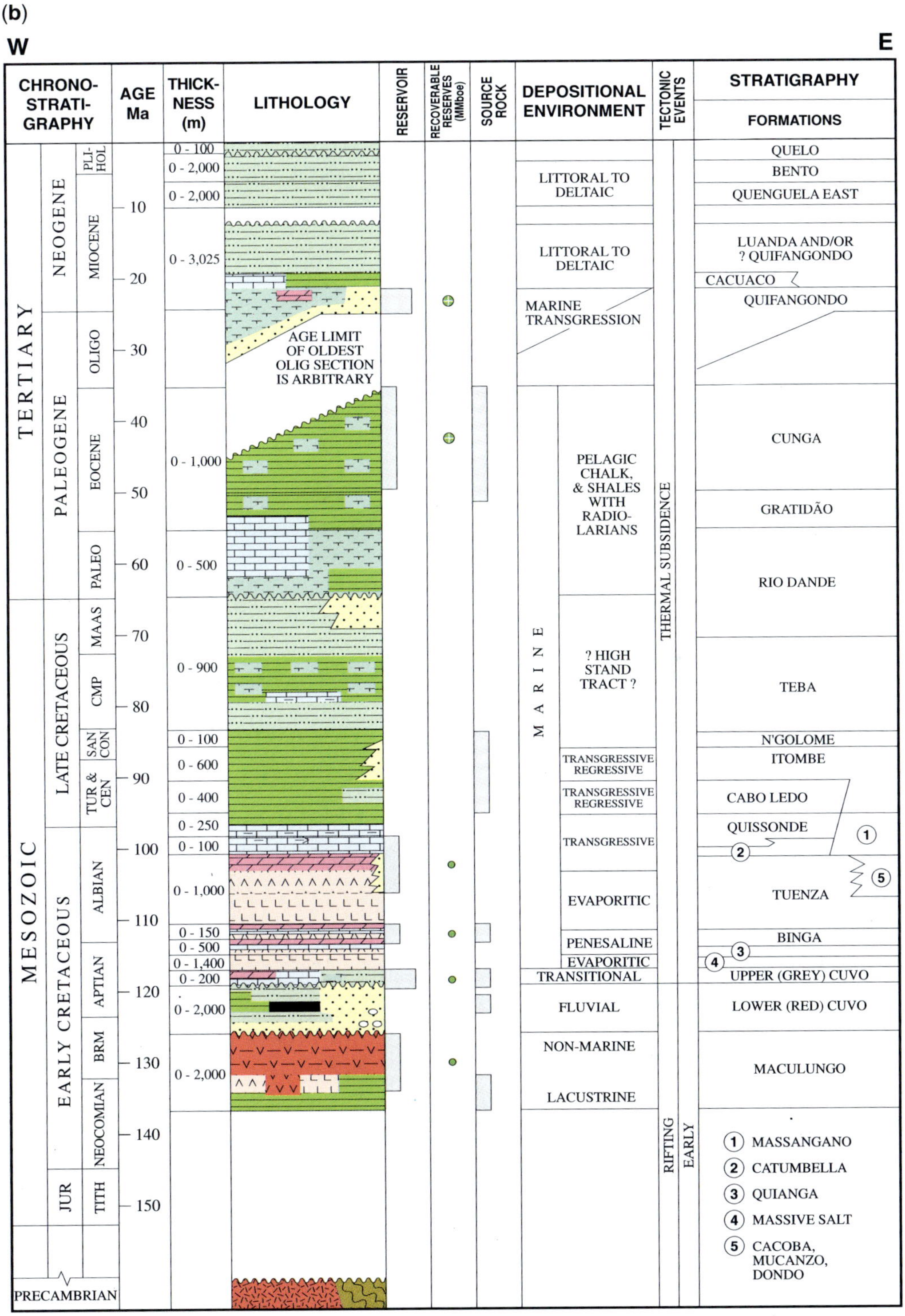

Fig. 5. (*continued*)

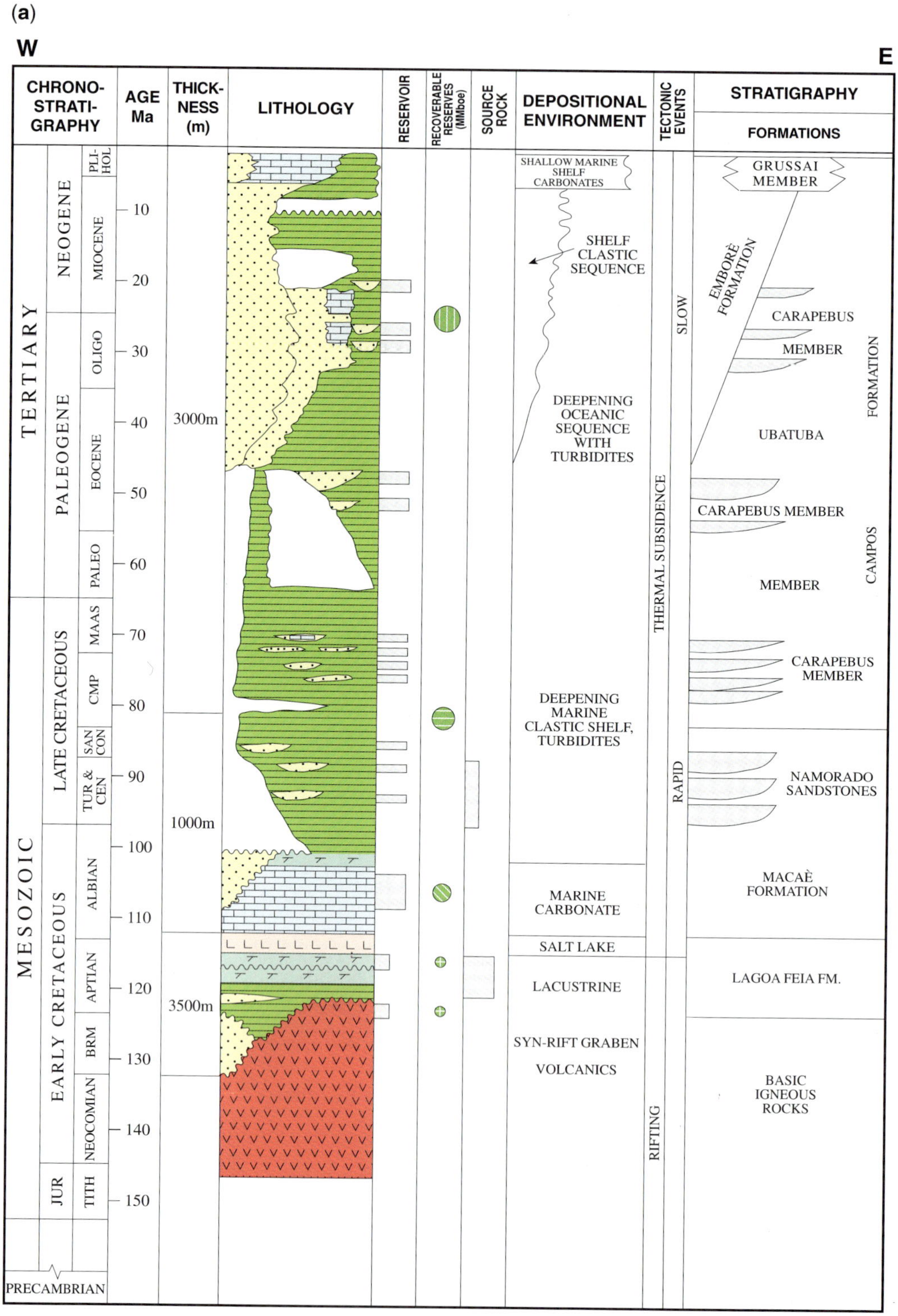

Fig. 6. Stratigraphy, depositional environment and hydrocarbon reserves of: (**a**) the Campos Basin; (**b**) the Kwanza Basin.

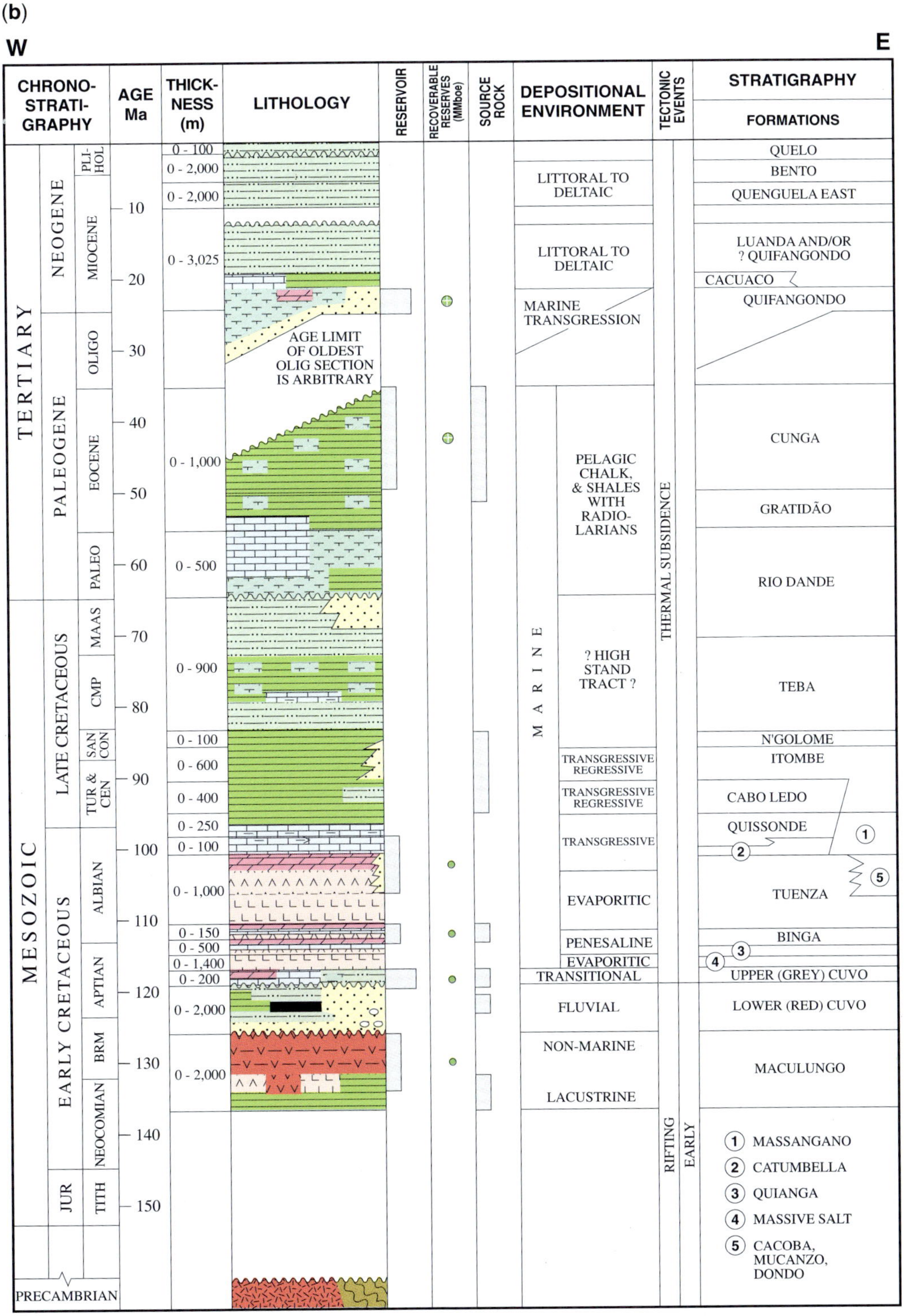

Fig. 6. *(continued)*

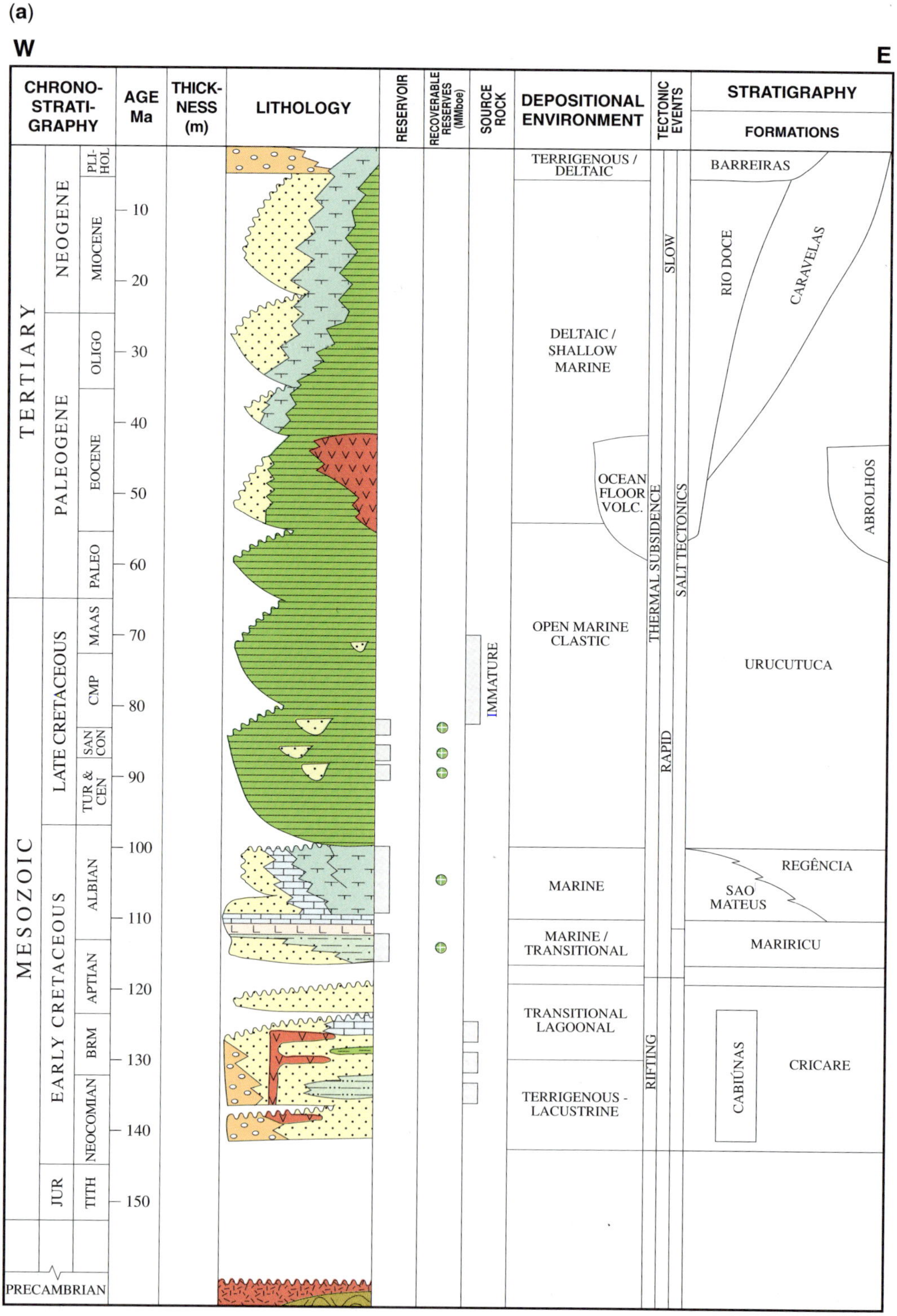

Fig. 7. Stratigraphy, depositional environment and hydrocarbon reserves of : (**a**) the Espirito Santos Basin; (**b**) the Lower Congo Basin.

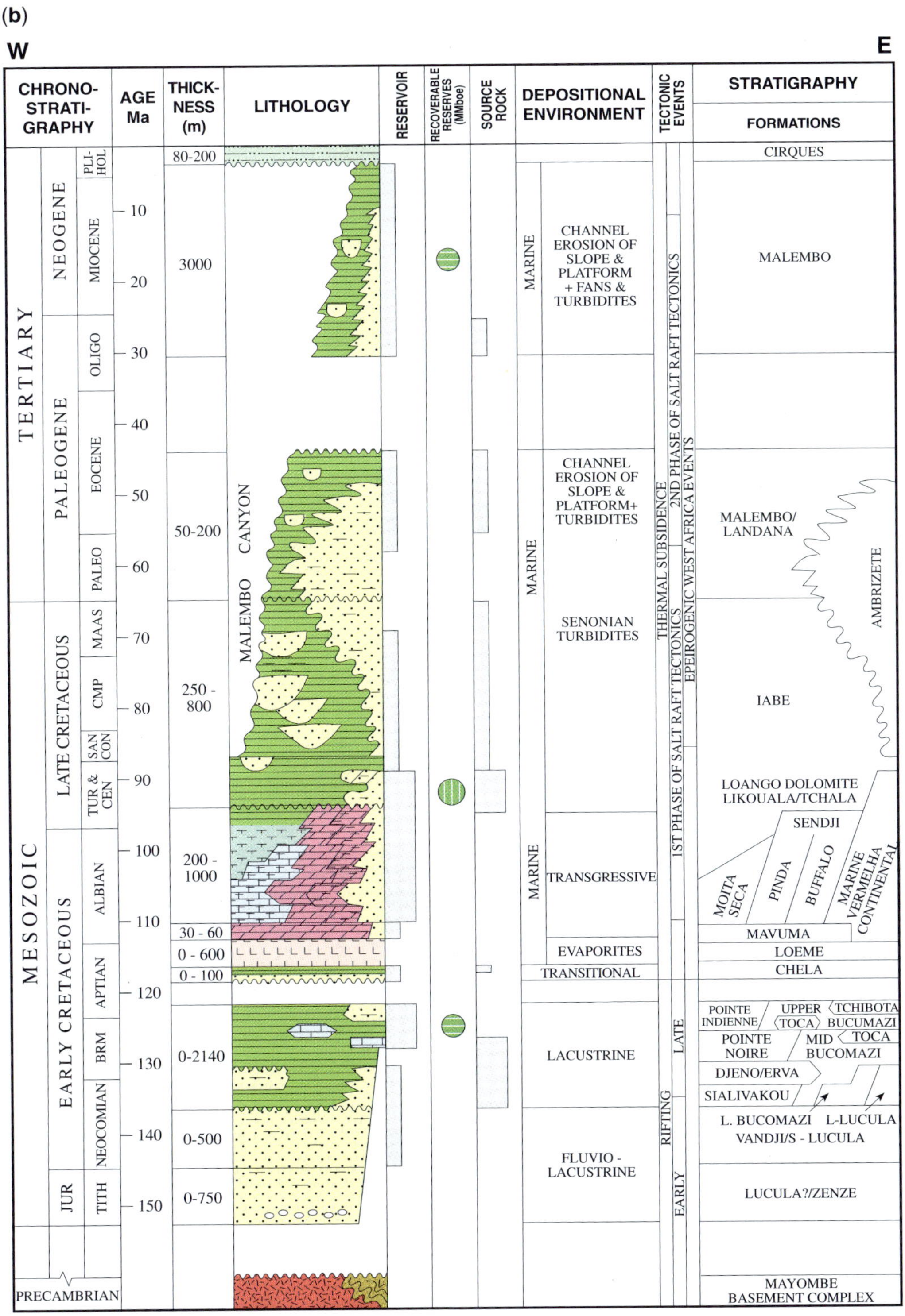

Fig. 7. (*continued*)

their thinness. This transgression was followed by the deposition of Aptian evaporites that included thick sequences of salt.

Subsequent deepening of the South Atlantic allowed marine waters to re-enter the basins, terminating evaporite deposition. The return to normal marine salinities was characterized by the deposition of Albian transgressive systems tract carbonates. There was further rifting at this time, possibly entirely thin-skinned driven by downslope sliding of the carbonate carapace to the salt, but more likely some of the rifting in the carbonates was soft-linked to renewed faulting in the basement. Some of the generated salt structures were active syn-depositionally and produced bathymetric highs on the seafloor that were characterized by the deposition of porous carbonate and clastic reservoirs.

The continued deepening of the Atlantic throughout the late Cretaceous and Tertiary provided an ever increasing gravitational head for downslope salt movement in the form of allochthonous salt 'rafts'. Their movement was strongly accentuated by up-dip upper Cretaceous and Tertiary sediment loading. Following development of a widespread Cenomanian unconformity, upper Cretaceous siliciclastic deposition yoked some of the rafts together only to be sometimes subsequently dismembered by accelerated Tertiary downslope movement. In a general way, the degree of salt rafting increases from north to south in the Lower Congo and Kwanza Basins.

Upper Oligocene withdrawal of the sea led to widespread erosion, which included the development of erosional channels subsequently filled with marine Miocene channel-fill sands and turbidites, related to the Late Drift Megasequence. The Congo Fan is the largest of these entities.

Sector IV

South America: Jequitinhonha–Cumuruxatiba, Almada–Camamu and Sergipe–Alagoas Basins. This sector includes, from south to north, the Jequitinhonha–Cumuruxatiba (Santos *et al.* 1994), the Almada–Camamu (Carvalho 1994; Netto *et al.* 1994) and the Sergipe–Alagoas (Schaller 1969; Van der Ven *et al.* 1989) Basins. Figure 8a from the Almada–Camamu Basin and Fig. 9a from the Sergipe–Alagoas Basin exemplify the stratigraphy and basin evolution. The Pre-Rift Megasequence is known from outcrop and wells. It has a different structural configuration from the equivalent section present to the south in Sector III. In the Sergipe–Alagoas Basin, Precambrian basement is overlain by Permo-Carboniferous glacial–fluvial–marine sequences. In the northern basins there are Jurassic pre-rift continental clastic sequences. Rifting was initiated during the Berriasian–Valanginian with deposition of thick conglomerates and sandstones in the grabens. Valanginian sediments are essentially lacustrine and infill the rift topography, followed by further sandstones and lacustrine shales reflecting late pulses of rifting and rift-flank uplift at the edge of the main zone of Atlantic opening. Restricted marine sequences, including bituminous shales and evaporites, accumulated in the Aptian with a more open marine environment in the Albian at the start of the drift phase, associated with evaporites and carbonates in the north. Thermal subsidence and oceanward tilting of the basins led to the deposition of progradational shelf–slope systems and the deposition of turbiditic sequences.

Africa: Gabon Coastal and Douala Basins. The Gabon Coastal (Brink 1974; Teisserene & Villemin 1990) and Douala (Belmonte 1966; Nguene *et al.* 1992; Pauken 1992) Basins constitute the central and northern part, respectively, of the series of West African basins developed between the Cameroon Volcanic Ridge and the Walvis Ridge. Their history is similar to that described above for the Lower Congo Basin. Figure 8b from the Gabon Coastal Basin and Fig. 9b from the Douala Basin exemplify the stratigraphy and basin evolution. Following development of a widespread Cenomanian unconformity, Turonian and younger organic-rich marine marls were deposited. The latter became mature for hydrocarbon generation only in the Gabon Coastal Basin where Maastrichtian through Tertiary deltaic loading, associated with the proto-Ogooue River, sufficed to bury them deeper than elsewhere.

Westward tilting of the basin, combined with Oligocene withdrawal of the sea, led to widespread erosion which in some places breached an Albian–Cenomanian reservoir interval and included the development of erosional channels subsequently filled with marine Miocene channel-fill sands and turbidites.

Continued deepening of the Atlantic throughout the late Cretaceous and Tertiary provided an ever increasing gravitational head for downslope salt movement which was strongly accentuated by up-dip sediment loading. The limited distribution of salt in the Douala Basin minimized this effect.

Sector V

South America: Potiguar Basin. This sector includes the Potiguar Basin, which extends 200 km inland and offshore to a water depth of 2000 m

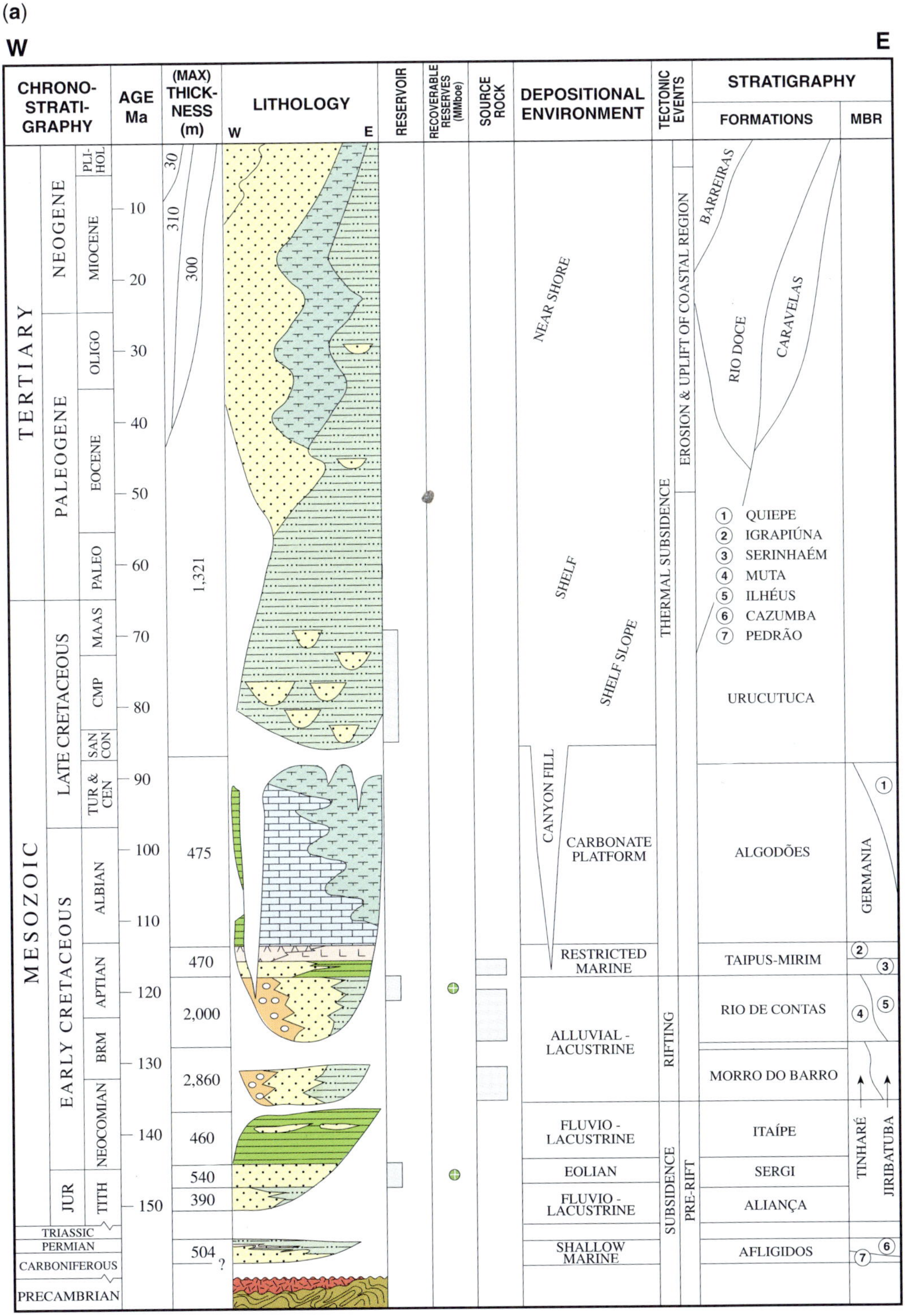

Fig. 8. Stratigraphy, depositional environment and hydrocarbon reserves of: (**a**) the Almada–Camamu Basin; (**b**) the Gabon Coastal Basin.

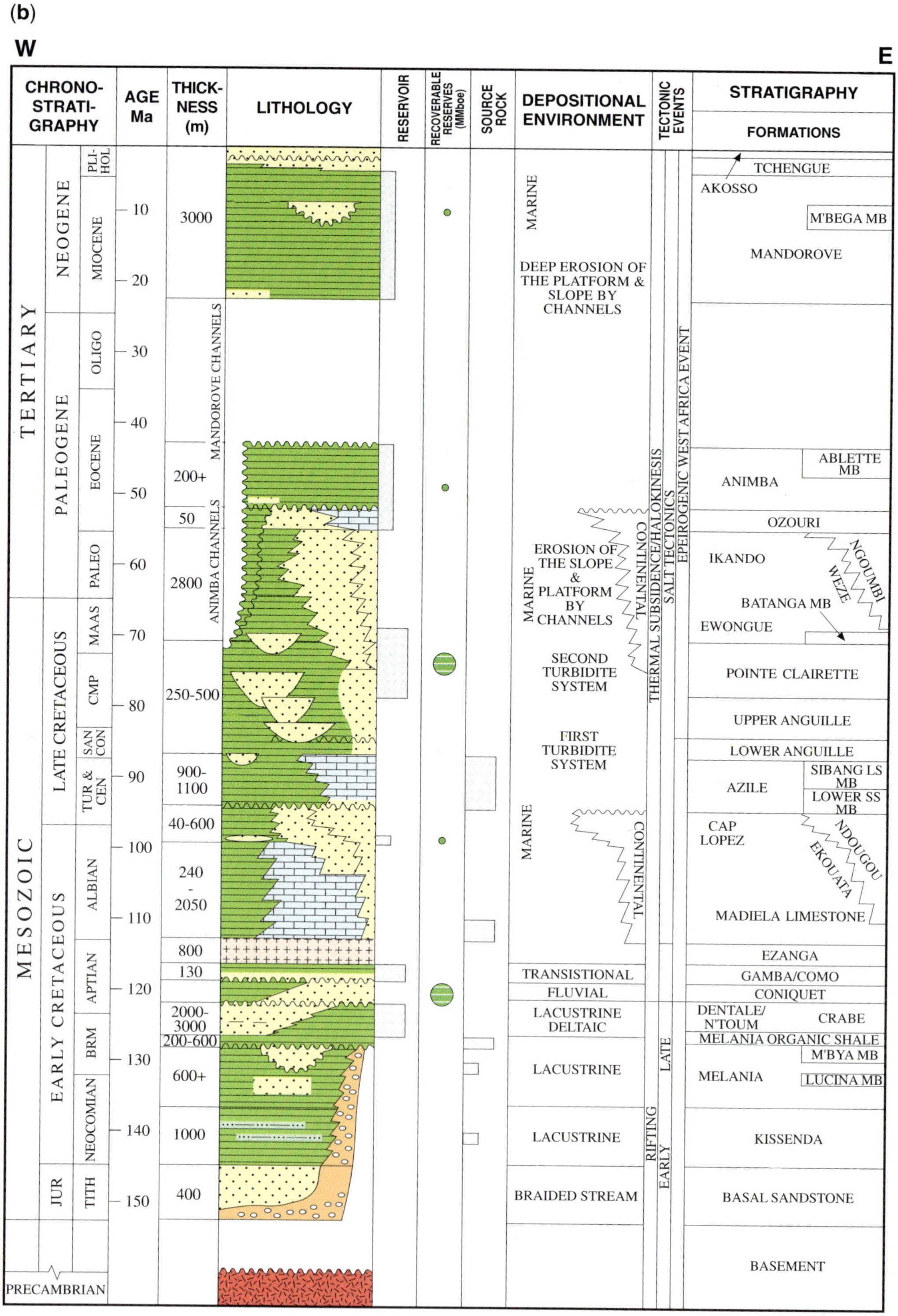

Fig. 8. (*continued*)

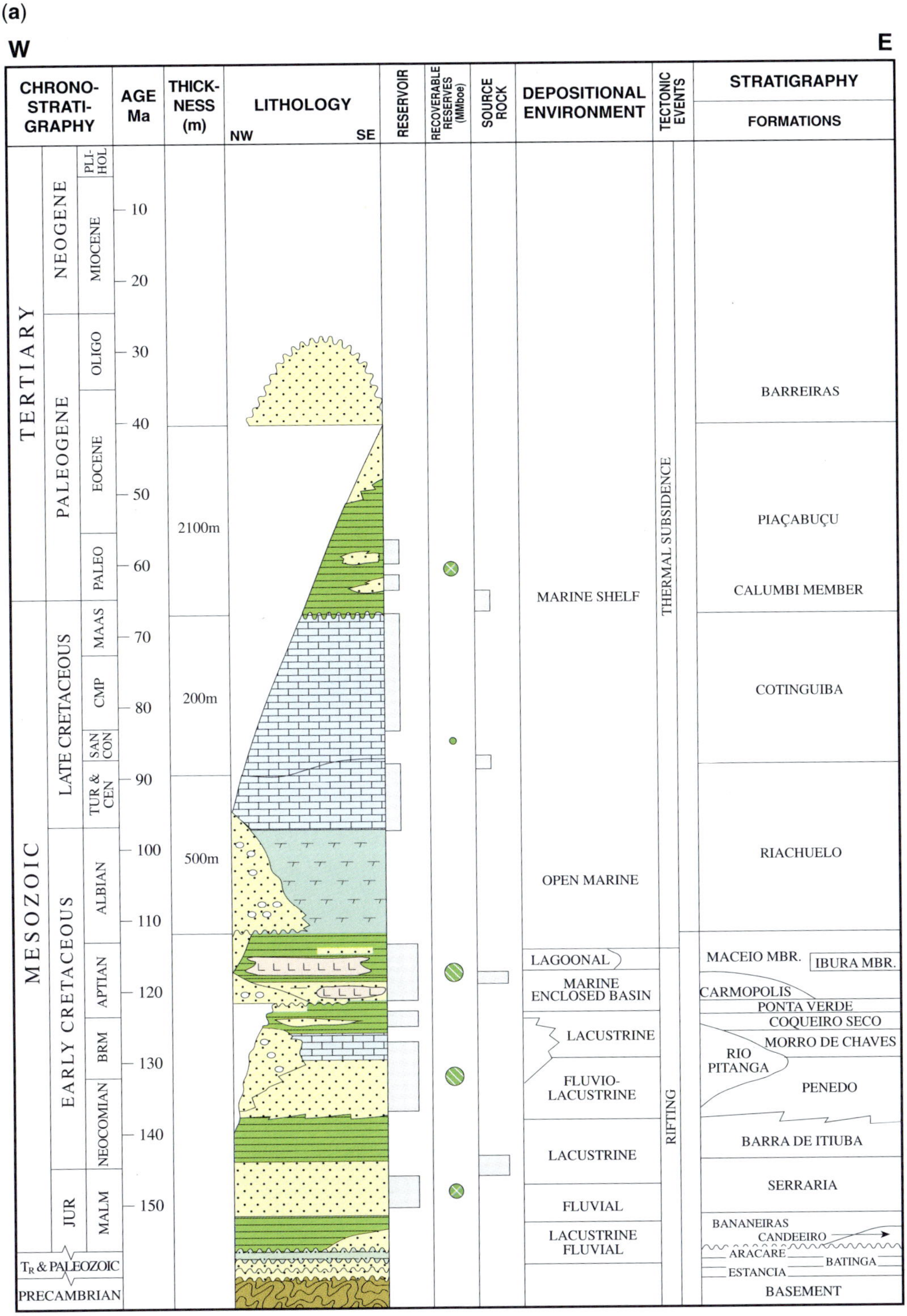

Fig. 9. Stratigraphy, depositional environment and hydrocarbon reserves of: (**a**) the Sergipe–Alagoas Basin; (**b**) the Douala Basin.

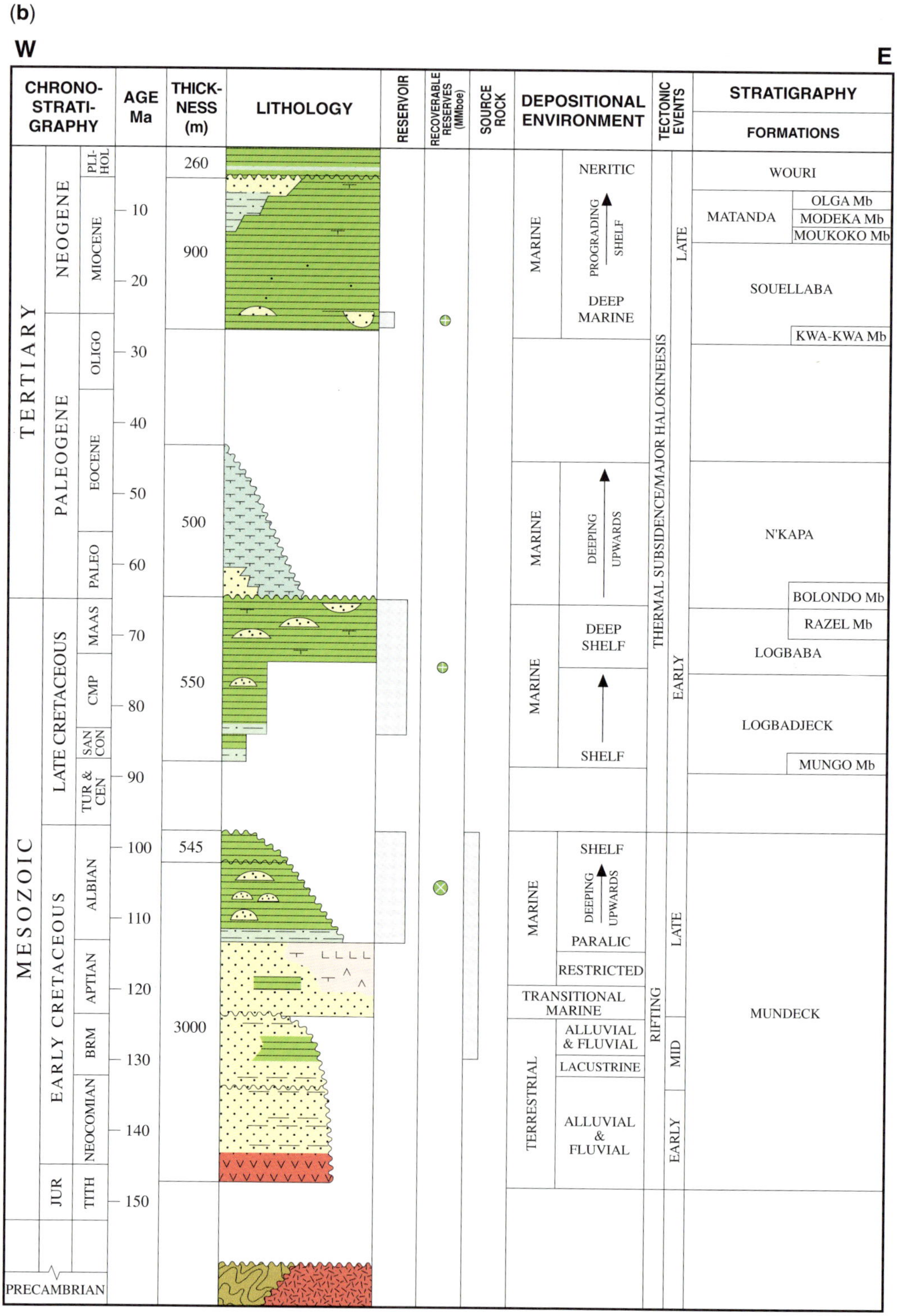

Fig. 9. (*continued*)

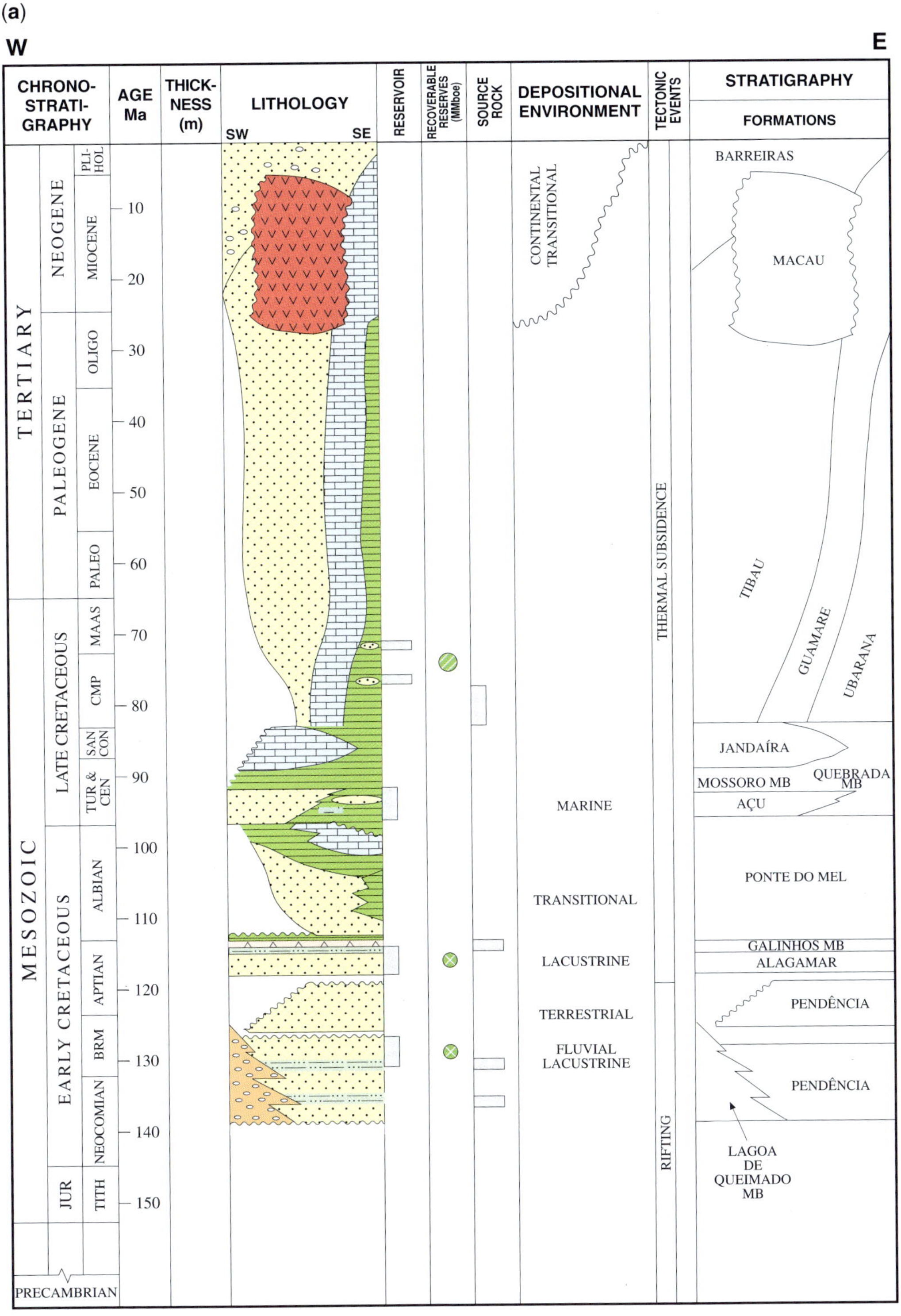

Fig. 10. Stratigraphy, depositional environment and hydrocarbon reserves of: (**a**) the Potiguar Basin; (**b**) the Niger Delta Basin.

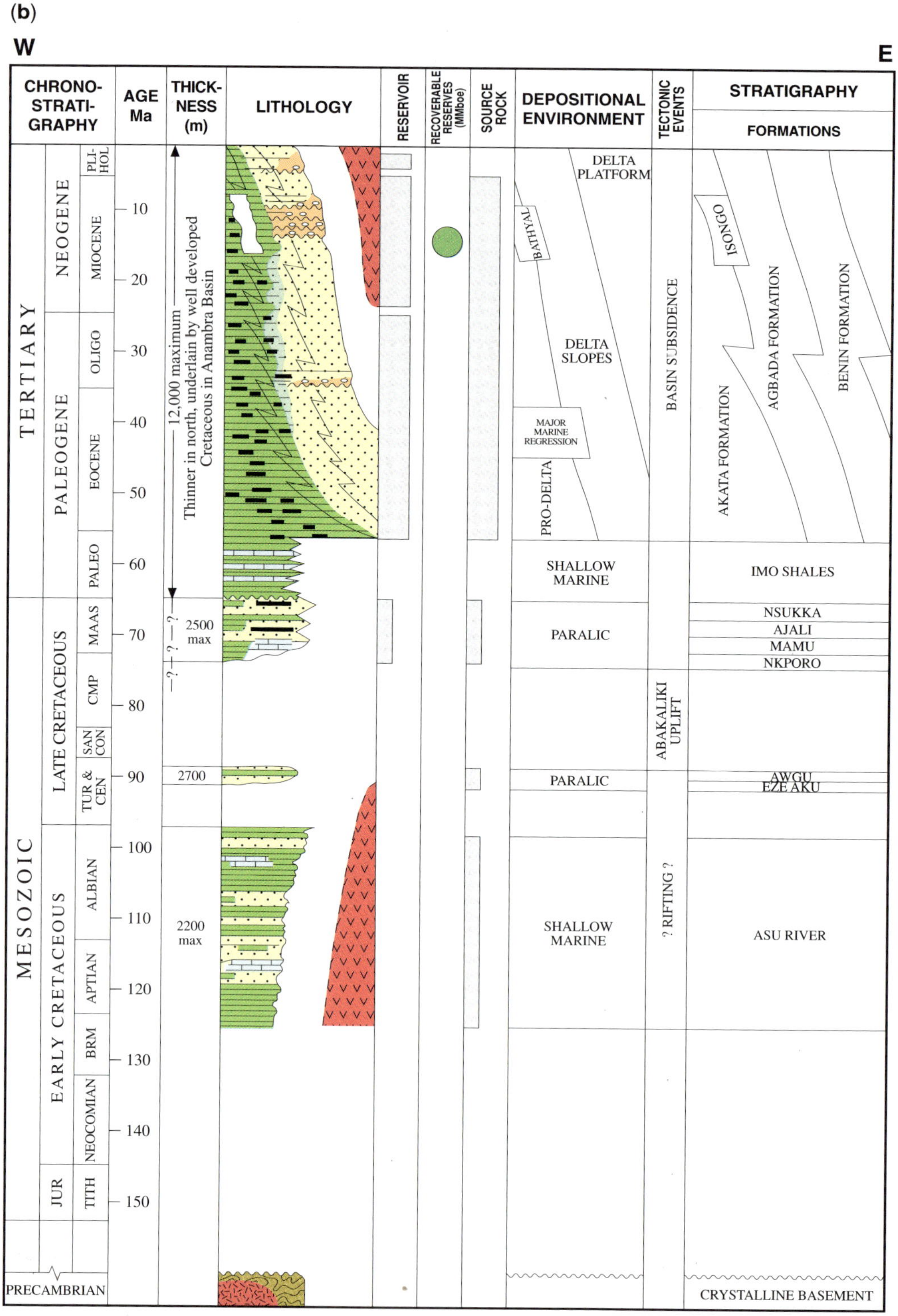

Fig. 10. (*continued*)

beneath the present continental shelf (de Matos 1991, 1992). Its tectonic history is similar to that of Sector III, but the basin contains no salt and only minor anhydrite. Figure 10a shows the stratigraphy and basin evolution. Rifting began in the onshore part of the basin during the Valanginian and resulted in the accumulation of 500–5000 m of syn-rift sediments, including lacustrine shales and turbidites into which progradational deltaic, fan–delta and alluvial plain sandstones and conglomerates were deposited. This onshore basin produced northeast trending half-grabens.

The offshore basin developed as a passive thermal subsidence basin following Aptian rifting along west- to northwest trending rifts. A regional post-rift Aptian unconformity is overlain by lacustrine carbonates and clastic sequences of Albian age. Deltaic clastics and subsequently thick turbidite deposits record the thermal subsidence in the offshore region. The onshore part of the Potiguar Basin remained relatively elevated.

Africa: Niger Delta. The Niger Delta consists of a coarsening, upward-regressive sequence of Tertiary clastics which prograded over the passive continental margin. Figure 10b shows the stratigraphy and basin evolution. The site of the Niger Delta was established at the initial opening of the South Atlantic between the middle Jurassic and the early Cretaceous. The Gulf of Guinea formed the site of a triple junction during continental breakup, with the northeast–southwest trending Benue Trough representing a failed arm. The point of intersection of the present Niger and Benue Rivers was a tectonically controlled graben (Anambra Basin) which accumulated a maximum thickness of 7000 m of clastic sediments by the late Cretaceous.

The present Niger Delta (Evamy *et al.* 1984; Bustin 1988) began to form in the Eocene, following the end-Palaeocene marine transgression during which the Imo Shales were deposited. The Eocene regression marked the beginning of the southward progradation of the Tertiary Niger Delta. This upward-coarsening regressive sequence is comprised of three diachronous units: pro-delta claystones at the base (Akata Formation), delta-front claystones and sands in the middle (Agbada Formation), and delta-top sands at the top (Benin Formation).

The build-up and distribution of sediments in the delta since Eocene times has been controlled by the degree of tectonic (thermal) subsidence and sediment supply. The sediments are thicker in the western half of the delta, but more extensive in the east. The building and southward progradation of the delta took place in the form of a succession of east–west elongated depositional belts, all of which include sediments attributable to the diachronous Akata, Agbada and Benin Formations which range in age from early Eocene to Quaternary.

Sector VI

South America: Marajó Basin, Pará–Maranhao Basin, Gurupi Graben System and the Barreirinhas Basin. This sector encompasses the family of northwest–southeast trending en-echelon basins which form the northeast coastal region of Brazil. In the west, the Marajó Basin defines the edge of the Foz de Amazonas Basin. To the east lies the Pará–Maranhão Basin, the Gurupi Graben System and the Barreirinhas and Ceará Basins (Pamplona 1969; Feijo 1994). The Marajó Basin, the Gurupi Graben System and the Barreirinhas Basin en-echelon set of basins forms an east–west trending sector, *c.* 400 km wide, between the western continuation of the Saint Paul and Romanche Fracture Zones. The en-echelon nature of the basins suggests that they formed under right lateral shear conditions between the precursors of these two oceanic fracture zones, and that they transfer spreading from the Cassipore Rift System in the northwest to the Ceará Rift System in the southeast.

In the Gurupi Graben System (Lima *et al.* 1994), rifting occurred during the Aptian and early–middle Albian, and the graben system is superimposed on, and modifies, a thermal subsidence basin formed at the outer edge of an earlier phase of rifting. Rift sediments are mainly fluvial/lacustrine sandstones and shales. There are no post-rift sediments; the graben system was held up by flexural forces caused by lithospheric bending due to subsidence along the Atlantic margin to the northeast.

During the late Albian–Recent there was post-rift thermal subsidence in the western part of the Barreirinhas Basin, but Cenomanian inversion and basin uplift in the east. The inversion was caused by transpression along the southern transform system, accompanied by the development of a compressive push-up sector at an offset in the transform.

Africa: Abidjan, Saltpond and Benin Embayment Basins. The basins of the transform margin of West Africa also lie between east–northeast-trending extensions of the oceanic fracture zones. The Côte d'Ivoire (Abidjan), Tano and Saltpond Basins, offshore Côte d'Ivoire and Ghana to the west are separated from the Benin Embayment (Ghana–Togo–Benin) by the Ivory Coast–Ghana Ridge which is continuous with the Romanche Fracture Zone. The Côte d'Ivoire Basin terminates abruptly at the Lagunes Fault Zone, parallel to the coastline, along which up to 10 km of displacement has accumulated. Rift sediments are Albian and are entirely clastic; as in Sector V there is no salt sequence in this sector. Rift traps are structural, with an increasing element of stratigraphic trapping up-section. The post-rift Cretaceous, Cenomanian to Maastrichtian, is also dominantly clastic, with carbonate development only in the lower Cenomanian. An intra-Senonian (*c.* base Campanian) unconformity separates progradational sediments below from gravity deposits above. Most petroleum discoveries have been in the Albian section, an exception being a Devonian reservoir in the Saltpond Field.

Sector VII

South America: Foz do Amazonas Basin and Amazon Cone. This sector comprises the area around the mouth of the Amazon, including the Amazon Cone. The Foz do Amazonas Basin (Brandão & Feijo 1994) underwent an early phase of rifting during the Triassic and a second phase during the Apto-Albian, related to seafloor spreading in the Central Atlantic. The thermal subsidence sequence began in the late Cretaceous, although the centre of subsidence and sedimentation has shifted eastwards from the continental shelf to the abyssal plain. From the Palaeocene–Miocene the clastic supply was limited and thick Tertiary carbonate banks built out across the platforms. Landward, these platforms are offset by fine, low-energy clastic sediments. From the Miocene–Recent, uplift of the Andes led to a massive influx of sediments ending the carbonate-dominated depositional regime. At this time there was relatively little thermal subsidence and the clastic wedge forming the Amazon Cone built outwards over the continent–oceanic break. Sediment thicknesses in the upper cone reach 14 km (Damuth & Kumar 1975).

Africa. There is no directly equivalent basin in Africa.

Petroleum systems

Three petroleum systems dominate the South Atlantic province. They are: (1) the Late and Middle Drift Megasequences of the Niger Delta (Haack *et al.* 1997); (2) the Middle and Early Drift Megasequences of West Africa (Cameroon to Angola) (Burwood 1997); (3) the Pre-Salt Megasequences of Brazil (Mello & Maxwell 1990) and West Africa from Gabon to Angola (Schoellkopf & Patterson 1997).

Information on the quality and richness of these and the other less important petroleum systems is provided by Burwood (1999), Davison (1999) and Schiefelbein *et al.* (1999). A comparison of Brazilian and West African oils has also been made by Mello *et al.* (1991).

There is a wide distribution of reservoirs throughout the stratigraphic column, including fractured basalts in the basement and Neocomian coquina shelly clastic sediments in the pre-salt section, overlain by several horizons of turbiditic sandstones in the middle and late Cretaceous, Eocene and Oligo-Miocene.

In Brazil, the oil reservoired in fractured basalt and in the Lagoa Feia Formation coquina was able to migrate by direct contact with the Neocomian source rocks. However, the greatest share of the hydrocarbon reserves is associated with late Cretaceous and Tertiary reservoirs, not with the Neocomian reservoirs that are close to the source. The oils reservoired in the Post-Salt marine sequence required pathways through the evaporites. Migration pathways developed as a consequence of thinning and evacuation of the salt layer during salt movement.

Rift-phase faults permitted the movement of oil from the source rocks to the base of the salt. The oil then migrated through the salt layer using welds caused by the evacuation of salt during salt movement and by the downthrow of the Post-Salt carapace onto the Pre-Salt succession during listric faulting. Gravity gliding continued throughout the late Cretaceous and Tertiary enabling further vertical oil migration into the less disturbed overlying turbiditic sands. These large Oligo-Miocene sheet-like turbidites make the largest reservoirs and constitute the most prolific plays.

In the Campos Basin, oil gravity varies from 26° API in the deeper accumulations to 19° API in the shallower accumulations, possibly indicating several pulses of oil migration. The heaviest oils may have reached the reservoirs earlier in the basin history, when the temperature was low enough for biodegradation. Further thin-skinned tectonics, caused by gliding above the salt, reactivated old listric faults and formed new faults, allowing new migration of oil. The lighter oils are regarded as mixtures of these remigrated heavier oils with more mature oils generated during a second migration. The latter reached the reservoirs when they were deeper and their temperatures were high enough to prevent biodegradation by bacteria.

In the Lower Congo Basin, the Post-Salt reservoirs have been charged by both the Pre-Salt Bucomazi source rocks and the Post-Salt Cenomanian–Turonian marine section. The

Aptian Chela sandstones have provided an up-dip conduit for migrating Pre-Salt hydrocarbons, and access to Post-Salt reservoirs depends on the existence of windows in the regional salt seal, non-sealing faults, and sandstone and/or silt-stone carrier beds.

Salt movement played a particularly important role in determining directions of migration. Firstly, it provided the 'holes' in the regional salt seal that allowed migration of Pre-Salt hydrocarbons into Post-Salt reservoirs. Secondly, it provided for localized depositional changes that enhanced reservoir quality on, and around, the growing salt structures, including the distribution of turbidites. Thirdly, it provided the structural relief conducive to the entrapment of hydrocarbons. In addition, maturation of Post-Salt upper Cretaceous source rocks has been provided by the Tertiary depositional load of the Congo River Delta and is responsible for the large hydrocarbon accumulation in the Oligocene and Miocene clastic reservoirs that are currently being discovered in Angola.

Geographical distribution of reserves

The distribution of productivity is irregular and large areas of the South Atlantic margin remain without significant discoveries. The reasons for this pattern are examined below. Table 1 details the reserves both by sector and megasequence.

Sector I (0.9% of total reserves)

Sector I basins in South America contain no proven discoveries of oil or gas due mainly to the lack of mature source rock. Possible source rocks include lower Cretaceous black shales in the eastern parts of the basins together with lacustrine shales of Permian age, recorded in wells in the southeastern part of the basin. In the southern African basins, the syn-rift sequence contains dry gas and oil-prone source rocks. These are locally developed in each basin, as Berriasian local rift flank uplift separated the basins causing restricted circulation in deep marine, poorly oxygenated conditions. In a Barremian low stand, dry to wet gas prone shales were deposited. The Aptian inversion was followed by a period of sediment starvation over much of the area, during which time organic-rich shales were deposited. Other episodes of sediment starvation during the Albian–Coniacian resulted in thin and immature source rocks.

Sector II (0.6% of total reserves)

Sector II contains no reserves in Brazil (Pelotas Basin). Gas is found, however, in the Southwest African Coastal Basin (Kudu Field with 3.2 TCF proven and probable reserves), where the source is believed to lie in late-rift Neocomian non-marine beds. Lack of source is a major factor for lack of reserves in the Pelotas Basin. Lack of an oil source is a major negative factor in the northern part of the Southwest African Basin.

Sector III (28.1% of total reserves)

The Campos Basin is the most important hydrocarbon province in Brazil with over 70% of the country's reserves. Turbidite sandstones are the most prolific reservoirs, comprising over 60 accumulations. Their successful exploration has been partly due to their distinctive seismic character, and to advances in seismic resolution which have enabled detailed mapping of facies changes and pinch-outs associated with reservoir lenses.

The Santos Basin is close to the main industrial centres of Brazil and stratigraphically it is analogous to the Campos Basin to the north. However, although some 80 exploration wells have been drilled, the structural framework of the basin and the nature of the sedimentary fill are less well known. The major problem in the basin is the paucity of mature source rocks. The geothermal gradient is low. The late Cretaceous source rocks, where sampled, are immature to only just mature and cannot have been in the oil window for > 10 Ma. The lower Cretaceous source rocks offer the best hope for success, but their distribution beneath the salt is not well documented. Since these lower Cretaceous source rocks are sealed by the Aptian salt, vertical migration requires the presence of structural or stratigraphic breaks between Pre-Salt and Post-Salt reservoirs. The best targets are middle Cretaceous oolitic grainstones and upper Turonian turbidites. However, the reservoirs are deep and poor porosity is an inherent risk.

Sector III basins in Africa are also prolific; most of the known reserves are concentrated in the Lower Congo Basin where deepwater discoveries are becoming as important as they are in the Campos Basin. The bulk of the discoveries are associated with the Middle and Late Drift Megasequence plays; however, 1.5 billion barrels of oil equivalent has been discovered in the rift succession. The Kwanza Basin has been less successful to date, but is less well explored, and the potential for deepwater discoveries remains significant, although the absence of the Tertiary Congo Delta sediment load mitigates against

Table 1. *Distribution of reserves in the basins of the South Atlantic margins*

	South America		Africa	
Megasequence	Age Range	MMboe	Age Range	MMboe
Sector VII	**Foz De Amazonas Basin**		No basins	
Late sag	Middle Miocene–Recent	147		
Middle sag	Albian–early Miocene			
Early sag	Aptian–Albian			
Late syn-rift				
Middle syn-rift				
Early syn-rift	Triassic–early Cretaceous			
Pre-rift				
Sector VI	**Barreirinhas System**		**Abidjan (Ivory Coast) Basin**	
Late sag	Late Cretaceous–Recent	22		
Middle sag				
Early sag	Middle Albian–Cenomanian	8	Cenomanian–early Senonian	
Late syn-rift	Early–middle Albian	135	Albian	200
Middle syn-rift				
Early syn-rift				
Pre-rift	Ordovician–Neocomian	4		
Sector V	**Potiguar Basin**		**Niger Delta and Anambra Basins**	
Late sag	Campanian–Recent	350	Eocene–Recent	53 540
Middle sag	Cenomanian–Santonian		Maastrichtian–Palaeocene	
Early sag	Albian		Turonian	118
Late syn-rift	Aptian	180	Aptian–Albian	
Middle syn-rift	Barremian–Aptian			
Early syn-rift	Neocomian–Barremian	111	Precambrian	
Pre-rift				
Sector IV	**Jequitinhonha–Cumuruxatiba, Almada–Camamu and Sergipe–Alagoas**		**Douala and Gabon Coastal Basins**	
Late sag	Turonian–Recent	375	Oligocene–Recent	88
Middle sag			Senonian–middle Eocene	1681
Early sag	Albian–Cenomanian	6	Albian–Cenomanian	
Late syn-rift	Aptian	350	Late Barremian–late Aptian	157
Middle syn-rift	Barremian–Aptian	1020		
Early syn-rift	Late Neocomian	206	Late Neocomian–Barremian	
Pre-rift	Late Palaeozoic–Neocomian	800	Late Jurassic–early Neocomian	
Sector III	**Campos Santos and Espirito Santo Basins**		**Lower Congo and Kwanza Basins**	
Late sag	Senonian–Recent	10 500	Miocene–Recent	1198
Middle sag			Albian–middle Eocene	9152
Early sag	Albian–Turonian	1200	Middle–late Albian	6
Late syn-rift	Aptian	200	Late Neocomian–late Barrem.	1450
Middle syn-rift	Barremian	32		
Early syn-rift	Neocomian		Late Jurassic–late Barremian	
Pre-rift	?	14		
Sector II	**Pelotas Basin**		**Southwest African Coastal Basin**	
Late sag	Turonian–Recent		Tertiary–Recent	
Middle sag			Late Cenomanian–Maastrict.	
Early sag	Aptian–Cenomanian		Aptian–early Cenomanian	5
Late syn-rift	Neocomian–Barremian		Late Hauterivian–Aptian	500
Middle syn-rift	Late Jurassic			
Early syn-rift	Middle Jurassic		Late Jurassic–mid Aptian	
Pre-syn-rift	Permian-early Triassic			

Table 1. *Contd.*

Sector I	**Salado and Colorado Basins**	**Outeniqua Basin**	
Late sag	Palaeocene		
Middle sag			
Early sag	Aptian–Maastrichtian	Barremian–Albian	189
Late syn-rift	Neocomian–Barremian	Oxfordian–Barremian	529
Middle syn-rift			
Early syn-rift			
Pre-rift		Devonian	8

maturation of a Post-Salt hydrocarbon source. Reserves in Sector III are fairly evenly split between South America and Africa, although the average reservoir age is younger on the Brazilian side.

Sector IV (5.5% of total reserves)

The Sergipe–Alagoas Basin is the most productive in the Brazilian part of this sector; the bulk of the reserves lies in the Rift Megasequence. In this sector, and in the related basins, the younger reservoirs lie principally in deeper water and have not yet been tested.

In Africa, Sector IV comprises the Douala and Gabon Coastal Basins, which have collectively yielded slightly lower reserves than their Brazilian counterparts. The late rift sub-salt sequence and the Middle Drift Megasequence (Senonian turbidites) contain the bulk of the reserves.

Sector V (64.3% of total reserves)

In Brazil, Sector V is represented by the Potiguar Basin which has a fair number of discoveries, but relatively modest reserves of oil and gas. The reserves are distributed between the Early and Late Rift, and the Late Drift Megasequences. Although the basin is sourced from the rift phase, it lies to the north of the Aptian salt basin and so lacks the trap types that depend on the existence of a salt layer.

On the African side, the Niger Delta Basin falls into this sector and it dominates the reserves of the entire province. It differs from the other basins in being much younger, in that the deltaic succession sediments considerably post-date the rift–drift transition. Prior to separation (de Matos 1999) the Potiguar Basin would have trended towards the Niger Delta Basin, and its African extension presumably underlies the Niger Delta. The existence of a Potiguar equivalent source beneath the Tertiary delta is an intriguing possibility that has been examined by Frost (1997).

Sector VI (0.4% of total reserves)

In the Sector VI basins of Brazil, exploration success has been limited and discoveries are small. The reservoir horizons are of poor quality due to excessive diagenesis and cementation. They are also highly segmented by tectonism. Source rocks are relatively poor in this sector of the Brazilian coastal basins.

In the Abidjan Basin, and in related basins of the African margin, the discoveries have also been limited in both number and size, but are of relatively greater importance to the countries concerned. The late rift (Albian) succession is dominant in terms of reserves, on both the African and South American sides of the ocean.

Sector VII (0.2% of total reserves)

To date, the Foz do Amazonas Basin, including the Amazon Cone, has been a major disappointment by comparison with the Niger Delta; only gas has been discovered, and the Pirapema Field (0.8 TCF) is the only discovery of any significance. The main problem is the uncertainty of any source rock. In the onshore parts of the underlying graben system, Eocene black shales and Miocene coals may provide some source potential. So far, the basin lacks the good Cretaceous source rocks which occur in the offshore Brazilian basins to the south. To date, Tertiary sediments have also been found to be poor in organic material. However, the early Neogene sediments are probably sufficiently rich in organic material in the northwest of the Amazon Cone to be able to source some structural and stratigraphic plays. Traces of hydrocarbons have been found in the more porous Palaeogene carbonates at the edge of the

Table 2. *Atlantic margin discovered reserves (proven and probable)*

			Africa			
	South America		With Niger Delta		Without Niger Delta	
Megasequence	MMboe	%	MMboe	%	MMboe	%
Late sag (drift)	11394	72.7	54826	79.7	1286	8.4
Middle sag (drift)			10833	15.7	10833	70.9
Early sag (drift)	1214	7.8	318	0.5	318	2.1
Sub-total	**12608**	**80.5**	**65977**	**95.9**	**12437**	**81.4**
Late syn-rift	865	5.5	2836	4.1	2836*	18.6
Middle syn-rift	1052	6.7				
Early syn-rift	317	2				
Pre-rift	818	5.2	8	0	8	0
Sub-total	**3052**	**19.4**	**2844**	**4.1**	**2844**	**18.6**
Totals	**15660**	**99.9**	**68821**	**100**	**15281**	**100**

Grand total: South America plus Africa = 84481 MMboe.
*Predominantly Cabinda (Angola).

carbonate bank, possibly sourced from Neogene shales and clays further offshore.

Conclusions

The Late Sag Megasequence comprises 78% of the total proven and probable discovered reserves of the South Atlantic basins (Table 2). The Niger Delta reserves account for 63% of the South American and African total of 84.5 billion oil-equivalent barrels. If the Niger Delta contribution is excluded, the Late Sag proportion of the total reserves reduces to 41%, and the South American and African reserve figures become grossly similar, 15.7 and 15.3 billion oil-equivalent barrels, respectively. The South American and African syn-rift and post-rift thermal sag reserve totals are also grossly similar with values of 2.2 and 12.6 billion barrels for South America v. 2.8 and 12.1 billion barrels, for Africa, respectively (Table 2). Not surprisingly, the main reserve difference between the two sides of the Atlantic is thus seen to be the Niger Delta.

The Niger Delta is an interesting anomaly for, among all the major, similar age deltas of Africa (Congo, Zambeze and Nile), the Niger Delta is the only prolific producer. Among the other Atlantic margin basins, there is a correlation between productivity and the presence of Aptian salt that reflects not only the occurrence of salt-generated structures, including the development of syn-depositional stratigraphic traps, but also the occurrence of prolific Pre- and Post-Salt hydrocarbon-generating source rocks. Most important, however, has been the tilting of the Aptian salt basins towards the Atlantic which resulted in *en masse* downslope salt movement and the creation of salt holes through which Pre-Salt generated hydrocarbons could move into Post-Salt reservoirs.

In contrast to the Pre-Salt source rocks, Post-Salt source rocks are generally regionally immature. Notable exceptions occur in offshore Gabon, where the Maastrichtian palaeo-Ogooue River delta load has insured local Tertiary maturation, and in offshore Angola, where Tertiary Congo River deposition seems to have produced the same effect for the recent deepwater giant Oligocene and Miocene discoveries. In both cases, the source rock is Upper Cretaceous in age. It remains to be demonstrated whether similar situations occur in the Brazilian salt basins.

The implications for future exceptions seem clear. Significant success will continue to occur in the Late Sag Megasequence where local Tertiary sediment loading has succeeded in depressing Upper Cretaceous source rocks into the oil window. The most productive part of the South Atlantic region is expected to be Sectors III and IV, characterized collectively by the occurrence of prolific Pre- and Post-Salt source rocks, salt-related structures, tectonic tilting and related Tertiary loading. With this in mind, Sector IV in Gabon appears especially interesting in the deepwater area, west of the mouth of the Ogooue River. Deepwater Niger Delta exploration is expected to continue to add reserves, which can be produced, to an already enormous reserve database.

The authors are greatly indebted to Nick Cameron for the invitation to contribute this paper, and for much help and encouragement in its preparation. We are also grateful to Petroconsultants S.A. for permission to use material originally issued in the Basin Monitor report series and data from the company's databases. M. J. Durham, D. N. Pratt, and R. C. Selley are thanked for their helpful reviews of the manuscript.

References

BELMONTE, Y. C. 1966. Stratigraphie du bassin sedimentaire du Cameroun. *Proceedings of the 2nd West African Micropaleontology Colloquium*, Ibadan, 7–24.

BRANDAO, J. A. S. L. & FEIJO, F. J. 1994. Bacia da Foz do Amazonas. *Boletin Geoçiencias da Petrobras*, **8**, 91–99.

BRINK, A. H. 1974. Petroleum geology of Gabon Basin: *AAPG Bulletin*, **58**, 216–235.

BURWOOD R. 1997. Source rock control for Angolan Atlantic margin Petroleum Systems. Hedberg Research Symposium, *Petroleum Systems of the South Atlantic Margin*, Rio de Janeiro, 16–19 November, extended abstract.

—— 1999. Angola: source rock control for Lower Congo Coastal and Kwanza Basin petroleum systems. *This volume*.

BUSTIN, R. M. 1988. Sedimentology and characteristics of dispersed organic matter in Tertiary Niger Delta; origin of source rocks in a deltaic environment. *AAPG Bulletin*, **72**, 277–298.

CAMPOS, C. W. M., PONTE, F. C. & MIURA, K. 1974. Geology of the Brazilian continental margin. *In*: BURK, C. A. & DRAKE, C. L. (eds) *Continental Margins*. Springer, New York 447–461.

CARVALHO, K. W. B. DE 1994. Geologia da bacia sedimentar do Rio Almada. *Boletin Geoçiencias da Petrobras*, **8**, 5–55.

DAMUTH, J. E. & KUMAR, N. 1975. Amazon Cone: Morphology, Sediments, Age, and Growth Pattern. *Bulletin of the Geological Society of America*, **86**, 863–787.

DAVISON, I. 1991. Brazil's many sedimentary basins offer attractive exploration targets. *Oil & Gas Journal*, **Aug. 5**, 52–55.

—— 1999. Tectonics and hydrocarbon distribution along the Brazilian South Atlantic margin. *This volume*.

DE MATOS, R. M. D. 1991. Crustal structure and evolution of the Potiguar Basin, Northeast Brazil. *AAPG Bulletin*, **75**, 630.

—— 1992. The Northeast Brazil rift system. *Tectonics*, **11**, 766–791.

—— 1999. History of the northeast Brazilian rift system: kinematic implications for the break-up between Brazil and West Africa. *This volume*.

DIAS, J. L., SAD, A. R. E., FONTANA, R. L. & FEIJO, F. J. 1994. Bacia de Pelotas. *Boletin Geoçiencias da Petrobras*, **8**, 235–245.

DUVAL, B., CRAMEZ, C. & JACKSON, M. P. A. 1992. Raft tectonics in the Kwanza Basin, Angola. *Marine and Petroleum Geology*, **9**, 389–404.

EVAMY, B. D., HAREBOURE, J., KAMERLING, P., KNAAP, W. E., MOLLOY, F. A. & ROWLANDS, P. H. 1984. Hydrocarbon habitat of Tertiary Niger Delta. *Memoir American Association of Petroleum Geologists*, **35**, 325–351.

FEIJO, F. J. 1994. Bacia de Barreirinhas. *Boletin Geociencias da Petrobras*, **8**, 103–109.

FROST, B. 1997. An alternative Cretaceous age Petroleum System model for the Niger Delta. Abstract, *'Oil and Gas Habitats of the South Atlantic'*, 24–26 February 1997. Geological Society, London.

GERRARD, I. & SMITH, G. C. 1982. Post-Paleozoic Succession and Structure of the Southwestern African Continental Margin. *In*: WATKINS, J. S. & DRAKE, C. L. (eds) *Studies in Continental Margin Geology*. Memoir American Association of Petroleum, **34**, 49–76.

GIBBONS, M. J., WILLIAMS, A. K., PIGGOTT, N. & WILLIAMS, G. M. 1983. Petroleum geochemistry of the Southern Santos Basin, offshore Brazil. *Journal of the Geological Society, London*, **140**, 423–430.

GOES, A. M. O. & FEIJO, F. F. 1994. Bacia do Parnaiba. *Boletin Geoçiencias da Petrobras*, **8**, 57–67.

GONCALVES, A., OLIVEIRA, M. A. M. DE & MOTTA, S. DE O. 1979. Geologia da Bacia de Pelotas e da Plataforma de Florianopolis. *Boletin tec. da Petrobras*, **22**, 157–174.

GUARDADO, L. R., GAMBOA, L. A. & LUCCHESI, C. F. 1990. Petroleum geology of the Campos Basin, Brazil, a model for a producing Atlantic-type Basin. *In*: EDWARDS, J. D. & SANTOGROSSI, A. (eds) *Divergent/Passive Margin Basins*. Memoir American Association of Petroleum Geologists, **48**, 3–79.

——, WOLFF, B. & BRANDAO, J. A. S. L. 1997. Campos Basin, Brazil, a model for producing Atlantic-type basins. *Offshore Technology Conference*, Houston, Texas, 5–8 May 1997, 457–462.

HAACK, R. C., SUNDARAMAN, P. & DAHL, J. 1997. Niger Delta Petroleum System. Hedberg Research Symposium, *Petroleum Systems of the South Atlantic Margin*, Rio de Janeiro, 16–19 November, extended abstract.

JUNGSLAGER, E. H. A. 1999. Petroleum habitats of the Atlantic margin of South Africa. *This volume*.

KEELEY, M. L. & LIGHT, M. P. R. 1993. Basin evolution and prospectivity of the Argentine continental margin. *Journal of Petroleum Geology*, **16**, 451–464.

KOUTSOUKOS, E. A. M., MELLO, M. R. & AZAMBUJA FILHO, N. C. DE 1991. Micropalaeontological and geochemical evidence of mid-Cretaceous dysoxic-anoxic palaeoenvironments in the Sergipe Basin, northeastern Brazil. *In*: TYSON, R. V. & PEARSON, T. H. (eds) *Modern and Ancient Continental Shelf Anoxia*. Geological Society, London, Special Publications, **58**, 427–447.

LETULLIER, A. 1995. *Southwest Africa Offshore and Nama–Kalahari Basins*. Non-exclusive report in

Africa Exploration Opportunities Series. Petroconsultants, Geneva, Switzerland.

LIGHT, M. P. R., MASLANYJ, M. P., GREENWOOD, R. J. & BANKS, N. L. 1993. Seismic sequence stratigraphy and tectonics offshore. Namibia: *In*: WILLIAMS, G. D. & DOBB, A. (eds) *Tectonics and Seismic Sequence Stratigraphy*. Geological Society, London, Special Publications, **71**, 163–191.

LIMA, H. P., ARANHA, L. G. F. & FEIJO, F. J. 1994. Bacias de Bragança–Viseu, São Luis e Graben de Ilha Nova. *Boletin Geoçiencias da Petrobras*, **8**, 111–116.

LUNDIN, E. R. 1992. Thin-skinned extensional tectonics on a salt detachment, northern Kwanza Basin, Angola. *Marine and Petroleum Geology*, **9**, 405–411.

——, ALBERT, K., LAURITZEN, O. & LORANGE, E. 1991. Tertiary uplift of the Kwanza Basin in Angola. Geologie Africaine 1st Stratigraphy & Paleogr. West African Sediment. Basins Colloq. (Libreville, Gabon). *Proceedings of Elf Aquitaine Memoirs*, **13**, 99–117.

MACEDO, J. M. 1989. Evolução tectonica da bacia de Santos e areas continentais adjacentes. *Boletin Geoçiencias da Petrobras*, **3**, 159–173.

MCHARGUE, T. R. 1990. Stratigraphic development of proto-South Atlantic rifting in Cabinda, Angola – a petroliferous lake basin. *In*: KATZ, B. J. (ed.) *Lacustrine Basin Exploration – Case Studies and Modern Analogues*. Memoir. American Association of Petroleum Geologists, **50**, 307–326.

MELLO, M. R. & MAXWELL, J. R. 1990. Organic geochemical and biological marker characterization of source rocks and oils derived from lacustrine environments in the Brazilian continental margin. *In*: KATZ, B. J. (ed.) *Lacustrine Basin Exploration – Case Studies and Modern Analysis*. Memoir. American Association of Geologists, **50**, 77–95.

——, MOHRIAK, W. U., KOUTSOUKOS, E. A. M. & FIGUEIRA, J. C. A. 1991. Brazilian and West African Oils: Generation, Migration, Accumulation and Correlation. *13th World Petroleum Congress*, Buenos Aires, Vol. 2, 153–164.

MILANI, E. J., FRANCA, A. B. & SCHNEIDER, R. L. 1994. Bacia do Parana. *Boletin Geoçiencias da Petrobras*, **8**, 69–82.

MILLIMAN, J. D. 1978. Morphology and structure of upper continental margin off southern Brazil. *AAPG Bulletin*, **62**, 1029–1048.

MOHRIAK, W. U., MELLO, M. R., DEWEY, J. F. & MAXWELL, J. R. 1990. Petroleum geology of the Campos Basin, offshore Brazil. *In*: BROOKS, J. (ed.) *Classic Petroleum Provinces*. Geological Society, London, Special Publications, **50**, 119–141.

MUNTINGH, A. 1993. Geology, prospects in Orange basin offshore western South Africa. *Oil & Gas Journal*, **Jan 25**, 106–109.

NETTO, A. S. T., FILHO, J. R. W. & FEIJO, F. J. 1994. Bacias de Jacuipe, Camamu e Almada. *Boletin Geoçiencias da Petrobras*, **8**, 173–184.

NGUENE, F. R., TAMFU, S., LOULE, J.-P. & NGASSA, C. 1992. Palaeoenvironments of the Douala and Kribi/Campo sub-basins in Cameroun, West Africa. *In*: *Géologie Africaine: Colloque de Géologie de Libreville*, Recueil de Communications, 129–136.

PAMPLONA, H. R. P. 1969. Litoestratigrafia da bacia cretacea de Barreirinhas. *Boletin tec. da Petrobras*, **12**, 261–290.

PAUKEN, R. J. 1992. Sanaga Sud Field, offshore Cameroon, West Africa. *In*: HALBOUTY, M. T. (ed.) *Giant Oil and Gas Fields of the Decade 1978–1988*. Memoir. American Association of Petroleum Geologists, **54**, 217–230.

PEES, S. T. & DELANEY, P. J. V. 1985. Oil, gas potential of Brazil's Pelotas Basin. *Oil & Gas Journal*, **Feb. 18**, 117–122.

PEREIRA, M. J. & MACEDO, J. M. 1990. A Bacia de Santos: perspectivas de uma nova provincia petrolifera na plataforma continental sudeste Brasileira. *Boletin Geoçiencias da Petrobras*, **4**, 3–11.

PONTE, F. C., FONSECA, J. D. R. & CAROZZI, A. V. 1980. Petroleum habitats in the Mesozoic–Cenozoic of the continental margin of Brazil. In: MIALL, A. D. (ed.) Facts and Principles of World Petroleum Occurrence. Memoir. Canadian Society of Petroleum Geologists, **6**, 857–886.

SANTOS, C. F., GONTIJO, R. C., ARAUJO, M. B. & FEIJO, F. J. 1994. Bacias de Cumuruxatiba e Jequitinhonha. *Boletin Geoçiencias da Petrobras*, **8**, 185–190.

SCHALLER, H. 1969. Revisao estratigraficas da bacia de Sergipe/Alagoas. *Boletin tec. da Petrobras*, **12**, 21–86.

SCHIEFELBEIN, C. F., ZUMBERGE, J. E., CAMERON, N. R. & BROWN, S. W. 1999. Petroleum systems in the South Atlantic margins. *This volume*.

SCHOELLKOPF, N. B. & PATTERSON, B. A. 1997. Petroleum Systems of Cabinda, Angola. Hedberg Research Symposium, *Petroleum Systems of the South Atlantic Margin*, Rio de Janeiro, 16–19 November, extended abstract.

TANKARD, A. J., ULIANA, M. A., WELSINK, H. J. *ET AL.* 1995. Structural and tectonic controls of basin evolution in southwestern Gondwana during the Phanerozoic. *In*: TANKARD, A. J., SUÁREZ SORUCO, R. & WELSKIN, H. J. (eds) *Petroleum Basins of South America*. Memoir. American Association of Petroleum Geologists, **62**, 5–52.

TEISSERENC, P. & VILLEMIN, J. 1990. Sedimentary basins of Gabon, geology and oil systems. *In*: EDWARDS, J. D. & SANTOGROSSI, A. (eds) *Divergent/Passive Margin Basins*. Memoir. American Association of Petroleum Geologists, **48**, 117–199.

URIEN, C. M., MARTINS, L. R. & ZAMBRANO, J. J. 1976. The geology and tectonic framework of southern Brazil, Uruguay and northern Argentina continental margin, their behavior during the South Atlantic opening. *Academy Brasileira Ciencias, Anais*, **48**, 365–376.

——, ZAMBRANO, J. J. & YRIGOYEN, M. R. 1995. Petroleum basins of southern South America: an overview. *In*: TANKARD, A. J., SUÁREZ SORUCO, R. & WELSINK, H. J. (eds) *Petroleum Basins of South America*. Memoir. American Association of Petroleum Geologists, **62**, 63–78.

VAN DER VEN, P. H., CAINELLI, C. & FERNANDES, G. J. F. 1989. Bacia de Sergipe–Alagoas: geologia e exploracao. *Boletin Geoçiencias da Petrobras*, **3**, 307–319.

VIEIRA, R. A. B., MENDES, M. P., VIEIRA, P. E., COSTA, L. A. R., TAGLIARI, C. V., BACELAR, L. A. P. & FEIJO, F. J. 1994. Bacias do Espirito Santo e Mucuri. *Boletin Geoçiencias da Petrobras*, **8**, 191–202.

WILLIAMS, B. G. & HUBBARD, R. J. 1984. Seismic stratigraphic framework and depostional sequences in the Santos Basin, Brazil. *Marine and Petroleum Geology*, **1**, 90–104.

ZALAN, P. V., NELSON, E. P., WARME, J. E. & DAVIS, T. L. 1985. The Piaui Basin: Rifting and wrenching in an equatorial Atlantic transform basin. *Soc. Econ. Paleontology and Mineralogy*, Special Publication, **37**, 177–192.

Tectonics and hydrocarbon distribution along the Brazilian South Atlantic margin

IAN DAVISON

Department of Geology, Royal Holloway, University of London, Egham, Surrey TW20 0EX, UK

Abstract: Break-up of the Brazilian margin produced a wide variety of structural styles depending on opening kinematics and thermal configuration of the lithosphere. Wide margins developed in Santos and Campos, where stretched continental crust extends for up to 600 km from the shoreline; whereas narrow margins may be only 40 km wide. The wide margins host the largest oil fields, such as those discovered in the Campos Basin. Wide margins have good hydrocarbon potential in the post-rift sequence, but narrow margins probably have the better unknown syn-rift plays which have still to be drilled in deepwater. Little is known of the narrow margins, such as the Jacuípe and Pernambuco–Cabo Basin area, as most of the prospective hydrocarbon structures lie in water depths > 1500 m.

Hydrocarbons are hosted by reservoirs ranging from fractured Neocomian volcanics to Miocene turbidites, with over 80% of the reserves trapped in the Campos Basin turbidite sandstones. These sandstones occur as constrained channel-fills on the upper platform and slope, and as unconstrained lobes up to 30 km in long dimension on the mid to lower continental slope which are situated in 500+ m water depths at the present day. The wide margins will continue to be the most productive in Brazil, but deepwater production technology has advanced enough to make the deeper parts of the narrower margins attractive exploration areas.

The Brazilian Atlantic margin tectonic and stratigraphic evolution, and their impact on hydrocarbon distribution, are reviewed in this paper. The Brazilian margin is clearly segmented with basement highs and igneous complexes (e.g. Abrolhos and Royal Charlotte Highs) separating individual basins (Fig. 1 and Enclosure 1). Most of the basins have been explored for hydrocarbons to varying degrees, although the deeper parts (> 1000 m water depth) of the basins are poorly explored. This paper attempts to summarize the evolution of the Brazilian margin and discuss the general implications for hydrocarbon potential, focusing on the differences between the structure and sedimentary patterns developed on wide margins compared to narrow margins (Davison 1997). Several recent papers have been written on the Brazilian margin and its hydrocarbon potential and these complement this paper (e.g. Szatmari *et al.* 1985; Figueiredo 1985; Guardado *et al.* 1989; Chang *et al.* 1988, 1992; Davison 1991; Mello *et al.* 1994).

Tectonics of the Brazilian margin

Afro-Brazilian Depression

Before Atlantic rifting began, a large gentle sag, called the Afro–Brazilian Depression (Ponte 1971) formed to the north of the Espírito Santo Basin and extended as far north as the Sergipe–Alagoas Basin (Fig. 1). It was filled with coarse-grained fluvial sandstones, occasional evaporites and red claystones of Late Jurassic age (Dom João stage of Petrobras). This early stage of sedimentary fill is exposed in the Recôncavo Basin and is not affected by syn-sedimentary faulting, and is therefore classed here as an early sag phase [cf. early rift phase of Chang *et al.* (1988, 1992)]. The total stratal thickness may reach up to 1000 m, but averages 600 m (Chang *et al.* 1992). Sedimentary facies and thickness variations are very regular over large distances (100 km). These sedimentary strata have also been recognized in Gabon, Congo and the Cabindá Basins (Matos 1992). The northernmost limit of these sediments is the Patos Shear Zone (Matos 1992; Enclosure 1). They may have extended further north than this but were subsequently eroded. The origin of this early sag basin is not fully understood. It may be due to ductile lithospheric stretching, without break-up of the upper crust, or may simply have been a sediment-filled topographic depression surrounded by mountainous areas developed during earlier orogenies. More work is required to resolve this.

Main phase of Atlantic rifting

Rifting south of the Walvis Ridge in Argentina and Namibia may have started in Triassic–Jurassic times. North of the Brazilian

DAVISON, I. 1999. Tectonics and hydrocarbon distribution along the Brazilian South Atlantic margin. *In:* CAMERON, N. R., BATE, R. H. & CLURE, V. S. (eds) *The Oil and Gas Habitats of the South Atlantic.* Geological Society, London, Special Publications, **153**, 133–151.

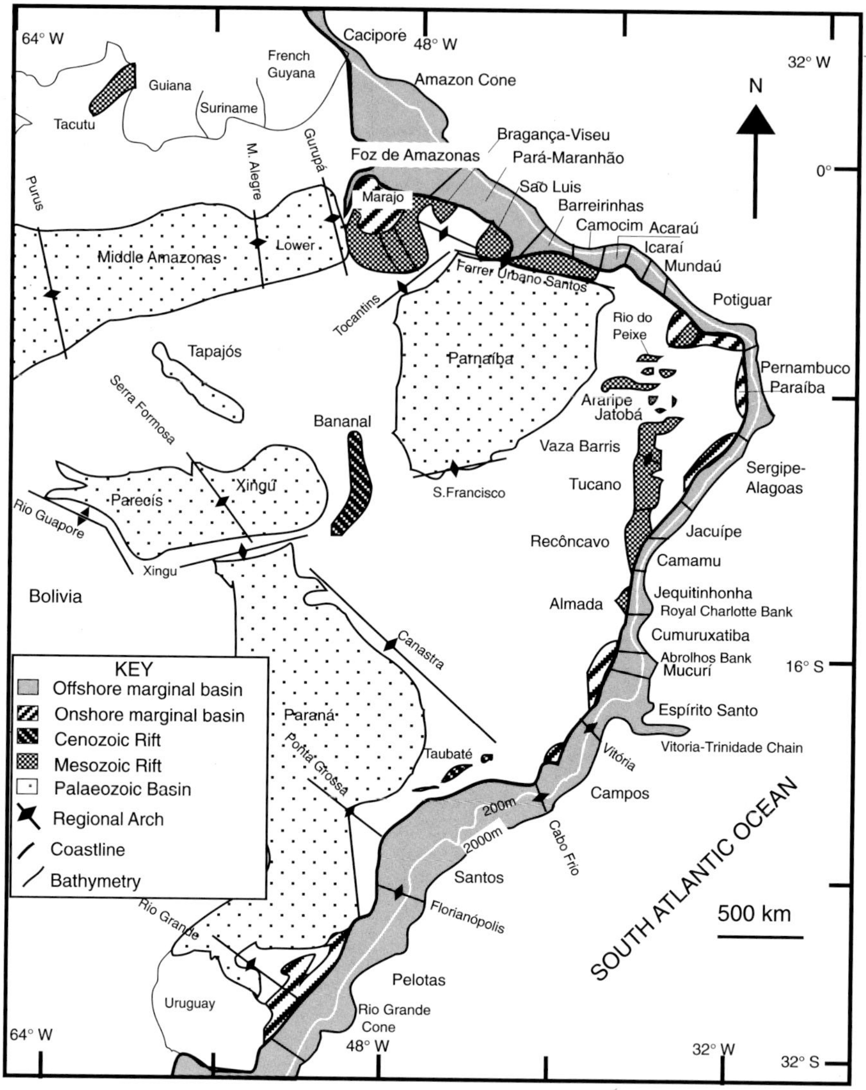

Fig. 1. Outline map of Brazilian Atlantic margin sedimentary basins.

margin rifting of the Central Atlantic began *c.* 180 Ma ago, some 50 Ma before South Atlantic rifting (Rabinowitz & La Breque 1979; Hay *et al.* 1994). Although the South Atlantic Ocean drifting is known to have started in the south and gradually spread northward, the earlier rifting did not proceed in such a simple manner. The oldest known rift sediments of Berriasian age (134–145 Ma) are situated in the Recôncavo to Espírito Santo section of the margin and the Potiguar Basin (Santos *et al.* 1994; Netto *et al.* 1994; Vieira *et al.* 1994; Fig. 2). This compares closely with the timing of African rifting along the conjugate Gabon–Angola margin (McHargue 1990; Teisserenc & Villemin 1990). Further south in the Campos Basin, rifting is interpreted to have commenced in the Hauterivian (Rangel *et al.* 1994; Fig. 2). The amount of syn-rift section in the Santos and Campos Basins is disproportionately small compared to the amount of stretching calculated from the thermal subsidence (Chang *et al.* 1992). This may be explained by some of the accommodation space created by rifting being filled with basalts which are highly rotated, but they are not seismically imaged as most of the seismic energy is reflected from the top basalt interface. Hence, rifting may have commenced earlier than the

Hautervian in the Santos and Campos Basins, but volcanic rocks rather than sediment were deposited. There may be a diachronous time span for the onset of rifting, up to 10 Ma in duration, with several individual rift basins opening simultaneously, which subsequently became linked by Hauterivian times. The absolute dating of the Brazilian stage boundaries is still not sufficiently precise to define this further, as the South Atlantic ostracod stratigraphy is difficult to tie into internationally recognized stage boundaries.

During the late Barremian a significant change in rift kinematics occurred when the Potiguar and Recôncavo–Tucano–Jatobá Rifts became inactive, and major rifting initiated along the Equatorial margin (Barreirinhas–Ceará; Fig. 2). The main rift phase is thought to have initiated in Aptian times along the Equatorial margin and the Benue Trough (Feijó 1994*b*), with the Barreirinhas Basin being the latest segment of the margin to subside in mid-Albian times (*c*. 112 Ma; Feijó 1994*b*). The Sergipe–Alagoas Basin also underwent important subsidence during the Aptian and salt was deposited during rifting in this basin (Feijó 1994*a*). The margin south of the Sergipe–Alagoas Basin experienced the transition from rift to drift stage during the Aptian. The base of the salt is almost flat in the southern basins (Fig. 3), and extension above the Aptian salt is mainly interpreted to be caused by downslope gravity sliding.

In summary, the main phase of rifting along the southern part of the Brazilian margin occurred between 120 and 135 Ma. Rifting had ceased by Aptian times from Pelotas to Jacuípe, but rifting was still active in the Alagoas and Pernambuco Basins until the early Albian (Dias 1994).

The main rifting phase (120–135 Ma) north of the Rio Grande Rise–Walvis Ridge system coincides with the age of the magnetic anomalies south of the ridge, where ocean spreading propagated from south to north (Nüller 1991). Rifting in each Brazilian basin probably spanned a 10–15 Ma period before oceanic spreading occurred.

Crustal thinning of at least 50% ($\beta = 1.5$) on the wide margins, was achieved in a 10–15 Ma period (Chang *et al.* 1992), which is equivalent to a strain rate of *c*. $1.26 \times 10^{-15}\ s^{-1}$. Tectonic subsidence outstripped sedimentation during the early rifting phase and individual deepwater lakes developed along most of the margin between the Campos and the Sergipe–Alagoas Basins, sometimes with more than one lake in each basin. It is still not known whether such deepwater lakes developed in the Santos and Pelotas Basins, as all the wells drilled into the syn-rift section are located in the shallow proximal part of the basins which are dominated by coarse clastic sediments and basalts. Important surface uplift occurred in the early Cretaceous period on the rift flanks in the Serra do Mar region (Gallagher *et al.* 1995).

Regional uplift also occurred during the post-Albian period in northeast Brazil, where Albian marine carbonates in the Araripe Basin are now at regional elevations of 800 m above sea level (Magnavita *et al.* 1994). This regional uplift may be due to a magmatic underplating event which has thickened the crust and produced sporadic magmatism along the Brazilian margin up to 500 km inland from the Atlantic margin (Fig. 4; Magnavita *et al.* 1994). However, there is no other evidence available to support this hypothesis and there is no hotspot present today (Grand *et al.* 1997).

Major fracture zones are traceable in oceanic crust almost to the present-day coastline (Sandwell & Smith 1995; Fig. 5) and these have been a major influence in the tectonic development of the margin. The clearest correlations are present in the central region of the Brazilian coastline, immediately south of the Sergipe– Alagoas Basin, where the Ascension Fracture Zone joins up almost directly to the African margin (Fig. 5; Karner & Driscoll 1999). The width of the stretched zone of continental crust is very variable, from 600 km in the Santos Basin to narrow zones of *c*. 40 km width in the Jacuípe and Pernambuco Basins (Davison 1997). Some of the lateral transitions between wide and narrow margins are very abrupt and occur across fracture zones such as the Rio Grande–Walvis Fracture Zone (Fig. 5).

Magmatism associated with South Atlantic break-up

Onshore. The Atlantic opening was facilitated by the Tristan Plume and possibly by a weak northeast Brazilian plume, although linear lithospheric breaks were already developed in the southern Bahia and Recôncavo Basins when plume activity began. It has been suggested that there was a minor Early Cretaceous plume underneath the Benue Trough–Atlantic junction (Standlee *et al.* 1992). Seismic tomographic studies do not indicate a present-day hotspot under northeast Brazil (Grand *et al.* 1997). There are several sites along the margin where important dyke swarms were intruded. The Ponta Grossa dykes adjacent to the Santos Basin have been dated at 131–134 Ma, using $Ar^{40/39}$ (Turner *et al.* 1994) and 120–125 Ma (Renne *et al.* 1996*b*); a tholeiitic dyke swarm (Rio Ceará Mirim magmatism) along the southern edge of the Potiguar Basin has been dated at 124–144 Ma using K–Ar and fission track ages (Sial 1976; Sial *et al.* 1981; Gomes 1981). It has been suggested that the Ponta Grossa dyke

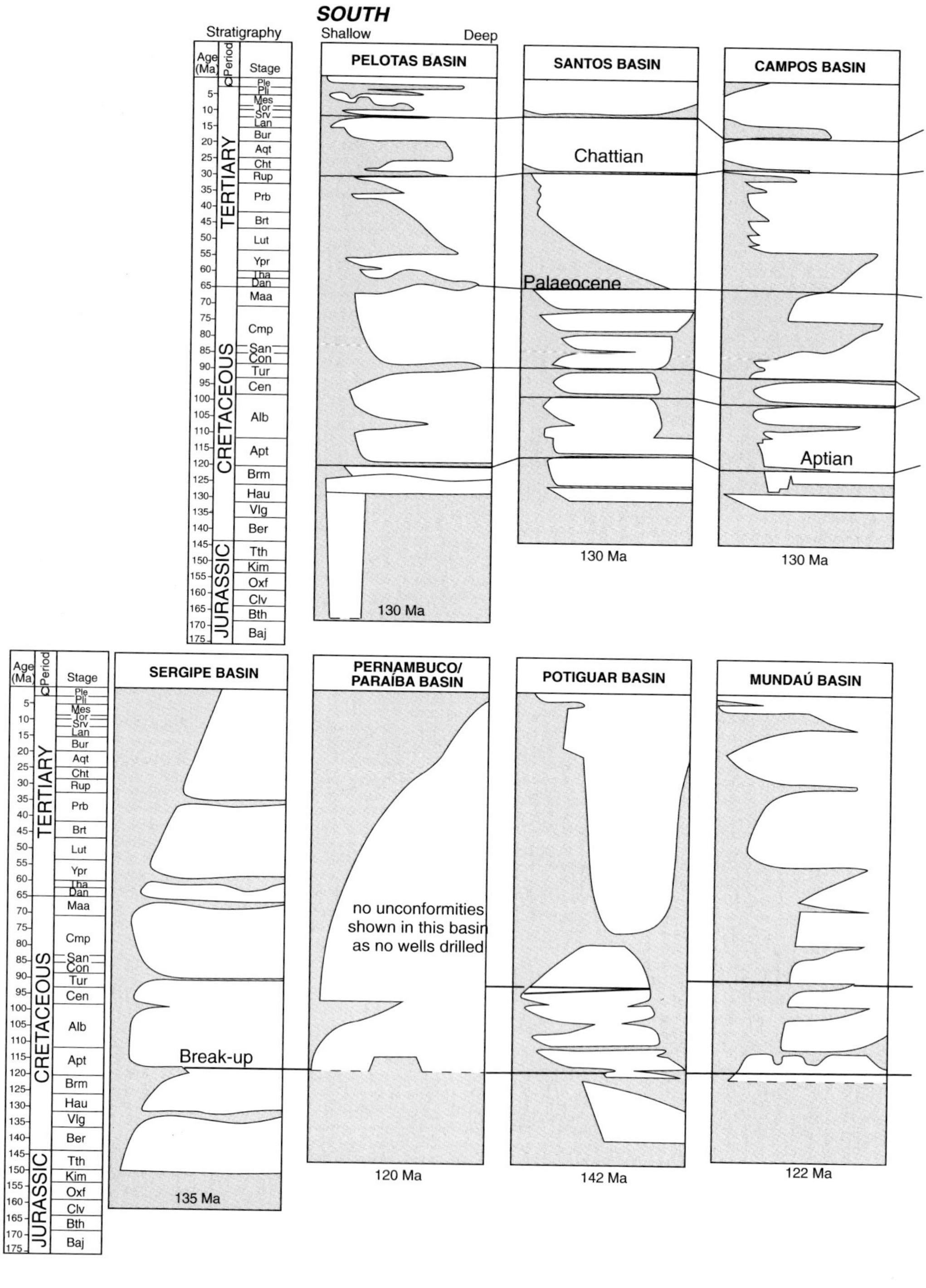

Fig. 2. Unconformity correlation (non-deposition and erosional events) and timing of initial rifting along the Brazilian margin. Data taken from papers in *Boletim de Geociências da Petrobras*, **8**, 1994. Sedimentary hiatuses are shaded grey. Timescale used: Gradstein *et al.* (1994).

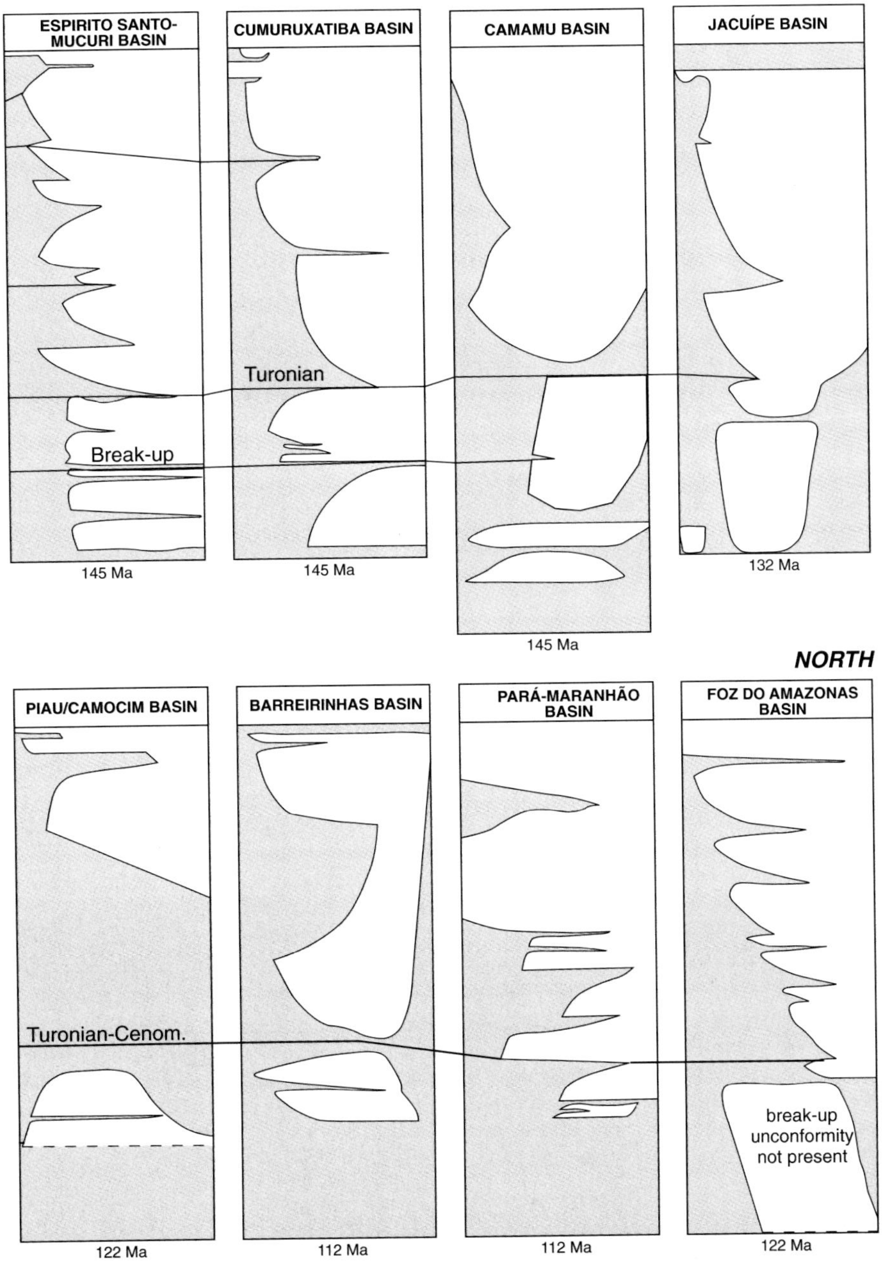

Fig. 2. (*continued*)

swarm was produced by differential opening between the crust either side of the São Paulo–Walvis Ridge with a clockwise rotation of southern South America interpreted to have produced the dyke swarm which strikes orthogonally to the margin (Chang & Kowsmann 1991). Hence, these dykes are thought to date the main rifting event which extended from 134 to 120 Ma.

The Tristan Plume was originally centred over the Paraná Basin and it produced the most extensive continental flood basalt province in the world. $Ar^{40/39}$ ages indicate the basalts were erupt-

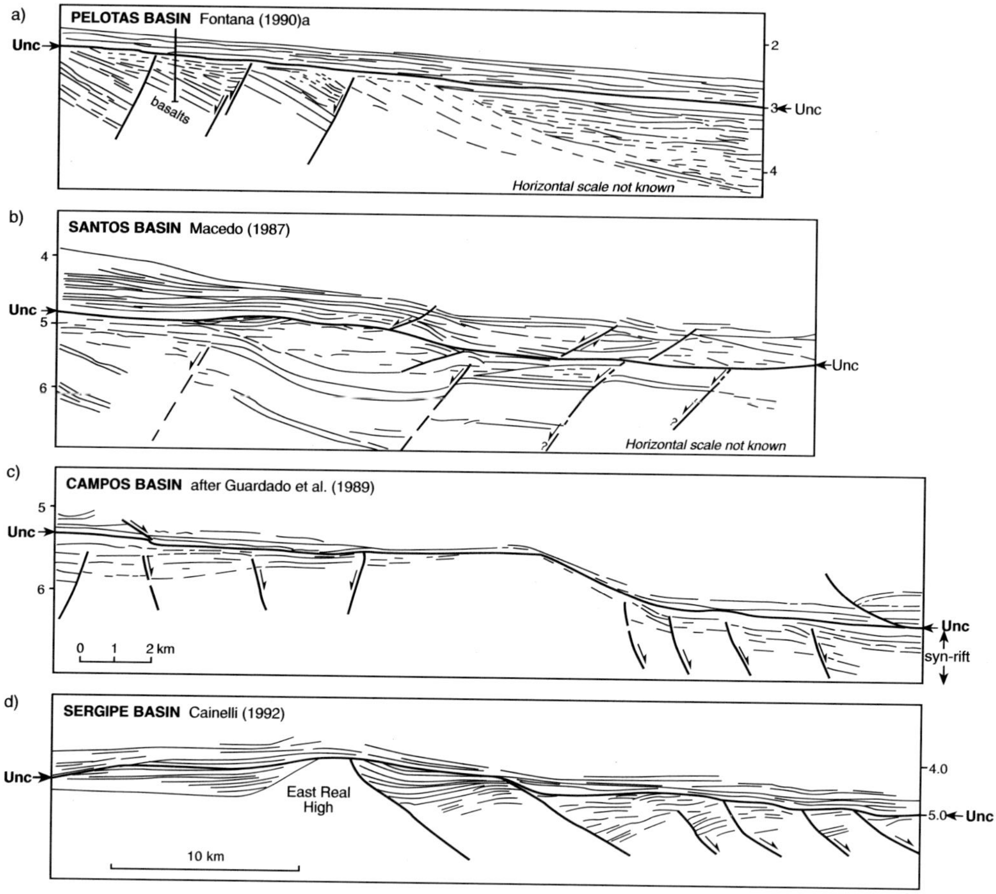

Fig. 3. Geoseismic interpretations of the Aptian unconformity at the base of the evaporite sequence. (**a**) Pelotas Basin (based on Fontana 1990*b*); (**b**) Santos Basin (based on Macedo 1987); (**c**) Campos Basin (based on Guardado *et al.* 1998); (**d**) Sergipe Basin (based on Cainelli 1992).

ed between 137 and 127 Ma (Turner *et al.* 1994; Renne *et al.* 1996*a*). The Etendeka volcanics in Namibia have similar ages (127–131 Ma; Turner *et al.* 1994; Renne *et al.* 1996*a*). The increase in acidic component and a decrease in magma volume in Namibia suggest the plume was centred under Brazil, and that the final split of the two continents occurred to the east of the plume head.

Offshore. The Paraná and Etendeka volcanic provinces are linked by the aseismic ridges of the Rio Grande Rise and the Walvis Ridge, and all of these volcanics are linked to the Tristan Plume (O'Connor & Duncan 1990). The age of the first volcanic activity offshore has been determined by K–Ar dating on basalts in the Santos, Campos and Espírito Santo Basins at *c.* 130–120 Ma (Fodor *et al.* 1983). However, these rocks are highly altered and the ages may reflect inherited argon. Basalt flows are rotated in the Pelotas, Santos and Campos Basins (Fig. 3), indicating rifting outlived plume volcanism in the south of Brazil. The plume post-dates rifting in northern Brazil which began in the Berriasian in the Recôncavo Basin. This indicates regional extension was occurring when the Paraná Plume impinged. It is possible that the Paraná Plume triggered the spreading process, allowing the Recôncavo Rift to propagate southward. Thick seaward-dipping wedge reflectors of probable early Cretaceous age have been recently interpreted on seismic sections in deepwater in the Sergipe–Alagoas and Jacuípe Basins (Mohriak 1995). Gravity data also suggests a large volcanic centre in the Jacuipe Basin (Enclosure 1).

Enclosure 1 (opp. p. 138): map of tectonics, sedimentation and hydrocarbon distribution along the South Atlantic margins. CD-Rom contains large map and individual maps of six African and eight Brazilian basins.

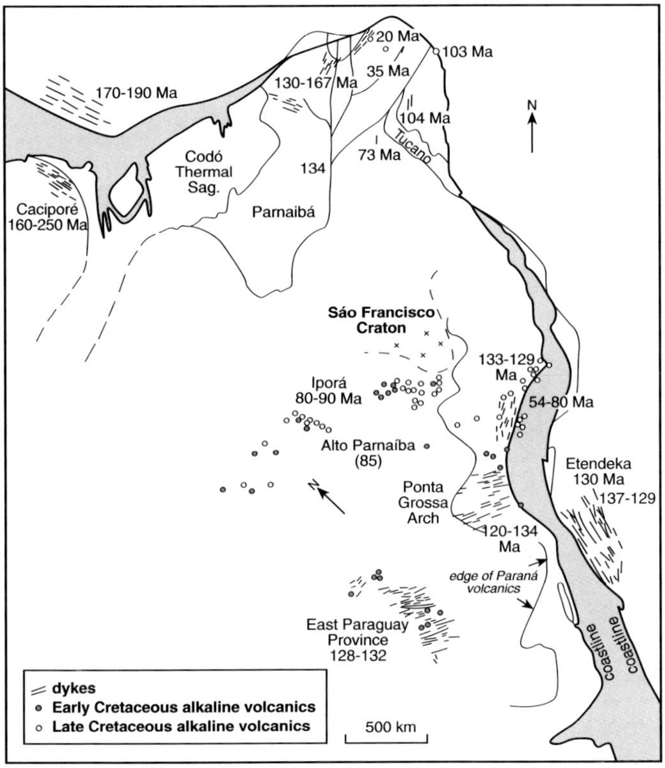

Fig. 4. Mesozoic magmatism along the Atlantic margin. Data from Sial *et al.* (1981) and Magnavita *et al.* (1994).

The volcanic centres of Ceará, Fernando de Noronha, Abrolhos, Royal Charlotte, Ferraz Ridge (Enclosure 1) and the alkaline magmatism of the states of São Paulo and Rio de Janeiro are Late Cretaceous–Tertiary in age (Fig. 4). Hence, the Brazilian margin has been the site of continued volcanic activity since South Atlantic break-up. This activity is asymmetric as it is largely confined to the Brazilian side of the Atlantic. The Abrolhos massif is the largest of these and has been dated at 40–50 Ma; again, these ages may be regarded as suspect as they were obtained using K–Ar techniques on altered basalts. The massif consists of basalts, diabase dykes, tuffs and volcanic breccia.

The Drift Phase

Break-up unconformity. The Break-up unconformity (Falvey 1974) can be correlated along many parts of the Brazilian margin. The unconformity generally appears as a low to high angle unconformity, with much of the fault-block topography infilled by syn-rift sediments before Aptian flooding surfaces produced a parallel-bedded unit overlying the unconformity surface (Fig. 3). This is well illustrated in the Pelotas Basin where the absence of the salt enables the syn-rift sediment wedges to be clearly imaged. The rifting took place rapidly during and after the igneous rocks were erupted (Fig. 3). The topography was infilled during the later rift stage and the Aptian unconformity surface is sub-parallel to the uppermost syn-rift sediments in the basins, but there is an angular truncation onto the eroded fault block crests (e.g. Fig. 3d). In the Santos and Campos Basins the unconformity is more subtle because there appears to have been less syn-rift rotation of fault blocks and the salt produces decoupling of the rift faults near to the mid-Aptian unconformity. However, there is still a slight angular unconformity of < 5° (Fig. 3b and c). In the Sergipe–Alagoas Basin there is an angular unconformity between the mid-Aptian rift strata and the overlying late Aptian drift sediments. Up to 700 m of rock may have been removed from the uplifted footwall blocks (calculated from truncation on Fig. 3d). In the intervening grabens the unconformity becomes conformable with the evaporites, indicating

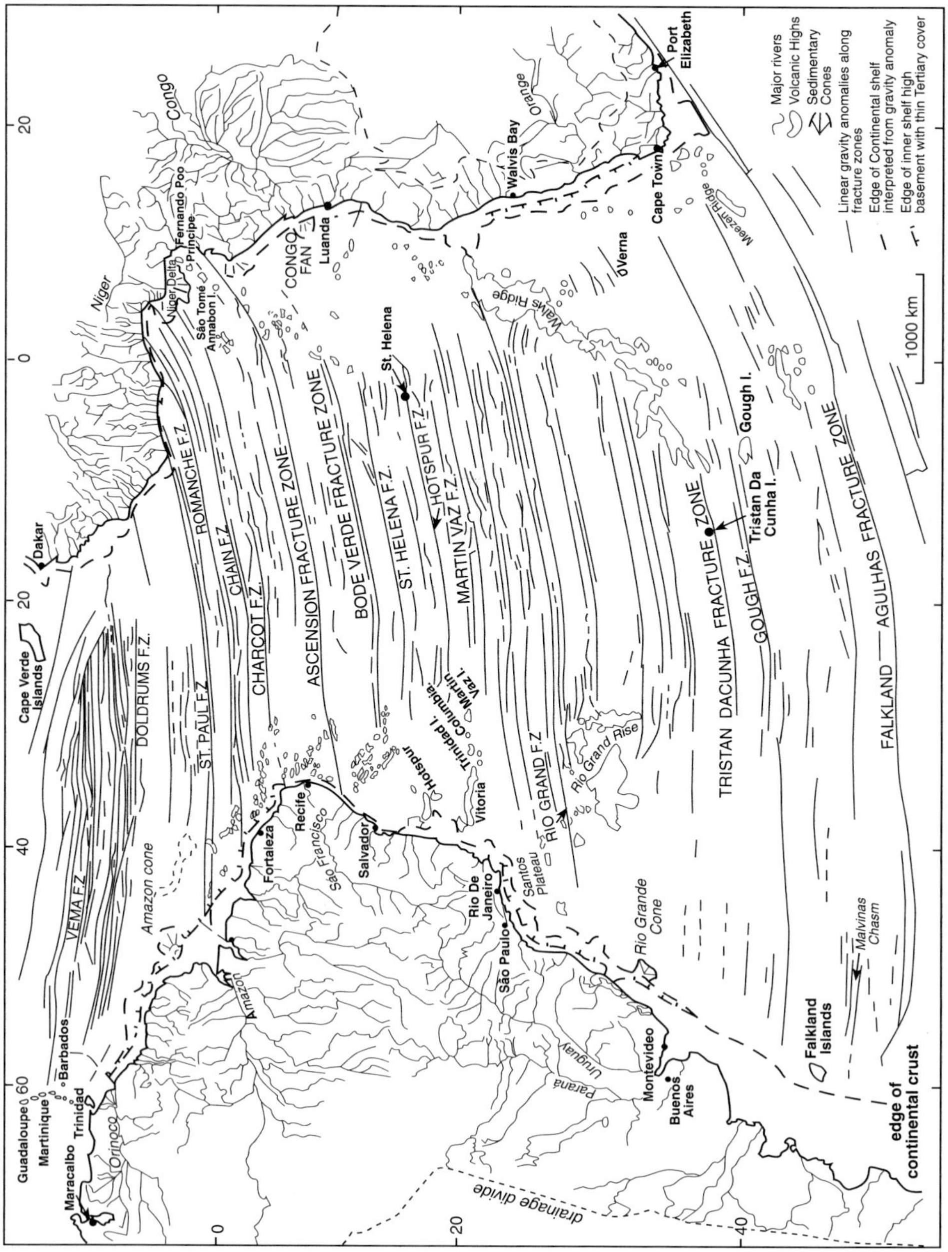

Fig. 5. Tectonic summary map of the South Atlantic. Fracture zone interpretation is based on the Seasat gravity map of Sandwell & Smith (1995).

infilled topography in the deeper graben. Uplift will be enhanced in these footwall blocks if the locus of rifting moved further offshore with time (Karner & Driscoll 1999).

All the above examples indicate that there has been a period of erosion of footwall crests after extension, but there has not been sufficient erosion to remove significant amounts of material from the intervening graben areas. The uplift and erosion occurs across most of the extended area and is not confined to the rift flanks (cf. Braun & Beaumont 1989). The break-up unconformity was followed by an increase in relative sea level which rapidly inundated the infilled peneplained topography, except for over the largest fault blocks such as the Badejo High in the Campos Basin which was not covered by salt deposits (Guardado *et a.* 1989). The subsidence after erosion was probably due to final crustal failure at the end of rifting which removed lateral flexural support from the central part of the plate, due to asthenospheric upwelling at a proto-ocean ridge. Oceanic spreading is believed to have started immediately after the salt deposition and so far autochthonous Aptian salt has not been proven on oceanic crust (W. Mohriak, pers. comm.). However, Karner & Driscoll (1999) suggest that drifting may have initiated in Barremian times.

Drift phase. The oldest magnetic anomaly in the south Atlantic ocean near Cape Town is probably M9 *c*. 130 Ma (Austin & Uchupi 1982). North of 33° S, seafloor spreading commenced around anomaly M4 (Mid-Hauterivian, 127 Ma), whereas north of the Walvis–São Paulo Ridge seafloor spreading is interpreted to have commenced at the Aptian–Albian boundary (113 Ma) (Chang *et al.* 1992). The drift phase of the south Atlantic is therefore thought to have occurred some 25 Ma after the onset of plume volcanism and 7–10 Ma after main phase of continental rifting. The last part of the margin to break apart was the Pernambuco Basin, which began rifting in Aptian times and by mid–late Albian times permanent contact of the North and South Atlantic Oceans had been established (Dias-Brito 1987). Brazil and Africa have moved northward relative to the Tristan hotspot some 2900 km between 130 Ma and magnetic anomaly 13, estimated at 34 Ma. This represents a northward drift of the South American and African plates relative to the hotspot of *c*. 30 mm/year.

The southern Atlantic spreading velocity of *c*. 15 mm/year continued from the start of spreading at 135 Ma to the Aptian–Albian boundary at 112 Ma, when the relative plate movement vector changed and the spreading velocity increased to 35 mm/year (Rabinowitz & La Breque 1979). This approximately coincides with the opening of the Benue Trough which has preserved Albian continental sediments (Akande *et al.* 1997). There was also a change in the plate velocity vectors in the latest Turonian–early Coniacian times (88–89.5 Ma) (Pereira 1992). This event coincides with a notable absence of Turonian sediments along most of the Brazilian margin (Fig. 2), followed by deep water anoxic shales immediately above the unconformity. A possible explanation for absence of sediments during the Turonian is that ocean currents were completely changed due to the change in plate movement direction and linking of the South to the North Atlantic. This may have caused widespread erosion of thin Turonian sediments along most of the Brazilian margin. Palaeocene, Chattian to Aquitanian and Serravalian age unconformities can also be correlated across the southern basins from Pelotas to Espírito Santo. These unconformities may be due to eustatic sea-level changes, but they are not easily correlated along the whole margin (Fig. 2).

Transpressional events have been responsible for erosional events of Maastrichtian–Oligocene age strata in the shallow parts of the Equatorial transform marginal basins. Very large flower structures developed from the Acaraú Basin to the Barreirinhas Basin, with the largest inversion events taking place in the Barreirinhas and Piauí–Camocim Basins where no Palaeocene or Eocene sediments are preserved in the shallow parts of the basin (Zalán 1984; Azevedo 1991; Feijó 1994*b*; Beltrami *et al.* 1994) (Fig. 2).

Stratigraphic evolution of the Brazilian Atlantic margin

The first syn-rift strata deposited along most of the margin were lacustrine shales and sandstones, with coarse alluvial fans deposited along the faulted basin borders. The early rift lakes were progressively filled with fluvio-deltaic sandstones. In contrast, during the Barremian stage, lacustrine shelly limestones, organic shales and coarse siliclastics were deposited between the Santos and Sergipe–Alagoas Basins. The carbonate banks were eventually covered by prograding clastics. Syn-rift carbonate banks of coquina shells are considered to be good reservoirs in the Campos Basin, but are sporadically developed and difficult to predict (Dias *et al.* 1988). Non-marine turbidite sandstones are found within the Barremian rift sequence in the south Bahia basins and these may constitute attractive reservoirs where they are not too deeply buried. The Aptian stage (Alagoas stage of Petrobras) was marked by the transitional continental to marine sequence. This sequence was deposited on a regional unconformity after the major phase of extension had mainly ceased. The unconformity is present along most of the Atlantic margin from the Pelotas Basin to the Potiguar Basin, although it is probably of slightly different age (early–Mid-Aptian; Fig. 2) and geometry (Fig. 3). Coarse clastic deposits of Aptian age were deposited in the shallow shelf areas and these gradually gave way to evaporites in deeper parts of the basin. The Aptian sequence, from the Recôncavo Basin northwards, consists of arid alluvial–fluvial coarse clastic deposits associated with local sabkhas, which were deposited on a peneplain above eroded and rotated earlier strata. Hence, the evaporites are interpreted to have been deposited in shallow water.

The salt basin matches closely with known coeval deposits on the African passive margin. Evaporites appear better developed on the African side in the northern part of the basin, and on the Brazilian side in the southern part of the basin (Leyden 1976; Enclosure 1). Although the evaporites are known to be Aptian along most of the margin, they were probably deposited in a series of individual basins along the margin rather than one simple large depression. Evaporite deposits may have reached original thicknesses of 2000 m, considering the size of the diapirs and their withdrawal synclines in the Santos and Campos Basins (Cobbold & Szatmari 1991; Chang *et al.* 1992), and they thin progressively northward, where only a few isolated diapirs are observed in the Sergipe–Alagoas Basin. Evaporites were deposited as far north as the Icaraí Sub-basin in Ceará, where thicknesses of 100 m have been penetrated by wells (Regali 1989; Azevedo 1991). Two separate layers of synrift Aptian salt have been identified in the Sergipe–Alagoas Basin (Uesugui 1987).

During the Albian stage the evaporites were followed by deposition of an extensive shallow marine carbonate platform with a restricted fauna, probably controlled by high salinities. Clastic fan-deltas were also deposited during this interval, and interfinger with the carbonate platforms. Further offshore, marls and shales were deposited on the continental slope. At the end of the Albian stage, fine-grained calcilutites, marls and shales were deposited in deeper waters. High-energy oolitic limestones and calcarenites were deposited on top of salt swells and can be good hydrocarbon reservoirs. During the Cenomanian stage, the palaeowater depth was in the region of 200–300 m over most of the marginal basins and a marked anoxic event occurred which produced Cenomanian black shales with high total organic carbon (TOC) levels in many of the offshore basins (Mello *et al.* 1989). The shales have abundant and diverse radiolaria and planktonic foraminifera, which can be correlated across the Atlantic to the Angola Basin where they are richer in organic content (Azevedo *et al.* 1987; Arai 1988). During this transgressive open marine sequence, Cenomanian turbiditic sandstones were deposited along most of the Brazilian margin. These form important reservoirs, especially in the Campos Basin.

From Campanian to Santonian times oxic conditions were widespread. A prograding clastic sequence with offlap geometry developed with local aggradational high-energy carbonate platforms developed particularly in the Sergipe–Alagoas area (Chang *et al.* 1992). The onset of the progradation varies along the margin from Turonian times in the Santos Basin, to Maastrichtian–Palaeocene age further north in the Campos Basin, and to Eocene age in the Sergipe–Alagoas Basin (Chang *et al.* 1992).

Very large turbidite fans (up to 300 km^2 in area), from Albian to Early Miocene age, poured into the Campos Basin during low sea-level stands and periods of tectonic movement (Bruhn & Walker 1995). Salt movements occurred throughout this period, with preferential turbiditic sand deposition in withdrawal basins adjacent to salt structures and on their landward side.

Source rocks

Good reviews of the source rock potential of the Brazilian basins are available (e.g. Mello 1988; Mello *et al.* 1988, 1994; Table 1). The Neocomian saline lacustrine source is the most important along the margin as this is believed to generate the Campos and Espírito Santo oils (Mello 1988). The Sergipe and Potiguar oils are believed to be sourced from Neocomian lacustrine freshwater and Aptian evaporite source rocks (Mello 1988). The source of the Santos Basin oils is not known as no published information is available. In the main depocentre of the Santos Basin the Neocomian source rocks are probably buried to 7 km, indicating they are now generating gas, and probably sourced the Merluza Field. However, the oils of Tubarão and adjacent fields in the southwest of the Santos Basin may be sourced from open marine shales of mid–late Cretaceous age.

In general, mature source rocks are present along most of the Brazilian Atlantic margin and began generating hydrocarbons in Late Cretaceous–Tertiary times. The Aptian evaporites may cause problems of migration from Neocomian source rocks into post-salt reservoirs.

Table 1. *List of source rock characteristics of the Brazilian Atlantic margin (after Mello 1988)*

Stratigraphic age	Source type	TOC range (%)	Hydrocarbon yield (kg HC $tonne^{-1}$)	Kerogen type	Other information
Early Neocomian	Lacustrine freshwater shales	1–6	4–37	Type I/II; 25–35% plant debris; lipid algal matter 40–60%	Sulphur < 0.1%, low $CaCO_3$ < 7%
Late Neocomian	Lacustrine saline	2–5	10–38	Type I lipid-rich algal and bacterial	$CaCO_3$ 2–45%, sulphur 0.5%
Aptian	Hypersaline marine	1.5–14	5–97	Type II amorphous 45–60%; herbaceous 15–25%; woody–coaly 10–25%	$CaCO_3$ 7–45%, sulphur 0.5–2.5%
Albian carbonates	Deep marine shales and carbonates	1–4	2–22	Type II amorphous 40–50%; herbaceous 15–25%; woody–coaly 20–30%	$CaCO_3$ up to 65%, sulphur 0.3–1.7%
Cenomanian–Campanian open marine	Open marine anoxic black shales	1–4	8–19	Type II mainly amorphous with minor herbaceous and woody material	$CaCO_3$ 11–40%, sulphur < 0.6%
Maastrichtian to Holocene	Open marine shelf shales	Up to 2 (locally richer around Amazon)	Poor	Type II and III amorphous organic matter	$CaCO_3$ 6–20%, sulphur < 0.7%

The migration pathway will depend on the formation of salt welds or faults cutting through thin salt. Where the salt is thickest in the Santos and Espírito Santo Basins this may cause a problem. Where the salt is thin, as in the Campos Basin, salt welds are common due to dissolution and mechanical salt withdrawal. Growth faults are developed above the salt due to gravity sliding and oil can migrate up these faults into post-salt reservoirs.

The Cenomanian black shales are not as well developed on the Brazilian margin compared to the African margin (Arai 1988; Mello & Maxwell 1991). This is reflected in the present-day organic productivity in the South Atlantic, which is highly asymmetric and biased to the African margin (Fig. 6). If the present-day productivity is similar to that in the past, open marine organic-rich shales would be expected only in the extreme northern and southern parts of Brazil.

Localized Tertiary source rocks may be developed in the Amazonas Delta where land plant material has been washed down from the Amazon Jungle. DSDP cores indicate there are background TOC levels of 1% in the fine-grained sediments on the Amazon Fan at the present day (Flood *et al.* 1995).

The only basin which does not appear to have good mature source rocks in the Neocomian or Aptian sequence is the Pelotas Basin, which may have been maintained above sea level as a volcanic highland until late in the rift phase. Reddened sub-aerial basalts and intercalated clastics have been recorded from this basin (Fontana 1987). However, the giant Kudu gas field on the opposing African margin (Enclosure 1) is apparently sourced by Aptian marine shales (Davies & van der Spuy 1990) and this remains a possibility for sourcing in the Pelotas Basin.

Reservoirs

Syn-rift (Berriasian–Aptian)

In the Sergipe–Alagoas and Bahia Basins early rift sandstones of the Upper Jurassic to Early Cretaceous Sergi Formation constitute good reservoirs. In the syn-rift sequence coarse fan sandstones on the margins of the basins may be good reservoirs and turbidite sandstones in the deeper parts of the rift basins are present (e.g. Recôncavo and Almada Basins). In the Campos Basin the most attractive syn-rift reservoirs are coquina limestone banks of the Lagoa Feia Formation which are of limited extent (Dias *et al.* 1988).

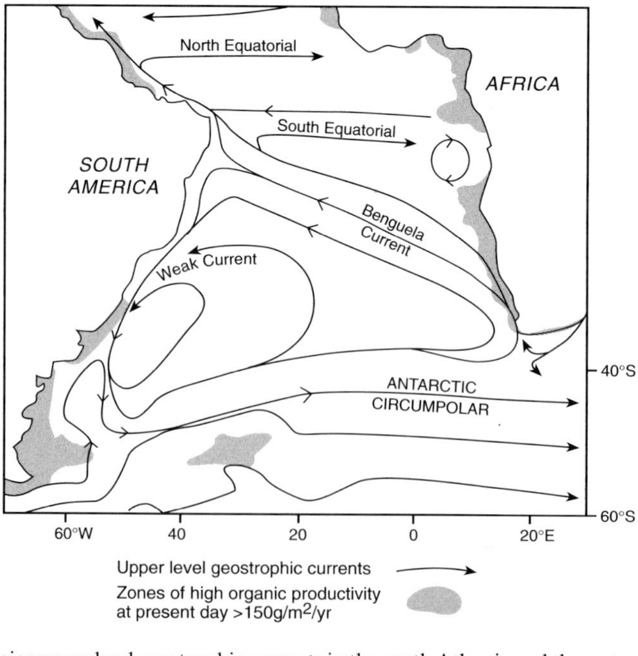

Fig. 6. Map of the major upper level geostrophic currents in the south Atlantic and the zones of high organic productivity at the present day. The present-day distribution of organic productivity is independent of the width of the margin and is mainly controlled by the ocean circulation patterns, such as the localized circular current which has induced a low productivity zone in the central part of the African margin around Angola. Compiled from data in Berger & Wefer (1996).

Albian

A shallow marine carbonate bank developed along most of the margin during the Albian, these Albian rocks can retain a reasonable porosity (*c.* 20%) and are producing reservoirs in the Campos Basin. Along the Equatorial margin carbonate production rates were higher and carbonates were deposited from the Cretaceous to the present day, reaching a thickness of 6 km in the Pará–Maranhão Basin. Elsewhere, they were eventually drowned by a prograding clastic wedge fed in by major rivers which managed to breach the uplifted flanks of the Atlantic. Albian turbidites produce oil in the Albacora Field in the Campos Basin.

Late Cretaceous–Tertiary

The rivers which delivered the detritus to the margin switched outlets so that localized depocentres of different ages are spread along the margins (e.g. Karner & Driscoll 1999). For example, the main depocentre in the Santos Basin developed in Cretaceous sediments with up to 7 km of Cretaceous strata imaged on seismic data, whereas the Cretaceous rocks were much thinner in the Campos Basin, with only 1–2 km developed. The main sedimentary input must have switched during the Tertiary so that sediments were preferentially deposited in the Campos Basin, whereas the Tertiary sedimentation was relatively thin in the Santos Basin (1 km).

Late Cretaceous turbidite sandstones in the Campos and Santos Basins were produced during an overall transgression, and they were probably deposited during temporary falls in sea level. However, comparison of the timing of the turbidite deposits in relation to the global sea-level curve indicates that local tectonics and sediment supply were probably more important than global sea-level changes (Bruhn & Walker 1995).

The size of the megaturbidite lobes will depend on the size of slope failures which occur along the shelf break, with the fan most likely to become unstable during sea-level falls or rapid depositional events. Hence, deepwater turbidite lobes should get larger through time as the shelf break prograding clastic wedges increases in size. This is supported by the Albacora Field data that indicate a maximum lobe size of 6.6 km in the Albian, increasing to 7.7 km in the Eocene, 19.6 km in the Oligocene and 20.3 km in the Miocene (Peres 1990; Guardado *et al.* 1989; Fig. 8). The turbidites in the Campos Basin are very clean and this has been attributed to ponding of sands in the São Tomé Low, which were winnowed before being deposited further downslope and then probably winnowed again by contourites (Peres 1990; Carminatti & Scarton 1991).

The Amazonas and Rio Grande Cones were deposited from Oligocene times, and this may correlate with a marked phase of erosion in the Andes. Both of these deltas probably contain sandstones in channels in the upper part of the fan and then as mouth lobes and overbank crevasse splay deposits in the mid and lower fan regions. Good quality sandstones have been encountered in shallow DSDP wells on the Amazonas Fan (Flood *et al.* 1995).

Hydrocarbon plays

Syn-rift plays

Narrow margins have good potential for rotated syn-rift fault-block plays, as these are relatively shallowly buried on narrow margins and most of the clastic sediment has bypassed the shelf (Fig. 7b). Syn-rift rotated fault blocks will also be an attractive target in the outer shelf area of wide margins, where the prograding clastic wedges or carbonate banks begin to thin oceanward, so that the syn-rift source and reservoir rocks are not too deeply buried (Fig. 7a). The presence of source rocks in close proximity to reservoirs is probable in most syn-rift settings and prediction of reservoir distribution is the main problem as the seismic data is poorer in the deep syn-rift section. Syn-rift plays are more attractive in the outer shelf area of wide margins and along unexplored narrow margins close to shore. These deeper rotated fault blocks may be attractive exploration prospects if early syn-rift sandstone reservoirs and deep lacustrine source rocks are present.

Albian carbonate and sandstone plays

Albian–Recent carbonate banks were developed along most of the Brazilian margin. These carbonates were structured by salt movement or gravity sliding on salt or overpressured shale of Aptian age. Sourcing of the carbonates was produced from Neocomian, Aptian or Albian shales. Top seals were produced by thick Cretaceous shales, which are present along most parts of the margin. The carbonate banks trapped clastic deposits on their landward side and sandstone reservoirs of Albian age could have been developed (Fig. 7a). In the Pará–Maranhão and Foz de Amazonas Basins, the carbonates were developed from Albian through to the Recent and top seals will be a problem in these areas.

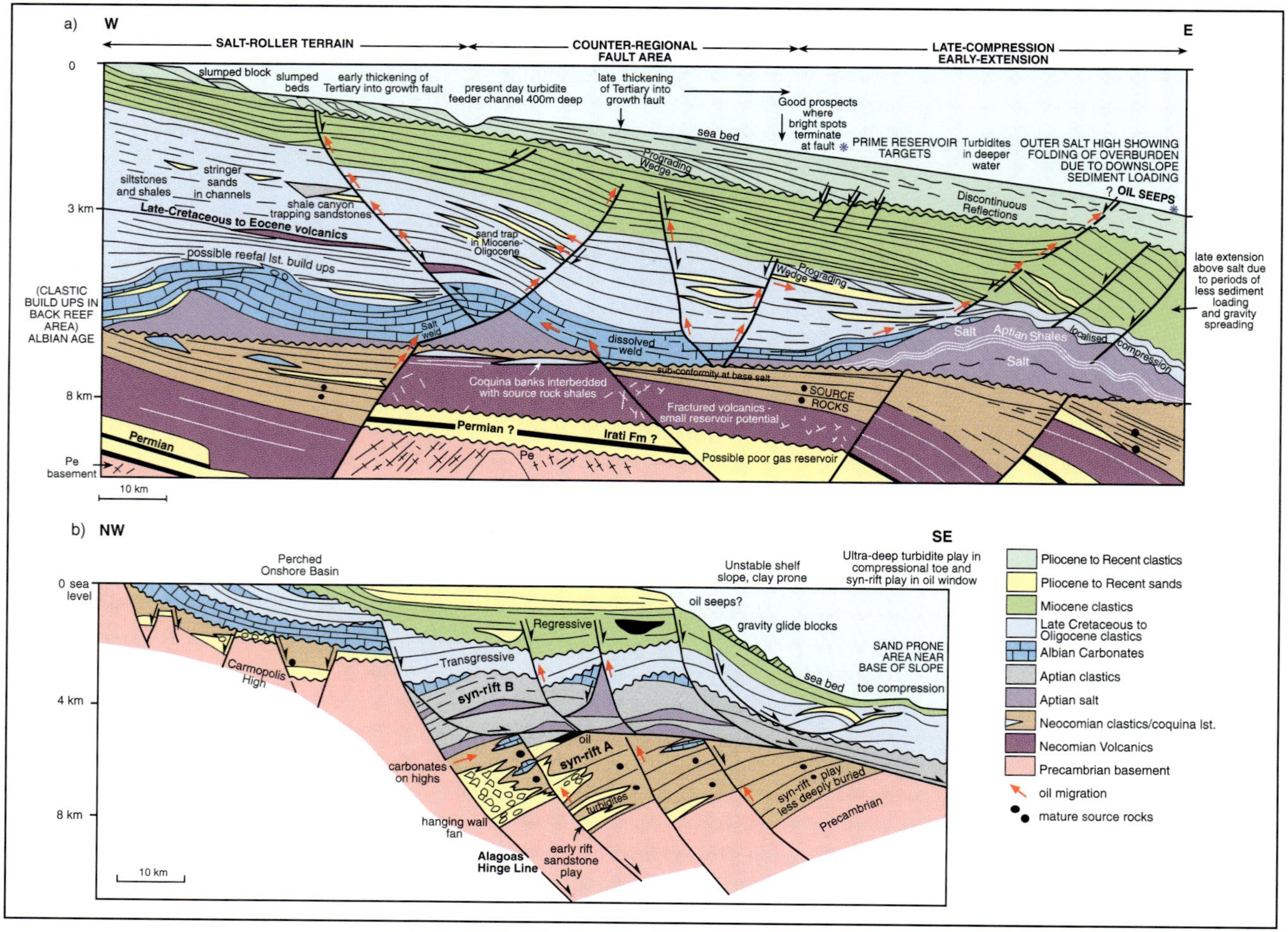

Fig. 7. Schematic cross-sections showing the main hydrocarbon play types along the Brazilian Atlantic margin. (**a**) Wide margin, deepwater Campos Basin, based on sections in Figueiredo & Martins (1990) and Mohriak *et al.* (1995), (**b**) narrow margin, Sergipe–Alagoas, partly based on Lana (1990).

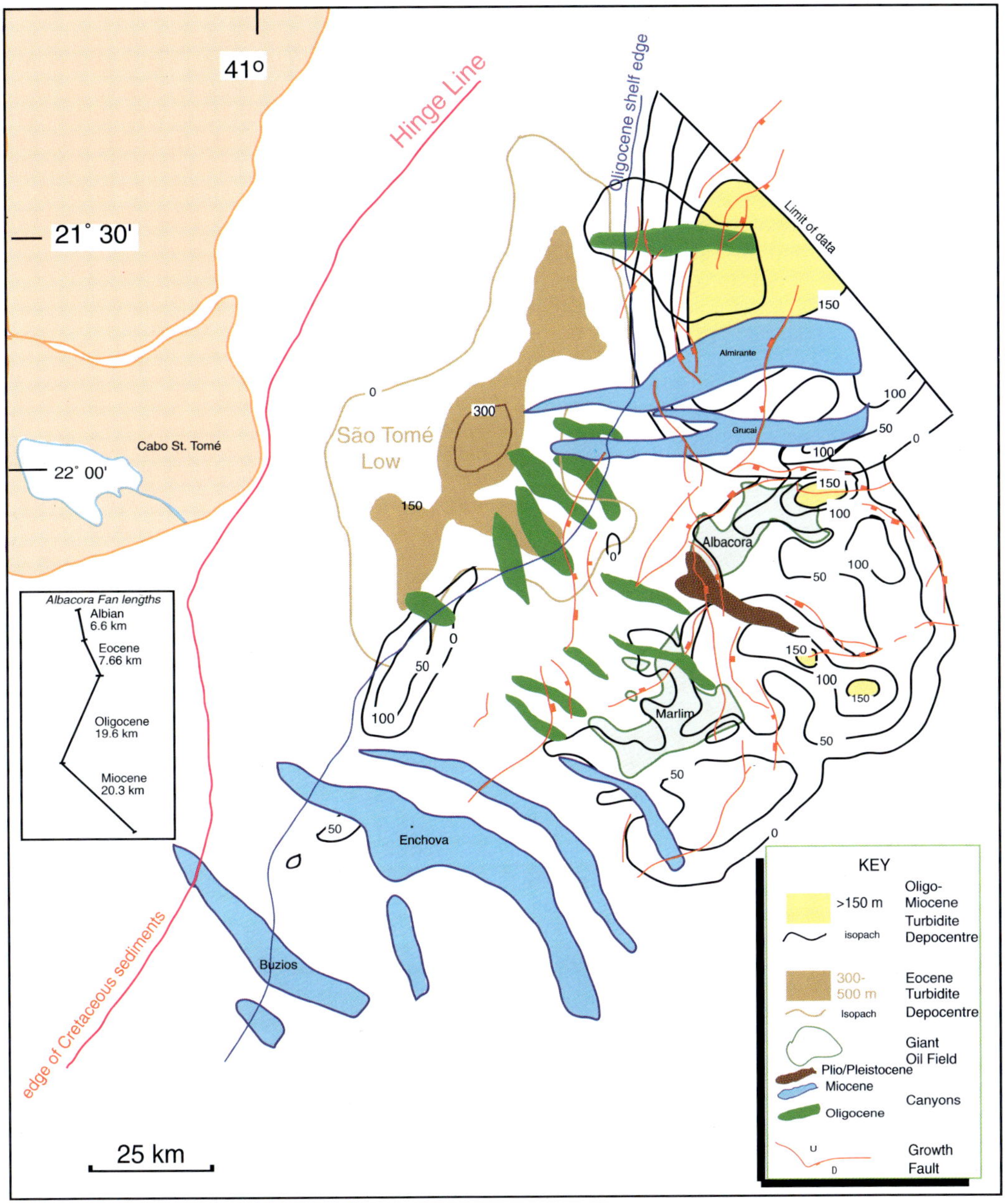

Fig. 8. Map of large turbidite lobes and feeder canyons known in deepwater > 500 m. Note the size of the lobes increased from Albian to Oligocene times in the Albacora Field (see inset). Canyon positions from Peres (1990); fault positions from Guardado *et al.* (1989).

Drift phase Late Cretaceous–Tertiary turbidite plays

The slope of the margin generally dictates how much sediment is lodged on it. As turbidites generally maintain their momentum on surface slopes of 2° or more they are unlikely to be deposited in depths < 3000 m in those narrow margin basins where the surface topography exceeds 2° down to the 3000 m isobath (e.g. Jacuípe, Pernambuco). The unconstrained sand lobes will thus be deposited in > 3000 m water. Hence, the narrow margins are not preferred sites for exploration of turbidites in < 2000 m of water. Turbidites tend to be trapped in lows produced by salt withdrawal on wide margins. These can either pond in faulted grabens formed by massive salt withdrawal, as in the northern Santos and southern Campos Basin (Demercian *et al.* 1993), or on the landward side of gently flexed rim synclines against salt swells and diapirs (Campos Basin). Attractive turbidite reservoir horizons can be developed at any stage from Albian to Miocene, and these may be revealed by bright spots on seismic sections.

A prerequisite for good deepwater turbidite plays is that the sandstones are clean, well-sorted and highly connected. Campos Basin turbidites are exceptionally good sandstone reservoirs. Their high quality has been attributed to wave washing of sandstones trapped in the São Tomé low in Eocene times (Fig. 8). They were then redeposited by turbidite flows further downslope and further winnowing of the sandstone lobes was produced by contourites (Carminatti & Scarton 1991).

Many compressional structural traps have been identified in deepwaters which are produced by gravity gliding further upslope (e.g. Barreirinhas, Azevedo 1991; Santos, Cobbold & Szatmari 1991; Rio Grande Cone, Pelotas, Fontana 1990*a*). These compressional structures may cause inversion of previous lows which preferentially accumulated thick turbidite sandstones. They also produce topographic highs at the sea bed which can preferentially trap turbidite sandstones behind the fold crests.

Salt welds are required to allow syn-rift oils to migrate up into Tertiary and Cretaceous reservoirs. Salt covered highs, such as the Cabo Frio area, are attractive because the salt provides effective detachment and structuring of the overlying reservoirs, and sufficient withdrawal to provide migration into Post-Salt reservoirs. Drift phase turbidite plays require active faulting after deposition of a thick claystone seal which is not breached. These faults can either be produced by gravity gliding of a prograding wedge above the Aptian salt (Campos and Santos Basins, Fig. 7a) or overpressured Tertiary claystones (Amazonas Cone and Rio Grande Cone in the Pelotas Basin, Fontana 1990*a*); or by differential compaction above large basement faults.

Tertiary age turbidite feeder channels have been identified on the shallow shelf areas in a number of other basins [Sergipe–Alagoas, Cainelli 1994; Northern Potiguar (Aracati), Gouyet 1988]. However, no significant sandstone discoveries have been made in the adjacent deepwaters as yet, except for a small Tertiary turbidite sandstone (1700 m water depth) in the southern part of the Sergipe–Alagoas Basin.

Conclusion

Large turbidite prospects have still to be drilled in the Campos Basin and in more untested basins to the north and south. These constitute prime targets in present-day water depths of 500–3000 m. The turbidite flows are largest in unconstrained lobes. It is this target which has still to be explored in ultradeep water off the narrow margins of Brazil where a next generation of major oil fields may be found in water depths > 3 km. Identification of large unexplored turbidite lobes with appreciable seismic impedance is the major task of the oil industry at present. However, attractive syn-rift plays remain to be explored in both shallow and deepwater.

Brian Frost, Luciano Magnavita and Dave G. Roberts are thanked for their useful reviews of this paper.

References

AKANDE, S. O., OJO, O. J., GEBHARDT, H. & ERDTMANN, B. D. 1997. Paleoenvironments, source rocks and thermal maturation studies of the Benue Trough Rift Basins, Nigeria. American Association Petroleum Geologists, Hedberg Conference, *Petroleum Systems of the South Atlantic Margin*, Rio de Janeiro, Nov. 1997, Abstract volume, 1–3.

ARAI, M. 1988. Geochemical reconnaissance of Mid-Cretaceous anoxic event in the Santos Basin. *Revista Brasileira Geociências*, **18**, 273–282.

AUSTIN, J. A. & UCHUPI, E. 1982. Continental to oceanic transition of Southwest Africa. *AAPG Bulletin*, **66**, 1328–1347.

AZEVEDO, R. P. 1991. *Tectonic evolution of Brazilian equatorial continental margin basins*. PhD Thesis, University of London, UK.

——, GOMIDE, J. & VIVIERS, M. C. 1987. Geohistoria da Bacia de Campos – Brasil; do Albiano ao Maestrichtiano. *Revista Brasileira Geociências*, **17**, 139–146.

BELTRAMI, C. V., ALVES, L. E. M. & FEIJÓ, F. J. 1994. Bacia do Ceará. *Boletim Geociências Petrobrás*, **8**, 117–126.

BERGER, W. H. & WEFER, G. 1996. Central themes of South Atlantic circulation. *In*: WEFER, G., BERGER, W. H., SIEDLER, G. & WEBB, D. J. (eds) *The South Atlantic: present and past circulation*. Springer, Berlin, 1–11.

BRAUN, J. & BEAUMONT, C. 1989. A physical explanation of the relation between flank uplifts and the break-up unconformity at rifted continental margins. *Geology*, **17**, 760–764.

BRUHN, C. H. L. & WALKER, R. G. 1995. High resolution stratigraphy and sedimentary evolution of coarse-grained canyon-filling turbidites from the Upper Cretaceous transgressive megasequence, Campos Basin Offshore Brazil. *Journal of Sedimentary Research*, **B65**, 426–442.

CAINELLI, C. 1992. *Sequence stratigraphy, canyons and gravity mass-flow deposits in the Piaçabuçu Formation, Sergipe Alagoas basin, Brazil.* PhD Thesis, University of Texas, USA.

—— 1994. Shelf processes and canyon/channel evolution controlling turbidite systems: examples from the Sergipe–Alagoas Basin, Brazil. *GCSSEPM, 15th Annual Research Conference, Proceedings, Submarine fans and turbidite systems*, 39–50.

CARMINATTI, M. & SCARTON, J. C. 1991. Seismic stratigraphic aspects of the Oligocene sequences in the Campos Basin, Offshore Brazil: an overview. *In*: LINK, M. & WEIMER, P. (eds) *Seismic Facies and Sedimentary Processes of Submarine Fans and Turbidite Systems*. Springer, Berlin, 241–246.

CHANG, H. K. & KOWSMANN, R. 1991. Significado dos diques do Arco de Ponta Grossa na abertura diferencial do Atlântico do Sul. *Simposio Nacional de Estudos Tectonicos*, Universidade Estadual Paulista, Abstracts volume, 77–78.

——, —— & FIGUEIREDO, A. M. F. 1988. New concepts of the development of the East Brazilian marginal basins. *Episodes*, **11**, 194–202.

——, ——, —— & BENDER, A. A. 1992. Tectonics and stratigraphy of the East Brazil rift system: an overview. *Tectonophysics*, **213**, 97–138.

COBBOLD, P. R. & SZATMARI, P. 1991. Radial gravitational gliding on passive margins. *Tectonophysics*, **188**, 249–289.

DAVIES, C. P. N. & VAN DER SPUY, D. 1990. Chemical and optical investigations into hydrocarbon source potential and thermal maturity of the Kudu 9A-2 and 9A-3 boreholes. *Communications Geological Survey Namibia*, **6**, 49–58.

DAVISON, I. 1991. Brazil's many sedimentary basins offer attractive exploration targets. *Oil and Gas Journal*, **August 5**, 52–55.

—— 1997. Narrow and wide margins of the Brazilian South Atlantic. *Journal of the Geological Society of London*, **154**, 471–476.

DEMERCIAN, S., SZATMARI, P., COBBOLD, P. & COELHO, D. F. 1993. Style and pattern of salt diapirs due to thin-skinned gravitational gliding, Campos and Santos basins, offshore Brazil. *Tectonophysics*, **228**, 393–344.

DIAS, J. L. 1994. Stratigraphy, sedimentology and volcanism of the Lower Cretaceous rift phase along the Eastern Brazilian continental margin. *International Sedimentological Congress*, Recife, **14**, S1-1–S1-2.

——, OLIVEIRA, J. Q. & VIEIRA, J. C. 1988. Sedimentological and stratigraphic analysis of the Lagoa Feia Formation, rift phase of the Campos Basin, offshore Brazil. *Revista Brasileira de Geociências*, **18**, 252–260.

DIAS-BRITO, D. 1987. A bacia de Campos no Meso Cretáceo: uma contribuição á paleoceanografia do Atlântico Sul primitivo. *Revista Brasileira Geociências*, **17**, 162–167.

FALVEY, D. A. 1974. The development of continental margins in plate tectonic theory. *Australian Petroleum Exploration Journal*, **14**, 95–106.

FEIJÓ, F. J. 1994*a*. Bacias de Sergipe e Alagoas. *Boletim Geociências Petrobrás*, **8**, 149–162.

—— 1994*b*. Bacia de Barreirinhas. *Boletim Geociências Petrobrás*, **8**, 103–110.

FIGUEIREDO, A. M. F. 1985. Geologia das Bacias Brasilieras. *Schlumberger WEC Conference*, Rio de Janeiro, 1–38.

—— & MARTINS, C. C. 1990. 20 anos de exploração da Bacia de Campos e o sucesso nas águas profundas. *Boletim Geociências Petrobrás*, **4**, 105–123.

FLOOD, R. D., PIPER, D. J. W., KLAUS, A. *ET AL.* 1995. Leg Synthesis. *Proceedings of Ocean Drilling Program Initial Reports*, **155**, 17–21.

FODOR, R. V., MCKEE, E. H. & ASMUS, H. E. 1983. K–Ar ages and the opening of the South Atlantic: basaltic rock from the Brazilian margin. *Marine Geology*, **54**, M1–M8.

FONTANA, R. L. 1987. *Desenvolvimento termo-mecánico da bacia de Pelotas e parte sul da plataforma de Florianópolis*. Masters Thesis, Universidade Federal de Ouro Preto, Brazil.

—— 1990*a*. Investigações geofisicas preliminares sobre o Cône do Rio Grande-Bacia de Pelotas-Brasil. *Acta Geologica Leopoldensia, São Leopaldo*, **13**, 161–170.

—— 1990*b*. Desenvolvimento termo-mecánico da Bacia de Pelotas e parte sul da plataforma de Florianopolis. *In*: RAJA GABAGLIA DE, G. P. & MILANI, E. J. (eds) *Origem e evolução de bacias sedimentares.* Gavea, Rio de Janeiro, 377–400.

GALLAGHER, K., HAWKESWORTH, C. J. & MANTOVANI, M. S. M. 1995. Denudation, fission track analysis and the long term evolution of passive margin topography: Application to the southeast Brazilian margin. *Journal South American Earth Sciences*, **8**, 65–78.

GOMES, J. R. C. 1981. *Projecto Radambrasil, Folhas SB 24–25, Jaguaribe e Natal, Geologia, geomorfologia, pedologia, vegetação e uso potencial da terra.* Ministerio de Minas e Energia, Secretaria Geral, Rio de Janeiro, 27–30.

GOUYET, S. 1988. *Evolution tectono-sédimentaire des marges Guyanaise et nord Bresilienne au cours de l'ouverture de l'Atlantique du sud.* PhD Thesis, Université de Pau, France.

GRADSTEIN, F. M., AGTERBERG, F. P., OGG, J. G., HARDENBOL, J., VEEN P., VAN, THIERRY, J. & HUANG, Z. 1994. A Mesozoic timescale. *Journal of Geophysical Research*, **99**, 24051–24074.

GRAND, S. P., HILST, VAN DER, R. D. & WIDIYANTORO, S. 1997. Global seismic tomography: a snapshot of convection in the Earth. *GSA Today*, **7**, 1–7.

GUARDADO, L. R., GAMBOA, L. A. P. & LUCCHESI, C. F. 1989. Petroleum geology of the Campos Basin, Brazil, a model for a producing Atlantic type Basin. *In*: EDWARDS, J. D. & SANTAGROSSI, P. A. (eds) *Divergent/Passive Margin Basins.* American Association of Petroleum Geologists Memoir, **48**, 3–80.

JAY, W. W., WOLD, C. N. & VOIGT, S. 1994. Equatorial gateways in the Atlantic meridional and zonal. *International Sedimentological Congress*, **14**, S9-4–S9-6.

KARNER, G. D. & DRISCOLL, W. 1999. Tectonic significance of syn- and post-rift sedimentary packages comprising the West African and eastern Brazilian margins. *This volume*

LANA, M. C. 1990. Bacia de Sergipe–Alagoas: uma hipótese de evolução tectono-sedimentar. *In*: RAJA GABAGLIA, G. P. & MILANI, E. J. (Coordinators) *Origem e Evolução das Bacias Sedimentares*. Gavea, Rio de Janeiro, 311–332.

LEYDEN, R. 1976. Salt distribution and models for the Eastern Brazil margin. *Anais Academia Brasileira Ciências*, **48**, (supplement), 159–168.

MACEDO, J. M. 1987. *Evolução estrutural da Bacia de Santos e áreas continentais adjacentes.* Masters Thesis, Universidade Ouro Preto, Brazil.

MAGNAVITA, L. P., DAVISON, I. & KUSZNIR, N. 1994. Uplift and erosion history of Recôncavo–Tucano–Jatobá Basin, North East Brazil. *Tectonics*, **13**, 367–388.

MCHARGUE, T. R. 1990. Stratigraphic development of proto-south Atlantic rifting in Cabinda Angola – a petroliferous lake basin. *In*: KATZ, B. (ed.) *Lacustrine Basin Exploration, Case Studies and Modern Analogs* AAPG Memoir, **50**, 307–326.

MATOS, R. M. D. 1992. The northeast Brazilian rift system. *Tectonics*, **11**, 776–791.

MELLO, M. R. 1988. *Geochemical and molecular studies of the depositional environments of source rocks and their derived oils from the Brazilian marginal basins.* PhD Thesis, University of Bristol, UK.

—— & MAXWELL, J. R. 1991. Organic geochemical and biological marker characterisation of source rocks and oils derived from lacustrine environments in the Brazilian continental margin. *In*: KATZ, B. (ed.) *Lacustrine Basins: Case Studies and Modern Analogs* AAPG Memoir, **50**, 77–95.

——, GAGLIANONE, P. C., BRASSELL, S. C. & MAXWELL, J. R. 1988. Geochemical and biological marker assessment of depositional environments using Brazilian offshore oils. *Marine and Petroleum Geology*, **5**, 205–223.

——, KOUTSOUKOS, E. A. M., MOHRIAK, W. U. & BACOCCOLI, G. 1994. Selected petroleum systems in Brazil. *In*: MAGOON, B., DOW, L. & WALLACE, G. (eds) *The Petroleum System; From Source to Trap.* AAPG Memoir, **60**, 499–512.

——, ——, HART, M. B., BRASSELL, S. C. & MAXWELL, J. R. 1989. Late Cretaceous anoxic events in the Brazilian continental margin. *Organic Geochemistry*, **14**, 529–542.

MOHRIAK, W. U. 1995. Salt tectonics structural styles: contrasts and similarities between the south Atlantic and the Gulf of Mexico. *GCSSEPM 16th Annual Research Conference*, Houston 4–6 Dec., 1995, 177–191.

——, MACEDO, J. M., CASTELLANI, R. T. *ET AL.* 1995. Salt tectonics and structural styles in the deep water province of the Cabo Frio region, Rio de Janeiro, Brazil. *In*: JACKSON, M. P. A., ROBERTS, D. G. & SNELSON, S. (eds) *Salt Tectonics: A Global Perspective*. AAPG Memoir, **65**, 273–304.

NETTO, A. S. T., FILHO, J. R. W. & FEIJÓ, F. J. 1994. Bacias de Jacuípe, Camamu e Almada. *Boletim Geociências Petrobrás*, **8**, 173–184.

NÜRNBERG, D. & MÜLLER, R. D. 1991. The tectonic evolution of the south Atlantic from late Jurassic to present. *Tectonophysics*, **94**, 27–53.

O'CONNOR, J. M. & DUNCAN, R. A. 1990. Evolution of the Walvis Ridge–Rio Grande Rise hot spot system: implications for African and South American plate motions over plumes. *Journal of Geophysical Research*, **95**, 17 474–17 502.

PEREIRA, M. J. 1992. Considerações sobre a estratigrafia do Cenomaniano – Santoniano em algumas bacias marginais Brasileiras e sua implicação na história e sedimentaçaõ da margem continental. *Boletim Geociências Petrobrás*, **6**, 171–176.

PERES, W. E. 1990. *Seismic-stratigraphic study of the Oligocene–Miocene shelf fed turbidite systems of the Campos Basin, Brazil.* PhD Thesis, University of Texas, Austin.

—— 1993. Shelf-fed turbidite system model and its application to the Oligocene deposits of the Campos Basin, Brazil. *AAPG Bulletin*, **77**, 81–101.

PONTE, F. C. 1971. Evolução paleontologica do Brasil oriental e Africa ocidental. *CEPG-IV, Petrobrás*, SETUP–Salvador Brazil.

RABINOWITZ, P. D. & LABREQUE, J. 1979. The Mesozoic South Atlantic and evolution of its continental margins. *Journal of Geophysical Research*, **B84**, 5973–6002.

RANGEL, H. D., MARTINS, F. A. L., ESTEVES, F. R. & FEIJÓ, F. J. 1994. Bacia de Campos. *Boletim Geociências Petrobrás*, **8**, 203–218.

REGALI, M. S. P. 1989. A idade dos evaporitos da plataforma continental do Ceará, Brasil, e sua relação com os outros evaporitos das bacias nordestinas. *Boletim de Instituto de Geociências*, Universidade de São Paulo, publicação especial, **7**, 139–143.

RENNE, P. R., GLEN, J. M., MILNER, S. C. & DUNCAN, A. R. 1996*a*. Age of Etendeka flood volcanism and associated intrusions in southwestern Africa. *Geology*, **24**, 659–662.

——, DECKART, K., ERNESTO, M., FERAUD, G. & PICCIRILLO, E. M. 1996*b*. Age of the Ponta Grossa Dike Swarm (Brazil) implications to Paraná flood volcanism. *Earth and Planetary Science Letters*, **144**, 199–211.

SANDWELL, D. T. & SMITH, W. H. F. 1995. *Map of Marine Gravity Anomaly from Satellite Altimetry*. Scripps Institute of Oceanography, La Jolla, California.

SANTOS, C. F., GONTIJO, R. C., ARAÚJO, M. B. & FEIJÓ, F. J. 1994. Bacias de Cumuruxatiba e Jequitinhonha. *Boletim Geociências Petrobrás*, **8**, 185–190.

SIAL, A. N. 1976. The post-Paleozoic volcanism of northeastern Brazil and its tectonic significance. *Anais Academia Brasileira Ciências*, **48**, (supplement), 299–311.

——, LONG, D. E., PESSOA, D. A. R. & KAWASHITA, A. K. 1981. Potassium–argon ages and strontium isotope geochemistry of Mesozoic and Tertiary basaltic rocks, northeastern Brazil. *Anais Academia Brasileira Ciências*, **53**, 112–115.

STANDLEE, L. A., BRUMBAUGH, W. R. & CAMERON, N. 1992. Controlling factors in the initiation of the South Atlantic Rift System. Proceedings 1e Colloque de Stratigraphie et de Paléogeographie des bassins sedimentaries ouest-africains. *Libreville, Gabon, 6–8 May. Bulletin des Centres de Recherches Exploration–Production Elf-Aquitaine, Memoire*, **13**, 141–152.

SZATMARI, P., MILANI, E., CONCEIÇÃO, J. C. LANA, M. C. & LOBO, A. P. 1985. How South Atlantic rifting affects Brazilian oil reserves distribution. *Oil and Gas Journal*, **83**, 107–114.

TEISSERENC, P. & VILLEMIN, J. 1990. Sedimentary basin of Gabon – Geology and Oil Systems. *In*: EDWARDS, J. D. & SANTAGROSSI, P. A. (eds) *Divergent/Passive Margin Basins*. AAPG Memoir, **48**, 117–199.

TURNER, S., REGELOUS, M., HAWKESWORTH, C. & MANTOVANI, M. 1994. Magmatism and continental break up in the South Atlantic: high precision ^{40}Ar–^{39}Ar geochronology. *Earth Planetary Science Letters*, **121**, 333–348.

UESUGUI, N. 1987. Posição estratigráfica dos evaporitos da Bacia de Sergipe–Alagoas. *Revista Brasileira Geociências*, **17**, 131–134.

VIEIRA, R. A. B., MENDES, M. P., VIEIRA, P. E., COSTA, L. A. R., TAGLIARI, C. V., BACELAR, L. A. P. & FEIJÓ, F. J. 1994. Bacias do Espírito Santo e Mucuri. *Boletim Geociências Petrobrás*, **8**, 191–202.

ZALÁN, P. V. 1984. *Tectonics and sedimentation of the Piauí–Camocin sub-basin, Ceará Basin, offshore northeastern Brazil.* PhD Thesis, Colorado School of Mines.

Petroleum habitats of the Atlantic margin of South Africa

ERIC H. A. JUNGSLAGER

Petroleum Licensing Unit, SOEKOR (Pty) Ltd, PO Box 307, Parow 7499, South Africa

Abstract: Petroleum system logic is applied along South Africa's western margin to improve understanding of the observed petroleum habitats and the evaluation of the area's potential for hosting profitable hydrocarbons. Three major groups of source rocks are identified and, consequently, the source-centred petroleum systems and complementary exploration plays form three major groups: (1) those sourced from syn-rift, Upper Jurassic–Neocomian, generally lacustrine source rock pods; (2) those sourced from Barremian–lower Aptian, generally marine, transitional succession source pods; (3) those possibly sourced from the Cenomanian–Turonian boundary, generally marine, drift succession source pods. These groups are further subdivided into individual petroleum systems in terms of specific individual source rocks, level of certainty, type of hydrocarbon and extent; and into individual exploration plays in terms of stratigraphic level of the reservoir and trap type. The geographic location of the known and actively generating oil and gas pods, whether in the syn-rift or transitional successions, is the major control on the observed distribution of petroleum habitats. The predominantly gas-bearing areas coincide with the main depocentre of the Orange Basin where the lower Aptian source rocks are in the gas window. More oil-bearing areas are predicted outside the depocentre, in deeper water, where the lower Aptian source rocks may become more oil prone and are possibly located in the oil window. These possibly more oil-prone areas coincide with major syn-sedimentary structures and hypothetical deepwater turbidite sandstones. They represent a new frontier for future exploration.

The aim of this paper is to review and explain the observed distribution of petroleum along the Atlantic margin of South Africa (Figs 1–13), and to identify possible new areas for future exploration (Figs 14–16). The concepts of 'petroleum system' (Perrodon 1983; Demaison & Huizinga 1994; Magoon & Dow 1994; Smith 1994; Magoon & Sanchez 1995) and exploration 'play' (White 1988; Allen & Allen 1990; Magoon & Sanchez 1995) are powerful tools that can help achieve this. With the paucity of geochemical biomarker fingerprinting data in this relatively poorly explored area, the possible relationships among the hydrocarbons, and between each hydrocarbon type and its source rocks, are inferred by applying petroleum system logic in the sense of Perrodon (1983), Demaison & Huizinga (1994) and Smith (1994), rather than by using the more rigorous approach of Magoon & Dow (1994).

General geological framework

An extensive literature dealing with the geological and petroleum aspects of this area already exists, e.g. McLachlan & McMillan (1979), Gerrard & Smith (1982), Dingle *et al.* (1983), Jaunich (1983*a*, *b*, *c*, *d*, *e*), Jungslager (1992, 1996), Light *et al.* (1992*a*, *b*, 1993*b*), Dingle (1993), Barton *et al.* (1993), Benson (1992), Broad & Mills (1993), Muntingh (1993), Muntingh & Brown (1993), Brown *et al.* (1995). To this must be added the literature on the conjugate continental margins, especially Argentina, the Falkland–Malvinas Plateau and Antarctica (Martin & Hartnady 1986; Uliana *et al.* 1989; Maslanyj *et al.* 1992; Light *et al.* 1993*a*, *b*; Keeley & Light 1993; Lawrence & Johnson 1995; Tankard *et al.* 1995; King *et al.* 1996; Richards *et al.* 1996*a*, *b*; Ben-Avraham 1997).

Figures 1, 2, 5 and 10 summarize the general geological framework and its relation to the known distribution of hydrocarbons. Briefly, the Orange Basin is delineated by the isopach map of the drift succession (Fig. 10) (Gerrard & Smith 1982). The underlying syn-rift succession displays generally isolated graben east of the medial hinge zone, and a large syn-rift wedge to the west between the medial hinge and the marginal ridge (Figs 2 and 5). The block-faulted syn-rift succession is followed by a ramp-like transitional succession which displays a gradual transgressive regime culminating in a major flooding event during early Aptian time. The lower Aptian shales are conveniently included in the transitional succession for the purposes of the petroleum systems analysis. These regional shales form a major source rock and a regional seal. They cap continental carrier beds below. Finally, a full-drift succession, with typical prograding Vailian sequences, is developed. This succession displays a largely structureless palaeoshelfal area in the east and a growth-faulted, toe-thrusted area to the west, above the Cenomanian–Turonian boundary source rocks. The Turonian Stage constitutes a regional source and seal capping shelfal carrier beds below (Fig. 2).

Jungslager, E. H. A. 1999. Petroleum habitats of the Atlantic margin of South Africa. *In:* Cameron, N. R., Bate, R. H. & Clure, V. S. (eds) *The Oil and Gas Habitats of the South Atlantic.* Geological Society, London, Special Publications, **153**, 153–168.

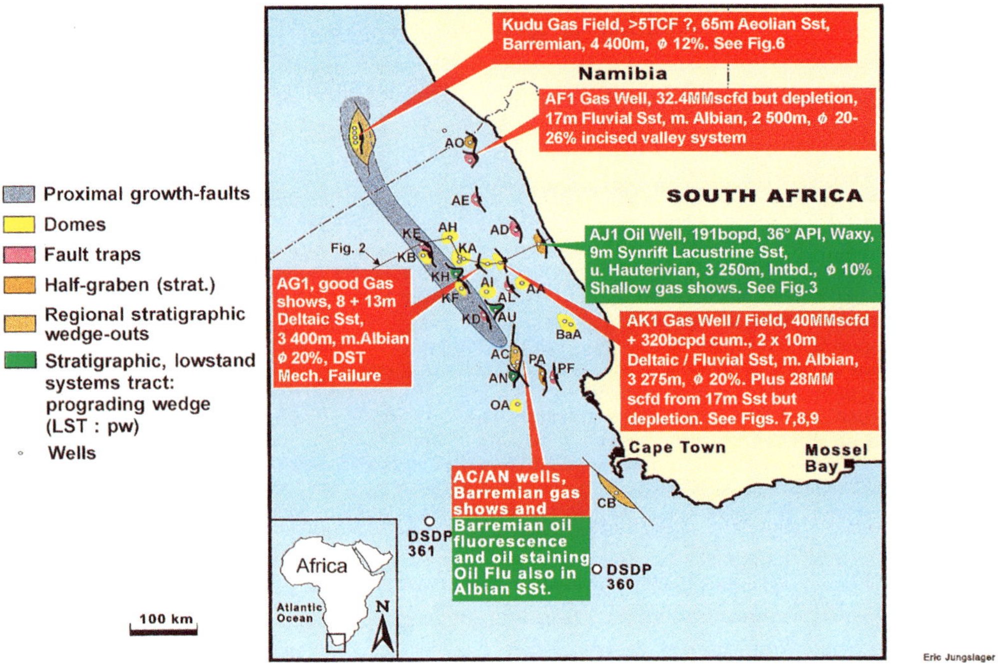

Fig. 1. The main petroleum discoveries and shows along South Africa's Atlantic margin. The types of traps tested historically are also shown. TCF, trillion cubic feet; MMscfd, million standard cubic feet per day; bopd, barrels of oil per day; bcpd, barrels of condensate per day; ϕ, porosity; Sst, sandstone; intbd., interbedded; cum., cumulatively; u, upper; m, middle.

Comparison of Fig. 2 with sections off West Africa and Brazil (e.g. Shannon & Naylor 1989; Carminatti & Scarton 1991) highlights a number of similarities among these areas: e.g. multiple hinge zones; multiple rifting events; a post-rift transitional interval (or late rift sag) between the active rift and the drift successions; a major flooding event during early Aptian time south of the Walvis–Rio Grande Ridge correlating with the evaporites of the Aptian Salt Basin north of the ridge; and syn-sedimentary growth faults and associated toe-thrusts. Notable differences are the absence of salt and carbonates in the study area.

Petroleum systems and complementary plays

Magoon & Dow (1994) define a petroleum system as 'a natural system that encompasses a pod of active source rock and all related oil and gas and which includes all the geological elements and processes that are essential if a hydrocarbon accumulation is to exist'. The genetic relationship among all the hydrocarbons of a petroleum system, and between them and their common source rock, is essential to the concept, and ideally is determined by geochemical analyses on samples of hydrocarbons and their source rocks in a petroliferous basin. Since the fingerprinting analyses are sparse in this poorly explored area, and only two liquid petroleum discoveries have to date been made (AJ1 and AK1 wells; Fig. 1) 'petroleum system logic' (Perrodon 1983; Demaison & Huizinga 1994; Smith 1994) is applied instead of the more rigorous approach of Magoon & Dow (1994), which may apply best to more mature petroleum provinces. Smith (1994) defines petroleum system logic as 'the thought process required to develop an integrated interpretation of the processes of petroleum generation, migration, and accumulation'. The more plausible genetic relationships between active source pods and their hydrocarbon accumulations are thus largely inferred here to develop a working model from strong circumstantial petroleum geological evidence using geochemical correlation data where available.

Magoon & Sanchez (1995) have proposed a new concept of exploration play (of undiscovered accumulations/prospects) which is complementary to a petroleum system and which 'shares the same pod of active source rock and at least some of the same plumbing system'. It differs from the traditional play concept centred on the reservoir-trap aspects regardless of whether the accumulations belong to the same petroleum system (i.e. originate from the same source pod).

Three major groups of source rocks are known and inferred in the area: (1) syn-rift source rocks of late Hauterivian age are known locally and syn-rift source rocks ranging from Late Jurassic to Neocomian age are inferred to occur regionally; (2) several Barremian age source rocks and an early Aptian age source rock are known; (3) a Cenomanian–Turonian boundary source rock is known (Barton *et al.* 1993; Muntingh 1993). They are directly related to the major stages of development of the margin and their resulting tectono-stratigraphic units: syn-rift, transitional and mature drift (Fig. 2).

Consequently, the source-centred petroleum systems and complementary exploration plays also form three major groups: (1) those sourced from mature pods of generally lacustrine, syn-rift source rocks, possibly ranging from Late Jurassic to Neocomian age; (2) those sourced from generally marine, mature source pods of Barremian to early Aptian age within the transitional interval; (3) those sourced from generally marine, mature source pods of Cenomanian–Turonian boundary age within the mature drift interval.

These major groups of petroleum systems and their complementary exploration plays are further subdivided in terms of specific individual source rocks where possible (e.g. Barremian v. lower Aptian individual source rock pods of the transitional interval), level of certainty (known, hypothetical or speculative), type of hydrocarbon (oil or gas) and extent.

Finally, individual trap-centred traditional plays are identified as targets for exploration within the limits of the above petroleum systems and their complementary plays on the basis of reservoir facies, stratigraphic level and type of trap. A systematic hierarchy is thus followed for analysing the distribution of petroleum in the basin fill.

In the following sections the petroleum systems, their complementary plays and traditional exploration plays are discussed, from the oldest generating source rocks upward through the stratigraphy.

Petroleum systems and complementary plays sourced from the syn-rift succession

Figures 2 and 5 illustrate the main tectonic and plate tectonic aspects of the syn-rift succession in the area during Late Jurassic–Neocomian pre-separation times. The presence of source rocks within a number of individual rift troughs

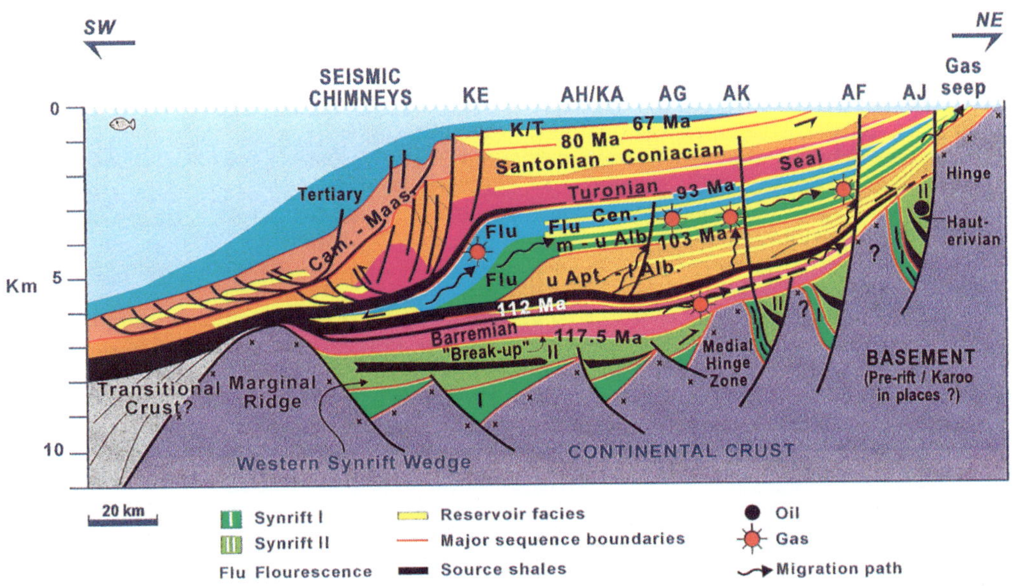

Fig. 2. Generalized composite stratigraphic cross-section through the Orange Basin depocentre, offshore South Africa, with the main petroleum aspects highlighted. The time scale of Haq *et al.* (1987*a*, *b*, 1988) is used. See Fig. 1 for location.

distributed over the south Gondwana area is clearly depicted. The occurrence of the source rocks is not continuous. This interval has only been targeted and significantly penetrated by three wells in the study area: AJ1, AO1 and PA1 (Fig. 1). These wells were stratigraphic tests drilled in the centre of three isolated half-graben. Three wells, BaA1 BaA2 and AC1, may have intersected an early syn-rift remnant, or possibly a pre-rift section. Two additional wells, AD1 and AF1, are interpreted to have intersected the uppermost portion of the syn-rift interval (Fig. 1).

The syn-rift fill is generally characterized by continental 'red beds' and igneous material. Only in the AJ half-graben have lacustrine sapropelic black shales and petroleum been encountered. All the above wells, except AC1, have tested relatively shallow, isolated graben at the extreme eastern margin of the basin (Figs 1 and 5). The only proven petroleum system related to the syn-rift succession is thus located in a single isolated trough.

The western syn-rift wedge (Fig. 2) has not been tested. The AC1 well intersected a localized igneous centre closely related to the medial hinge and a major fracture zone (Fig. 5) but failed to sample the wedge. There is no evidence to date of syn-rift-sourced hydrocarbons outside the syn-rift succession.

The individual petroleum systems and plays identified and inferred for the syn-rift succession are as follows.

Proven, late Hauterivian, syn-rift-sourced, lacustrine oil system and complementary play within the AJ graben. This petroleum system was proven in the AJ1 well within the syn-rift fill of an isolated half-graben (Figs 1–5) (Barton *et al.* 1993; Muntingh 1993). Although the find was non-commercial (Fig. 1) it is of particular significance since it confirms the presence of rich oil-prone lacustrine source rocks in the area of interest. The characteristics of the source rocks and the associated oil are shown in Figs 1 and 4. Thermal modelling indicates that the source rocks have only entered the top of the oil window, mainly as a result of the low geothermal gradient (2.5° C 100 m^{-1}). The source rocks are imaged seismically as relatively high amplitude, continuous reflectors (Fig. 3). Similar reflectors are seen in Fig. 3 toward the base of the AJ half-graben, possibly indicating additional, more mature source rocks, and thus possibly a hypothetical syn-rift petroleum system sourced from Upper Jurassic–Valanginian source rocks, encountered elsewhere in southwest Gondwana (Fig. 4).

Although Barton *et al.* (1993) have cautioned that igneous rocks may also be similarly imaged on seismic data, a high resolution airborne magnetic survey over the AJ graben has shown that large amounts of igneous material are probably absent when compared with a similar survey over the igneous PA half-graben (Fig. 1). If additional and more mature source rocks are indeed present, the AJ graben holds possibly larger oil volumes

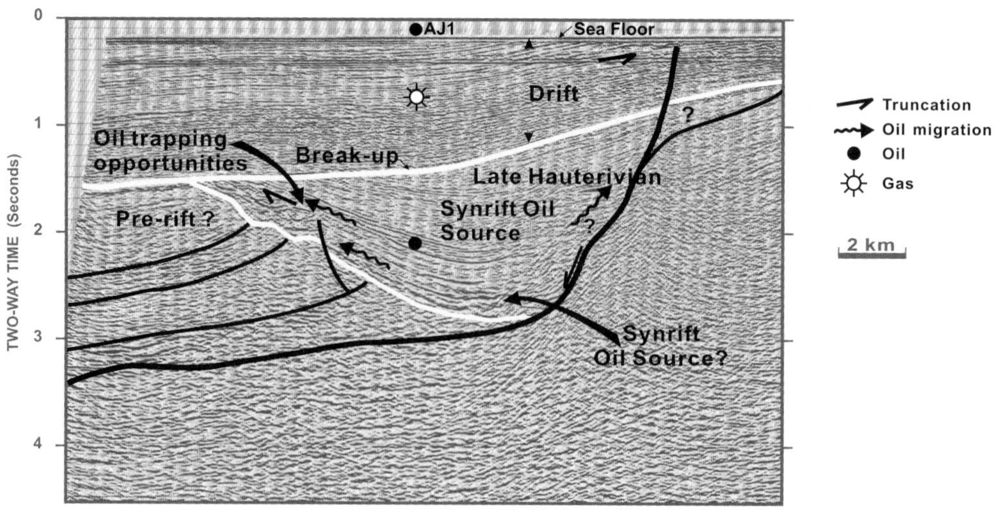

Fig. 3. Two-dimensional seismic line across the AJ half-graben (see Fig. 5 for location). The lacustrine source rocks intersected in the AJ1 well are the relatively continuous high amplitude reflections at *c.* 2.1 s two-way time. Possible additional source rocks may be imaged deeper in the undrilled part of the half-graben.

than intersected in the AJ1 well, especially in the up-dip play along the passive limb of the half-graben where structural trapping has been mapped, and where lacustrine or other continental sandstones are predicted (Fig. 3).

Hypothetical, Upper Jurassic–Neocomian, syn-rift-sourced oil system within a trend of graben, down-dip of AJ graben and intra-rift tilted fault-block play. Hypothetical petroleum systems and plays similar to those in the AJ graben may be developed in a trend of more deeply buried graben developed to the west and down-dip from the isolated AJ trough and east of the medial hinge line (Figs 2 and 5) where: (1) regular, high amplitude seismic reflectors similar to those in the AJ graben are seen on seismic profiles; (2) higher thermal maturity is expected; (3) large intra-rift 'buried-hill' structural traps have been mapped which may contain syn-rift reservoir facies forming a play to target in future exploration. Gas, rather than oil, may, however, be expected where these graben are deeply buried, depending on the geothermal gradient (e.g. Fig. 2).

Speculative, Upper Jurassic, syn-rift-sourced gas system within the western syn-rift wedge and its pinch-out play. Hypothetical syn-rift-sourced petroleum systems are predicted within the western syn-rift wedge west of the medial hinge line (Figs 2 and 5). The western syn-rift wedge has not been intersected. Seismic data, however, indicate a syn-rift origin (Fig. 14), with source shales of Late Jurassic–Neocomian age thus possibly developed (Figs 4 and 5).

In the portion of the margin underlying the main post-rift depocentre (Figs 4 and 10) any hydrocarbons generated would presently be within the gas window (Fig. 2). The main play is the pinch-out of the western syn-rift wedge between the medial hinge and the 'break-up' unconformity (Figs 2 and 5). Such a play would be located at depths > 5 km with associated concerns regarding poro-perm quality in continental sandstones (Fig. 2). The pinch-out traps against the medial hinge could, however, potentially be very large (Fig. 5).

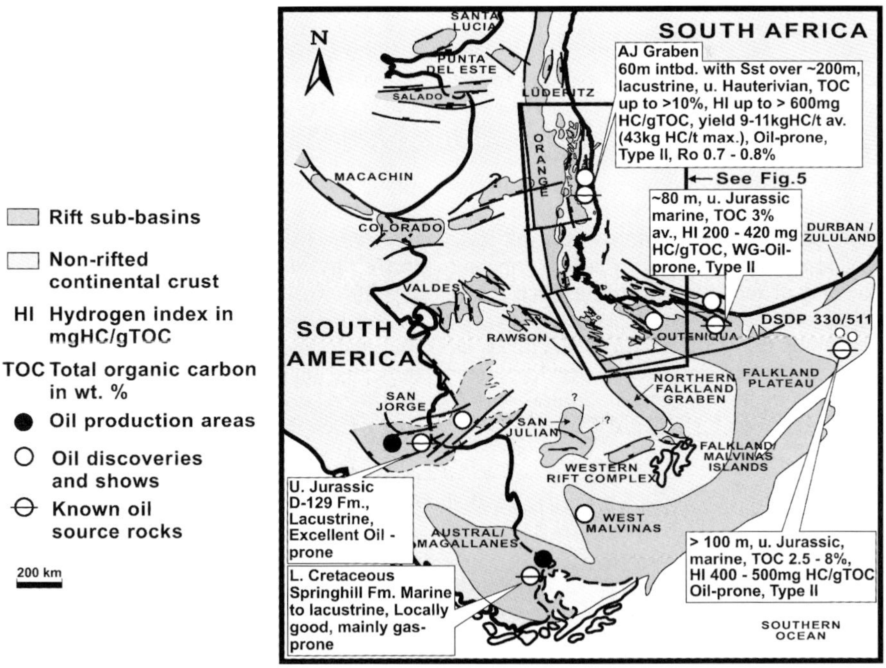

Fig. 4. Distribution of Late Jurassic to Early Cretaceous (Neocomian) rift basins within southwest Gondwana. Significant petroleum aspects are highlighted. The plate reconstruction is after De Wit *et al.* (1988). Other sources are: Ben-Avraham *et al.* (1997), Broad & Mills (1993), Burden (1992), Deroo *et al.* (1983), Keeley & Light (1993), King *et al.* (1996), Lawrence & Johnson (1995), Light *et al.* (1993a,b), Malan (1993), McMillan *et al.* (1997), Muntingh (1993), Richards *et al.* (1996*a*, *b*), Roux (1997), Tankard *et al.* (1995) and Uliana *et al.* (1989).

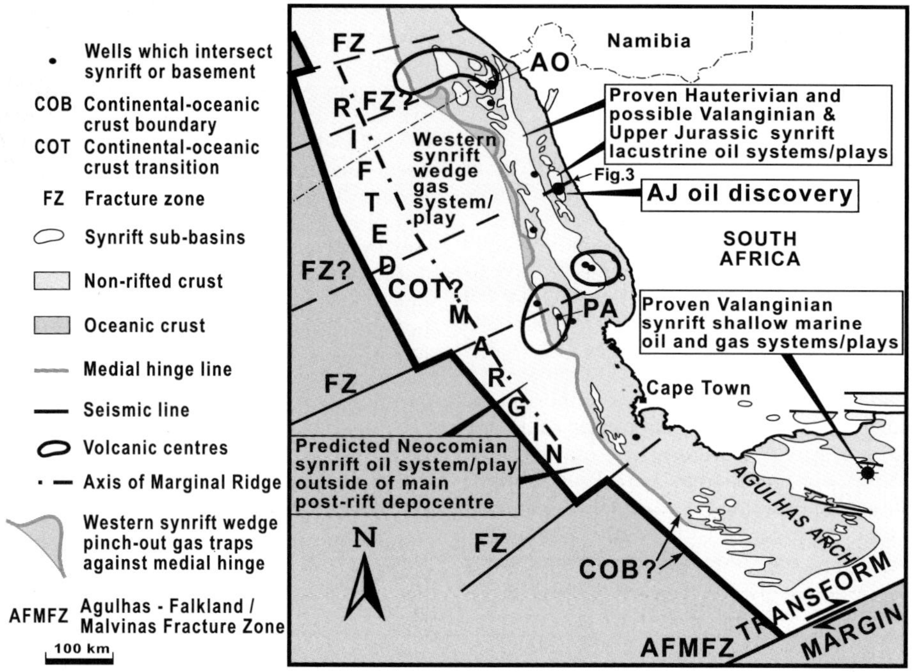

Fig. 5. Distribution of known and predicted syn-rift oil and gas systems and plays. Major tectonic elements are shown. See also Fig. 4.

Hypothetical, Upper Jurassic–Neocomian syn-rift oil system within the southern portion of the western syn-rift wedge and a tilted fault-block play. In the southern portion of the margin, outside of the main post-rift depocentre (Figs 5 and 10), the syn-rift succession is likely to be buried within the oil window (Fig. 10). By analogy with the producing Valanginian, syn-rift shallow marine sandstone play of the adjacent Bredasdorp Basin immediately to the east of the Agulhas Arch (Fig. 5; Broad & Mills 1993), the likelihood of oil accumulations may be increased within possible similar reservoir facies capping similar tilted fault-blocks on the western side of the arch.

Petroleum systems and plays sourced from the transitional succession

A restricted seaway was developed in the South Atlantic from at least the time of the end of active rifting to the start of continental drifting around the end of the Hauterivian Stage, and culminated in a period of major anoxia during the early Aptian stage. Several Barremian source rocks are developed, in addition to the culminating lower Aptian shales. These conditions were abruptly terminated by oxygenation in middle Aptian time (Bolli *et al.* 1978; Gerrard & Smith 1982; Katz & Pheifer 1982; Dingle *et al.* 1983; Clifford 1986; Herbin *et al.* 1987; De Wit *et al.* 1988; Jacquin & de Graciansky 1988; Benson 1990; Davies & van der Spuy 1990; McMillan 1990). It is suggested that this anoxic area of the Southern Atlantic be called the 'Aptian Anoxic Basin', by analogy with its northern correlative the 'Aptian Salt Basin' of West Africa (Fig. 11).

The excellent quality oil-prone source rocks that were developed in this restricted seaway, as indicated by the DSDP 361 drill site (Herbin *et al.* 1987), are predicted to extend to the east, relatively close to the more distal drilling that has taken place, whereafter they rapidly increase in terrestrial, more gas-prone character, as the limits of the ancient restricted seaway are reached. The main characteristics of these source rocks are given in Fig. 12. Thermal modelling (e.g. Fig. 13), together with the lateral change in organic matter type, results in the interpreted distribution of active source kitchens shown in Fig. 12.

When Figs 1, 2, 10 and 12 are compared it becomes evident why the distribution of hydrocarbon habitats follows the observed pattern. It is directly controlled by the distribution of the active source kitchens/pods of the major source rocks in the area, that is to say, the lower Aptian shales. Migration pathways, such as faults, unconformities and carrier beds, play a relatively secondary role.

The presence of a major gas charge in the main depocentre of the Orange Basin (Fig. 10), where the Barremian and lower Aptian source shales are buried within the gas window, is evidenced in the transitional and full-drift successions by gas discoveries, significant gas shows, gas seeps and direct seismic indicators, such as interpreted gas chimneys and bottom simulating reflectors, and AVO anomalies (Figs 1, 2, 10, 12 and 16). The following petroleum systems and plays are sourced from these Barremian and lower Aptian source rocks.

Proven, lower Aptian (and Barremian)-sourced gas system within the transitional succession in the main Orange Basin depocentre and stratigraphic Barremian aeolian sandstone play (Kudu-type). This play is confirmed by the Kudu Gas Field offshore southern Namibia, in the northern part of the Orange Basin depocentre (Figs 1, 2, 6, 10 and 16). This gas field has recently been declared commercially viable; proven reserves *c.* 1.8 TCF (trillion cubic feet) and an ultimate resource possibly > 5 TCF have been quoted by the operator. Descriptions of the gas field were published by Namcor prior to the licensing of the acreage and can be found in Benson (1990), Davies & van der Spuy (1990, 1993), McLachlan (1990), McMillan (1990), Marot (1990), Van Rijswijck & Steyn (1990) and Wickens & McLachlan (1990). The most striking features of this gas accumulation are: that it is stratigraphically trapped in continental aeolian sandstones; that no gas–water contact was encountered; that the reservoir is some 1600 psi overpressured; that the porosities average 12% and the permeabilities 43 mD at a depth of *c.* 4400 m; that the deliverability of the reservoir is substantial, despite the depth of burial; that the reservoir sands are closely associated with basalts; that the gas is *c.* 96% methane; and that the main reservoir is overlain by a gas-charged, but tight, shallow marine sandstone, itself overlain by some 200 m of spent source rocks.

Aeolian coastal dune fields are often geographically extensive depositional systems, such as portrayed by the present-day Namib Desert. Ancient deposits also reflect this, such as the Rotliegend of southern North Sea gas province (Glennie

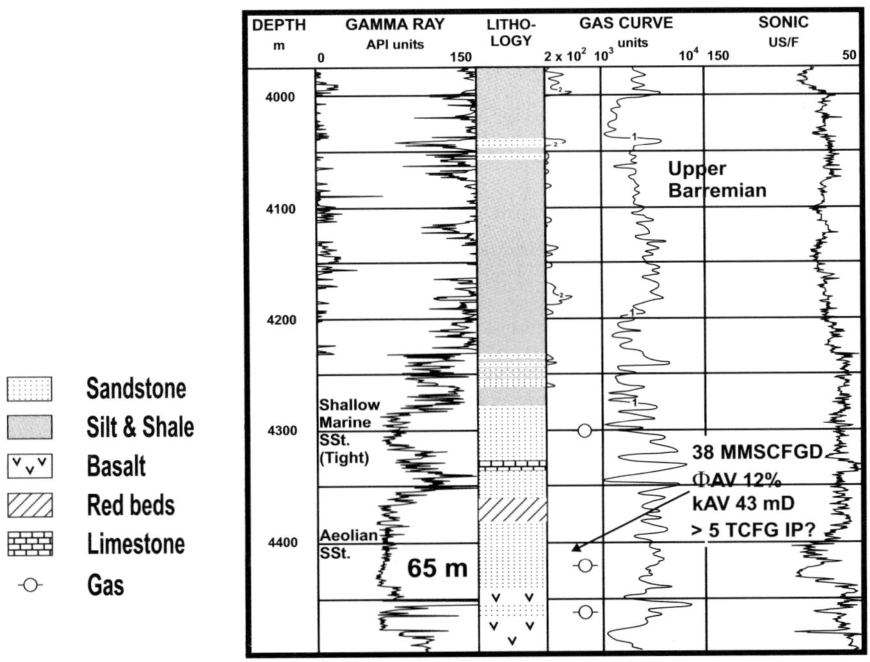

Fig. 6. Electric and gas logs of the reservoir interval in the Kudu 3 well, Kudu Gas Field, offshore southern Namibia. DST flow rate is in millions of standard cubic feet of gas per day (MMSCFGD). The average porosity (ΦAV) of the main Barremian aeolian reservoir sandstone is 12% and the average permeability (kAV) is 43 millidarcies (mD). See Figs 1 and 9 for location. Adapted from Marot (1990), van Rijswijck & Steyn (1990) and Wickens & McLachlan (1990).

1990). It is suggested on this basis that the Kudu-type aeolian play extends southward into the South African portion of the Orange Basin, as already suggested by Light *et al.* (1992*a*, 1993*b*).

It may not be inconceivable that the Kudu stratigraphic trap extends regionally, although probably not continuously, into the main depocentre area off South Africa where ample evidence of a major gas charge is evidenced, including a large thermogenic methane seep near the mouth of the Orange River (Figs 1, 2 and 10). This play has not been targeted by the drill off South Africa, principally as a result of the likelihood of gas which was seen historically less desirable than oil (Fig. 1). Several large prospects have, however, been mapped and are largely defined by high seismic amplitude anomalies. The traps have a large stratigraphic component and prediction of porosity is difficult at depths *c.* 4000 m.

Hypothetical, Barremian–lower Aptian-sourced gas system within the transitional succession in the main depocentre and the Barremian shallow marine sandstone play. This play has repeatedly yielded significant gas shows from the Kudu location in the north to the AC/AN well locations in the south (Figs 1, 10, 12 and 16). However, the sandstones are invariably tight, cemented by carbonates and secondary quartz. It does, however, confirm that a major gas charge is operative at this stratigraphic level throughout the basin and that these sandstones are major regional carrier beds (Fig. 2). A potential regional pinch-out play has been delineated to the northeast of the AC/AN wells along a basement high and is associated with a wet gas 'Sniffer' anomaly. Porosity may be developed in this more proximal area where the thick Aptian shales may still form a seal.

Speculative, Barremian–lower Aptian-sourced oil–gas system within the drift succession in hypothetical deepwater turbidite sandstone plays trapped over the marginal ridge. This hypothetical petroleum system and play is strategically located where the Aptian oil pod is likely to be best developed (Fig. 12). In addition, it coincides with the position of the marginal ridge (Gerrard & Smith 1982) which is likely to have been a focus of migration as shown in Figs 2 and 12. Lower Cretaceous turbidites that may be draped over, or ponded against, the ridge form the targets of this future deepwater play.

Proven, Barremian–lower Aptian-sourced gas system within the stratigraphically trapped middle Albian–Cenomanian sandstones of the drift succession in the main depocentre. This gas play has been confirmed in a series of discoveries off South Africa (Figs 1, 2, 7–10 and 16), the best of which is the AK1 well (Fig. 1). It is the most

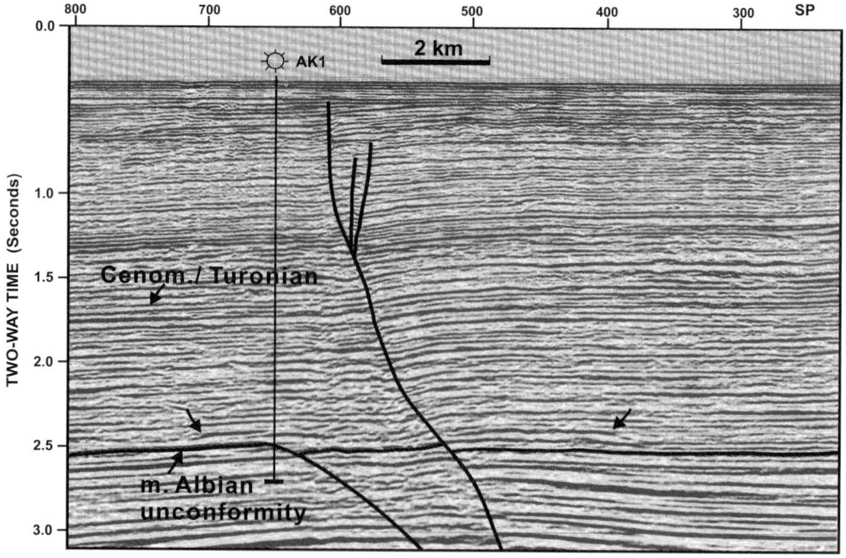

Fig. 7. Two-dimensional seismic line across the AK gas-bearing structure. See Figs 9 and 10 for location. Higher seismic amplitudes at the level of the gas reservoirs are noticeable at 2.4 s two-way time where the well was drilled. A similar anomaly is seen on the downthrown side of the master fault at the same level between shotpoints 400 and 500. This may represent an extension of the gas field. Arrows indicate the anomalies. Positive AVO (amplitude variation v. offset) anomalies are also associated with the gas sands in well AK1 (not shown).

significant play in the South African portion of the Orange Basin to date. The more important features of this play type exemplified by the AK1 discovery are, *inter alia*, its stratigraphic trapping; the excellent poro-perm qualities and the extremely good flow rates for relatively thin sandstones (Figs 1 and 8); seismic amplitude anomalies associated with the gas-bearing sandstones (Fig. 7), especially AVO anomalies; and the absence of gas–water contacts indicating possibly large stratigraphically trapped systems.

Although this play type can be confidently mapped seismically over the entire main depocentre area between the AK1 location and the Namibian border, individual traps are difficult to identify mainly because of the almost complete absence of structure. It is suggested, from sequence stratigraphic mapping, that these reservoirs formed within broad compound incised valley systems, and that they may be best developed off-structure along the valleys' thalwegs which appear to follow the trend of underlying graben systems (Fig. 5). The future development of this play will most likely depend on the application of sophisticated seismic attribute technology.

Speculative, Barremian–lower Aptian-sourced oil–gas systems within the Upper Cretaceous growth-fault and toe-thrust belt in the northern part of the margin. Plays are hypothetical turbidite sandstones. The trend and extent of this structural play (Fig. 14) are shown in Figs 10 and 16. The structural features seen on the seismic lines are quite clearly growth faults, such as those encountered in the Gulf of Mexico (Lopez 1990) and the Niger Delta (Shannon & Naylor 1989). These are not 'slumps' as previously described (Gerrard & Smith 1982; Dingle 1983; Dingle & Robson 1992). A number of these roll-over structures display possible gas escape features. Complementary toe-thrusts are developed down-dip of the extensional growth faults. These toe-thrust structures are situated over the active oil pod (Fig. 12) and may host deep-sea fan turbidite systems akin to those described in the Tertiary of the Campos Basin of Brazil, where they host giant reserves of oil (Carminatti & Scarton 1991; Dias *et al.* 1991). It has also been shown that similar deep-sea sands in the Gulf of Mexico display excellent reservoir properties (Peterson & Cooke 1995; Craig & Hyde 1997). This play is predicted to be a choice target for future deep water exploration off South

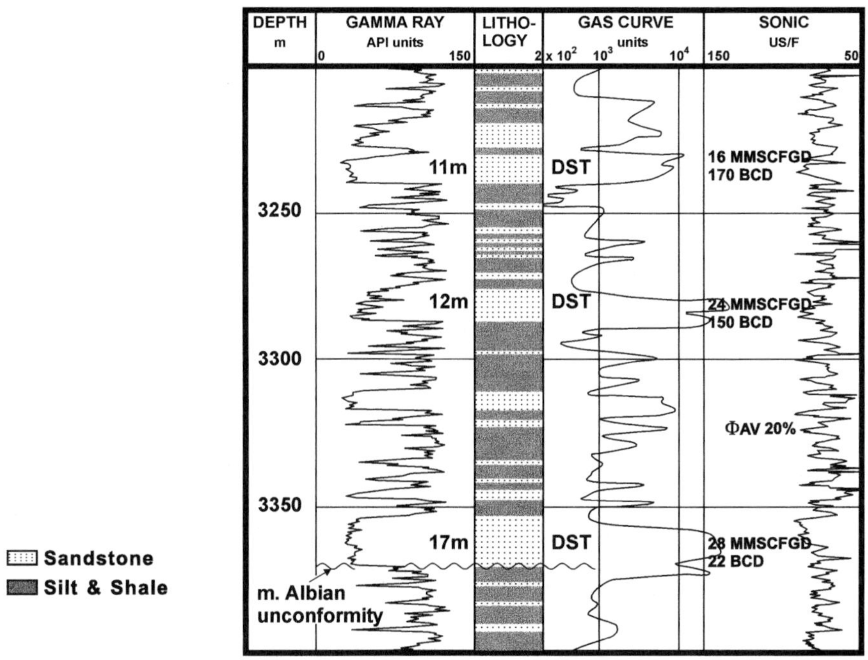

Fig. 8. Electric and gas logs over the gas reservoirs in well AK1. DST flow rates are in millions of standard cubic feet of gas per day (MMSCFGD) and barrels of condensate per day (BCD). The average porosity (ΦAV) of these middle Albian fluvio-deltaic sandstones is 20%.

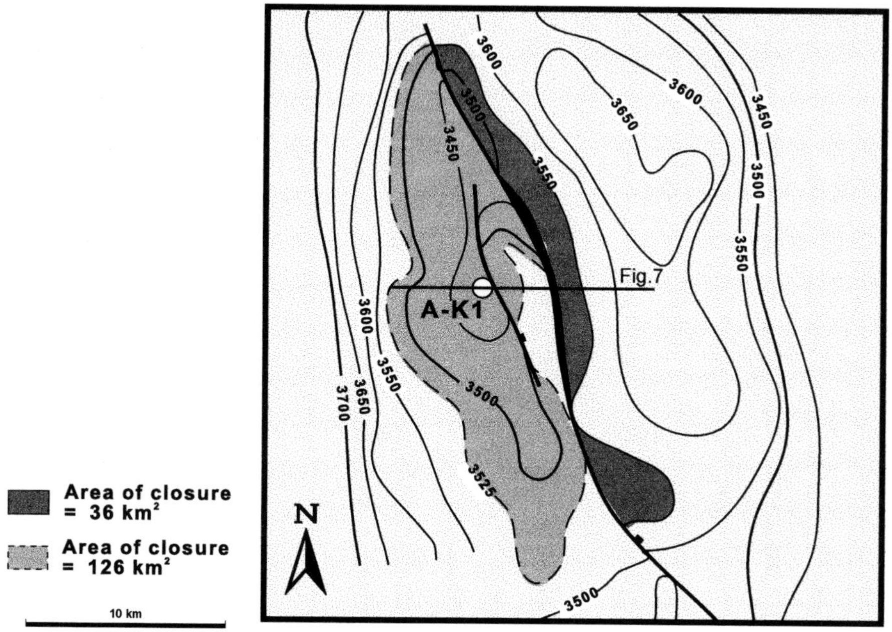

Fig. 9. Structural depth map in metres at the level of the middle Albian second-order sequence boundary at the base of the gas sands in well AK1. See Figs 7 and 8.

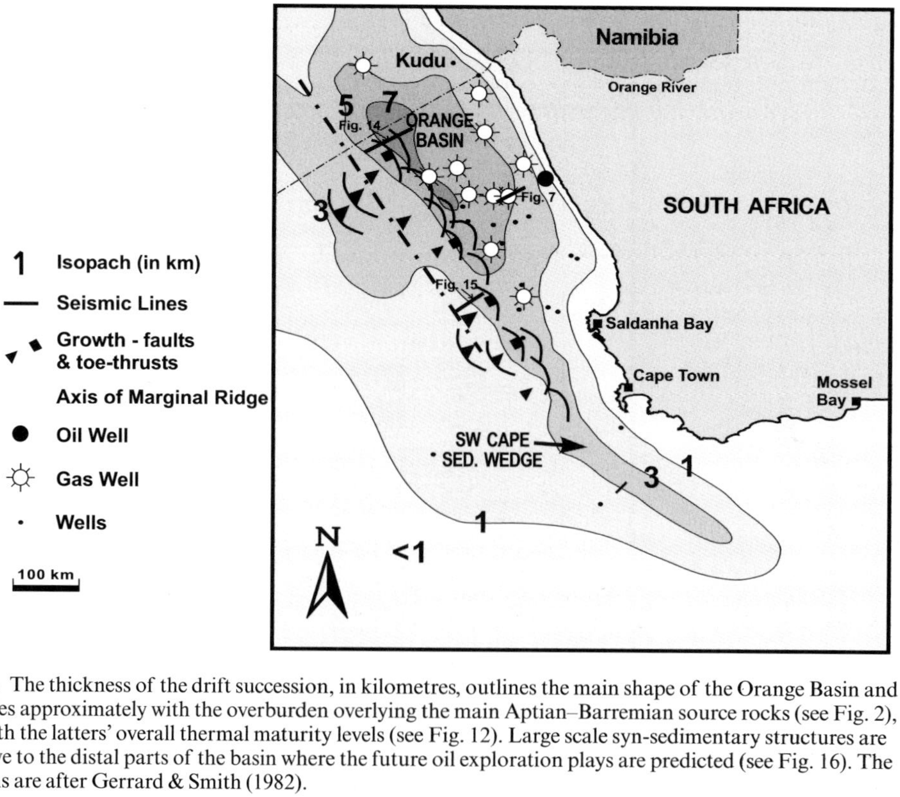

Fig. 10. The thickness of the drift succession, in kilometres, outlines the main shape of the Orange Basin and coincides approximately with the overburden overlying the main Aptian–Barremian source rocks (see Fig. 2), and thus with the latters' overall thermal maturity levels (see Fig. 12). Large scale syn-sedimentary structures are exclusive to the distal parts of the basin where the future oil exploration plays are predicted (see Fig. 16). The isopachs are after Gerrard & Smith (1982).

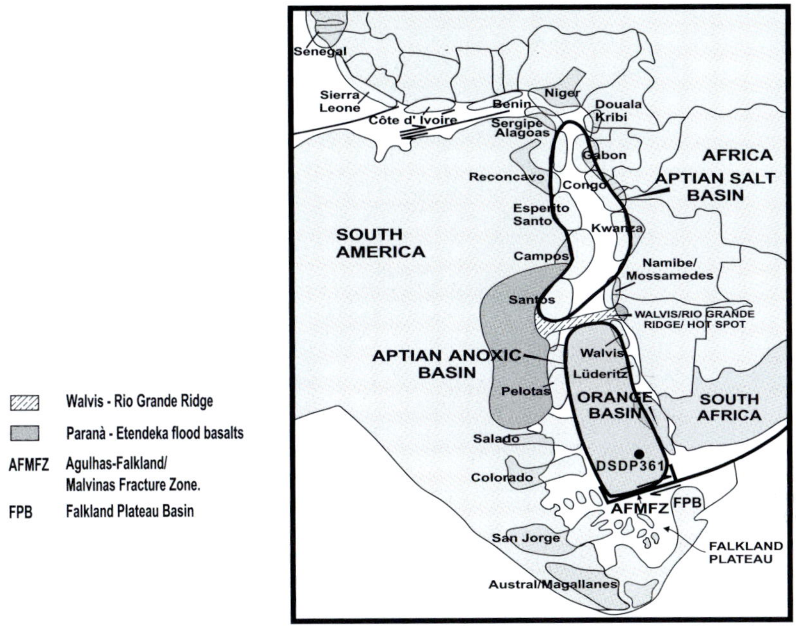

Fig. 11. Anoxia was developed in the southern proto-South Atlantic, here called the 'Aptian Anoxic Basin', on newly formed oceanic crust between major tectonic barriers to circulation (Walvis–Rio Grande Igneous Ridge, the Falkland Plateau, and the South American and Southwest African continental margins) during the Barremian, and especially early Aptian stages. Anoxia reached its maximum distribution during the early Aptian, as depicted here. Resulting source rocks reaching excellent oil-prone quality (e.g. DSDP 361, Fig. 12) are thus possibly located in and beyond the distal parts of the local coastal basins. See also Fig. 4. Sources are: Bolli *et al.* (1978), Herbin *et al.* (1987), Jacquin & Graciansky (1988) and Katz & Pheifer (1982).

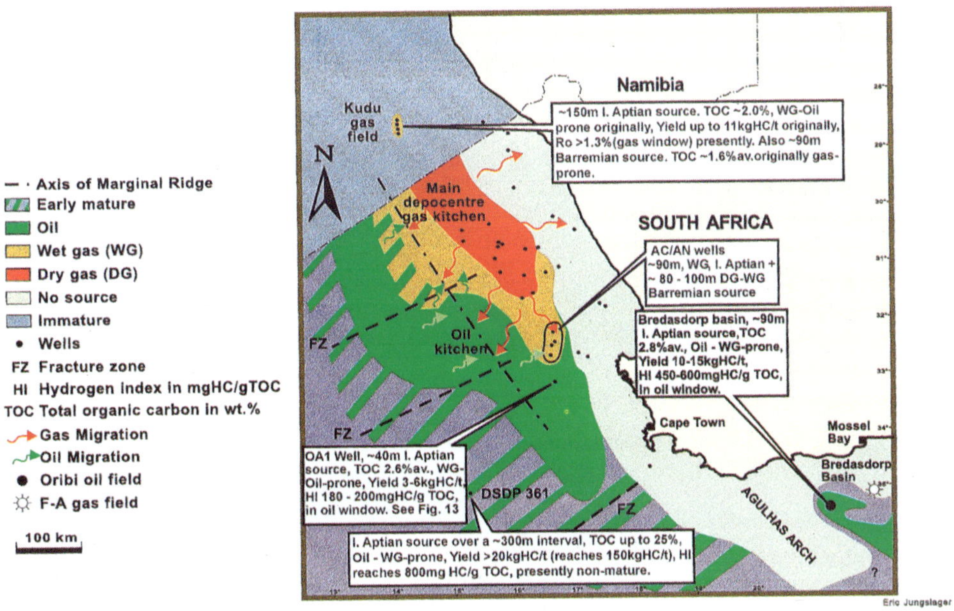

Fig. 12. The active marine source rock kitchen/pods at the lower Aptian level as observed and predicted to be at work along South Africa's Atlantic margin (cf. with Figs 1, 10 and 16). Data sources are: Davies & van der Spuy (1990, 1993); Herbin *et al.* (1987) and Muntingh (1993). The lower Aptian oil pod charging the Oribi Oil Field in the Bredasdorp Basin (Burden & Davies 1997*a*, *b*) is shown for comparison.

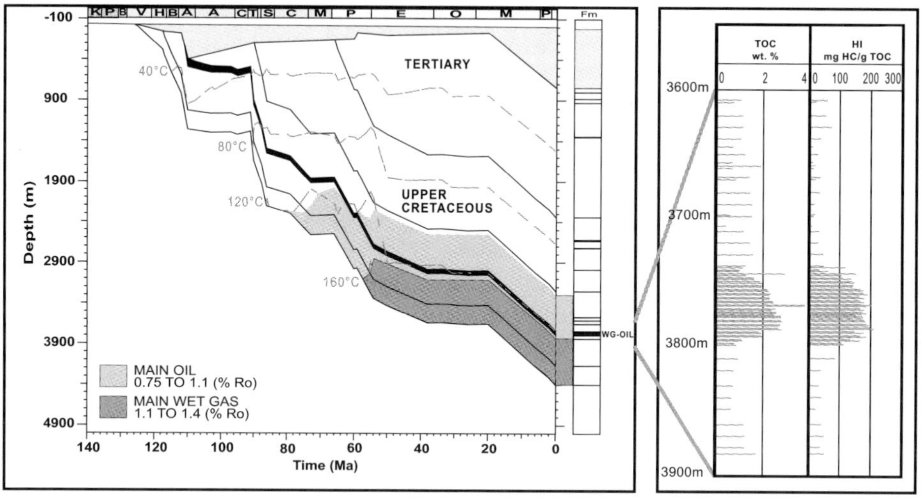

Fig. 13. Geohistory reconstruction and thermal modelling for the OA1 well site. See Figs 1 and 2 for location.

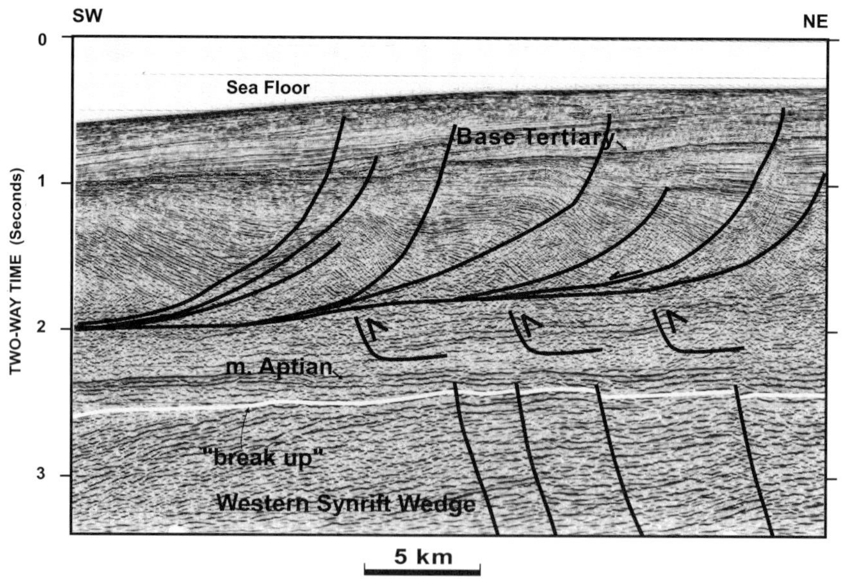

Fig. 14. Two-dimensional seismic line illustrating growth faults and toe-thrusts in the uppermost Cretaceous distal frontier. See Fig. 10 for location.

Africa's western coast. The challenge however, will be, to locate the sandstone bodies. Naturally, three-dimensional seismic data will be required to gain confidence in the prediction and presence of elongated channel–levee systems.

Speculative, Barremian–lower Aptian-sourced oil system within the Upper Cretaceous and Tertiary toe-thrust belts in the central and southern portions of the margin. Plays are hypothetical turbidite sandstones. Figure 15 shows large toe-thrust

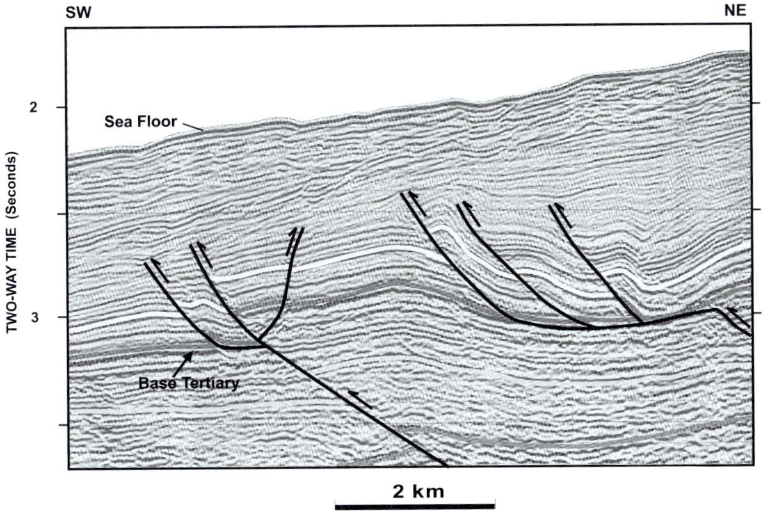

Fig. 15. Two-dimensional seismic line illustrating toe-thrusts in the Upper Cretaceous and Tertiary southern frontier area. See Fig. 10 for location.

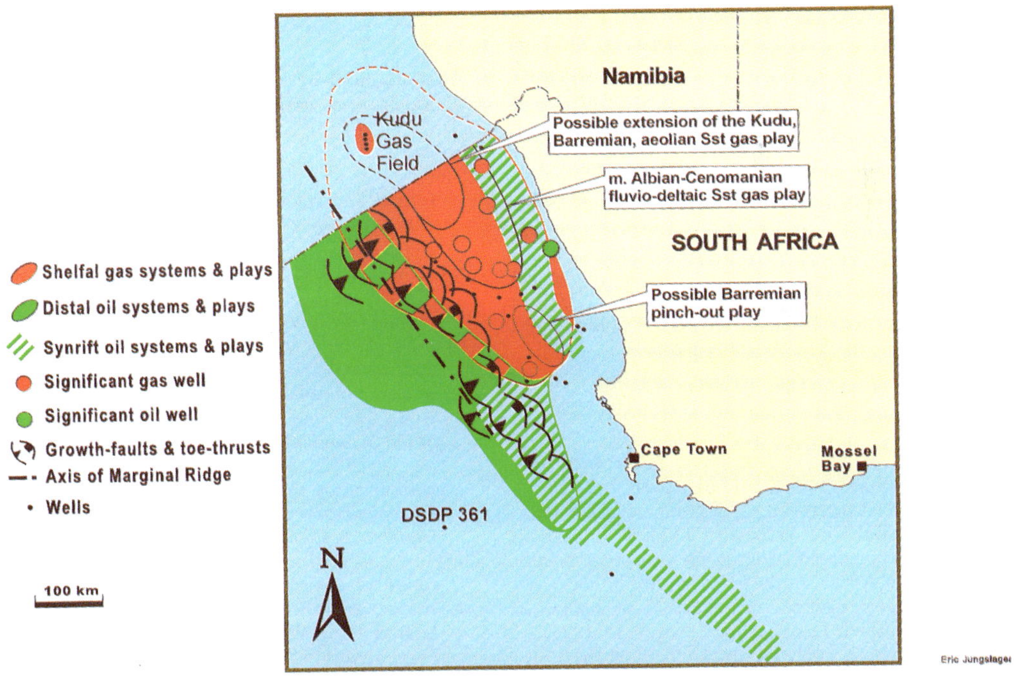

Fig. 16. Outline of the main known and predicted petroleum systems and complementary exploration plays along South Africa's Atlantic margin.

structures that form an additional deepwater play in the area. Here, the Tertiary section so prolific in the Aptian Salt Basin may become prospective as it thickens significantly southward. This play is hypothetical, but its position relative to the likely Aptian oil source pods is favourable (Fig. 12).

Petroleum systems and complementary plays sourced from the drift succession

Potential source rocks have been encountered at the Cenomanian–Turonian boundary. They are of poor, predominantly terrestrial, gas-prone quality and are immature where drilled. They may, however, improve to oil-prone quality down-dip. This is speculative, but potentially important for the unexplored growth-fault and toe-thrust plays.

Conclusions

The observed distribution of petroleum (Figs 1–10), when viewed in relation to the major source rocks and their active source pods (Figs 4, 5 and 11–13), indicates a very close geographic relationship. This is the main basis for predicting a possible oil system and complementary plays west of previous drilling and in close proximity to the predicted lower Aptian oil pod (Figs 14–16). It can also be argued that only a limited portion of the entire sedimentary fill, and one which happens to largely coincide with an active gas charge, has been explored by the drill to date, whilst the most promising areas for oil remain undrilled. This has led, it is believed, to the premature, and possibly erroneous, perception that the area is gas prone as a whole.

SOEKOR (Pty) Ltd are thanked for allowing publication. LASMO are thanked for sponsoring the colour figures. Chris P. N. Davies of SOEKOR E & P and David van der Spuy (Consultant) are acknowledged for aspects of the Aptian source kitchens diagram (Fig. 12) and for the geohistory diagram for the OA1 well site (Fig. 13). Comments by Dave Roberts, Nigel Banks and Jennifer Miles helped improve this paper.

References

ALLEN, P. A. & ALLEN, J. R. 1990. *Basin Analysis – Principles and Applications*. Blackwell.

BARTON, K. R., MUNTINGH, A. & NOBLE, R. D. P. 1993. Geophysical and geological studies applied to hydrocarbon exploration on the west coast margin of South Africa. *Extended Abstracts of the Third International Congress of the Brazilian Geophysical Society*, Rio de Janeiro, Brazil, Nov. 7–11, 1993.

BEN-AVRAHAM, Z., HARTNADY, C. J. H. & KITCHIN, K. A. 1997. Structure and tectonics of the Agulhas–Falkland fracture zone. *Tectonophysics*, **282**, 83–98.

BENSON, J. M. 1990. Palynofacies characteristics and palynological source rock assessment of the Cretaceous sediments of the northern Orange Basin (Kudu 9A-2 and 9A-3 boreholes). *Communications of the Geological Survey of Namibia*, **6**, 31–39.

—— 1992. Palynological expression of depositional systems tracts from the Cretaceous of the Orange Basin and Eastern Cape Basin. Extended abstracts. *Southwestern African Continental Margin: Evolution and Physical Characteristics*, 4–11. Geological Society of Namibia and Geological Survey of Namibia, 'Miniconference', Windhoek, April 1992.

BOLLI, H. M., RYAN, W. B. F., FORESMAN, J. B. *ET AL.* 1978. *Initial Reports of the Deep Sea Drilling Project*, **40**.

BROAD, D. S. & MILLS, S. R. 1993. South Africa offers exploratory potential in variety of basins. *Oil and Gas Journal*, **Dec. 6**, 38–44.

BROWN, L. F., BROWN, L. F. JR, BENSON, J. M., BRINK, G. J. *ET AL.* 1995. *Sequence Stratigraphy in Offshore South African Divergent Basins, An Atlas on Exploration for Cretaceous Lowstand Traps by SOEKOR (Pty) Ltd.* AAPG Studies in Geology, **41**.

BURDEN, P. L. A. 1992. Soekor, partners explore possibilities in Bredasdorp basin off South Africa. *Oil and Gas Journal*, **Dec. 21**, 109–112.

—— & DAVIES, C. P. N. 1997*a*. South Africa Offshore E&P – 1: Exploration to first production on Block 9 off South Africa. *Oil and Gas Journal*, **Sept. 8**, 92–97.

—— & —— 1997*b*. South Africa Offshore E&P – 2: Oribi field is South Africa's first offshore crude oil production. *Oil and Gas Journal*, **Sept. 15**, 63–65.

CARMINATTI, M. & SCARTON, J. C. 1991. Sequence stratigraphy of the Oligocene turbidite complex of the Campos basin, offshore Brazil – an overview. *In*: WEIMER, P. & LINK, M. H. (eds) *Seismic Facies and Turbidite Systems*. Springer, 214–246.

CLIFFORD, A. C. 1986. African oil – past, present and future. *In*: HALBOUTY, M. (ed.) *Future Petroleum Provinces of the World*. AAPG Memoir, **40**, 339–372.

CRAIG, M. J. K. & HYDE, S. T. 1997. Deepwater Gulf of Mexico more profitable than previously thought. *Oil and Gas Journal*, **March 10**, 41–50.

DAVIES, C. P. N. & VAN DER SPUY, D. 1990. Chemical and optical investigations into the hydrocarbon source potential and thermal maturity of the Kudu 9A-2 and 9A-3 boreholes. *Communications of the Geological Survey of Namibia*, **6**, 49–58.

—— & —— 1993. The Kudu wells: Results of a biomarker study related to burial history modeling. *Communications of the Geological Survey of Namibia*, **8**, 45–56.

DEMAISON, G. & HUIZINGA, B. J. 1994. General classification of petroleum systems using three factors: charge, migration and entrapment. *In*: MAGOON, L. B. & DOW, W. G. (eds) *The Petroleum System – From Source to Trap*. AAPG Memoir, **60**, 73–89.

DEROO, G., HERBIN, J. P. & ROUCACHE, J. 1993. Organic geochemistry of Upper Jurassic – Cretaceous sediments from site 511, leg 71, Western Atlantic. *Initial Reports of the Deep Sea Drilling Project*, **71**.

DE WIT, M. J. & RANSOME, I. G. D. (eds) 1992. *Inversion Tectonics of the Cape Fold Belt, Karoo and Cretaceous Basins of Southern Africa.* Balkema, Rotterdam.

——, JEFFERY, M., BERGH, H. & NICOLAYSON, L. 1988. *Geological map of some sectors of Gondwana reconstructed to their disposition* c. *150 Ma*. Bernard Price Institute of Geophysical Research, University of Witwatersrand, South Africa and American Association of Petroleum Geologists, Tulsa, scale 1:10 00 0000, two sheets.

DIAS, J. L., SCARTON, J. C., ESTEVES, F. R., CARMINATTI, M. & GUARDADO, L. R. 1991. Aspectos da Evolucao Tectono-Sedimentar e a Ocorrencia de Hidrocarbonetos na Bacia de Campos. *In*: RAJA GABAGLIA, G. P. & MILANI, E. J. (eds) *Origem e Evolucao de Bacias Sedimentares*, 2nd Edicao. Petroleo Brasileiro S. A., Rio de Janeiro, 333–360.

DINGLE, R. V. 1983. Slump structures on the outer continental margin of South-Western Africa. *In*: BALLY, A. W. (ed.) *Seismic Expression of Structural Styles*. AAPG Studies in Geology, **15**, 24–30.

—— 1993. Structural and sedimentary development of the continental margin off southwestern Africa. *Communications of the Geological Survey of Namibia*, **8**, 35–43.

—— & ROBSON, S. H. 1992. Southwestern Africa continental rise: structural and sedimentary evolution. *In*: POAG, C. W. & GRACIANSKY, P. C. DE (eds) *Geologic Evolution of Atlantic Continental Rises*. Von Ostrand, 62–76.

——, SIESSER, W. G. & NEWTON, A. R. 1983. *Mesozoic and Tertiary Geology of Southern Africa*. Balkema, Rotterdam.

GERRARD, I. & SMITH, G. C. 1982. Post-Paleozoic succession and structure of the southwestern African continental margin. *In*: WATKINS, J. S. & DRAKE, C. L. (eds) *Studies in Continental Margin Geology*. AAPG Memoir, **34**, 49–74.

GLENNIE, K. W. 1990. *Introduction to the Petroleum Geology of the North Sea*. Blackwell.

HAQ, B. U., HARDENBOL, J. & VAIL, P. R. 1987*a*. The new chronostratigraphic basis of Cenozoic and Mesozoic sea level cycles. *In*: ROSS, C. A. & HAMAN, D. (eds) *Timing and Depositional History of Eustatic Sequences: Constraints on Seismic Stratigraphy*. Special Publication of the Cushman Foundation for Foraminiferal Research, **24**, 7–31.

——, —— & —— 1987b. Chronology of fluctuating sea levels since the Triassic. *Science*, **235**, 1156–1166.

——, —— & —— 1988. Mesozoic and Cenozoic chronostratigraphy and eustatic cycles of sea-level change. *In*: WILGUS, C. K., HASTINGS, B. S., POSAMENTIER, H., VAN WAGONER, J., ROSS, C. A. & KENDELL, C. G. ST C. (eds) *Sea-Level Changes: An Integrated Approach*. Special Publication of the Society of Economic Paleontologists and Mineralogists, **42**, 71–108.

HERBIN, J. P., MULLER, C., GRACIANSKY, P. C. DE, JACQUIN, T., MAGNIEZ-JANNIN, F. & UNTERNEHR, P. 1987. Cretaceous anoxic events in the South Atlantic. *Revista Brasileira de Geociencias*, **17**, 92–99.

JACQUIN, T. & GRACIANSKY, P. C. DE 1988. Cyclic fluctuations of the anoxia during Cretaceous time in the South Atlantic Ocean. *Marine and Petroleum Geology*, **5**, 359–369.

JAUNICH, S. 1983*a*. The South-Western African Continental Margin – Part 1. *In*: BALLY, A. W. (ed.) *Seismic Expression of Structural Styles*. AAPG Studies in Geology, **15**, 70–107.

—— 1983*b*. The South-Western African Continental Margin, synsedimentary faulting. *In*: BALLY, A. W. (ed.) *Seismic Expression of Structural Styles*. AAPG Studies in Geology, **15**, 108–117.

—— 1983*c*. The South-Western African Continental Margin, velocity anomalies. *In*: BALLY, A. W. (ed.) *Seismic Expression of Structural Styles*. AAPG Studies in Geology, **15**, 118–122.

—— 1983*d*. The South-Western African Continental Margin, buried paleotopography. *In*: BALLY, A. W. (ed.) *Seismic Expression of Structural Styles*. AAPG Studies in Geology, **15**, 6–9.

—— 1983*e*. Tertiary intrusions on the South-Western African Margin. *In*: BALLY, A. W. (ed.) *Seismic Expression of Structural Styles*. AAPG Studies in Geology, **15**, 10–14.

JUNGSLAGER, E. H. A. 1992. Sequence stratigraphy of the Orange Basin, western shelf of South Africa. *In*: CURNELLE, R. (ed.) *Geologie Africaine – Compte-rendu des Colloques de géologie de Libreville*, 6–8 Mai 1991. Bulletin des Centres de Recherches Exploration–Production Elf Aquitaine, Mémoire, **13**, 170.

—— 1996. Structure, stratigraphy and petroleum aspects of South Africa's Western Margin (abstract). *In*: JARDINÉ, S., KLASZ, I. DE & DEBENAY, J.-P. (eds) *Géologie de L'Afrique et de L'Atlantique Sud*. Bulletin des Centres de Recherches Exploration–Production Elf Aquitaine, Mémoire, **16**, 713.

KATZ, B. J. & PHEIFER, R. N. 1982. Characteristics of Cretaceous organic matter in the Atlantic. *In*: WATKINS, J. S. & DRAKE, C. L. (eds) *Studies in Continental Margin Geology*. AAPG Memoir, **34**, 617–628.

KEELEY, M. L. & LIGHT, M. P. R. 1993. Basin evolution and prospectivity of the Argentine continental margin. *Journal of Petroleum Geology*, **16**, 451–464.

KING, E. C., LIVERMORE, R. A. & STOREY, B. C. 1996. Weddell Sea tectonics and Gondwana break-up: an introduction. *In*: STOREY, B. C., KING, E. C. & LIVERMORE, R. A. (eds) *Weddell Sea Tectonics and Gondwana Break-up*. Geological Society, London, Special Publications, **108**, 1–10.

LAWRENCE, S. & JOHNSON, M. 1995. Shelf north of Falklands may be new S. Atlantic petroleum province. *Oil and Gas Journal*, **March 6**, 52–55.

LIGHT, M. P. R., MASLANYJ, M. P. & BANKS, N. L. 1992*b*. New geophysical evidence for extensional tectonics on the divergent margin offshore Namibia. *In*: STOREY, B. C., ALABASTER, T. & PANKHURST, R. J. (eds) *Magmatism and the Causes of Continental Break-up*. Geological Society, London, Special Publications, **68**, 257–270.

——, KEELEY, M. L., MASLANYJ, M. P. & URIEN, C. M. 1993*a*. The tectono-stratigraphic development of Patagonia, and its relevance to hydrocarbon exploration. *Journal of Petroleum Geology*, **16**, 465–482.

——, MASLANYJ, M. P., GREENWOOD, R. J. & BANKS, N. L. 1993*b*. Seismic sequence stratigraphy and tectonics offshore Namibia. *In*: WILLIAMS, G. D. & DOBB, A. (eds) *Tectonics and Seismic Sequence Stratigraphy*. Geological Society, London, Special Publications, **71**, 163–191.

——, BANKS, N. L., GREENWOOD, J., MASLANYJ, M., HORN, I. W., DAVIDSON, K. & SIMON, P. 1992*a*. Sismostratigraphie, stratigraphie sequentielle et tectonique de l'offshore de la Namibie. *In*: CURNELLE, R. (ed.) *Géologie Africaine – Compte rendu des Colloques de géologie de Libreville*, 6–8 Mai 1991. Bulletin des Centres de Recherches Exploration–Production Elf Aquitaine, Mémoire, **13**, 79–98.

LOPEZ, J. A. 1990. Structural styles of growth faults in the U.S. Gulf Coast Basin. *In*: BROOKS, J. (ed.) *Classic Petroleum Provinces*. Geological Society, London, Special Publications, **50**, 203–219.

MCLACHLAN, I. R. 1990. Introductory note on the history of the Kudu prospect. *Communications of the Geological Survey of Namibia*, **6**, 5.

—— & MCMILLAN, I. K. 1979. Microfaunal biostratigraphy, chronostratigraphy and history of Mesozoic and Cenozoic deposits on the coastal margin of South Africa. *In*: ANDERSON, A. M. & VAN BILJON, W. J. (eds) *Some Sedimentary Deposits and Associated Ore Deposits of South Africa.* Geological Society, South Africa, Special Publication, **6**, 161–181.

MCMILLAN, I. K. 1990. Foraminiferal biostratigraphy of the Barremian to Miocene rocks of the Kudu 9A-1, 9A-2 and 9A-3 boreholes. *Communications of the Geological Survey of Namibia*, **6**, 23–29.

——, BRINK, G. J., BROAD, D. S. & MAIER, J. J. 1997. Late Mesozoic sedimentary basins off the south coast of South Africa. *In*: SELLEY, R. C. (ed.). *African Basins. Sedimentary Basins of the World*, **3**. Elsevier, 319–376.

MAGOON, L. B. & DOW, W. G. 1994. The petroleum system. *In*: MAGOON, L. B. & DOW, W. G. (eds) *The Petroleum System – From Source to Trap*. AAPG Memoir, **60**, 3–24.

—— & SANCHEZ, R. M. O. 1995. Beyond the Petroleum System. *AAPG Bulletin*, **79**, 1731–1736.

MALAN, J. 1993. Geology, potential of Algoa, Gamtoos basins of South Africa. *Oil and Gas Journal*, **Nov. 15**.

MAROT, J. E. B. 1990. Petrography and porosity of the gas-bearing sandstones intersected by the Kudu 9a-2 and 9a-3 boreholes. *Communications of the Geological Survey of Namibia*, **6**, 41–48.

MARTIN, A. K. & HARTNADY, C. J. H. 1986. Plate tectonic development of the South West Indian Ocean: a revised reconstruction of East Antarctia and Africa. *Journal of Geophysical Research*, **91**, 4767–4786.

MASLANYJ, M. P., LIGHT, M. P. R., GREENWOOD, R. J. & BANKS, N. L. 1992. Extension tectonics offshore Namibia and evidence for passive rifting in the South Atlantic. *Marine and Petroleum Geology*, **9**, 590–601.

MUNTINGH, A. 1993. Geology, prospects in Orange basin offshore western South Africa. *Oil and Gas Journal*, **Jan. 25**, 104–109.

—— & BROWN, L. F., JR 1993. Sequence stratigraphy of petroleum plays, post-rift Cretaceous rocks (Lower Aptian to Upper Maastrichtian), Orange Basin, Western Offshore, South Africa. *In*: WEIMER, P. & POSAMENTIER, H. W. (eds) *Siliciclastic Sequence Stratigraphy – Recent Developments and Applications*. AAPG Memoir, **58**, 71–98.

PERRODON, A. 1983. Dynamics of oil and gas accumulations. Bulletin des Centres de Recherches Exploration–Production Elf Aquitaine, Mémoire, **5**.

PETERSON, R. H. & COOKE, D. W. 1995. Deep-water northern Gulf of Mexico hydrocarbon plays. *Gulf Coast Association of Geological Societies Transactions*, **XLV**, 481–486.

RICHARDS, P. C., GATLIFF, R. W., QUINN, M. F. & FANNIN, N. G. T. 1996*a*. Petroleum potential of the Falkland Islands offshore area. *Journal of Petroleum Geology*, **19**, 161–182.

——, ——, ——, WILLIAMSON, J. P. & FANNIN, N. G. T. 1996*b*. The geological evolution of the Falkland Islands continental shelf. *In*: STOREY, B. C., KING, E. C. & LIVERMORE, R. A. (eds) *Weddell Sea Tectonics and Gondwana Break-up*. Geological Society, London, Special Publications, **108**, 105–128.

ROUX, J. 1997. Potential outlined in Southern Outeniqua basin off South Africa. *Oil and Gas Journal*, **July 21**, 87–91.

SHANNON, P. M. & NAYLOR, D. 1989. Niger Delta. *In*: *Petroleum Basin Studies*. Graham & Trotman, 169–180.

SMITH, J. T. 1994. Petroleum system logic as an exploration tool in a frontier setting. *In*: MAGOON, L. B. & DOW, W. G. (eds) *The Petroleum System – From Source to Trap.* AAPG Memoir, **60**, 2549.

TANKARD, A. J., ULIANA, M. A., WELSINK, H. J. ET AL. 1995. Structural and tectonic controls of basin evolution in Southwestern Gondwana during the Phanerozoic. *In*: TANKARD, A. J., SUAREZ SORUCO, R. & WELSINK, H. J. (eds) *Petroleum Basins of South America.* AAPG Memoir, **62**, 5–52.

ULIANA, M. A., BIDDLE, K. T. & CERDAN, J. 1989. Mesozoic extension and the formation of Argentine sedimentary basins. *In*: TANKARD, A. J. & BALKWILL, H. R. (eds). *Extensional Tectonics and Stratigraphy of the North Atlantic Margins.* AAPG Memoir, **46**, 599–614.

VAN RIJSWIJCK, P. G. & STEYN, A. 1990. A short note on the results of the geophysical logging and testing at the Kudu 9A-2 and 9A-3 boreholes. *Communications of the Geological Survey of Namibia*, **6**, 7.

WHITE, D. A. 1988. Oil and gas play maps in exploration and assessment. *AAPG Bulletin*, **64**, 1158–1178.

WICKENS, H. DE V. & MCLACHLAN, I. R. 1990. The stratigraphy and sedimentology of the reservoir interval of the Kudu 9A-2 and 9A-3 boreholes. *Communications of the Geological Survey of Namibia*, **6**, 9–22.

Petroleum systems in the South Atlantic margins

C. F. SCHIEFELBEIN[1], J. E. ZUMBERGE, N. R. CAMERON & S. W. BROWN

GeoMark Research, Inc., 9748 Whithorn Drive, Houston, TX 77095, USA
[1] Present address: Geochemical Solutions International, 2203 Timberloch Place, Suite 235, Woodlands, Texas 77380, USA

Abstract: Results obtained from the detailed geochemical analysis of more than 150 crude oils were used to : (1) identify a number of different petroleum systems present in the South Atlantic margins; and (2) establish geochemical criteria useful in distinguishing lacustrine-derived oils from marine-derived oils. The multiplicity and type of effective source rocks were determined by using the geochemical characteristics of the crude oils to establish the number of compositionally distinct oil families, and by inferring palaeoenvironmental conditions of source rock deposition and possible age. Using multivariate statistical techniques and bulk geochemical parameters, isotopic compositions and biomarker distributions, a number of general petroleum systems were identified, differing from each other in terms of source palaeoenvironment and age. The oils are separated into two broad groups: lacustrine and marine. Further detailed examination of each broad group, sometimes using more specific geochemical criteria, typically resulted in the establishment of sub-groups of oils according to specific source environment, such as lacustrine fresh v. saline conditions. Areas where oils of mixed provenance occur are also identified. The geochemical criteria used in discriminating between lacustrine and marine oil types were determined based on geochemical differences that exist between known marine-sourced (mainly post-salt) oils from the Kwanza and Gabon (Port Gentil area) Basins and known lacustrine-sourced (pre-salt) oils from the Congo Basin.

In areas where substantial oil production has been established, a regional oil study is an excellent way of identifying, evaluating and comparing the various petroleum systems in a province by first establishing the number of compositionally distinct oil types or families. A genetic classification of the different oil types is proposed. This is collectively labelled as a 'Petroleum System', being aware that a complete definition of the term implies the knowledge of many other different factors controlling the occurrence of oil in nature, such as: maturity of source rock, expulsion, migration, accumulation and retention (Demaison & Huizinga 1991). Geochemical data of oils are interpreted in such a way that oil–oil correlations are made and source rock inferences are proposed. The source inferences are possible since the geochemical characteristics of the oil reveal important information on the source age and the palaeoenvironmental conditions of its deposition. Also, a regional oil study approach is particularly useful because the generated and expelled oil can be considered to be representative of the average of a given source rock section, which can exhibit extreme variations in its geochemical characteristics (Moldowan *et al.* 1992). Previous investigators have reported the occurrence of both lacustrine (Pre-Salt) and marine (mainly Post-Salt) oil in the salt basins of Brazil and West Africa (Mello *et al.* 1988*a*, *b*; Teisserence & Villemin 1990; Burwood *et al.* 1995 and refs therein).

Approach

In order to better understand the petroleum systems of the South Atlantic margins, more than 150 oils from the major Brazil and West Africa coastal basins (Ivory Coast, Nigeria, Gabon, Congo and Angola) have been geochemically analysed. Crude oil locations are shown in Fig. 1. The detailed analytical program included measurements of bulk parameters (API gravity, and %S, V and Ni content), whole oil gas chromatography, stable carbon isotope composition of C15+ hydrocarbon fractions, and terpane, sterane and aromatic biomarker distributions using gas chromatography/mass spectrometry (GC/MS). The analytical data were compared employing multivariate statistical techniques such as hierarchical cluster (HCA) and principal component (PCA) analyses using the commercial software program 'Pirouette'. The primary objective of the statistical analyses is to reduce the dimensionality of the information to a few important components that best explain the variation in the data. The geochemical data used in the statistical analyses are primarily source dependent. The 20 variables used were the pristane/phytane ratio, the ^{13}C values of the saturated and aromatic hydrocarbon fractions, and 15 biomarker ratios, including the distribution of the 14β, 17β-C_{27}, -C_{28} and -C_{29} steranes (based on the abundance of the 20R stereoisomers measured from *m/z* 218), total steranes/total hopanes, C_{26}-tricyclic/C_{27}-

SCHIEFELBEIN, C. F., ZUMBERGE, J. E., CAMERON, N. R. & BROWN, S. W. 1999. Petroleum systems in the South Atlantic margins. *In:* CAMERON, N. R., BATE, R. H. & CLURE, V. S. (eds) *The Oil and Gas Habitats of the South Atlantic.* Geological Society, London, Special Publications, **153**, 169–179.

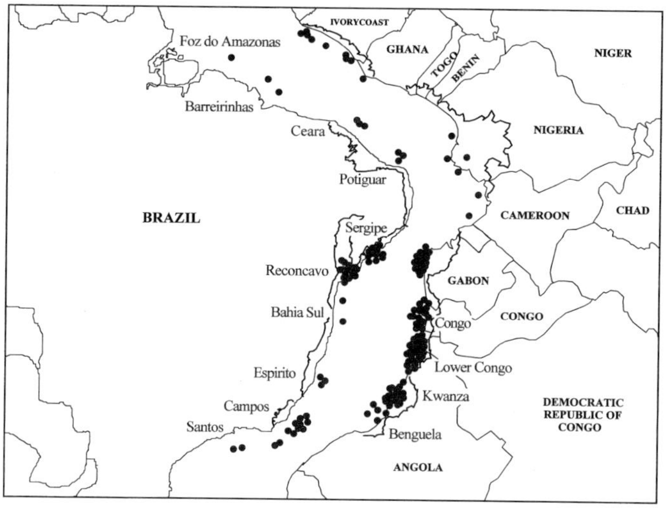

Fig. 1. Crude oil location map.

pentacyclic terpanes, gammacerane/C_{30}-hopane, C_{31}-hopanes/C_{30} hopane, C_{29}-hopane/C_{30}-hopane and oleanane/C_{30}. Ratios based on the distribution of tricyclic and tetracyclic terpanes were also used: C_{19}-Tri/C_{23}-Tri, C_{23}-Tri/C_{24}-Tri, C_{26}-Tri/C_{25}-Tri and C_{24}-Tetra/C_{26}-Tri. Finally, the distribution of the triaromatic dinosteroids and triaromatic 3- and 4-methylsteroids measured from *m*/*z* 245 (Moldowan *et al.* 1995) were used and the abundance of an unknown doublet relative to a C_{27} diasterane (measured from *m*/*z* 259) was also included.

With this approach, a number of general petroleum systems were identified, differing from each other in terms of age and type of depositional environment of the source rocks. The oils are separated into two broad groups: lacustrine and marine. Areas where oils of mixed provenance occur are also identified. Further detailed examination of each petroleum system, sometimes using more specific geochemical criteria, typically resulted in the establishment of subgroups of oils presumably derived from specific source environments, such as fresh v. saline lacustrine depositional conditions.

Results and discussion

Selected results of the analyses of the oils are shown in Figs 2–8. The cross-plot of the saturated/aromatic (sat/aro) hydrocarbon ratio v. sulphur content (Fig. 2) indicates that, in general, many of the oils from the South Atlantic margin have low sulphur contents and are enriched in saturated hydrocarbons (sat/aro > 2.0), consistent with an origin from organic matter deposited in a lacustrine environment. Exceptions are oils from the Kwanza and Gabon (Port Gentil area) Basins that have increased concentrations of sulphur and lower sat/aro ratios (sat/aro < 2.0). Although the majority of these high sulphur oils originated from marine source rocks (Teisserence & Villemin 1990; Burwood *et al.* 1990), these parameters may also be affected by post-generative alteration processes, such as bacterial alteration and water washing.

Figure 3 is a cross-plot of the pristane/phytane isoprenoid ratio v. the stable carbon isotopic composition of the C15+ aromatic hydrocarbons. The areas indicated for oils originating from source rocks deposited in marine or lacustrine environments are arbitrarily defined based on relationships observed for the South Atlantic margin oils. The isoprenoid ratios for the lacustrine-derived oils from the Congo Basin are > 1.5, while the majority of the marine oils from the Kwanza and Gabon Basins have isoprenoid ratios < 1.5. Many Lower Congo Coastal Basin oils from Post-Salt reservoirs have intermediate isoprenoid ratios, possibly indicative of lacustrine and marine oil mixing in the reservoir. The majority of the samples from Nigeria have elevated isoprenoid ratios associated with an origin from marine organic matter deposited in a deltaic environment. The isoprenoid ratios for most of the Brazilian oils from the Campos, Santos and Reconcavo Basins are elevated, consistent with a lacustrine origin. The isotopic compositions of the South Atlantic margin oils varies considerably, from *c.* –31 to –21 permil.

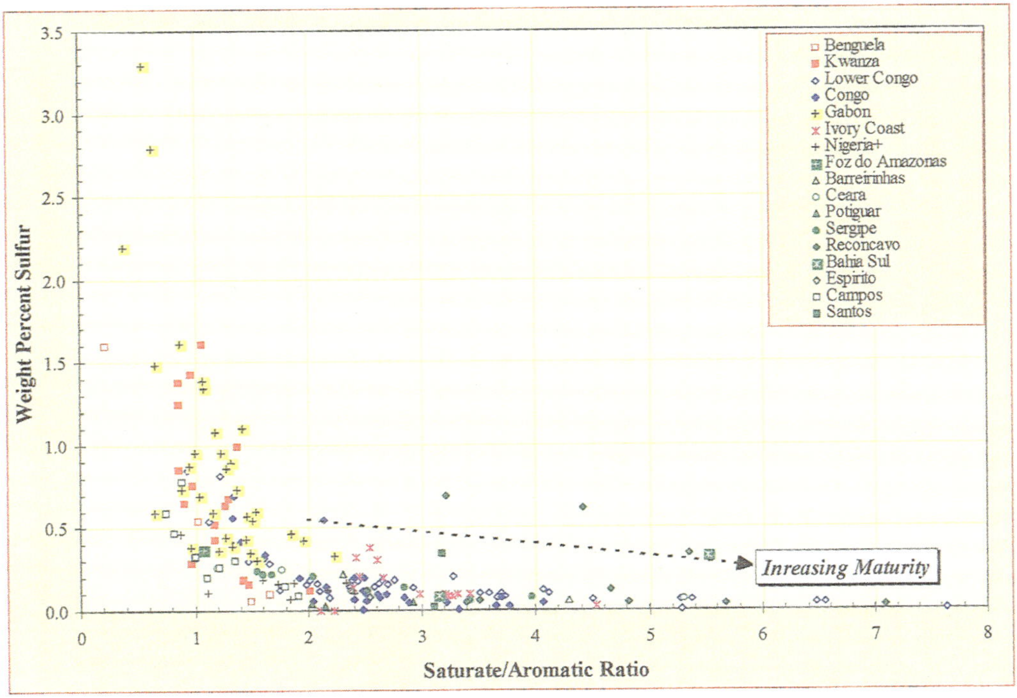

Fig. 2. Saturate/aromatic ratio v. API gravity for the South Atlantic margin oils.

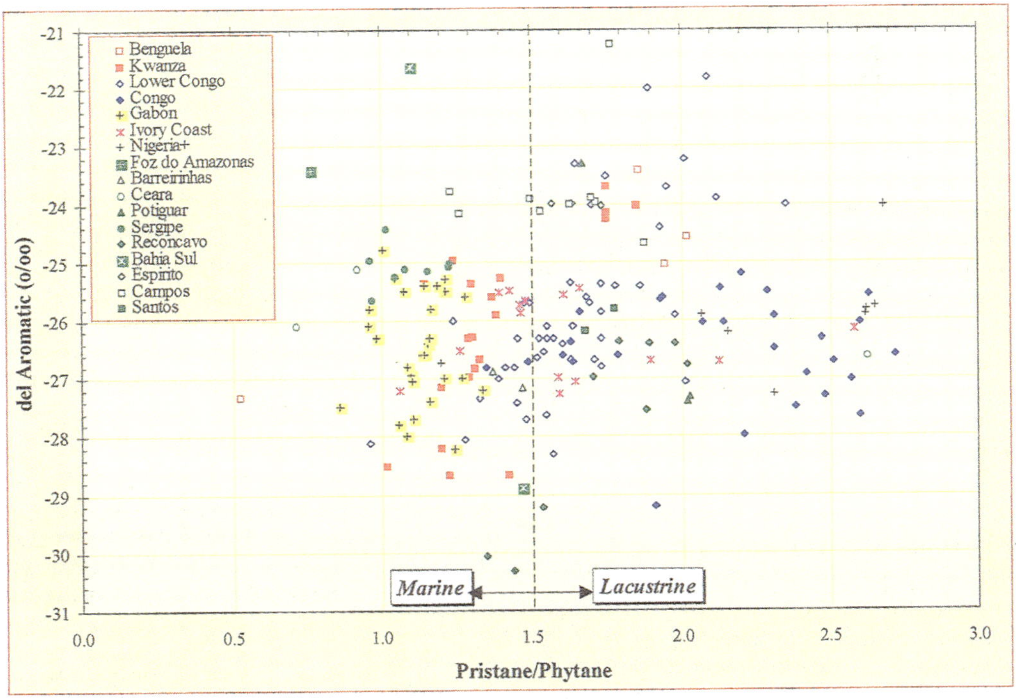

Fig. 3. Pristane/phytane isoprenoid ratio v. the stable carbon isotopic composition of the C15+ aromatic hydrocarbons for the South Atlantic margin oils.

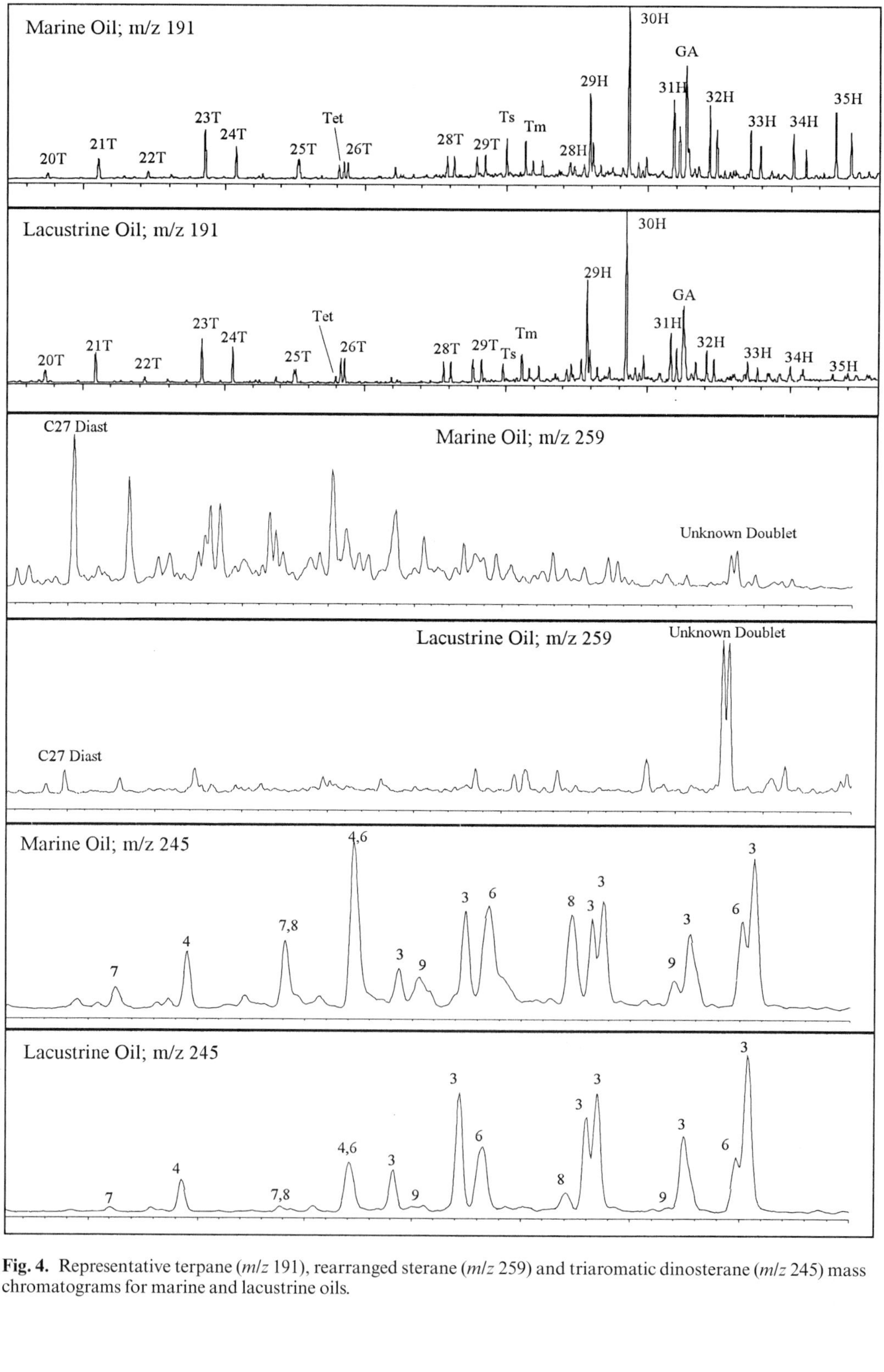

Fig. 4. Representative terpane (*m/z* 191), rearranged sterane (*m/z* 259) and triaromatic dinosterane (*m/z* 245) mass chromatograms for marine and lacustrine oils.

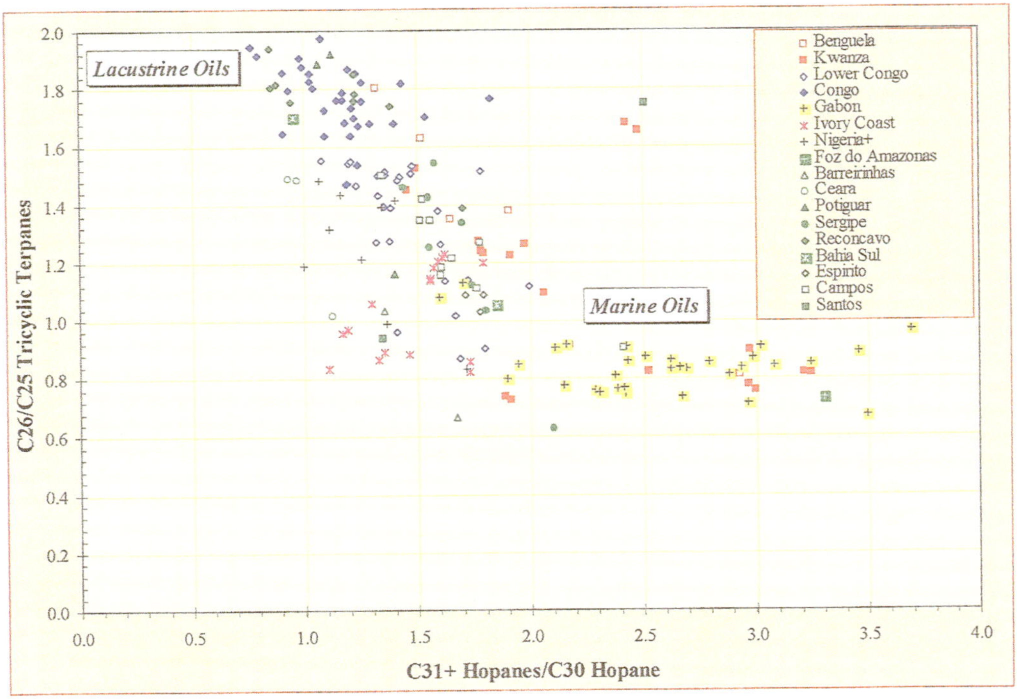

Fig. 5. Extended hopanes (C_{31+})/C_{30} hopane v. C_{26}/C_{25} tricyclic terpane ratios for the South Atlantic margin oils.

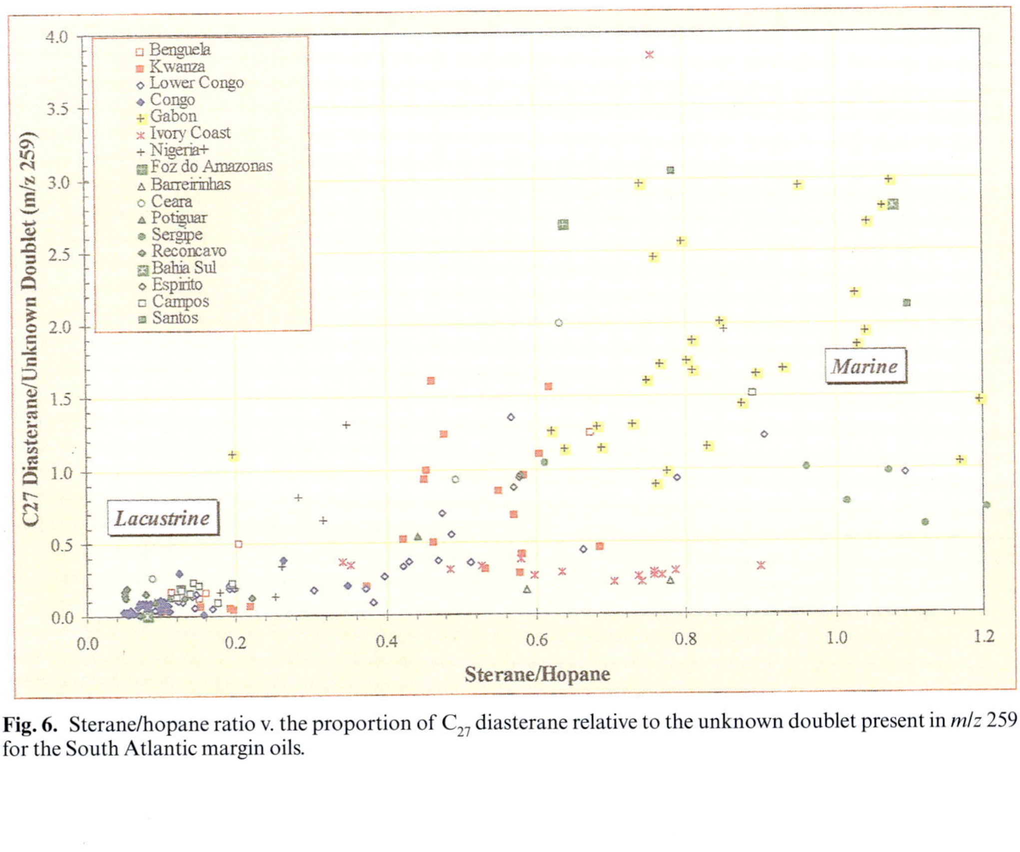

Fig. 6. Sterane/hopane ratio v. the proportion of C_{27} diasterane relative to the unknown doublet present in *m/z* 259 for the South Atlantic margin oils.

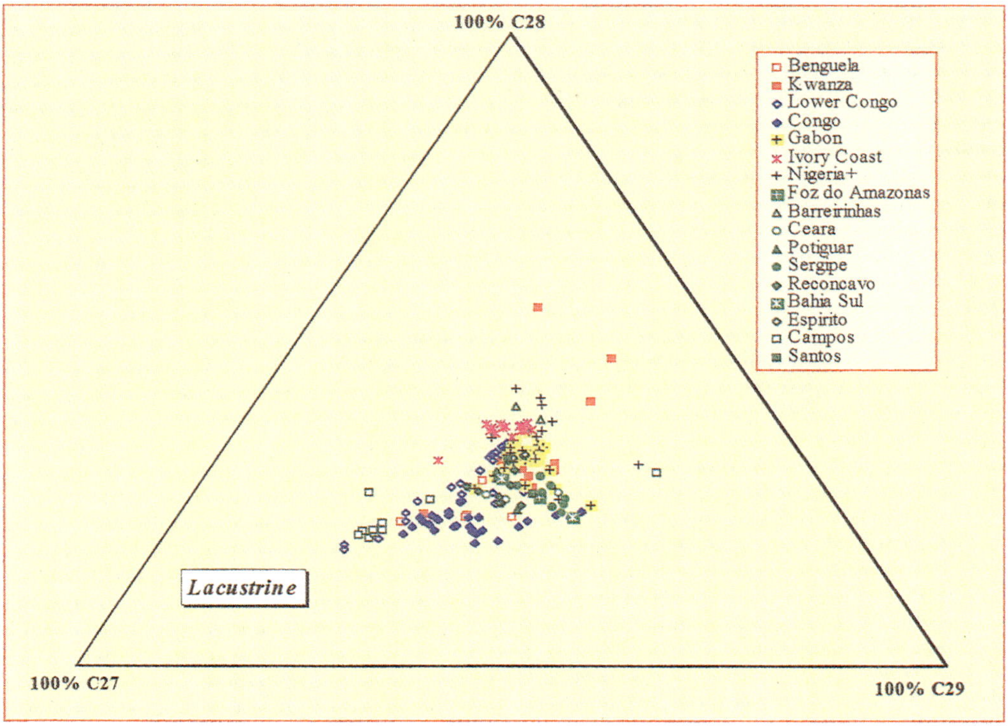

Fig. 7. αββ Sterane compositions (*m/z* 218) for the South Atlantic margin oils.

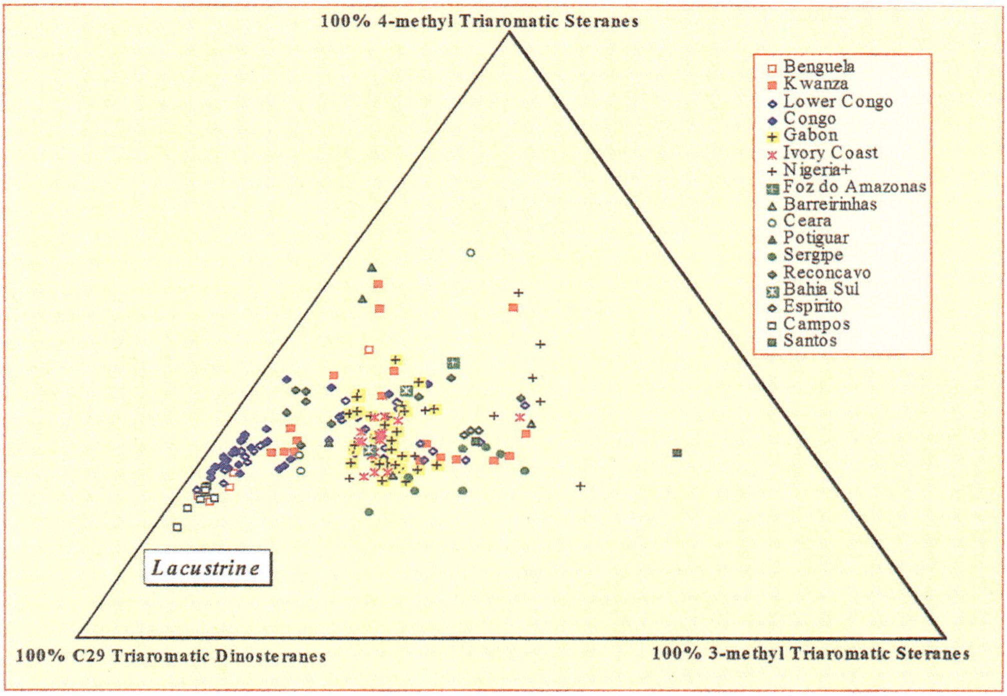

Fig. 8. Triaromatic dinosterane compositions (*m/z* 245) for the South Atlantic margin oils.

Terpane (m/z 191), rearranged sterane (m/z 259) and triaromatic dinosterane (m/z 245) mass chromatograms representing marine and lacustrine oil from the South Atlantic margins are shown in Fig. 4.

Figure 5 is a cross-plot of the proportion of extended hopanes (C_{31+}) relative to C_{30} hopane v. the proportion of C_{26} relative to C_{25} tricyclic terpanes. In general, lacustrine-derived oils have very low proportions of extended hopanes and C_{26}/C_{25} tricyclic terpane ratios > 1.2. Marine-derived oils have higher proportions of extended hopanes and C_{26}/C_{25} tricyclic ratios < 1 (see m/z 191, Fig. 4).

Figure 6 is a cross-plot of the proportion of steranes relative to hopanes v. the proportion of a C_{27} diasterane relative to an unknown doublet (from m/z 259; Fig. 4). In general, lacustrine-derived oils have very low proportions of steranes and contain high concentrations of the unknown doublet relative to the C_{27} diasterane. Marine-derived oils have higher proportions of steranes and C_{27} diasteranes.

Figure 7 is a ternary diagram depicting the relative proportion of αββ steranes (m/z 218). Lacustrine-derived oils from the Congo Basin appear to be enriched in C_{27} steranes, while the marine-derived oils from the Kwanza and Gabon Basins have more diverse αββ sterane compositions.

Figure 8 is a ternary diagram depicting the relative proportion of triaromatic dinosteranes (m/z 245; Moldowan *et al.* 1995). Lacustrine-derived oils from the Congo Basin appear to be enriched in triaromatic dinosteranes (labelled as 3 in Fig. 4), while the marine-derived oils from the Kwanza and Gabon Basins have more diverse compositions and contain higher relative abundances of 3-methyl (labelled as either 4 or 6 in Fig. 4) and 4-methyl (labelled as either 7 or 9 in Fig. 4) triaromatic steranes.

With the help of multivariate statistical analyses (HCA and PCA), oils from the South Atlantic margin can be separated into two broad groups, either marine or lacustrine. The marine group in the top portion of the cluster analysis dendrogram shown in Fig. 9 contains several sub-clusters representing either pure marine oils, mixed lacustrine–marine oils, marine transitional oils, or oils originating from Tertiary source rocks deposited in a deltaic environment. The lacustrine oils in the bottom cluster of the dendrogram are separated into three sub-groups, primarily reflecting differences in water chemistry (e.g. fresh v. saline). These separations are possible because the geochemical characteristics of oil reveal important information on the source age and the paleoenvironmental conditions of its deposition (Didyk *et al.* 1978; Seifert &

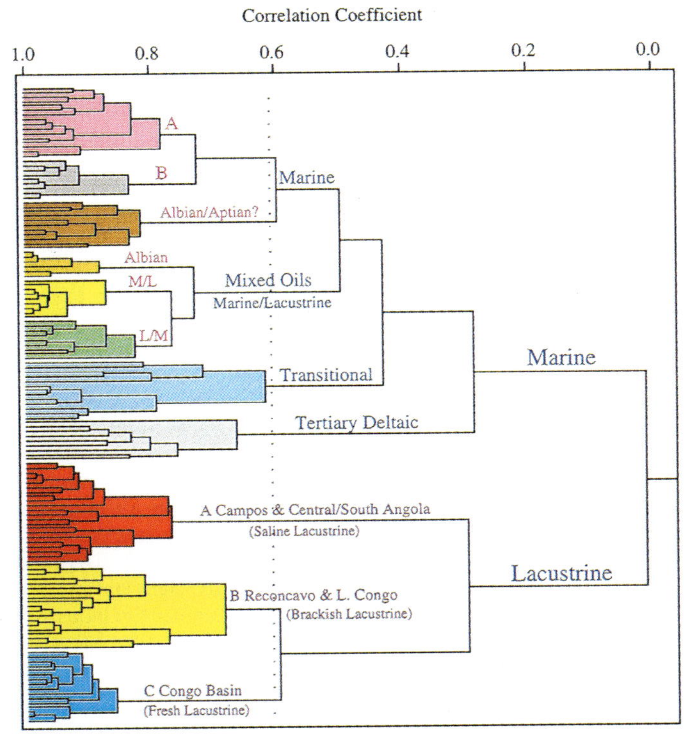

Fig. 9. Cluster analysis dendorgram for the South Atlantic margin oils.

Moldowan 1978, 1981; Moldowan *et al.* 1985; Peters *et al.* 1986; Powell 1986; Zumberge 1987; ten Haven *et al.* 1988; Mello *et al.* 1988*a*, *b*; Volkman 1988; Pu *et al.* 1991). For example, constraints on the age of the source rock can be made by the presence or absence of oleanane, a specific biomarker derived from angiosperms and associated with a Tertiary-aged source due to the evolution of higher land plant organic matter (Ekweozor & Udo 1987). When an oil contains significant concentrations of oleanane (Ol/C_{30}-hopane > 0.2), a Cenozoic source age can be confidently inferred (Moldowan *et al.* 1994).

The South Atlantic margin oils originating from source rocks deposited in a marine environment (marine oils) are characterized by variable to high sulphur content, generally lower sat/aro and pristane/phytane ratios, high proportions of steranes relative to hopanes, high proportions of extended hopanes, and variable $\alpha\beta\beta$ sterane and triaromatic dinosterane compositions. Marine oils have low C_{26}/C_{25} tricyclic terpane ratios and abundant diasteranes relative to the unknown doublet (from *m/z* 259). Marine oils can also be identified based on the presence of the 24-*n*-propylsteranes, a highly characteristic biomarker for marine algae as detected by MRM analyses (Moldowan *et al.* 1990).

The South Atlantic margin oils originating from source rocks composed mainly of algal kerogen (Type I) deposited in a lacustrine environment (lacustrine oils) are distinguished by low sulphur contents, C_{26}-tricyclic/C_{25}-tricyclic ratios >> 1, low proportions of steranes relative to hopanes, and low abundances of extended (C_{31+}) hopanes. Lacustrine oils are enriched in C_{27} $\alpha\beta\beta$ steranes and C_{29} triaromatic dinosteranes. Another important lacustrine classifier is the abundance of an unknown compound detected in the *m/z* 259 mass chromatogram (with a *m/z* 414 parent) relative to regular steranes. This compound is present in significant concentrations in lacustrine oils from West Africa, offshore Brazil, Utah and the Far East, but it is only a minor constituent in marine and terrestrial oils (relative to steranes). In general, lacustrine oils from the South Atlantic margins have pristane/phytane ratios > 1.5 and *c.* < 3.0. Classification based on stable carbon isotopes is difficult because the isotopic composition of lacustrine oils can vary between –20 to –30 permil (Burwood *et al.* 1995).

The South Atlantic margin oils classified as 'Tertiary deltaic' originated from mainly terrigenous kerogen (Type III) deposited in fluvial deltaic or paralic environments. Tertiary deltaic oils are characterized by the absence of 24-*n*-propylsteranes, high abundances of low molecular weight (C_{19} and C_{20}) tricyclic terpanes and C_{24} tetracyclic terpanes, a predominance of C_{29} $\alpha\beta\beta$ steranes, elevated pristane/phytane ratios, positive canonical variables and abundant oleanane.

Conclusions

The PCA loadings and scores plots shown in Fig. 10a and b, respectively, can be used to separate the oils according to source type. The distribution of oil types is shown in Fig. 11.

The South Atlantic margin oils originating from source rocks deposited in marine or transitional marine environments (positive PC1 and variable PC2 values; Fig. 10b) are primarily from the Ivory Coast and the Gabon (Port Gentil area), Kwanza and Benguela Basins in West Africa and the Santos, Espirito and Sergipe Basins offshore Brazil. Certain oils from offshore Ivory Coast and the Lower Congo Coastal and Kwanza Basins displayed intermediate geochemical characteristics (mixed oils), possibly suggesting commingling of marine-derived and lacustrine-derived oils (Connan *et al.* 1987; ten Haven 1996) and/or the presence of a mixed kerogen assemblage (Burwood *et al.* 1995).

The South Atlantic margin oils derived from Tertiary source rocks deposited in a deltaic environment (positive PC1 and negative PC2 values; Fig. 10b) are from Nigeria, Benin, Equatorial Guinea, Cameroon and the Foz do Amazonas Basin of northern Brazil. Tertiary oils may also be present offshore the Ivory Coast and have been reported offshore northern Angola (Connan *et al.* 1987).

Lacustrine oils are widespread in West Africa and offshore Brazil and are thought to originate from organic-rich Pre-Salt (Neocomian–Barremian) source rocks (Mello *et al.* 1988*a*, *b*; Burwood *et al.* 1995). Lacustrine oils have negative PC1 and variable PC2 values (Fig. 10b). Lacustrine oils from West Africa are present in the Congo, Lower Congo Coastal and Benguela Basins. Brazilian oils with lacustrine affinities are present in the Reconcavo, Campos, Potiguar, Ceara and Bahia Sul Basins.

Certain geochemical characteristics of lacustrine oils can often be used to separate the oils according to more specific source environments. For example, the relative abundance of gammacerane may reflect conditions of increasing palaeosalinity (ten Haven *et al.* 1988). Carbon isotope values for the lacustrine oils are variable and may reflect differences in algal speciation, water chemistries and/or carbon budget (Sofer 1984; Katz & Mertani 1989). The pristane/phytane isoprenoid ration may also reflect environmental conditions (ten Haven *et al.* 1988).

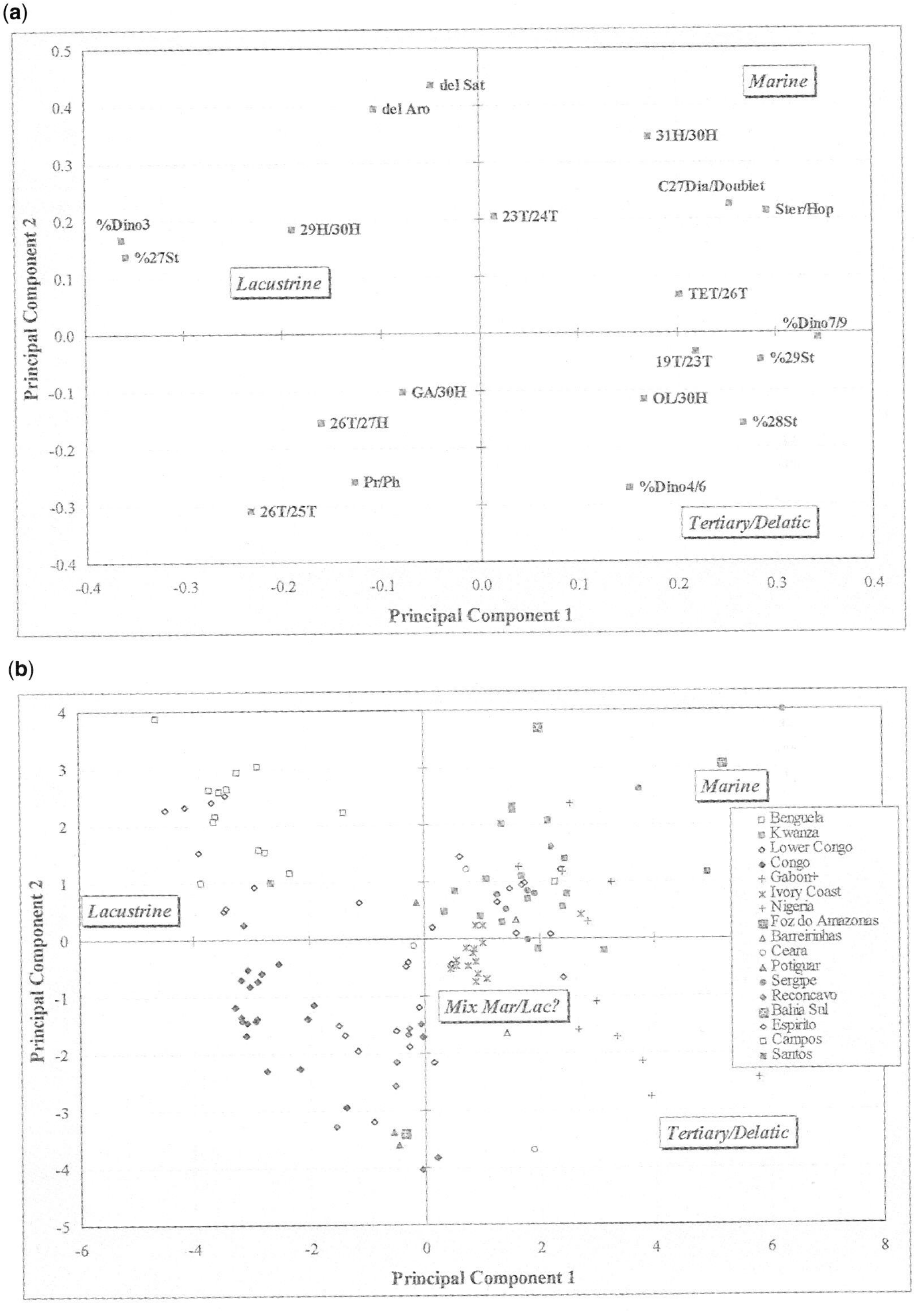

Fig. 10. Principal component (**a**) loadings and plot (**b**) scores for the South Atlantic margin oils.

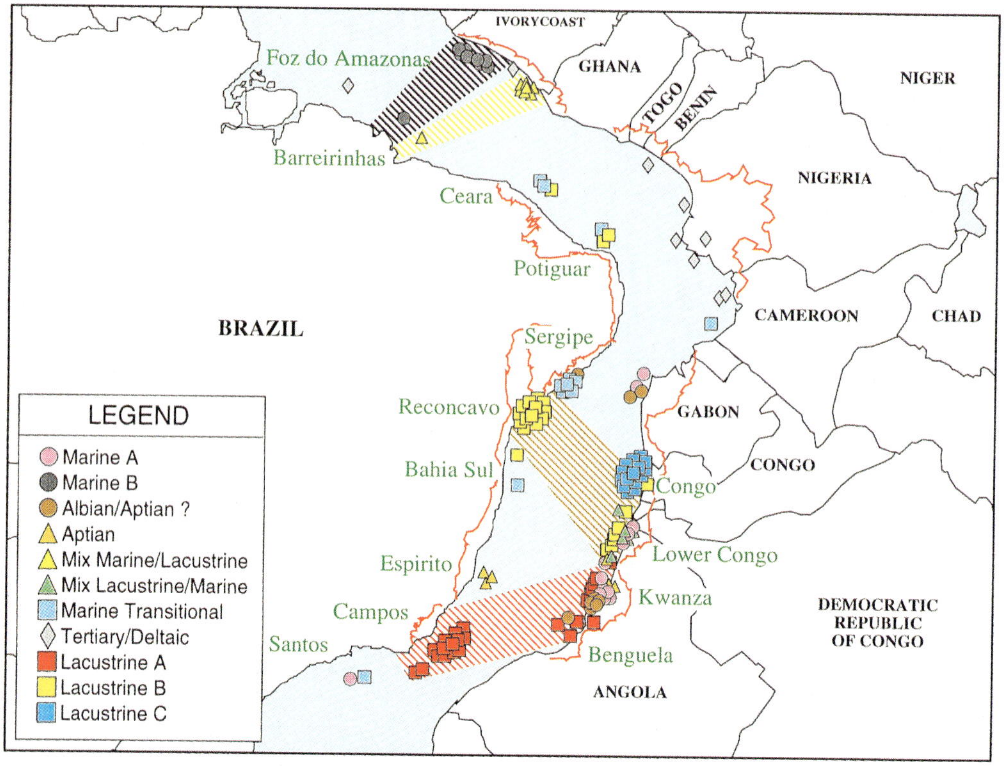

Fig. 11. Oil family location map for the South Atlantic margin oils.

Alternatively, the differences in oil chemistries may be related to facies variations within a single, long-term lake system. Lacustrine oils from the Campos Basin offshore southern Brazil are geochemically similar to lacustrine oils from offshore central and southern Angola (Lower Congo and Benguela Basins), implying an origin from algal source rocks deposited in similar saline lacustrine environments. In addition, oils from the Reconcavo and Lower Congo Basins are geochemically similar, implying an origin from algal source rocks deposited in similar brackish/saline lacustrine environments. None of the Brazilian oils included in this study could be correlated to the West African oils from the Congo Basin, thought to originate from algal source rocks deposited in fresh, deepwater lacustrine environments.

References

BURWOOD, R., JACOBS L. & PAULET, J. 1990. Kerogen pyrolysis–carbon isotope technology: Application to source-oil correlation problems. *Reviews of Palaeobotany and Palynology*, **65**, 367–377.

——, DE WITTE, S. M., MYCKE, B. & PAULET, J. 1995. Petroleum geochemical characterisation of the Lower Congo Coastal Basin Bucomazi Formation. *In*: KATE, B. (ed.) *Petroleum Source Rocks*. Springer, New York, 235–263.

CONNAN, J., LEVACHE, D., SALVATORI, T., RIVA, A., BURWOOD, R. & LEPLAT, P. 1987. Petroleum geochemistry of Angola: An Agelfi Study. *Abstract, 13th International Meeting EAOG*, Venice, 89–90.

DEMAISON, G. & HUZINGA, B. J. 1991. Genetic classification of petroleum systems. *AAPG Bulletin*, **75**, 1626–1643.

DIDYK, B. M., SIMONEIT, B. R. T., BRASSELL, S. C. & EGLINTON, G. 1978. Organic geochemical indicators of paleoenvironmental conditions of sedimentation. *Nature*, **272**, 216–222.

EKWEOZOR, C. M. & UDO, O. T. 1987. The oleananes: Origin, maturation, and limits of occurrence in Southern Nigeria sedimentary basins. *Organic Geochemistry*, **13**, 131–140.

TEN HAVEN, H. 1996. Applications and limitations of Mango's light hydrocarbon parameters in petroleum correlation studies. *Organic Geochemistry*, **24**, 957–976.

——, DE LEEUW, J., SINNINGHE DAMSTE, J., SCHENCK, P., PALMER, S. & ZUMBERGE, J. 1988. Application of biological markers in the recognition of paleo hypersaline environments. *In*: KELTS, K., FLEET, A.

& TALBOT, M. (eds) *Lacustrine Petroleum Source Rocks*. Blackwell, 123–130.

KATZ, B. & MERTANI, B. 1989. Central Sumatra – A geochemical paradox. *Proceedings of the Indonesian Petroleum Society, 18th Annual Conventions*, 403–425.

MALLO, M. R., GAGLIANONE, P. C., BRASSELL, S. C. & MAXWELL, J. R. 1988*a*. Geochemical and biological marker assessment of depositional environments using Brazilian offshore oils. *Marine and Petroleum Geology*, **5**, 205–223.

——, TELNAES, N., GAGLIANONE, P. C., CHICARELLI, M. I., BRASSELL, S. C. & MAXWELL, J. R. 1988*b*. Organic geochemical characterization of depositional paleoenvironments of source rocks and oils in Brazilian marginal basins. *Organic Geochemistry*, **13**, 31–45.

MOLDOWAN, J. M., SEIFERT, W. K. & GALLEGOS, E. J. 1985. Relationship between petroleum composition and depositional environment of petroleum source rocks. *AAPG Bulletin*, **69**, 1173–1180.

——, DAHL, J., HUIZINGA, B. J., FAGO, F. J., HICKEY, L. J., PEAKMAN, T. M. & TAYLOR, D. W. 1994. The molecular fossil record of oleanane and its relation to angiosperms. *Science*, **265**, 768–771.

——, ——, FAGO, F. J. *et al.* 1995 Correlation of biomarkers with geologic age. Abstract 17 International EAOG Meeting San Sebastian, 418–420.

——, FAGO, F. J., LEE, C. Y., *et al.* 1990. Sedimentary 24-*n*-propylcholestanes, molecular fossils diagnostic of marine algae. *Science*, **247**, 309–312.

——, LEE, C. Y., SUNDARARAMAN, P., *et al.* 1992. Source correlation and maturity assessment of select oils and rocks from the Central Adriatic Basin (Italy and Yugoslavia). *In*: MOLDOWAN, J. M., ALBRECHT, P. & PHILP, R. P. (eds) *Biological Markers in Sediments and Petroleum*. Prentice Hall, New Jersey, 370–401.

PETERS, K. E., MOLDOWAN, J. M., SCHOELL, M. & HAMPKINS, W. B. 1986. Petroleum isotopic and biomarker composition related to source rock organic matter and depositional environment. *Organic Geochemistry*, **10**, 17–27.

POWELL, T. 1986. Petroleum geochemistry and depositional setting of lacustrine source rocks. *Marine and Petroleum Geology*, **3**, 200–219.

PU, F., PHILP, R. P., ZHENZI, L., ZINKE, Y. & GUANGGUO, Y. 1991. Biomarker distributions in crude oils and source rocks from different sedimentary environments. *Chemical Geology*, **93**, 61–78.

SEIFERT, W. K. & MOLDOWAN, J. M. 1978. Application of steranes, terpanes, and monoaromatics to the maturation, migration, and source of crude oils. *Geochimica et Cosmochimica Acta*, **42**, 77–95.

—— & —— 1981. Paleoreconstruction by biological markers. *Geochimica et Cosmochimica Acta*, **45**, 783–794.

SOFER, Z. 1984. Stable carbon isotope compositions of crude oils: Application to source depositional environments and petroleum alteration. *AAPG Bulletin*, **42**, 1141–1149.

TEISSERENCE, P. & VILLEMIN, J. 1990. Sedimentary basin of Gabon – Geology and oil systems. *In:* EDWARDS, J. D. (ed.) *Divergent/Passive Margin Basins*. AAPG Memoir, **48**, 117–199.

VOLKMAN, J. K. 1988. Biological marker compounds as indicators of the depositional environment of petroleum source rocks. *In*: KELTS, K., FLEET, A. & TALBOT, M. (eds) *Lacustrine Petroleum Source Rocks*. Geological Society, London, Special Publications, **40**, 103–122.

ZUMBERGE, J. 1987. Prediction of source rock characteristics based on terpane biomarkers in crude oils: A multivariate statistical approach. *Geochimica et Cosmochimica Acta*, **51**, 1625–1637.

Angola: source rock control for Lower Congo Coastal and Kwanza Basin petroleum systems

R. BURWOOD

Fina Exploration Ltd, Fina House, 1 Ashley Avenue, Epsom, KT18 5AD, UK

Abstract: The purpose of this paper is to provide an overview of petroleum occurrence and provenance for the 1000 km West African Atlantic Margin from Cabinda to mid-Angola. Over this margin the Lower Congo Coastal and Kwanza provinces cumulatively account for reserves of *c.* 6 gigabarrels oil recoverable (GBOR). These are dominantly reservoired in Pinda carbonate traps of the former basin. However, with production from a range of aggradational wedge, carbonate platform and pre-salt reservoirs, a diversity in oil character presupposes complex hydrocarbon habitats charged by multiple sourcing.

Each of these two major Atlantic margin salt basins constitutes a different, source rock driven, hydrocarbon habitat. As classic passive margin pull-apart basins, Early Cretaceous initiated rift events (Pre-rift, Syn-rift I, II, etc.) evolved into the drift phase opening of the southern Atlantic. A striking feature of this progression was widespread evaporite deposition of the Aptian Loeme salt. This separates two distinct sedimentary and tectonic domains of the Pre- and Post-Salt.

The core Lower Congo Coastal habitat is dominated by the Pre-Salt Bucomazi Formation sourced 'poly' petroleum system. These lacustrine, often super-rich, sediments reveal considerable organofacies variations between their basin fill (Syn-rift I) and sheet drape (Syn-rift II) development, accounting for the compositional diversity in their progenic petroleums. Of crucial impact is a cognate diversity in their kerogen kinetic behaviour. This controls the conditions and timing of generation and realization of charge potential.

With the Lower Congo Coastal habitat extending southwards towards the Ambriz Spur, the Bucomazi facies proper appears restricted to the northern and deeper proto-lake trend. Over the more weakly subsident margins such troughs host inferior sheet drape potential. Elsewhere, Upper Cretaceous–Palaeogene marine clastic Iabe Formation sourced petroleum systems are hydrocarbon productive, yielding petroleums of unique, and/or mixed Pre-Salt hybrid, source provenance.

A less attractive Pre-Salt source system is reflected in the Kwanza habitat. Consequential effects of salt movement have proved important for halokinetic induced structuration, Cretaceous block rafting and pseudograben (Tertiary trough) source deposition. Marginal reserves of Infra-Cuvo and Cuvo derived petroleums are known. With limitations in charge potential at this level, the modest Kwanza reserves (0.1 GBOR) have a Post-Salt provenance deriving primarily form the Eocene Margas Negras and subordinate Aptian Middle Binga sourced petroleum systems. Cases of multi-sourced hybrid petroleums abound. In addition to the organic-rich Albian micrites, Upper Cretaceous black shales provide unproven possible offshore potential.

The Angolan Atlantic margin hosts several basins collectively seen as part of a greater West African Aptian Salt Basin (Edwards & Bignell 1988). A schematic chrono-stratigraphy for the main features, summarizing tectonic and source reservoir information, is illustrated in Fig. 1. Although sharing essentials of a common chrono-tectonic development from rift inception in the earliest Cretaceous, the component Lower Congo Coastal and Kwanza Basins have disparate structural and sedimentary styles. The latter constraint influences both source rock deposition and reservoir fabric considerations. Moreover, the basins are not contiguous with transform fault delineated regional highs (e.g. Ambriz, Cabo Ledo) separating intermediate domains.

Nevertheless, Aptian salt (Loeme Formation) forms an important regional marker separating an essentially non-marine, lacustrine syn-rift depositional regime (Pre-Salt) from the drift phase (Post-Salt) marine conditions attendant to the opening of the southern Atlantic Ocean. In the Lower Congo Coastal area, extensive and deep lacustrine depocentres have hosted accumulation of thick, source quality, Pre-Salt sediments of the Bucomazi Formation (Burwood *et al.* 1995). South of the Zaire River canyon, and certainly the Ambriz Spur, such time-equivalent Pre-Salt sediments are not so richly and/or necessarily widely developed. Herein lies one important reason for the present apparent disproportionate prospectivity (Fig. 2) in favour of the Lower Congo Coastal Basin over the Kwanza Basin.

Elsewhere, marine Post-Salt candidate sources, showing patchy development in the Albian and

BURWOOD, R. 1999. Angola: source rock control for Lower Congo Coastal and Kwanza Basin petroleum systems. *In:* CAMERON, N. R., BATE, R. H. & CLURE, V. S. (eds) *The Oil and Gas Habitats of the South Atlantic.* Geological Society, London, Special Publications, **153**, 181–194.

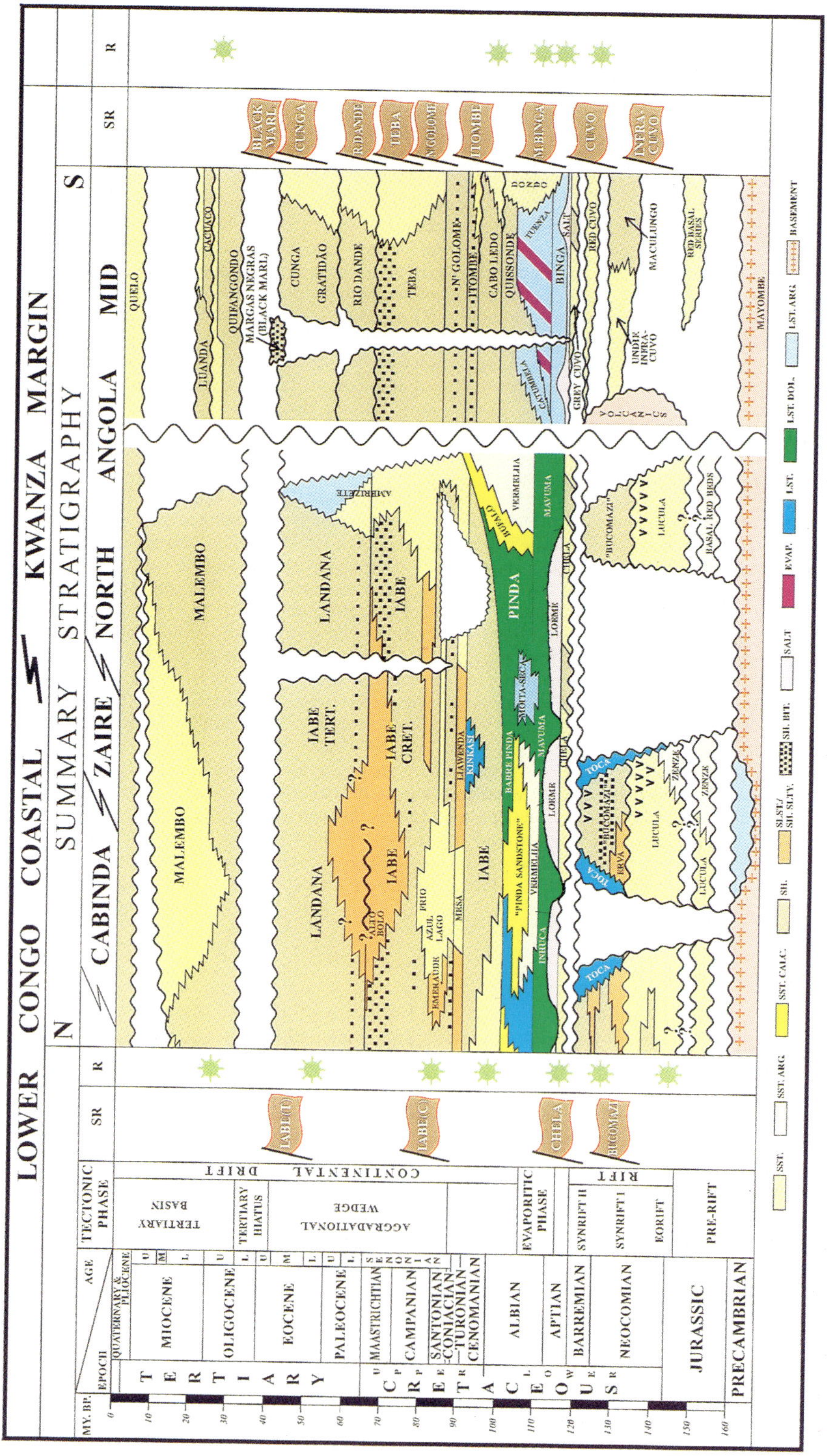

Fig. 1. Schematic stratigraphy for the Lower Congo Coastal and Kwanza Basins summarizing major tectonic phases and local source–reservoir relationships.

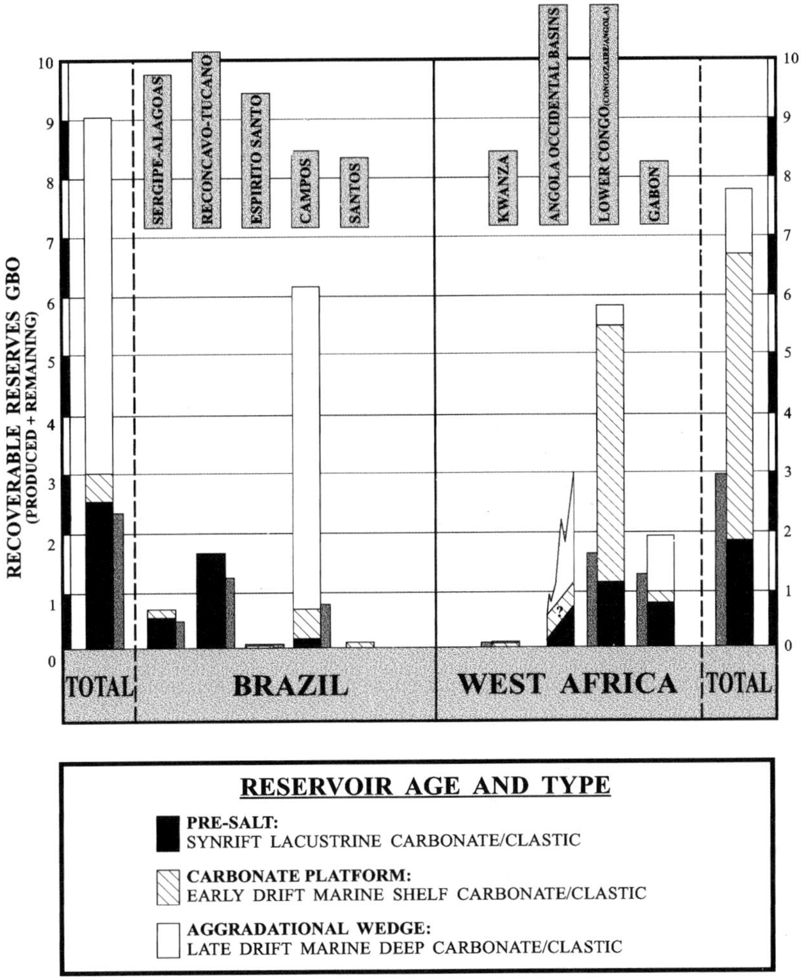

Fig. 2. South Atlantic margin: comparative estimates of cumulative crude oil reserves (bor, barrels oil recoverable) for the productive West African and Brazilian basins.

more ubiquitously as Late Cretaceous black shales [Iabe (Cretaceous), Itombe–Teba Formations, etc.], are found over the margin but are often burial-constrained with respect to maturity (Burwood *et al.* 1990*a*). Widespread Palaeogene candidate source development [Iabe (Tertiary), Rio Dande–Cunga Formations] share a similar limitation. Halokinetic block-rafting of the Late Cretaceous section, particularly in the Kwanza area (Brognon & Verrier 1966), resulted in Tertiary troughs that hosted rich Eocene, and locally productive, source rock developments of the Margas Negras Formation (Burwood *et al.* 1990*a*).

In the present enquiry an understanding of the operative hydrocarbon habitat has been approached from a 'petroleum system' vantage (Magoon & Dow 1994), aided by contemporary geochemical data acquisition practices and interpretative concepts such as 'geochemical inversion' (Bissada *et al.* 1992). Here, the petroleum system concept can be used as an effective model to account for discovered hydrocarbon accumulations and by implication be applied to evaluation of non- or under-explored acreage.

For synthesis of such an approach, key objectives to be addressed include:

- crude oil characterization leading to the recognition of diversity/multiplicity in generic families, hence petroleum system(s);
- source rock candidature, characterization and quantification of petroleum potential;
- source–oil assignments confirming hydrocar-

bon provenance and the existence, plus any diversity, in operative petroleum system(s);

- evaluation of considerations relating to oil generation and emplacement history, mixed-source aggregate fluids and any post-emplacement/in-reservoir alteration processes;
- quantification of basinal generation and charge volumetrics as input to prospect evaluation and risk analysis.

In the body of this contribution *de novo* data is summarized primarily in support of the first two of these concerns. This permits realization of the subsequent considerations, revealing that the 'Aptian Salt Basin' is a complex hydrocarbon habitat where unique oils, plus mixed aggregates derived from hybrid petroleum systems, are encountered. This naturally has implications regarding the diversity of play types that could be

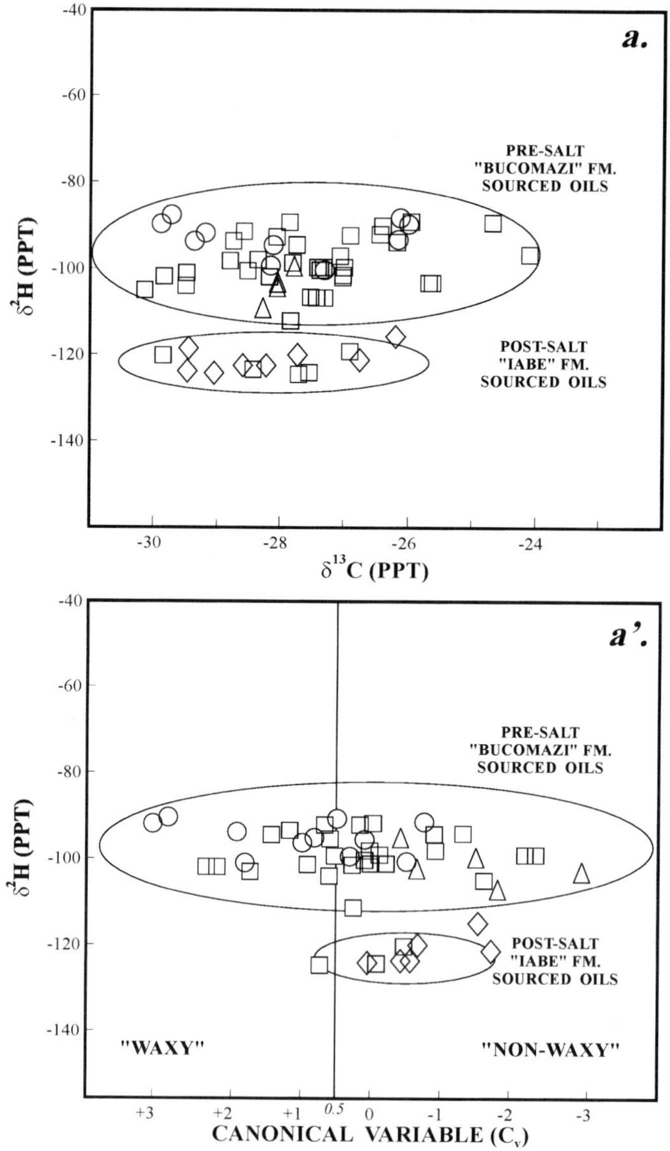

Fig. 3. Differentiation of: (**a**) Lower Congo Coastal and (**b**) Kwanza Basin oils into lacustrine (Pre-Salt) and marine (Post-Salt) depositional source-style provenance in terms of stable isotope (^2H, ^{13}C)-derived parameters. For definition of the canonical variable (C_v) see Appendix. Reservoirs identified as: ○, Pre-Salt; ⓥ, Pre-Salt volcanics; □, early drift carbonate platform; Δ, late drift marine clastics; ◇, Late Palaeogene – Neogene clastics.

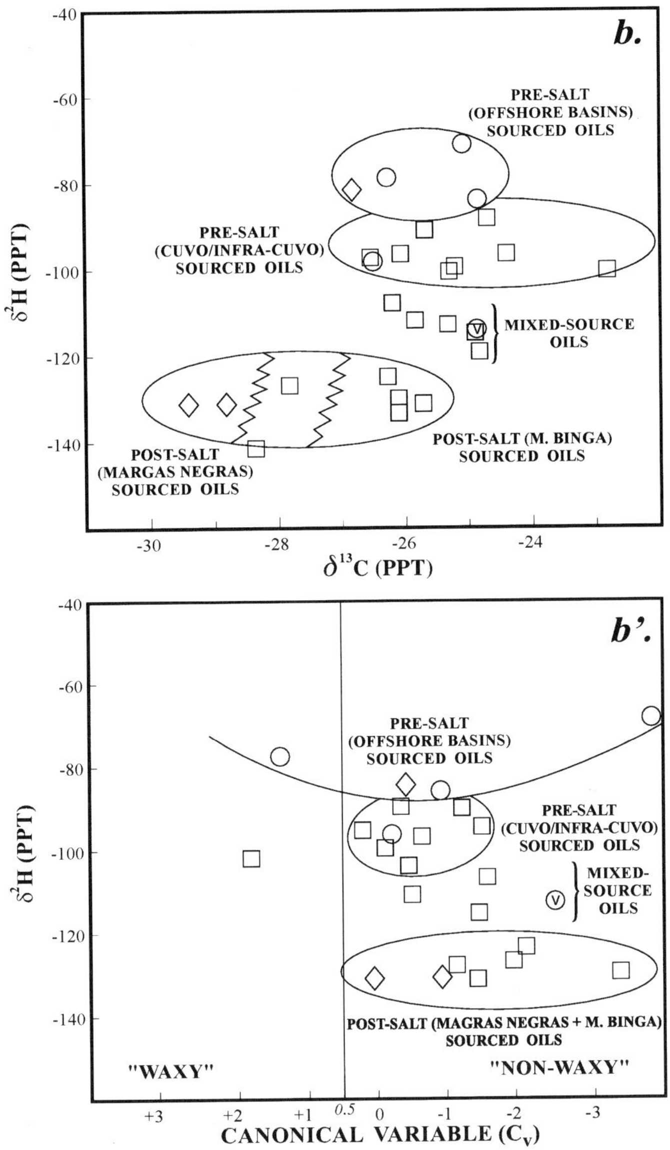

Fig. 3. (*continued*)

anticipated and the formulation of exploration strategies.

Petroleum characteristics

Angolan margin oil type and quality can be highly disparate, ranging from average gravity crudes of variable waxiness to heavier in-reservoir biodegraded (Cegonha) or condensate-like (Puaca) fluids and gas accumulations (Livuite). A multiplicity of origins is implicated.

In this context, oil data inversion analysis has provided an invaluable insight into the character and identity of the contributory source rocks. Stable isotope procedures provide a robust test for initial differentiation of Pre-Salt (lacustrine sourced) from the Post-Salt (marine sourced) oils (Fig. 3a, b). Whereas oils of lacustrine source rock provenance are habitually enriched in ^{2}H, note the wide ^{13}C signature range (8 ppt), implying large source organofacies variations and the ability to recognize hybrid, dual-origin oils. For the Lower Congo Coastal Basin Pre-Salt origin oils, note also the greater abundance of a 'waxy' connotation (canonical variable parameter; Sofer 1984). This again implies a variation in source rock organofacies between this habitat and that of the Kwanza Basin. These gross segregations can be further refined through application of biomarker signatures to source palaeoenvironmental interpretations (cf. Schiefelbein *et al.* 1999). Including sterane/hopane and tricyclic terpane ratios (differentiation of marine v. non-marine influences; Fig. 4); gammacerane (hypersalinity/water column stratification) and 18[α]oleanane (Tertiary terrestrial input; Fig. 5) abundances, these data help to identify viable source–oil correlation assignments.

Candidate source rock systems

Over the Angolan margin, 12 discrete candidate source rock systems have been recognized, quantified as to potential and typed according to generic signature (Figs 6a–f and 7a–f). Each system is denoted as to its operative domain as: LCCB, Lower Congo Coastal Basin; NAOB, Northern Angolan Offshore Basins; KB, Kwanza Basin, KOB, Kwanza Offshore Basins. The contributory source rocks vary in age from Early Cretaceous (Neocomian?–Barremian) to Eocene and are productive either individually or in combination, as hybrid petroleum systems. Due to its complexity, and extent, the Bucomazi Formation should be treated as a poly-system, its diverse organofacies resulting in a spectrum of progenic hydrocarbons.

For the description of each candidate source rock system a comprehensive data set was assembled for a type section setting where possible. This was aimed at providing source quantification information in the form of typical effective thickness, hydrocarbon proneness (GOPR value) and richness (formation mean S2 value), the latter being translated into petroleum potential values either in metric ($m^3\,km^{-3}$) or traditional (bbl/acre ft) units. As an aid to source–oil correlation assignments, the formation range and source potential weighted mean kerogen pyrolysate carbon isotope signature is quoted, together with pertinent diagnostic biomarker information where appropriate. For purposes of conducting

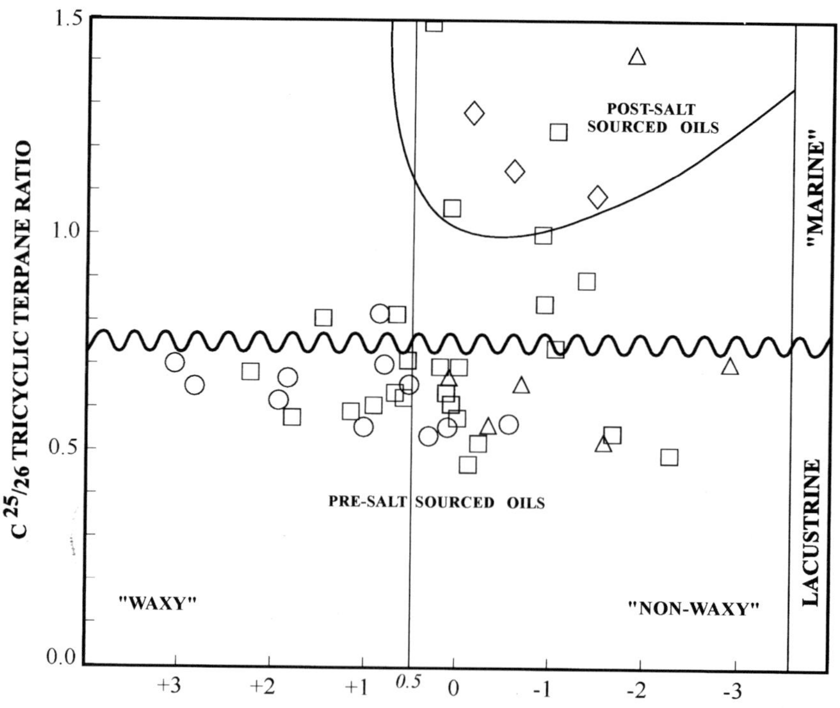

Fig. 4. Differentiation of Lower Congo Coastal Basin oils into lacustrine (non-marine) and marine source depositional-style genera using biomarker (C_{25}/C_{26} tricyclic triterpane) ratio and stable carbon isotope (C_v) derived parameters. Reservoir identity as in Fig. 3.

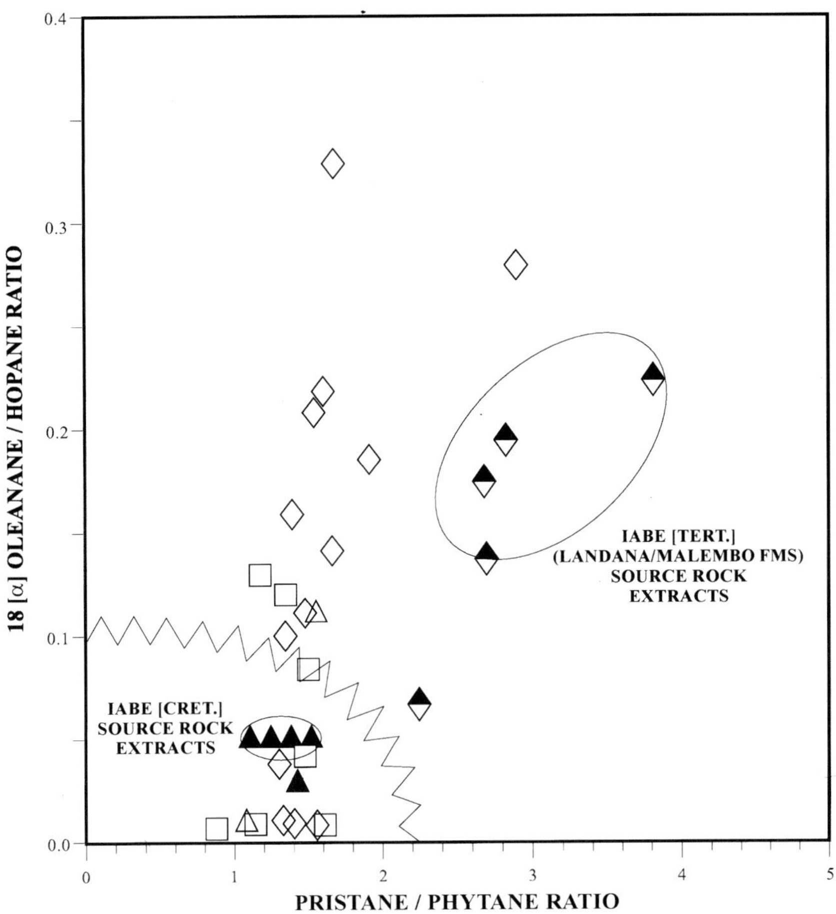

Fig. 5. Differentiation of Lower Congo Coastal Basin Post-Salt marine depositional source-style in terms of the 18[α]oleanane/hopane ratio segregating those oils of exclusive Iabe (Cretaceous) provenance from analogues showing a greater/lesser Iabe (Tertiary) source influence. Reservoir identity as in Fig. 3, source rock extracts denoted as: ⬘, Iabe (Tertiary); ▲, Iabe (Cretaceous).

kinetic-controlled basin modelling, a typical kerogen kinetic data set (E_a, A), as determined by the Rock-Eval 5/Optkin procedure, is also included. A more detailed treatment of the Bucomazi and Chela Formations in this way has been previously published (Burwood *et al.* 1995). A glossary as to the source evaluation parameters employed and petroleum system designation nomenclature is given in the Appendix.

Petroleum systems – discussion

Individual petroleum systems, with respect to their source rock component, are allocated to operative domains in the Lower Congo Coastal Basin (LCCB); Northern Angolan Offshore Basins (NAOB); Kwanza Basin (KB) and Kwanza Offshore Basins (KOB), as illustrated in Fig. 8. Due to the multiplicity in trap opportunities (cf. Fig. 1), numerous variations on the reservoir theme exist in completion of petroleum system definition. The illustrated systems, although reflecting a dominant reservoir/reserves style, should not be seen as exclusive.

To the north (Lower Congo Coastal Basin proper), Bucomazi Formation sourcing dominates in the form of the poly (!*) petroleum system (Fig. 6*a*, *b*). Here, it is important to distinguish between the basin fill organic-rich zone as opposed to the kinetically more labile, but less subsident, sheet drape (Late Bucomazi Formations sediments, the activation energy contrast having substantial implications for

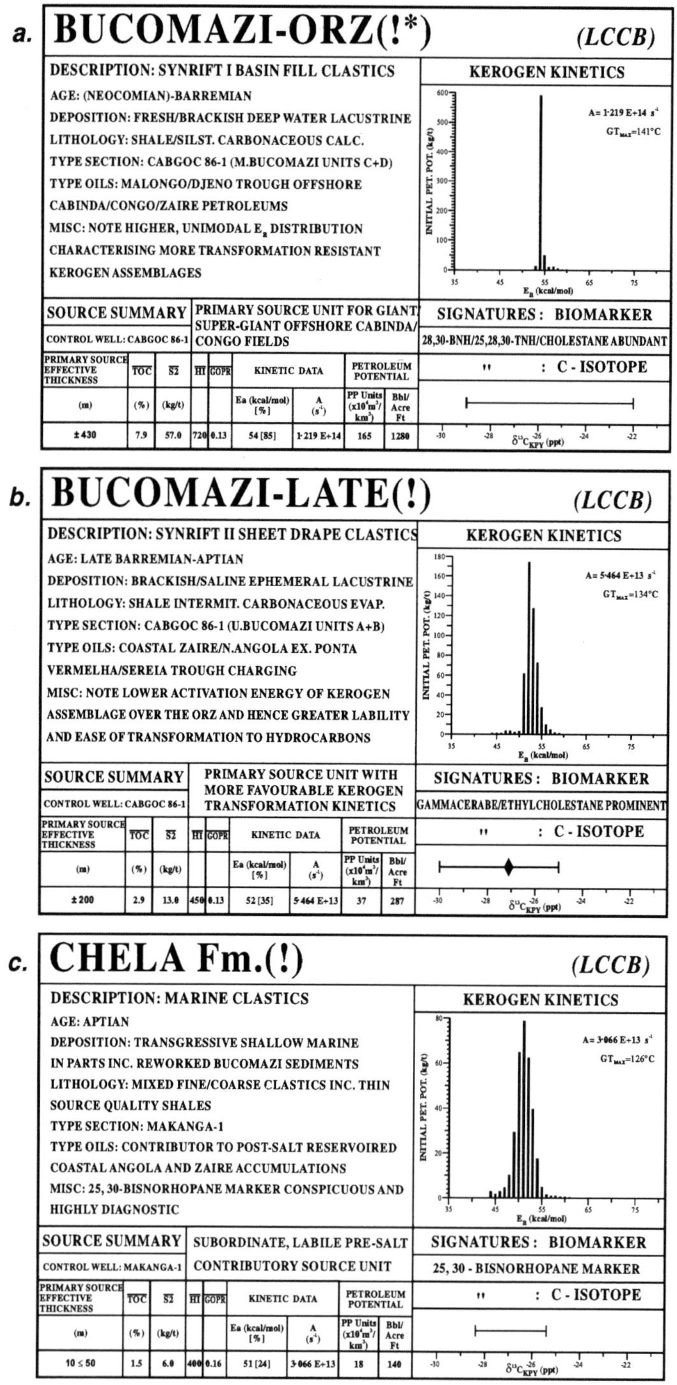

Fig. 6. Summary data compendia for the source rock components of petroleum systems operative in the Lower Congo Coastal and Northern Angola Offshore Basins. See Appendix for petroleum system nomenclature and definition of source evaluation parameters employed.

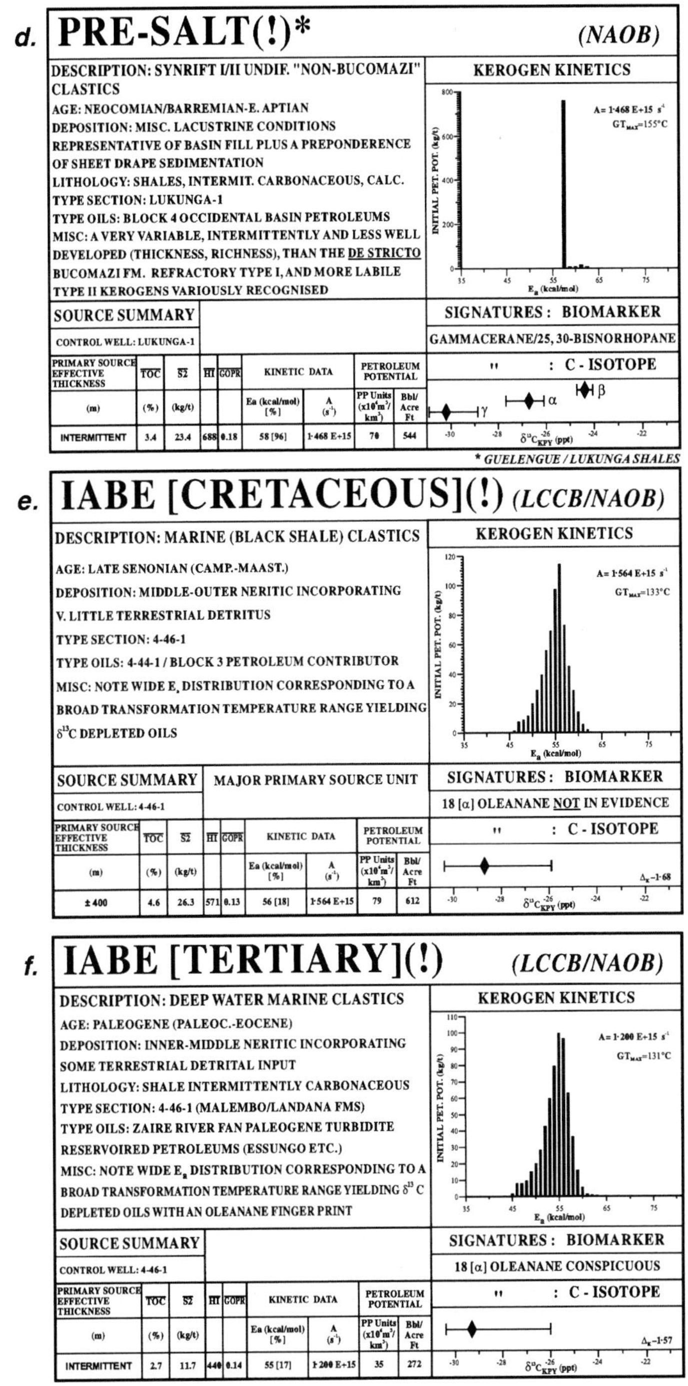

Fig. 6. (*continued*)

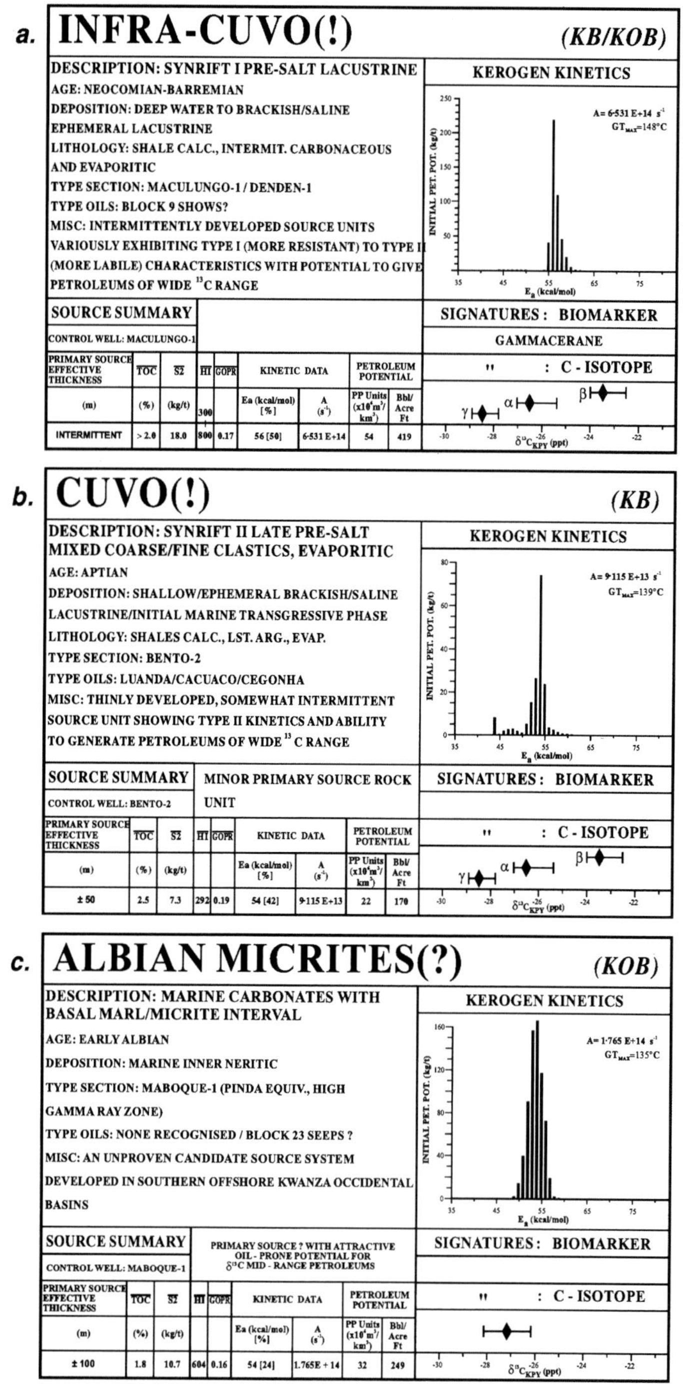

Fig. 7. Summary data compendia for the source rock components of petroleum systems operative in the Kwanza and Kwanza Offshore Basins. See Appendix for petroleum system nomenclature and definition of source evaluation parameters employed.

d. M. BINGA Fm.(!) *(KB)*

DESCRIPTION: SHALLOW MARINE CARBONATES /MARLS/MICRITES

AGE: APT./ALBIAN

DEPOSITION: TRANSGRESSIVE RESTRICTED (LITTORAL/ PARALIC) MARINE HOSTING THIN ORGANIC-RICH MICRITES

TYPE SECTION: TOBIAS-6 (M.BINGA/SUB-LITHOGRAPHIC ZONE)

TYPE OILS: CABO LEDO HIGH FIELDS (TOBIAS, GALINDA)

MISC: THINLY DEVELOPED "SLEEPER" SOURCE UNIT CHARGING ca 150 MBOIP

KEROGEN KINETICS

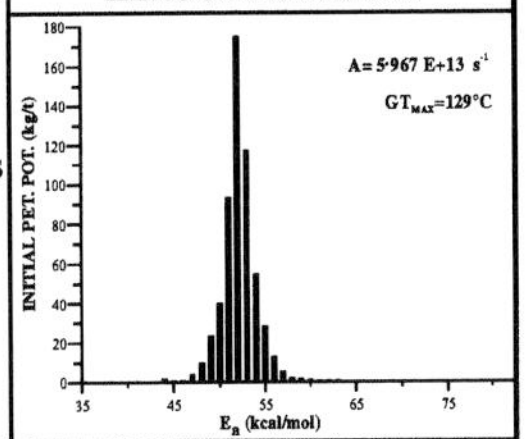

SOURCE SUMMARY — THINLY DEVELOPED, HIGH GAMMA RAY SEDIMENTS SUPPORTING SUPER-RICH LABILE KEROGEN ASSEMBLAGES

CONTROL WELL: TOBIAS-6

PRIMARY SOURCE EFFECTIVE THICKNESS	$\overline{TOC}$	$\overline{S2}$	$\overline{HI}$	$\overline{GOPR}$	KINETIC DATA		PETROLEUM POTENTIAL	
(m)	(%)	(kg/t)			Ea (kcal/mol) [%]	A (s^{-1})	PP Units ($x10^6m^3/km^3$)	Bbl/ Acre Ft
5 ≤ 10	6.3	39.0	614	0.15	52 [30]	5·967 E+13	117	907

SIGNATURES : BIOMARKER

GAMMACERANE - RICH

" : C - ISOTOPE

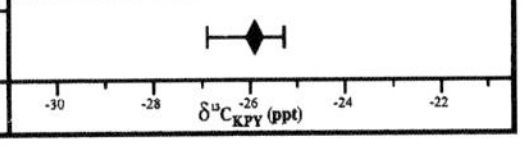

e. MARGAS NEGRAS(!) *(KB)*

DESCRIPTION: MARINE "BASAL BASIN FILL" CLASTICS

AGE: EOCENE

DEPOSITION: TERTIARY TROUGH FILL ATTENDANT TO CRETACEOUS SECTION BLOCK RAFTING (HALOKINETICS)

LITHOLOGY: SHALE, CARBONACEOUS

TYPE SECTION: FUNDA-3/BLACK MARL FM.

TYPE OILS: QUENGUELA TROUGH PETROLEUMS (QUENGUELA NORTH, LEGUA, ETC.)

MISC: A "MICRO-HYDROCARBON HABITAT" FACILITATED BY BLOCK RAFTING DEPOSITION OF LABILE KEROGEN ASSEMBLAGES

KEROGEN KINETICS

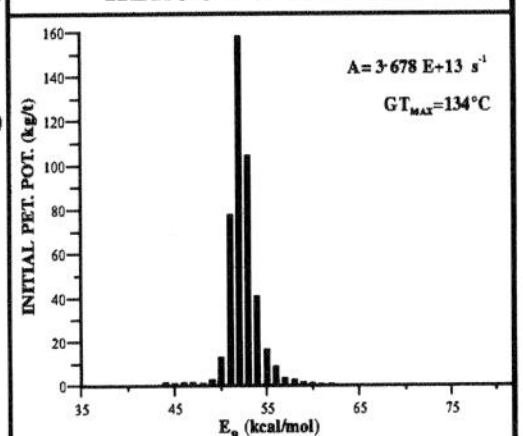

SOURCE SUMMARY — PRIMARY SOURCE CHARGING "MICRO-HABITAT" OF 221 MBOIP

CONTROL WELL: FUNDA-3

PRIMARY SOURCE EFFECTIVE THICKNESS	$\overline{TOC}$	$\overline{S2}$	$\overline{HI}$	$\overline{GOPR}$	KINETIC DATA		PETROLEUM POTENTIAL	
(m)	(%)	(kg/t)			Ea (kcal/mol) [%]	A (s^{-1})	PP Units ($x10^6m^3/km^3$)	Bbl/ Acre Ft
5 < 50	3.6	19.0	524	0.18	52 [36]	3·678 E+13	57	442

SIGNATURES : BIOMARKER

" : C - ISOTOPE

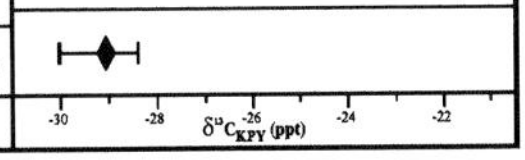

f. TEBA-ITOMBE(?) *(KB/KOB)*

DESCRIPTION: MARINE (BLACK SHALE) CLASTICS

AGE: LATE CRETACEOUS (CENO.-MAAST.) PACKAGE

DEPOSITION: MIDDLE-OUTER NERITIC. MINIMAL TERRESTRIAL ORGANO-DETRITAL INPUT

TYPE SECTION: MACULUNGO-1/CUNGA-1

TYPE OILS: NONE REOGNISED TO DATE

MISC: POTENTIALLY V.ATTRACTIVE SOURCE SYSTEM APPARENTLY SUFFERS FROM INSUFFICIENT BURIAL OVERBURDEN VS REQUIRED GT_{MAX} OF ca 133°C

KEROGEN KINETICS

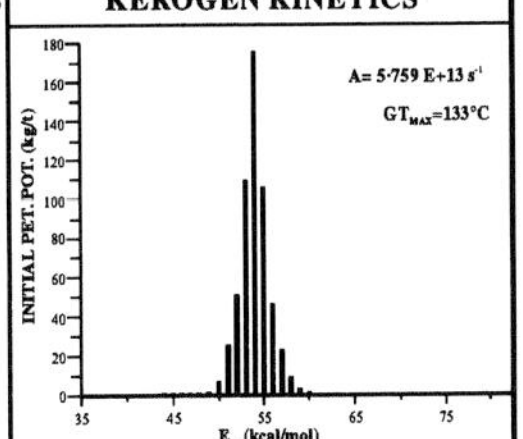

SOURCE SUMMARY — EXCELLENT UN-REALISED OIL-PRONE SOURCE POTENTIAL

CONTROL WELL: COMPOSITE A/B

PRIMARY SOURCE EFFECTIVE THICKNESS	$\overline{TOC}$	$\overline{S2}$	$\overline{HI}$	$\overline{GOPR}$	KINETIC DATA		PETROLEUM POTENTIAL	
(m)	(%)	(kg/t)			Ea (kcal/mol) [%]	A (s^{-1})	PP Units ($x10^6m^3/km^3$)	Bbl/ Acre Ft
± 600	4.8	17.1	358	0.18	54 [31]	2·202 E+14	51	398

SIGNATURES : BIOMARKER

" : C - ISOTOPE

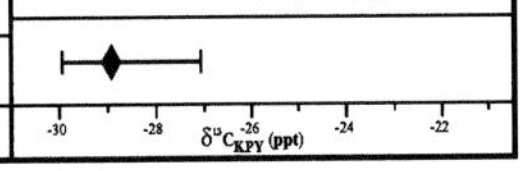

Fig. 7. (*continued*)

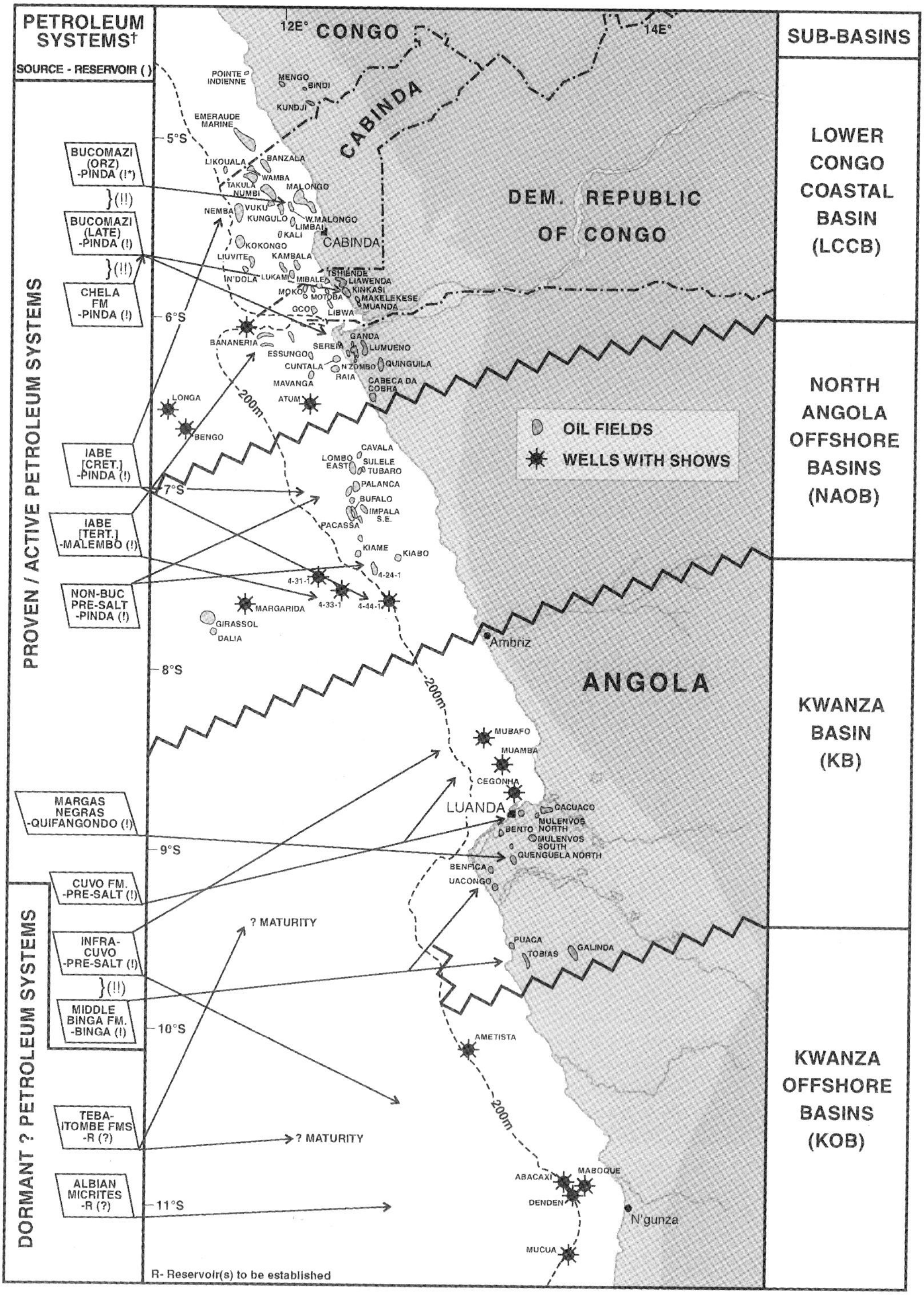

Fig. 8. Angolan Atlantic margin with location of major and/or significant oil accumulations and distribution of the contributory petroleum systems in terms of their source rock component(s). Petroleum system† definition reflects a dominant, but not necessarily exclusive, reservoir style.

generation timing and volumetrics. In this way the hybrid provenance of the super-giant Malongo trend oils (organic-rich zone dominated) can be differentiated from the more modest Late Bucomazi derived coastal Zaire and northwest Angola plays (Tshiende, N'Zombo Fields, etc.). The latter oils also often show a prominent 25,30-bisnorhopane signature indicative of a source contribution from the minor, curiosity Chela Formation (Fig. 6c) system (Burwood *et al.* 1990*b*).

Additionally, a conspicuous feature of these Lower Congo Coastal oils is their wide range of carbon isotopic signatures and variable, often high, waxiness (Fig. 3a), this being a feature of the disparate organofacies of the poly-petroleum system. According to thermal maturation history, and kerogen kinetic control, much scope exists for co- and/or sequential generation over a broad stratigraphic section. The type section in CABGOC 86-1 is over 750 m in extent (Burwood *et al.* 1995), whilst McHargue (1990) reports up to 1800 m of Bucomazi being present in the deepest depocentres. This provides opportunity for mixing of charges of equally disparate characteristics.

In the more occidental offshore Lower Congo, the marine deposited organic-rich Iabe (Cretaceous) and Palaeogene Iabe (Tertiary) (Fig 6e, f) are prospective (e.g. Nemba, Mavanga, Essungo, Bufalo Fields, etc.) These oils are typically depleted in ^{2}H and show the 'non-waxy' source characteristics associated with dominantly neritic marine-deposited sediments (Fig. 3a) 18[α]Oleanane abundance, being prominent in the Iabe (Tertiary) derived oils, helps differentiate between the relative source contributions (cf. Fig. 5). The opportunity for hybrid Bucomazi–Iabe petroleums exists (e.g. Raia, Atum, Cabeca da Cobra Fields, etc.).

Southwards, Northern Angola Offshore Basin (NAOB) Pre-Salt source potential is less extensive and more thinly developed, but still remains prospective as witnessed by charge to the Kiame and Kiabo Fields. Enriched in ^{2}H, the progenic oils are of the ^{13}C heavier variety and fit the 'non-waxy' Pre-Salt derived genera (Fig. 3a). This is taken to imply a provenance from more ephemeral lacustrine-deposited sediments, not unlike those seen in the Late Bucomazi Formation. The Pre-Salt (non-Bucomazi) system nevertheless shows rather refractory kerogen assemblages (Fig. 6d), requiring a forceful generative regime for realization of potential. Elsewhere, sourcing from the Iabe formations is more prevalent, with unique Iabe (Cretaceous) oils found at 4-44-1, Cavala and Impala, etc. Similarly, evidence for hybrid Pre-Salt–Iabe (Cretaceous) (e.g. Sulele, Palanca, Pacassa), and indications of Iabe (Tertiary) influenced oils (e.g. Margarida), have become apparent (Fig. 5).

South of the Ambriz Spur in the Kwanza Basin (KB), Pre-Salt (Infra-Cuvo and Cuvo Formations) potential becomes more tenuous (Fig. 7a, b). Shows ascribable to such sourcing (e.g. Mubafo, Cegonha through Ametista to Mucua), in addition to marginal onshore production (Luanda, Cacuaco Fields), set the scene. However, even the level of original oil in place estimates (< 10 MB001P) would imply that there is a paucity of meaningful source rock development, or that some other impediment to effective generation or to emplacement and accumulation exists.

The main Kwanza onshore production is dominated by two very localized and curiosity source systems (Burwood *et al.* 1990*a*). Thinly developed micrites, with labile kerogen assemblages, of the Albian Middle Binga Formation provide charge to the Tobias and Galinda Fields (Fig. 7d). Elsewhere, Quenguela North and Mulenvos North oils are Margas Negras Formation derived, these Eocene sediments being unconformably deposited on the Aptian Loeme–Cuvo Formations within halokinetic block-rafted Tertiary troughs (Fig. 7e). Instances of hybrid oils, being aggregates from these two sources and the Pre-Salt, have been widely recognized (e.g. Bento, Benfica and Uacongo Fields).

For the overall Kwanza area (KB/KOB), the most attractive potential is considered to be associated with marine Late Cretaceous and Palaeogene sediments. The Teba–Itombe (and Paleogene Cunga–Rio Dande) packages show sustained oil-prone development but suffer from regional immaturity (Fig. 7f). Additionally, intermittent onshore Middle Cretaceous source indications (Quissonde Formation) are more impressively developed in the offshore as the time-equivalent Albian Micrites (Fig. 7c). At appropriate maturity these oil-prone sediments would be prospective yielding progenic oils with distinct carbon isotope signatures.

Conclusions

A combination of detailed source rock evaluation and oil data inversion analysis has led to the recognition and assignment of 12 candidate petroleum systems over the greater Angolan margin. To account for the generic variety in oil types, the concepts of 'poly' and 'hybrid' petroleum systems have been introduced. Although of a distribution restricted to the core of the Lower Congo Coastal Basin, the Bucomazi Formation is unquestionably the premier regional source rock and responsible for a larger part of the cumulative proven reserves. Lesser, but still attractive, source potential is also observed for the non-Bucomazi Pre-Salt and marine deposited Late Cretaceous–Palaeogene Iabe Formation sediments over the sub-Lower Congo margin. Whereas proven reserves in the Kwanza Basin can be ascribed to two marginal petroleum systems (Middle Binga and Margas Negras Formations), the Albian micrites and Late Cretaceous shales have considerable, as yet unrealized, potential in the offshore province.

References

BISSADA, K. K., ELROD, L. W., DARNELL, L. M., SZYMEZYK, H. M. & TROSTLE, J. L. 1992. Geochemical inversion – a modern approach to inferring source-rock identity from characteristics of accumulated oil and gas. *Proceedings of the Indonesian Petroleum Association*, **21**, 165–199.

BROGNON, G. & VERRIER, G. 1966. Oil and geology in Kwanza Basin Angola. *AAPG Bulletin*, **50**, 108–158.

BURWOOD, R., JACOBS, L. & PAULET, J. 1990*a*. Kerogen pyrolysis – carbon isotope technology: Application to source-oil correlation problems. *Review on Paleobotany and Palynology*, **65**, 367–377.

——, CORNET, P. J., JACOBS, L. & PAULET, J. 1990*b*. Organofacies variation control on hydrocarbon generation: A Lower Congo Coastal Basin (Angola) case history. *Organic Geochemistry*, **16**, 325–338.

——, DE WITTE, S. M., MYCKE, B. & PAULET, J. 1995. Petroleum geochemical characterisation of the Lower Congo Coastal Basin Bucomazi Formation. *In*: KATZ, B. J. (ed.) *Petroleum Source Rocks*. Springer, 235–263.

EDWARDS, A. & BIGNELL, R. 1988. Hydrocarbon potential of West African salt basin. *Oil and Gas Journal*, **86**, 71–74, (12 Dec), 55–58 (19 Dec).

MCHARGUE, T. R. 1990. Stratigraphic development of proto-South Atlantic rifting in Cabinda, Angola – a petroliferous lake basin. *In*: KATZ, B. J. (ed.) *Lacustrine Basin Exploration Case Studies and Modern Analogs*. AAPG Memoir, **50**, 307–326.

MAGOON, L. B. & DOW, W. G. 1994. The petroleum system. *In*: MAGOON, L. B. & DOW, W. G. (eds) *The Petroleum System – From Source to Trap*. AAPG Memoir, **60**, 3–23.

SCHIEFELBEIN, C. F., ZUMBERGE, J. E., CAMERON, N. R. & BROWN, S. W. 1999. Petroleum systems in the South Atlantic margins. *This volume*.

SOFER, Z. 1984. Stable carbon isotope compositions of crude oils: application to source depositional environments and petroleum alteration. *AAPG Bulletin*, **68**, 31–49.

Appendix

Source evaluation parameters

PP units	Petroleum Potential ($\times 10^6$ m$^3_{(OE)}$ km$^{-3}_{(rock)}$) calculated from S2 value (OE, oil equivalent of 35° API fluid; rock, source rock with density of 2.55 g cm^{-3}). Conversion factor to bbl/acre ft ($\times 7.758 \times 10^{-6}$)
GOPR	Gas–oil Proneness Ratio (0.0–1.0). Pyrolysis, gas chromatographic measure of kerogen HC product, zero and unity being exclusively oil prone and gas prone, respectively
$\delta^{13}C_{KPY}$	Kerogen Pyrolysate (Simulated Oil) Carbon Isotope Signature, formation mean values are source richness weighted against S2 and calculated thus: $\delta^{13}C_{(pyrolysate)} = [\Sigma\delta^{13}C_{(pyrolysate)} \times S2/\Sigma S2]$ Values reported v. NBS22 at –29.8 ppt (PDB)
E_a/A	Kerogen transformation kinetics determined on bitumen-free sediments using the Rock-Eval 5/Optkin procedure
$GT_{(max)}$	Real (geological) time optimum 50% kerogen transformation temperature at heating rate 1°C Ma^{-1}.
C_v	Canonical variable $[(-2.53\delta^{13}C_{sat} + 2.22\delta^{13}C_{arom}) - 11.65]$

Petroleum system nomenclature (after Magoon & Dow 1994)

Known/simple (proven source–oil correlation)	(!)
Known/complex (poly-system)	(!*)
Known/complex (proven hybrid system)	(!!)
Hypothetical (prognosed with confidence)	(.)
Hybrid/part proven (major source known)	(!.)
Speculative	(?)

Upper Cretaceous palaeoenvironments and benthonic foraminiferal assemblages of potential source rocks from the western African margin, Central Atlantic

ANN HOLBOURN[1], WOLFGANG KUHNT[2], ABDERRAZZAK EL ALBANI[2], THOMAS PLETSCH[2], FLORIAN LUDERER[2] & THOMAS WAGNER[3]

[1]*Department of Palaeontology, Natural History Museum, Cromwell Road, London SW7 5BD, UK*

[2]*Geologisch–Paläontologisches Institut, Christian-Albrechts-Universität zu Kiel, Olshausenstr. 40, D-24118 Kiel, Germany*

[3]*Fachbereich 5, Geowissenschaften, Universität Bremen, Klagenfurter Str. 5, 28359 Bremen, Germany*

Abstract: Upper Cretaceous benthonic foraminiferal assemblages, clay minerals, kerogen types and carbonate microfacies are examined in organic-rich sediments from western African coastal basins along a latitudinal transect extending from southern Morocco to Nigeria. Samples from onshore sections of the Tarfaya Basin, southern Morocco, from DSDP Site 369, from offshore commercial wells on the Casamance Shelf, Senegal, from DSDP Site 367, from ODP Site 959 and from onshore sections of the Benue Trough in Nigeria, are analysed. Organic-rich sediments from these sites are characterized by smectite-rich clay mineral assemblages, high amounts of marine organic matter and a typical benthonic foraminiferal biofacies dominated by buliminids and bolivinids. Benthonic foraminifera provide robust proxy indicators of palaeoproductivity and organic matter flux to the seafloor. The palaeobiogeographic distribution of benthonic foraminiferal biofacies can be used to identify areas and stratigraphic intervals of hydrocarbon accumulation. Lower Turonian benthonic foraminiferal assemblages exhibit extremely low diversity and are strongly dominated by a *Gabonita* biofacies, in contrast to Coniacian to Maastrichtian assemblages, which display higher diversity and contain numerous species of *Buliminella, Praebulimina, Afrobolivina, Bolivina* and *Orthokarstenia*. The palaeobathymetric distribution of these high productivity benthonic foraminiferal biofacies reflects the establishment of a productivity driven oxygen minimum zone along the western African shelf, associated with an upwelling system active off coastal Africa during most of the Late Cretaceous. The extent of the high productivity equatorial belt appears to have shifted latitudinally during the Late Cretaceous as sea level and circulation patterns changed. The zone of highest productivity was broadest in the Early Turonian, when it coincided with the maximum sea-level rise and highest atmospheric temperature in the Late Cretaceous.

The Late Cretaceous was generally characterized by high atmospheric CO_2 concentrations, warm climates without major ice-caps, flatter equator to pole temperature gradients than today's and elevated sea levels (Barron & Washington 1985; Berner 1991; Cerling 1991; Frakes *et al.* 1992; Freeman & Hayes 1992; Barron *et al.* 1995; Huber *et al.* 1995; Bush & Philander 1997). The focus of deepwater formation switched from the poles to the low latitudes as extensive, shallow epicontinental seas, with high evaporation rates, became the sources of warm, oxygen-depleted bottom waters (Brass *et al.* 1982; Hay 1988; Barron & Peterson 1990; Spicer & Corfield 1992; Barron *et al.* 1993; Hay 1995*a*, *b*). At times of highest sea level, in the Early Turonian and Early Campanian, the widespread flooding of continental blocks triggered major changes in palaeocirculation and palaeoproductivity, which led to the global accumulation of organic-rich sediments (Eicher & Diner 1991). These hydrographic changes had a major impact on the nutrient and oxygen budgets of coastal basins along the western African margin, where general circulation models predict that a major cell of upwelling was active during most of the Late Cretaceous (Kruijs & Barron 1990; Bush & Philander 1997), which led to increased surface water productivity and to the establishment of intensified oxygen minimum zones on the outer shelf. Environmental conditions along the western African margin were, therefore, exceptionally favourable for the accumulation of thick sequences of organic-rich sediments and coastal basins became the prime targets for the trapping of potential hydrocarbon resources during the Late Cretaceous.

As the composition of benthonic foraminiferal

HOLBOURN, A., KUHNT, W., EL ALBANI, A., PLETSCH, T., LUDERER F. & WAGNER, T. 1999. Upper Cretaceous palaeoenvironments and benthonic foraminiferal assemblages of potential source rocks from the western African margin, Central Atlantic. *In:* CAMERON, N. R., BATE, R. H. & CLURE, V. S. (eds) *The Oil and Gas Habitats of the South Atlantic.* Geological Society, London, Special Publications, **153**, 195–222.

assemblages is strongly dependent on the productivity of surface waters, and on the oxygenation of bottom water masses, high productivity benthonic foraminifera in fossil assemblages may serve as proxies of original phytodetritus flux rates to the seafloor and bottom water oxygenation. Both these factors are also influential in the formation of source rocks and benthonic foraminiferal palaeobiogeographic distribution patterns may, therefore, be used to trace the distribution of potential source rocks in sedimentary basins. Benthonic foraminifera have radiated to occupy most niches in surficial sediments of marine ecosystems, where they represent an essential link in the oceanic food chain (Gooday 1994). Recent studies by Lutze & Colbourn (1984), Altenbach & Sarnthein (1989) and Gooday & Turley (1990) have indicated that food and oxygen are the most influential factors regulating the distribution of benthonic foraminifera. Tyson & Pearson (1991), Barmawidjaja *et al.* (1992) and Jorissen *et al.* (1992) also suggested that the depth of the redox potential discontinuity below the sediment surface has a profound influence on the distribution of benthonic foraminifera within sediments, while Linke & Lutze (1993) considered distribution to be a dynamic process reflecting the need to optimize food acquisition. Loubere (1994, 1996) also showed that benthonic foraminiferal assemblage composition can be quantitatively related to surface ocean productivity and to bottom water oxygen content. Modern benthonic foraminiferal communities in areas of increased organic matter export flux rates, which result in an enhanced food supply and dysoxic conditions at the seafloor, are characterized by high numbers of small buliminids and bolivinids regarded as opportunistic infaunal detritus feeders (Lutze & Colbourn 1984; Lutze *et al.* 1986; Corliss & Chen 1988; Hermelin 1992; Jian *et al.* 1999; Kuhnt *et al.* 1999).

Sedimentological data, generated from X-ray diffractometry of the clay fraction, carbonate microfacies analysis and Rock-Eval pyrolysis, are used as independent indicators of siliciclastic and carbon flux to the seafloor, bottom water oxygenation and sea-level changes. The composition of clay mineral assemblages provide information on the proximality or distality from the sediment source, as detrital minerals such as illite, chlorite and kaolinite are preferentially deposited in the most proximal part of sedimentary basins, whereas smectite is selectively deposited in more distal environments on account of its higher buoyancy (Chamley 1989). Variations in the degree of sorting, maturity and grain size of sediments reflect fluctuations in the intensity of the terrigenous flux, which may also be related to sea-level highstands or lowstands. Observations of fine laminations in sediments with a high organic matter content indicate strong dysoxia, or even anoxia, in the original environment of deposition. Measurements of the 'total organic content' (TOC) in sediments provide evaluation of organic carbon accumulation at the seafloor and the degree of oxicity/anoxicity at the sediment–water interface at the time of deposition. Finally, kerogen analyses allow the distinction of lacustrine (Type I) and marine (Type II) organic matter, which are both derived from planktonic organisms, and organic matter originating from land plants (Type III).

The palaeobathymetric, biostratigraphic and palaeobiogeographic distribution of Upper Cretaceous benthonic foraminifera in organic-rich sediments from coastal basins along the western African margin from southern Morocco to Nigeria are investigated. The study is concentrated on two main time slices: the Early Turonian and the Early Campanian, which correspond to the two most prominent highstands of the Late Cretaceous. Samples from onshore sections of the Tarfaya Basin in southern Morocco, from DSDP Site 369 on the continental slope off southern Morocco, from offshore commercial wells on the Casamance Shelf off Senegal, from DSDP Site 367 in the Cape Verde Basin, from ODP Site 959 on the Côte d'Ivoire–Ghana Marginal Ridge and from onshore sections of the Benue Trough in Nigeria are examined. Samples from these locations represent a latitudinal transect along the east Atlantic margin, which may be used to trace the position of the main upwelling systems during the Late Cretaceous.

Location and geological setting

Tarfaya Basin

The Tarfaya Basin is a Mesozoic basin extending along the coast of southern Morocco between latitudes 28° N and 24° N (Fig. 1), which has remained tectonically stable since the Cretaceous. A palaeoenvironmental evolution of the basin during the Late Cretaceous is presented in El Albani *et al.* (1999). Bituminous marls were deposited in this basin during the Turonian in an open shelf setting, in water depths of the order of 200–300 m (Kuhnt *et al.* 1990). At the more distal site of Well S75 the Turonian sediment (40 m thick) consists almost exclusively of organic matter and biogenic carbonate. The carbonate content can reach up to 90% and is principally composed of planktonic foraminifera and

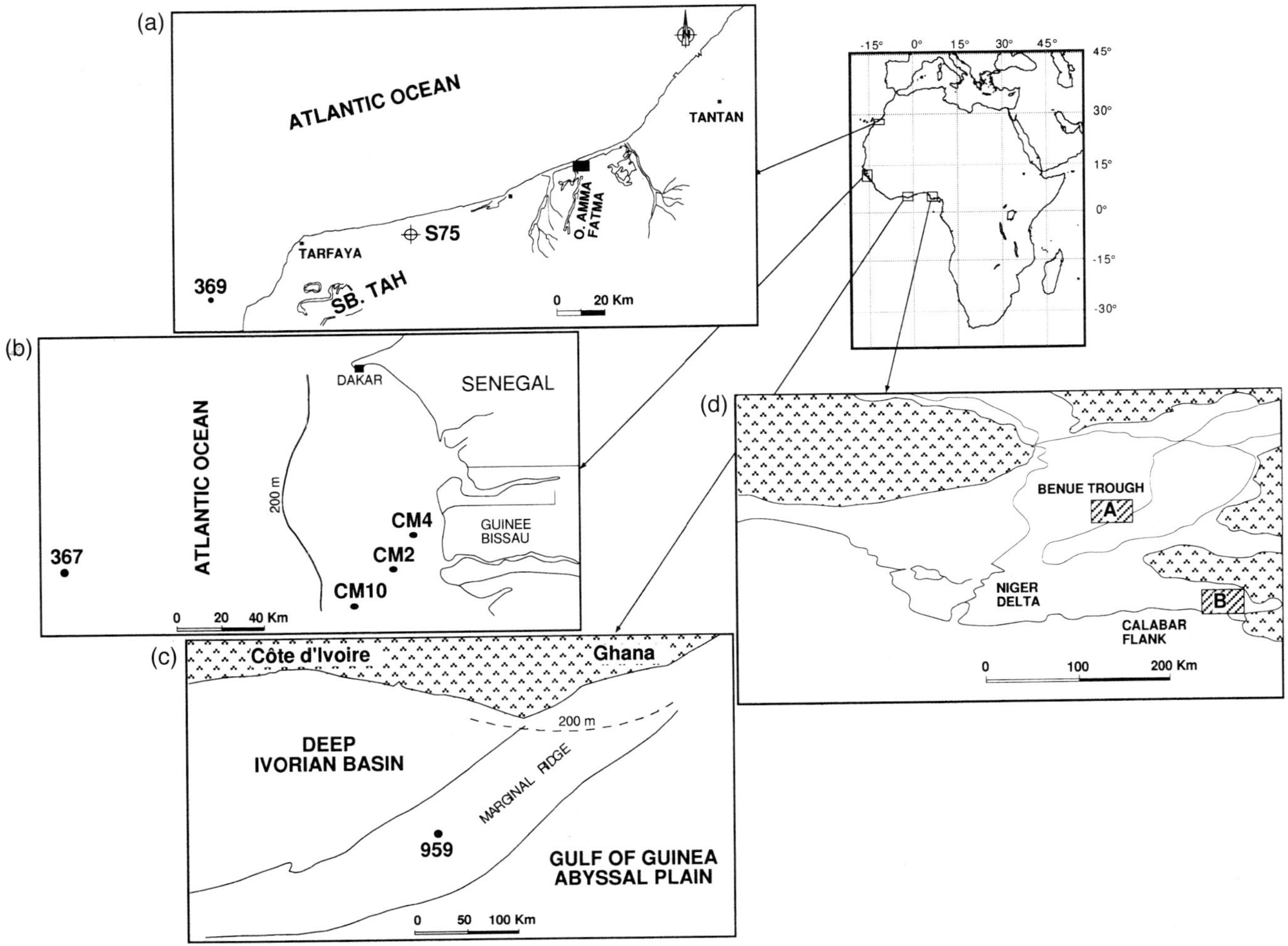

Fig. 1. Location map for (a) the Tarfaya Basin, DSDP Site 369, (b) the Casamance Shelf off Senegal, DSDP Site 367, (c) ODP Site 959, (d) the Central Benue Trough and Calabar Flank in Nigeria.

calcareous nannoplankton. At the more proximal onshore location of Oued Amma Fatma the Lower Turonian succession (*c.* 17 m thick) is intercalated with coarse bioclastic limestones, frequently displaying erosive bases and hummocky cross-stratification, which are probably storm generated. The Lower Campanian sediments from Sebkha Tah represent a 20 m thick outcrop section, which consists of alternating dark and light marl layers. The organic matter is mostly of marine origin (kerogen Type II) and the maturity is low with T_{max} values between 400 and 420 °C (Kuhnt *et al.* 1990; Wagner & Pletsch 1999).

Casamance Shelf, Senegal

Several exploration wells were drilled on the Casamance Shelf, offshore Senegal. The most seaward of the wells, CM10, is situated close to the shelf edge while Wells CM2 and CM4 are in more proximal locations (Fig. 1). Thick Turonian–Maastrichtian sedimentary sequences were recovered at Well CM2 (over 1600 m), Well CM4 (over 1200 m) and Well CM10 (over 1200 m). The Turonian at Well CM2 is *c.* 150 m thick and consists of dark brown and bituminous clays. The Coniacian–Santonian section is *c.* 350 m thick and corresponds to dark claystones intercalated with calcareous and, occasionally, sandy layers. Similar lithological sequences were recovered at Well CM4, where the Turonian interval is nearly 100 m thick and the Coniacian–Santonian section just under 200 m. At Well CM10 the Turonian interval is *c.* 150 m thick and the Coniacian-Santonian section, which corresponds to dark claystones and calcareous claystones with some lighter clayey calcareous layers, is 100 m thick. The Campanian–Maastrichtian successions at the three sites consist of dark claystones with silty, sandy or calcareous intervals, and show a marked increase in silt and sand contents towards the top of the successions (upper 400 m at Well CM2 and Well CM4 and upper 150 m at Well CM10). In the wells studied the maturity of the organic matter is still low with T_{max} values between 422 and 429 °C (Herbin *et al.* 1986; Kuhnt *et al.* 1990). However, effective source rocks of the same age may be located in deeper settings along the continental margin.

DSDP Site 367

Hole 367 is situated at the base of the continental rise in the Cape Verde Basin in a water depth of 4768 m. It is located at 12° 29.2′ N, 20° 02.8′ W (Fig. 1), *c.* 330 km west on a direct transect from Wells CM2–CM10. Upper Cretaceous abyssal clay from the Plantagenet Formation (lithologic Unit 3) was recovered at this site in Cores 367-15R to -17R (Lancelot *et al.* 1977). This lithologic unit was spot-cored and consists of greyish blue–green and reddish brown silty clay, with occasional mottled zones, faint laminations and slight bioturbation. A Campanian–Maastrichtian age was determined for Cores 367-16R and -15R, based on the occurrence of deepwater agglutinated foraminifera index species (Holbourn *et al.* 1999). The maturity of the organic matter is extremely low with T_{max} values between 396 and 403 °C (Herbin *et al.* 1986).

DSDP Site 369

Hole 369 was drilled on the continental slope off southern Morocco at 26° 35.5′ N, 14° 59′ W, in a water depth of 1752 m (Fig. 1). Campanian–Maastrichtian sediments (lithologic Unit 2) were recovered in Core 369-36R to -39R at this site (Lancelot *et al.* 1977). Lithologic Unit 2 was spot-cored and comprises Sub-unit 2a consisting of argillaceous nannofossil chalk and limestone, with minor porcellanite and chert (Cores 369-36R and -37R) and Sub-unit 2b composed of argillaceous chalk (Cores 369-38R and -39R).

ODP Site 959D

Hole 959D is situated on the northern flank of the Côte d'Ivoire–Ghana Marginal Ridge at 3° 37.656′ N, 2° 44.149′ W in a water depth of 2090.7 m, on a small plateau close to the top of the ridge as shown in Fig. 1 (Mascle *et al.* 1996). The Lower Turonian sediments (Core 959D-68R) in the lower part of lithologic Unit 4a consist of calcareous sandstone, including thin conglomeratic beds with phosphatic pebbles and glauconitic clasts. Intercalated are black fissile shales with planktonic foraminifera of the *W. archaeocretacea* Zone (Fig. 2a). These shales have high TOC values and correspond to the global anoxic event of the Cenomanian–Turonian boundary event. The Upper Turonian–Lower Coniacian sediments (Core 959D-67R), from the upper part of lithologic Unit 4a, comprise sandy limestone and sandy dolomite, which are heavily bioturbated at some levels. A phosphatic hardground in Section 959D-66R-CC marks the boundary between the Lower Coniacian and the Middle Coniacian (Fig. 2b). The Middle Coniacian–Lower Santonian sediments in Core 959D-66R above this hardground, and at the base of Core 959D-65R, are

Fig. 2. Organic-rich sediments from ODP Hole 959D. (**a**) Section 159-959D-68R-1, 50–68 cm. Black fissiles shales with phosphatic and glauconitic clasts corresponding to the CTBE. (**b**) Section 159-959D-66R-CC, 13.2–30.8 cm. Phosphatic hardground marking boundary between Lower and Middle Coniacian. (**c**) Section 159-959D-66R-6, 99-116 cm. Phosphatic conglomerate below black massive claystone and laminated dark brown claystone. (**d**) Section 159-959D-66R-1, 108–125 cm. Massive black claystone and finely laminated dark brown claystone.

characterized by sequences of massive, black claystone alternating with finely laminated dark brown claystone, corresponding to the lowermost part of lithologic Unit 3 (Fig. 2c, d). Phosphatic hardgrounds are present in Core sections 959D-66R-4 and -66R-7, and a phosphatic conglomerate occurs in Core section 959D-66R-6, 111–118 cm (Fig. 2c), probably formed from the erosion of another hardground. The maturity of the organic matter is low: pyrolitic and clay mineral data suggest a moderate overprint of the organic matter which hardly reached the top of the oil window (Wagner & Pletsch 1999).

Nigeria (Central Benue Trough and Calabar Flank)

Lower Turonian sediments from the Central Benue Trough (Fig. 1) consist of marginal clastic sequences on the edge of the basin and black, laminated organic-rich calcareous marls and limestones, overlain by fine-grained sandstones, lumachelle beds and limestones; hardgrounds and tempestites occur in the central part of the basin. The palaeoenvironments of the Central Benue Trough have been interpreted as coastal areas to the north ranging to a deep outer shelf with shelf slope transition to the south (Kuhnt *et al.* 1990). The Calabar Flank, situated at the eastern edge of the Gulf of Guinea (Fig. 1), has been considered to be part of the South Atlantic continental margin, southeast of the Benue Trough (Nyong & Ramanathan 1985). Lower Turonian continental margin sequences deposited close to the shelf edge along the Calabar Flank consist of dark grey fissile shales intercalated with some limestones and some lumachelle beds, which were probably storm generated. The maturity of the organic matter is generally low. Occasional high thermal maturity is related to local volcanic heat flow (Ojoh 1988).

Methods

For micropalaeontological studies, the samples were dried, weighted, soaked in distilled water and wet-sieved through a 63 µm (#230 mesh) screen; then the residue was dried. Consolidated samples were first treated with a buffered 5% hydrogen peroxyde solution to help break them up before sieving. In rare cases where the hydrogen peroxyde treatment did not lead to complete disintegration of the clay, the dried samples were soaked in a concentrated anionic tenside solution (REWOQUAT of REQO Chemie, Steinau an der Straße, Germany), which usually disintegrated even slightly silicified samples. Generally, the complete residue was picked for benthonic foraminifera. In a few exceptional samples, with extremely high faunal content, samples were split with a standard Otto-splitter. Fragments of tubular species were counted as one individual, indeterminable fragments or extremely deformed specimens were picked but not included in the counts. Faunal occurrences only were recorded for cutting samples. The foraminiferal slides are housed in the micropalaeontology collections of the Geologisch-Paläontologisches Institut at the Christian Albrechts University in Kiel. Electron micrographs were made on a Camscan SEM at the Geologisch-Paläontologisches Institut at the Christian Albrechts University in Kiel.

The Shannon–Weaver information function (Shannon & Weaver 1949) was used to evaluate benthonic foraminiferal diversity. This measure of heterogeneity, which is frequently used as an index of biodiversity, takes into account the number of species and the equitability of their distribution within a sample. It is defined by the equation:

$$H(S) = -\sum_{i=1}^{S} p_i \ln p_i$$

where S is the number of species and p_i the proportion of the ith species (p = per cent divided by 100).

Clay mineralogy was analysed using a Philips X-ray diffractometer (PW 1390) equipped with a copper anticathode (40 kV, 25 mA). X-ray diffractions were performed on oriented mounts. Six successive steps were followed for the preparation of the oriented mounts: (1) grinding of the rock; (2) splitting in distilled water; (3) decarbonation in hydrochloric acid (5*N*) followed by successive centrifugations to remove excess acid and to defloculate clay minerals; (4) micro-homogenization using a mixer; (5) sampling of the clay after decantation; (6) preparation of oriented mounts after centrifugation. Three diffractograms were run for each sample set (natural sample, glycolated sample and natural sample dried at 490 °C over 2 h). The analytical and interpretative procedures concerning the diffractograms are described by Holtzapffel (1985).

For DSDP Site 367, Casamance and Nigeria samples, TOC values and kerogen types were determined with a Rock-Eval II plus TOC instrument at the Institut Français du Pétrole in Paris, following the methodology described by Espitalié *et al.* (1985). The type of organic matter was determined using the hydrogen index (HI)/oxygen index (OI) ratio, plotted in modified Van Krevelen diagrams (Espitalié *et al.* 1977). TOC measurements for ODP Site 959 and Tarfaya samples were performed on dried, homogenized samples at Bremen University using a Leco CS-300 device. Calculation of hydrogen indices for these samples used the s2 records from Rock-Eval pyrolysis and TOC data from Leco measurements.

Distribution of benthonic foraminifera, organic carbon and terrigenous flux

Tarfaya Basin

Oued Amma Fatma. Lower Turonian benthonic foraminiferal assemblages from Oued Amma Fatma show extremely low diversity (Shannon–Weaver index between 0.0 and 0.9) and marked variations in abundance. They are strongly dominated by a *Gabonita* biofacies composed of *Gabonita levis, Gabonita obesa* and *Praebulimina* sp. Planktonic foraminifera are present in very high numbers and ostracods are abundant in a few samples. TOC values fluctuate markedly between 0.93 and 7.25, and the organic matter is of planktonic marine origin (Type II). A summary of the benthonic foraminiferal, clay mineral and Rock-Eval results is presented in Fig. 3. Changes in the number of buliminids per gram show a negative correlation with fluctuating TOC values (Fig. 4). Upper bathyal to outer shelf depths are suggested by the foraminiferal assemblages and also by the clay mineral assemblages, which are dominated by smectite (up to 80%), indicative of a relatively distal setting with a reduced clastic flux. Marked changes in depositional energy are evident from the lithological succession, which is composed of dark calcareous claystones intercalated with coarse bioclastic limestones, often with erosive bases and hummocky cross-stratification. The high numbers of planktonic foraminifera, and the presence of radiolarians and diatoms, in the assemblages suggest a high surface water productivity. Observations of thin sections confirm that the sediment contains abundant organic matter, often laminated with numerous planktonic foraminifera and buliminids. Strongly dysoxic conditions, hostile to benthonic life, were probably induced intermittently at the seafloor by fluctuations in organic matter export flux rates, or changes in bottom water circulation, and are reflected by marked variations in benthonic foraminiferal abundance and TOC values.

Well S75. Lower–Middle Turonian benthonic foraminiferal assemblages from this exploration well exhibit extremely low diversity [$H(S) < 1.0$] and are also dominated by a *Gabonita* biofacies, comprised of *Gabonita levis, Gabonita obesa* and *Praebulimina* sp. The sediment contains a high proportion of calcium carbonate (up to 80%). A summary of the benthonic foraminiferal and Rock-Eval results is presented in Fig. 5. Fluctuations in the accumulation rates of benthonic foraminifera show a negative correlation with TOC values, which reach up to 12.86% (Fig. 6). An outer shelf, or upper bathyal, environment is indicated by the benthonic foraminiferal assemblages; however, a slightly more distal depositional environment than Oued Amma Fatma in the main depocentre of the basin was proposed by Kuhnt *et al.* (1997), on account of the higher sediment accumulation rate. High productivity in surface water is suggested by the presence of radiolaria, the very high numbers of planktonic foraminifera and the dominance of heterohelicids. The marked variations in TOC and benthonic foraminiferal accumulation rates reflect either changing oxygenation levels at the sediment–water interface, in response to cyclic fluctuations in carbon flux rates to the seafloor, or changing bottom water circulation. A detailed cyclostratigraphic analysis of the sedimentary succession from this well by Kuhnt *et al.* (1997) showed that these cyclic changes occurred on a Milankovitch scale and allowed a high resolution stratigraphy to be established for the lower part of the *W. archaeocretacea* Zone.

Sebkha Tah. Lower Campanian benthonic foraminiferal assemblages from the Sebkha Tah section show relatively high diversity. The Shannon–Weaver index fluctuates *c.* 2.0, except near the top of the section where it drops just < 2.0 in a few samples. Faunal abundance increases between Sample TAH36A and Sample TAH51 (reaching over 589 specimens per gram in Sample TAH36A), except in Samples TAH36, TAH39A, TAH39 and TAH41, where the number of specimens drops markedly (as low as 29 specimens per gram in Sample TAH39 with the highest TOC value of 5.32%). Buliminids and bolivinids represent up to 61% of the assemblage in samples with the highest abundance of benthonic foraminifera and elevated TOC values. A summary of the benthonic foraminiferal, clay mineral and Rock-Eval results is presented in Fig. 7. High surface water productivity is suggested from the abundance of planktonic foraminiferal tests, which include high numbers of heterohelicids, and by the presence of radiolarians. The composition of the assemblages indicates an outer shelf, or upper bathyal, environment with a high organic carbon export flux to the seafloor. Observations of thin sections confirm that the sediment contains abundant

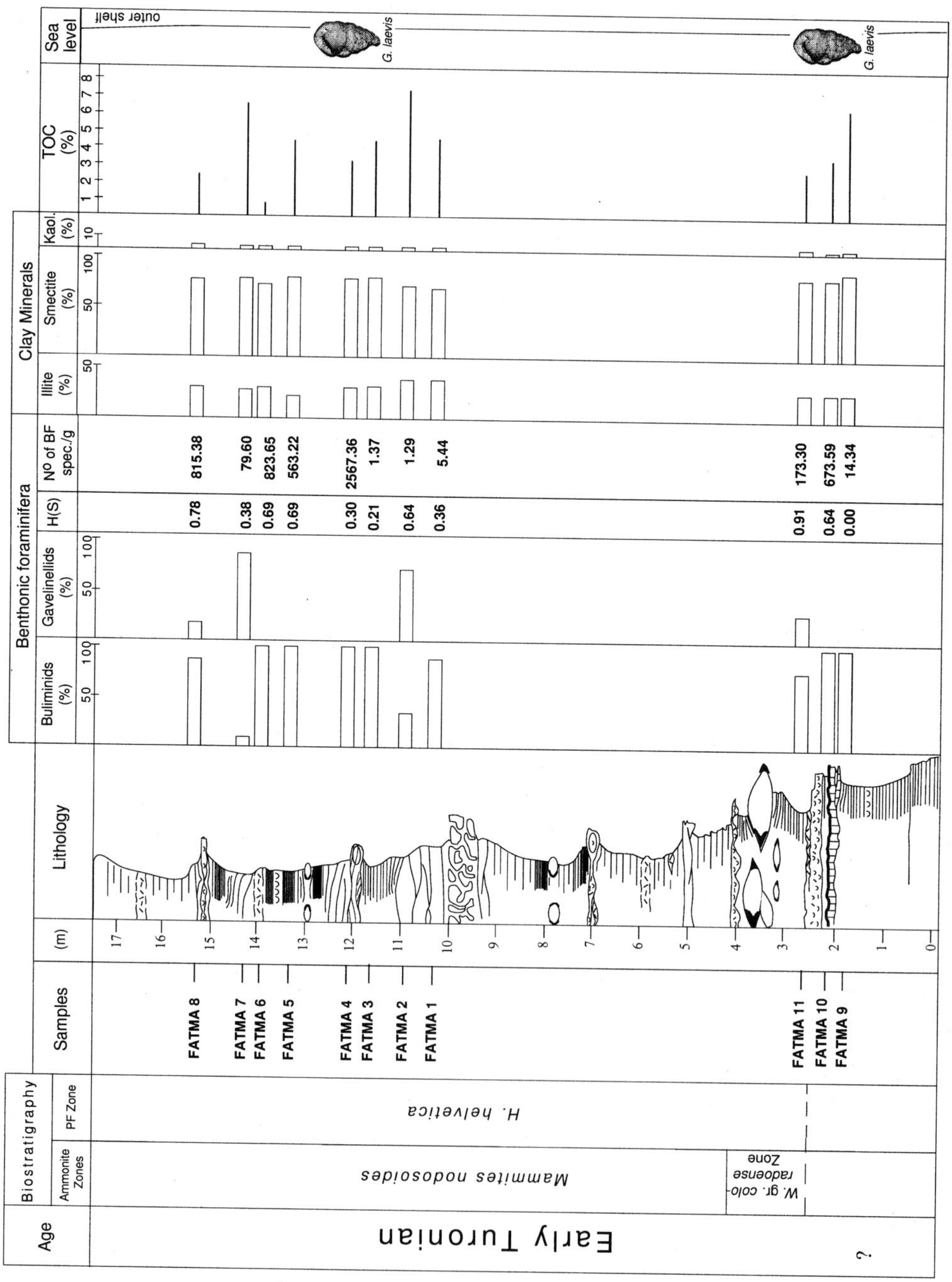

Fig. 3. Summary of benthonic foraminiferal, clay mineral and geochemical results for the Oued Amma Fatma section.

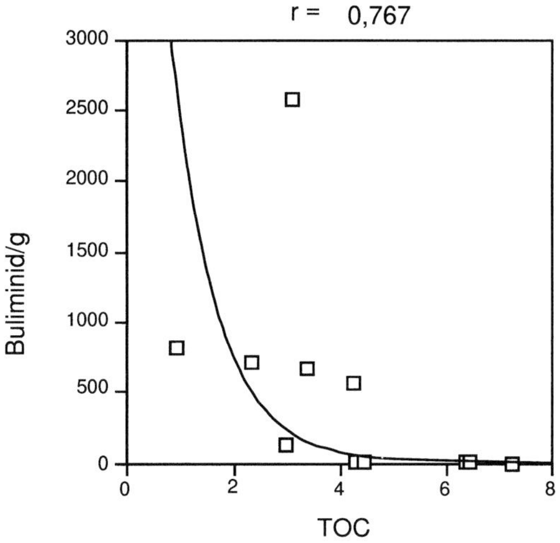

Fig. 4. Regression plot between TOC and the number of buliminid tests in samples from the Oued Amma Fatma section.

organic matter (Fig. 8b). The dominance of smectite in the clay mineral assemblages, and the abundance of Type II organic matter derived from planktonic marine organisms, also indicate a relatively distal environment with a reduced clastic influence. Fluctuations in benthonic foraminiferal abundance and assemblage composition probably indicate changing rates of organic matter export flux related to fluctuations in surface water productivity, or changes in bottom water circulation, affecting the preservation of organic matter. The relatively high diversity of the benthonic foraminiferal assemblages suggests that conditions at the sediment–water interface never became as severely dysoxic as in the Early Turonian.

Casamance Transect, off Senegal

Upper Cenomanian–Maastrichtian benthonic foraminiferal assemblages from four offshore wells along a transect across the Casamance margin, extending from the inner shelf to the abyss, were examined. A detailed study of the benthonic foraminiferal assemblages is presented in Holbourn *et al.* (1999). A summary of the benthonic foraminiferal, clay mineral and Rock-Eval results for Wells CM2, CM4 and CM10 on the Casamance Shelf is presented in Figs 9–11. The Upper Cretaceous assemblages from the most proximal wells (Wells CM2 and CM4) contain coarsely agglutinated foraminifera with organically cemented walls such as *Ammotium nkalagum, Kutsevella* and *Ammobaculites* spp. The benthonic foraminiferal assemblages and the illite–kaolinite-rich clay mineral assemblages are diagnostic of a restricted, shallow inner shelf environment receiving a high terrigenous flux (Fig. 8e, f). At these two wells, sea-level highstands are characterized by elevated TOC values (up to 6.7%) and abundant planktonic foraminifera in the Early Turonian, and by a typical buliminid benthonic foraminiferal biofacies in the Early Campanian. The Turonian–Maastrichtian sediments from the distal well at the edge of the continental shelf (Well CM10) contain smectite-rich clay mineral assemblages and diversified buliminid associations, which include many species characteristically found in equatorial basins (Holbourn *et al.* 1999). The assemblages from Well CM10 contrast markedly in composition with the organically cemented flysch-type deepwater agglutinated foraminifera assemblages from the deep basin at DSDP Site 367. The distribution patterns of organic-rich, partly laminated sediments and benthonic foraminiferal assemblages diagnostic of high organic carbon flux rates in Well CM10 (Fig. 8d) provide evidence that a productivity driven oxygen minimum zone was established on the outer shelf of the Casamance margin for most of the Late Cretaceous. During sea-level highstands at the Cenomanian–Turonian boundary and in the Early Campanian the oxygen minimum zone intensified and expanded to reach more proximal environments of the inner shelf.

Côte d'Ivoire–Ghana Transform Margin (ODP site 959)

A relatively low diversity Lower Turonian assemblage (Shannon–Weaver index between 0.97 and 1.63) dominated by *Bolivina anambra, Praebulimina* sp. 1, *Praebulimina* sp. 2, *Praebulimina* sp. 3 or by *Gavelinella* spp. is found in Core 959D-68R. The Middle Coniacian–Lower Santonian assemblage from Core 959D-66R and the base of Core 959D-65R is generally abundant and diverse (Shannon–Weaver index ≥ 2.5) but displays marked drops in abundance and diversity, which appear related to changes in TOC values, in Samples 959D-66R-7 (21–25 cm), 959D-66R-6 (15–17 cm), 959D-66R-1 (110–114 cm), 959D-65R-7 (31–35 cm) and 959D-65R-6 (28–32 cm) (Shannon–Weaver index between 1.09 and 1.85) Buliminids are abundant in the assemblage but differ markedly from that in Core 959D-68R, including *Praebulimina robusta, Praebulimina prolixa longa, Praebulimina fang, Neobulimina subregularis* and *Buliminella* cf. *gabonica*. A marked change in faunal composition occurs in Sample 959D-65R-5 (30–34 cm) with a TOC value of

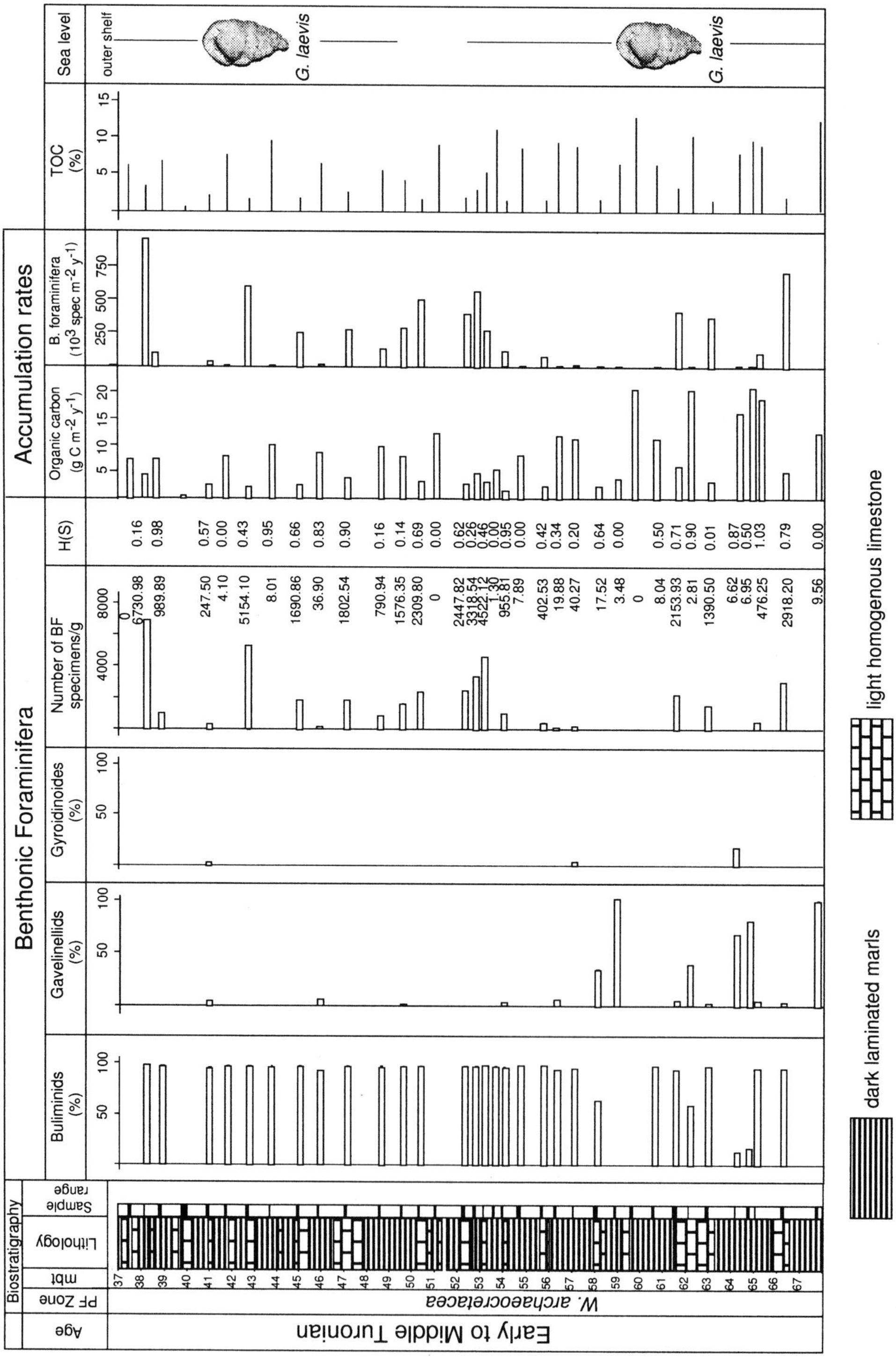

Fig. 5. Summary of benthonic foraminiferal and geochemical results for Well S75.

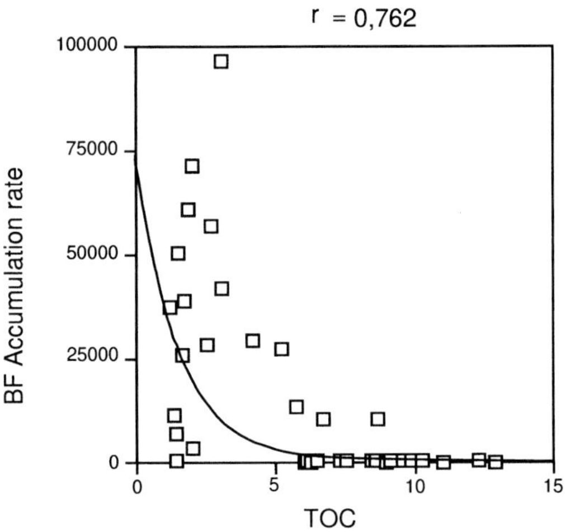

Fig. 6. Regression plot between TOC and benthonic foraminiferal (BF) accumulation rate in samples from Well S75.

2.35%, where numerous nodosariids and agglutinated tests are recorded together with a few buliminids. A summary of the benthonic foraminiferal, clay mineral and Rock-Eval results is presented in Fig. 12. A detailed study of the benthonic foraminiferal assemblages is presented in Holbourn & Kuhnt (1998).

Lower Turonian and Middle Coniacian–Lower Santonian benthonic foraminiferal assemblages from ODP Site 959 are dominated by two different buliminid associations, which, together with the relatively high TOC values, suggest poorly oxygenated bottom waters in an outer shelf or upper slope setting. The distinct composition of the two buliminid associations found in Hole 959D suggests that endemism was stronger during the Early Turonian when circulation was probably more restricted in the Deep Ivorian Basin. The Early Turonian association in Core 959D-68R represents a typical 'African' association, previously only recorded from Nigeria (Petters 1982). In contrast, the Middle Coniacian–Early Santonian buliminid association in Core 959D-66R and at the base of Core 959D-65R has widespread distribution in equatorial Atlantic basins, which indicates more open connections between the Deep Ivorian Basin and other equatorial Atlantic basins. The clay mineral assemblages indicate a high detrital flux during the deposition of Core 959-68R, whereas minerals indicating rapid erosion of rock substrates are completely absent from Core 959-67R upsection, suggesting drowning of terrestrial source areas in the vicinity of Site 959 after the Turonian. The benthonic foraminiferal data provide evidence that an intensified oxygen minimum zone became established off the Côte d'Ivoire–Ghana Transform Margin in the Early Turonian and in the Middle Coniacian–Early Santonian. Intermittent increases in benthonic foraminiferal diversity in Core 959D-66R and at the base of Core 959D-65R, and corresponding decreases in TOC, probably reflect episodic improvements in bottom water oxygenation as periodic contraction–expansion of the oxygen minimum zone led to changes in its depth range and intensity.

Nigeria

Nkalagu Quarry, Central Benue Trough. The Upper Turonian–Coniacian benthonic foraminiferal assemblages from the Nkalagu Quarry in Central Nigeria show very low diversity (Shannon–Weaver index between 0.48 and 1.10) and are strongly dominated by agglutinated taxa with coarse, organically cemented walls such as *Ammotium nkalagum, Kutsevella* and *Ammobaculites* sp. These assemblages are similar in composition to assemblages from Wells CM2 and CM4 on the Casamance Shelf and are diagnostic of shallow, restricted conditions on the inner shelf. The clay mineral assemblages are dominated by kaolinite, illite and chlorite, which point to a high detrital input from the continent. Analysis of thin sections shows that the sediment contains abundant angular grains of quartz, mica and plagioclase feldspar, indicating a very proximal sediment source. TOC values are low, between 0.25 and 0.51%. The organic matter is of Type III, indicating a terrestrial origin. A small proportion of bolivinids and buliminids is present in the basal sample (Sample 26/2–7), which contains no chlorite, suggesting a reduced detrital influence at the time of deposition. This is also supported by the microfacies analysis, which reveals that this sample contains finer quartz grains and a higher proportion of organic matter. A summary of the benthonic foraminiferal, clay mineral and Rock-Eval results is presented in Fig. 13.

Lokpanta Section, Central Benue Trough. The Lower Turonian benthonic foraminiferal assemblages from the Lokpanta Section exhibit low diversity and are strongly dominated by a *Gabonita* biofacies, which consists of endemic species not documented in the literature. The abundance of planktonic foraminifera, the high TOC values, ranging from 1.71 to 5.21, and the Type II organic matter suggest that the depositional setting was a middle or outer shelf with a relatively high organic matter export flux to the seafloor derived from primary production in sur-

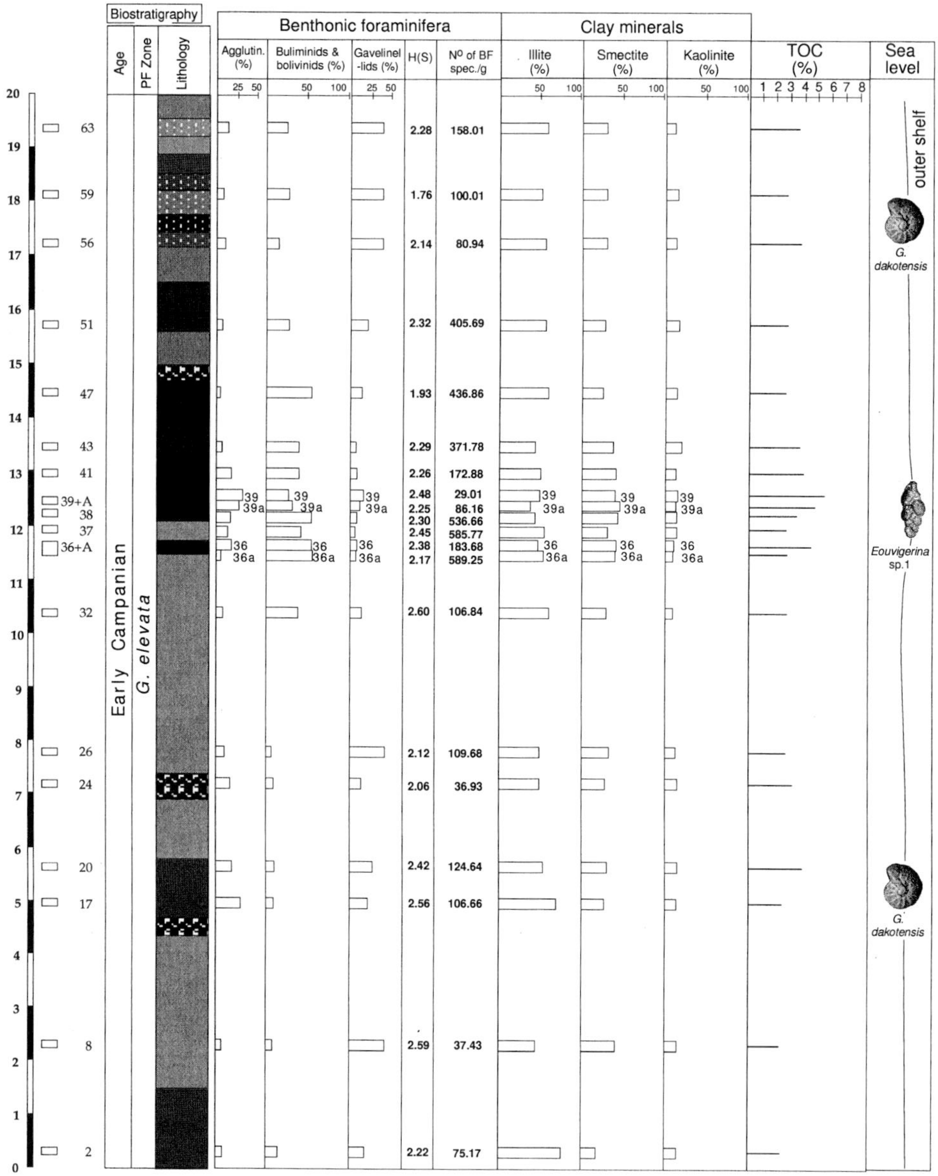

Fig. 7. Summary of benthonic foraminiferal, clay mineral and geochemical results for the Sebkha Tah section.

face water. Studies of thin sections show that the sediment contains abundant laminated organic matter with numerous planktonic and benthonic foraminifera and calcispheres (Fig. 8a, c). A detrital influence is indicated by the presence of kaolinite and illite in the clay mineral assemblages; however, the absence of chlorite and the presence of smectite in one sample point to a relatively distal middle or outer shelf setting. A summary of the benthonic foraminiferal, clay mineral and Rock-Eval results is presented in Fig. 14.

Ngbanocha Section, Central Benue Trough. The Lower Turonian benthonic foraminiferal assemblages from the Ngbanocha Section contain mainly gavelinellids, which may reflect either extremely poor oxygenation at the seafloor at the time of deposition or represent a preservational

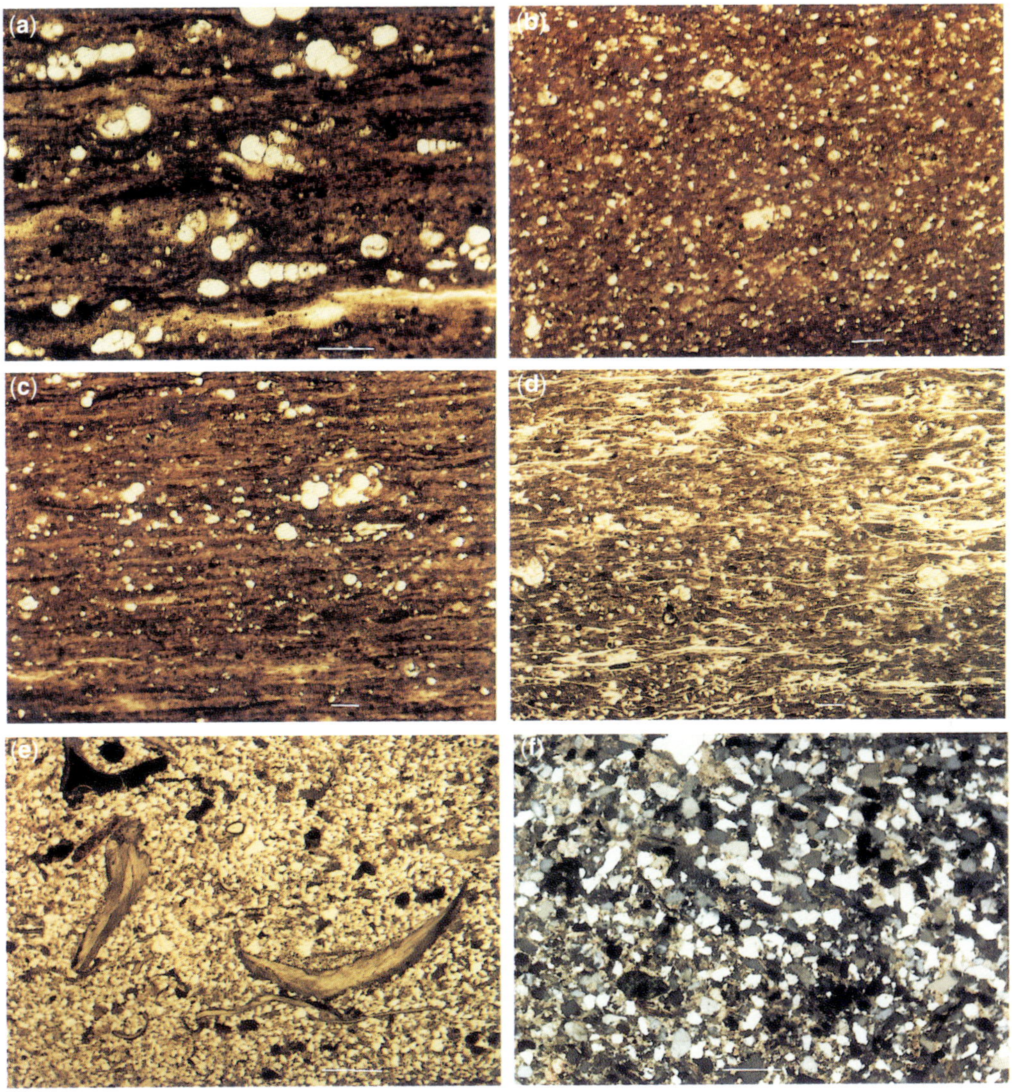

Fig. 8. (**a**) Laminated organic-rich lower Turonian sediment from the Lokpanta section, central Nigeria including numerous benthonic and planktonic foraminifera (Sample 27/2-4.) (**b**) Organic-rich lower Campanian sediment from the Sebkha Tah section including numerous benthonic and planktonic foraminifera (Sample TAH 39). (**c**) Laminated organic-rich lower Turonian sediment from the Lokpanta section, central Nigeria, including numerous benthonic and planktonic foraminifera (Sample 27/2-4.) (**d**) Partly laminated organic-rich Maastrichtian sediment from Well CM10, Casamance Shelf, including numerous benthonic and planktonic foraminifera (Sample CM10-C2, T1). (**e**) Poorly sorted Coniacian to Santonian proximal sediment from Well CM2, Casamance Shelf, with shell debris and angular quartz grains (Sample CM2-C3, 1972.8–1973.6 m). (**f**) Poorly sorted Coniacian to Santonian proximal sediment from Well CM2, Casamance Shelf, with angular quartz grains (Sample CM2-C3, 1981.6–1982 m). Scale bar = 1000 μm.

bias. Sample 27/2–16 also contains a few agglutinated forms, typical of restricted, proximal settings. Planktonic foraminifera are very abundant, suggesting an open marine environment with a high surface water productivity. The dominance of detrital clay minerals, such as illite, kaolinite and chlorite, and the absence of smectite in the clay assemblages indicate a high clastic flux. Observations of abundant grains of angular quartz, mica and plagioclase feldspar in thin sections also point to a high detrital input. Laminated organic matter is present in thin sections, but

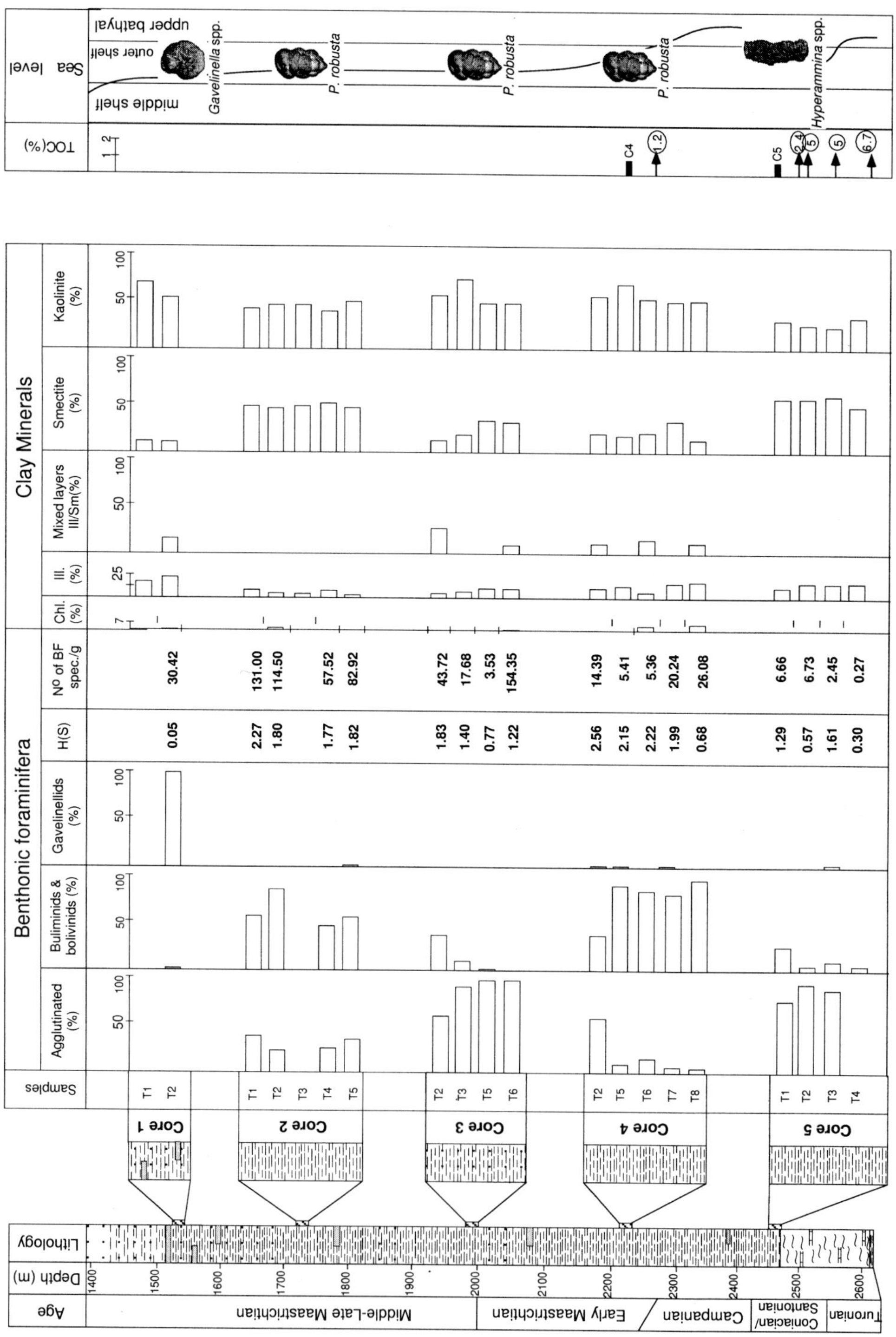

Fig. 9. Summary of benthonic foraminiferal, clay mineral and geochemical results for Well CM10. Biostratigraphy in Holbourn *et al.* (1999).

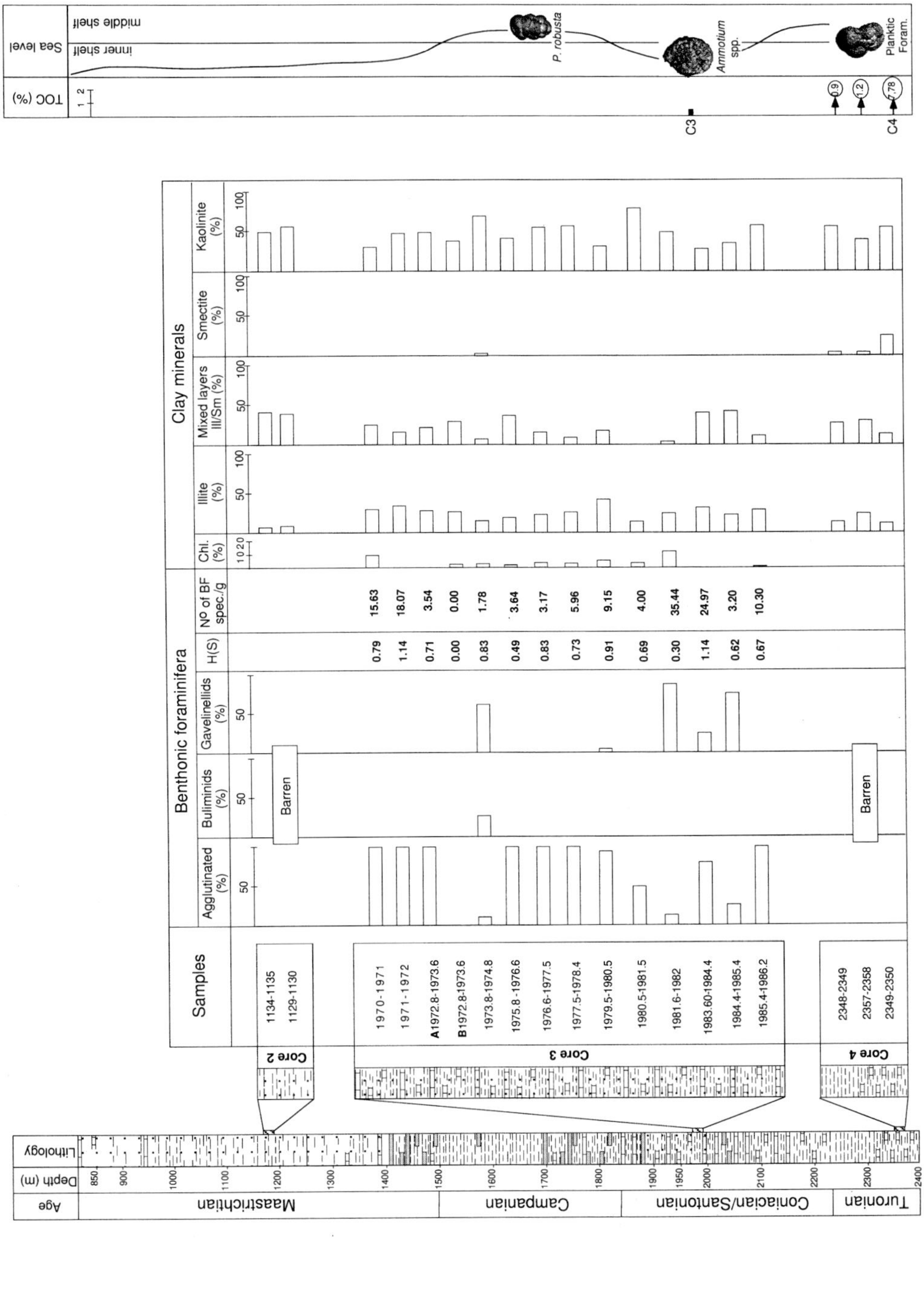

Fig. 10. Summary of benthonic foraminiferal, clay mineral and geochemical results for Well CM2. Biostratigraphy in Holbourn *et al.* (1999).

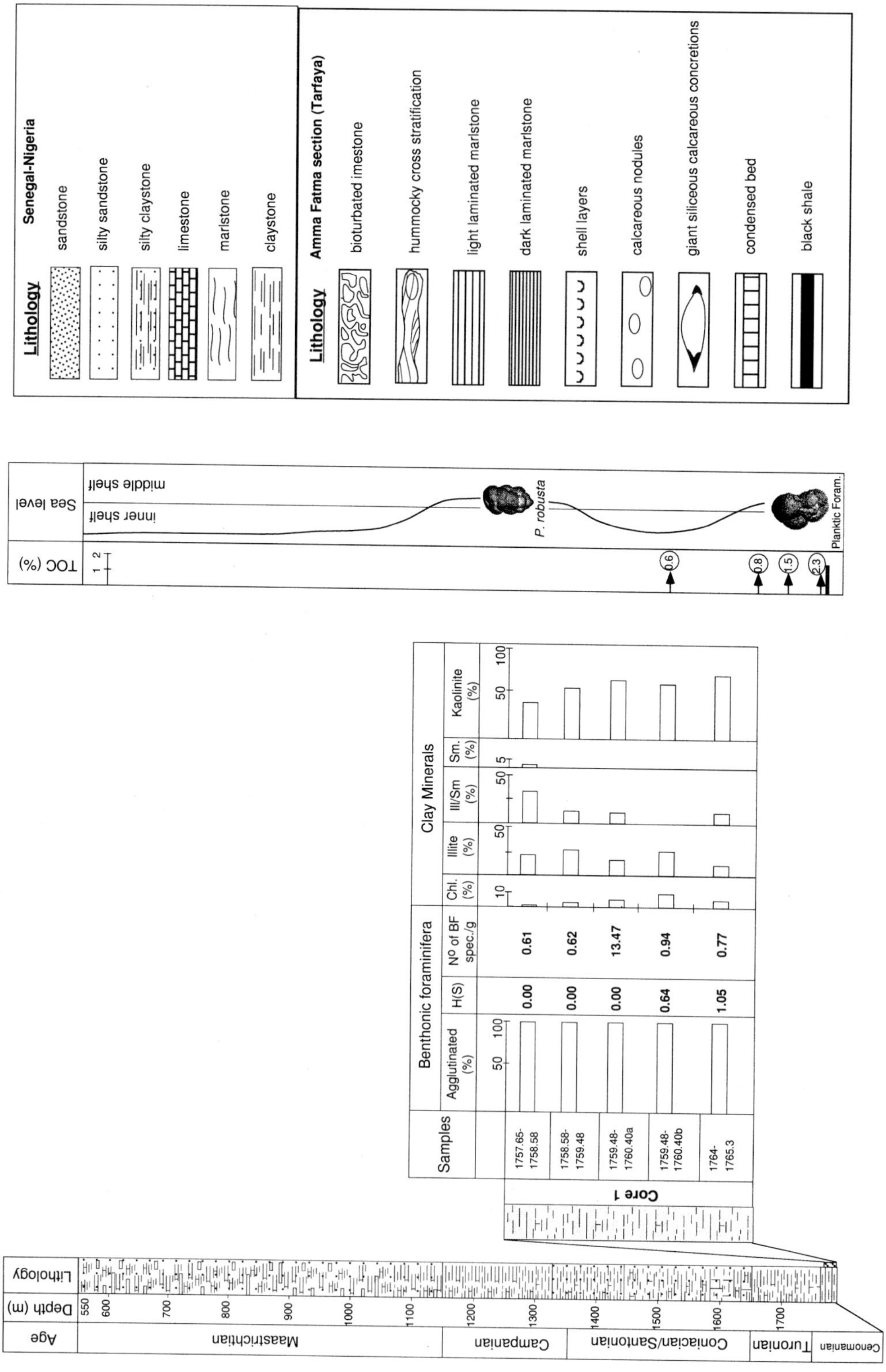

Fig. 11. Summary of benthonic foraminiferal, clay mineral and geochemical results for Well CM4. Biostratigraphy in Holbourn *et al.* (1999).

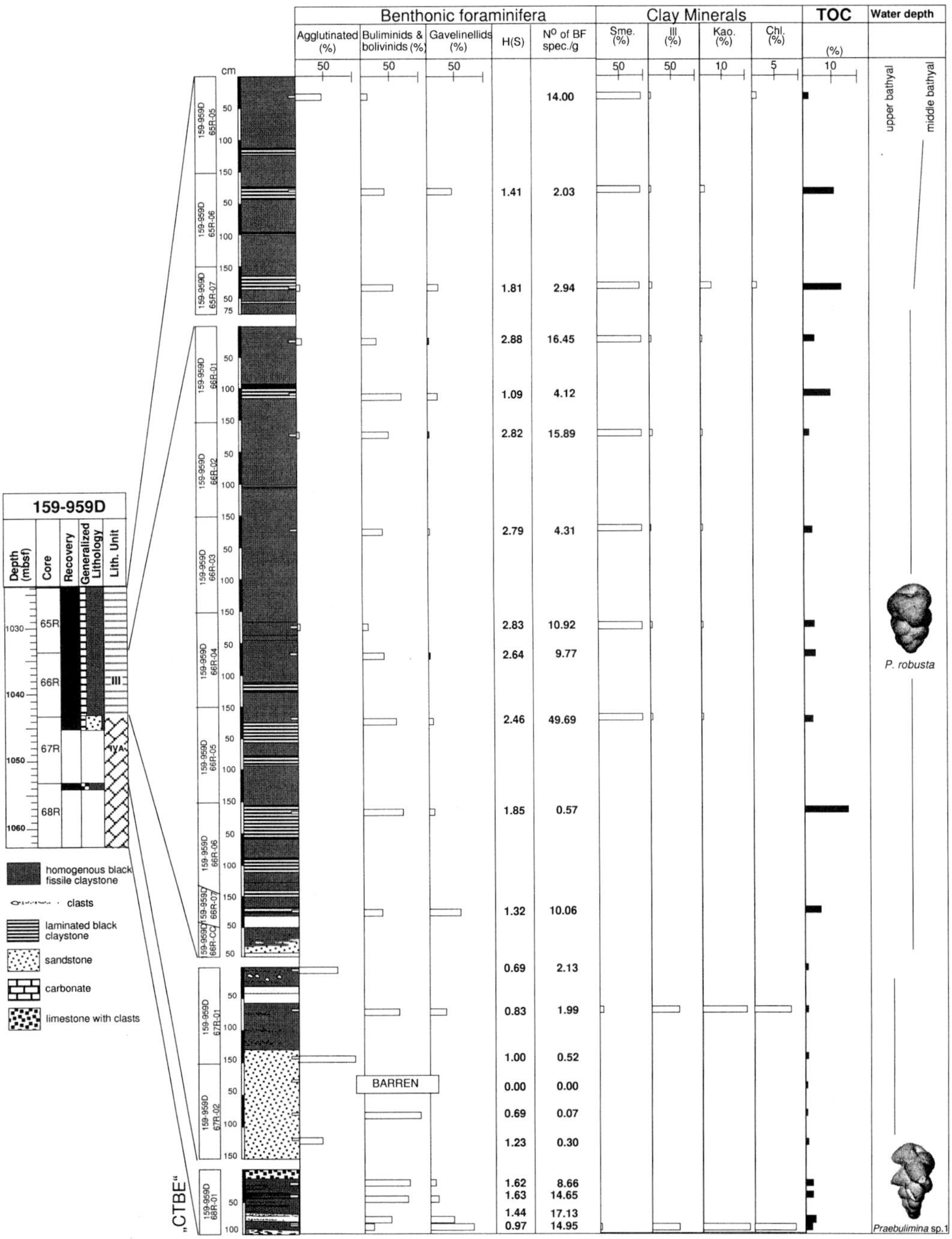

Fig. 12. Summary of benthonic foraminiferal, clay mineral and geochemical results for ODP Hole 959 (Section 959D-68R-1 to 65R-5). Biostratigraphy in Holbourn & Kuhnt (1998).

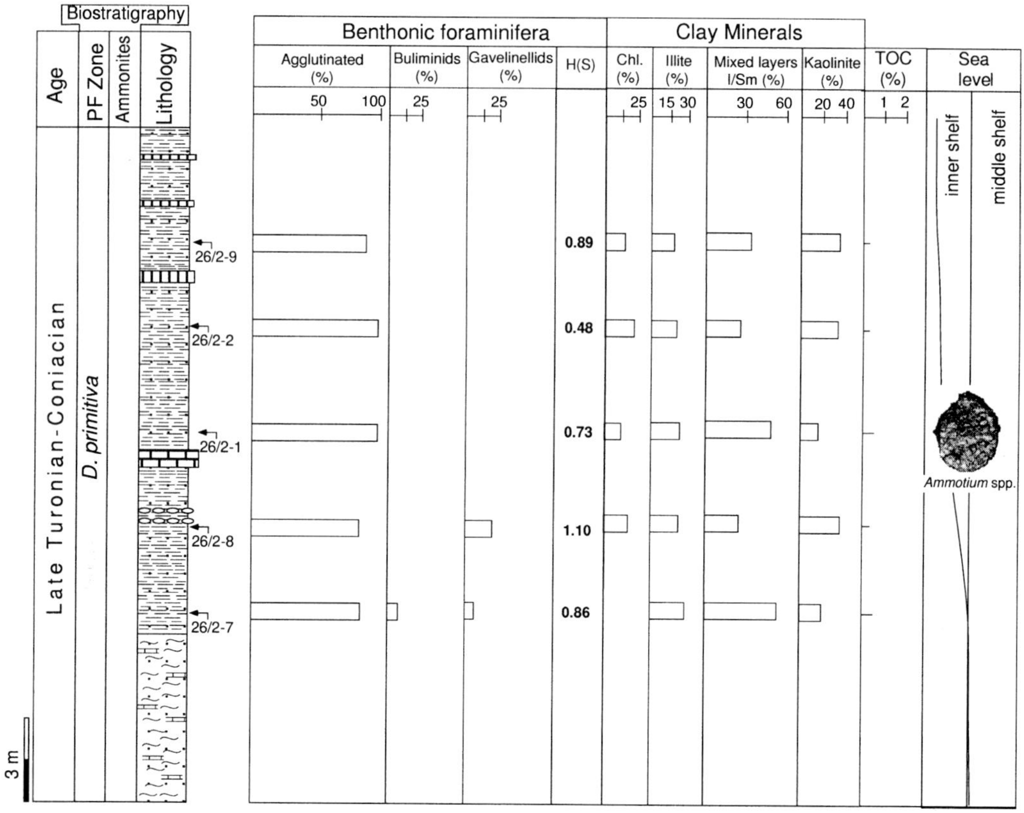

Fig. 13. Summary of benthonic foraminiferal, clay mineral and geochemical results for the Nkalagu Quarry, Central Nigeria.

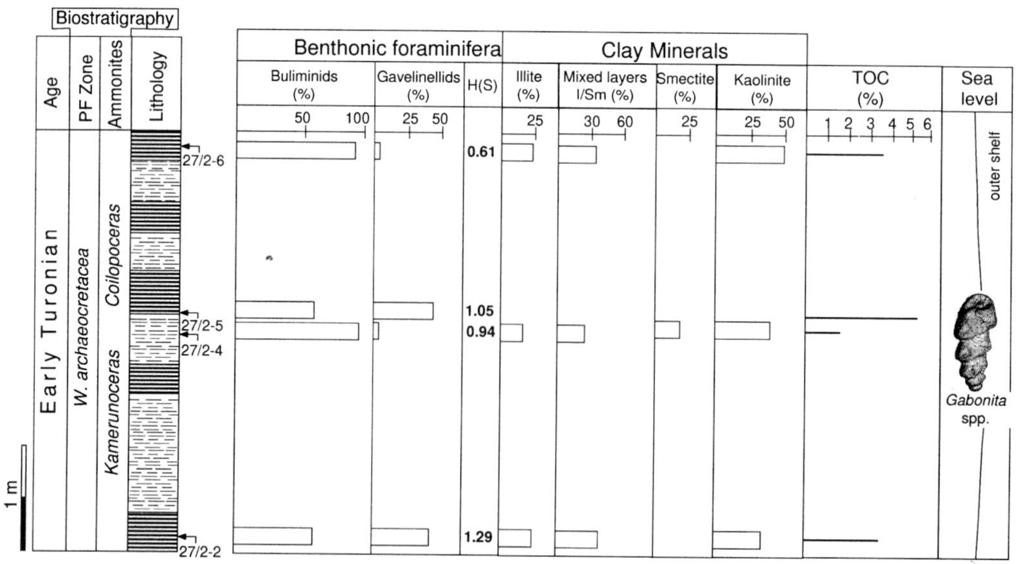

Fig. 14. Summary of benthonic foraminiferal, clay mineral and geochemical results for the Lokpanta Section, Central Nigeria.

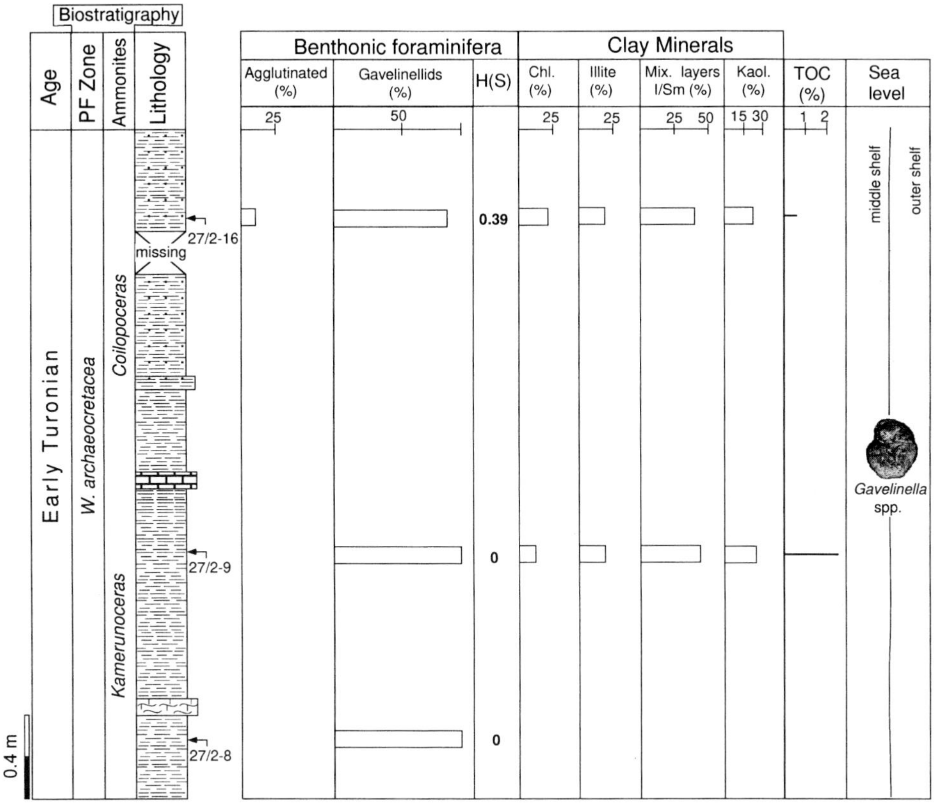

Fig. 15. Summary of benthonic foraminiferal, clay mineral and geochemical results for the Ngbanocha Section, Central Nigeria.

generally appears to have been diluted by the clastic flux. TOC values vary between 0.48 and 2.67 and the organic matter is predominantly of terrestrial origin (Type III). The environment is interpreted as an inner or middle shelf under a strong clastic influence, based on the foraminiferal, clay mineral and Rock-Eval data. A summary of the benthonic foraminiferal, clay mineral and Rock-Eval results is presented in Fig. 15.

Calabar Flank. The Lower Turonian benthonic foraminiferal assemblages from the Calabar Flank show extremely low diversity and are strongly dominated by gavelinellids. As in the Ngbanocha Section, the absence of other calcareous tests of benthonic foraminifera may be indicative of low oxygenation conditions at the seafloor or may be due to the selective preservation of thicker corrosion-resistant gavelinellids tests. The presence of planktonic foraminifera and radiolarians, and the relatively high hydrogen index values, indicating a mixture of Types II and III organic matter, suggest an open marine environment with a relatively high surface water productivity, such as a middle or outer shelf setting. Abundant grains of quartz, mica and plagioclase feldspar observed in thin sections point to a significant detrital flux. The organic matter content of the sediment appears to have been high, and low TOC values may reflect poor preservation of the original organic matter content of the sediment. The presence of a few agglutinated tests, typical of shallow, proximal settings, indicate an environment which was probably still relatively shallow and in close proximity to the continent. A summary of the benthonic foraminiferal, clay mineral and Rock-Eval results is presented in Fig. 16.

Temporal and spatial distribution of high productivity benthonic foraminiferal assemblages

Productivity proxies

In Pleistocene–Holocene sediments palaeoproductivity or carbon flux proxies include the

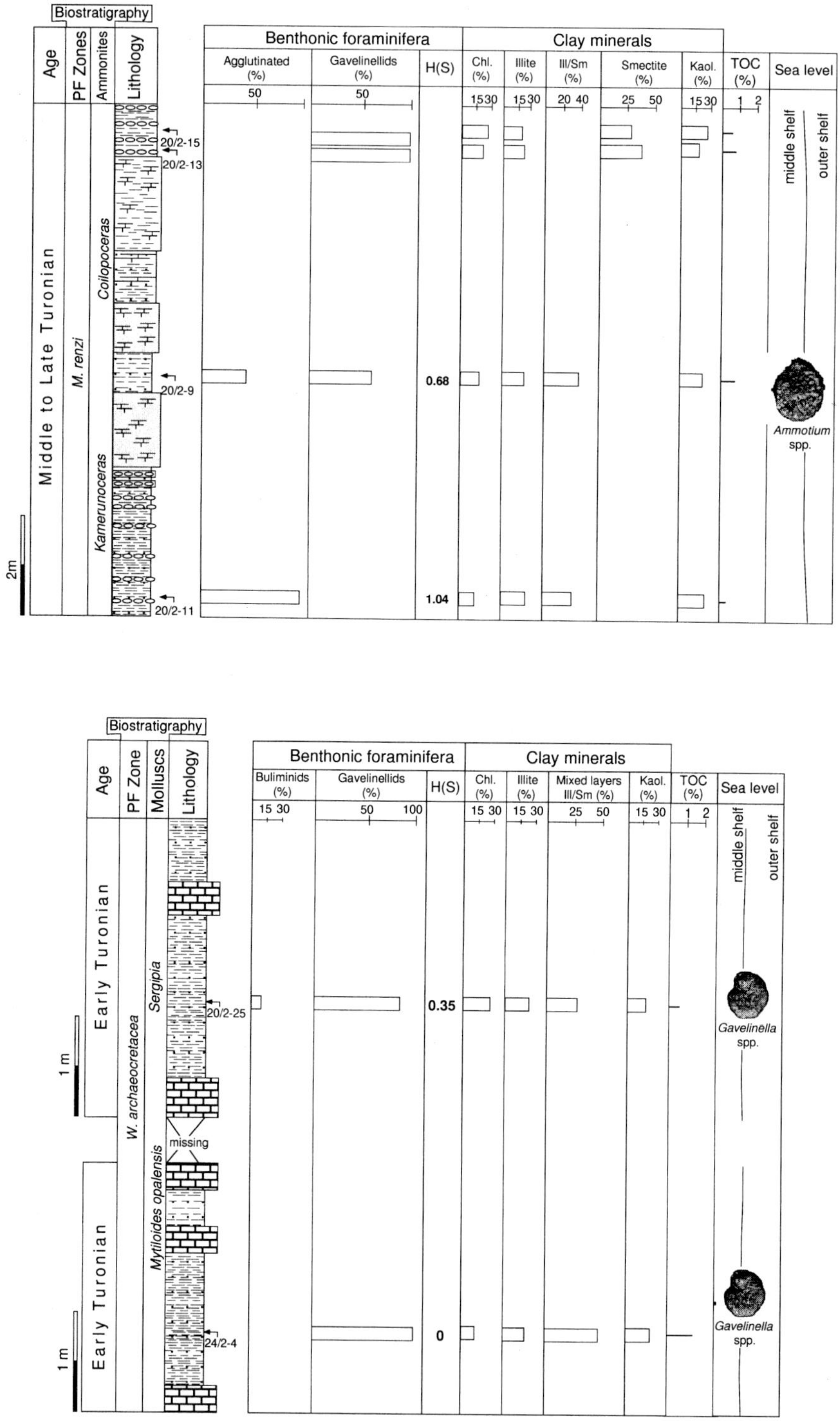

Fig. 16. Summary of benthonic foraminiferal, clay mineral and geochemical results for the Calabar Flank, western Nigeria.

accumulation rate of organic matter (Sarnthein *et al.* 1992), the accumulation rate of benthonic foraminiferal tests (Herguera & Berger 1991) and the quantitative composition of benthonic foraminiferal communities (Loubere 1996). However, the application of these proxies to Cretaceous sediments is strongly limited by several factors. Firstly, the stratigraphic resolution of biostratigraphic, magnetostratigraphic and radiometric dating of Cretaceous sediments does not allow a sufficiently precise calculation of organic matter or foraminiferal test accumulation. Secondly, different palaeoceanographic conditions probably prevailed in the Late Cretaceous (for instance, warm, saline deepwater), which led to higher flux rates of organic matter to the seafloor than in the modern ocean, characterized by cold, oxygenated deepwaters. Finally, the application of quantitative methods, such as transfer functions to Upper Cretaceous assemblages, is extremely limited, as no analogue situation exists even for the comparatively slow evolving benthonic faunas.

For this study the benthonic foraminifera and organic matter distribution within Well S75 provides the unusual possibility to, at least partly, overcome some of these problems. Firstly, a high resolution cyclostratigraphic framework is available which allows accumulation rates with a resolution of 20 ka to be calculated (Kuhnt *et al.* 1997) and, secondly, the preservation of foraminiferal tests and organic matter is sufficiently good to allow reasonable accumulation rates to be computed for these components (Fig. 5). If transfer functions for modern oceans are used (Sarnthein *et al.* 1992), palaeoproductivity in the order of 150–200 gC/ym^2 is calculated, a value that is in the range of modern high productivity areas. Foraminiferal test accumulation rates are strongly fluctuating, probably as a response to the negative influence of dysoxic conditions on the foraminiferal populations. The foraminiferal assemblages are dominated by buliminids and bolivinids, as in present day high productivity–low oxygen environments (Kuhnt *et al.* 1999; Jian *et al.* 1998). The monospecific gavelinellids, which are found in samples characterized by a lower carbon flux, may be suspension feeders, which occupied ecological niches equivalent to those of present day *Cibicidoides* or may, alternatively, represent a preservational bias.

Palaeobathymetric distribution

High productivity benthonic foraminiferal assemblages are characteristic of outer shelf and upper bathyal environments studied from the Tarfaya Basin, southern Morocco, the Casamance Shelf, off Senegal, the Côte d'Ivoire–Ghana Marginal Ridge and the Benue Trough, Nigeria, during the Late Cretaceous. The palaeobathymetric distribution of these assemblages reflects the establishment of a productivity driven oxygen minimum zone along the western African shelf, which was probably associated with an upwelling system active off coastal Africa during most of the Late Cretaceous (Kruijs & Barron 1990; Bush & Philander 1997). High surface water productivity leading to an enhanced organic matter export flux to the seafloor, and to depleted oxygen resources at the seafloor, favoured the establishment of benthonic foraminiferal communities adapted to low oxygen levels and able to exploit vast food resources derived from the phytodetritus flux.

The intensity, depth and lateral extent of the oxygen minimum zone was probably controlled by the surface productivity and the resulting organic matter flux to the seafloor. During sea-level highstands the zone of high surface water productivity and associated oxygen minimum zone extended to more proximal environments of the inner shelf on the Casamance Shelf and in the Tarfaya and Agadir Basins (Wiedmann *et al.* 1982). Out into the basin, the organic matter flux became reduced as particles sank to greater depths. Campanian– Maastrichtian benthonic foraminiferal assemblages from DSDP Site 369 at the foot of the continental slope, offshore southern Morocco, contain some buliminids and bolivinids; however, the assemblages are more diversified and differ in composition from the assemblages of the Tarfaya Basin, reflecting a reduced phytodetritus flux to the seafloor, although surface productivity was probably still high. Consequently, oxygen was less depleted and diversified benthonic foraminiferal communities were able to flourish. At DSDP Site 367 in the Cape Verde Basin, situated in an even more distal position from the continental margin and in much greater water depths below the Carbonate Compensation Depth (CCD), the Campanian–Maastrichtian benthonic foraminiferal assemblages contain organically cemented cosmopolitan deepwater agglutinated foraminifera, which are typical of oligotrophic, abyssal environments. The extremely low TOC values (0.1–0.2%) recorded at this site (Lancelot *et al.* 1977) provide evidence that accumulation rates of organic carbon were considerably lower than in the Tarfaya Basin.

At Site 959 on the Côte d'Ivoire–Ghana Marginal Ridge, high productivity benthonic foraminiferal assemblages and high TOC values provide evidence that an oxygen minimum zone was established on the outer shelf or upper slope during the Early Turonian and Middle

Coniacian–Early Santonian. The sudden increase in the proportion of deepwater agglutinated foraminifera in Core 959D-65R, and the disappearance of all calcareous taxa in Core 959D-64R, suggest that a rapid subsidence below the CCD occurred in the Late Santonian–Early Campanian (Holbourn & Kuhnt 1998). The remaining Upper Cretaceous succession contains mainly cosmopolitan deepwater agglutinated foraminifera with organic cement (Kuhnt *et al.* 1998, 1999) which indicates that the site remained below the CCD until the Late Maastrichtian, at depths where the phytodetritus flux to the seafloor was greatly reduced.

Biostratigraphic distribution

The composition of high productivity benthonic foraminiferal assemblages varies in time and highlights important evolutionary trends in the Late Cretaceous. Lower Turonian assemblages

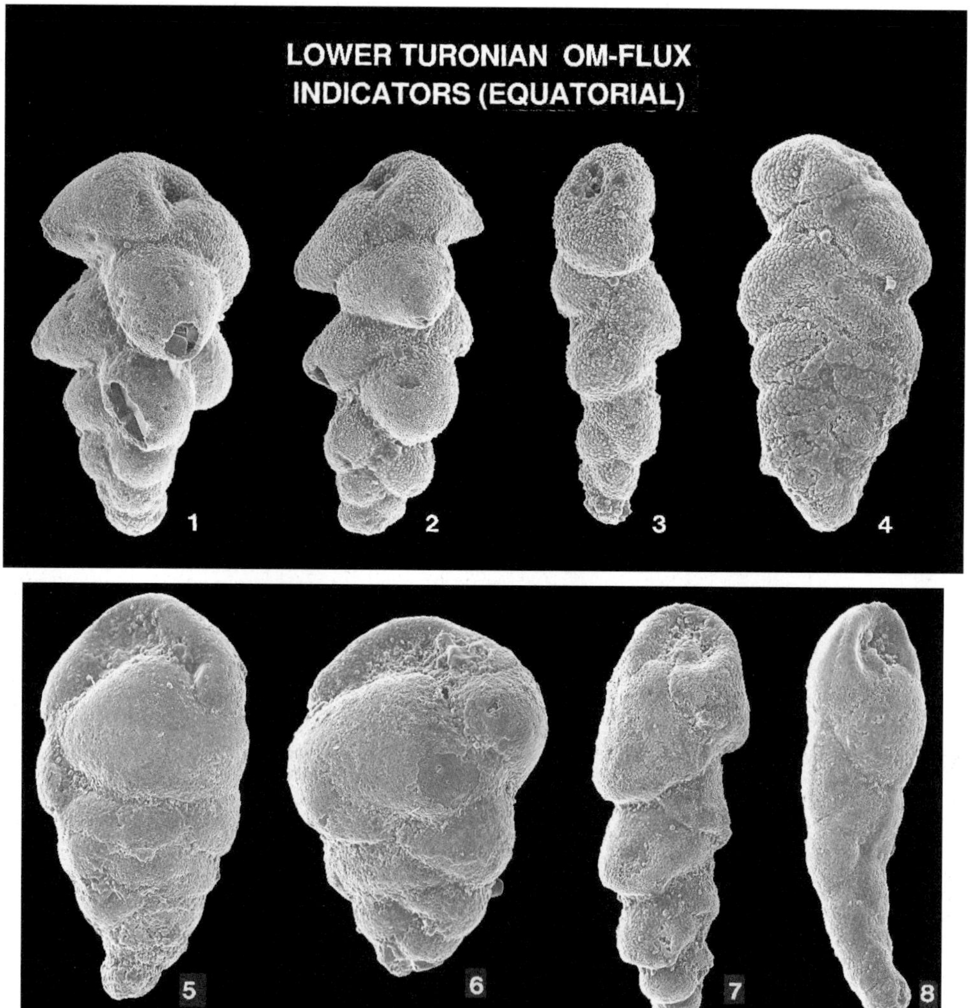

Fig. 17. Length of specimens in microns. 1. *Praebulimina* sp. 1, (340), Côte d'Ivoire–Ghana Marginal Ridge, Sample 959D-68R-1, 40–44 cm; 2. *Praebulimina* sp. 1, (290), Côte d'Ivoire–Ghana Marginal Ridge, Sample 959D-68R-1, 40–44 cm; 3. *Praebulimina* sp. 2, (330), Côte d'Ivoire–Ghana Marginal Ridge, Sample 959D-68R-1, 40–44 cm; 4. *Bolivina anambra* Petters, (230), Benue Trough, Nigeria Sample 27/2-4; 5. *Gabonita levis* (de Klasz, Marie and Rérat), (365), Amma Fatma, Tarfaya, Sample 4; 6. *Gabonita obesa* (de Klasz, Marie and Rérat), (300), Amma Fatma, Tarfaya, Sample 4; 7. *Gabonita* sp. 1, (380), Benue Trough, Nigeria, Sample 27/2-4; 8. *Gabonita* sp. 2, (225), Benue Trough, Nigeria, Sample 27/2-4.

show extremely low diversity and are dominated by *Gabonita levis, Gabonita obesa* and *Bolivina anambra*. In fact, these two species of *Gabonita* provide useful stratigraphic markers for the Early Turonian in western African basins. The number of buliminid and bolivinid species is very low in sediments older that Turonian, where they represent only a minor component of assemblages. Buliminids and bolivinids only started to diversify later in the Turonian, when it can be speculated that they could exploit many of the ecological niches left vacant after the major extinctions of the Cenomanian–Turonian boundary event. The Early Turonian biofacies probably comprises only the early ancestors of the main lineages, which evolved later during the Late Cretaceous. The higher diversity of Upper Cretaceous assemblages from Atlantic equatorial basins is exemplified in assemblages from the Casamance Shelf and the Côte d'Ivoire–Ghana Marginal Ridge, which include many species of buliminids and bolivinids. Extreme environmental conditions at the Cenomanian–Turonian boundary event in equatorial Atlantic basins probably restricted the early recolonization of the seafloor to a few opportunistic species able to cope with severe oxygen depletion, whereas improving conditions in the Middle and Late Turonian favoured the radiation of more specialized taxa which could occupy niches previously filled by different groups. The evolution of buliminids and bolivinids during the Late Cretaceous provides useful stratigraphic markers in western African equatorial basins, however, their full stratigraphic potential is yet to be exploited.

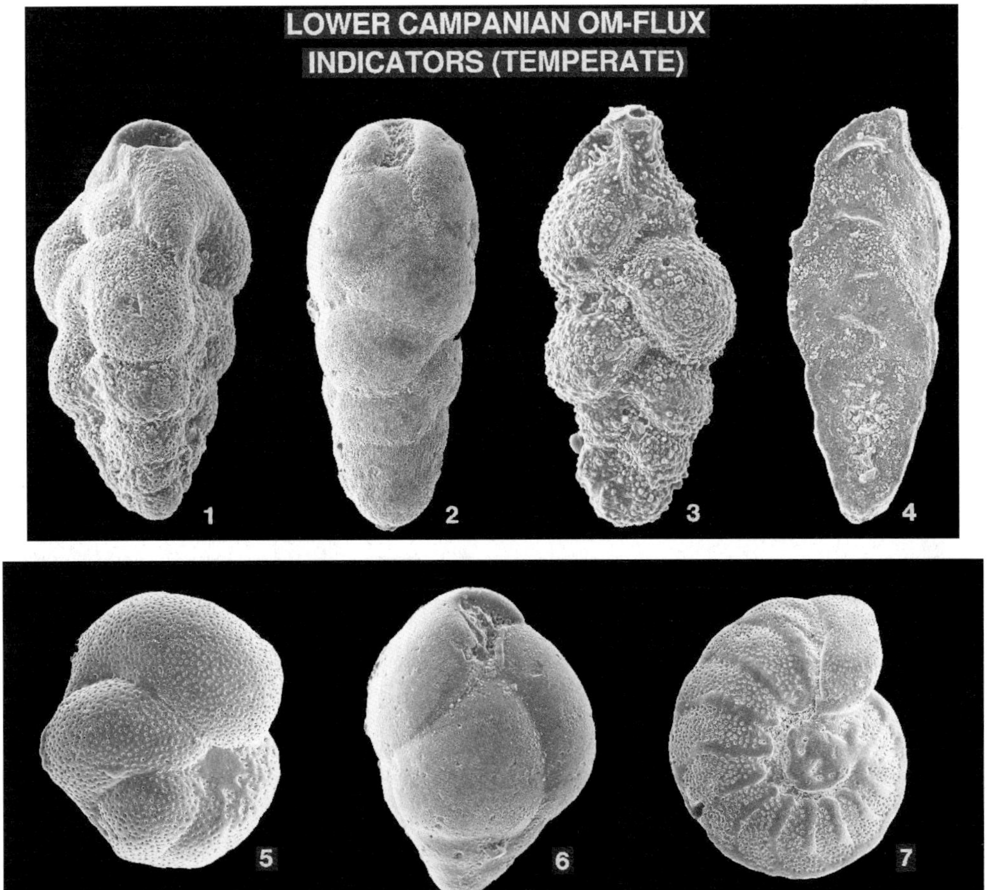

Fig. 18. Length of specimens in μm. 1. *Pyramidina* sp. 1, (280), Sample TAH 39; 2. *Praebulimina aspera* (Cushman and Parker), (685), Sample TAH 36; 3. *Eouvigerina aspera* Marss, (275), Sample TAH 39; 4. *Loxostomum eleyi* (Cushman), (310), Sample TAH 39a; 5. *Cibicides beaumontianus* d'Orbigny, (450), Sample TAH 20; 6. *Praebulimina reussi* (Morrow), (280), Sample TAH 63; 7. *Gavelinella dakotensis* (Fox), (615), Sample TAH 36a.

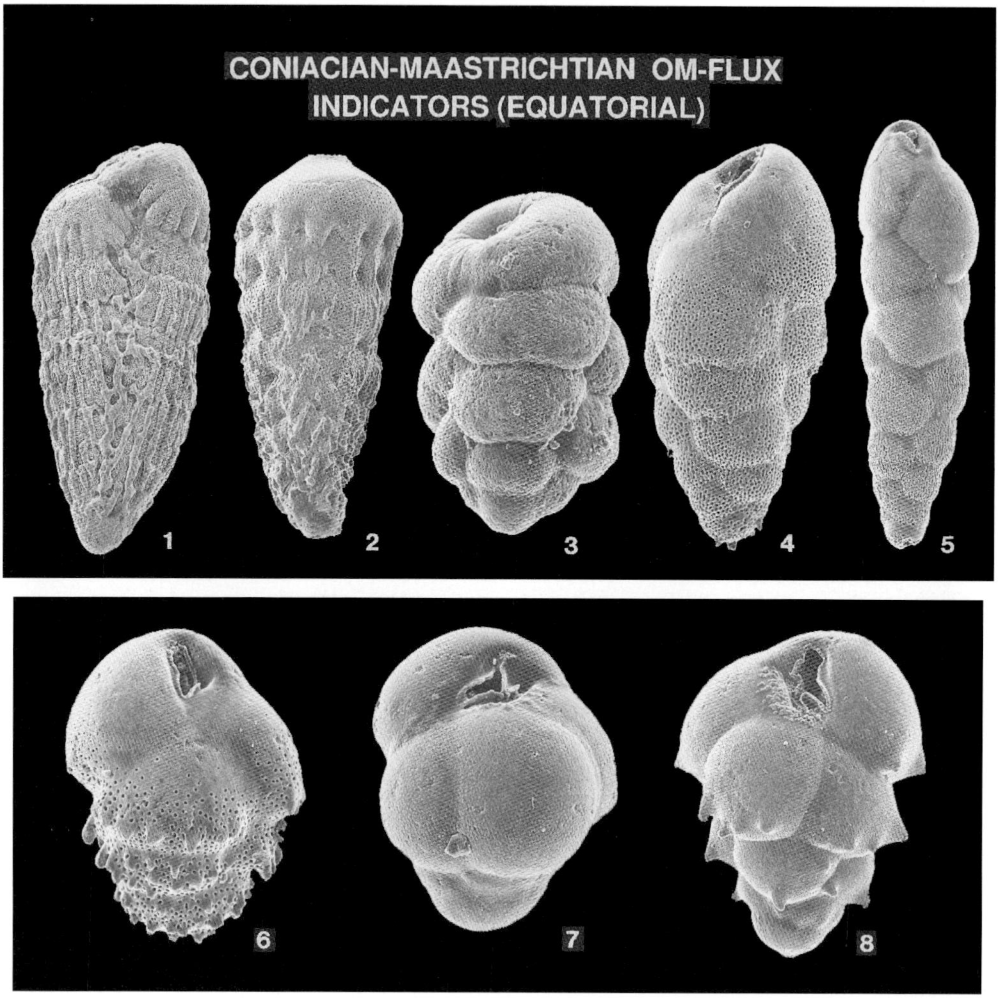

Fig. 19. Length of specimens in μm. 1. *Afrobolivina afra* Reyment (1295), Casamance Shelf, Senegal, Sample CM10-C4, T8; 2. *Orthokarstenia* cf. *dentata* (Chenouard, de Klasz and Meijer), (635), Casamance Shelf, Senegal, Sample CM10-C4, T2; 3. *Praebulimina exigua robusta* (de Klasz, Magné and Rérat, (240), Casamance Shelf, Senegal, Sample CM10-C5, T2. 4. *Praebulimina prolixa longa* (de Klasz, Magné and Rérat), (360), Casamance Shelf, Senegal, Sample CM10-C2, T2; 5. *Neobulimina subregularis* (de Klasz, Magné and Rérat), (510), Casamance Shelf, Senegal, Sample CM10-C2, T2; 6. *Praebulimina arkadelphiana* (Cushman and Parker), (210), Casamance Shelf, Senegal, Sample CM10-C2, T1; 7. *Buliminella quadrilobata* (de Klasz, Magné and Rérat), (200), Casamance Shelf, Senegal, Sample CM10-C2, T2; 8. '*Buliminella*' cf. *mamelligera* (de Klasz and Rérat), (320), Casamance Shelf, Senegal, Sample CM10-C2, T1.

Latitudinal distribution

The palaeobiogeographic distribution of high productivity benthonic foraminiferal assemblages shows that a *Gabonita* biofacies dominated by *Gabonita levis* and *Gabonita obesa* (Fig. 17, 5 and 6) was widely distributed in coastal basins of western Africa from Nigeria to southern Morocco (Kuhnt & Wiedmann 1995). In equatorial basins on the Côte d'Ivoire–Ghana Marginal Ridge and in the Benue Trough, where a high detrital input is evident, more endemic buliminid biofacies were established (Fig. 17, 1–4, 7, 8), probably due to a more restricted circulation in semi-enclosed basins. The *Gabonita* biofacies was widely distributed in Atlantic equatorial basins (Koutsoukos 1992) and was also recorded in Israel (Hamaoui 1965). It appears to be restricted to a broad equatorial belt of highest productivity on the outer shelf of coastal basins, receiving a low terrigenous input. The palaeogeographic distribution of this high productivity biofacies seems to have contracted latitudinally in the Early Campanian. Lower Campanian assemblages from the Tarfaya Basin (Fig. 18) show more similarity with assemblages from higher latitudes, such as the Paris Basin (Marie 1941), the Western Interior Basin (MacNeil & Caldwell 1981) and New Jersey slope (Nyong & Olsson 1984), than with assemblages from Equatorial Africa, which comprise numerous species of *Buliminella, Praebulimina, Afrobolivina, Bolivina* and *Orthokarstenia* (Fig. 19). A corresponding Lower Turonian assemblage (*Tappanina* biofacies) in the higher latitudes characterized areas outside the equatorial belt of highest productivity (Eicher & Worstell 1970; Kuhnt & Wiedmann 1995).

The zone of highest productivity was broadest in the Early Turonian, coinciding with the maximum sea-level rise and highest atmospheric temperature in the Late Cretaceous. An important mechanism for deepwater formation in the Late Cretaceous was probably the increased evaporation of extensive, shallow epicontinental seas in the low latitudes, leading to the formation of warm, dense, saline bottom and intermediate water, flowing in a reverse direction from the tropical regions towards the poles, which was a major agent of heat transport towards the poles (Hay 1995*a*, *b*). During times of sea-level rise and elevated atmospheric temperature, the formation of warm saline water enriched in nutrients intensified at low latitudes in extensive epicontinental seas. Eicher & Diner (1991) suggested that two contrasting circulation modes may have existed during the Late Cretaceous: a saline- and a temperature-driven mode. Late Cretaceous oceans may have been considerably more unstable than modern day oceans and extremely sensitive to relatively minor climatic or hydrographic changes, which may have triggered rapid switching from one dominant mode to another and led to major fluctuations in marine productivity. Such fluctuations may have rhythmically occurred on a Milankovitch scale (Kuhnt *et al.* 1997). Late Cretaceous general circulation models also predict that coastal upwelling may have been enhanced (Bush & Philander 1997) as a result of higher atmospheric CO_2 and warmer temperatures. During sea-level rises, intensified equatorial upwelling would have brought to the surface waters with a high nutrient content over a wide

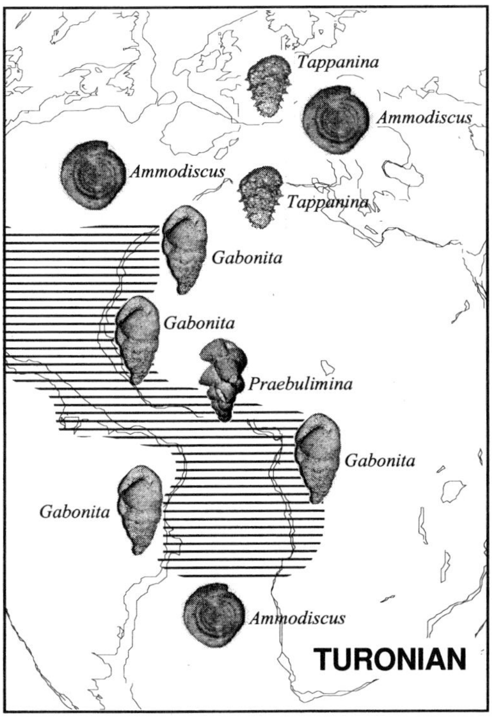

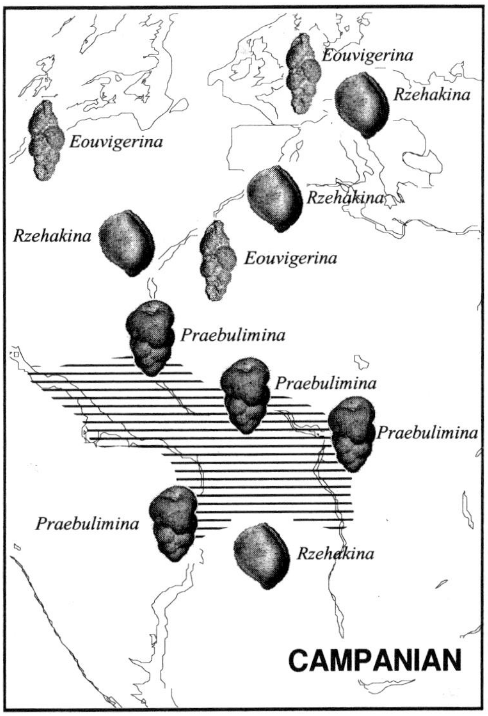

Fig. 20. Palaeobiogeographic distribution of high productivity benthonic foraminifera during the Early Turonian and Early Campanian transgression showing contraction of the high productivity equatorial belt in the Early Campanian.

latitudinal zone. The maximum extent of this zone of highest productivity was probably in the Early Turonian when maximum sea-level rise occurred and the highest atmospheric CO_2 and temperature prevailed.

Conclusions

The reconstruction of paleogeographic distribution patterns of enhanced productivity and resulting high carbon fluxes provides useful information to recognize areas of potential source rock formation. Benthonic foraminifera, which can be studied even in weathered sections and cuttings, record phytodetritus/organic matter flux rates at the time of deposition and may be better proxy-indicators for paleoproductivity than organic matter accumulation rates, which only record the preserved part of the primary production and are strongly biased by changing preservation conditions. This combined data set of benthonic foraminiferal distribution and organic matter accumulation patterns points to substantially enhanced primary production along the West African margin in *c.* 1000 km wide band north and south of the Cretaceous equator (Benue, Côte Ivoire, Casamance and Tarfaya Basins). In other areas, the formation of potential source rocks is restricted to a short time-span at the Cenomanian–Turonian boundary, and organic matter accumulation may be largely controlled by favourable preservation conditions.

Organic-rich sediments from outer shelf and upper bathyal environments are characterized by benthonic foraminiferal assemblages dominated by buliminids and bolivinids. Lower Turonian assemblages exhibit extremely low diversity and are strongly dominated by a *Gabonita* biofacies. In contrast, Coniacian–Maastrichtian assemblages display much higher diversity and contain numerous species of *Buliminella, Praebulimina, Afrobolivina, Bolivina* and *Orthokarstenia*. The extent of the high productivity equatorial belt appears to have shifted latitudinally during the Late Cretaceous as sea level and circulation patterns changed. The deposition of organic-rich sediments was especially favoured during two main time slices in the Early Turonian and Early Campanian, when major transgressive pulses occurred on a global scale (Haq *et al.* 1988). These two periods coincided with major changes in palaeocirculation and palaeoproductivity, which affected the nutrient and oxygen balance of coastal basins along the western African margin and led to the establishment of an expanded oxygen minimum zone and to the extensive deposition of organic-rich deposits along the western African margin of the central Atlantic. The intensity, depth and latitudinal extent of the oxygen minimum zone appear to have reached a maximum in the Early Turonian, when the highest sea level and warmest temperature occurred for the whole of the Cretaceous. The distribution of high productivity benthonic foraminiferal assemblages in the Early Turonian and Early Campanian is shown in Fig. 20.

We are especially grateful to Haydon Bailey and Malcolm Hart for reviewing the manuscript, and to Ivan de Klasz for taxonomic advice and critical comments. We extend our thanks to the staff of the SEM Lab and Photographic Unit at the Geologisch-Paläontologisches Institut of the Christian Albrechts University in Keil, to Anya Kirch for help with assembling the photographic plates and to Ingo Bohnert for draughting some of the figures. We thank the Deutsche Forschungsgemeinschaft (Grant Ku649-2 'Equatorial Atlantic Gateways') for financial support. A.E.L.H. acknowledges support of a Royal Society European Exchange Fellowship at Kiel University and A.E.A. of an Alexander von Humboldt Fellowship at Kiel University.

References

ALTENBACH, A. V. & SARNTHEIN, M. 1989. Productivity record in benthic foraminifera. *In*: BERGER, W. H., SMETACEK, V. S. & WEFER, G. (ed) *Productivity in the Oceans, Present and Past.* Dahlem-Konferenzen, 255–269.

BARMAWIDJAJA, D. M., JORISSEN, D. J., PUSKARIC, S. & VAN DER ZWAAN, G. J. 1992. Microhabitat selection by benthic foraminifera in the northern Adriatic Sea. *Journal of Foraminiferal Research*, **20**, 246–247.

BARRON, E. J. & PETERSON, W. H. 1990. Mid-Cretaceous ocean circulation: results from model sensitivity studies. *Paleoceanography*, **5**, 319–338.

—— & WASHINGTON, W. M. 1985. Warm Cretaceous climates: High atmospheric CO_2 as a plausible mechanism. *In*: SANDQUIST, E. T. & BROECKER, W. S. (eds) *Carbon Cycle and Atmospheric CO_2: Natural Variations Archean to Present*. American Geophysical Union, Geophysical Monography, **32**, 546–553.

——, FAWCETT, P. J. & PETERSON, W. H. 1995. A 'simulation' of mid-Cretaceous climate. *Paleoceanography*, **10**, 953–962.

——, PETERSEN, W. H., POLLARD, D. & THOMPSON, S. 1993. Past climate and the role of ocean heat transport: Model simulation for the Cretaceous. *Paleoceanography*, **8**, 785–798.

BERNER, R. A. 1991. A model for atmospheric CO_2 over Phanerozoic time. *American Journal of Science*, **291**, 339–376.

BRASS, G. W., SOUTHAM, J. R. & PETERSON, W. H. 1982. Warm saline bottom water in the ancient ocean. *Nature*, **296**, 620–623.

BUSH, A. B. G. & PHILANDER, S. G. H. 1997. The Late Cretaceous: Simulation with a coupled atmosphere–ocean general circulation model. *Paleoceanography*, **12**, 495–516.

CERLING, T. E. 1991. Carbon dioxide in the atmosphere: Evidence from Cenozoic and Mesozoic paleosols. *American Journal of Science*, **291**, 377–400.

CHAMLEY, H. 1989. *Clay Sedimentology*. Springer, 1–623.

CORLISS, B. H. & CHEN, C. 1988. Morphotype patterns of Norwegian Sea deep-sea benthic foraminifera and ecological implications. *Geology*, **16**, 716–719.

EICHER, D. L. & DINER, S. R. 1991. Environmental factors controlling Cretaceous limestone–marlstone rhythms. *In*: EINSELE, G., RICKEN, W. & SEILACHER, A. (eds) *Cycles and Events in Stratigraphy*. Springer, 79–93.

—— & WORSTELL, P. 1970. Cenomanian and Turonian Foraminifera from the Great Plains, United States. *Micropaleontology*, **16**, 269–324.

EL ALBANI, A., KUHNT, W., LUDERER, F., HERBIN, J. P. & CARON, M. 1999. Palaeoenvironmental evolution of the Late Cretaceous sequence in the Tarfaya Basin (southwest of Morocco). *This volume*.

ESPITALIÉ, J., DEROO, G. & MARQUIS, F. 1985. La pyrolyse Rock-Eval et ses applications. *Revue de l'Institut Français du Pétrole*, **40**, 563–579, 755–784, **41**, 73–89.

——, MADEC, M., TISSOT, B., MENNIG, J. H. & LEPLAT, P. 1977. Source rock characterization method of petroleum exploration. *Offshore Technology Conference*, Houston, **2935**, 1–6.

FRAKES, L. A., FRANCIS, J. E. & SYKTUS, J. I. 1992. *Climate Models of the Phanerozoic*. Cambridge University Press, 1–274.

FREEMAN, K. H. & HAYES, J. M. 1992. Fractionation of carbon isotopes by phytoplankton and estimates of ancient CO_2 levels. *Global Biogeochemical Cycles*, **6**, 185–198.

GOODAY, A. J. 1994. The biology of deep-sea foraminifera: A review of some advances and their applications in palaeoceanography. *Palaios*, **9**, 14–31.

—— & TURLEY, C. M. 1990. Responses by benthic organisms to inputs of organic material to the ocean floor: a review. *Philosophical Transactions of the Royal Society of London*, **A331**, 119–138.

HAMAOUI, M. 1965. Sur la présence de Gabonella (Foraminiferida) en Israel. *Palaeontographia Italiana*, **29**, 189–231.

HAQ, B. U., HARDENBOL, J. & VAIL, P. R. 1988. Chronology of fluctuating sea levels since the Triassic (250 million years to the Present). *Science*, **235**, 1156–1167.

HAY, W. W. 1988. Paleoceanography: A review for the GSA Centennial. *Bulletin of the Geological Society of America*, **100**, 1934–1956.

—— 1995*a*. Cretaceous paleoceanography. *Geologica Carpathica*, **46**, 257–266.

—— 1995*b*. Paleoceanography of marine organic carbon-rich sediments. *In*: HUC, A. Y. & SCHNEIDERMANN, N. (eds) *Paleogeography, Paleoclimate and Source Rocks*. AAPG, 21–59.

HERBIN, J. P., MONTARDET, L., MÜLLER, C., GOMEZ, R., THUROW, J. & WIEDMANN, J. 1986. Organic rich sediments at the Cenomanian–Turonian boundary in oceanic and coastal basins in the North Atlantic and Tethys. *In*: SUMMERHAYES, C. & SHACKLETON, N. J. (eds) *North Atlantic Paleoceanography*. Geological Society, London, Special Publications, **22**, 389–422.

HERGUERA, J. & BERGER, W. H. 1991. Paleoproductivity from benthic foraminiferal abundance: Glacial to postglacial change in the west-equatorial Pacific. *Geology*, **19**, 1173–1176.

HERMELIN, J. O. R. 1992. Variations in the benthic foraminiferal fauna of the Arabian Sea: a response to changes in upwelling intensity? *In*: SUMMERHAYES, C. P., PRELL, W. L. & EMEIS, K. C. (eds) *Upwelling Systems: Evolution Since the Miocene*. Geological Society, London, Special Publications, **64**, 151–166.

HOLBOURN, A. E. L. & KUHNT, W. 1998. Turonian–Santonian benthic foraminiferal assemblages from Site 959D (Côte d'Ivoire–Ghana Transform Margin, Equatorial Atlantic): Indication of a Late Cretaceous oxygen minimum zone. *Ocean Drilling Program, Scientific Results*, **159**, in press.

——, ——, EL ALBANI, A., LY, A., GOMEZ, R. & HERBIN, J. P. 1999. Palaeoenvironments and palaeobiogeography of the Late Cretaceous Casamance Transect (Senegal, NW Africa): Distribution patterns of benthic foraminifera, organic carbon and terrigenous flux. *Neues Jahrbuch für Geologie und Paläontologie*, in press.

HOLTZAPFFEL, T. 1985. Les minéraux argileux, préparation, analyse diffractométrique et détermination. *Annales de la Société Géologique du Nord*, **12**, 1–136.

HUBER, B. T., HODELL, D. A. & HAMILTON, C. P. 1995. Middle–Late Cretaceous climate of the southern high latitudes: Stable isotopic evidence for minimal equator-to-pole thermal gradients. *GSA Bulletin*, **107**, 1164–1191.

JIAN, Z., HESS, S., KIENAST, M., KUHNT, W., ERLENKEUSER, M. & SARNTHEIM, M. 1999. Deepwater variations in the northern and southwestern China Sea, the record of benthic foraminifera and stable isotopes. *In*: SARNTHEIN, M. & WANG, P. X. (eds) *Response of West Pacific Marginal Seas to Global Climatic Change*. Marine Geology, in press.

JORISSEN, F. J., BARMAWIDJAJA, D. M., PUSKARIC, S. & VAN DER ZWAAN, G. J. 1992. Vertical distribution of benthic foraminifera in the northern Adriatic Sea: The relation with the organic flux. *In*: VAN DER ZWAAN, C. G., JORISSEN, F. J. & ZACHARIASSE, W. J. (eds) *Approaches to Palaeoproductivity Reconstructions*. Marine Micropaleontology, **19**, 131–146.

KOUTSOUKOS, E. A. M. 1992. Late Aptian to Maastrichtian foraminiferal biogeography and paleoceanography of the Sergipe Basin, Brazil. *Palaeogeography, Palaeoclimatology, Palaeoecology*, **92**, 295–324.

KRUIJS, E. & BARRON, E. 1990. Climate model prediction of paleoproductivity and potential source rock distribution. *In*: HUC, A. Y. (ed.) *Deposition of Organic Facies*. AAPG Studies in Geology, **30**, 195–216.

KUHNT, W. & WIEDMANN, J. 1995. Cenomanian–Turonian Source Rocks: Paleobiogeographic and Paleoenvironmental Aspects. *In*: HUC, A. Y. (ed.) *Paleogeography, Paleoclimate and Source Rocks*. AAPG Studies in Geology, **40**, 213–232.

——, HESS, S. & JIAN, Z. 1999. Quantitative composition of benthic foraminiferal assemblages as a proxy indicator for organic carbon flux rates in the South China Sea. *In*: SARNTHEIN, M. & WANG, P. X. (eds) *Response of West Pacific Marginal Seas to Global Climate Change.* Marine Geology, in press.

——, MOULLADE, M. & KAMINSKI, M. 1998. Upper Cretaceous, K/T Boundary and lowermost Paleocene agglutinated foraminifera from Hole 959D. *Ocean Drilling Program Scientific Results*, **159**, in press.

——, NEDERBRAGT, A. & LEINE, L. 1997. Cyclicity of Cenomanian–Turonian organic-carbon-rich sediments in the Tarfaya Atlantic Coastal Basin (Morocco). *Cretaceous Research*, **18**, 587–601.

——, HERBIN, J. P., THUROW, J. & WIEDMANN, J. 1990. Distribution of Cenomanian–Turonian organic rich facies in the western Mediterranean and along the adjacent Atlantic margin. *In*: HUC, A. Y. (ed.) *Deposition of Organic Facies*. AAPG Studies in Geology, Tulsa, Oklahoma, **30**: 133–160.

LANCELOT, Y., SEIBOLD, E. *ET AL.* 1977. *Initial Reports of the Deep Sea Drilling Project*. US Government Printing Office (Washington, DC), 1–1259.

LINKE, P. & LUTZE, G. F. 1993. Microhabitat preferences of benthic foraminifera – a static concept or a dynamic adaptation to optimize food acquisition? *Marine Micropaleontology*, **20**, 215–234.

LOUBERE, P. 1994. Quantitative estimation of surface ocean productivity and bottom water oxygen concentration using benthic foraminifera. *Paleoceanography*, **9**, 723–737.

—— 1996. The surface of ocean productivity and bottom water oxygen signals in deep water benthic foraminiferal assemblages. *Marine Micropaleontology*, **28**, 247–261.

LUTZE, G. & COLBOURN, W. 1984. Recent benthic foraminifera from the continental margin of northwest Africa: community structure and distribution. *Marine Micropaleontology*, **22**, 361–401.

——, PFLAUMANN, U. & WEINHOLZ, P. 1986. Jungquartäre Fluktuationen der bentischen Foraminiferenfaunen in Tiefsee-Sedimenten vor NW-Afrika: Eine Reaktionauf Produktivitätsänderungen im Oberflächenwasser. *Meteor Forschungsergebnisse, Reihe C*, **40**, 163–180.

MASCLE, J., LOHMAN, G. P., CLIFT, P. D. *ET AL.* 1996. *Proceedings of the ODP Initial Reports*. College Station, TX (Ocean Drilling Program), **159**.

MCNEIL, D. H. & CALDWELL, W. G. E. 1981. Cretaceous Rocks and their foraminifera in the Manitoba Escarpment. *Geological Association of Canada, Special Paper*, **21**, 1–439.

MARIE, P. 1941. Les foraminifères de la Craie à *Belemnitella mucronata* du Bassin de Paris. *Mémoires du Muséum d'Histoire Naturelle*, **12**, 1–296.

NYONG, E. & OLSSON, R. K. 1984. A paleoslope model of Campanian to Lower Maastrichtian foraminifera in the North American Basin and adjacent continental margin. *Marine Micropaleontology*, **8**, 437–477.

—— & RAMANATHAN, R. 1985. A record of oxygen-deficient paleoenvironments in the Cretaceous of the Calabar Flank, southeastern Nigeria. *Journal of African Earth Sciences*, **3**, 455–460.

OJOH, K. A. 1988. Evolution géodynamique des bassins albo-santonien du sud-ouest du fossé de la Benoue (Nigeria). Thèse Aix-Marseille III, 1–241.

PETTERS, S. W. 1982. Central West African Cretaceous–Tertiary benthic foraminifera and stratigraphy. *Palaeontographica Abt. A*, **179**, 1–104.

SARNTHEIN, M., PFLAUMANN, U., ROSS, R., TIEDEMANN, R. & WINN, K. 1992. Transfer functions to reconstruct ocean paleoproductivity: a comparison. *In*: SUMMERHAYES, C. P., PRELL, W. L. & EMEIS, K. C. (eds) *Upwelling Systems: Evolution Since the Early Miocene*. Geological Society, London, Special Publications, **64**, 411–437.

SHANNON, C. E. & WEAVER, W. 1949. *The Mathematical Theory of Communication*. University of Illinois Press, 1–125.

SPICER, R. A. & CORFIELD, R. M. 1992. A review of terrestrial and marine climates in the Cretaceous with implications for modelling the 'Greenhouse Earth'. *Geological Magazine*, **129**, 169–180.

TYSON, R. V. & PEARSON, T. H. 1991. Modern and ancient continental shelf anoxia: an overview. *In*: TYSON, R. V. & PEARSON, T. H. (eds) *Modern and Ancient Continental Shelf Anoxia*. Geological Society, London, Special Publications, **58**, 1–24.

WAGNER, T. & PLETSCH, T. 1999. Tectono-sedimentary controls on Cretaceous black shale deposition along the opening Equatorial Atlantic Gateway (ODP Leg 159). *This volume*.

WIEDMANN, J., BUTT, A. & EINSELE, G. 1982. Cretaceous stratigraphy, environment and subsidence history of the Moroccan Continental Margin. *In*: VON RAD, U., HINZ, K., SARNTHEIN, M. & SEIBOLD, E. (eds) *Geology of the Northwest African Continental Margin*. Springer, 366–395.

Palaeoenvironmental evolution of the Late Cretaceous sequence in the Tarfaya Basin (southwest of Morocco)

A. EL ALBANI[1], W. KUHNT[1], F. LUDERER[1], J. P. HERBIN[2] & M. CARON[3]

[1]*Geologisch-Palaeontologisches Institut, Universität zu Kiel, Olshausenstr. 40, 24118 Kiel, Germany*
[2]*Institut Français du Pétrole, Reuil Malmaison, France*
[3]*Institut de Géologie, boulevard Perolles, Fribourg, Switzerland*

Abstract: The Late Cretaceous sedimentary evolution of the Tarfaya Basin is reviewed based on late Albian–Campanian outcrops and two cored exploration wells (S13 and S75) with a stratigraphic range from the early Cenomanian to the late Turonian. Standard sedimentological and micropalaeontological methods were used, together with clay mineralogy and organic geochemistry. Two distinct depositional environments were found. The first is late Albian–early/middle Cenomanian in age and dominated by terrigenous silici- and bioclastic sediments containing primarily terrestrial organic matter with a low organic content. Abundant illite and chlorite, eroded from crystalline rocks outcropping in the Anti-Atlas mountain and Mauritanides is present, as is abundant kaolinite reworked from soils surrounding the Tarfaya Basin. The second environment comprises late Cenomanian–Campanian sediments characterized by pelagic marls and limestones with a high organic content and a abundant pelagic macro- and microfauna. These sediments, which were probably deposited in an upwelling system, are characterized by higher contents of smectite.

The Tarfaya coastal basin, located in southwest Morocco, extends along the Atlantic coast in a north-northeast–south-southwest direction (Fig. 1), and has been the subject of several regional geological studies. These include Lehmann (1965), Choubert *et al.* (1966, 1972), Wiedmann *et al.* (1978, 1982) and Ranke *et al.* (1982).

The basin is limited by the Anti-Atlas to the north, the Mauritanides to the south, the Reguibat in the east and the Atlantic Ocean in the west. The evolution of the basin is closely connected with the geological history of the African craton and the opening of the Atlantic (Ranke *et al.* 1982) Mesozoic and Cenozoic sediments cover the basement, which is a complex of igneous and metamorphic rocks of Precambrian and Palaeozoic age. The post-Triassic subsidence of the basin is related to the opening of the Atlantic Ocean (Wiedmann *et al.* 1982). Important is the activation of the Zemmour Fault that separates the Anti-Atlas and the Tindouf Basin to the east (Choubert *et al.* 1966). With the activation of the fault, steady subsidence of the Tarfaya Basin commenced in the Triassic followed by stepwise subsidence in the Jurassic and the Cretaceous. Widespread carbon-rich marls and limestones (oil shales) were deposited in the Tarfaya Basin during the Late Cretaceous (Wiedmann *et al.* 1982; Leine 1986).

The Cretaceous and Cenozoic sections are poorly exposed, but may be examined along the shoreline and around Sebkhas and Oueds (Fig. 1). Because of the almost horizontal stratification it is possible, despite the poor outcrops, to reconstruct the Late Cretaceous sequence by combining individual outcrops. The oldest sediments are in the eastern part (Tassegdelt) and the youngest are in the west (Sebkha Tah) of the basin (Fig. 1). Exploration Wells S13 and S75, drilled by ONAREP and Shell in the 1970s and early 1980s, record an expanded Cenomanian and Turonian succession in the centre of the basin. The accompanying oil shales, which extend offshore, occupy an area of *c.* 170 000 km^3. Similar sedimentary successions are present in most of the northwest African coastal basins, e.g. the Agadir Basin (Morocco), the Casamance Basin (Senegal), the Benue Basin (Nigeria) (Kuhnt *et al.* 1990; Kuhnt & Wiedmann 1995; Holbourn *et al.* 1999).

Materials and methods

This study is based on field observations between 1993 and 1997, microfacies analyses of thin sections, clay mineralogy, micropalaeontological investigations and organic geochemistry. The objective is to combine these data into a sequence stratigraphic and palaeoceanographic framework for the Late Cretaceous of the Tarfaya Basin.

Continuous sedimentological descriptions were prepared for a composite section derived from nine outcrops. The stratigraphic ranges of each section are illustrated in Fig. 2. Micropalaeontological samples were collected at *c.* 2 m intervals for the outcrop sections and at 1 m intervals for the well

EL ALBANI, A., KUHNT, W., LUDERER, F., HERBIN, J. P. & CARON, M. 1999. Palaeoenvironmental evolution of the Late Cretaceous sequence in the Tarfaya Basin (southwest of Morocco). *In:* CAMERON, N. R., BATE, R. H. & CLURE, V. S. (eds) *The Oil and Gas Habitats of the South Atlantic.* Geological Society, London, Special Publications, **153**, 223–240.

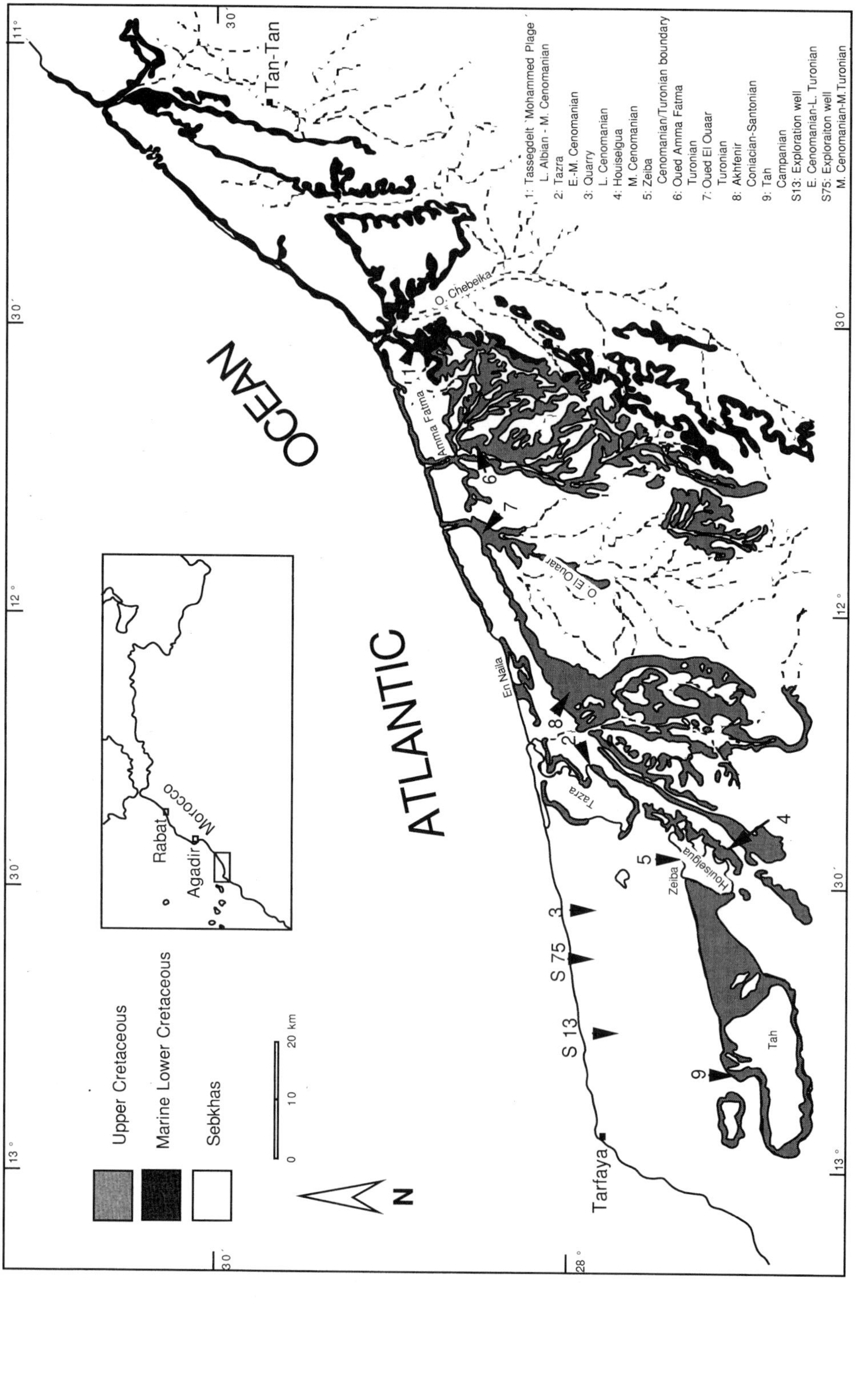

Fig. 1. Location map for the studied area in the Tarfaya coastal basin.

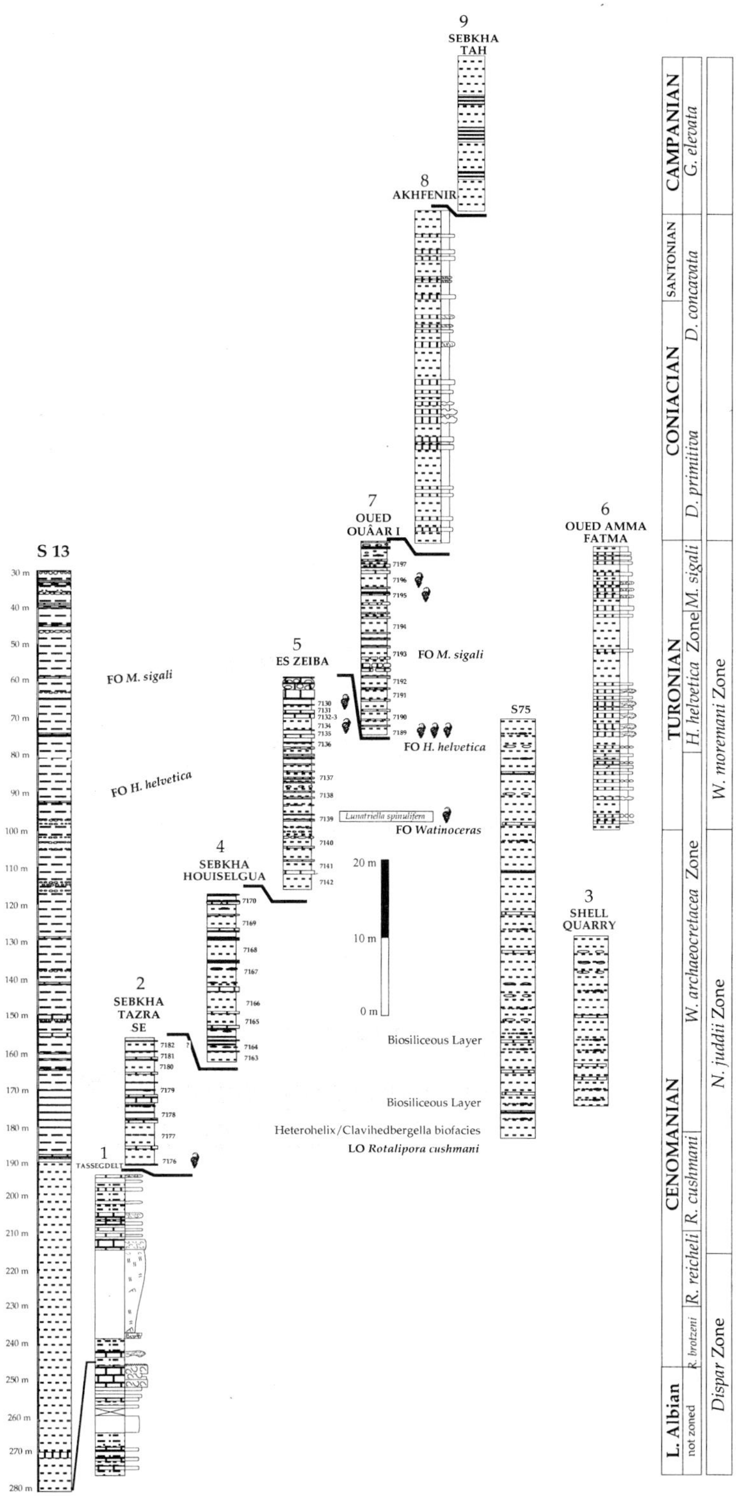

Fig. 2. Complete composite stratigraphic section of the Late Cretaceous sequence.

cores. For the Cenomanian–Turonian boundary, 10 cm sampling was used; 20 cm sampling was employed for the early Campanian. For the microfacies analyses, 150 samples were collected using a 1 m sample interval.

Clay mineralogy was performed on 200 samples (1 m resolution) using a Philips X-ray diffractometer (PW 1390). The analytical and interpretative procedures follow those of Holtzapfel (1985). The analyses were carried out in the University of Lille (France).

The organic matter of the outcrop sections was determined using Rock-Eval pyrolysis on the same samples as those analysed by X-ray diffraction. This method has been described by Espitalié *et al.* (1985). The 'total organic content' (TOC), the origin of organic matter and the carbonate content of the exploration Wells S13 and S75 was examined on a 1 m scale over the core and on a centimetre scale across the extinction zone of *Rotalipora cushmani* (between 2 m below and 2 m above the low occurrence of *Rotalipora cushmani*). The results are published in Kuhnt *et al.* (1990, 1997) and Luderer & Kuhnt (1997). These analyses were carried out with the Rock-Eval II analyser and the carbonate module at the Institute Français du Pétrole in Paris.

Biostratigraphic framework

The biostratigraphic correlation of the individual sections is mainly based on the Tethyan standard planktonic foraminiferal zonation (Caron 1985). The position of the studied sections within the Tethyan standard zonation is given in Fig. 3. Zonations for Wells S13 and S75 are given in Kuhnt *et al.* (1986, 1990, 1997) and Luderer & Kuhnt (1997). Five informal units are recognized, which are described below.

Late Albian–early/middle Cenomanian terrigenous sandstones and marls

This part of the profile is observed in the Tassegdelt section (82 m thickness, for locations see Fig. 1), in the basal part of Well S13 (120 m) and along the coast north of the Oued Amma Fatma ('Mohammed Plage'). These outcrops are located 100 and 120 km to the northeast of Tarfaya. Three lithological units were distinguished.

Unit 1. This unit represents the basal part of the section. It is composed of yellowish to greenish silty marls with intercalations of centimetre to decimetre thick silty limestones.

Unit 2. This unit is composed of fine-grained and bioturbated sandstones, which occupy the middle part of the outcrop (Fig. 4a). These sandstones are covered by a thick distinct bioclastic layer (7 m in thickness) which can be followed laterally over several kilometres and which is composed of monospecific oyster shells. Ammonites [*Desmoceras latidorsatum* and *Mortoniceras (Durnovarites)*] are found on a hardground surface on top of this bed and indicate a late Albian–early Cenomanian age (*dispar* Zone). The top of Unit 2 is a silty marl facies with intercalations of bioclastic carbonates (gastropods and bivalves) and a medium-grained, strongly bioturbated sandstone with ripple marks.

Unit 3. This unit corresponds to the top of the section. It is characterized by alternating silty marls and massive limestone layers which are occasionally bioclastic.

Late Cenomanian kerogen-rich limestones

The late Cenomanian units were sampled in a quarry situated 40 km northeast of Tarfaya, exposing 22 m of late Cenomanian sediments. The section consists of dark grey, laminated, kerogenous chalks, alternating with non-laminated lighter coloured limestones. The light layers are sometimes developed as nodular limestones (Fig. 4b). The late Cenomanian age of this section is defined by ammonites (*Neoptychites* sp. and *Vascoceras* cf. *cauvini*). Intercalations of concretions and lenses of chert and siliceous limestones are present. These siliceous layers are usually < 10 cm thick. The kerogenous chalks locally contain some fish debris.

The expanded sequences of late Cenomanian–early/middle Turonian age are documented in the exploration wells drilled by ONAREP and Shell. The lithology corresponds to the quarry section and is characterized by cyclic variations in TOC and carbonate content. The correlation of the quarry section with the exploration wells is based on the last occurrence of *Rotalipora cushmani*. Micropalaeontological analysis proves that most of the quarry section corresponds to the *Whiteinella archaeocretacea* Zone.

Turonian nodular limestones and laminated bituminous chalk

Turonian sediments are either documented in the Oued Amma Fatma (77 m) section, which is located 100 km northeast of Tarfaya, or within the exploration Wells S13 and S75. Three lithological units are distinguished.

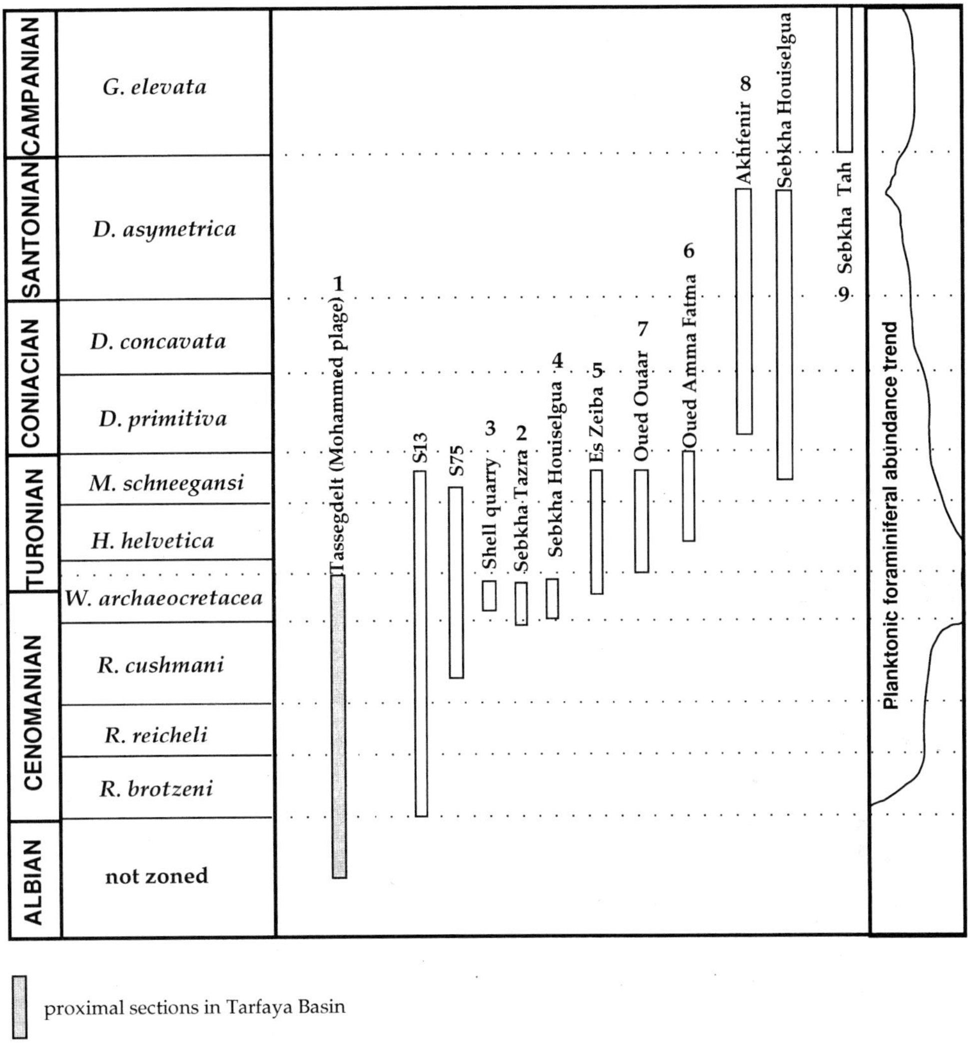

Fig. 3. Biostratigraphic correlation between proximal and distal parts of the Tarfaya Basin.

Unit 1. This unit represents the basal part of the outcrop section and is mainly composed of black laminated marls alternating with lighter limestone. Two types of limestone are present: (1) carbonate nodules which are aligned parallel to the stratification (Fig. 4c) – the nodules contain undeformed ammonites, molluscs and brachiopods; (2) undulating discontinuous, bioclastic beds showing hummocky cross-stratifications (Fig. 4d).

These sediments, based on the occurrence of the ammonites *Puzosia* sp., *Puzosia* cf. *curvatisulcata, Kamerunoceras* cf. *eschii, Benuites* cf. *reymenti, Watinoceras* gr. *coloradoense* and *Mammites nodosoides*, and the foraminiferid *Helvetoglobotruncana helvetica* are early to middle Turonian in age.

Unit 2. This unit occupies the middle part of the outcrop and is composed of bioclastic limestone with intercalations of marls.

Unit 3. This unit corresponds to the top of the section and is dominated by alternating marls and limestone (mudstone to wackestone).

Coniacian–Santonian marls and shell beds

The Coniacian–Santonian interval was sampled in the Akhfenir section situated 80 km northeast of Tarfaya (Fig. 1). The outcrop is a 91 m thick sequence of alternating brown marls and limestones, with rare shells of astartes and inoce-

Fig. 4. (**a**) Bioturbated sandstones (late Albian–middle Cenomanian), (**b**) dark grey, laminated, kerogenous chalks, alternating with non-laminated lighter coloured limestones (late Cenomanian–early Turonian), (**c**) centimetre plus carbonate nodules aligned parallel to the stratification (early Turonian); (**d**) undulating discontinuous beds showing hummocky cross-stratification (early Turonian). Bar scales = 30 cm.

ramids. Two main types of limestone are identified: carbonate nodules and laminated limestones.

The top of the section consists of condensed layers of thick and monospecific inoceramid shells. The Coniacian–Santonian age of these sediments is based on the ammonite fauna (*Forresteria petrocariense, Peroniceras* gr. *tridorsatum*) and a planktonic foraminiferal fauna (*Dicarinella concavata, Marginotruncana pseudolinneiana, Marginotruncana sinuosa, Marginotruncana undulata* and *Marginotruncana tarfayaensis*).

Campanian bituminous marls (oil shales)

The Campanian unit was sampled in the Sebkha Tah, 35 km southeast of Tarfaya (Fig. 1). The outcrop section (42 m in total) is mainly composed of black laminated marls alternating with light brownish laminated marls without a macrofauna. The early Campanian age is defined by planktonic foraminifera *Globotruncana elevata* and *Globotruncana fornicata*.

Facies and microfacies characterization

Macro- and microfacial analysis of the outcrop sections (Fig. 2) reveals eight distinct facies types.

Facies Type 1

Facies Type 1 is represented by the bed that forms the top of the previously described late Albian–early/middle Cenomanian Unit 2 and is composed of thick and monospecific oyster shells. The base is erosive. The microfacies analysis reveals a packstone to grainstone texture, dispersed angular quartz grains in the matrix and no microfauna (Fig. 5a). The oysters are fragmented. A proximal environment within the zone of the fairweather wavebase, 20–30 m below the surface, is indicated.

Facies Type 2

This facies consists of ochre-coloured, bioturbated silty sandstones containing dislocated lamellibranchs. It forms thick continuous layers and is often associated with lenticular sands, which are *c.* 1 cm thick. The microfacies analysis reveals thin-graded sandstone layers containing bioclastic fragments. These have a very low carbonate content and are devoid of microfauna. The sandstone is composed of graded angular quartz grains (70%, silt fraction size), feldspar (20%) and muscovite (5%) (Fig. 5b). The characteristics of this facies indicate a shoreface depositional setting.

Facies Type 3

Facies Type 3 consists of calcareous bioclastic layers intercalated within marls. It contains fragmented single species lamellibranch shells and no microfauna. The size of the shells gradually decreases towards the top of the layer (Fig. 5c). The carbonates are several decimetres thick and laterally very consistent. The abundance of fragmented lamellibranchs, the absence of bioturbation and the presence of marls suggest storm swell settings in a shoreface environment (Aigner 1985).

Facies Type 4

Facies Type 4 is composed of greenish to yellowish, silty clays with intercalations of silty limestones and lenticular sands with plane parallel or rippled laminations (Fig. 5d). The rare fauna mainly comprises ostracods (*Cythereis fahrioni bigarandis*), bivalve and fish debris. A transitional environment between shoreface and offshore is indicated.

Facies Type 5

Facies Type 5 is composed of grey nodular limestones with frequent hummocky cross-stratification. These layers are several centimetres thick and have erosive bases. Microscopic examination shows a wackestone texture. The numerous faecal pellets are rounded and mixed with abundant foraminifera, many of which are buliminids. Well preserved radiolarians, diatoms and abundant unorientated bivalve fragments are present locally (Fig. 6a). An offshore environment within the storm swell influenced zone is indicated.

Facies Type 6

This facies is composed of well-stratified, 1 dm thick limestone layers. The bottom of each layer is bioturbated. Bivalve shells (astartes and inoceramids), gastropods and ammonites are well preserved. The microfacies analysis reveals a wackestone to packstone texture (Fig. 6b). The floral and faunal content varies and includes

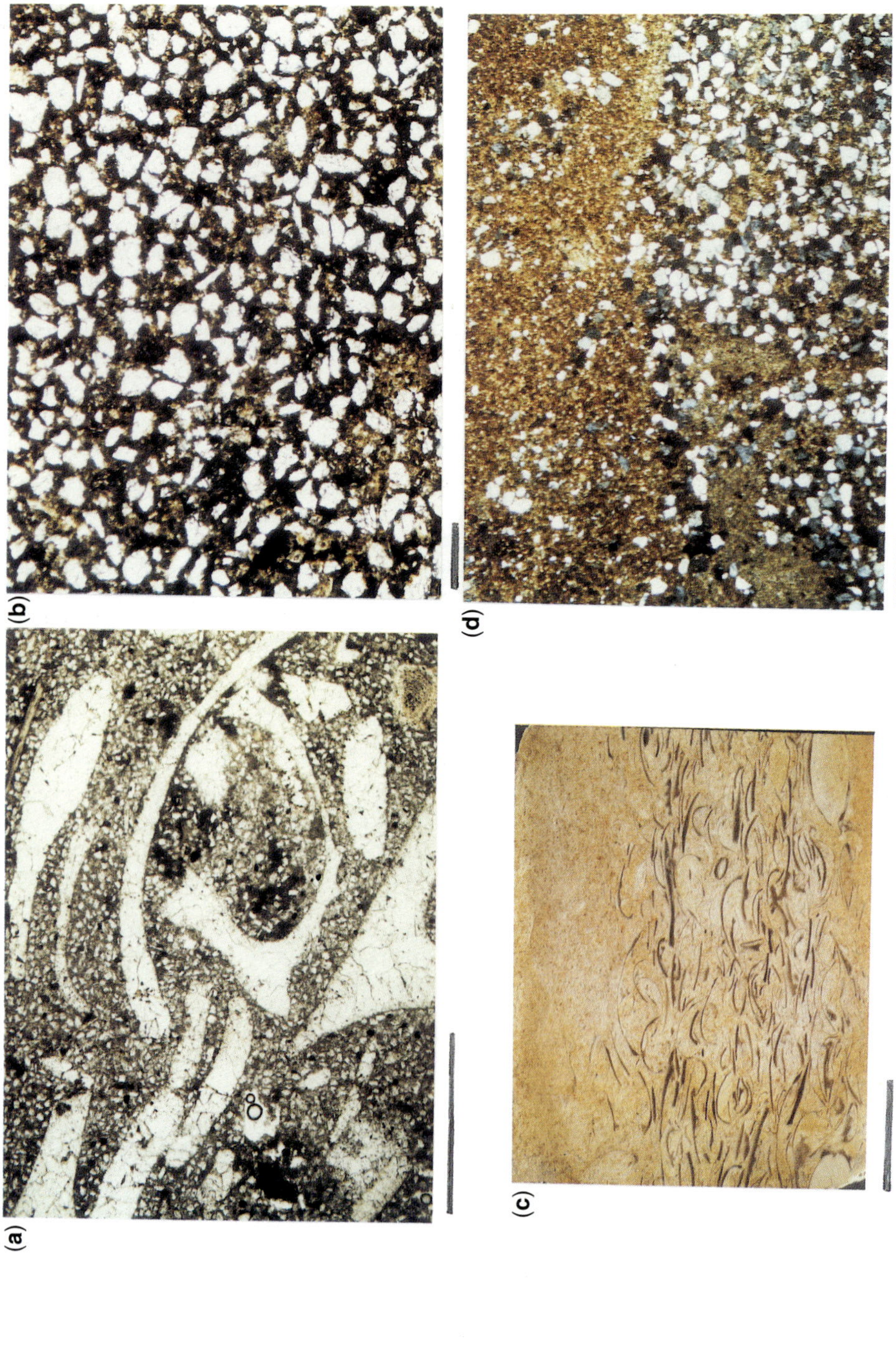

Fig. 5. **(a)** Facies Type 1: monospecific dislocated oyster shells. The microfacies analysis reveals a packstone to grainstone texture, dispersed angular quartz grains in the matrix and no microfauna (bar scale = 2 mm). **(b)** Facies Type 2. Sandstone grains (quartz). These have a very low carbonate content and are devoid of microfauna. The sandstone is composed of graded angular quartz grains (Bar scale = 2 mm). **(c)** Facies Type 3. Monospecific bivalve shells (± 1 cm in size) without microfauna. The size of the shells gradually decreases towards the top of the layer (bar scale = 1 cm). **(d)** Facies Type 4. Bioturbated silty clays (bar scale = 2 mm).

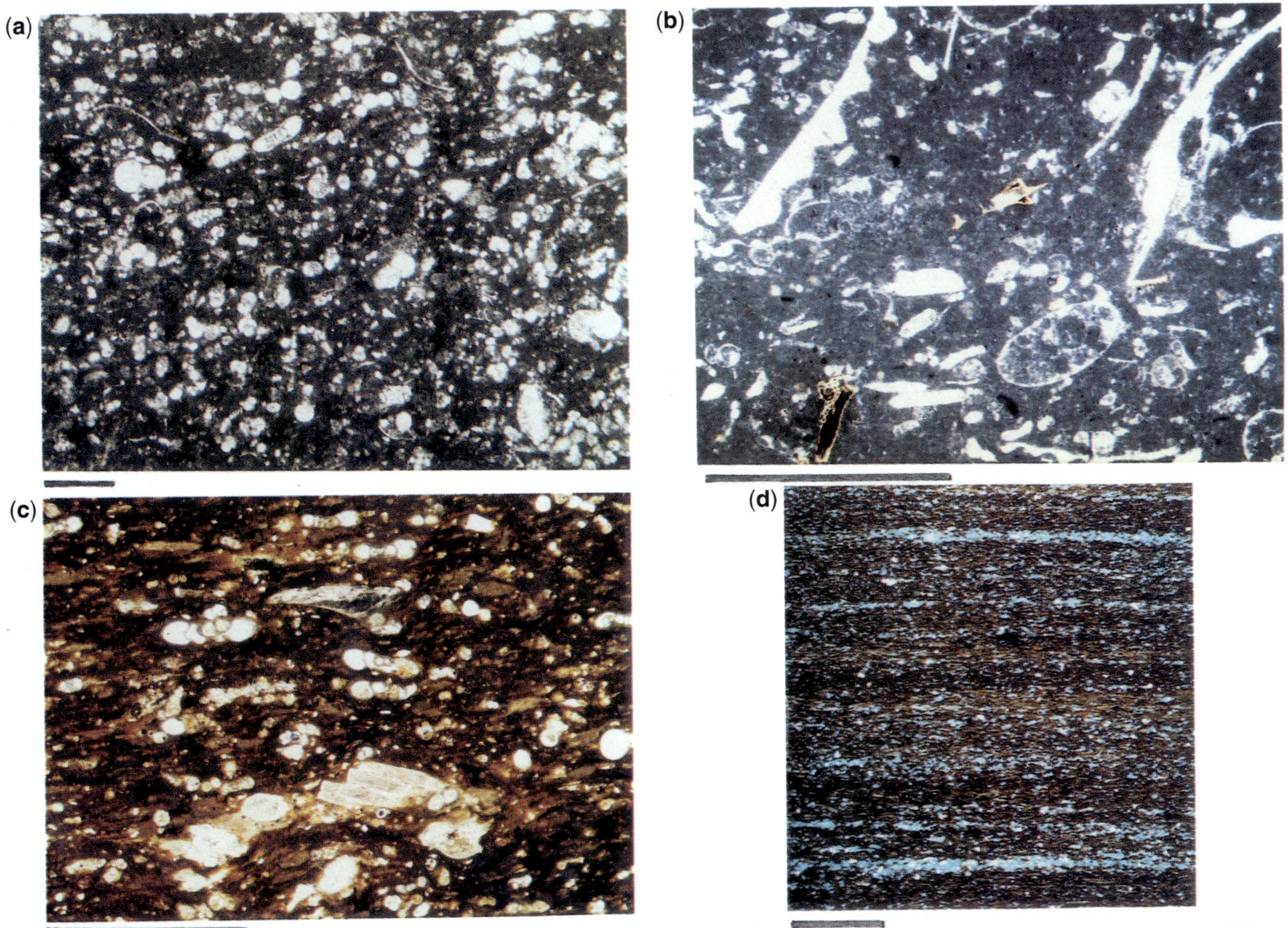

Fig. 6. (**a**) Facies Type 5. Faecal pellets mixed with abundant foraminifera, many of which are buliminids. Well preserved radiolarians, diatoms and abundant unorientated bivalve fragments (up to several millimetres in size) are present locally. (**b**) Facies Type 6. Lamellibranch (astartes and inoceramids) and gastropod shells. The microfacies analysis reveals a wackestone to packstone texture. The floral and faunal content varies and includes bivalves filled with round faecal pellets showing geopetal structures, foraminifera, radiolarians and diatoms, *Pithonella*/calcispheres and coccoliths. (**c**) Facies Type 7. Laminated marls with occasional complete astartes, inoceramid shells. The microfacies analysis reveals a grouping of varied species including benthonic and planktonic foraminifera, radiolarians and '*Chaetoceras*' diatoms, *Pithonella*/calcispheres, coccoliths and phosphatic remains. (**d**) Facies Type 8. Finely laminated, black limestones. The sediments contain a high amount of organic matter without apparent macrofauna. The thin dark laminae are mainly composed of clays mixed with organic matter. Planktonic foraminifera, radiolarians, '*Chaetoceras*' diatoms form the lighter laminae. Bar scales = 2 mm.

bivalves filled with round faecal pellets showing geopetal structures, foraminifera, radiolarians and diatoms, *Pithonella*/calcispheres and coccoliths. Some of the shells show partial to total epigenesis due to biogenic silica (opal). The matrix is mainly composed of round or flattened faecal pellets that approach 1 mm in size and contain organic remains. This facies was part of a low energy, offshore environment, as shown by the presence of complete shells and numerous faecal pellets. The varied planktonic fauna and abundant benthonic organisms indicate an open marine environment with a high oxygen and nutrient supply.

Facies Type 7

Facies Type 7 is composed of beige coloured laminated marls with occasional complete shells of astartes, inoceramids or ammonites. Partly nodular carbonated layers, with a mudstone to wackestone texture, are also present (Fig. 6c). The microfacies analysis reveals a grouping of varied species including benthonic and planktonic foraminifera, radiolarians and '*Chaetoceras*' diatoms (I. Brodie, pers. comm.), calcispheres/ *Pithonella*, coccoliths and phosphatic remains. Fine quartz grains are scattered in the micritic matrix. Biogenic silica is present. This facies indicates an offshore environment out of reach of the influence of storm waves. The calcareous layers are the result either of autocyclic factors, such as chemical changes in the water, or of allocyclic factors such as eustatic or climatic variations. The high number and variety of organisms might be due to upwelling systems (Kuhnt *et al.* 1990).

Facies Type 8

This facies is composed of marls and black limestones forming very thin laminations and showing some thalassinoid bioturbation. The sediments contain high levels of organic matter without apparent macrofauna, except for fish remains. The thin dark laminae are mainly composed of clays mixed with organic substances (Fig. 6d). In the lighter laminae, planktonic foraminifera, radiolarians, '*Chaetoceras*' diatoms, phosphatic fish remains, calcipheres/ *Pithonella* and coccoliths are visible. Very fine quartz grains and pyrites are scattered in the matrix. This facies implies a deep open marine, highly nutritive, anoxic environment deprived of hydrodynamic agitation. The abundance of organic matter indicates an extended oxygen minimum zone.

Two primary facies groupings

Stage I

From the late Albian–middle Cenomanian the sedimentary sequence is characterized by bioclastic layers intercalated in silty marls and occasionally bioturbated sandstones. In the early and middle Cenomanian, bioclastic beds are more frequent (Fig. 7). They correspond either to proximal storm deposits or to a continuous current activity, hampering the deposition of small particles, thus leading to the accumulation of shells (Bloos 1982).

Stage II

During the late Cenomanian–Campanian the sedimentary record is characterized by marls and limestones with a high organic content, and by a remarkable lateral continuity of nodular limestone and bioclastic beds. These carbonate-rich sediments are typical of basin sediments with a poor terrigenous input. The change in lithology is due to a high sea-level stand (Fig. 7).

The presence of fine angular quartz grains in the dark laminated marls indicates that strong sandstorms were blowing from the land to the ocean, starting or amplifying the upwelling. Stage II is characterized by high bioproductivity linked to an upwelling system. Exceptionally large Globotruncanidae species are present, their large size caused by favourable ecological environments rich in nutrients (Lehmann 1965).

A zonation for the different facies types and their depositional environments is suggested in Fig. 8.

Clay mineralogy, geochemistry and the hydrocarbon implications

The results of the clay analysis and the geochemical characterization of the organic matter are shown according to their stratigraphic position in Fig. 9. The Van Krevelen diagram (Fig. 10) presents the Rock-Eval pyrolysis data.

Within the early part of the section (late Albian–middle Cenomanian), the clay minerals are mainly composed of illite, irregular mixed layers of illite–smectite, kaolinite and chlorite. A smectite-rich interval occurs in the early–middle Cenomanian. In this interval the TOC is very low (TOC < 1 %). the 'Hydrogen index' (HI) is also low (HI < 100 mgHC/gTOC), as is T_{max} (420° < T_{max} < 428 °C). Immature Type III organic matter is indicated.

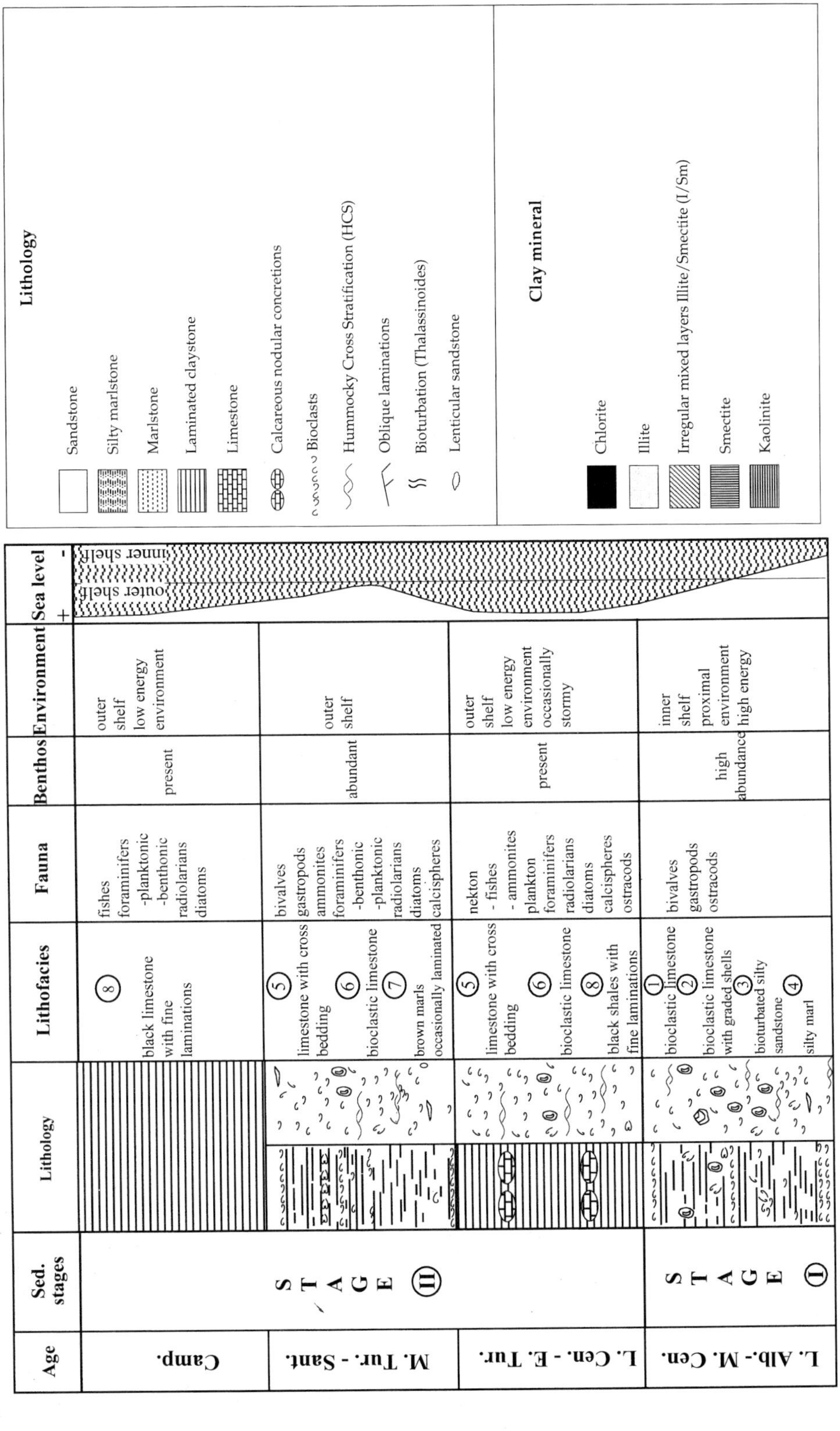

Fig. 7. Biofacies and lithofacies distribution in Stage I, Stage II and sea-level fluctuations.

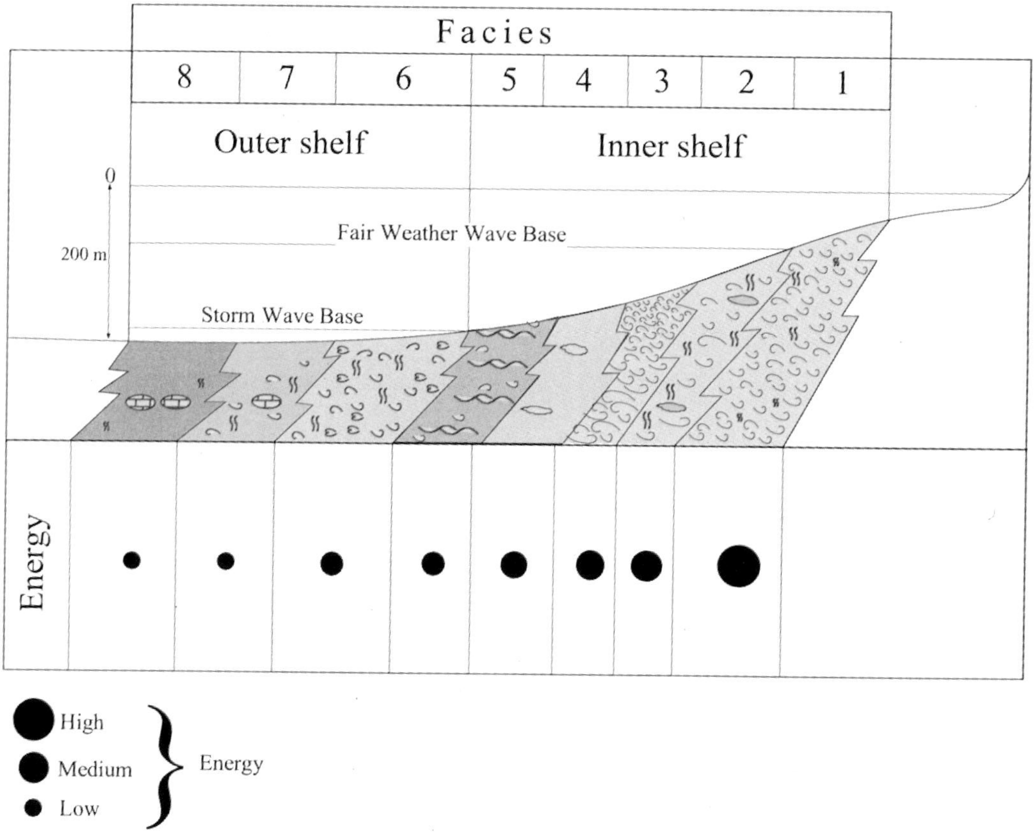

Fig. 8. Zonation of the lithofacies types and their depositional environments.

In the late Cenomanian, the clay mineral assemblage includes illite and kaolinite. The relative proportions of these two minerals fluctuate considerably. The $CaCO_3$ content is high, reaching up to 80%. In contrast to the previous section, the TOC varies from 2 to 16% and the HI is up to 750 mgHC/gTOC. Type II kerogens are present. T_{max} is low and rarely above 415 °C. The Turonian stage clay mineral assemblage is composed of illite, irregular mixed layers of illite–smectite, smectite and kaolinite. Their relative proportions vary considerably. The base and the top of the section are enriched in smectite, whereas the central part is enriched in kaolinite and in irregular mixed layers of illite–smectite. Only the sediments from the early Turonian show high rates of TOC, up to 16%. The organic matter is marine (HI < 800 mgHC/gTOC).

The Coniacian–Santonian interval contains the same clay minerals but shows a great variety in abundances. At the bottom of the section smectite is missing. The clays are mainly composed of kaolinite, illite and irregular mixed layers of illite–smectite in low proportions. At the top, smectite rates rise up to 80%. The levels of organic matter are generally low, except in three samples that show enrichment in irregular mixed layers of illite–smectite.

In the Coniacian–Santonian interval, organic matter is associated with smectite-rich layers. The TOC reaches 4% and the HI is 520 mgHC/gTOC, which corresponds to marine organic matter (Type II). T_{max} is low (< 420 °C).

In the Campanian, the clays are essentially composed of illite and irregular mixed layers of illite–smectite with minor proportions of kaolinite. The resumption of organic sedimentation is indicated by high TOC values (up to 4%). HI is 542 mgHC/gTOC and indicates marine planktonic organic matter. T_{max} is *c.* 425 °C. The $CaCO_3$ percentages are high and can reach 80%.

In the Late Cretaceous series from the Tarfaya Basin, the clay minerals are mainly composed of illite, kaolinite and smectitic minerals, including

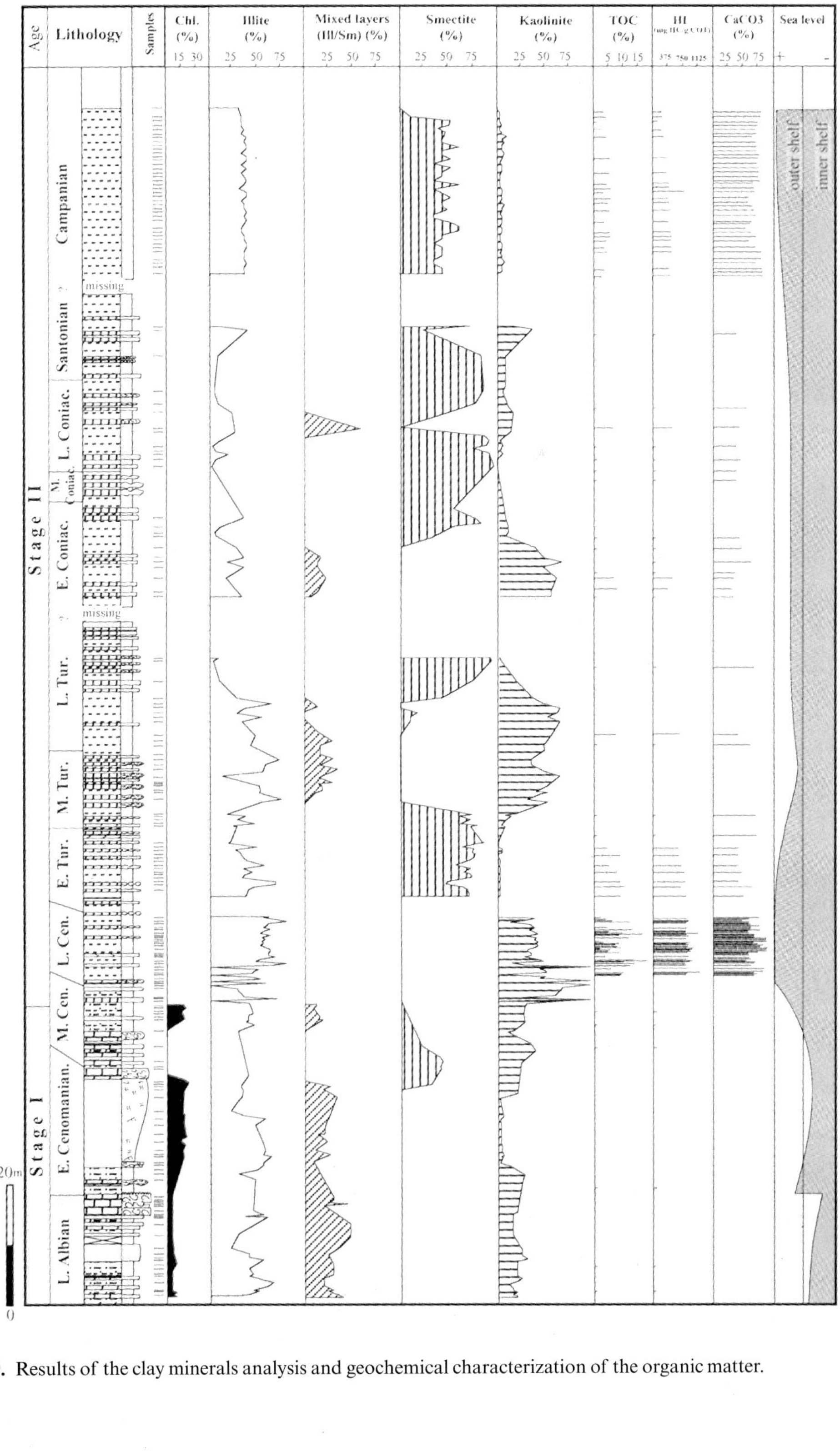

Fig. 9. Results of the clay minerals analysis and geochemical characterization of the organic matter.

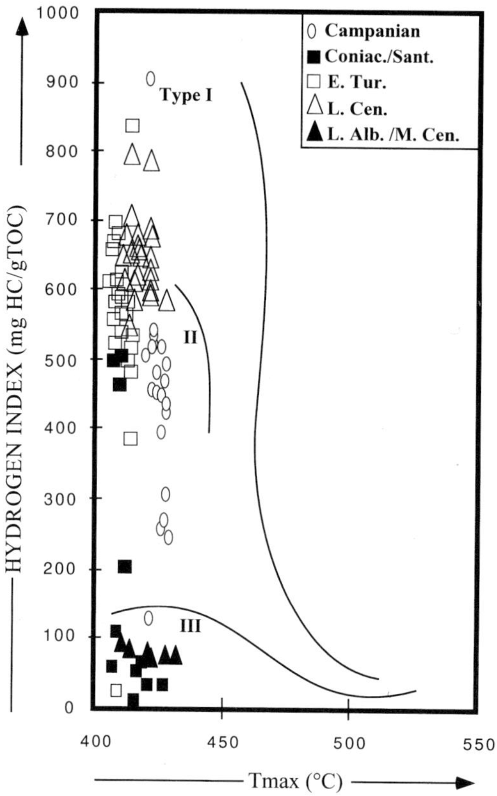

Fig. 10. Van Krevelen diagram for the Late Cretaceous.

smectites and irregular mixed layers of illite–smectite. A lesser amount of chlorite is present, especially in the late Albian–early/middle Cenomanian.

As with the facies analysis, the Late Cretaceous can also be divided, on the basis of the above information, into two stages:

(1) Stage I: late Albian–early/middle Cenomanian in age and characterized by chlorite, illite and kaolinite, and by continental organic matter (Fig. 11);
(2) Stage II: late Cenomanian–Campanian in age and characterized by abundant smectitic minerals and an abundance of marine organic matter (Fig. 11).

The high amount of primary clay minerals (illite and chlorite) in the late Albian and early–middle Cenomanian are attributed to the erosion of crystalline rocks in the Anti-Atlas and Mauritanides. The kaolinite is considered to be derived from soils in the same region. The lithofacies and microfacies results indicate proximal sedimentation, a finding that accords with Chamley (1989). A low sea-level stand is evident (Fig. 11; Chamley *et al.* 1990; Deconinck 1992; Deconinck & Chamley 1995). These results are consistent with the previously discussed occurrence of Type III kerogens.

In the early–middle Cenomanian the clay mineral assemblage is mainly composed of primary minerals and kaolinite. A smectite layer records a sudden rise in sea level, as does the facies analysis.

In Stage II the clay mineral assemblage shows a high proportion of smectite. A possible source is the reworking of smectites from submerged soil profiles during a marine transgression (Chamley *et al.* 1990; Deconinck 1992). The considerable illite and kaolinite supply during the middle–late Turonian reflects the tectonic activity that occurred in the High Atlas area (Emery & Uchupi 1984).

The two intervals of rich Type II organic matter indicate the conjunction of favourable circumstances. These include the physiography of the basin, the palaeoproductivity, the anoxicity and a high preservation potential. Pederson & Calvert (1990) considered that the presence of organic matter depended directly on productivity, whereas Tyson (1987) proposed that the preservation process is mainly important in the accumulation of organic matter. Summerhayes (1987) suggested that a high rate of sedimentation favoured the preservation of organic matter, whereas Vail *et al.* (1987), Van Wagoner *et al.* (1990). and Loutit *et al.* (1988) assumed that a low rate of sedimentation reduced the dilution of organic matter in the sediments.

High productivity leads to the rapid consumption of seawater oxygen. This induces an upwards shifting of the oxygen minimum zone (Fig. 12), which reduces the benthonic fauna (Fahrbach *et al.* 1981; Guillen & Calienes 1981; Packard *et al.* 1983; Owens *et al.* 1991). However, the oxygen level in the oceans depends not only on primary productivity, but also on other factors, such as the development of the oxygen minimum zone, the sedimentation rate, the depth of the water column, the intensity of currents and the topography of the oceanic floor.

Between these two periods of accumulation of organic matter, the TOC and the HI are very low. Important negative values of $\delta^{13}C$ recorded in a carbonate concretion from the Coniacian–Santonian interval suggests that organic matter was present during the deposition and that it was altered later (El Albani *et al.* 1997).

Conclusions

The sedimentological data indicate that the Late Cretaceous sediments in the Tarfaya Basin are

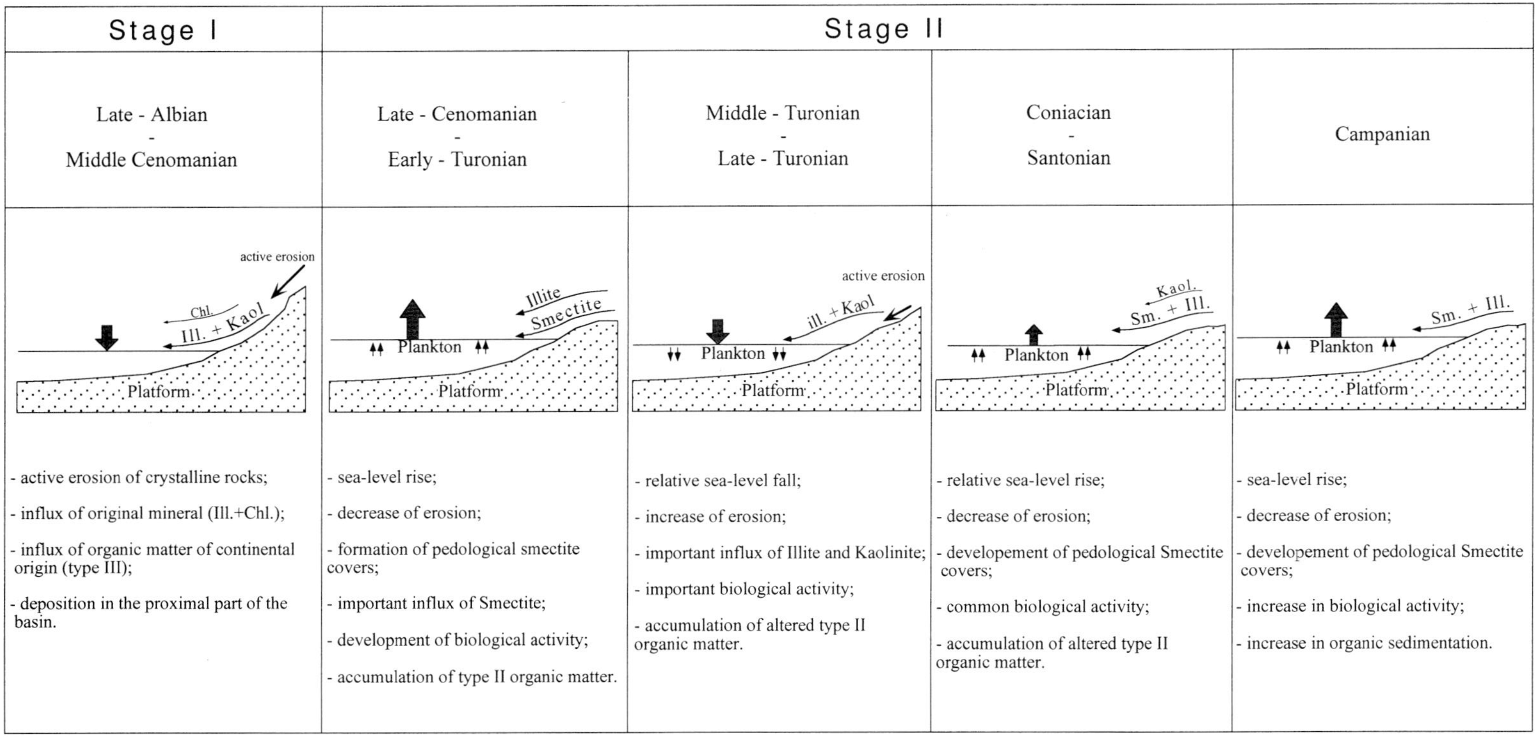

Fig. 11. Relationship between clay mineral assemblages and the palaeogeographic evolution of the Tarfaya Basin between the late Albian and the Campanian.

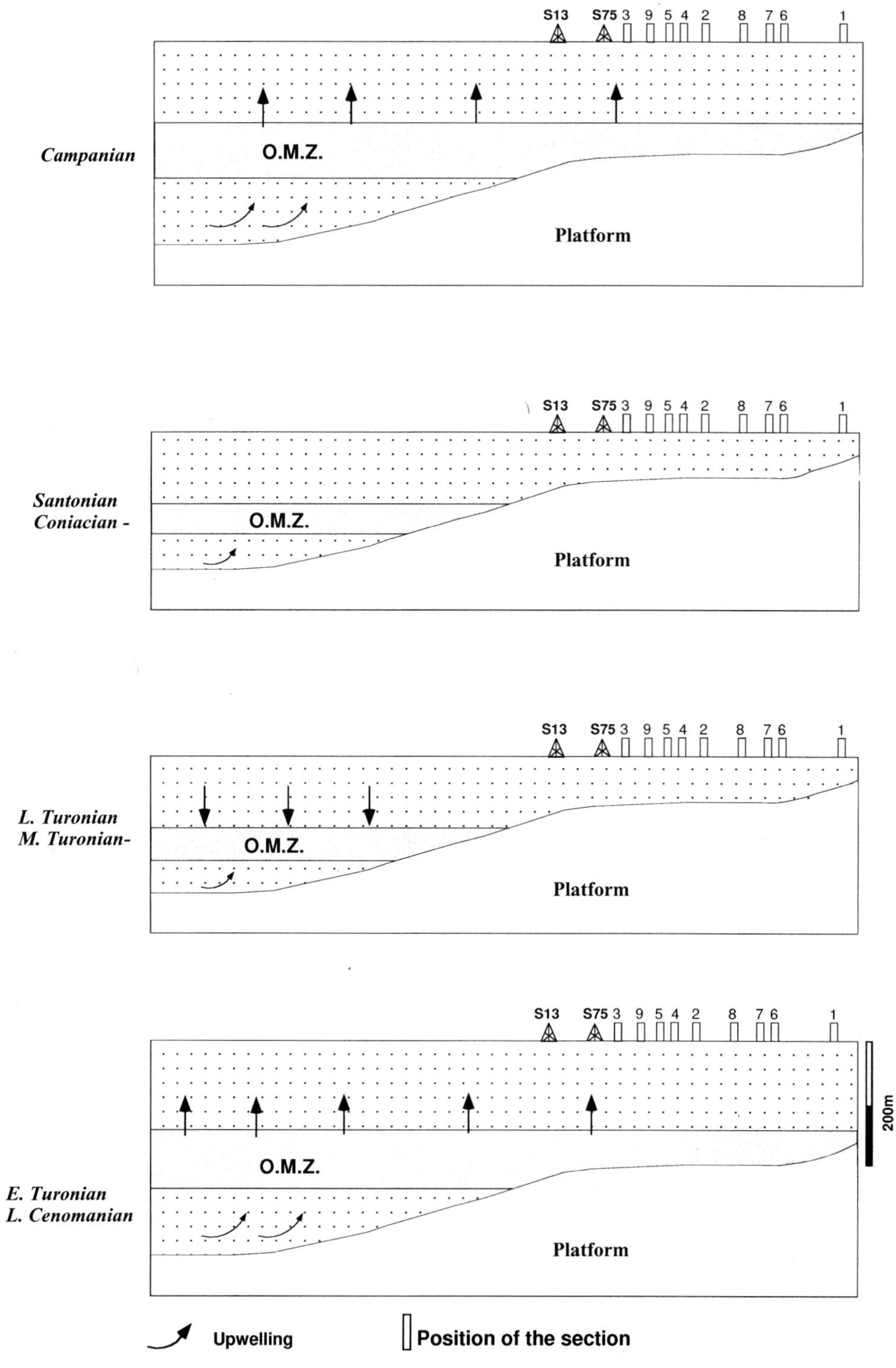

Fig. 12. Sea level and oxygen minimum zone (OMZ) changes between the late Cenomanian and the Campanian.

characterized by marine environments. From the late Cenomanian the sedimentary setting is deeper and marked by the deposition of thin sediments with a very diversified fauna. Two distinct evolutionary stages are recognized. **Stage I** (late Albian–middle Cenomanian) is characterized by the abundance of a proximal sedimentation corresponding to a shoreface environment. Siliciclastic supply and the development of coarse bioclastic sediments are dominant in this environment. The clay mineral assemblage, illite, chlorite and kaolinite from the Anti-Atlas and mauritanides indicate active erosion of the surrounding relief. These minerals characterize deposition in the proximal part of the basin. During this phase, most of the organic matter belongs to Type III (continental origin). **Stage II** (late Cenomanian–Campanian) is characterized by a sharp decrease in detrital supply, linked to a rise in sea level. The establishment of an upwelling system induced the conditions needed to deposit rich Type II source rocks. The Tarfaya oil shale deposit is favourably situated close to the Atlantic coast at shallow depth and contains large reserves of moderate grade oil shales, which are ideally suited to open-put mining. Overburden ratios in the western part of the basin are relatively low and the oil shales slightly thicker, and have a better average oil yield than elsewhere. According to 'Shell Exploration', oil shale reserves in this area are more than sufficient for a 50 000 barrel/day operation. The oil shale is immature, but it would become a world-class source with additional burial offshore.

The Deutsche Forschungsgemeinschaft (grant Ku-649/3) supported this study. The Alexander von Humboldt Stiftung provided the finances to AEA. We would like to thank the many colleagues with whom we have discussed the ideas presented in this paper, in particular, D. Vachard, M. Lopez, F. Robaszynski, J. Thurow, A. Strasser, F. Amedro. Finally, we wish to thank the referees for their constructive reviews.

References

AIGNER, T. 1985. Storm depositional systems. *Lecture Notes in Earth Sciences*, **3**.

BLOOS, G. 1982. Shell bed in the Early Lias of South Germany-facies and origin. *In*: EINSELE, G. & SEILACHER, A. (eds) *Cyclic and Event Stratification*. Springer, 223–239.

CARON, M. 1985. Cretaceous planktic foraminifera. *In*: BOLLI, H. M., SAUNDERS, E. & PERCH-NIELSEN, K. (eds) *Plankton Stratigraphy* Cambridge University Press, 17–86.

CHAMLEY, H. 1989. *Clay Sedimentology*. Springer, Berlin.

——, DECONINCK, J. F. & MILLOT, G. 1990. Sur l'abondance des minéraux smectitiques dans les sédiments marin communs déposés lors des périodes de haut niveau marin du Jurassique supérieur au Paléogène. *Compte Rendue Academie Sciences, Paris*, **311**, 1529–1536.

CHOUBERT, G., FAURE MURET, A. & HOTTINGER, L. 1966. Aperçu géologique du Bassin côtier de Tarfaya (Stratigraphie). *In*: CHOUBERT, G., FAURE MURET, A., HOTTINGER, L., VIOTTI, C. & LECOINTRE, G. (eds) *Le Bassin côtier de Tarfaya (Maroc Méridional)*. Notes et Mémoire Service Géologique du Maroc, **175/I**, 7–106.

——, —— & —— 1972. La série stratigraphique de Tarfaya Maroc sud occidental et le problème de la naissance de l'Atlantique. *Notes et Mémoire Service Géologique du Maroc*, **31**, 29–40.

DECONINCK, J. F. 1992. *Sédimentologie des argiles dans le Jurassique–Crétacé d'Europe occidentale et du Maroc*. Habilitation Thesis, Lille.

—— & CHAMLEY, H. 1995. Diversity of smectite origins in late Cretaceous sediments: Example of chalks from Northern France. *Clay Minerals*, **30**, 365–379.

EL ALBANI, A., CARON, M., DECONINCK, J. F. *ET AL*. 1997. Origin of nodules in early Turonian organic-rich sediments in Tarfaya Basin (Southwest of Morocco). *Compte Rendue Academie Sciences* Paris, **324**, serie IIa, 9–16.

EMERY, K. O. & UCHUPI, E. 1984. *The Geology of the Atlantic Ocean*. Springer, **1**.

ESPITALIÉ, J., DEROO, G. & MARQUIS, F. 1985. La pyrolyse Rock-Eval et ses applications. *Revue Institut Français du Pétrole*, **40**, 563–579, 755–784, **47**, 73–89.

FAHRBACH, E., BROCKMANN, N., LOSTAUNAU, N. & URQUIZO, W. 1981. The northern Peruvian upwelling system during the ESACAN experiment. *In*: RICHARDS, F. A. (ed.) *Coastal Upwelling*. Coastal Estuarine Sciences Series, Washington, 134–145.

GUILLEN, O. & CALIENES, R. 1981. Upwelling off Chimbote. *In*: RICHARDS, F. A. (ed.) *Coastal Upwelling*. Coastal Estuarine Sciences Series, Washington, 312–326.

HOLBOURN, A., KUHNT, W., EL ALBANI, A., PLETSCH, T., LUDERER, F. & WAGNER, T. 1999. Upper Cretaceous palaeoenvironments and benthonic foraminiferal assemblages of potential source rocks from the Western African margin, Central Atlantic. *This volume*.

HOLTZAPFEL, T. 1985. Les minéraux argileux, préparation, analyse diffractométrique et détermination. *Annales Societe géologique Nord, Mémoire*, **12**.

KUHNT, W. & WIEDMANN, J. 1995.. Cenomanian–Turonian source rocks: paleobiogeographic and paleoenvironmental aspect. *In*: HUC, A. Y. (ed.) *Paleogeography, Paleoclimates and Source Rocks*. AAPG Studies in Geology, **40**, 213–232.

——, NEDERBRAGT, A. & LEINE, L. 1997. Cyclicity of Cenomanian–Turonian organic-carbon-rich sediments in the Tarfaya Atlantic Coastal Basin (Morocco). *Cretaceous Research*, **18**, 587–601.

——, HERBIN, J. P., THUROW, J. & WIEDMANN, J. 1990. Distribution of Cenomanian–Turonian organic facies in the western Mediterranean and along the adjacent Atlantic margin. *In*: Huc, A. Y. (ed.) *Deposition of Organic Facies*. AAPG Studies in Geology, **30**, 133–160.

——, Thurow, J., Wiedmann, J. & Herbin, J. P. 1986. Oceanic anoxic conditions around the Cenomanian/Turonian Boundary and the Response of the Biota. *Mitteilungen aus dem Geologisch-Palaeontologischen Institut der Universität Hamburg*, Scope/Unep Sonderbland, **60**, 205–246.

Lehmann, R. 1965. Résultats d'une étude des Globotruncanidés du Crétacé supérieur de la province de Tarfaya (Maroc méridional). *Colloque International de Micropaléontologie*, Dakar, **32**, 113–117.

Leine, L. 1986. Geology of the Tarfaya oil shale deposit, Morocco. *Geologie en Mijnbouw*, **65**, 57–74.

Loutit, T. S., Hardenbol, J., Vail, P. R. & Baum, G. R. 1988. Condensed sections: the key to age determination and correlation of continental margin sequences. *In*: Wilgus, C., Hastings, B. S., Kendall, C. G., Posamentier, H. W. & Ross, C. A. (eds) *Sea level Change and Integrated Approach*. Special Publication SEPM, **42**, 183–213.

Luderer, F. & Kuhnt, W. 1997. A high resolution record of the *Rotalipora* extinction in laminated organic-carbon rich limestones of the Tarfaya Atlantic coastal Basin (Morocco). *Annales de la Société géologique du Nord*, **T. 5**, 199–205.

Owens, N. J. P., Law, C. S., Mantoura, P., Burkill, H. & Llewellyn, C. A. 1991. Methane flux to the atmosphere from the Arabian sea. *Nature*, **354**, 293–296.

Packard, T. T., Garfield, P. C. & Codispoti, L. A. 1983. Oxygen consumption and dentrification below the Peruvian upwelling. *In*: Suess, E. & Thiede, J. (ed.) *Coastal Upwelling, its Sedimentary Record, Part A, Response of the Sedimentary Regime to Present Coastal Upwelling*. NATO Conference Series, 147–173.

Pedersen, T. F. & Calvert, S. E. 1990. Anoxia vs. productivity: What controls the formation of organic-carbon-rich sediments and sedimentary rocks? *AAPG Bulletin*, **74**, 454–466.

Ranke, U., von Rad, U. & Wissmann, G. 1982. Stratigraphy, facies and tectonic development of the on- and offshore Aaiun–Tarfaya Basin – A review. *In*: von Rad, U., Hinz, K., Sarnthein, M. & Seibold, E. (eds) *Geology of the Northwest African Continental Margin*. Springer, 86–105.

Summerhayes, C. P. 1987. Organic rich Cretaceous sediments from the north Atlantic. *In*: Brooks, J. & Fleet, A. J. (ed.) *Marine Petroleum Source Rocks*. Geological Society, London, Special Publications, **26**, 301–316.

Tyson, R. V. 1987. The genesis and palynofacies of marine petroleum source rocks. *In*: Brooks, J. & Fleet, A. J. (ed.) *Marine Petroleum Source Rocks*. Geological Society, London, Special Publications, **26**, 47–67.

Vail, P. R., Colin, J. P., Du Chene, R., Kuchly, J., Mediavilla, F. & Trifilieff, V. 1987. La stratigraphie sequentielle et son application aux corrélations chronostratigraphiques dans le Jurassique du bassin de Paris. *Bulletin Société Géologique France*, **7**, 1301–1321.

Van Wagoner, J. C., Mitchum, R. M., Campion, K. M. & Rahmanian, V. D. 1990. Siliciclastic sequence stratigraphy in well logs, cores and outcrops. *AAPG Bulletin*, **7**.

Wiedmann, J., Butt, A. & Einsele, G. 1978. Vergleich von marokkanischen Kreide-Küstenaufschlüssen und Tiefseebohrungen (DSDP): Stratigraphie, Paläoenvironment und Subsidenz an einem passiven Kontinentalrand. *Geology Rundschau*, **67**, 454–508.

——, —— & —— 1982. Cretaceous stratigraphy, environment, and subsidence history at the Moroccan continental margin. *In*: von Rad, U., Hinz, K., Sarnthein, M. & Seibold, E. (ed.) *Geology of the Northwest African Continental Margin*. Springer, 366–395.

Tectono-sedimentary controls on Cretaceous black shale deposition along the opening Equatorial Atlantic Gateway (ODP Leg 159)

THOMAS WAGNER[1] & THOMAS PLETSCH[2]

[1]*Fachbereich 5, Geowissenschaften, Universität Bremen, Klagenfurter Str. 5, 28359 Bremen, Germany*
[2]*Geologisch–Paläontologisches Institut, Christian-Albrechts-Universität zu Kiel, Olshausenstr. 40, 24118 Kiel, Germany*

Abstract: The Cretaceous Equatorial Atlantic Gateway between the Central and South Atlantic basins is of interest not only for palaeoceanographic and palaeoclimatic studies, but also because it provided particularly favourable conditions for the accumulation and preservation of organic-rich sediments. Deposition of carbonaceous sediments along the Côte d'Ivoire–Ghana Transform Margin (Ocean Drilling Program Leg 159) was intimately linked to the plate tectonic and palaeoceanographic evolution of this gateway. Notably, the formation of a marginal basement ridge on the southeastern border of the transform margin provided an efficient shelter of the landward Deep Ivorian Basin against erosive and potentially oxidizing currents. Different subsidence histories across the transform margin were responsible for the development of distinct depositional settings on the crest and on both sides of the basement ridge. Whereas the southern, oceanward flank of the basement ridge was characterized by rapid, continuous deepening since the latest Albian–early Cenomanian, marine sedimentation on the northern, landward flank was interrupted by a period of uplift and erosion in the late Albian, and rapid subsidence started after the early Coniacian.

Organic-rich sediments occur throughout almost the entire Cretaceous section, but hydrogen-rich marine black shales were exclusively recovered from core sections above an uplift-related unconformity. These black shales formed when separation of Africa and South America was sufficient to allow permanent oceanic midwater exchange after the late Albian. Four periods of black shale accumulation are recorded, some of them are correlated with the global oceanic anoxic events: in the latest Albian–earliest Cenomanian, at the Cenomanian–Turonian boundary, during the middle Coniacian–early Campanian, and in the mid-Maastrichtian. These periods were characterized by increased carbon fluxes to the seafloor, induced by enhanced palaeoproductivity and intensified supply of terrestrial organic matter. Black shale deposition appears to be intimately linked to periods of rising or maximum eustatic sea level and to the expansion of the oxygen minimum zone, as indicated by foraminiferal biofacies. Intervals between black shale units, in contrast, indicate a shrinking oxygen minimum zone and enhanced detrital flux rates, probably related to lowering sea level.

Upper Cretaceous detrital limestones with high porosities may provide excellent hydrocarbon reservoirs, although their areal extent appears to be limited. Palaeogene porcellanites, capped by Neogene pelagic marls and clays, extend over a wider area and may provide another target for hydrocarbon exploration.

Cretaceous sedimentary deposits, in particular from past epicontinental shelf seas and along the continental margins of former ocean basins, are known to be replete with hydrocarbon source rocks (Bois *et al.* 1980). An ongoing debate has developed on the formation and distribution of these organic-rich sediments (e.g. Demaison & Moore 1980; Calvert 1987; Calvert & Pedersen 1993; Tyson 1987; Parrish 1995). Non-actualistic models of ocean circulation and productivity were proposed for this period of Earth's history, as the boundary conditions for the accumulation of organic matter were obviously different from today (e.g. Thiede & van Andel 1977; Summerhayes 1987; Zimmerman *et al.* 1987; Bralower & Thierstein 1987; Hay 1995*a*, *b*; Hay *et al.* 1997). Many of these models were based upon studies which were carried out on sediments from the Atlantic and its marginal basins. However, published work on the spatial distribution of organic-rich sediments in the Atlantic displays a conspicuous gap for the area which formed the Cretaceous gateway between the Central and the South Atlantic (Kuhnt *et al.* 1990). Ocean Drilling Project (ODP) Leg 159 opened the possibility to drill the Côte d'Ivoire–Ghana Transform Margin, one of the most prominent bathymetric features in the eastern Equatorial Atlantic (Fig. 1).

The Equatorial Atlantic is of particular interest for the palaeoceanographic and palaeoclimatic evolution of the Atlantic Ocean because it is there that the first connection between the formerly separated Central and South Atlantic

WAGNER, T. & PLETSCH, T. 1999. Tectono-sedimentary controls on Cretaceous black shale deposition along the opening of the Equatorial Atlantic Gateway (ODP 159). *In:* CAMERON, N. R., BATE, R. H. & CLURE, V. S. (eds) *The Oil and Gas Habitats of the South Atlantic.* Geological Society, London, Special Publications, **153**, 241–265.

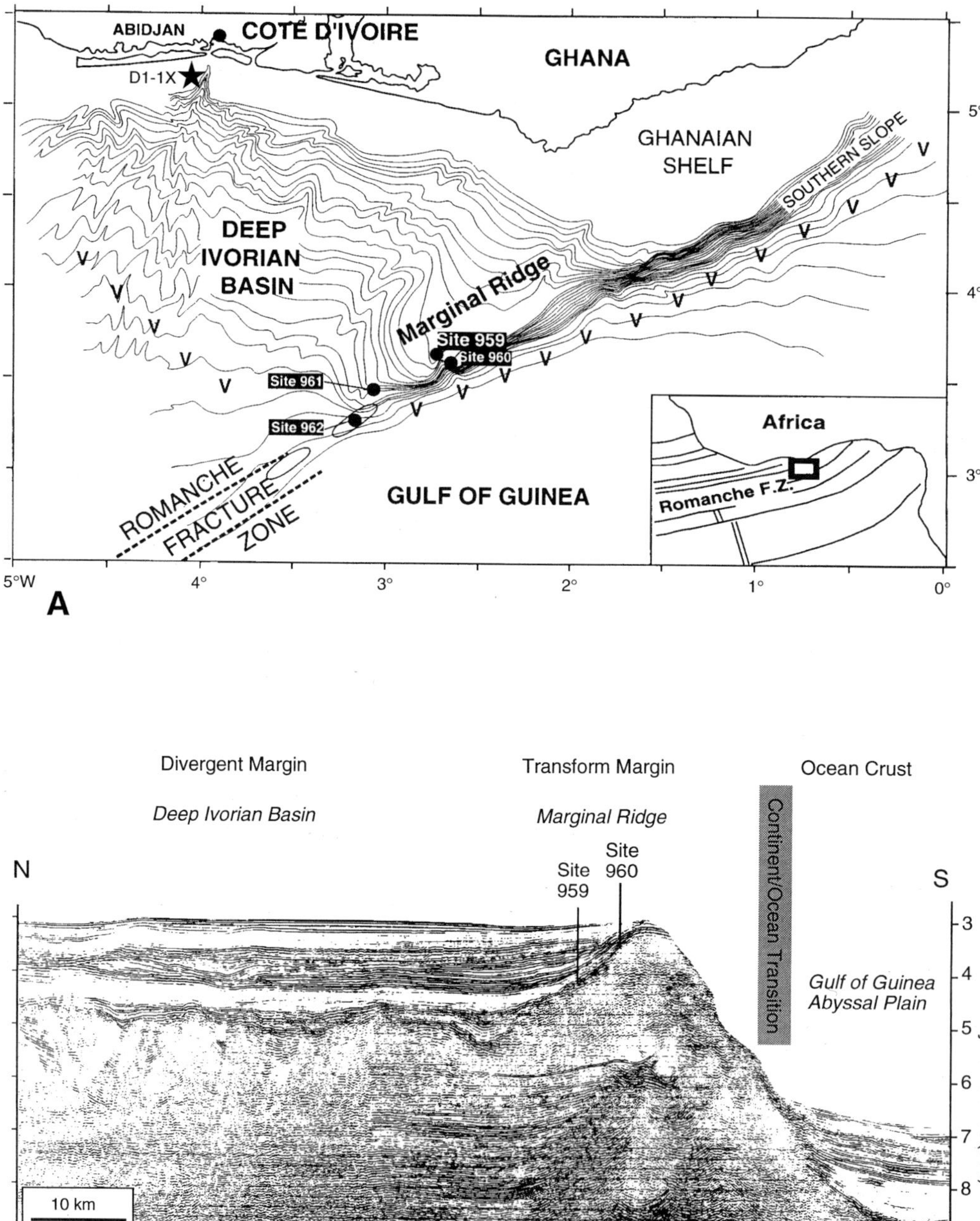

Fig. 1. (**a**) Location of drill sites of ODP Leg 159 along the Côte d'Ivoire–Ghana Transform Margin. Note position of exploration Well D1-1X off Abidjan. (**b**) Migrated multichannel seismic line MT02 across the Deep Ivorian Basin. From Mascle *et al.* (1996).

basins opened during the Cretaceous. The position of the continental margins parallel to the equatorial fracture zones suggests that a large part of this opening history was accomplished by lateral movements of the South American and African plates. Lateral or transform tectonics allow for the rapid opening and closure of marine passages which, in spite of their restricted width, may be relatively deep. However, this also makes transform margins complex to study. Little is known about their subsidence histories, sediment fluxes, palaeoenvironments and palaeoceanographic changes, compared to sedimentary basins on other continental margin types (e.g. Mitchell & Reading 1986; Einsele 1992). This is partly related to the fact that many ancient transform margin basins were involved in later compressional tectonic events which have complicated the original sedimentary geometries, or which have partly destroyed the palaeoenvironmental record due to thermal alteration. Transform margins are characterized by a great variety of basin formation mechanisms and common temporal and spatial changes in subsidence over short distances. Notably, the distribution of uplift with contemporaneous production of detritus and the formation of depressions that accommodate this debris strongly depend on the direction of lateral motion and on basin morphology itself (Christie-Blick & Biddle 1985). As a result of their susceptibility to changes in tectonic movements and basin configuration, the stratigraphic sequence and regional extent of sedimentary facies in transform margin basins are hardly predictable. Consequently, distribution of organic-rich deposits along transform margins is expected to be patchy with tectonically induced barriers which may serve as short, vertical pathways for hydrocarbon migration (Mann *et al.* 1997). Because of its large areal extension, and the lack of post-Cretaceous orogenic deformations, investigation of the Côte d'Ivoire–Ghana Transform Margin may help both in the development of a depositional model for the Cretaceous Equatorial Atlantic in general and in a better understanding of the distribution of potential source rocks along transform margin basins.

Presented here are organic geochemical, sedimentological and mineralogical data of Cretaceous and lower Palaeogene sediments recovered during ODP Leg 159 to assess their depositional conditions and their spatial distribution along the Côte d'Ivoire–Ghana Transform Margin. Although Leg 159 drill sites were not specifically chosen to investigate the hydrocarbon potential of the Côte d'Ivoire–Ghana Transform Margin and its adjacent basins, they may provide a first-order estimate of what is to be expected in the distal part of the Deep Ivorian Basin, which is actively explored for hydrocarbons. This study will focus on the quality of Cretaceous sediments as hydrocarbon source rocks at ODP Site 959, where the thickest and most complete Cretaceous section was recovered. Included are estimates of the thermal maturity and some preliminary considerations on potential hydrocarbon reservoirs and seals.

Regional geology of the Côte d'Ivoire–Ghana Transform Margin

Located on the intersection of the modern Romanche Fracture Zone with the Ghanaian shelf, the Côte d'Ivoire–Ghana Transform Margin has recorded lateral plate movements which started when South America and Africa separated, giving way to the Equatorial Atlantic Gateway. The time of separation is fairly speculative due to the lack of marine magnetic anomalies and biostratigraphic information from the adjacent oceanic crust (Rabinowitz & LaBreque 1979; Nürnberg & Müller 1991). The plate margins on both sides of the early Equatorial Atlantic were still affected by the lateral motion between the African and South American plates, whereas passive continental margins had developed around the Central and South Atlantic by the earliest Cretaceous (Mascle *et al.* 1988; Zalan *et al.* 1985). Based on regional plate reconstructions, it is now accepted that break-up and rifting along this segment started between 120 and 115 Ma (Fig. 2; Klitgord & Schouten 1986; Cande *et al.* 1988; Hay *et al.* 1998). Recent micropalaeontological and sedimentological data support this timing (Moullade *et al.* 1993, 1998; Pletsch *et al.* submitted).

The Côte d'Ivoire–Ghana Transform Margin is characterized by a steep, northeast striking slope which dips towards the Gulf of Guinea Abyssal Plain in the south and extends towards the southwest along the marginal basement ridge (Fig. 1a). Geophysical surveys, dredges and submersible dives along the margin in the northern Gulf of Guinea have shown that it consists of three main morphostructural elements which formed during the Early Cretaceous opening (Basile *et al.* 1996; Mascle *et al.* 1997). The Marginal Ridge, which is underlain by Pan-African basement, separates the oceanic Gulf of Guinea from the landward Deep Ivorian Basin in the north. Its southern slope corresponds to the transition from continental to oceanic crust (Fig. 1b; Mascle *et al.* 1996). Due to its elevated topography, the Marginal Ridge is only covered by a thin post-tectonic sediment blanket.

In spite of these studies, difficulties have remained for reconstructing the early history of the Côte d'Ivoire–Ghana Transform Margin, mainly due to the lack of age control of the seismic lines. Based on average spreading rates, Clift *et al.* (1997) proposed that the intra-continental transform regime was replaced by an active ocean–continent system at *c.* 105 Ma [middle Albian according to the timescale of Gradstein *et al.* (1995)]. Intense deformation along the margin ceased in the latest Albian–early Cenomanian

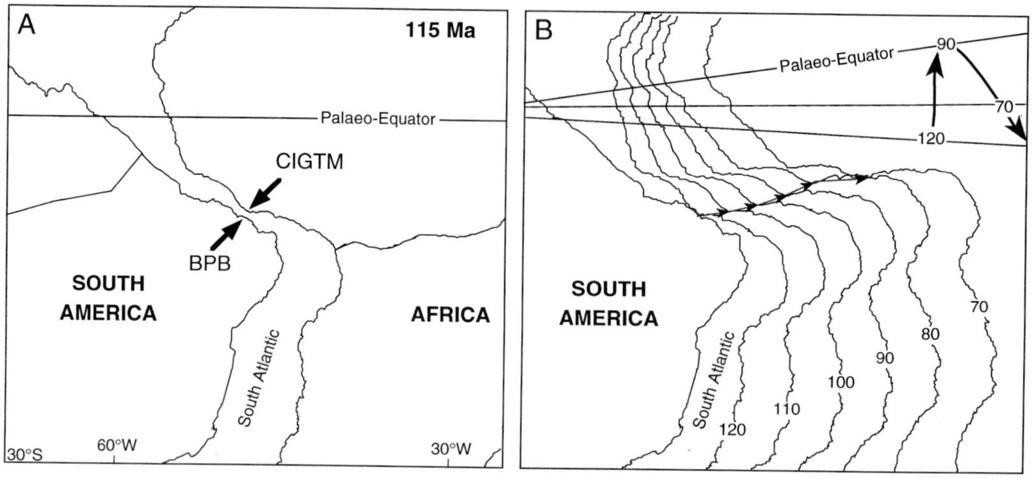

Fig. 2. Plate reconstructions: (**a**) initial separation (115 Ma, late Aptian); (**b**) movement of Africa with respect to South America being kept fixed. Based on a revised set of palaeomagnetic data and new calculations of plate margin deformation by Hay *et al.* (1998). CIGTM, Côte d'Ivoire–Ghana Transform Margin; BPB, Barreirinhas–Piauí Basin.

(99 Ma). Uplift of the Marginal Ridge soon after continent–ocean transform movement, induced by flexural unloading and rebound at the edge of the margin, was estimated to be in the order of 390 m to 1.4 km (Clift *et al.* 1997; Clift pers. comm.). Apatite fission track data suggest a 90 Ma age (late Turonian) for thermal rejuvenation of the margin due to ridge–transform intersection (Bouillin *et al.* 1997). This margin development fits well with a schematic three-stage model initially proposed by Mascle & Blarez (1986) and recently modified by Pletsch *et al.* (submitted; Fig. 3). The stages probably correspond to the tectonic evolution from an early Cretaceous intracratonic rift and transform basin without, or with only very limited, current exchange, to a Late Cretaceous passive margin with deepwater exchange comparable to modern oceanographic conditions. The early opening stage (until the middle Albian) was characterized by the formation of isolated rift or strike-slip basins with limited interconnection (Fig. 3a). Following an uplift event in the middle–late Albian, complete separation of the continental margins led to a continuous mid-water connection, but a strong differentiation in subsidence and basin morphology persisted along the margin (Fig. 3b). Finally, during the oceanic opening stage (middle Coniacian and afterwards), there was persistent subsidence along the margin and obstacles had sufficiently subsided to allow continuous deepwater circulation (Fig. 3c).

Materials and methods

For Site 959, organic carbon was measured on homogenized, calcium carbonate-free samples using a Leco CS-300 elemental analyser at Bremen University. Rock-Eval pyrolysis was performed on bulk samples using a Rock-Eval II device at the Alfred-Wegener-Institut in Bremerhaven, according to the procedures described by Espitalié *et al.* (1985). 'Hydrogen indices' (HI) and 'oxygen indices' (OI) were calculated using 'total organic carbon' (TOC) values obtained from Leco analysis. HI/OI ratios were plotted in Van Krevelen diagrams for kerogen typing. Shipboard organic geochemical data were used for ODP Sites 960–962.

Bulk rock and clay mineralogy were analysed by X-ray diffraction at the University of Lille, France, at the University of Kiel, Germany, and onboard *JOIDES Resolution*, according to different routine methods (Pletsch 1998; Mascle *et al.* 1996). Clay mineral results are reported as semiquantitative proportions to avoid discrepancies related to the use of different techniques and instrumental settings. Detailed clay mineralogical results will be published elsewhere.

ODP Leg 159 drill sites

ODP Leg 159 drilled 13 holes at four sites, eight of which penetrated Cretaceous sediments. A documentation of the recovered sequence and a

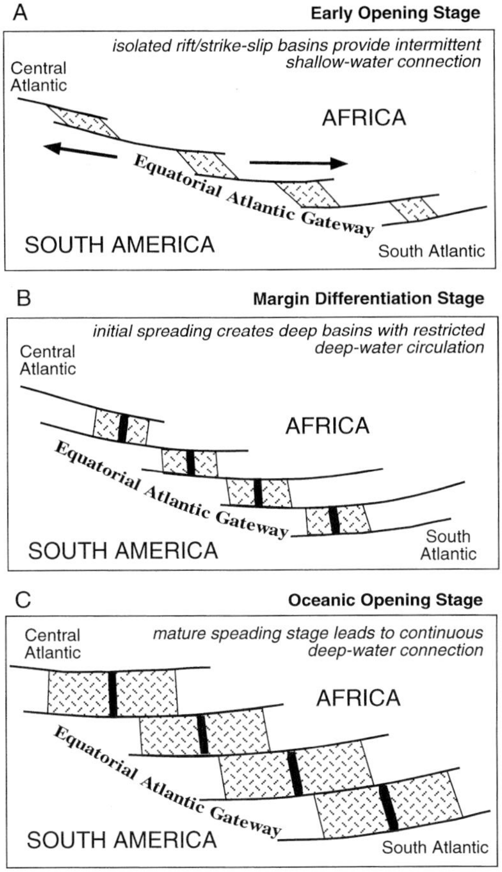

Fig. 3. Idealized cartoon of the opening Equatorial Atlantic Gateway and the possible relationship between plate tectonic situation, basin formation mechanisms and the development of progressively deeper circulation.

preliminary interpretation of the geologic history of the study area was published in the Initial Reports of ODP Leg 159 (Mascle *et al.* 1996). On the basis of a recent compilation and synthesis by Moullade *et al.* (1998), a subdivision of the stratigraphic succession into six 'depositional phases' (DP) was employed to facilitate the correlation between the drill sites (Table 1; Pletsch *et al.* submitted). Although it was tried to place depositional phase boundaries close to changes in lithofacies, organofacies, or biofacies, the boundaries were defined on the basis of known or tentative biostratigraphic ages rather than on changes in lithology. Figure 4 summarizes the stratigraphic range and dominant lithofacies of Cretaceous to Palaeocene deposits.

Late Aptian–middle Albian (DP 1)

DP 1 comprises poorly sorted siliciclastic mudstones, siltstones and sandstones (Figs 4 and 5a). Laminated siltstones at the base of Hole 960A, devoid of marine microfossils, were interpreted as freshwater, probably lacustrine, deposits, whereas sedimentary structures and rare marine microfossils in the overlying section at Site 960, and at Sites 959 and 961, suggest deposition in marginal marine, probably deltaic, environments (Mascle *et al.* 1996; Strand 1998; Pletsch *et al.* submitted). In contrast to the generally low textural maturity of these sediments, the compositional maturity of sandstones is fairly high, which, in conjunction with frequent plant material and non-fluorescing (terrigenous) amorphous organic matter, points to recycling of older sedimentary rocks (Strand 1998; Oboh-Ikuenobe *et al.* 1997). Clay minerals in DP 1 sediments are very diverse, but smectite is missing. Diagenetic minerals, including pyrite, ferroan and magnesian carbonates, and vermicular kaolinite occur in the open pore space of sandstones as well as in the more compacted mudstone interbeds (Figs 7a, b).

The organic matter is of kerogen types III or IV and shows low S2 yields, suggesting a poor source potential for liquid hydrocarbons (Figs 5b and 6). There is a pronounced variability in the thermal maturity of sedimentary organic matter, as indicated by fluctuations in the 'thermal alteration index' (TAI; obtained from palynological residues; Oboh-Iluenobe *et al.* 1997) and T_{max} records. Notably, thermal maturity of DP 1 is lowest at the base of Hole 959D, as is indicated by T_{max} values < 430 °C.

Late Albian–early Cenomanian (DP 2)

DP 2 was subdivided into the lower DP 2A and the upper DP 2B, although both yield upper Albian–lower Cenomanian microfossils. The subdivision is based on the assumption that the differences in tectonic deformation between DP 2A and DP 2B are related to a time-equivalent uplift event. At Site 959, DP 2A is lithologically comparable to DP 1, apart from the occurrence of late Aptian–middle Albian foraminifers that suggest reworking of older marine strata during late Albian times (Holbourn & Moullade 1998). The overlying DP 2B is composed of coarse-grained detrital limestones with large, uncemented and partly interconnected voids and fossil moulds. Siliciclastic components include common rock fragments, feldspar and silicified shale clasts of up to 6 mm in diameter. The latter are similar to *in situ* facies found in DP 1 sediments of Site 960.

Table 1. *Cretaceous depositional phases of ODP Leg 159*

DP	Age range	Lithology (drill site of occurrence)
1	Aptian - m. Albian	tectonically deformed irregular alternation of siliciclastic sandstones, siltstones, and mudstones (959, 960, 961)
2	l. Albian - l. Cenomanian	A: as DP 1 (959); tectonically deformed mixed calcareous/siliciclastic sand-, silt-, and mudstones (962) B: hardly deformed coarse detrital limestones (959); chert and porcellanite (962)
3	e. - m. Turonian	black shale with intercalated detrital limestone (959); glauconite clay (962)
4	l. Turonian - e. Coniacian	dolomitic marlstone (959), black shale and coarse detrital limestone (960); ?glauconite clay (962)
5	m. Coniacian - e. Campanian	black shale (959); phosphatic crust (960); ?glauconite clay (962)
6	m. Campanian - e. Palaeocene	black claystone (959); ?glauconite clay (962)

e early, m. middle, l. late

At Site 962, carbonate-rich intervals with open marine microfossil assemblages were observed which document pelagic sedimentation south of the Marginal Ridge (Fig. 8a). This is further suggested by upward increasing proportions of smectite in clay mineral assemblages. The overlying cherts and porcellanites were attributed to DP 2B.

Sedimentary organic matter in DP 2A is similar to DP 1, suggesting a dominance of Type III and IV kerogen of low to intermediate thermal maturity (Figs 5b, 6, 8b and 9). Comparison of Rock-Eval data obtained from Site 962 and 959, however, indicates a slightly higher hydrocarbon potential for gas and oil at the seaward position, considering fairly high HI (> 300 mgHC/gTOC) and S2 yields (approaching 10 kgHC/ton rock) from Type II kerogen at the top of DP 2A. The occurrence of predominantly marine kerogen in upper Albian–lower Cenomanian sediments of Site 962 constitutes the oldest recovery of potential hydrocarbon source rocks at ODP Leg 159 drill sites.

Few elemental and Rock-Eval data are available from DP 2B at Site 962. They show an uphole decrease in the proportion of hydrogen-rich kerogen. T_{max} values indicate the presence of immature organic matter, although this trend is not reproduced by TAI data (Oboh-Ikuenobe *et al.* 1997).

Early–middle Turonian (DP 3)

Turonian sediments comprise crudely laminated black shales which are intercalated with coarse-grained detrital limestones similar to those of DP 2B. Black shales at Site 959 lie on top of detrital limestones of DP 2B. Smectite appears for the first time at Site 959, but is mixed with various other clay minerals, including chlorite. Benthic foraminiferal assemblages in the black shales record an expansion of the oxygen minimum zone during a period of increased surface water productivity (Holbourn *et al.* 1999). In contrast to Site 959, only a few metres of carbonate-free glauconite-bearing smectite clay were recovered at Site 962 (Figs 4 and 8a). Several glauconite hardgrounds occur in this interval, one of them being interpreted to represent the Cenomanian–Turonian Boundary Event (Pletsch *et al.* submitted).

TOC contents rise to *c.* 4% in the black shales (Figs 5b and 6), the organic matter being composed of thermally immature Type II kerogen. HI up to 500 mgHC/gTOC and S2 yields > 10 kgHC/ton rock suggest a fairly good oil potential. This organic character is similar to those reported from other early Turonian source rocks in the Atlantic and Tethys Oceans (e.g. Arthur *et al.* 1987; Schlanger *et al.* 1987; Herbin *et al.* 1987; Kuhnt *et al.* 1990; Kuhnt & Wiedmann 1995).

Late Turonian–early Coniacian (DP 4)

At Site 959, several metres of upper Turonian–lower Coniacian dolomitic marlstones and detrital limestones were recovered. These deposits commonly contain silt- and sand-sized quartz,

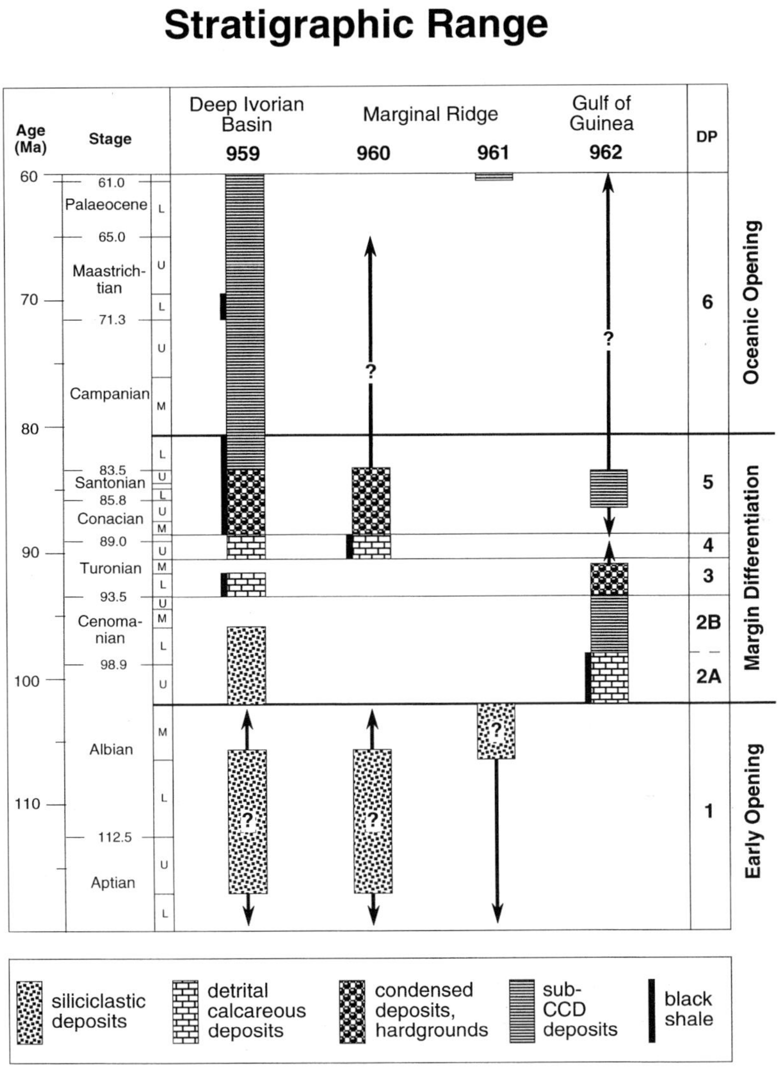

Fig. 4. Stratigraphic range and dominant lithofacies of Cretaceous–Palaeogene sediments recovered during ODP Leg 159 according to the biostratigraphic synthesis of Moullade *et al.* (1998). Timescale of Gradstein *et al.* (1995). The figure does not account for periods of non-deposition below the substage level.

and clay mineral assemblages which are similar to those of the underlying black shales. Benthic biofacies suggest an increasing detrital flux and a reduction of oxygen deficiency within the oxygen minimum zone (Holbourn *et al.* 1999). In striking contrast to Site 959, on top of black shales similar to those of DP3 at 959, *c.* 145 m of coarse-grained packstones and grainstones with abundant stromatoporoid, bivalve, echinoderm and calcareous algal debris were deposited at Site 960, located just 5 km to the south of Site 959 (Fig. 1). Compared to the detrital limestones of DP 2B at Site 959, they contain slightly more siliciclastic components, including abundant silicified shale clasts. Organic geochemical data of the DP 4 dolomitic marlstones reveal intermediate TOC contents with immature kerogen Type II/III (Figs 5b and 6).

Middle Coniacian–early Campanian (DP 5)

Sediments of DP 5 at Site 959 differ from the underlying deposits in the absence of coarse-

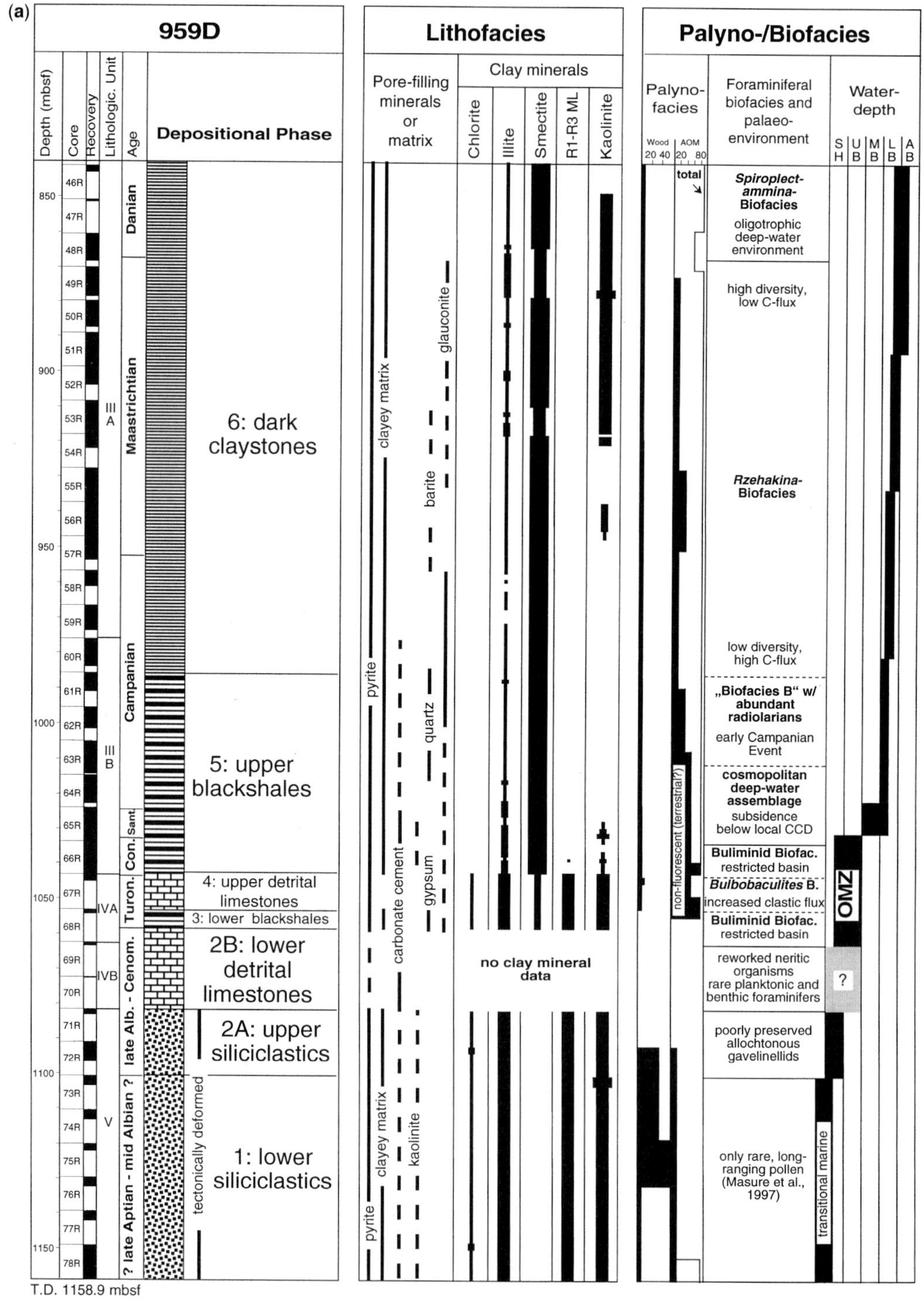

Fig. 5. (**a**) Summary of lithofacies and biofacies of the Cretaceous section in Hole 959D. The width of the bars for clay mineral species relates to the relative proportion of characteristic reflections in the diffractograms of the glycolated mounts. R1–R3 ML, illite–smectite mixed-layer minerals with long- ($R = 0$) or short-range ($R = 3$) crystal order; SH, shelf; UB, upper bathyal; MB, middle bathyal; LB, lower bathyal; AB, abyssal; OMZ, oxygen

(b)

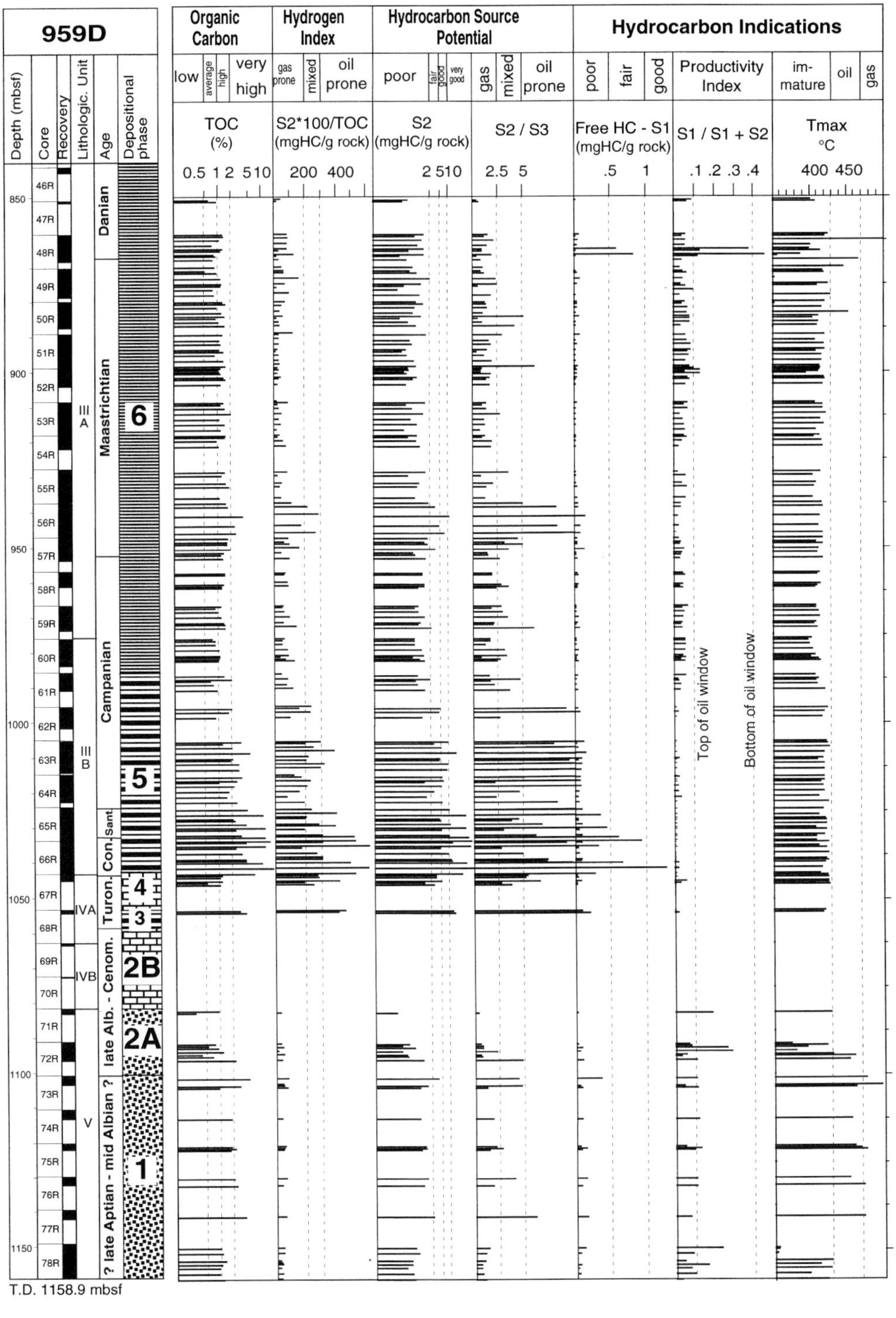

minimum zone; CCD, calcite compensation depth. Lithologic units are based on the revised subdivision of Janik *et al.* (1998). Palynofacies data by Oboh-Ikuenobe *et al.* (1997). (**b**) Rock-Eval pyrolysis results of Hole 959. Calculated parameters according to Espitalié (1985).

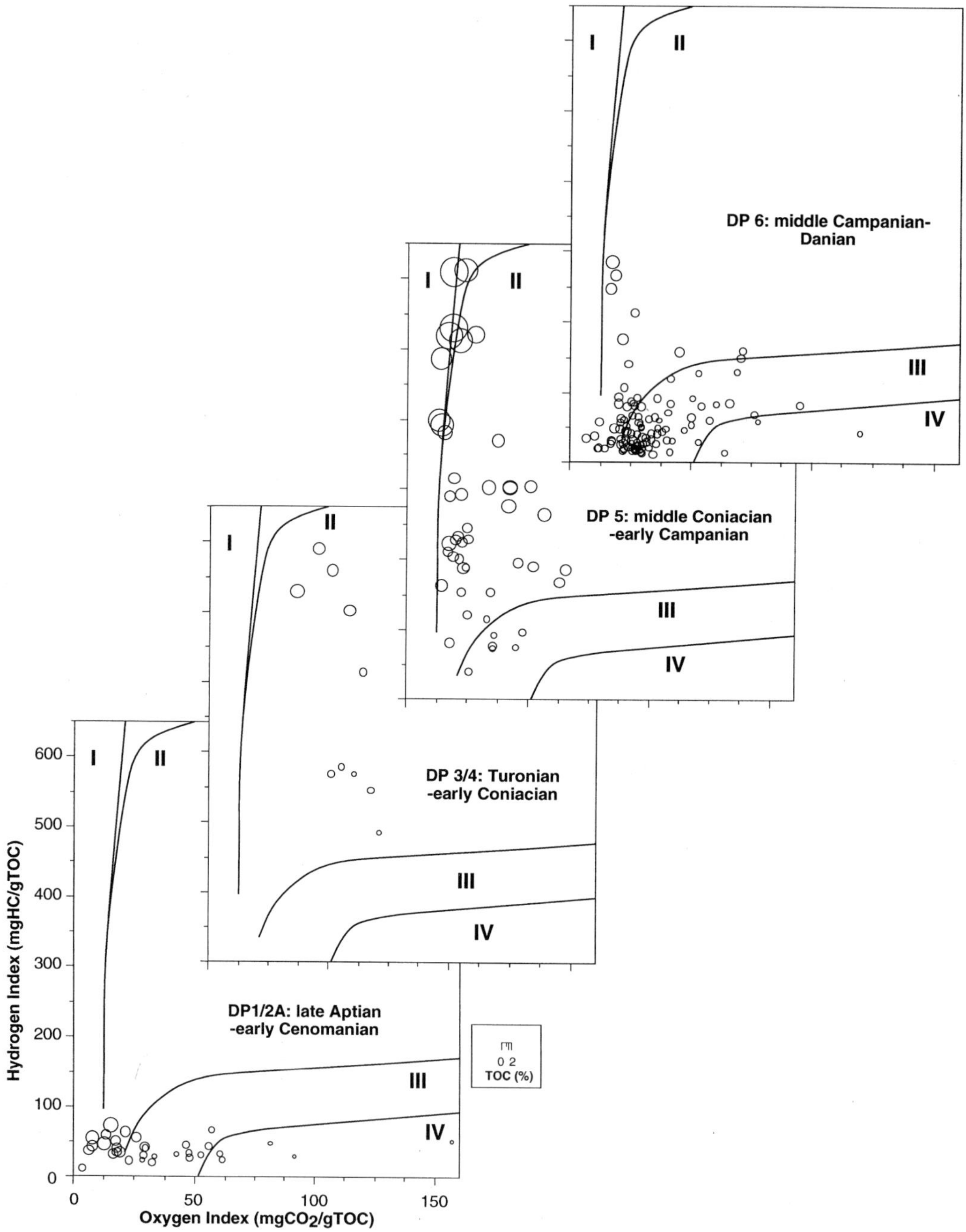

Fig. 6. Modified Van Krevelen diagrams for samples from different intervals of Hole 959D.

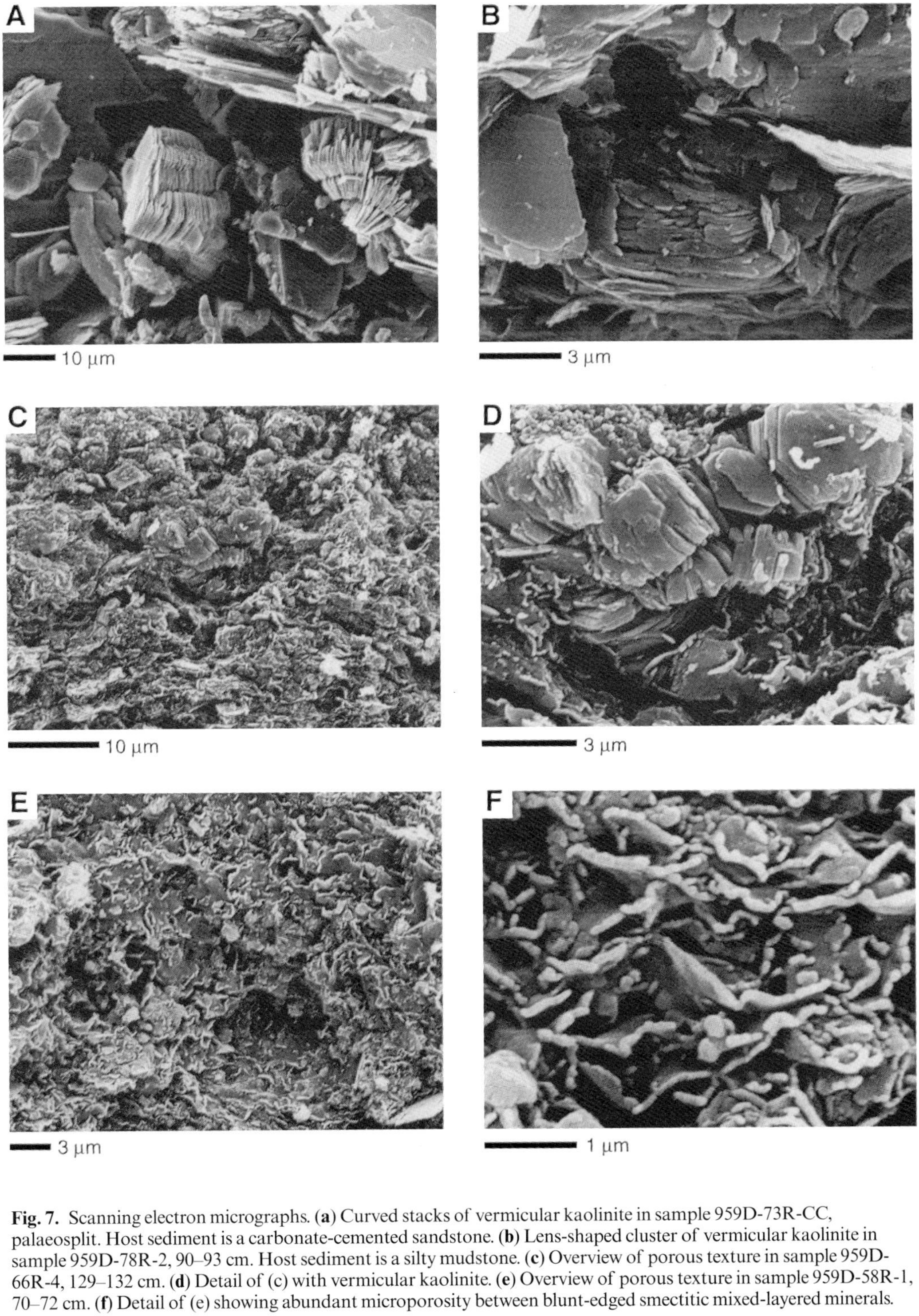

Fig. 7. Scanning electron micrographs. (**a**) Curved stacks of vermicular kaolinite in sample 959D-73R-CC, palaeosplit. Host sediment is a carbonate-cemented sandstone. (**b**) Lens-shaped cluster of vermicular kaolinite in sample 959D-78R-2, 90–93 cm. Host sediment is a silty mudstone. (**c**) Overview of porous texture in sample 959D-66R-4, 129–132 cm. (**d**) Detail of (c) with vermicular kaolinite. (**e**) Overview of porous texture in sample 959D-58R-1, 70–72 cm. (**f**) Detail of (e) showing abundant microporosity between blunt-edged smectitic mixed-layered minerals.

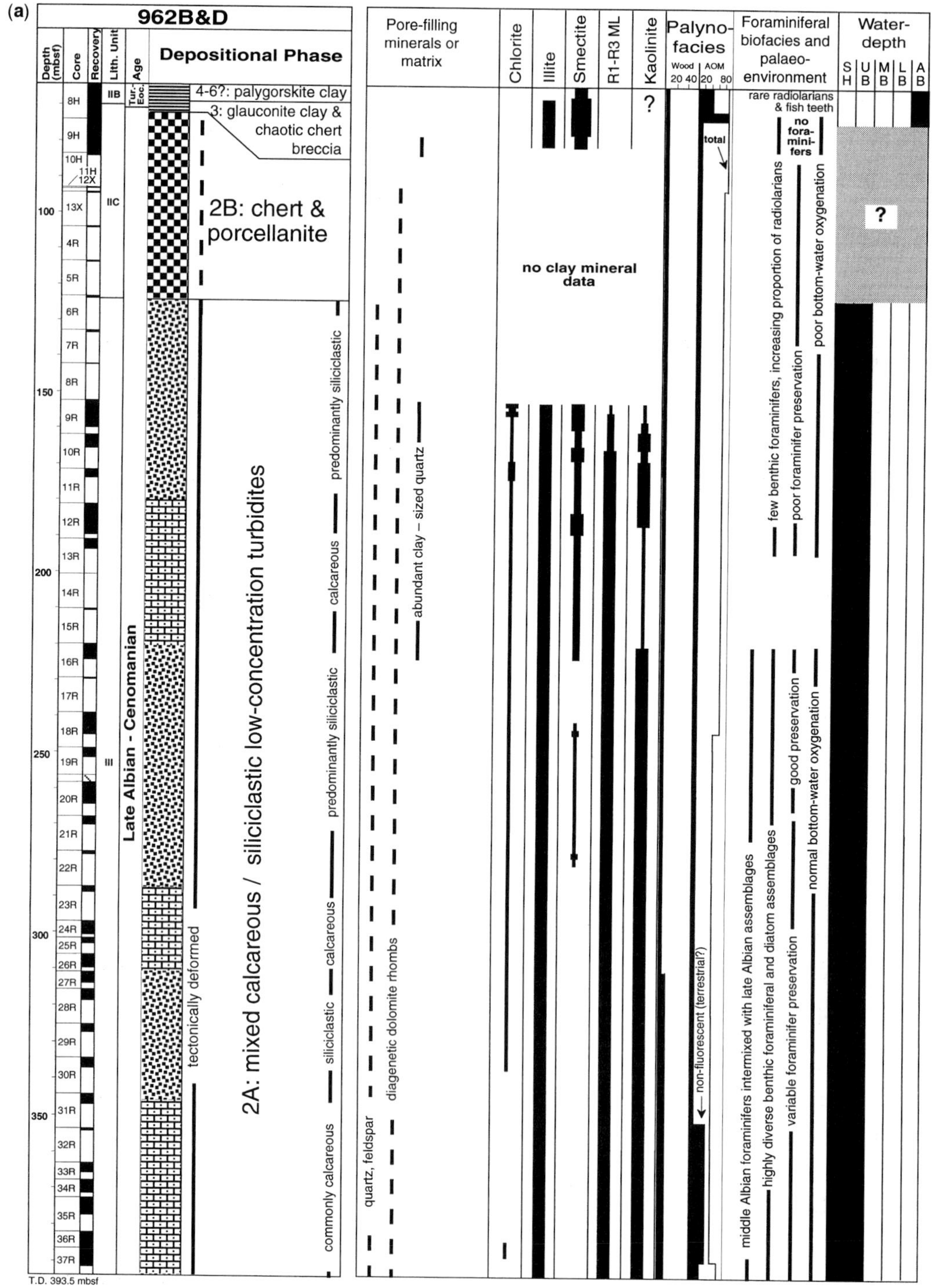

Fig. 8. (**a**) Summary of lithofacies, organofacies and biofacies of the Cretaceous section in Holes 962B and 962D. For explanations see Fig. 3. Palynofacies data by Oboh-Ikuenobe *et al.* (1997). (**b**) Rock-Eval pyrolysis results of Holes 962B and 962D. Calculated parameters according to Espitalié (1985).

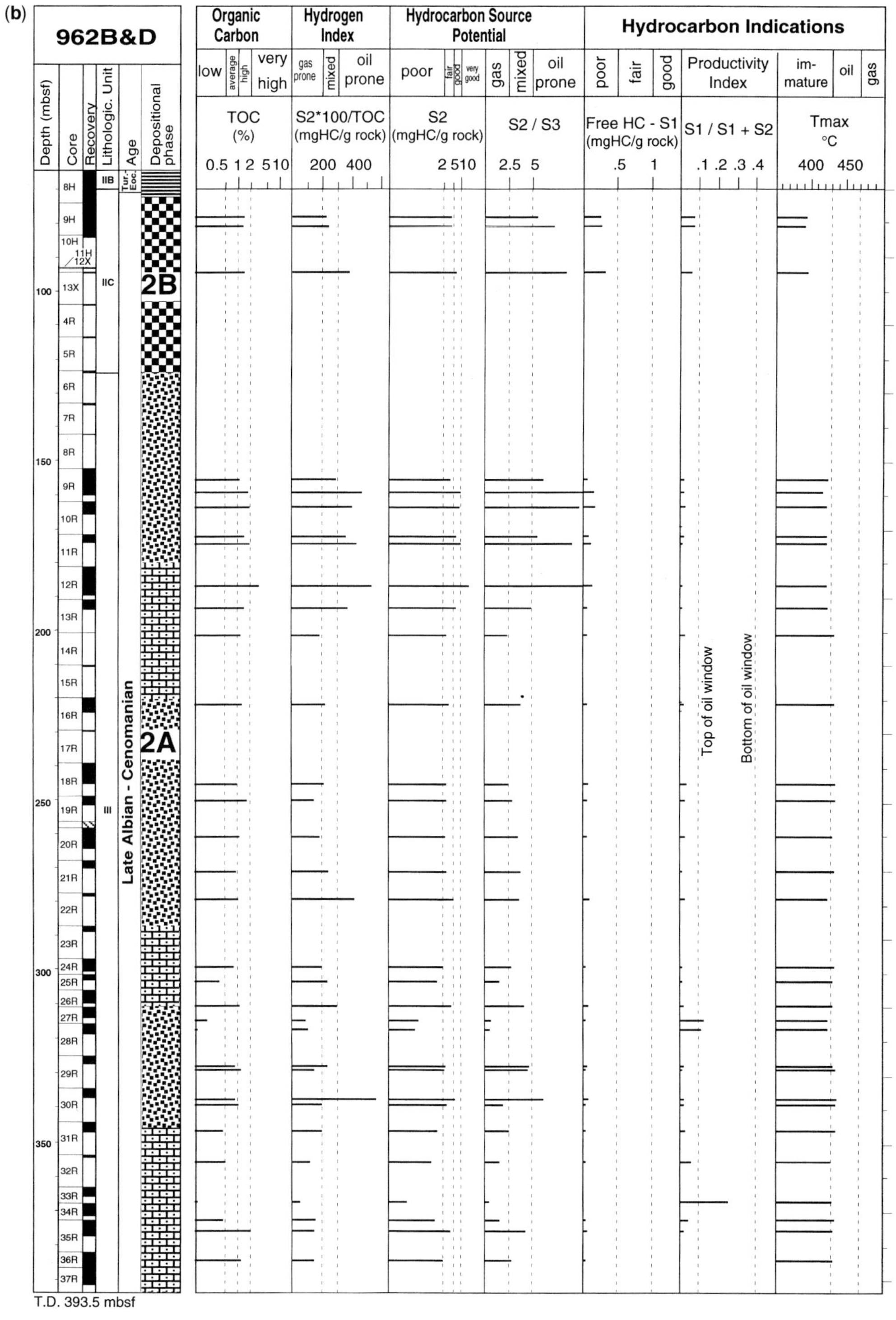

Fig. 8. (*continued*)

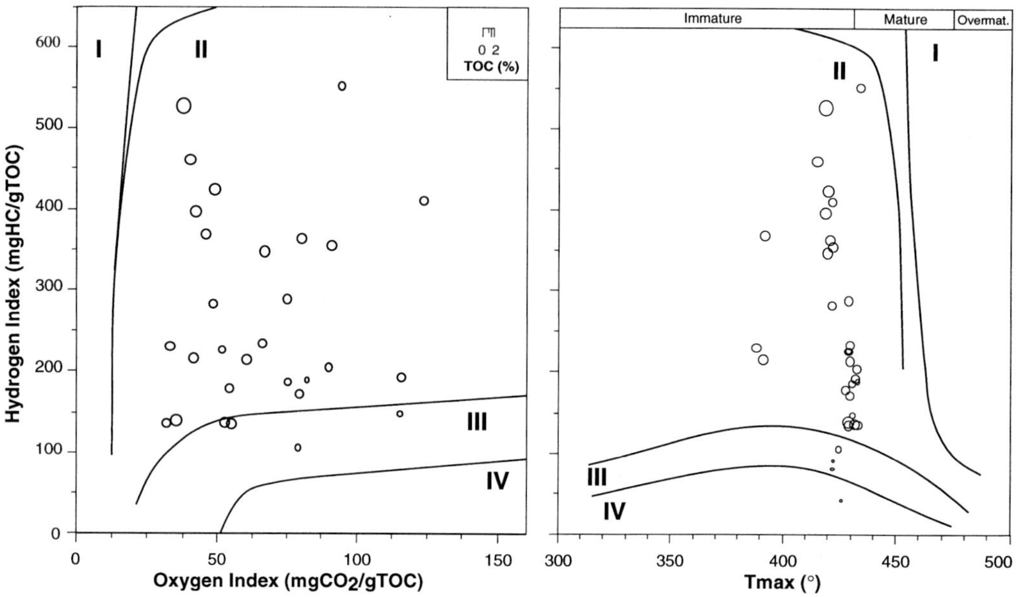

Fig. 9. Modified Van Krevelen and HI v. T_{max} diagrams for upper Albian–lower Cenomanian samples from Holes 962B and 962D.

grained extrabasinal debris, and notably in the complete lack of sand-sized siliciclastic material. In addition, clay mineral assemblages are composed mainly of smectite and are monotonous compared to those of the underlying section (Fig. 5a). This was interpreted as a shut-off of detrital sources other than hemipelagic input of clay-sized and fine silt-sized material, and very local reworking of authigenic mineral crusts, such as those recovered at Site 960 (Pletsch *et al.* 1999). Vermicular kaolinite was found in close association with organic-rich intervals, probably induced by the influence of organic acids generated by the black shales (Fig. 7c, d; Curtis 1987). Hardgrounds at the base of DP 5 correspond to a pronounced maximum in gamma-ray intensity in the uranium spectral band and provide excellent markers for correlation between Sites 959 and 960 (Mascle *et al.* 1996; Fig. 11). Uranium probably has its source in the abundant phosphate and organic matter of adjacent black shale units. Elevated potassium spectral intensities, detected at Site 960, may originate from the overlying glauconite clay. Highest TOC concentrations of *c.* 17% were measured in laminated marls in the lower part of DP 4. Kerogen preserved in these black shales generates considerable amounts of hydrocarbon (S2 up to 100 kgHC/ton rock), is thermally immature and of oil-prone Type I/II (Figs 5b and 6). Intercalated intervals, in contrast, consist of a structureless marly matrix with frequent authigenic mineral pebbles and shell fragments. These pebbly intervals still contain up to 4% TOC of immature kerogen Type II (Figs 5b and 6). Benthic foraminiferal biofacies in the black shales indicate a renewed intensification and/or increasing thickness of the oxygen minimum zone. Above the hardground interval, *c.* 50 m of carbonaceous marls and silicified claystones with generally > 2% TOC, accumulated during the middle Coniacian–early Campanian (Fig. 5).

Middle Campanian–late Palaeocene (DP 6)

Middle Campanian–late Palaeocene deposits at Site 959 consist of black, virtually carbonate-free claystones and minor siltstones (Figs 4 and 5). Intense bioturbation, which has destroyed most sedimentary structures, indicates the return of oxygenated bottom and pore water conditions during that time. Smectite continues to be the most abundant clay mineral with only minor amounts of illite. Kaolinite becomes enriched in the upper part of DP 6, namely in the mid-Maastrichtian and in the latest Maastrichtian to earliest Palaeocene. Peak contents in kaolinite roughly correlate with the Cretaceous–Tertiary boundary. Remarkably, clay-sized and fine silt-sized quartz is particularly abundant where kaolinite is rare or absent, but different to the underlying silicified claystones, no indications for a diagenetic quartz growth were observed in the SEM analyses (Fig. 5e, f).

The mid-Maastrichtian interval is characterized by a pronounced maximum in TOC, in S2 yields, in the proportion of non-fluorescent amorphous organic matter (Oboh-Ikuenobe *et al.* 1998), and in the abundance of the benthic foraminiferal genera *Rzehakina* (Kuhnt *et al.* 1998; Fig. 5b). The latter was interpreted to reflect temporarily enhanced marine organic carbon flux rates to the seafloor (Kuhnt *et al.* 1998). Organic matter from this interval is of immature, hydrocarbon-enriched Type II kerogen (Figs 5b and 6). Except for this mid-Maastrichtian interval, TOC contents remain between 1 and 1.5% throughout most of DP 6. Corresponding HI, S2 and T_{max} values persist at low levels, typical for hydrocarbon-depleted, immature kerogen Types III–IV (Figs 5b and 6). Higher maturity levels around the Cretaceous–Tertiary boundary are associated with pebbly mudstone beds which suggests reworking of older, organic-rich strata. Two samples close to the Cretaceous–Tertiary boundary contain considerable amounts of free hydrocarbons that have probably migrated from older strata.

Tectono-sedimentary evolution

The development of Cretaceous depositional environments along the Côte d'Ivoire–Ghana Transform Margin was intimately linked to the temporal and spatial development of water depth and exposure to oceanic currents. Since both the northern and the southern side of the marginal ridge experienced different subsidence histories, ODP Leg 159 deposits may serve to reconstruct the complex evolution of depositional settings and its associated palaeocirculation. Therefore, the sedimentary records of Sites 959–962 were tied into the three-stage model which outlines the general opening history of the Equatorial Atlantic Gateway (Fig. 3).

Early opening stage (middle Albian or earlier)

The oldest deposits of the Leg 159 drill sites are exclusively siliciclastic and record the transition from terrestrial to marine environments. At Site 960 this is represented by the change in sedimentary structures and compositions from those typical of freshwater lakes to those of a marine delta. As indicated by the presence of reworked marine microfossils, marine sedimentation must have started by the late Aptian–early Albian, or even earlier, at the other ODP sites, where the transition in the palaeoenvironment was not recovered (Fig. 10a). Currently, no lithologic or palaeontologic marker is available to refine the correlation between these contrasting facies. It is therefore not known whether the lacustrine environment at Site 960 co-existed with marginal marine environments at the other drill sites or whether it is older than these.

The abundance of plant debris and kerogen of Types III or IV during this early stage is attributed to the proximity of the depositional environments to terrestrial source areas. Considering the palaeogeographic position of ODP Leg 159 close to the palaeoequator (Fig. 2b), a dense tropical vegetation with intense weathering and run-off might have prevailed on the adjacent continents. Organic geochemical and palynological data suggest that contribution of marine organic matter was minor during that time. The pronounced accumulation of marine organic matter reported from other parts of the Atlantic and Tethys basins during the Aptian–middle Albian (e.g. Arthur *et al.* 1990; Erbacher *et al.* 1996) did not occur in the Deep Ivorian Basin either due to strong dilution with terrigenous material or because it was still inhibited by the incomplete or non-existent opening to oceanic midwater circulation north of the Côte d'Ivoire–Ghana Transform Margin. It is suspected that during the early opening stage the Deep Ivorian Basin was a very shallow marine depression which was separated from the initial Equatorial Atlantic Gateway by the Marginal Ridge. The Marginal Ridge must have been locally emerged and was probably wider than today to accommodate the lacustrine environments. Unfortunately, no data are available from more distal localities towards the axis of the gateway, but the recent discovery of a calcareous deepwater facies of earliest Cretaceous age along the Romanche Fracture Zone (Bonatti *et al.* 1996) strongly supports the assumption that deeper marine conditions existed, at least temporarily, south of the Côte d'Ivoire–Ghana Transform Margin.

Margin differentiation stage (late Albian–early Coniacian)

A pronounced differentiation of palaeowater depths and depositional environments along the Côte d'Ivoire–Ghana Transform Margin took place during the mid-Cretaceous. Up to this time, the Equatorial Atlantic Gateway was probably a morphologically complex strait with several discontinuous deeper basins aligned along the major transform fault trend (Fig. 3a). Uplift

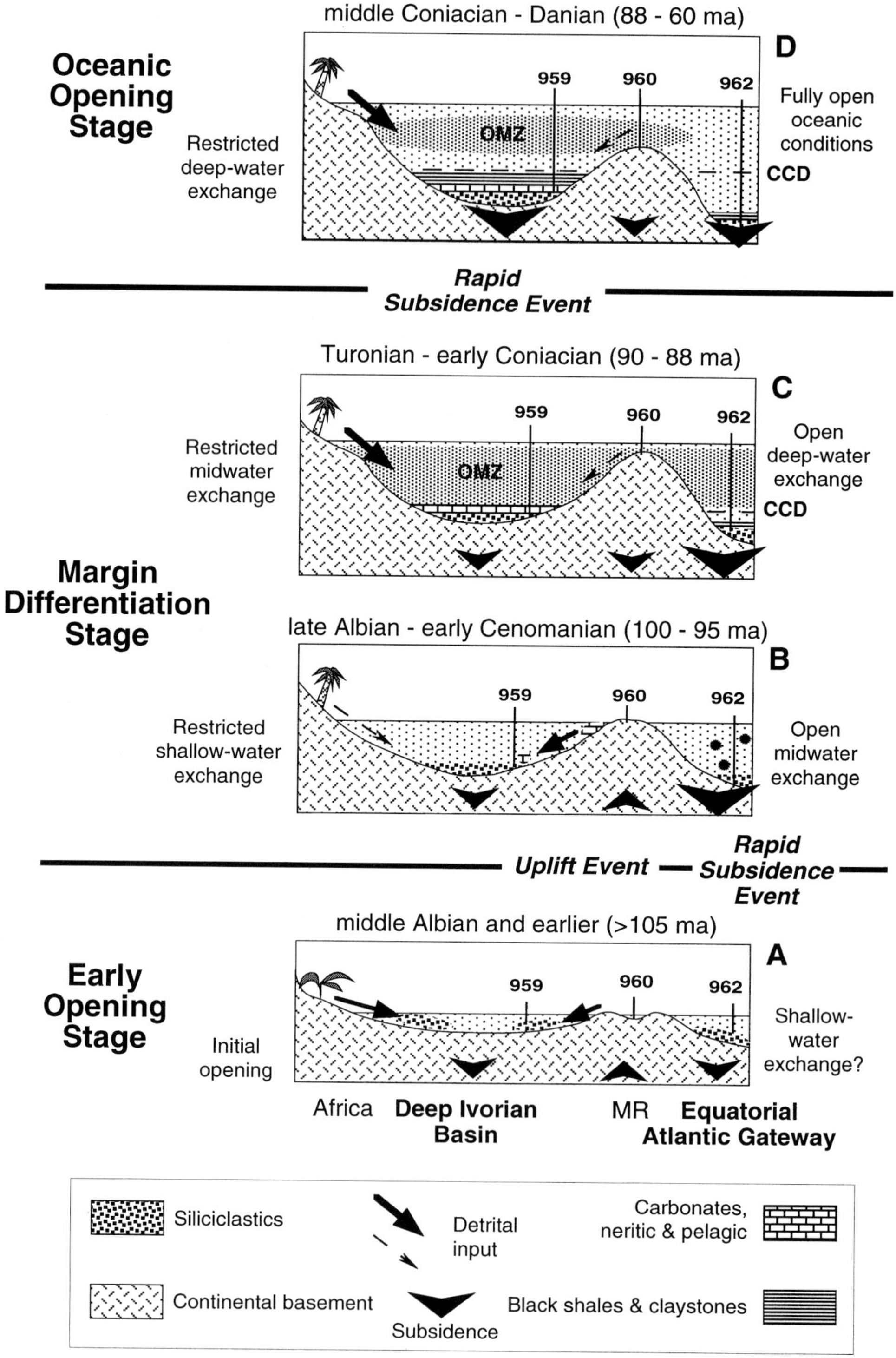

Fig. 10. Sketch of the depositional environments on a hypothetical transect from the African continent over the Deep Ivorian Basin and the Marginal Ridge (MR), to the Equatorial Atlantic Gateway. CCD, Calcite compensation depth; OMZ, oxygen minimum zone.

adjacent to Sites 959 and 962, as evidenced by the erosion of previously deposited marine sediments, by the formation of an unconformity and by abrupt changes in sediment composition, physical properties and tectonic deformation, had started in the middle–late Albian and probably culminated during the early Cenomanian (Fig. 10b). The presence of large feldspars and rock fragments in sediments overlying the unconformity at Site 959 may indicate a continued unroofing of deeper crustal levels adjacent to this site (Pletsch *et al.* 1999). Considering that the separation between Africa and South America was fairly advanced by the middle Albian (Fig. 2; Hay *et al.* 1998), it is assumed that the uplifted area was the Marginal Ridge, because the West African and Brazilian margins were too far removed from the Côte d'Ivoire–Ghana Transform Margin to have acted as siliciclastic sediment sources. The deeper level of unroofing on the Marginal Ridge supports the idea of a source for the silicilastics along the Côte d'Ivoire–Ghana Transform Margin itself.

While the morphologically dammed, landward Deep Ivorian Basin was still dominated by terrestrial organic matter supply, the onset of kerogen Type II deposition during the late Albian–early Cenomanian probably marks the first establishment of a midwater circulation on the southern slope of the Côte d'Ivoire–Ghana Transform Margin. Deepening on the oceanward side of the Marginal Ridge became extremely rapid with the seafloor subsiding to below the local calcite compensation depth during the Cenomanian, which at that time was at a water depth of *c.* 2000 m (Pletsch *et al.* submitted; Fig. 10b). At Site 962, an opening to erosive deepwater currents and a full deepwater connection between Central and South Atlantic is signalled by the deposition of highly condensed, carbonate-free clays from this time.

The Cretaceous Atlantic was particularly prone to the expansion of the oxygen minimum zone, favouring anoxia of the water column down to the seafloor, related to its marginal position with respect to the world ocean. This, in turn, allowed for an enhanced preservation of organic matter when marine productivity was increased during the Cenomanian–Turonian Boundary Event. Significant contribution of terrestrial organic matter and clay minerals point to an additional terrigenous input and intensified nutrient leaching as a stimulus for enhanced palaeoproductivity and organic carbon flux rates. The abundance of glauconite and pyrite in the thin Turonian–Santonian interval at Site 962 indicate that, despite the lack of TOC-rich strata, deepwaters were oxygen deficient as a result of the widespread expansion of the oxygen minimum zone (Fig. 10c). Connection to oxic deepwater circulation was probably not achieved before the spreading ridge had passed the Côte d'Ivoire–Ghana Transform Margin later in the Turonian (Fig. 3c; Clift *et al.* 1997). It is assumed that other obstacles in the Equatorial Atlantic Gateway behaved in a similar way, to allow a full, oxic deepwater connection between the Central and South Atlantic south of the margin after the early Turonian.

During the same period, the crest of the Marginal Ridge and the Deep Ivorian Basin remained at outer neritic to upper bathyal water depths. The occurrence of Cenomanian–Turonian Boundary Event facies at these sites, in spite of their restricted positions, implies the establishment of an open midwater exchange with the adjacent Central and South Atlantic basins (Fig. 10c). Deepwater exchange through the Deep Ivorian Basin, as indicated by the invasion of cosmopolitan deepwater faunas, was established only after the middle Coniacian, *c.* 8 Ma later than on the southern slope of the Côte d'Ivoire–Ghana Transform Margin. Rapid subsidence of the Deep Ivorian Basin since that time correlates with strong condensation on the Marginal Ridge in the middle Coniacian. Similar small-scale differences in the development of water depths and sedimentary environments have been observed in other marine transform settings (e.g. Schwentke & Kuhnt 1990; Reicherter *et al.* 1994) and are considered to be typical for the evolution of transform margins.

Oceanic opening stage (middle Coniacian and later)

The oceanic opening stage was heralded by a dramatic reduction in sedimentation rates and is characterized by relatively smooth changes in palaeowater depth. While the hemipelagic claystones in the landward Deep Ivorian Basin still record a distant input of terrigenous organic matter and very fine-grained siliciclastic material, the oceanward slope of the Côte d'Ivoire–Ghana Transform Margin is characterized by extreme condensation and/or erosion throughout the latest Cretaceous. Sediment starvation partly resulted from enhanced current winnowing, especially on the Marginal Ridge (Fig. 10d). The virtual absence of coarse-grained components and the dominance of smectite in clay mineral assemblages testify to the deep submergence of the Marginal Ridge and to tectonic quiescence during that time. Thus, by the latest

Cretaceous, the evolution of the Côte d'Ivoire–Ghana Transform Margin became similar to that of other mature passive continental margins.

Aspects of regional petroleum geology

Potential hydrocarbon source rocks

Siliciclastic sediments, deposited during the early opening stage (DP 1), typically contain hydrogen-poor organic matter with low oil potential (Fig. 11). Variable, but occasionally high, T_{max} records suggest reworking of fossil, previously heated organic matter from more proximal environments, or from deeper uplifted strata. Volatile hydrocarbons, generated during this earlier period of burial, were probably lost when overlying reservoirs became eroded during middle Albian–Cenomanian tectonic uplift. These previously deeply buried organic-rich units were observed at Sites 960 and 961, and probably extended over a larger area. Longer-chain volatile hydrocarbons, detected below the unconformity at the bottom of Hole 959D, may witness these deeper levels of hydrocarbon generation (Mascle *et al.* 1996).

The oldest marine black shales recovered during ODP Leg 159 are upper Albian–lower Cenomanian cherts, rich in marine-originated kerogen from oceanward Site 962 (Fig. 11). The damming effect of the Marginal Ridge, the lack of a continuous midwater circulation and the strong dilution of marine organic matter with terrigenous material probably prevented the formation of contemporaneous marine black shales in the Deep Ivorian Basin.

In contrast to the Lower Cretaceous situation, marine black shales of Turonian age were exclusively recovered on top of the Marginal Ridge and on its northern slope. These deposits correlate with a pronounced increase in marine productivity, known as the Cenomanian–Turonian Boundary Event, or Oceanic Anoxic Event II (e.g. Arthur *et al.* 1987; Schlanger *et al.* 1987 and refs therein). The absence of Cenomanian–Turonian Boundary Event black shales at Site 962 is probably related to erosion by deepwater currents, similar to many other pelagic settings of the Cretaceous Atlantic (e.g. de Graciansky *et al.* 1982, 1987; Zimmerman *et al.* 1987). More favourable conditions for the widespread accumulation of black shale facies probably existed in the Deep Ivorian Basin due to its sheltered position behind the Marginal Ridge, as was the case in many other shelf-slope settings along the West African and Brazilian margins (Kuhnt *et al.* 1990; Koutsoukos *et al.* 1991; Mello *et al.* 1989, 1995; Holbourn *et al.* 1999). The highest TOC contents and S2 yields in ODP Leg 159 deposits were observed in the middle Coniacian–lower Campanian black shale interval. Similar lithologies were deposited over extended areas along the West African margin, although more limited to restricted marginal basins (de Graciansky *et al.* 1987; Ly & Kuhnt 1994; El Albani *et al.* 1999). Black, partly laminated pyritic shales of Turonian–Campanian age reported from exploration well D1-1X, a few kilometres off the Ivorian coast (Saint-Marc & N'da 1997; Fig. 1) may indicate the extension of this source rock facies as far as the northern margin of the Deep Ivorian Basin. Biofacies data indicate that the formation of the mid-Maastrichtian black shales corresponds to a renewed pulse in marine palaeoproductivity, causing an expansion of the oxygen minimum zone, much like it is proposed for the older source rock units (Fig. 11; Holbourn *et al.* 1999).

The remarkable discrepancy between the condensed, organic-lean clays at the deep oceanward drill site and the recurring thick black shale intervals that accumulated on the landward slope of the Côte d'Ivoire–Ghana Transform Margin seems to be characteristic of many Upper Cretaceous transects in the South Atlantic (Herbin *et al.* 1987; Zimmermann *et al.* 1987). Atlantic bottom waters were sufficiently oxygenated to prevent any significant accumulation of organic matter in the deep basins, whereas sustained organic matter production, accumulation and preservation occurred along basin margins.

It is important to note that terrestrial organic matter contributed a considerable share to the organic carbon fraction of deposits on the Côte d'Ivoire–Ghana Transform Margin, not only in intervals between the black shales but, in particular, during enhanced accumulation of marine organic matter. This feature is in agreement with organic geochemical results reported from many other Late Mesozoic black shale unit from the Atlantic Ocean (Emeis 1987).

Thermal maturity

Besides organic maturity indicators, such as the T_{max} value from Rock-Eval pyrolysis and the TAI determined on palynomorphs, illite–smectite mixed-layer minerals are of particular interest with regard to hydrocarbon generation because they provide supplementary evidence for the degree of thermal alteration of sediments. This mineral group displays a stepwise increasing crystal order over a continuous range of intercalation levels from fully expandable smectite to nonexpandable illite (Reynolds 1980). The percentage of illitic layers (%I) in these minerals and the steps of crystal order (i.e. fully expandable R0-type 'Smectite' and less expandable R1- or R3-Type 'illite–smectite' clay) increase mainly as a result of increasing thermal alteration in the temperature range of the oil window (Pollastro 1993).

Elevated T_{max} and TAI values, and the consistent lack of highly expandable clay minerals reported from the tectonically deformed siliciclastic sediments below the unconformity of Leg 159 have fuelled speculations about an intense and pervasive thermal diagenetic overprint of these sediments (Mascle *et al.* 1996). Further studies on the TAI of palynomorphs and clay minerals in the 0.5 µm size fraction (Oboh-Ikuenobe *et al.* 1997; Holmes 1998) seem to support the idea of pervasive heating of the sediments below the unconformity. Notably, Holmes (1998) documented that these sediments exclusively contain R1 and minor R3 illite–smectite clay, whereas smectite is absent. This was explained as a post-depositional transformation at burial temperatures in the range of 100–180 °C. Similarly, assemblages containing R1-ordered illite–smectite clay at the bottom of Hole 962D are replaced upsection by more expandable clay, and by exclusively R0-ordered smectite at the top of DP 2A over an interval of 175 m. A palaeogeothermal gradient between 343 and 457 °C/km was calculated to bring the bottom of Hole 962D in the temperature range of 120–140 °C, where all R0 clay is reacted to R1 clay. This extreme geothermal gradient was interpreted to result from the same heating event either during, or shortly after, deposition of DP 2A, i.e. in late Albian–earliest Cenomanian times (Holmes 1998). Most of the findings reported by Holmes (1998) were reproduced in our studies of the slightly coarser (> 2 µm) size fraction. However, with the larger set of data on the expandability of illite-smectite mixed-layer minerals, and especially on T_{max} values, a different interpretation emerges.

At the temperatures proposed by Holmes (1998), kerogen should have obtained a thermally mature signature, because organic matter reacts more sensitively than clay minerals when exposed to short periods of intense heating (Smart & Clayton 1985). However, T_{max} values from that interval scatter *c.* 430 °C and never exceed 434 °C (Figs 8b and 9), which indicates the dominance of marginally mature organic matter. Moreover, eight samples from the bottom of Hole 959D consistently yielded T_{max} values of 427 °C or less in Type III kerogen, which is difficult to reconcile with downward increasing thermal alteration.

In view of the striking discrepancy in palaeotemperature estimates derived from clay minerals and organic matter, it is assumed that the sediments acquired their diagenetic/maturity level prior to deposition. Hence, the postulated thermal overprint most probably results from mixing with reworked, previously altered material. Deeply buried sediments on the Marginal Ridge may have been eroded during the middle–late Albian and deposited at Sites 959 and 962 to create the apparent thermal overprint signature observed at these basinal sites. This interpretation is supported by an apatite fission track age of 105 Ma which pre-dates the depositional age of DP 2 (Clift *et al.* 1997, 1998). According to the available mineralogic and organic geochemical data, maximum temperatures at the bottom of Holes 959D and 962D did not exceed *c.* 80 °C. Additional data (i.e. vitrinite reflectance data) are required to substantiate this interpretation.

The presence of smectite in sediments overlying the unconformity (DP 2B–6), as well as estimates based on oxygen isotopes from carbonate cements and veins, indicate maximum palaeotemperatures < 100 °C (Holmes 1998; Marcano *et al.* 1998; Pletsch *et al.* 1999). Similarly, none of the diagenetic minerals in post-unconformity sediments (different kaolinite polytypes, barite, carbonate minerals and microquartz) imply a pervasive heating of the host sediments. Rock-Eval data generally confirm the burial temperatures mentioned above, although some overprint of the autochthonous maturity signal by admixture of reworked organic matter must be considered (Oboh-Ikuenobe *et al.* 1997; 1998; Wagner unpublished data).

According to this interpretation, organic-rich sediments of the Côte d'Ivoire–Ghana Transform Margin hardly reached the top of the oil window as they have not been buried to sufficient depth. Consequently, no significant hydrocarbon generation is expected to have occurred. This is corroborated by the very low production indices throughout the entire core section, which suggest that these sediments have retained most of their hydrocarbon generative potential (Figs 5b and 8b). On the northern margin of the Deep Ivorian Basin, however, Turonian–Coniacian strata were encountered at *c.* 1.8 km depth in exploration well D1-X1, off Abidjan (Fig. 1a; Saint-Marc & N'da 1997). Further, seismic line MT02 (Fig. 1a), which provides a cross-section through the distal part of the Deep Ivorian Basin, displays the quasi-continuous extension of post-rift deposits in a landward direction. Much of the interval that contains the potential source rocks at Site 959 is buried about 1.5 times deeper in the adjacent southern Deep Ivorian Basin. Considering that equivalent potential hydrocar-

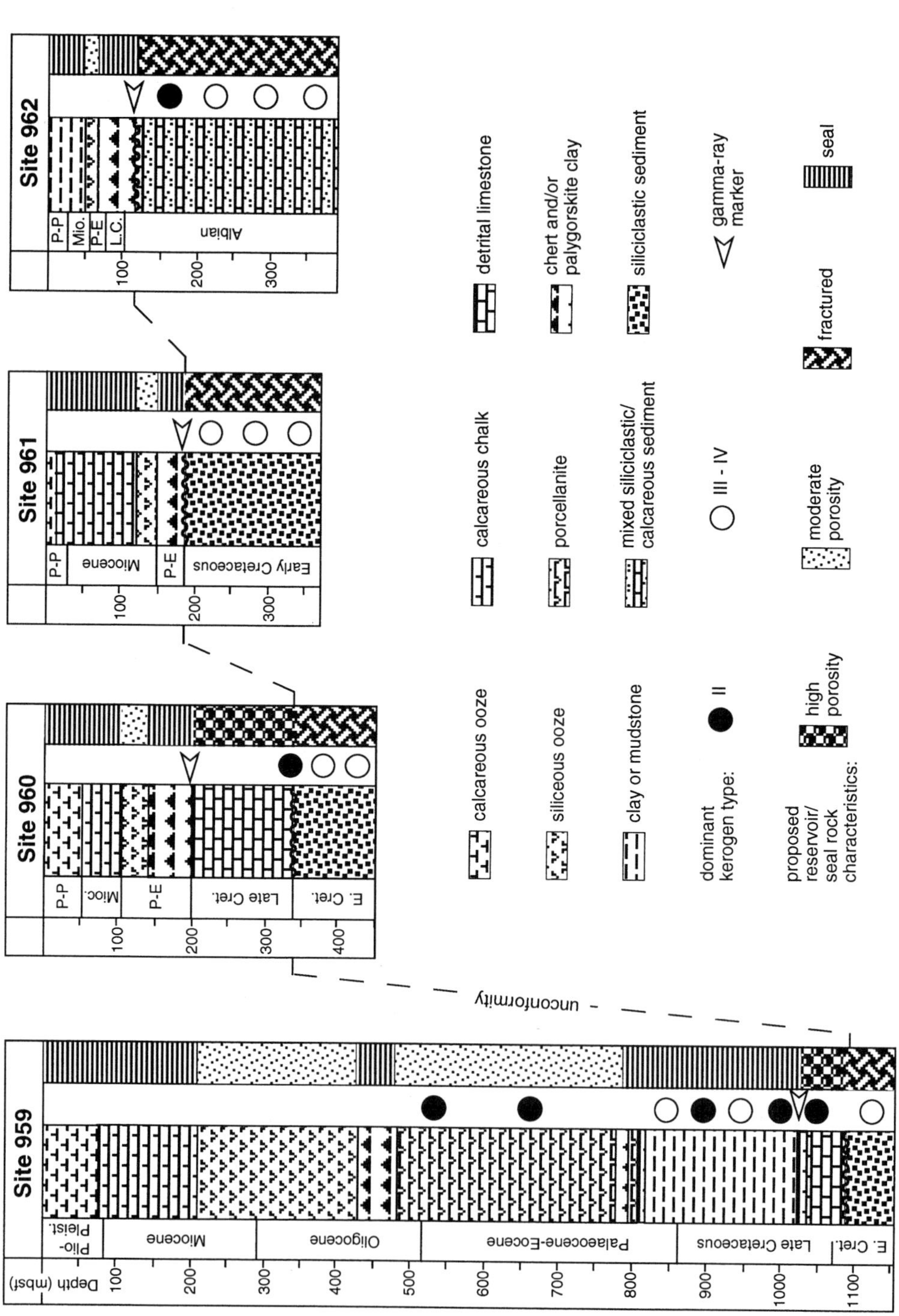

Fig. 11. Summary of the stratigraphic section, source rock potential, estimated reservoir and seal characteristics, and prominent gamma-ray log markers of lithofacies recovered during ODP Leg 159. See text for discussion.

source rocks were buried to even greater depths in the central Deep Ivorian Basin, source rocks equivalent to those found at the ODP Leg 159 drill sites may have generated hydrocarbons downdip of the elevated Côte d'Ivoire–Ghana Transform Margin.

Reservoirs and seals

Reservoirs in sediments of the early opening stage are probably limited both in volume and quality due to intense syn-depositional faulting and fracturing and the generally low porosity of texturally immature sandstones (Fig. 11). The potential for secondary porosity to develop is low because unstable components such as feldspars are scarce. In contrast, the highly porous and permeable detrital limestones of late Albian–Coniacian age have excellent reservoir characteristics but their spatial distribution is poorly constrained by the ODP Leg 159 drill sites. The extreme thickness of this facies on top of the present ridge, and its almost complete lack some 5 km northwards, may indicate that much of the calcareous debris had bypassed Site 959 and accumulated further downslope in the central Deep Ivorian Basin. Since these detrital deposits are covered by clay-rich hemipelagic sediments with low permeability of DP 5 and 6 they might provide an interesting exploration target.

Pore space in the laminated black shale facies is limited to microfossil tests and to shelter porosity below larger shelly debris, whereas some intergranular porosity is preserved in the pebbly facies (Fig. 7c). Porosity strongly decreases in an interval that comprises the upper part of DP 5 and the bottom of DP 6 (Janik *et al.* 1998). These sediments are intensively silicified and riddled with mineralized faults and veins (Holmes 1998). Extensional tectonic events as late as the middle Eocene may have allowed to fill potential reservoirs when corresponding source rocks had been buried down to the oil window in these settings (Clift *et al.* 1997; Marcano *et al.* 1998; Pickett & Allerton 1998).

Upper Eocene and Oligocene porcellanites, deposited along the Marginal Ridge when deformation had largely ceased, reveal partly very high porosities and may also have served as hydrocarbon reservoirs. These siliceous facies typically grade into less porous hemipelagic sediments when approaching the continental shelf, but may extend over large areas of the Deep Ivorian Basin. On the Côte d'Ivoire–Ghana Transform Margin this unit is capped by an undeformed blanket of Cenozoic pelagic marls which may act as an effective barrier to the loss of hydrocarbons.

Conclusions

The evolution of the Cretaceous Equatorial Atlantic Gateway is not only of palaeoceanographic interest but it also provided particularly favourable conditions for the accumulation of organic-rich sediments. The occurrence of hydrogen-rich marine black shales is restricted to Cretaceous sections above an uplift-related unconformity which formed when separation of Africa and South America was sufficient to allow permanent oceanic midwater exchange during the late Albian. Four periods of black shale accumulation are recorded, some of which correlate with the global oceanic anoxic events: in the latest Albian–earliest Cenomanian, at the Cenomanian–Turonian boundary, during the middle Coniacian–early Campanian and in the early Maastrichtian. Biofacies indicate that these periods were controlled by elevated marine organic carbon flux rates, induced by a combination of enhanced palaeoproductivity and supply of terrestrial organic matter. Black shale deposition appears to be intimately linked to periods of rising or maximum eustatic sea levels and to the expansion of the oxygen minimum zone. Intervals between black shale units, in contrast, indicate a shrinking oxygen minimum zone and enhanced detrital flux rates in correspondence to lowering sea levels.

With regard to the thermal maturation of the organic matter, contrasting models are currently discussed, postulating a total range of maximum paleotemperatures between 180 and 80 °C. Pyrolytic and clay mineralogic data presented in this study favour a moderate thermal overprint of the organic matter in black shales which hardly reached the top of the oil window. Hence, intense hydrocarbon generation from organic-rich Cretaceous units is not expected to have occurred along the Marginal Ridge.

Intercalated or overlying detrital limestones and porcellanites of Cretaceous and early Cenozoic age, respectively, have retained high porosities and may provide excellent hydrocarbon reservoirs.

We would like to thank Nick Cameron, Andy Fleet and Jonathan Turner for critical and helpful comments on a former version of this manuscript and P. Clift, J. Erbacher, A. Holbourn, M. A. Holmes, W. Kuhnt, F. Oboh-Ikuenobe, and E. Söding, who shared their data and expertises with us. Technical assistance by R. Henning, J. Kim, S. Kolonic, J. Funk, U. Langrock, H. Orthey (Bremen), by P. Recourt, P. Vanderaveroet, D. Malengros, and N. Pernot (Lille), and by A. Kirch, U. Schmölke, U. Schuldt and A. Frantz (Kiel) is gratefully acknowledged. Funding was provided by the German Research Council (DFG-grant Ku 649/2).

References

ARTHUR, M. A., DEAN, W. E. & STOW, D. A. V. 1984. Models for the deposition of Mesozoic–Cenozoic fine-grained organic-carbon-rich sediment in the deep sea. *In*: STOW, D. A. V. & PIPER, D. J. W. (eds) *Fine-Grained Sediments: Processes and Products*. Geological Society, London, Special Publications, **15**, 527–560.

——, SCHLANGER, S. O. & JENKYNS, H. C. 1987. The Cenomanian–Turonian oceanic anoxic event, II: Palaeoceanographic controls on organic-matter production and preservation. *In*: BROOKS, J. & FLEET, A. J. (eds) *Marine Petroleum Source Rocks*. Geological Society, London, Special Publications, **26**, 401–420.

——, JENKYNS, H. C., BRUMSACK, H. J. & SCHLANGER, S. O. 1990. Stratigraphy, geochemistry, and paleoceanography of organic carbon-rich Cretaceous sequences. *In*: GINSBURG, R. N. & BEAUDOIN, B. (eds) *Cretaceous Resources, Events and Rhythms, Background and Plans for Research*. NATO ASI Series, Series C: Mathematical and Physical Sciences, Kluwer, **304**, 75–120.

BASILE, C., MASCLE, J., SAGE, F., LAMARCHE, G. & PONTOISE, B. 1996. Pre-cruise and site-surveys: a synthesis of marine geological and geophysical data on the Côte d'Ivoire–Ghana Transform Margin. *In*: MASCLE, J., LOHMAN, G. P., CLIFT, P. D. ET AL. (eds) *Proceedings of the Ocean Drilling Program, Initial Reports*. College Station, TX (Ocean Drilling Program), **159**, 47–60.

BOIS, C., BOUCHÉ, P. & PELET, R. 1980. Histoire géologique et répartition des reserves d'hydrocarbures dans le monde. *Revue de l'Institut Français de Pétrole*, **35**, 242–270.

BONATTI, E., LIGI, M., BORSETTI, A. M., GASPERINI, L., NEGRI, A. & SARTORI, R. 1996. Lower Cretaceous deposits trapped near the equatorial Mid-Atlantic Ridge. *Nature*, **380**, 518–520.

BOUILLIN, J. P., POUPEAU, G., LABRIN, E., ET AL. 1997. Fission track study: heating and denudation of marginal ridge of the Ivory Coast–Ghana transform margin. *Geo-Marine Letters*, **17**, 55–61.

BRALOWER, T. J. & THIERSTEIN, H. R. 1987. Organic carbon and metal accumulation rates in Holocene and mid-Cretaceous sediments: palaeoceanographic significance. *In*: BROOKS, J. & FLEET, A. J. (eds) *Marine Petroleum Source Rocks*. Geological Society, London, Special Publications, **26**, 345–369.

CALVERT, S. E. 1987. Oceanographic controls on the accumulation of organic matter in marine sediments. *In*: BROOKS, J. & FLEET, A. J. (eds) *Marine Petroleum Source Rocks*. Geological Society, London, Special Publications, **26**, 137–151.

—— & PEDERSEN, T. F. 1993. Geochemistry of recent oxic and anoxic marine sediments: implications for the geological record. *Marine Geology*, **113**, 67–88.

CANDE, S. C., LABREQUE, J. L. & HAXBY, W. B. 1988. Plate kinematics of the South Atlantic: Chron 34 to present. *Journal of Geophysical Research*, **93**, 13479–13492.

CHRISTIE-BLICK, N. & BIDDLE, K. T. 1985. Deformation and basin formation along strike-slip faults. *In*: BIDDLE, K. T. & CHRISTIE-BLICK, N. (eds) *Strike-Slip Deformation, Basin Formation, and Sedimentation*. Society of Economic Paleontologists and Mineralogists, Special Publications, **37**, 1–34.

CLIFT, P. D., CARTER, A. & HURFORD, A. J. 1998. Apatite fission track analysis of Sites 959 and 960 on the Transform continental margin of Ghana, West Africa. *In*: MASCLE, J., LOHMAN, G. P. & MOULLADE, M. (eds) *Proceedings of the Ocean Drilling Program, Scientific Results*. College Station, TX (Ocean Drilling Program), **159**, 35–42.

——, LORENZO, J., CARTER, A., HURFORD, A. J. & ODP LEG 159 SCIENTIFIC PARTY 1997. Transform tectonics and thermal rejuvenation on the Côte d'Ivoire–Ghana margin, west Africa. *Journal of the Geological Society of London*, **154**, 483–489.

CURTIS, C. 1987. Mineralogical consequences of organic matter degradation in sediments: inorganic/organic diagenesis. *In*: LEGGET, J. K. & ZUFFA, G. G. (eds) *Marine Clastic Sedimentology*. Graham & Trotman, 108–123.

DE GRACIANSKY, P. C., BROSSE, E., DEROO, G. ET AL. 1982. Les formations d'âge Crétacé de l'Atlantique Nord et leur matière organique: paléogéographie et milieux de dépôt. *Revue de l'Institut Français de Pétrole*, **37**, 275–336.

——, ——, ——, ET AL. 1987. Organic-rich sediments and palaeoenvironmental reconstructions of the Cretaceous North Atlantic. *In*: BROOKS, J. & FLEET, A. J. (eds) *Marine Petroleum Source Rocks*. Geological Society, London, Special Publications, **26**, 317–344.

DEAN, W. E., ARTHUR, M. A. & STOW, D. A. V. 1984. Origin and geochemistry of Cretaceous deep-sea black shales and multicolored claystones, with emphasis on Deep Sea Drilling Project Site 530, southern Angola Basin. *In*: HAY, W. W. & SIBUET, J. C. (eds) *Initial Reports of the Deep Sea Drilling Project*. US Government Printing Office (Washington, DC), **75**, 819–844.

DEGENS, E. T., EMEIS, K. C., MYCKE, B. & WIESNER, M. G. 1986. Turbidites, the principal mechanism yielding black shales in the early deep Atlantic Ocean. *In*: SUMMERHAYES, C. P. & SHACKLETON, N. J. (eds) *North Atlantic Palaeoceanography*. Geological Society, London, Special Publications, **21**, 361–376.

DEMAISON, G. J. & MOORE, G. T. 1980. Anoxic environments and oil source bed genesis. *American Association of Petroleum Geologists Bulletin*, **64**, 1179–1209.

EINSELE, G. 1992. *Sedimentary Basins: Evolution, Facies and Sediment Budget*. Springer.

EL ALBANI, A., KUHNT, W., LUDERER, F., HERBIN, J. P. & CARON, M. 1999. Palaeoenvironmental evolution of the Late Cretaceous sequence in the Tarfaya Basin (southwest of Morocco). *This volume*.

EMEIS, K. C. 1987. Cretaceous black shales of the south Atlantic Ocean: The role and origin of recycled organic matter. *Mitteilungen des Geologisch-Paläontologischen Instituts der Universität Hamburg, SCOPE/UNEP Sonderband*, **62**, 209–232.

ERBACHER, J., THUROW, J. & LITTKE, R. 1996. Evolution of radiolaria and organic matter variations: A new approach to identify sea-level changes in mid-Cretaceous pelagic environments. *Geology*, **24**, 499–502.

ESPITALIÉ, J., DEROO, G. & MARQUIS, P. 1985–1986. La pyrolyse Rock-Eval et ses applications. *Révue de l'Institut Français de Pétrole*, **40**, 563–581, 755–784, **41**, 73–89.

GRADSTEIN, F. M., AGTERBERG, F. P., OGG, J. G. HARDENBOL, J., VAN VEEN, P., THIERRY, J. & HUANG, Z. 1995. A Triassic, Jurassic and Cretaceous time scale. *In*: BERGGREN, W. A., KENT, D. V., AUBRY, M. P. & HARDENBOL, J. (eds) *Geochronology, Time Scales and Global Stratigraphic Correlation*. Society of Economic Paleontologists and Mineralogists, Special Publication, **54**, 95–126.

HAY, W. W. 1995*a*. Cretaceous paleoceanography. *Geologica Carpathica*, **46**, 257–266.

—— 1995*b*. Paleoceanography of marine organic-carbon-rich sediments. *In*: HUC, A. Y. (ed.) *Paleogeography, Paleoclimate, and Source Rocks*. AAPG Studies in Geology, **40**, 21–59.

——, DECONTO, R. & WOLD, C. N. 1997. Climate: Is the past the key to the future? *Geologische Rundschau*, **86**, 471–491.

——, ——, —— ET AL. 1998. An alternative global Cretaceous paleogeography. *In*: BARRERA, E. & JOHNSON, C. (eds) *The Evolution of Cretaceous Ocean/Climate Systems*. Geological Society of America, Special Publication, in press.

HERBIN, J. P., MULLER, C., DE GRACIANSKY, P. C., JACQUIN, T., MAGNIEZ-JANNIN, F. & UNTERNEHR, P. 1987. Cretaceous anoxic events in the South Atlantic. *Revista Brasileira de Geociências*, **17**, 92–99.

HOLBOURN, A. E. L. & MOULLADE, M. 1998. Lower Cretaceous benthic foraminiferal assemblages form ODP Leg 159 (Côte d'Ivoire–Ghana Transform Margin, Equatorial Atlantic): Biostratigraphic, paleoenvironmental and paleobiogeographic significance. *In*: MASCLE, J., LOHMAN, G. P. & MOULLADE, M. (eds) *Proceedings of the Ocean Drilling Program, Scientific Results*. College Station, TX (Ocean Drilling Program), **159**, 347–362.

HOLBOURN, A., KUHNT, W., EL ALBANI, A., PLETSCH, T., LUDERER, F. & WAGNER, T. 1999. Upper Cretaceous palaeoenvironments and benthonic foraminiferal assemblages of potential source rocks from the western African margin, Central Atlantic. *This volume*.

HOLMES, M. A. 1998. Thermal diagenesis of Cretaceous sediment recovered during ODP Leg 159 to the Côte d'Ivoire–Ghana margin. *In*: MASCLE, J., LOHMAN, G. P. & MOULLADE, M (eds) *Proceedings of the Ocean Drilling Program, Scientific Results*. **159**. College Station, TX (Ocean Drilling Program), **159**, 53–71.

JANIK, A. G., HOOD, J. A. & ASK, M. V. 1998. Physical properties data at Hole 959D of Leg 159 – Comparison of core and log measurements and a proposed revision of lithological units. *In*: MASCLE, J., LOHMAN, G. P. & MOULLADE, M. (eds) *Proceedings of the Ocean Drilling Program, Scientific Results*. College Station, TX (Ocean Drilling Program), **159**.

KLITGORD, K. D. & SCHOUTEN, H. 1986. Plate kinematics of the Central Atlantic. *In*: TUCHOLKE, B. E. & VOGT, P. R. (eds) *Geology of North America: The Western Atlantic Region*. Geological Society of America, DNAG Series, 351–378.

KOUTSOUKOS, E. A. M., MELLO, M. R. & DE AZAMBUJA FILHO, N. C. 1991. Micropalaeontological and geochemical evidence of mid–Cretaceous dysoxic–anoxic palaeoenvironments in the Sergipe Basin, northeastern Brazil. *In*: TYSON, R. V. & PEARSON, T. H. (eds) *Modern and Ancient Continental Shelf Anoxia*. Geological Society, London, Special Publications, **58**, 427–447.

KUHNT, W. & WIEDMANN, J. 1995. Cenomanian–Turonian source rocks: paleobiogeographic and paleoenvironmental aspects. *In*: HUC, A. Y. (ed.) *Paleogeography, Paleoclimate, and Source Rocks*. AAPG Studies in Geology, **40**, 213–231.

——, MOULLADE, M. & KAMINSKI, M. A. 1998. Upper Cretaceous, K/T boundary, and lowermost Paleocene agglutinated foraminifera from Hole 959D (ODP Leg 159, Côte d'Ivoire–Ghana Transform Margin). *In*: MASCLE, J., LOHMAN, G. P. & MOULLADE, M. (eds) *Proceedings of the Ocean Drilling Program, Scientific Results*. College Station, TX (Ocean Drilling Program), **159**, 389–412.

——, HERBIN, J. P., THUROW, J. & WIEDMANN, J. 1990. Distribution of Cenomanian–Turonian organic facies in the Western Mediterranean and along the adjacent Atlantic margin. *In*: HUC, A. Y. (ed.) *Deposition of Organic Facies*. AAPG Studies in Geology, **30**, 133–160.

LESPINASSE, M., LEROY, J. L., PIRONON, J. & MOIRON, M. C. 1998. The fluids from the marginal ridge of the Côte d'Ivoire–Ghana transform margin (Leg ODP 159, Hole 960A) as thermal indicators. *In*: MASCLE, J., LOHMAN, G. P. & MOULLADE, M. (eds) *Proceedings of the Ocean Drilling Program, Scientific Results*. College Station, TX (Ocean Drilling Program), **159**, 49–52.

LY, A. & KUHNT, W. 1994. Late Cretaceous benthic foraminiferal assemblages of the Casamance shelf (Senegal, NW Africa). Indication of a Late Cretaceous oxygen minimum zone. *Revue de Micropaléontologie*, **37**, 49–74.

MANN, U., HANTSCHEL, T., SCHAEFER, R. G., KROSS, B., LEYTHAEUSER, D., LITTKE, R. & SACHSENHOFER, R. F. 1997. Petroleum migration: mechanisms, pathways, efficiencies and numerical simulations. *In*: WELTE, D. H., HORSFIELD, B. & BAKER, D.R. (eds) *Petroleum and Basin Evolution*. Springer, 403–520.

MARCANO, M. C., LOHMAN, K. C. & PICKETT, E. A. 1998. Uplift and exposure history of the Côte d'Ivoire–Ghana Transform Margin: geochemistry of pore-filling and fracture vein calcites. *In*: MASCLE, J., LOHMAN, G. P. & MOULLADE, M. (eds) *Proceedings of the Ocean Drilling Program, Scientific Results*. College Station, TX (Ocean Drilling Program), **159**, 71–80.

MASCLE, J. & BLAREZ, E. 1986. Evidence for transform margin evolution for the Côte d'Ivoire–Ghana continental margin. *Nature*, **326**, 378–381.

——, —— & MARINHO, M. 1988. The shallow structure of the Guinea and Ivory Coast–Ghana transform margins: their bearing on the Equatorial Atlantic evolution. *Tectonophysics*, **155**, 193–209.

——, LOHMAN, G. P., CLIFT, P. D. ET AL. 1996. *Proceedings of the Ocean Drilling Program, Initial Reports*. College Station, TX (Ocean Drilling Program), **159**.

——, ——, —— & ODP 159 SCIENTIFIC PARTY. 1997. Development of a passive transform margin: Côte d'Ivoire–Ghana transform margin – ODP Leg 159 preliminary results. *Geo-Marine Letters*, **17**, 4–11.

MELLO, M. R., KOUTSOUKOS, E. A. M., HART, M. B., BRASSELL, S. C. & MAXWELL, J. R. 1989. Late Cretaceous anoxic events in the Brazilian continental margin. *Organic Geochemistry*, **14**, 529–542.

——, TELNAES, N. & MAXWELL, J. R. 1995. The hydrocarbon source potential in the Brazilian Marginal basins: a geochemical and paleoenvironmental assessment. *In*: HUC, A. Y. (ed.) *Paleogeography, Paleoclimate, and Source Rocks*. AAPG Studies in Geology, **40**, 233–272.

MITCHELL, A. H. G. & READING, H. G. 1986. Sedimentation and tectonics. *In*: READING, H. G. (ed.) *Sedimentary Environments and Facies*. Blackwell, 471–519.

MOULLADE, M., MASCLE, J., BENKHELIL, J., COUSIN, M. & TRICART, P. 1993. Occurrence of marine mid-Cretaceous sediments along the Guinean slope (Equamarge II cruise): their significance for the evolution of the central Atlantic African margin. *Marine Geology*, **110**, 63–72.

——, WATKINS, D. K., OBOH-IKUENOBE, F. E. ET AL. 1998. ODP Leg 159, Equatorial Atlantic: Mesozoic biostratigraphic, paleoenvironmental and paleobiogeographic synthesis. *In*: MASCLE, J., LOHMAN, G. P. & MOULLADE, M. (eds) *Proceedings of the Ocean Drilling Program, Scientific Results*. College Station, TX (Ocean Drilling Program), **159**, 481–490.

NÜRNBERG, D. & MÜLLER, R. D. 1991. The tectonic evolution of the South Atlantic ocean from Late Jurassic to present. *Tectonophysics*, **191**, 27–53.

OBOH-IKUENOBE, F. E., YEPES, O. & ODP LEG 159 SCIENTIFIC PARTY 1997. Palynofacies of sediments from the Côte d'Ivoire–Ghana transform margin: Preliminary correlation with some regional events in the Equatorial Atlantic. *Palaeogeography, Palaeoclimatology, Palaeoecology*, **129**, 291–314.

——, —— & GREGG, J. M. 1998. Palynostratigraphy, palynofacies and thermal maturation of Cretaceous–Paleocene sediments from the Côte d'Ivoire–Ghana Transform Margin. *In*: MASCLE, J., LOHMAN, G. P. & MOULLADE, M. (eds) *Proceedings of the Ocean Drilling Program, Scientific Results*. College Station, TX (Ocean Drilling Program), **159**, 277–318.

PARRISH, J. T. 1995. Paleogeography of Corg-rich rocks and the preservation versus production controversy. *In*: HUC, A. Y. (ed.) *Paleogeography, Paleoclimate, and Source Rocks*. AAPG **40**, 1–20.

PICKETT, E. A. & ALLERTON, S. 1998. Structural observations from the Côte d'Ivoire–Ghana Transform Margin. *In*: MASCLE, J., LOHMAN, G. P. & MOULLADE, M. (eds) *Proceedings of the Ocean Drilling Program, Scientific Results*. College Station, TX, (Ocean Drilling Program), **159**, 3–12.

PLETSCH, T. 1998. Origin of Lower Eocene palygorskite claystones on the Côte d'Ivoire–Ghana Transform Margin (Leg 159, eastern Equatorial Atlantic). *In*: MASCLE, J., LOHMAN, G. P. & MOULLADE, M. (eds) *Proceedings of the Ocean Drilling Program, Scientific Results*. College Station, TX (Ocean Drilling Program), **159**, 141–156.

——, WAGNER, T., ERBACHER, J. ET AL. submitted. Cretaceous depositional environments along the opening Equatorial Atlantic Gateway (Côte d'Ivoire–Ghana Transform Margin, ODP Leg 159). *Palaeogeography, Palaeoclimatology, Palaeoecology*.

POLLASTRO, R. 1993. Considerations and applications of the illite/smectite geothermometer in hydrocarbon-bearing rocks of Miocene to Mississippian age. *Clays and Clay Minerals*, **41**, 119–133.

RABINOWITZ, P. D. & LABREQUE, J. 1979. The Mesozoic South Atlantic and evolution of its continental margins. *Journal of Geophysical Research*, **85**, 5973–6002.

REICHERTER, K., PLETSCH, T., KUHNT, W., MANTHEY, J., HOMEIER, G., WIEDMANN, J. & THUROW, J. 1994. Mid-Cretaceous paleogeography and paleoceanography of the Betic Seaway (Betic Cordillera, Spain), *Palaeogeography, Palaeoclimatology, Palaeoecology*, **107**, 1–33.

REYNOLDS, R. C. 1980. Interstratified clay minerals. *In*: BRINDLEY, G. W. & BROWN, G. (eds) *Crystal Structures of Clay Minerals and their X-Ray Identification*. Mineralogical Society Monograph, London, **5**, 249–303.

SAINT-MARC, P. & N'DA, V. 1997. Biostratigraphie et paléoenvironnements des dépôts crétacés au large d'Abidjan (Golfe de Guinée). *Cretaceous Research*, **18**, 545–565.

SCHLANGER, S. O., ARTHUR, M. A., JENKYNS, H. C. & SCHOLLE, P. A. 1987. The Cenomanian–Turonian Oceanic Anoxic Event. I. Stratigraphy and distribution of organic carbon-rich beds and the marines $\delta^{13}C$ excursion. *In*: BROOKS, J. & FLEET, A. J. (eds) *Marine Petroleum Source Rocks*. Geological Society, London, Special Publications, **26**, 371–399.

SCHWENTKE, W. & KUHNT, W. 1992. Subsidence history and continental margin evolution of the Western Pyrenean and Basque Basins. *Palaeogeography, Palaeoclimatology, Palaeoecology*, **95**, 297–318.

SMART, G. & CLAYTON, T. 1985. The progressive illitization of interstratified illite–smectite from the Carboniferous sediments from northern England and its relationship to organic maturity indicators. *Clay Minerals*, **20**, 455–466.

STRAND, K. 1998. Sedimentary facies and sediment composition in response to tectonics of the Côte d'Ivoire–Ghana Transform Margin. *In*: MASCLE, J., LOHMAN, G. P. & MOULLADE, M. (eds) *Proceedings of the Ocean Drilling Program, Scientific Results*. College Station, TX (Ocean Drilling Program), **159**, 113–123.

SUMMERHAYES, C. P. 1987. Organic-rich Cretaceous sediments from the North Atlantic. *In*: BROOKS, J. & FLEET, A. J. (eds) *Marine Petroleum Source Rocks*. Geological Society, London, Special Publications, **26**, 301–316.

THIEDE, J. & VAN ANDEL, T. H. 1977. The deposition of anaerobic sediments in the Late Mesozoic Atlantic Ocean. *Earth and Planetary Science Letters*, **33**, 301–309.

TYSON, R. V. 1987. The genesis and palynofacies characteristics of marine petroleum source rocks. *In*: BROOKS, J. & FLEET, A. J. (eds) *Marine Petroleum Source Rocks*. Geological Society, London, Special Publications, **26**, 47–67.

WATKINS, D. K., SHAFIK, S. & SHIN, I. C. 1998. Calcareous nannofossils from the Cretaceous of Leg 159. *In*: MASCLE, J., LOHMAN, G. P. & MOULLADE, M. (eds) *Proceedings of the Ocean Drilling Program, Scientific Results*. College Station, TX (Ocean Drilling Program), **159**, 319–334.

ZALAN, P. V., NELSON, E. P., WARME, J. E. & DAVIS, T. L. 1985. The Piaui Basin: rifting and wrenching in an Equatorial Atlantic transform basin. *In*: BIDDLE, K. T. & CHRISTIE-BLICK, N. (eds) *Strike-Slip Deformation, Basin Formation, and Sedimentation*. Society of Economic Paleontologists and Mineralogists, Special Publication, **37**, 177–192.

ZIMMERMANN, H. B., BOERSMA, A. & MCCOY, F. W. 1987. Carbonaceous sediments and palaeoenvironment of the Cretaceous South Atlantic Ocean. *In*: BROOKS, J. & FLEET, A. J. (eds) *Marine Petroleum Source Rocks*. Geological Society, London, Special Publications, **26**, 271–286.

Miocene benthonic foraminiferal morphogroups in an oxygen minimum zone, offshore Cabinda

R. C. PREECE[1], M. A. KAMINSKI[1] & T. W. DIGNES[2]

[1]*Postgraduate Unit of Micropalaeontology, Department of Geological Sciences, University College London, Gower Street, London WC1E 6BT, UK*
[2]*Chevron Overseas Petroleum Inc., 6001 Bollinger Canyon Road, San Ramon, CA 94583-2324, USA*

Abstract: Two hundred and seventy-eight benthonic foraminiferal species were recorded (126 agglutinating and 152 calcareous) from CABGOC 128-3, offshore Angola. This assemblage is comprised of a cosmopolitan fauna named the Agua Salada Fauna, known from the Caribbean, in the Vienna Basin, West Africa, Libya and Borneo. The fluctuating relative abundance of benthonic foraminiferal morphogroups in combination with 'total organic carbon' (TOC) data is suggested to reflect variation in the intensity of the oxygen minimum zone as it steadily increases upwards. An infaunal agglutinated 'opportunistic' (*r*-selected) group moves from antiphase into phase with TOC upwards in CABGOC 128-3. A morphologically comparative calcareous group displays the opposite trend, moving from phase into antiphase with TOC data. Planispiral agglutinated foraminifera emulate the steady increase in TOC upwards in CABGOC 128-3. This group includes *Cyclammina cyclops*, previously only recorded in the Arctic. Elongate, flattened calcareous morphologies undergo a sudden expansion towards the top of the well and are apparently unaffected by TOC. This overall trend is supported by concurrent poor preservation amongst distinct calcareous groups susceptible to dissolution, as expected in an area of dysaerobic conditions. The disappearance of all agglutinated faunas in the late Miocene is considered to be a reflection of oceanographic change.

Miocene sediments of equatorial upwelling regions contain a distinctive assemblage of opportunistic (*r*-selected) agglutinated foraminfera termed the Agua Salada Fauna. First identified from the Oligocene–Miocene Agua Salada Formation of Venezuela, the Agua Salada Fauna is dominated by agglutinated foraminiferal morphogroups considered to be typical of modern dysaerobic environments (Kaminiski *et al.* 1995; Kaminski & Kuhnt 1995), and identified by the presence of *Valvulina flexilis*. The Agua Salada Fauna, first named by Renz (1948), has received little attention over the last 50 years. Originally examined by Cushman & Renz (1941), and mentioned by Akers (1954), the Agua Salada Fauna was then re-evaluated by Blow some years later (Blow 1959). There has been some confusion owing to changes of lithological nomenclature in the type area of the Agua Salada Fauna (Renz 1959). However, since this early work the fauna has only received fleeting reference (Basou 1976; Diaz de Gamero 1997, 1985*a*, *b*) until the present day (Giffuni 1991; Moreno-Vaquez 1995; Kaminski *et al.* 1998).

The fauna has been identified from diverse sites. Kaminski *et al.* (in press) identified the fauna from the Miocene of the Gulf of Mexico and compared it favourably with types from the Smithsonian Institution, Libya, and the Sabah Basin. Basou (1976) identified elements of the Agua Salada Fauna from the Guinea Bay, offshore Sekonda Takoradi, Ghana, while Akers (1954), in his paper relating the ecological aspects and stratigraphic significance of *Cyclammina cancellata*, refers to the Agua Salada Fauna of Venezuela and *C. cancellata's* bearing within it. Moy (1968) recovered elements of the fauna from Oligo-Miocene sediments of Libya and Hungary. Similarly, Ujiie (1970, 1979) identified elements of the fauna from the Miocene Sandakan Formation, North Borneo.

Geological overview

Cabinda is a small detached enclave of Angola that produces most of that country's oil. The Cabinda Basin, north of Angola, is a post-rift Tertiary sequence represented exclusively by Neogene sediments. To the south (Zaire–Kwanza Basin), there is a complete Tertiary sequence grading down to Palaeocene sediments (Brun *et al.* 1984). The area's mixed carbonate–clastic system renders log analysis problematic and hinders understanding in the area (Brown 1993). Typical fields are fault-bounded rafts with one or more rotated fault blocks, separated through gravity sliding on the underlying salt decollement (McHargue 1990; Spaw & Koehler 1981).

The CABGOC 128-3 well lies offshore Angola in *c.* 100 meters water depth at 5°48′S, 11°47′E (Fig. 1). The well recovered uppermost

PREECE, R. C., KAMINSKI, M. A. & DIGNES, T. W. 1999. Miocene benthonic foraminiferal morphogroups in an oxygen minimum zone, offshore Cabinda. *In:* CAMERON, N. R., BATE, R. H. & CLURE, V. S. (eds) *The Oil and Gas Habitats of the South Atlantic.* Geological Society, London, Special Publications, **153**, 267–282.

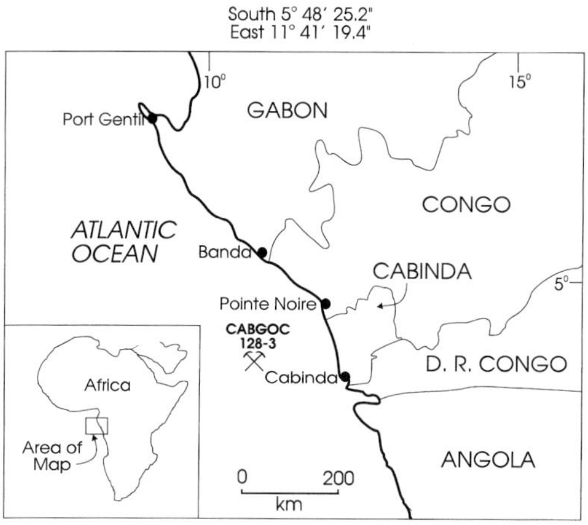

Fig. 1. Location map of CABGOC 128-3.

Oligocene–mid-Miocene lower slope laminated, fissile, silty shales containing the Agua Salada Fauna. The well is largely barren of planktonic foraminifera; similarly, calcareous nannofossil recovery is poor leading to extremely ambiguous age constraints.

The sedimentary basins of West Africa are bordered by the St Paul Fracture Zone to the north and the Walvis Ridge to the south. The African continent separated from South America in the late Jurassic–early Cretaceous, resulting in the production of a series of marginal basins: the Côte d'Ivoire–Ghana Basin; the Nigeria–Cameroon Basin; and the Gabon–Congo–Angola basins along the western African margin.

The geological history of these basins is intimately linked with the first period of rifting (Vale 1960). Alluvial and lacustrine sediments were deposited in grabens at the initial rift. The various basins then record slightly differing environments throughout the Mesozoic (Brun *et al.* 1984). The Gabon–Congo–Angola basins are characterized by Aptian lacustrine deposits corresponding to the initial period of rifting. This is succeeded by a transgressive sequence of carbonates of Albian–Turonian age, followed by a Cenomanian–Miocene clastic series.

Orogenic movements initiated in the Albian affected the Tertiary series, principally the Miocene of the Gabon and Angola basins. The movements resulted in local unconformities, isolated to individual subsiding basins accumulating Tertiary sediments. A following period of drift encouraged the development of a carbonate bank, this fragmented and eventually began to 'slide' on a Cretaceous salt decollement. The Albian–Turonian sequence of carbonates is succeeded by a Cenomanian–Miocene clastic series. The Eocene is a regressive phase recognized in all the basins. The upper Oligocene is only identified locally (Brice *et al.* 1982).

Oceanographic overview

It is only during the past decade that physical oceanographical investigations performed over the continental margins on eastern oceanic boundaries have improved the understanding of coastal upwelling (Smith 1992). Despite the difference in their size, the tropical regions of the Atlantic and Pacific are primarily influenced by similar dominant dynamics – remote wind stress. In both oceans, equatorial undercurrents impinge their eastern boundaries and act in the formation, in the case of the Atlantic, of the Guinea and Angola Domes (Picaut 1985). However, published temporal and spatial data remain sparse offshore the Cabinda area.

The dominant oceanographic system in the South Atlantic is the South Atlantic subtropical gyre. On its eastern side it introduces northward flowing, cold, upwelled water to within 100 nautical miles of the west coast of South Africa and Namibia (Siesser 1980). This gyre is not the rectangular system often depicted, but an anticyclonic triangular gyre, confined to the southwestern Atlantic (Gordon & Bosley 1992; Peterson & Stramma 1991; Stramma 1991; Wacongne & Piton 1992). Its northeastern limb extends diagonally across the ocean from South Africa to Cabo San Rogue. The northeastern South Atlantic circulates as a cyclonic gyre, implying that organic carbon and opal on the Walvis Ridge and southeastern Angola Basin could be advected from productive waters to the north. The South Atlantic surface circulation pattern and the southern hemisphere Neogene climate were both affected by the evolution of the southern Atlantic system. This, in turn, was largely driven by variations in the Antarctic ice mass. This system has gradually strengthened since the mid-Miocene and has gradually migrated northward from the Cape Basin in the south to the Walvis Ridge and Angola Dome to the north, in a series of small-scale cyclic north–south oscillations (Diester-Hass *et al.* 1990, 1992; Hay & Brock 1992). These oscillations were the product of varying Antarctic continental ice volumes.

The late Miocene was a time of major oceanographic reorganization marked by two important events. Firstly, a carbon isotope shift marks the onset of Messinian desiccation; secondly, Mediterranean isolation followed. Early in the Miocene, as the Arabian plate impinged on the Eurasian plate, the Mediterranean became cut-off from the Tethys and Pacific Oceans. This forced the Mediterranean into a landlocked state that persisted until the late Miocene, when a connection was re-established between the Red Sea and the Indian Ocean in the south. This triggered the Messinian Salinity Crisis and the drying out of the Mediterranean, culminating in the deposition of massive evaporites.

The late Miocene is characterized by a sharp cooling, resulting in the strengthening of the thermocline and intensification of equatorial upwelling (Berger *et al.* 1981). There is evidence to suggest that the shift is intimately linked to changes in deep and bottom water production, and, in turn, a product of the isolation of the Mediterranean where it is a source of deepwaters. Simultaneously, the North Atlantic was becoming a source of deepwater, forcing a fall of the CCD and facies changes.

The Messinian Salinity Crisis, as a result of the isolation of the Mediterranean Basin and the formation of massive evaporite deposits, caused a substantial decrease in the salt content of the ocean and, in turn, facilitated ice build-up in the Antarctic. The pulsing discharge of heavy brine waters from the Mediterranean to the world's oceans led to oceanic stratification and a brief period of extreme carbonate dissolution in the latest Miocene.

Carbonate dissolution far above the CCD is a common feature of the south Atlantic system and is considered to be related to sea-level fluctuations and Antarctic ice extent (Diester-Hass 1988). Varying organic matter input, as a result of sea-level fluctuations, in turn a product of glacio-eustatic variation (the only major mechanism capable of forcing rapid worldwide sea-level fluctuations), effects carbonate dissolution.

Foraminiferal overview

Micropalaeontological investigations of the South Atlantic, and West Africa in particular, are reviewed by Jones (1996), while Reyment (1966) and refs therein) provides an overview of earlier ideas. Hanse (1965) reviewed the microfauna of Angola, the benthonic fauna listed is very similar to the calcareous component found in the Cabinda Basin. The most comprehensive micropalaeontological report on West Africa, however, is still that of Kogbe & Me'hes (1986).

The majority of the pertinent foraminiferal work on West Africa has been completed by a small group of individuals. Seiglie & Frost (1979) identified the significance of mid-Tertiary larger foraminifera common to West Africa and the Caribbean, in addition, they reported observations upon species of smaller benthonic foraminifera and their cosmopolitan nature across the widening Atlantic. Cenozoic West African Tertiary agglutinated foraminifera became the focus of attention in the 1980s (Seiglie & Baker 1983; Seiglie *et al.* 1986) and, more recently, Haman *et al.* (1993) reported a new species of *Virgulinella* from the Miocene of Cabinda. Finally, Cretaceous foraminiferal zonation of the Cabinda area is comparatively well studied with reference to sea level (Seiglie & Baker 1982), and facies and morphogroups (Dignes 1995, 1996).

Materials and methods

One hundred and seventy-eight continuous 60 ft composite ditch cutting samples spanning the Neogene were processed for foraminifera, and picked and counted for benthonic species. Preservation was often poor, many species being heavily resilicified and internally filled with diagenetic pyrite. Following Jones & Charnock (1985) the agglutinated data were divided into seven morphogroups: A = unilocular/multilocular, tubular of branching; B^1 = unilocular, globular; B^2 = unilocular, multilocular, flattened; B^3 = multilocular, planispiral/trochospiral (most lituolids); B^4 = unilocular/multilocular, irregular, conical; C^1 = multilocular, elongate mixed growth; C^2 = multilocular, elongate, quinqueloculine/milioline (C^1 and C^2 were combined for the purposes of this study); and D = multilocular, trochospiral, conical. Following Bernhard (1986), the calcareous forms were divided into seven morphogroups, where: 1 = elongate, flattened, ovate to flattened in apertural view, parallel to subparallel sides, biserial, lanceolate and palmate; 2 = tapering or flaring, circular in apertural view, tapering sides, triserial and multiserial; 3 = cylindrical group, circular in apertural view, parallel sides, uniserial; 4 = flattened, planispiral, discoidal, round peripheries; 5 = biumblicate, planispiral, flattened; 6 = lenticular, biconvex, angled periphery, planispiral and coiled biserial; 7 = planoconvex, spiroconvex, umbiliconvex and trochospiral. CABGOC 128-3 contained no morphologies appropriate to Bernhard group 8. Of these groups, only agglutinated groups B^4, C^1 and C^2, and calcareous groups 1 and 2 proved to be significant. For the purposes of this study, groups C^1 and C^2 are presented together as one group.

Opportunistic foraminiferal species are considered, in this study, to be members of agglutinated morphogroup C. Kaminski (1988*a*) and Kaminski *et al.* (1988, 1995) identified modern members of this 'deepwater agglutinating foraminifera' group in the Panama Basin as infaunal opportunists (*r*-selected), capable of recolonization and tolerant of low O_2. They were characterized as small, tapered and elongate morphotypes such as *Reophax* and '*Textulara*'. The full foraminifera dataset will be published elsewhere (Preece in prep.).

'Total organic carbon' (TOC) data were measured by Chevron Overseas Petroleum Inc., by high-temperature combustion using a Leco instrument.

Results

Analysis of 178m 60 ft composite ditch cutting samples from CABGOC 128-3 revealed a diverse assemblage of agglutinated and calcareous benthonic foraminifera known collectively as the Agua Salada Fauna. This included the distinctive variant species of *C. cylops*, previously only described from the Arctic (McNeil 1988). Planktonic foraminiferal recovery was extremely poor, similarly calcareous nannofossils were rare and age constraints at CABGOC 128-3 remain ambiguous.

Two hundred and seventy eight benthonic foraminiferal species were recorded within mud chippings spanning a depth of 570–10 160 ft. The agglutinating component comprises 126 species, including various members of the genera *Cyclammina*, *Haplophragmoides*, *Recurvoides*, *Reticulophragmium*, *Eggerella*, *Karreriella*, *Reophax*, *Textularia* and *Valvulina*. *Cyclammina* species include a variation of *C. cyclops* previously documented as an Arctic species (McNeil 1988).

Through applying the agglutinating morphogroups of Jones & Charnock (1985) and the calcareous groups of Bernhard (1986), a distinctive cyclicity emerges. Specific calcareous genera appear to reflect this in coincident calcite dissolution. An antiphase relationship between opportunistic infaunal species and TOC values is recorded.

The uppermost 26 samples were disregarded, as they contained low total abundance counts in comparison to the majority of samples in CABGOC 128–3 (Fig. 2). Diversity varies greatly, decreasing markedly toward the top of the well (Fig. 3). Downhole, the ratio of agglutinated taxa:calcareous taxa fluctuates greatly (Fig. 4). Three cycles of increasing dominance by agglutinated taxa are recognized in the well. TOC data also displays three maxima; the TOC profile is displayed on all of the morphogroup graphs.

This pattern of a peaked maxima is mirrored to the greatest extent by agglutinated morphogroup C, the sediment-dwelling infaunals (Figs 5 and 10). The relative frequency of multilocular elongate genera (including *Alveolvalvulina*, *Ammobaculites*, *Bulbobaculites*, *Eggerella*, *Gravellina*, *Jarvisella*, *Karreriella*, *Martinottiella*, *Reophax*, *Subreophax*, *Textularia* and *Valvulina*) exhibit three repetitive cycles. This group, by far the most abundant overall in CABGOC 128-3, records a low relative abundance at the base of the well but becomes dominant in the three peaks

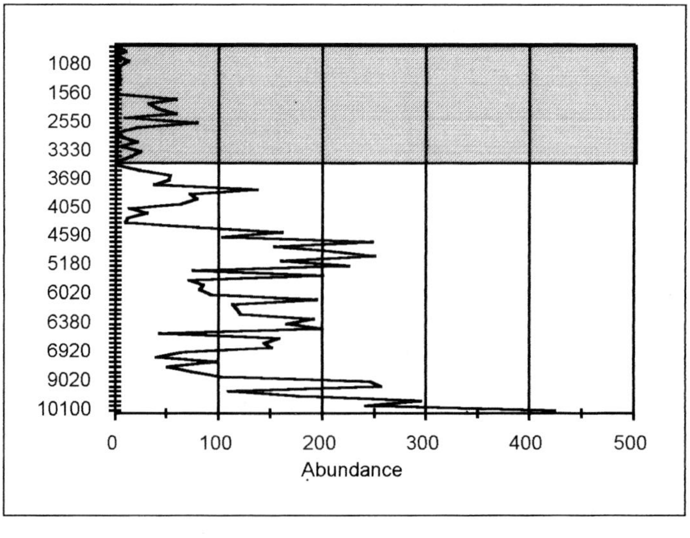

Fig. 2. Foraminiferal abundance curve at CABGOC 128-3 (the shaded area indicates samples of low total abundance which are not considered further in this study).

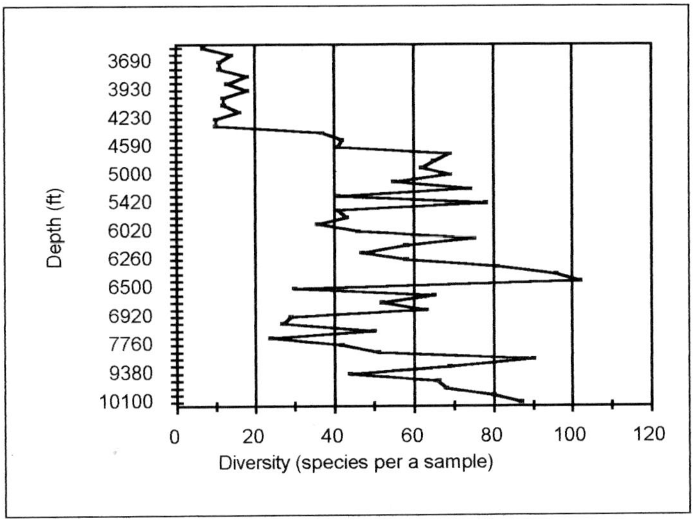

Fig. 3. Foraminiferal diversity at CABGOC 128-3 (species per sample).

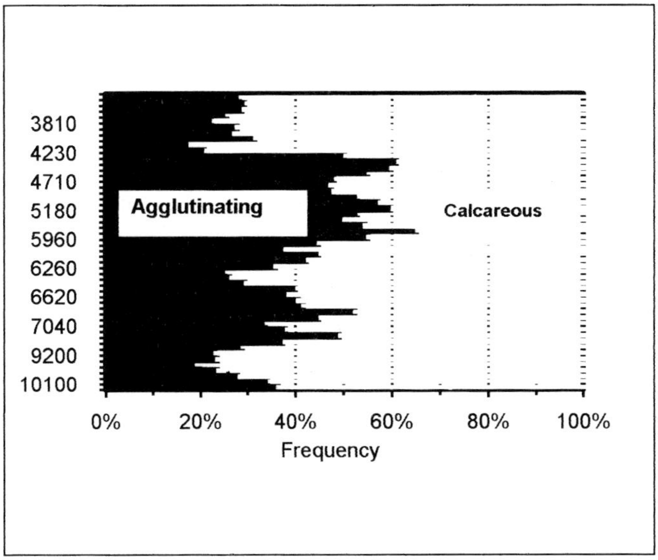

Fig. 4. Relative abundance of agglutinating to calcareous taxa at CABGOC 128-3.

increasing in dominance upwell. The cyclic nature morphogroup C is in antiphase with the TOC curve at the base of CABGOC 128-3, migrating into phase upwell.

Group B^3 (Figs 6 and 10), the multilocular planispiral group, consists of the genera *Cribrostromoides*, *Cyclammina*, *Discamminoides*, *Haplophragmoides*, *Reticulophragmium* and *Recurvoides*. This epifaunal group displays an increasing intensity 'sawtooth' pattern upwell, in phase with TOC values. This group exhibits a prominent decline at the mid Miocene–?upper Miocene transition.

Most calcareous genera show a constant frequency within CABGOC 128-3. However, morphogroup 1 (Figs 7 and 10), the elongate flattened, biserial, lanceolate and palmate forms (i.e. *Bolivina* and *Plectofrondicularia*), display a fluctuating low abundance until a sudden expansion coincident with morphogroup C's decline.

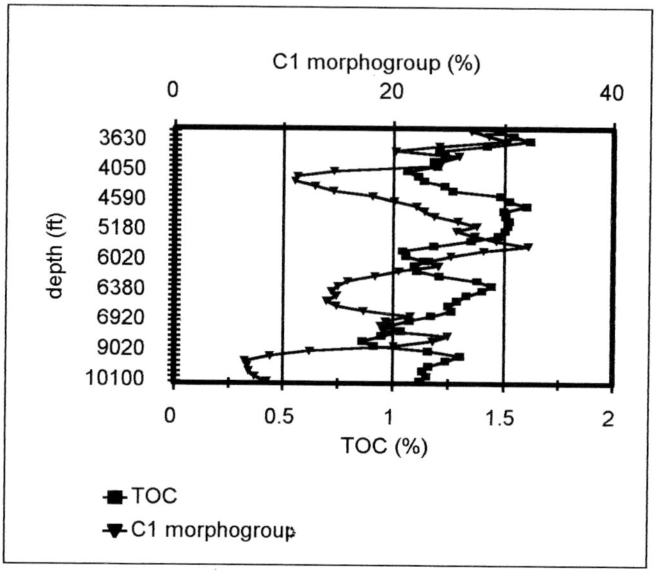

Fig. 5. Profile of morphogroup C (sediment-dwelling infaunal group) and TOC at CABGOC 128-3.

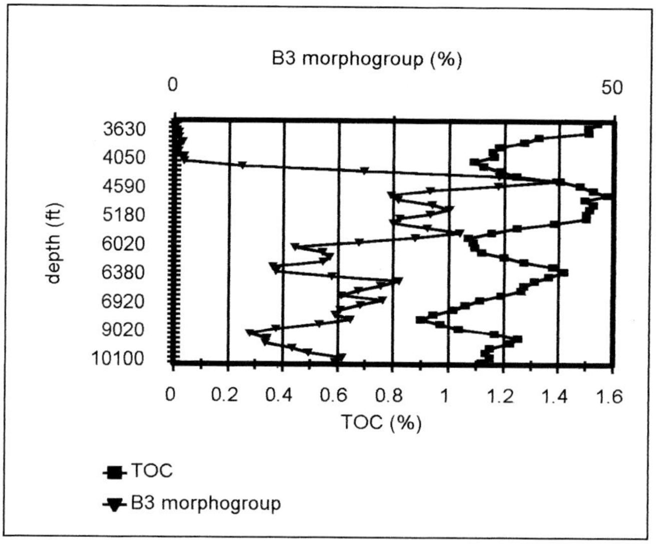

Fig. 6. Profile of morphogroup B^3 (epifaunal agglutinated group) and TOC at CABGOC 128-3.

Morphogroup 2 (Figs 8 and 10), the tapering triserial and multiserial *Bulimina* and *Uvigerina* group, exhibit a triple pattern of increasing abundance similar to that of morphogroup C. This fluctuating frequency profile is an antiphase with agglutinated morphogroup C, and hence migrates from a phase relationship with TOC values into antiphase upwell. The magnitudes of this group's profile and that of morphogroup C are comparable.

Species preserved solely as pyrite internal moulds (*Globobulimina perversa*, *G. ovata*, *Protoglobobulimina pupoides*, *Chilostomella ovidea*, *Vaginulopsis spinulosa*, *Valvulineria inequalis*, *V. palmerae*, *V. venezuela*, *Virgulina exilis* and *V. pontoni*) display a gradual fluctuating decline

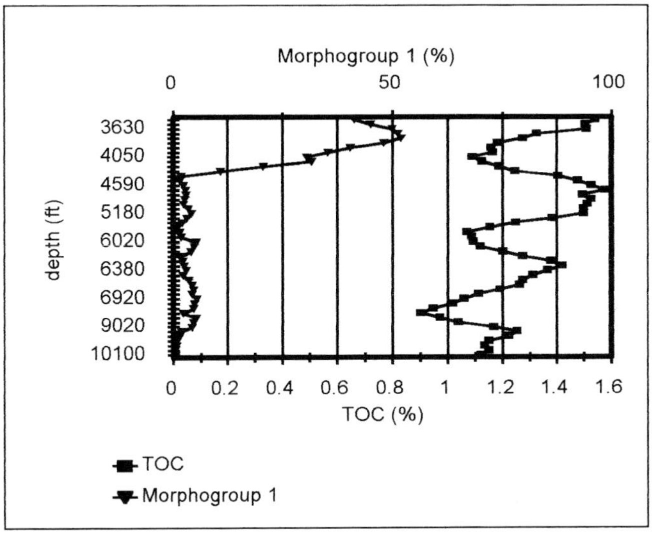

Fig. 7. Profile of morphogroup 1 (elongate flattened, biserial, lanceolate and palmate forms) and TOC at CABGOC 128-3.

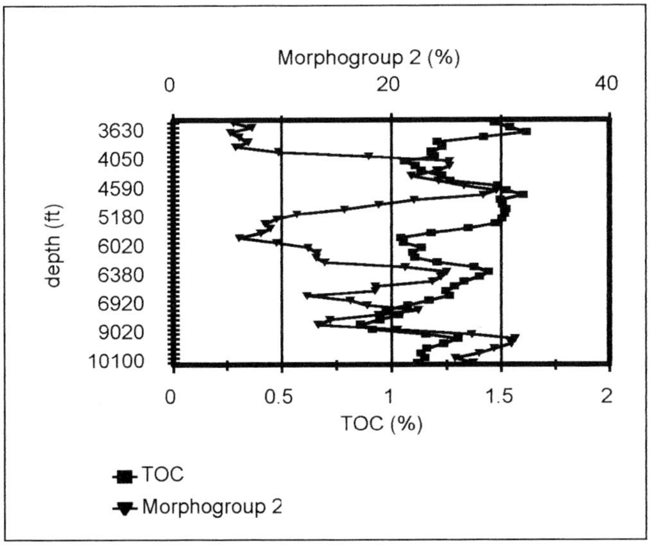

Fig. 8. Profile of morphogroup 2 (tapering triserial and multiserial group) and TOC at CABGOC 128-3.

upwards in CABGOC 128-3 (Fig. 9), until an almost complete decline at the top of the well.

Discussion

Morphogroup analysis reveals a distinctive cyclicity in CABGOC 128-3 well. This is postulated as reflecting the periodic intensification of the local upwelling system, through the introduction of nutrient-rich waters to the surface, facilitating phytoplankton blooms, hence heightening the oxygen minimum zone upon their deposition and decay (Gooday 1988). High TOC values are interpreted as being indicative of increased surface productivity at CABGOC 128-3. The introduction of nutrient-rich upwelled waters to the continental margin expedites phytoplankton pro-

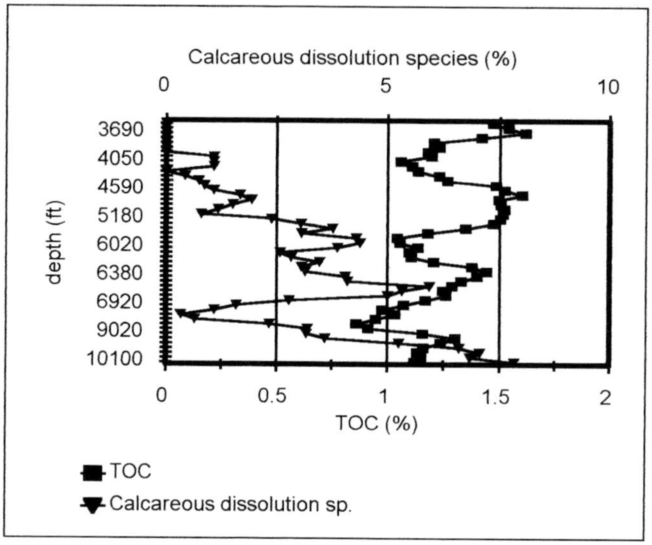

Fig. 9. Profile of genera susceptible to carbonate dissolution at CABGOC 128-3.

duction. This is essentially a function of upwelling rate and the nutrient concentration. Particulate organic matter descends through the water-column, utilizing oxygen as it is oxidized, thereby expanding the oxygen minimum zone (Demaison & Moore 1980; Southam *et al.* 1982). Hence, TOC values may reflect the intensity of the oxygen minimum zone at CABGOC 128-3. However, the response of the benthonic foraminiferal fauna to the upwelling proxy appears to shift with time upwell in CABGOC 128-3 from a food-limited environment to an oxygen-limited one (Jorissen *et al.* 1995).

Morphogroup analysis for ecological and palaeoecological interpretation is a useful tool. It is independent of species level taxonomy and is thus relatively elementary to translate from one worker to another. This approach permits comparison of assemblages of differing ages. The power of this relatively simple tool is demonstrated by the plethora of studies (e.g. Severin 1983; Katz & Thunell 1984; Jones & Charnock 1985; Bernhard 1986; Corliss & Chen 1988; Nagy 1992; Tyszka 1994). Here, the study is based on test shape, although successful work based upon wall composition and structure has also been conducted, e.g. Murray (1973).

Opportunism and morphogroup analysis has been extensively studied on modern foraminiferal faunas, however, these have concentrated mainly on calcareous forms. Hermelin & Shimmield (1990) and Hemerlin (1992) isolated specific assemblages indicative of oxygen minimum zone conditions in the Arabian Sea, considered to be *r*-selected. Sen Gupta & Machain-Castillo (1993) identified opportunistic species dominating low-oxygen environments; these species were observed as always being present in low numbers until adverse conditions predominated when they essentially 'bloomed'. In particular, they noted the predominance of elongate morphotypes (*Bolivina*) in low-oxygen conditions, but found no modern characteristic morphology indicative of poor oxygen levels. Similarly, Gooday (1993) identified modern abyssal calcareous opportunists. However, these were characteristically small, trochospiral with thin, transparent tests. These calcareous species are interpreted as being opportunistically adapted to seasonal food pulses (Smart *et al.* 1994). Infaunal taxa typified areas of intense upwelling off northwest Africa where food flux is more constant through time (Linke & Lutze 1993). Observed microhabitat preferences in living benthonic foraminifera showed a flexible approach towards food acquisition among specific faunas and were considered as opportunistic in life habit.

Kaminski *et al.* (1988, 1995) noted deepwater agglutinating foraminifera in modern seasonally dysaerobic environments to be dominated by tapered, elongate morphotypes. These were considered as infaunal and opportunistic in life habit and correlate to morphogroup C in this study. Similarly, Koutsoukos *et al.* (1990) noted elongate, tapered agglutinates in the latest Cenomanian–earliest Turonian of northeast Brazil and the Anglo–Paris Basin to proportionally increase in numbers with decreasing oxygen levels in an oxygen minimum zone. Tyszka (1994) described Jurassic calcareous benthonic foraminiferal assemblages to decrease with interpreted falls in O_2, while agglutinated, opportunistic faunas increased.

Through applying morphologic and taxonomic parameters upon benthonic foraminifera, Kaiho (1994) erected a dissolved oxygen index for palaeoenvironmental interpretation, although indexes have been developed using other aspects of the meiofauna, e.g. Wignall & Myers (1988).

The pattern of three maxima is demonstrated clearly by morphogroup C, the sediment-dwelling infaunals (Figs 5 and 10). Modern-day representatives of this group are regarded as opportunistic species (Kaminski & Kuhnt 1995; Kaminski *et al.* 1995) and are often interpreted as indicators of increased productivity in upwelling regions. However, at the base of CABGOC 128-3 they are in exact antiphase with TOC data, moving into phase towards the top of the well. This is regarded as a dynamic shift in environmental forcing from food acquisition to oxygen minimum zone limitation (Sen Gupta & Machain-Castillo 1993). Jorissen *et al.* (1995) define this interplay between food availability and O_2 concentration conceptually. This group is always present throughout CABGOC 128-3, increasing in dominance upwell. Towards the base of the well, abundant food supply is not coupled with low oxygen and the group is out competed by calcareous group 2. This group's opportunistic life habit becomes apparent upwards as oxygenation levels gradually fall, as expressed through increasing TOC values, and the group steadily migrates into phase with TOC. A sudden expansion of the oxygen minimum zone in the uppermost section of CABGOC 128-3 is expressed through morphogroup C's highest values.

The pattern displayed by calcareous morphogroup 2 (Figs 8 and 10) also displays three maxima, the peaks occurring between those of agglutinating group C, and hence in phase with TOC data at the base of the well, moving out of phase towards the top. To the authors' knowledge this is the only report of this juxtaposition of calcareous to agglutinated morphogroup data in combination with TOC values. Towards the base of CABGOC 128-3, morphogroup 2 maximize their food acquisition potential. As the oxygen minimum zone intensifies up well they become less dominant in a dynamic environmental forcing shifting away from food acquisition towards O_2 limitation (Sen Gupta & Machain-Castillo 1993; Jorissen *et al.* 1995) and migrate out of phase with TOC data. They suffer a dramatic decline in the uppermost part of the well coincident with group C's expansion and the sudden expansion and intensification of the oxygen minimum zone.

Group B^3 (Figs 6 and 10), the epifaunal group, displays a gradual 'sawtooth' increase in dominance upwards in CABGOC 128-3. This group includes the distinctive variant species of *C. cyclops*, previously only described from the Arctic (McNeil 1988). This epifaunal group displays synchronized increases in frequency with TOC, thus implying an opportunistic life habit. They do not appear to be greatly affected by the gradual shift from food acquisition control to O_2 limitation, as experienced by morphogroups C and 2 in the Cabinda Basin. The sudden decline of this group in the uppermost part of the well (late Miocene) is coincident with the dramatic turnover experienced by the other two groups and it is postulated that this is in response to the sudden expansion and intensification of the oxygen minimum zone.

Morphogroup 1 (Figs 7 and 10) displays a sudden rise towards the top of CABGOC 128-3, having maintained a previously relatively low frequency. The genera within this morphogroup are typically small in size, have high surface area: volume ratios and are highly perforate. These are characteristics of forms adapted to dysoxic conditions (Corliss 1985; Sen Gupta & Machain-Castillo 1993). This 'bloom' is considered to reflect the sudden pronounced expansion of the oxygen minimum zone associated with late Miocene oceanographic reorganization. This group appears to be unaffected by TOC fluctuations.

Carbonate dissolution is a common phenomenon associated with high productivity zones. In CABGOC 128-3, the gradual intensification of the oxygen minimum zone is mirrored by the relative decrease in species preserved solely as internal moulds (Fig. 9). These species live infaunally, close to the redox boundary (Corliss & Emerson 1990), and are thus well adapted to low oxygen conditions. However, the sudden expansion and intensification of the oxygen minimum zone towards the top of the well severely affected this group, resulting in their disappearance. Post-mortem dissolution of other taxa is impossible to quantify.

Sedimentation rate is naturally also a controlling factor upon TOC accumulation. Lithologically, CABGOC 128-3 displays consistent laminated, fissile, silty shales and, hence, sedimentation rate is not considered to have varied greatly. The deepwater setting of CABGOC 128-3 is considered to negate any eustatic fluctuations experienced more acutely in shallower water. The nutrient content of the water mass is viewed as one of the primary factors that is capable of controlling TOC formation through the intensification of the local upwelling system. The lack of

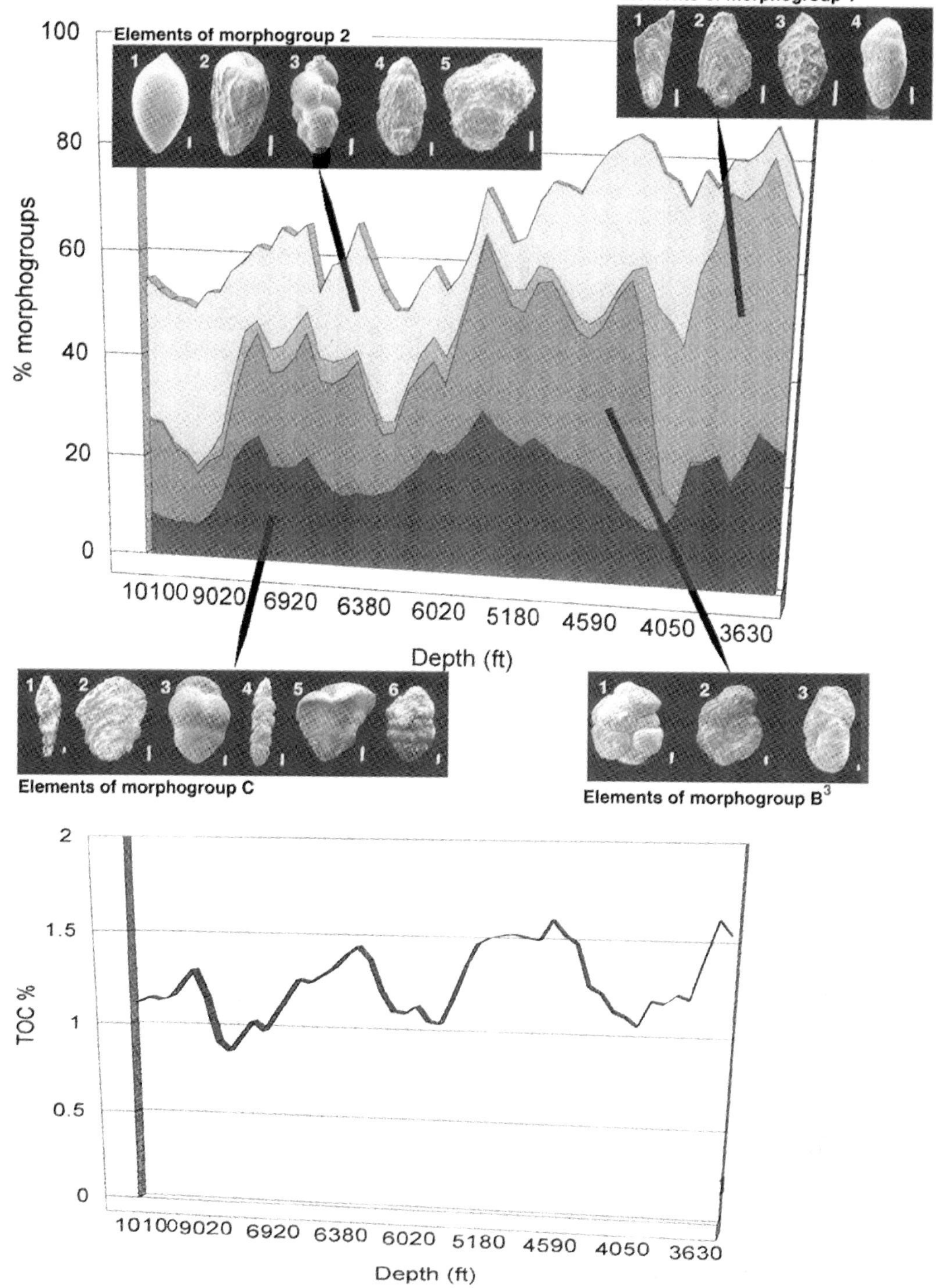

Fig. 10. Composite profile of morphogroup and TOC data at CABGOC 128-3, showing elements of morphogroups. Elements of morphogroup C: 1, *Reophax subfusiforms* Earland 1933; 2, *Pavontina styrica* Schubert 1914; 3, *Eggerella* cf. *bradyi* Cushman 1911; 4, *Karrerulina apicularis* Cushman 1911; 5, *Textularia* cf. *abbreviata*

age control in CABGOC 128-3 has rendered any sequence study impractical.

Elements of the discussed morphogroups are displayed in a composite cross-plot with TOC in Fig. 10. These data suggest that in the Cabinda Basin, a gradual intensification and expansion of the oxygen minimum zone is reflected through benthonic foraminiferal morphogroups. The groups display a migration away from food-limiting factors towards O_2 limitation. Other factors, such as pH, salinity, facies, food type and amount, etc., in addition to O_2, may have some control on relative foraminiferal abundance, either individually or in combination. It is hoped that further investigations into the Agua Salada Fauna will elucidate this quandary.

Conclusions

The TOC profile at CABGOC 128-3 is considered to be a proxy for the local upwelling intensity in the Cabinda Basin. Upwelling forces phytoplankton proliferation and subsequent high accumulation rates of organic carbon. Oxidization of this particulate organic matter eventually leads to local dysoxic conditions.

Morphogroup analysis of a Miocene benthonic foraminiferal assemblage, coupled with TOC data, is presented as an approach for modern micropalaeontological and oceanographic study. At CABGOC 128-3 a gradually developing oxygen minimum zone is identified through morphogroup analysis. This undergoes a sudden expansion and intensification towards the top of the well in the late Miocene.

The elongate, agglutinated morphogroup (group C), considered to be opportunistic, is found to be in antiphase with TOC data at the base of CABGOC 128-3, becoming in phase towards the top. The morphologically comparative calcareous group (group 2) is in antiphase with group C and hence moves from in phase into antiphase up CABGOC 128-3 with regard to TOC. This relationship is unprecedented within the foraminiferal literature to date, but is considered to be a reflection of an environmental shift migrating away from food as a primary limiting factor to O_2.

Epifaunal, planispiral, agglutinated taxa (group B^3) display a gradual increase in dominance up CABGOC 128-3. This group reflects TOC highs within its profile and, hence, it is inferred to be more opportunistic than previously assumed. Its sudden decline towards the top of the well is coincident with the sudden expansion and intensification of the oxygen minimum zone.

Small perforate calcareous taxa with a high surface area: volume ratio, regarded as being indicative of dysoxic conditions, exhibit a sudden increase in dominance towards the top of CABGOC 128-3. This is associated with a pronounced expansion of the oxygen minimum zone.

Calcite dissolution, as a result of acid pore waters in low oxygen conditions, is expressed by the preservation of susceptible species as internal pyrite moulds.

The dramatic decline of foraminiferal abundance within the upper Miocene of CABGOC 128-3 is regarded as a reflection of the late Miocene oceanographic reorganization. A refined stratigraphy for this area must be established before more precise timing can be determined. However, this remains problematic as planktonic foraminiferal and calcareous nannofossil recovery is extremely poor.

Thanks to Jim Davey, Toby Stiles and Sorin Filipescu for help in the preparation of photographic materials and technical advice. TOC data is provided by S. Northam of Chevron Overseas Petroleum Inc. R. C. P. acknowledges a NERC CASE studentship with Chevron Overseas Petroleum Inc., many fruitful discussions with numerous UCL postgraduates and the constructive criticisms of Ray Bate and two anonymous reviewers.

d'Orbgny 1846; 6, *Valvulina flexilis* Cushman & Renz 1941. Elements of morphogroup B^3: 1, *Haplophragmoides globigerinoides* Haeusler 1882; 2, *Reticulophragmium venezuelanum* Maync 1957; 3, *Cyclammina* cf. *cyclops* McNeil 1988. Elements of morphogroup 1: 1, *Plectofrondicularia longistriata* Le Roy 1939; 2, *Plectofrondicularia vaughi* Cushman 1927; 3, *Bolivina bryamensis* Cushman 1923; 4, *Bolivina tenuisstriata* Cushman & Ellisor 1939. Elements of morphogroup 2: 1, *Guttulina* sp. 1; 2, *Bulimina falconensis* Renz 1948; 3, *Uvigerina modeloensis* Cushman & Kleinpell 1934; 4, *Uvigerina peregrina* Cushman 1923; 5, *Uvigerina spinosa* Boersma 1984. Bar = 100 μm.

Appendix 1. Benthonic foraminiferal data for morphogroup analysis at CABGOC 128-3. (Overall totals include species not members of the four listed morphogroups.)

	Morphogroup C 26 Genera 64 Species	**Morphogroup B3** 9 Genera 23 Species	**Morphogroup 1** 2 Genera 16 Species	**Morphogroup 2** 11 Genera 42 Species		
Well depth (ft)	Total per sample	Total per sample	Total per sample	Total per sample	Overall Agglutinating Total	Overall Calcareous Total
3570	5	0	9	0	5	19
3630	9	0	22	6	9	45
3690	23	0	24	3	23	30
3750	10	0	19	2	11	28
3810	30	2	76	7	32	105
3870	18	1	36	6	20	53
3930	11	1	40	6	13	66
3990	25	0	28	1	26	38
4050	3	0	4	3	3	11
4110	4	1	11	10	5	26
4230	2	0	5	4	2	10
4290	1	2	3	2	4	6
4350	9	36	2	23	45	35
4470	28	74	0	27	104	57
4590	16	42	3	29	56	48
4710	55	62	4	90	121	127
4830	37	36	4	43	66	88
4950	46	58	9	52	100	111
5000	61	70	5	43	126	124
5120	44	61	3	27	102	59
5180	63	74	7	33	135	90
5300	22	16	4	3	39	37
5420	33	38	7	19	91	109
5540	22	25	0	9	43	29
5780	32	30	2	5	59	26
5900	27	32	0	7	53	29
6020	19	12	2	5	30	63
6080	53	22	6	30	81	113
6140	27	15	12	19	39	75
6200	32	34	1	11	71	47
6260	19	20	1	19	39	82
6320	30	20	5	36	42	149
6380	24	12	5	53	39	127
6440	26	33	1	52	64	136
6500	7	12	2	8	17	27
6620	27	48	1	43	80	78
6800	14	16	8	16	35	109
6860	29	34	10	26	67	84
6920	16	16	1	6	46	18
6980	10	11	2	10	20	21
7040	11	15	10	22	25	73
7160	12	9	0	12	19	33
7760	18	18	3	9	41	34
8180	30	12	3	9	52	50
9020	41	23	22	50	65	181
9200	25	30	4	65	55	201
9380	6	12	2	32	19	91
9920	14	22	2	54	51	125
9980	21	24	4	91	39	256
10 040	17	55	0	66	76	166
10 100	39	81	2	118	152	272

Appendix 2. Benthonic foraminifera species of morphogroups.

C Morphogroup
Agglutinated sp. 2
Agglutinated sp. 6
Agglutinated sp. 7
Alveovalvulina suteri
Ammobaculites agglutans
A. sp. 1
A. sp. 2
A. sp. 3
A. sp. 4
Ammomargulina sp. 1
Bigenerina sp. 1
Bulbobaculites sp. 1
Eggerella cf. *bradyi*
E. forestensis
E. karamatensis
Gaudryina thalmanni
Gravellina narivaensis
Jarvisella karamatensis
Karrerulina apicularis
K. cf. *coniformis*
K. conversa
K. horrida
K. microgranulosa
K. cf. *microgranulosa*
K. sp. 1
K. sp. 2
Liebusella pozonensis
Martinottiella cf. *cyclostomata*
M. sp. 1
Pavontina styrica
Pavopsammia flabellum
Pavonitina adanula
Pseudogaudryina alazanensis
Reophax acosta
R. dupex
R. subfusiforms
R. sp. 1
R. sp. 2
R. sp. 4
R. sp. 5
Spiroplectammina barrowi
Subreophax scalaris
S. sp. 1
S. sp. 2
Textularia cf. *abbreviata*
T. crassisepta
T. earlandi
T. isidroensis
T. kugleri
T. leuzingeri
Textularia mississippiensis
T. panamensis
T. pozonensis
T. tatumi
T. teasi
T. sp. 1
T. sp. 2
Textulariella barretti
Textulariaopsis sp. 1
Valvulina flexilis
Vernuillina sp. 1
Vulvulina jacuraensis
V. pachyeilus
V. spinosa

B3 Morphogroup
Budashevella multicameratus
Cribrostromoides sp. 1
Cyclammina cancellata
C. cf. *cyclops*
C. placenta
C. prussila
Discamminoides tobleri
Haplophragmoides carinatum
H. emaciatum
H. globigerinoides
H. narivaensis
H. cf. *narivaensis*
H. obliquicameratus
Recurvoides contortus
R. higginsi
R. obsoletum
Reticulophragmium rotundirosata
R. venezuelanum
R. sp. 1
R. sp. 2
R. sp. 4
Spiropsammia primula
S. sp. 1

Morphogroup 1
Bolivina advena
B. alata
B. alazanensis
B. bryamensis
B. caudriae
B. cf. *cochei*
B. inconspicua
B. marginata var. *multicostata*
B. pisciformis
B. pozonensis
B. suteri
B. tenuisstriata
B. tongi
B. sp. 1
Plectofrondicularia longistriata
P. vaughi

Morphogroup 2
Bulimina affinis
B. cf. *alazanensis*
B. falconensis
B. inflata
B. inflata var. *alligata*
Protoglobobulimine pupoides
Buliminella cf. *basistriata* var. *nuda*
Gumbelina cubensis
Guttulina irregularis
G. jarvisi
G. sp. 1
Pseudoglandulina comatula
P. incisa
P. laevigata
P. sp. 1
Siphogenerina lamellata
Siphogenerina smithi
Trifarina bradyi
Uvigerina altacostata
U. auberiana
U. cf. *beccarii*
U. capayana
U. carapitana
U. cf. *hannai*
U. hispida
U. laviculata
U. lirettensis
U. mantaensis
U. modeloensis
U. peregrina
U. pilulata
U. pigmae
U. proboscida
U. rugosa
U. rustica
U. spinosa
U. sp. 1
Uvigerinella sparsicostata
Virgulina exilis
V. pontoni
Virgulinella ossamagnifica

References

AKERS, W. H. 1954. Ecologic aspects and stratigraphic significance of the foraminifera *Cyclammima cancellata* Brady. *Journal of Paleontology*, **28**, 132–152.

BASOU, I. A. 1976. Benthic foraminifera in shelf sediments of the Gulf of Guinea. *Biologiya Morya*, **1**, 22–29.

BERGER, W. H. ET AL. 1981. The deep-sea record: Major steps in Cenozoic ocean evolution. *In*: WARME, J. E. ET AL. (eds) *The Deep Sea Drilling Project: A Decade of Progress*. Society of Economic Paleontologists and Mineralogists, Special Publication, **32**, 489–504.

BERNHARD, J. M. 1986. Characteristic assemblages and morphologies of benthic foraminifera from anoxic, organic-rich deposits: Jurassic through Holocene. *Journal of Foraminiferal Research*, **16**, 207–215.

BLOW, W. H. 1959. Age, correlation and biostratigraphy of the Upper Tocuyo (San Lorenzo) and Pozon Formations, Eastern Venezuela. *Bulletin of American Paleontology*, **39**, 67–251.

BRICE, S. E. ET AT. 1982. Tectonics and sedimentation of the South Atlantic rift sequence: Cabinda, Angola. *In*: WATKINS J. S. & DRAKE, C. L. (eds) *Studies in Continental Margin Geology. AAPG Memoir*, **34**, 5–18.

BROWN, D. 1993. West Africa challenges Explorers. *AAPG Explorer*, **14**, 12, 13, 15.

BRUN, L. ET AL. 1984. Répartition stratigraphique et paléoécologique des principales espèces de Bolivinitidae (foraminifères) du Tertiaire du Golfe de Guinée. *Benthos '83; Second International Symposium of Benthic Foraminifera*, 91–104.

CORLISS, B. H. 1985. Microhabitats of benthic foraminifera within deep-sea sediments. *Nature* **314**, 435–438.

—— & CHEN, C. Morphotype patterns of Norwegian Sea deep-sea benthic foraminifera and ecological implications. *Geology*, **16**, 716–719.

—— & EMERSON, S. 1990. Distribution of Rose Bengal stained benthic foraminifera from the Nova Scotia Continental Margin and Gulf of Maine. *Deep Sea Research*, **37**, 381–400.

CUSHMAN, J. A. & RENZ, H. H. 1941. New Oligocene–Miocene foraminifera from Venezuela. *Contributions from the Cushman Laboratory for Foraminiferal Research*, **17**, 1–27.

DEMAISON, G. J. & MOORE, G. T. 1980. Anoxic environments and oil source bed genesis. *AAPG Bulletin*, **64**, 1179–1209.

DIAZ DE GAMERO, M. L. 1977. *Estratigrafia y Micropaleontologia del Olioceno y Mioceno inferior del centro de la cuenca de Falcon, Venezuela*. GEOS 22-3-60, Escuela de Geologia y Minas, UCV, Caracas.

—— 1985*a*. Micropaleontologia de la formacion Agua Salada, Falcon Nororiental. *In*: ESPEJO, A. ET AL. (eds) *Estratigrafia paleontologia y sedimntacion. VI Congresso Geologico Venezolano*. Sociedad Venezolana de Geologos, **1**, 384–453.

—— 1985*b*. Estratigrafia de Falcon Nororiental. *In*: ESPEJO, A. ET AL. (eds) *Estratigrafia paleontologia y sedimntacion. VI Congresso Geologico Venezolano*. Sociedad Venezolana de Geologos, **1**, 454–502.

DIESTER-HASS, L. 1988. Sea level changes, carbonate dissolution and history of the Benguela Current in the Oligocene–Miocene off Southwest Africa (DSDP Site 362, Leg 40). *Marine Geology*, **79**, 213–242.

—— ET AL. 1990. Miocene history of the Benguela current and Antarctic ice volumes: Evidence from rhythmic sedimentation and current growth across the Walvis Ridge (Deep Sea Drilling Project Sites 362 and 532). *Paleoceanography*, **5**, 685–707.

—— ET AL. 1992. The Benguela Current and associated upwelling on the southwest African Margin: a synthesis of the Neogene–Quaternary sedimentary record at DSDP Sites 362 and 532. *In*: PRELL, C. P. & EMEIS, K. C. (eds) *Upwelling Systems: Evolution Since the Early Miocene*. Geological Society, London, Special Publications, **64**, 331–342.

DIGNES, T. W. 1995. Albian to Cenomanian foraminferal zones of Angola (Cabinda), west Africa: relationships to time, facies, and morphogroups. *In*: KAMINSKI, M. A. ET AL. (eds) *Proceedings of the Fourth International Workshop on Agglutinated Foraminifera*. Grybowski Foundation Special Publication, **3**, 210.

—— 1996. Albian to Cenomanian foraminiferal zones of Angola (Cabinda): relationships to time, facies and morphogroups. *In*: JARDINE, S., DE KLASZ, I. & DEBENAY, J. P. *Géologie de L'Afrique et de L'Atlantide Sud – Compte – rende des colloques de géologie D'Angers*, 16–20 Juillet 1994, 705.

GIFFUNI, G. I. S. A. 1991. Biostratigraphy of a Middle Miocene–Pliocene sequence from Cumarebo area, Falcon State, Northwestern Venezuela. *AAPG Bulletin*, **75**, 580.

GOODAY, A. J. 1988. A response by benthic foraminifera to the deposition of phytodetritus in the deep sea. *Nature*, **332**, 70–73.

—— 1993. Deep-sea benthic foraminiferal species which exploit phytodetritus: characteristic features and controls on distribution. *Marine Micropalaeontology*, **22**, 187–205.

GORDON, A. L. & BOSLEY, K. T. 1992. Cyclonic gyre in the tropical South Atlantic. *Deep Sea Research*, **38**, 323–343.

HAMAN, D. ET AL. 1993. *Virgulinella ossamagnifica* n. sp., from the Miocene of Cabinda, Equatorial West Africa. *Journal of Foraminiferal Research*, **23**, 205–208.

HANSE, A. 1995. Les microfaunes en Angola. *Memoires Bureau Recherche Geologique Miniere*, **32**, 327–334.

HAY, W. H. & BROCK, J. C. 1992. Temporal variation in intensity of upwelling off southwest Africa. *In*: SUMMERHAYES, C. P., PRELL, W. L. & EMEIS, K. C. (eds) *Upwelling Systems: Evolution Since the Early Miocene*. Geological Society, London, Special Publications, **63**, 463–497.

HERMELIN, J. O. R. 1992. Variations in the benthic foraminiferal fauna of the Arabian Sea a response to changes in upwelling intensity? *In*:

SUMMERHAYES, C. P. ET AL. (eds) *Upwelling Systems: Evolution Since the Early Miocene.* Geological Society, London, Special Publications, **64**, 151–166.

—— & SHIMMIELD, G. B. 1990. The importance of the Oxygen Minimum Zone and sediment geochemistry in the distribution of recent benthic foraminifera in the Northwest Indian Ocean. *Marine Geology*, **91**, 1–29.

JONES, R. W. 1996. *Micropalaeontology in Petroleum Exploration.* Oxford Science.

—— & CHARNOCK, M. A. 1985. Morphogroups of agglutinating foraminifera, their life positions and feeding habits and potential applicability in (palaeo)ecological studies. *Revue de Paléobiologie*, **4**, 311–320.

JORISSEN, F. J. ET AL. 1995. A conceptual model explaining benthic foraminiferal microhabitats. *Marine Micropaleontology*, **26**, 3–15.

KAIHO, K. 1994. Benthic foraminiferal dissolved-oxygen index and dissolved-oxygen levels in the modern ocean. *Geology*, **22**, 719–722.

KAMINSKI, M. A. 1988*a*. Cenozoic deep-water agglutinated foraminifera in the North Atlantic (PhD. Thesis). *Woods Hole Oceanographic Institute/Massachusetts Institute of Technology Technological Report*, **88**.

—— & KUHNT, W. 1995. Tubular agglutinated foraminfera as indicators of organic carbon flux. *In*: KAMINSKI, M. A. ET AL. (eds) *Proceedings of the Fourth International Workshop on Agglutinated Foraminifera.* Grzybowski Foundation Special Publication, **3**, 141–144.

—— BERGGREN, W. A. & PREECE, R. C. in press. The Agua Salada Fauna: A Neogene deep-sea agglutinated foraminiferal fauna adapted to dysaerobic conditions of the low latitude upwelling zones, in press.

—— ET AL. 1995. Response of deep-water agglutinated foraminifera to dysoxic conditions in the California Borderland basins. *In*: KAMINSKI, M. A., GEROCH, S. & GASINSKI, M. A. (eds) *Proceedings of the Fourth International Workshop on Agglutinated Foraminifera.* Grzybowski Foundation Special Publication, **3**, 131–140.

—— ET AL. 1988. Life history and recolonization among agglutinated foraminifera in the Panama Basin. *Abhandlungen des Geologischen B–A*, **41**, 229–243.

KATZ, M. E. & THUNELL, R. C. 1984. Benthic foraminiferal biofacies associated with Middle Miocene to Early Pliocene oxygen-deficient conditions in the eastern Mediterranean. *Journal of Foraminiferal Research*, **14**, 187–202.

KOGBE, C. A. & ME'HES, K. 1986. *Micropaleontology and Biostratigraphy of the Coastal Basins of West Africa.* Pergamon.

KOUTSOUKOS, A. M. ET AL. 1990. Latest Cenomanian–earliest Turonian oxygen tolerant benthonic foraminifera: a case-study from the Sergipe Basin (N.E. Brazil) and the western Anglo–Paris Basin (southern England). *Palaeogeography, Palaeoclimatology, Palaeoecology*, **77**, 145–177.

LINKE, P. & LUTZE, G. F. 1993. Microhabitat preferences of benthic foraminifera – a static concept or a dynamic adaptation to optimise food acquisition? *Marine Micropaleontology* **20**, 215–234.

MCHARGUE, T. R. 1990. Stratigraphic development of Proto-South Atlantic rifting in Cabinda, Angola – a petroliferous lake basin. *In*: KATZ, B. J. (ed.) *Lacustrine Basin Exploration. AAPG Memoir*, **50**, 307–326.

MCNEIL, D. H. 1988. *Cyclammina cyclops*, n. sp., in the Eocene Richards Formation, Beaufort Sea area of Arctic Canada. *Journal of Foraminiferal Research*, **18**, 114–123.

MORENO-VAQUEZ, J. 1995. Neogene biofacies in eastern Venezuela and their calibration with seismic data. *Marine Micropaleontology*, **26**, 287–302.

MOY, R. L. 1968. *Benthonic foraminifera from the Oligo-Miocene of Libya and Hungary.* PhD Thesis, University of Wales (Aberystwyth).

MURRAY, J. W. 1973. *Distribution and Ecology of Living Benthic Foraminiferids.* Heinemann.

NAGY, J. 1992. Environmental significance of foraminiferal morphogroups in Jurassic North Sea deltas. *Palaeogeography, Palaeoclimatology, Palaeoecology*, **95**, 111–134.

PETERSON, R. G. & STRAMMA, L. 1991. Upper-level circulation in the South Atlantic Ocean. *Progress in Oceanography*, **26**, 1–73.

PICAUT, J. 1985. Major dynamics affecting the eastern tropical Atlantic and Pacific oceans. *CalCOFI Report*, **26**, 41–50.

PREECE, R. C. in prep. *Neogene benthic foraminifera from equatorial upwelling zones.* PhD Thesis, University College London, in prep.

RENZ, H. H. 1948. Stratigraphy and fauna of the Agua Salada Group, state of Falcon, Venezuela. *Geological Society of America, Memoir*, **32**.

RENZ, H. H. 1959. San Lorenzo Formation (Agua Salada Group) of Eastern Falcon, Venezuela, a valid lithostratigraphic name. *AAPG Bulletin*, **43**, 2500–2502.

REYMENT, R. A. 1966. Brief review of the stratigraphic sequences of West Africa (Angola to Senegal). *Proceedings of the 2nd West African Micropaleotonology Colloquium*, Ibadan, 1965, 162–175.

SEIGLIE, G. A. & BAKER, M. B. 1982. Foraminiferal zonation of the Cretaceous off Zaire and Cabinda, West Africa and its geological significance. *In*: WATKINS, J. S. & DRAKE, C. L. (eds) *Studies in Continental Margin Geology.* AAPG Memoir, **34**, 651–658.

—— & —— 1983. Some west African Cenozoic agglutinated foraminiferas with inner structures: taxonomy, age and evolution. *Micropalaeontology*, **29**, 391–403.

—— & FROST, S. H. 1979. Significance of Middle Tertiary large foraminifera common to West Africa and Caribbean. *AAPG Bulletin*, **63**, 525.

—— ET AL. 1986. *Alveolvalvulinidae*, n. fam., and Neogene diversification of agglutinated foraminifers with inner structures. *Micropalaeontology*, **32**, 169–181.

SEN GUPTA, B. K. & MACHAIN-CASTILLO, M. L. 1993. Benthic foraminifera in oxygen poor habitats. *Marine Micropaleontology*, **20**, 183–200.

SEVERIN, K. P. 1983. Test morphology of benthic foraminifera as a discriminator of biofacies. *Marine Micropaleontology*, **8**, 65–76.

SIESSER, W. G. 1980. Late Miocene origin of the Benguela Upswelling System off northern Namibia. *Nature*, **208**, 283–285.

SMITH, R. L. 1992. Coastal upwelling in the modern ocean. *In*: SUMMERHAYES, C. P., PRELL, W. L. & EMEIS, K. C. (eds) *Upwelling Systems: Evolution Since the Early Miocene*. Geological Society, London, Special Publications, **63**, 9–28.

SOUTHAM, J. R. ET AL. 1982. Dynamics of anoxia. *Palaeogeography, Palaeoclimatology, Palaeoceanography*, **40**, 183–198.

SPAW, R. H. & KOEHLER, R. P. 1981. Geology and reservoir distribution, Pinda Formation, offshore Zaire and southern offshore Cabinda. *AAPG Bulletin*, **65**, 996.

STRAMMA, L. 1991. Geostrophic transport of the South Equatorial Current in the Atlantic. *Journal of Marine Research*, **49**, 281–294.

TYSZKA, J. 1994. Response of Middle Jurassic benthic foraminiferal morphogroups to dysoxic/anoxic conditions in the Pieniny Klippen Basin, Polish Carpathians. *Palaeogeography, Palaeoclimatology, Palaeoceanography*, **110**, 55–81.

UJIIE, H. 1970. Miocene foraminiferal faunas from the Sandakan Formation, North Borneo. *Contributions to the Geology and Palaeontology of Southeast Asia*, **85**, 165–185.

— 1979. New species and subspecies of benthonic foraminifera from the Miocene Sandakan Formation, North Borneo. *Contributions to the Geology and Palaeontology of Southeast Asia*, **182**, 87–101.

VALE, F. S. 1960. Bacia sedimentar de Cabinda. *Boletin Serviciole Geologia Mineria*, **2**, 31–44.

WACONGNE, S. & PITON, B. 1992. The near surface circulation in the north eastern corner of the South Atlantic Ocean. *Deep Sea Research*, **39**, 1273–1298.

WIGNALL, P. B. & MYERS, K. J. 1988. Interpreting benthic oxygen levels in mudrocks: a new approach. *Geology*, **16**, 452–455.

Non-marine ostracod assemblages of the Pre-Salt rift basins of West Africa and their role in sequence stratigraphy

RAYMOND H. BATE

Lacustrine Basin Research Limited, Little Lower Ease, Cuckfield Road, Ansty, West Sussex RH17 5AL, UK

Abstract: Early Cretaceous ostracods living within the rift lakes of Central Gondwana reflect, through their morphology, abundance and species diversity, depositional environments and changes in water chemistry that occurred within the lakes at that time. By studying the indigenous ostracod assemblages it is possible to interpret past environmental changes with respect to both water depth and water chemistry, and to utilize this information to estimate the probability of source and reservoir sediment deposition. The alternation of siliclastic and carbonate sequences throughout rift lacustrine sediments can be explained by changes in the depositional environments and related water-level movements. Such changes being brought about as a direct result of tectonism leading to increases or decreases in precipitation as a consequence of river capture or regional changes in climate. Two cycles of sedimentary infill are recognized, each initiated by an onlapping, fining-upwards transgressive systems tract within which a maximum flooding surface is developed. This is followed by a coarsening-upwards highstand systems tract in which a prograding sequence retains the diverse ostracod fauna of the transgressive systems tract below. A coarsening-upwards lowstand systems tract, having a low diversity ostracod fauna, brings each cycle to a close. An understanding of the cyclic sequence stratigraphy of these basins suggests that the break-up of Gondwana commenced as early as the late Barremian. Commercially, these rift basin lakes contain excellent hydrocarbon source rocks, but it is the interpretation of the lacustrine faunas that provides the essential information as to lake size and hydrocarbon potential.

Commercial interest in the capacity of lacustrine sediments to generate hydrocarbons has attracted oil company interest for many years now, particularly in China, Indonesia, Brazil and West Africa, and, more recently, offshore Vietnam and Mongolia. Although many lacustrine basins develop as sag basins, the marginal basins of Brazil and West Africa developed within the geologically important rift (for which the present-day East African Rift is an essential analogue) that preceded the splitting of Gondwana. As a result of sustained oil company exploration, on both sides of the South Atlantic, many major oil fields have been discovered whose hydrocarbons are directly sourced from the Lower Cretaceous syn-rift lake sediments; even though the reservoirs may be either syn-rift lacustrine or post-rift marine. Two examples can be quoted: firstly, the giant late-rift Rabi-Kounga Field, onshore Gabon (Boeuf *et al.* 1991), that has been a major success for Shell, and secondly, Petrobras' giant Campos Basin fields, offshore Brazil, where organic-rich lacustrine shales of the Lagoa Feia Formation have sourced syn-rift bioclastic limestones and post-rift marine Cretaceous carbonates, and Cretaceous and Tertiary turbidites (Abrahao & Warme 1990)

Interpretation of syn-rift lake sediments is difficult, mainly owing to the poor quality seismic detail produced as a consequence of the fragmented overlying salt succession. Because of this, new techniques, based upon the application of ostracod diversity and shell ornamentation, have been developed to facilitate the interpretation of lake size, sediment deposition and environmental history. These techniques rely upon a precise interpretation of the observed diversity and ornamentation changes through time, and require a sound knowledge of lacustrine faunas.

Lower Cretaceous ostracods living within the rift lakes of central Gondwana reflect, through their morphology, abundance and species diversity, depositional environments and changes in water chemistry that occurred within the lakes at that time. Consequently, ostracods represent perhaps the most important group of freshwater microfossils that are able to provide, through these changes, information leading to the recognition of past environmental conditions. Although in China calcareous algae (charophytes) have provided supportive environmental data, this has not been possible here as these microfossils are almost totally missing from the West African rift sequence, for reasons not yet understood.

By studying the indigenous ostracod assemblages it is possible, therefore, to interpret past environmental changes and to utilize this information to estimate lake size, water chemistry (e.g. salinity), and an assessment as to the probability of source and reservoir sediment deposition.

An analysis of the lacustrine ostracod faunas

BATE, R. H. 1999. Non-marine ostracod assemblages of the Pre-Salt rift basins of West Africa and their role in sequence stratigraphy. *In:* CAMERON, N. R., BATE, R. H. & CLURE, V. S. (eds) *The Oil and Gas Habitats of the South Atlantic.* Geological Society, London, Special Publications, **153**, 283–292.

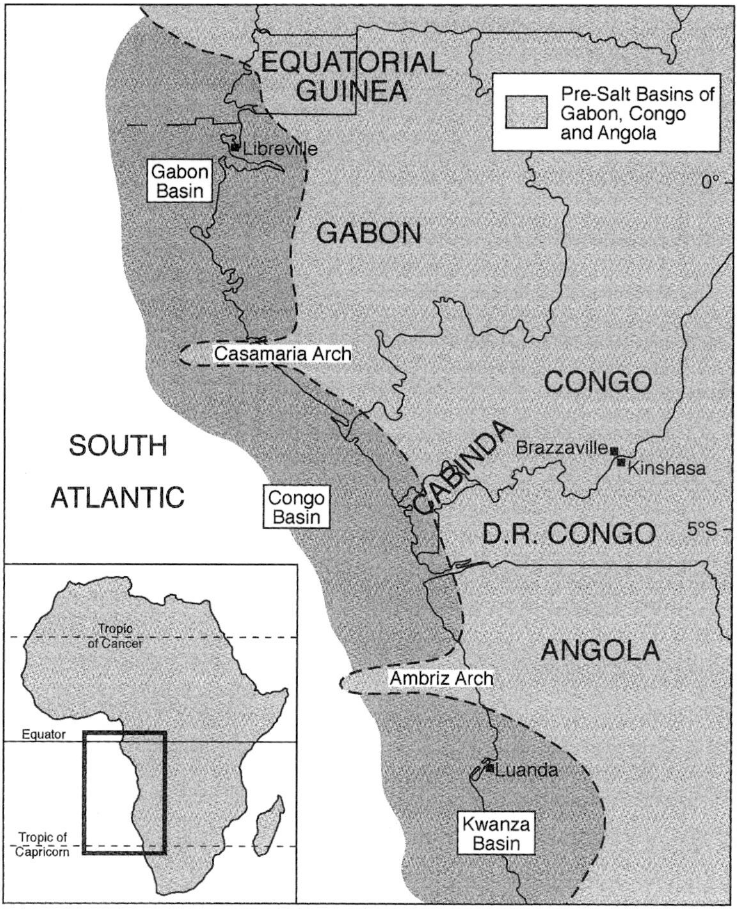

Fig. 1. Area of study along the West African South Atlantic margin.

from the Lower Cretaceous rift of Central Gondwana indicates that the intra-rift basins have experienced similar environmental histories; the sediment infill of each basin containing a comparable sequence of ostracod faunas whether that basin be in West Africa or coastal Brazil. Consequently, it is possible to recognize environmental changes that have affected the entire region, not just one basin, and this has important implications for the entire South Atlantic region as a whole. On break-up, the north–south oriented rift basins, once sited in close proximity to each other, were pulled apart and separated by the developing ocean. As a consequence they are now to be found attached either to the eastern seaboard of South America (Brazil) or the western seaboard of West Africa (Gabon to Angola).

That it is possible to interpret the phases of environmental change experienced by these rift lakes owes much to the researches of Carbonnel (1982) and Carbonnel & Peypouquet (1979, 1983), who observed that water chemistry plays a key role in controlling shell ornamentation in ostracods (especially lacustrine/euryhaline species). They were able to demonstrate the relationship and effect that the ionic equilibrium of Mg^{2+} and Ca^{2+} have on ostracod shell morphology. For example, with increasing salinity, Ca^{2+} ions decrease, by combining with carbonates, and smooth-shelled ostracods tend to dominate as a result. Where Mg^{2+} is low and dissolved Ca^{2+} is high, and where there is an abundance of organic detritus, ornamented species tend to predominate. As ostracods cannot store the minerals that combine to form their shells, at every moult shell-forming minerals have to be extracted from the lake water. For those ostracods in which the genetic make-up provides for shell strengthening, through noding or reticulation, the availability of minerals in solution will determine the extent to which this shell form is reproduced. The higher energy oligohaline environments require shell strengthening and chemical availability makes this possible. As lower energy saline environments reduce the availability of shell-forming minerals, smooth-shelled ostracods appear as a result. It is perhaps too simplistic to state that there is insufficient material available to produce an ornamented shell, but that is basically what appears to be the case. Thus, ostracods, as aquatic animals, interpret lake environments more precisely than do land-derived organisms that may be brought into the system by rivers sourced many hundreds of miles away.

From the concepts given above, it is possible to interpret past environmental conditions where species diversity (rise or fall) is known to reflect changes in lake water level (transgressions, highstands, regressions, lowstands), while changes in ornamentation reflect changes in lake water chemistry. This study of these Lower Cretaceous ostracod assemblages has shown that two distinct faunal cycles existed; each cycle commencing with a transgressive phase, followed by a highstand and terminating in a lowstand. These faunal cycles are the product of time, tectonism and climate change. Lambiase (1990, p. 269), referring to the published article of Brice *et al.* (1982), states that two depositional sequences, separated by an unconformity, represent the sediment infill of these West African continental rift basins. The faunal cycles recognized coincide precisely with the rift phases that provided the accommodation space required for lake development and sediment infill (Karner & Driscoll this volume). Logically, there is a time lag between rifting and the production of accommodation space; the initiation of lacustrine conditions and the onset of sequential ostracod faunal changes.

Sequence stratigraphy

The sequence terminology applied here is that normally associated with marine sequences and, as the effects of a rise and fall in water level on sediment deposition and faunal development are broadly comparable, it is considered unnecessary to introduce separate terminologies for lake systems.

Each lake sequence, for which faunal evidence has identified the depositional environment at its base, is seen to have been initiated by a basal swamp into which fluvial clastic sediments were deposited. High energy deposited fluvial sands rarely contain a fauna, either because the organic remains disintegrate or because the rapid deposition of sediment tends to smother everything attempting to live in that environment. Under such conditions, the precise depositional environment is difficult to determine. However, where low energy sediments exist, the shallow water environment of these basal sediments is identified by the presence of the ostracod *Darwinula* and, not infrequently, the co-occurrence of conchostracans as inhabitants of ephemeral pools. The early lakes developed as stable, balanced (or underfilled) lake systems that enlarged, through a transgressive phase, as a result of increasing precipitation. This transgressive phase continued until outflow (through either evaporation and/or overspill) equalled inflow and the system again became balanced. At that time the system reached highstand, when evaporation became equal to precipitation, while the final phase, that of a lowstand, occurred when evaporation was greater than precipitation, at which time the lake waters almost certainly became saline.

As the drying up of a lake (evaporation greater than precipitation) initiates a rise in salinity and the development of a stressed environment, carbonate precipitation is frequently triggered and species diversity (although not species abundance) becomes restricted and smooth-shelled faunal populations develop. Such stressed conditions were particularly prevalent at the end of Neocomian–early Barremian times and again at the top of the Pre-Salt succession during the early Aptian. From this it can be seen that the onset of lake shrinkage and an increase in water salinity coincides precisely with the end of each of the two cycles of rifting, possibly related to the infilling of available accommodation space and the lowering of surface relief. Each transgressive phase, on the other hand, relates to the early stages of rifting after increased accommodation space has been provided and increased terrain relief possibly triggered the initiation of increased precipitation.

Using sequence stratigraphic terminology it is thus possible to recognize lowstand, transgressive and highstand systems tracts, while a maximum flooding surface, because of the considerable input of sediment during periods of increased precipitation, is not so much a surface as an interval of sediment deposition associated with species diversity.

For the first lake cycle, although exhibiting freshwater conditions at its base and during the transgressive phase, appears to become rapidly saline with most of the ostracods encountered being smooth-shelled. In West Africa, the lowstand phase is frequently removed by erosion and the next cycle of lake development lies unconformably on older sediments of the transgressive highstand sequence.

The second cycle commenced with a dramatic period of lake expansion when most of the older Neocomian–early Barremian lakes flooded and coalesced to form very large lake systems. It was also during this second transgressive phase that the major source rocks developed as a direct consequence of the size of the lakes and as a result of the organic productivity generated. These factors effectively combined to produce the considerable quantities of hydrocarbons known to exist. Clearly, knowledge of the size and age of these Pre-Salt lakes is essential when assessing the hydrocarbon potential of a specific region.

As for the first cycle, the transgressive phase of the second cycle ended with a balanced lake system in which highstand sedimentation was established. The highly diverse fauna that developed earlier also steadily declined as precipitation fell away and lowstand saline conditions returned. This finally marked the end of the syn-rift sedimentation.

The sediment infill of a lake involves the input of fluvial clastics through the agency of axial or hinge margin deltas; turbidity flows extending into deeper parts of the basin and re-worked beach sands that form around the lake margins. Downslope the coarser clastics give way to finer-grained silts and eventually grade into deepwater muds. Carbonates develop on highs (over accommodation zones) or at lake margins during both transgressive and saline phases. Recognition of highstand, lowstand and transgressive systems tracts thus provides important data for the understanding of sediment deposition at any point in time. For example, during the transgressive phase of lake development not only will deepwater muds develop under anoxic conditions but there will be a significant enlargement of deltas as the load capacity of the rivers increases; turbidity flows will also become more prevalent. As lake levels rise, the growth of aquatic plants can initiate carbonate sedimentation as a result of carbon dioxide extraction from the water. However, carbonate development is also a feature of lowstand systems tracts where water salinity increases and the carbonate branch point is reached. It should be stressed, however, that many lake carbonates are largely bioclastic deposits and can form within all system tracts other than where water chemistry reaches particularly high pH levels.

The sedimentary sequence that infilled the West African Rift consists of two cyclic events that themselves may be subdivided into three systems tracts: the base of each cycle commencing with a transgressive systems tract, followed by a highstand systems tract and ending with a lowstand systems tract. Each systems tract, and the environmental conditions under which sedimentation took place, may be identified through its contained fauna, providing a timetable for the deposition of both source and reservoir rocks. The cycles and the sequence stratigraphic phases that go to make up this rift sequence are summarized in Table 1, in which the ostracod zonation used is essentially the Elf zonation (see Grosdidier *et al.* 1996). Credit for the establishment of the lacustrine ostracod faunas of coastal Brazil and of West Africa must go to Krömmelbein whose many papers are nicely summarized in Krömmelbein (1966).

Rift environmental events

The stratigraphical sections encountered in the West African rift system are dateable throughout as Lower Cretaceous (Neocomian–Aptian) in age. Syn-rift sediments of Jurassic age, comparable with the lacustrine sediments of the interior basin of the Democratic Republic of Congo (Lualaba Series) (Grekoff 1957), have not been recorded, although pre-rift Jurassic sediments (M'vone Formation) are known from Gabon. Consequently, the lacustrine events noted here are entirely of Lower Cretaceous age.

Although the lithostratigraphic terminology used throughout the different countries of West Africa is not the same, nevertheless, the same sequence of depositional environments recognizable reflect the co-occurrence of environmental conditions throughout the entire region. These sequential events are illustrated on Fig. 2 here for Gabon, Congo and Angola (Cabinda).

Ostracod zonation

Changes in the ostracod faunas through time are affected both by the longevity of the genera/species comprising the assemblages and by the environmental changes that may produce conditions lethal to some. As the entire West African rift underwent a synchronous sequence of events, the effects of environmental change had no major effect on the stratigraphical application of these faunas. Nowhere, for example, has it been possible to demonstrate a transgressive period of lake deepening that was concurrent with a lake lowstand in which saline conditions predominated. An axial connection between all the major lakes would appear, therefore, to have been in existence. Consequently, time-equivalent faunas are correlatable throughout the syn-rift sequences and may

Table 1. *Cyclical sequence stratigraphy of lacustrine syn-rift sediments (see Fig. 3 for ostracod zonation)*

Second cycle	**Top of last syn-rift sequence**
Phase 7	Lowstand systems tract, saline lacustrine (AS12) sediments of Aptian age unconformably overlie Barremian, highstand clastic and carbonate sediments. Early Aptian sediments may be removed as a result of structural movements. Low diversity ostracod faunas present but with abundant individuals. Dated as Zones AS11 and 12.
Phase 6	Coarsening-upwards highstand systems tract, late Barremian lacustrine sediments with abundant ostracods. Dated as Zone AS10.
Phase 5	Fining-upwards transgressive systems tract sequence terminating in a maximum flooding surface as late Barremian lakes enlarge after rifting, development of accommodation space and increased precipitation. Ostracod species diversity increases, while clastic and carbonate reservoirs develop along with anoxic, organic-rich source rocks. Dated as Zones AS 8 and 9.
First cycle	**Top of first syn-rift sequence**
Phase 4	Early Barremian sediments will be missing where an unconformable boundary exists between early Barremian and Neocomian sediments. In terms of ostracod zonation, it is usually Zones AS5–7 that are removed. Early Barremian sediments, when preserved, indicate lowstand sedimentation with probable saline lake waters. Clastic and carbonate reservoir rocks may occur, along with organic-rich source rocks. These lowstand systems tract conditions bring the first cycle to a close. Dated as AS6 and 7.
Phase 3	Neocomian, coarsening-upwards highstand systems tract. Dated as Zone AS5.
Phase 2	Neocomian lakes deepen and ostracod diversity increases. A maximum flooding surface may be preserved at the top of this fining-upwards transgressive system tract sequence. Dated as Zones AS3 and 4.
Basal unit	
Phase 1	Lowstand systems tract sag basin swamps of Neocomian age, gradually become shallow lakes as rifting develops; fluvially-derived clastics frequently infill basin. Lowstand source and reservoir rock potential, although lakes not as large as those developed during late Barremian times. Dated as Zone AS2.

be used to date and correlate sections within all the countries of West Africa from Gabon to Angola. In West Africa, the current ostracod zonation scheme used is that established by Elf, with some minor modification in the Aptian. The zonation used is prefixed AS (Ante Saliferes) and numbered 2–12 (an AS1 Zone was not established but would be relevant to the topmost Jurassic), and can be used to date the two cycles of lake development and, consequently, of rifting.

Syn-rift I – first cycle

The initial phase of basin sag (lowstand), in which fluvial sediments frequently predominate, occurs at the base of the syn-rift succession in which the faunas indicate shallow, freshwater conditions, that have been dated as Zone AS2. Subsequent lake development commences in AS3 Zone times, continues through AS4 Zone times and reaches highstand during AS5 Zone times. The end of the first depositional lake cycle, as well as that of rifting, coincide at AS6 and 7 Zone times with a lowstand and the development of saline conditions. At that time, the lakes shallowed and infilled with sediment; subsequent aerial exposure frequently removing these later deposited sediments, producing a significant unconformity as a result.

Continuous sedimentation is restricted to the deep lacustrine depocentres where sedimentation was continuous throughout the Lower Cretaceous. Dating of these sediments is provided by the short stratigraphical range of the genera concerned (as illustrated in Fig. 3) where Zones AS2–4 are dominated by *Cypridea* spp. Geologically, Zones AS2–5 represent the Neocomian, ending with an *Ilhasina* spp.–dominated fauna, while Zones AS6 and 7 (dominated by *Reconcavona* spp. and *Paracypridea* spp.) represent the onset of the Barremian.

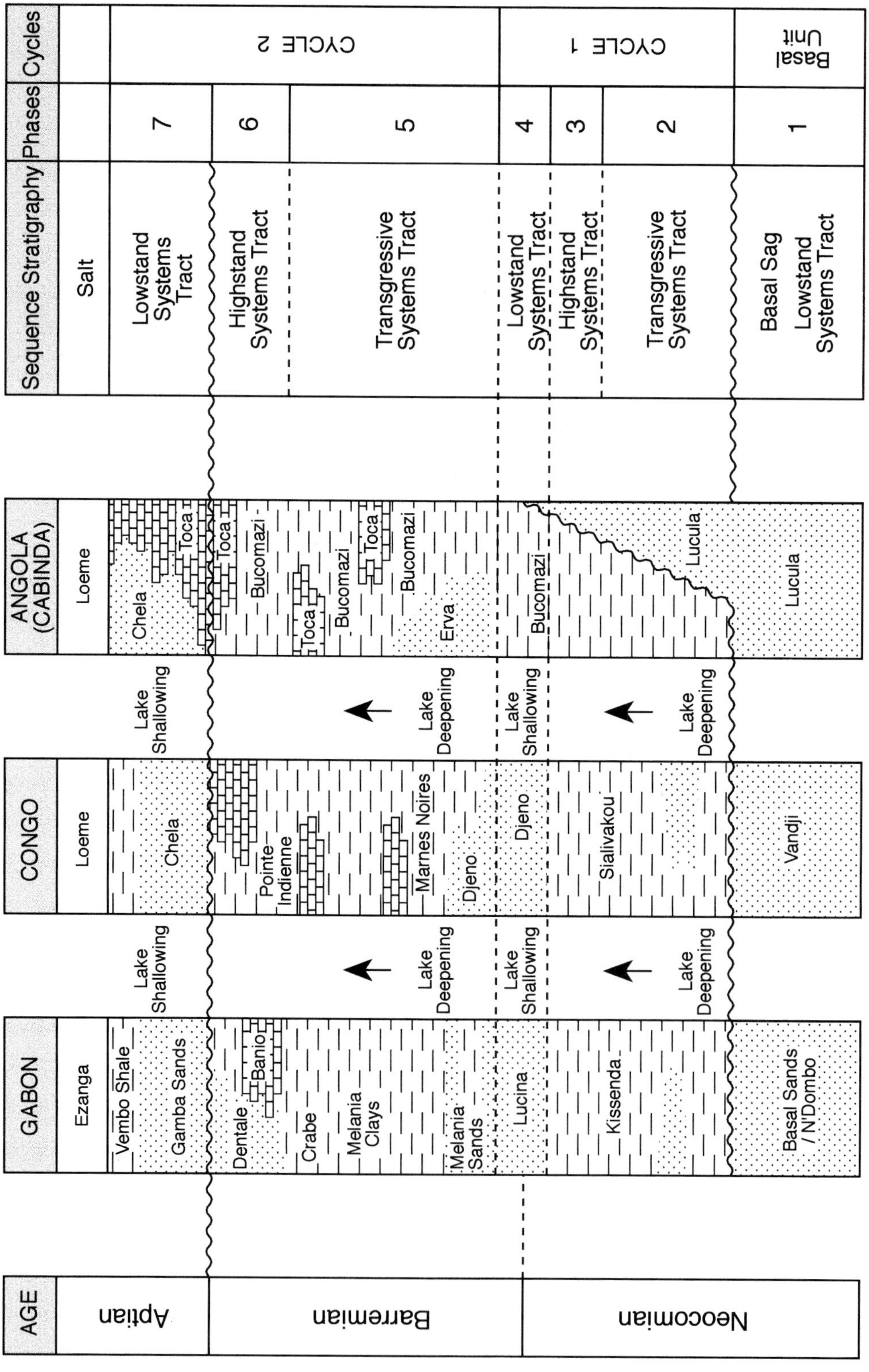

Fig. 2. Generalized rift sequence stratigraphy for Gabon, Congo and Angola (Cabinda).

CYCLES / PHASES	AGE	ASSEMBLAGES	ENVIRONMENT	ZONE	MORPHOLOGY
End of Second Cycle **Phase 7** AS11-AS12 Lowstand Systems Tract	APTIAN	*Pattersoncypris angulata symmetrica*	Saline lacustrine	AS12	
		Pattersoncypris angulata salitrensis	Saline lacustrine ~~~ Unconformity ~~~	AS11	
Phase 6 AS10 Highstand Systems Tract	BARREMIAN	*Hourcqia africana africana; Paracypridea cf.langdoni; Damonella tinkoussouensis; Reconcavona ? ultima; Cypridea loango;*	Highstand lacustrine	AS10	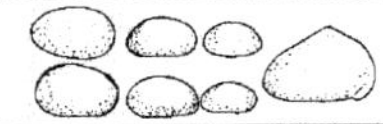
Maximum Flooding Surface **Phase 5** AS8-AS9		*Hourcqia africana confluens* *Cypridea kouilouensis; Petrobrasia zairensis; P.diversicostata; P.congolensis;P.pontanigrensis; Theriosynoecum papillaris; Cypridea hollensis; Cypridea sylvesterbradleyi; C.sostensis; C.mataensis; C.devexa; C.fida; C.mira; C.nannorostrata; C.onusta; C.matinversa; Petrobrasia gutta; Bisulcocypris postangularis Salvadoriella ? pusilla*	Maximum flooding surface Transgressive lacustrine	AS9	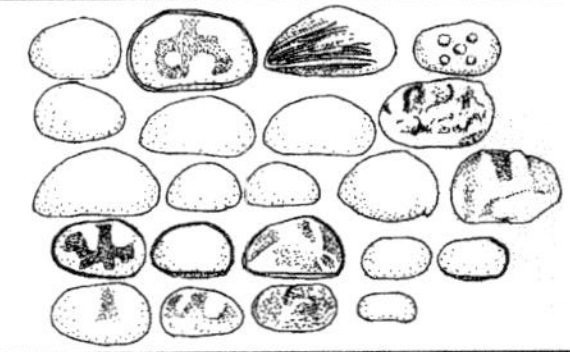
Transgressive Systems Tract **Start of Second Cycle**		*Cypridea remboueensis; C.bibullata tribullata Reconcavona cancellata; R.centronodosa Salvadoriella redunca redunca; Paracypridea quadrirugosa; Petrobrasia gibbosa*	Transgressive lacustrine	AS8	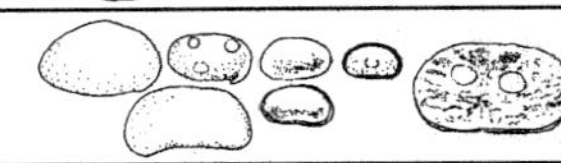
End of First Cycle **Phase 4** AS6-AS7		*Reconcavona swaini; R.triebeli; Cypridea bapounou*	Restricted lacustrine	AS7	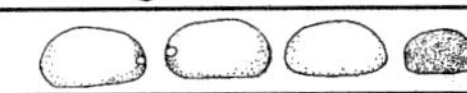
Lowstand Systems Tract		*Reconcavona striatula; R.gastracantha; R.swaini; Paracypridea elegans; P.similis; P.langdoni; P.obovata*	? Saline lacustrine	AS6	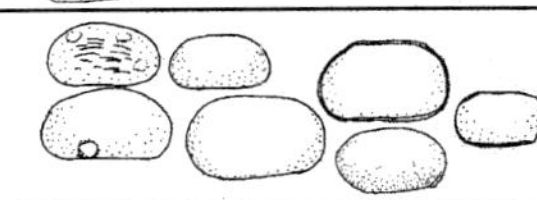
Phase 3 AS5 Highstand Systems Tract	NEOCOMIAN	*Ilhasina torosa; I.amphotera; I.remanei; Cypridea vulgaris; Paracypridea maacki; P.impar; P.bicallosa.*	Highstand lacustrine	AS5	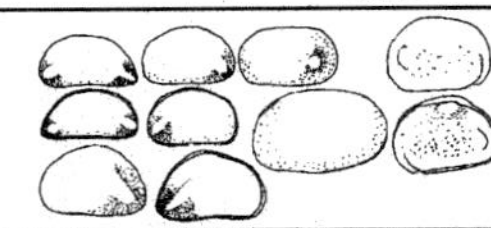
Phase 2 AS3-AS4 Transgressive Systems Tract		*Cypridea hadronodosa; C.salvadoriensis; C.subtilis; C.vulgaris; C.ellipsoidea; C.sellata; C.grekoffi; Tucanocypris camposi; T.apiculata*	Maximum flooding surface	AS4	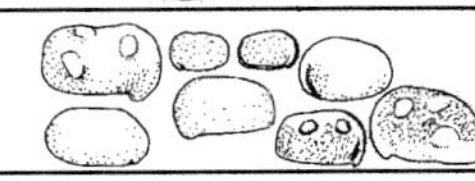
Start of First Cycle		*Cypridea ambigua; C.sellata; C.trindosa*	Fluvial/lacustrine	AS3	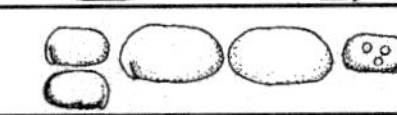
Phase 1 AS2 Basal Unit		*Cypridea kegeli; C.ambigua; C.primaria;* *C.acuta; C.armata; Darwinula spp.*	Fluvial/lacustrine	AS2	

Fig. 3. Ostracod zonation of the Pre-Salt section, West Africa.

Syn-rift II – second cycle

The second period of tectonism, coupled with increased water supply, provided the mechanism by which the major lacustrine transgressive phase (Zones AS8 and 9) of the late Barremian was both initiated and made possible. It was during this period that the major ostracod diversity developed and in which *Cypridea* spp., in association with *Petrobrasia* spp., dominated. Although there was a diverse ostracod fauna throughout the freshwater transgressive phase, at the end of this period, faunal diversity declined and in AS10 Zone times a period of highstand returned to be typified by a dominance of smooth-shelled ostracods. Lowstand conditions of the early Aptian (Zones AS11 and 12), represented by a further decline in species diversity in which all the ostracods are smooth-shelled, brought this final phase of syn-rift sedimentation to a close as the lakes infilled and dried up. In some cases submergence of the rift margins was not completed and residual surface relief remained.

The pre-salt Aptian sediments developed throughout West Africa appear not to be as complete as those encountered in Brazilian Atlantic basins. Consequently, some future revision of the AS10–12 zonation might become necessary should a more complete succession be demonstrated. However, the sequence of environmental changes observed, together with their associated ostracod faunas, indicate that a major break in deposition is unlikely to exist. While *Hourcqia africana* (of which the subspecies *confluens* in Zone AS9 pre-dates the AS10 subspecies *africana*) introduces the onset of highstand conditions, *Pattersoncypris* spp. characterize the Aptian saline lowstand (Zones AS11 and 12) of both West Africa and Brazil.

Minor differences occur within the ostracod assemblages of the West African Basins due to the fact that some species are locally restricted, or at least appear to be so at the present time. Although the new species encountered in West Africa have not yet been recorded from the coastal basins of Brazil, the fact that the ostracod faunas are essentially identical does allow for a correlation of the zonal schemes that presently exist. As labelling of the zones by means of a numerical system stands in the way of promoting a single unified scheme, it is hoped that, under the auspices of IGCP 381, the use of ostracod species/subspecies names as zonal indices will lead to the acceptance of a single scheme for all the South Atlantic basins. A provisional comparison of the two zonal schemes is presented in Table 2.

Table 2.

	West Africa (Grosdidier *et al.* 1996)	Brazil (Moura 1972)
Aptian	AS12	
	AS11	
Barremian	AS10	
	AS9	RT009(3)-RT008(1)
	AS8	RT007(4)-RT007(1)
	AS7	RT006(2)-RT005(5)
	AS6	RT005(4)-RT005(2)
Neocomian	AS5	RT005(1)-RT004(3)
	AS4	RT004(2)-RT003(2)
	AS3	RT003(1)-RT002(2)
	AS2	RT002(1)
Jurassic		RT001

Polymorphism and diversity

As observed during marine transgressions, species diversity increases during periods of water-level rise. This is precisely the situation in lacustrine environments when the lakes deepen (precipitation is greater than evaporation) and/or the waters transgress onto the surrounding land surface. During these transgressive cycles ostracods diversify and their populations contain a significant number of ornamented species. Many freshwater ostracods, however, are always smooth and a diverse fauna of smooth species could be supportive of a transgressive phase – although, in all probability, still indicative of increasingly saline conditions. Ornamented ostracods exist both throughout periods of lake water deepening and during lake recessions, when alkaline waters predominate. A substantial drying up of the lake, normally accompanied by arid or semi-arid land conditions, will produce conditions in which ornate ostracods either cannot survive or lose their ornamentation.

Quantifying the species/subspecies diversity present within a section often only establishes the diversity present within that part of the lake system. This is because the sections encountered are not always representative of the basin as a whole. It is possible, however, through the interpretation of data from a large number of sections (wells), to establish average figures for a particular dateable horizon and the systems tract to which it belongs. The number of species/subspecies potentially inhabiting a systems tract is given in Table 3 and comparable figures may be gleaned from Moura (1972).

Table 3.

Zone	Species/Subspecies Diversity	Systems Tract
AS12	1–< 2	LST
AS11	1–< 2	LST
AS10	5–< 12	HST
AS9	10–< 20	TST
AS8	7–< 10	TST
AS7	3–< 8	LST
AS6	7–< 9	LST
AS5	7–< 12	HST
AS4	4–< 8	TST
AS3	3–< 5	TST
AS2	3–< 6	LST

Polymorphism describes the occurrence, within a single species, of different morphological characteristics that develop in response to changing environmental conditions. With respect to freshwater/saline lacustrine ostracods, a species might be either ornamented or smooth-shelled, depending upon the water chemistry of the lake or other water body in which it is living. Morphological variations have been recorded in both *Cypridea sylvesterbradleyi*, from the late Barremian lake sediments of Gabon and Angola (Bate 1994), and in *Schlerocypris* gp. *clavularis* from the Recent sediments of Lake Bogoria, East Africa (Carbonel & Peypouquet 1983). Polymorphism, in the strict sense, is rarely observed in the lacustrine ostracods of these West African Rift lakes. It is the replacement of an ornamented species by a smooth-shelled, related species (or vice versa) that occurred here. During periods of lowstand, where the lake waters become increasingly saline, changes in shell morphology bring about the replacement of ornamented genera by smooth genera, the change occurring at a more genetically complex level.

Application to oil exploration

Commercially, the Cretaceous West African lakes are known to contain excellent hydrocarbon source rocks (Burwood *et al.* 1995) that have fed reservoir sands/carbonates of both syn- and post-rift sequences. Importantly, it is the size of these lakes that has, to a considerable extent, determined the hydrocarbon reserves available. Consequently, in order to be able to estimate the exploration potential of a particular area, it is essential that both the size and the environmental conditions experienced by the lakes are fully understood. The examination and interpretation of the indigenous lacustrine faunas and floras provides much of this information. With respect to applying these observations to hydrocarbon exploration, it is during periods of maximum lake development and deepening that the major source sediments are deposited. Also, potential reservoir sediments deposited prior to and during periods of lake expansion will be sealed by the deposition of transgressive organic-rich muds, thereby presenting the explorationist with syn-rift, as well as post-rift, targets. As far as developing high-quality organic source rocks is concerned, saline conditions also provide excellent conditions for the development of an anoxic bottom layer by restricting water turn over and preventing the introduction of oxygenated water to the lake bottom. The fact that these lakes are almost certainly smaller than their freshwater counterparts could reduce their hydrocarbon potential, although it is possible that this factor could be offset by the source rocks being richer or just at the right level of maturity.

Opening of the South Atlantic

As a consequence of recognizing the changing environmental events through which the West African lakes passed, it has been noted that the major period of lake deepening consistently occurred during the late Barremian, after the necessary accommodation space had been provided following emplacement of the Atlantic Hinge. To explain this ubiquitous lake expansion event it is necessary to consider the strong probability that a major climate change occurred at that time. The change envisaged is for a period of increased precipitation to have commenced at about the beginning of the late Barremian (ostracod Zones AS8 and 9), thereby enabling the major rivers of that time to flood the rift, something they had been unable to do during the Neocomian–early Barremian. That there should have been such a dramatic change in climate at the centre of Gondwana requires explanation and the one proposed here concerns the onset of continental break-up at a much earlier time than has been suggested previously. Although marine Aptian sediments containing foraminifera having Tethyan affinities have been recorded from the Sergipe–Alagoas Basin (Koutsoukos & Bengtson 1993), more interestingly, marine carbonates dated as Berriasian and resembling peri-Tethyan Maiolica-type pelagic limestones have been found near the intersection of the Mid-Atlantic Ridge and the Romanche Transform Fault (Bonatti *et al.* 1996). Whether the marine waters entered this equatorial region via the Benue Trough, prior to separation of Africa from South

America, is not yet clear. What is indicated, however, is that the incoming of Tethyan marine waters into a proto-South Atlantic occurred prior to the late Barremian. This would, in all probability, have led to a significant change in climate and, following the period of desiccation that occurred at the close of the Neocomian–early Barremian, is considered to have been responsible for the expansion of the late Barremian lake system in West Africa.

Dingle (1996), from his knowledge of South African marine Cretaceous ostracods, observed that the Walvis Ridge was not breached until Cenomanian–Turonian times, and it is only after this time that South African ostracods were freely able to migrate northwards into the developing South Atlantic. On the other hand, the marine Albian ostracods present in the Pinda (Angola) and Madiela (Gabon) Formations already indicate a northern opening of the South Atlantic through their genetic affinity with European Albian ostracods. This, together with the record of marine Berriasian-dated limestones (Bonatti *et al.* 1996) and Aptian foraminifera having Tethyan affinities (Koutsoukos & Bengtson 1993) within the equatorial region of the South Atlantic, provides supportive evidence that a Tethyan connection was established prior to the breaching of the Walvis Ridge. Although the increased precipitation and resultant lake expansion identified throughout late Barremian times does not necessarily support either a northern or a southern opening of the South Atlantic, the faunal evidence available does support a prior northern opening.

I would like to thank my co-editor, Nick Cameron, for his continued interest in this work and for kindly reading through the manuscript. This paper has benefitted considerably as a result.

References

ABRAHAO, D. & WARME, J. E. 1990. Lacustrine and associated deposits in a rifted continental margin – Lower Cretaceous Lagoa Feia Formation, Campos Basin, offshore Brazil. *In*: KATZ, B. J. (ed.) *Lacustrine Basin Exploration – Case Studies and Modern Analogs*. AAPG Memoir, **50**, 287–305.

BATE, R. H. 1994. On *Cypridea sylvesterbradleyi* Bate sp.nov. *Stereo-Atlas of Ostracod Shells*, **21**, 103–106.

BOEUF, M. A. G., CLIFF, W. J. & HOMBROEK, J. A. R. 1991. Discovery and development of the Rabi–Kunga field: a giant oil field in a rift basin, onshore Gabon. *World Petroleum Congress*, Buenos Aires, **2**, 33–49.

BONATTI, E., LIGI, M., BORSETTI, A. M., GASPERINI, L., NEGRI, A. & SARTORI, R. 1996. Lower Cretaceous deposits trapped near the equatorial Mid-Atlantic Ridge. *Nature*, **380**, 518–520.

BRICE, S. E., COCHRAN, M. D., PARDO, G. & EDWARDS, A. D. 1982. Tectonics and sedimentation of the South Atlantic rift sequence – Cabinda, Angola. *In*: WATKINS, J. S. & DRAKE, C. L. (eds) *Studies in Continental Margin Geology*. AAPG Memoir, **34**, 5–18.

BURWOOD, R. G., DE WITTE, S. M., MYCKE, B. & PAULET, J. 1995. Petroleum geochemical characterisation of the Lower Congo Coastal Basin Bucomazi Formation. *In*: KATZ, B. (ed.) *Petroleum Source Rocks, Casebooks in Earth Sciences*. Springer, 235–263.

CARBONEL, P. 1982. Les Ostracodes, traceurs des variation hydrologiques dans des systèmes de transition eaux–eaux salées. *Mémoire Société De France*, 117–128.

—— & PEYPOUQUET, J.-P. 1979. Les Ostracodes des series du Bassin de l'Omo. *Bulletin De l'Institut De Géologie Du Bassin D'Aquitaine, Bordeaux*, **25**, 167–199.

—— & —— 1983. Ostracoda as indicators of ionic concentrations and dynamic variations: methodology (Lake Bogoria, Kenya). *In*: *Applications of Ostracoda*. University of Houston Geoscience, 264–276.

DINGLE, R. V. 1996. Cretaceous ostracoda of the SE Atlantic and SW Indian Ocean: A stratigraphical review and atlas. *In*: JARDINE, S., DE KLASZ, I. & DEBENAY, J.-P. (eds) *Géologie de l'Afrique et de l'Atlantique Sud*. Elf Aquitaine Memoire, **16**, 1–11.

GREKOFF, N. 1957. Ostracodes du Bassin du Congo. I. Jurassique superieur et Cretace inferieur du nord du bassin. *Annales du Musee Royal du Congo Belge, Series 8*, **19**, 1–89.

GROSDIDIER, E., BRACCINI, E., DUPONT, G. & MORON, J.-M. 1996. Non-marine Lower Cretaceous biozonation of the Gabon and Congo Basins. *In*: JARDINE, S., DE KLASZ, I. & DEBENAY, J.-P. (eds) *Geologie de l'Afrique et de l'Atlantique Sud*. Elf Aquitaine Memoire, **16**, 67–82.

KARNER, G. D. & DRISCOLL, N. W. 1999. Tectonic significance of syn and post-rift sedimentary packages comprising the West African and Eastern Brazilian margins. *This volume*.

KOUTSOUKOS, E. A. M. & BENGSTON, P. 1993. Towards an integrated biostratigraphy of the Upper Aptian–Maastrichtian of the Sergipe Basin, Brazil. *Documents Des Laboratoires De Géologie De Lyon*, **125**, 241–262.

KROMMELBEIN, K. 1966. On 'Gondwana Wealden' Ostracoda from NE Brazil and West Africa. *Proceedings of the 2nd West African Micropalaeontological Colloquium*, 113–119.

LAMBIASE, J. J. 1990. A model for tectonic control of lacustrine stratigraphic sequences in continental rift basins. *In*: KATZ, B. J. (ed.) *Lacustrine Basin Exploration – Case Studies and Modern Analogs*. AAPG Memoir, **50**, 265–276.

MOURA, J. A. 1972. Algumas especies e subspecies novas de ostracodes da bacia reconcavo/tucano. *Bulletin tech. Petrobras*, **15**, 245–263.

Walvis Ridge barrier: its influence on palaeoenvironments and source rock generation deduced from ostracod distributions in the early South Atlantic Ocean

R. V. DINGLE

British Antarctic Survey, Madingley Road, Cambridge CB3 0ET, UK (Present address: Geological Institute, University of Copenhagen, Øster Valdgade 10, 1350 Copenhagen K, Denmark)

Abstract: The early South Atlantic was subdivided latitudinally by the proto-Walvis Ridge, which separated early Cretaceous open marine conditions in the south from non-marine environments to the north. Changes in the palaeogeography, which can be monitored by the distribution of ostracod assemblages, determined the establishment of a wide variety of depositional environments, including evaporitic and total organic carbon-rich facies. Limited northward marine influxes across the barrier are postulated for mid-Aptian time, while large-scale southward migrations of marine ostracods from the northern sector occurred in early Cenomanian and/or Turonian times. The latter established cosmopolitan faunas in the southern Atlantic and effectively marked the breakdown of the Walvis Ridge as a physical barrier to north–south migrations, although possible oceanographic factors militated against longitudinal faunal movements. Variations in the composition of ostracod assemblages during so called 'oceanic anoxic events' and the production of potential source rocks is briefly reviewed.

Opening of the South Atlantic began in late Cretaceous time (M9–10, 131 Ma, Valanginian–Hauterivian boundary) after an earlier phase of intracontinental rifting, which may have extended from the late Jurassic (e.g. Sibuet *et al.* 1984; Nürnberg & Müller 1991). For the first 30–40 Ma, sedimentary and oceanographic environments were significantly different either side of a progressively enlarging shallow marine and subaerial volcanic edifice (proto-Walvis Ridge), which exerted a major influence on marine migrations between the Austral and Equatorial regions of western Gondwana (Reyment 1980). These changes can be documented using Ostracoda and, in a classic example of palaeobiogeographic reconstruction, Krömmelbein (e.g. 1966, 1967, 1976) first established the congruity of the non- and shallow marine Neocomian–early Albian faunas between the coastal basins of Gabon–Congo and Brazil (Sergipe) in the northern sector of the South Atlantic. Tambareau (1982) and Dingle (1982, 1988) summarized the regional distribution of ostracods within Gondwana during its early break-up.

In the present contribution evidence from early–mid-Cretaceous ostracods for changing palaeoenvironments in the northern and southern sectors of the South Atlantic will be discussed, and the time at which the Walvis Ridge ceased to have a significant influence on faunal movements between the two sectors will be reviewed.

Faunas and palaeonvironments

Marine and non-marine early–mid-Cretaceous ostracod-bearing strata are found both north and south of the Walvis Ridge. In the northern South Atlantic they occur in the narrow conjugate coastal basis of Brazil and west-central Africa, while in the southern sector they occur offshore and in scattered coastal basins of southwest and southern Africa. No detailed account has been published of the faunas from the Argentine offshore areas and the closest well-documented sequences occur in the Andean Neuquén Basin (e.g. Malumián & Báez 1976) (Fig. 1a).

Non-marine sequences

During the rift phase of the South Atlantic opening, continental and freshwater sequences were deposited in narrow, elongate valleys along the length of the proto-ocean (Reyment & Tait 1972). Fluvial and lacustrine environments occurred from ?Hauterivian to at least Barremian time north of the Walvis Ridge (upper Coruripe and Cocobeach sequences) (Fig. 1), and in the Sergipe Basin extended into the lower part of the Aptian Muribeca Formation (e.g. Reyment & Tait 1972; Koutsoukos *et al* 1991*a*, *b*, Grosdidier *et al.* 1996). The ostracod assemblages in these sediments are diverse and abundant, with a total of at least 248 species, of which 192 occur in Brazil, 187 in Gabon and 131 (53%) are common

DINGLE, R. V. 1999. Walvis Ridge barrier: its influence on palaeoenvironments and source rock generation deduced from ostracod distributions in the early South Atlantic Ocean. *In:* CAMERON, N. R., BATE, R. H. & CLURE, V. S. (eds) *The Oil and Gas Habitats of the South Atlantic.* Geological Society, London, Special Publications, **153**, 293–302.

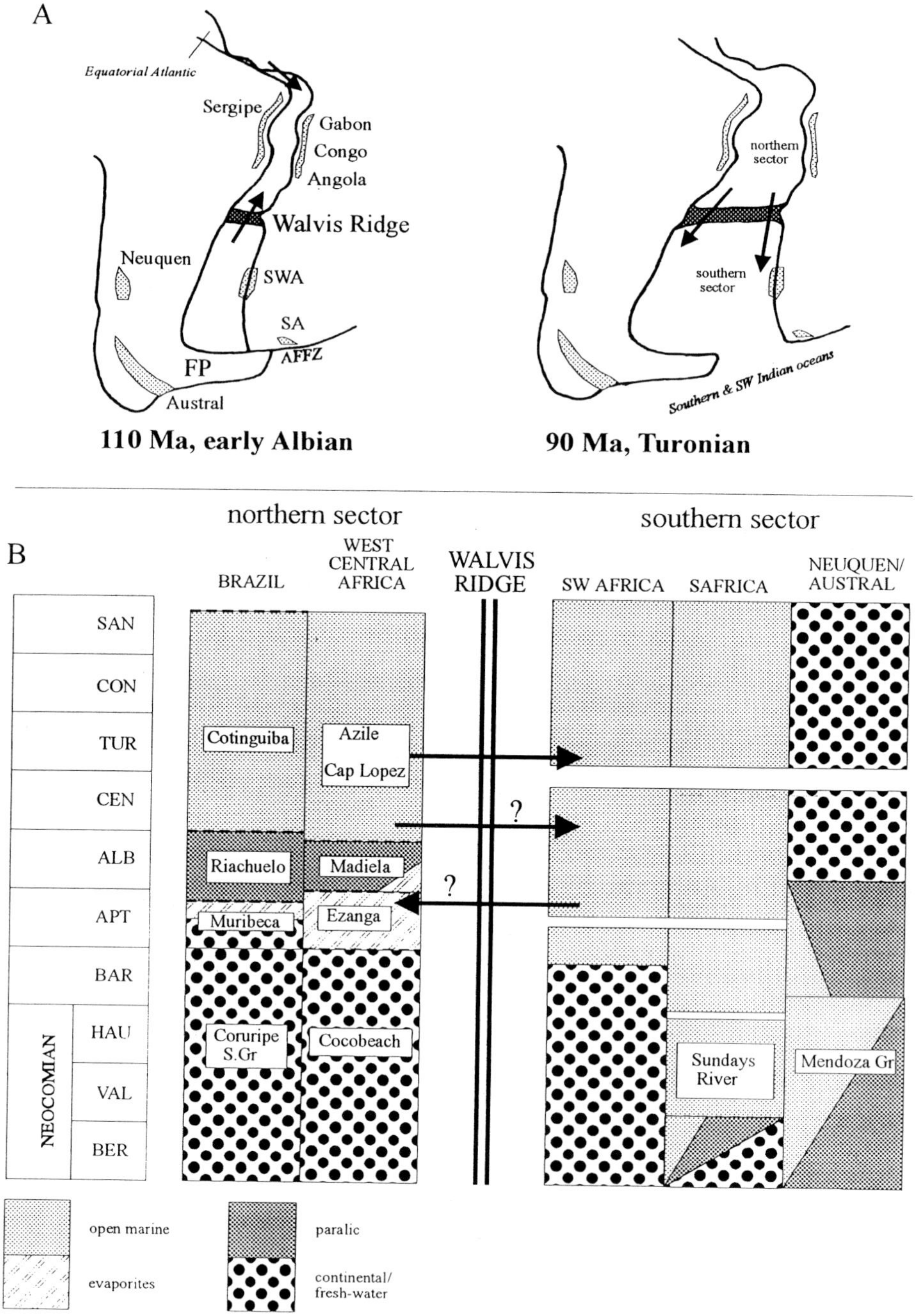

Fig. 1. Mid-Cretaceous palaeogeographies and stratigraphy of South Atlantic. (**a**) Location of sediment basins adjacent to the South Atlantic: 110 Ma reconstruction relates to late Aptian–Albian events and 90 Ma reconstruction relates to Cenomanian–Turonian events. Continental outlines and positions based on the reconstructions of Nürnberg & Müller (1991). Arrows show postulated faunal migration routes. SWA, basin off southwestern Africa; SA, basins off southern Africa; AFFZ, Agulhas–Falkland Fracture Zone; FP, Falkland Plateau. (**b**) Summary stratigraphies of main sediment basins north and south of the Walvis Ridge barrier. Only formations mentioned in text are identified. Nomenclature for west central Africa relates to the Gabon Basin. Arrows show postulated faunal migration events. Summarized from Reyment & Tait (1972), Brenner & Oertli (1976), Malumian & Baez (1976), Grosdidier (1979), McMillan (1990), Koutsoukos *et al.* (1991*b*), Duarte-Morais (1996) and Dupont (1996).

to both areas (Tambareau 1982). They have been used to effect a detailed biostratigraphy (e.g. Krömmelbein 1966; Grosdidier *et al.* 1996) and have been interpreted as indicating warm, non-seasonally humid conditions, that gave way in late Barremian time to a warm, seasonally arid climate (Grosdidier *et al.* 1996). Tambareau (1982) identified a high degree of endemism, and of the 131 species common to both sides of the South Atlantic, 88 (67%) belong to genera restricted to the southern hemisphere. Of the 33 genera recorded, 19 (58%) are restricted to the southern hemisphere, the most important of which were *Reconcavona* (27 spp.), *Petrobrasia* (19 spp.), *Cypridea* (*Sebastianites*) (17 spp.) and *Paracypridea* (16 spp.).

South of the Walvis Ridge off southwest Africa, continental sandstones, shales and lavas of Barremian age are overlain by marine sediments, but no freshwater faunas have been reported. Further southeast, along the rifted sector bounded by the Agulhas–Falkland Fracture Zone, the continental and paralic rift infill is of ?late Jurassic–early Valanginian age, but a preliminary report lists only two species (*Cypridea* sp. and *Ilyocypris* sp.) (McLachlan *et al.* 1976). From Argentina, Musacchio (e.g. 1979) has reported 35 species of non-marine ostracods from ?Berriasian–Aptian sediments in the Neuquén Basin. These occur in nine genera, three of which are restricted to the southern hemisphere, but none of which occur to the north of the Walvis Ridge (*Rayosoana, Huillicythere* and *Wolburgiopsis*).

Marine sequences

Neocomian–Barremian. Marine conditions were established earliest south of the Walvis Ridge, when the southern sector was connected to the proto-Indian Ocean around and across the Falkland Plateau (e.g. Dingle 1982). In the basins of southern Africa, a distinctive late Jurassic–Barremian fauna was characterized by progonocytherids (*Majungaella*, *Amicytherideal Fastigatocythere*, *Progonocythere*), cytherurids (*Paranotacythere*, *Procytherura*), schizocytherids (*Sondagella*), cytherididis (*Rostrocytheridea*) and platycopids (*Cytherella*, *Cytherelloidea*) (Brenner & Oertli 1976; Dingle 1982; Valicenti & Stephens 1984). A total of 54 species of ?Berriasian–Barremian ostracods have been recorded from the southeast South Atlantic margin (Dingle 1996).

In the Neuquén and Austral Basins of Argentina, the late Jurassic–Valanginian shallow-water marine ostracod faunas are very similar to those from southern Africa (e.g. Malumian & Baez 1976; Musacchio 1979; Ballent *et al.* 1997), and Tambareau (1982) compiled an inventory of 44 species. Seven of these are common to southern Africa: *Sondagella theloides*, *S. colchesterensis*, *Majungaella nematis*, *Cytherelloidea agyroides*, *Procytherura maculata*, *Rostrocytheridea ornata* and '*Progonocythere*' *reticulata*. This late Jurassic–Barremian austral Gondwana fauna has been termed the South Gondwana fauna (Dingle 1982, 1988). As discussed by Musacchio (1979), Tambareau (1982), Dingle (1982) and Musacchio & Simeoni (1995), it contains elements that are locally endemic (e.g. 50% of the Neuquén Hauterivian fauna), but over large areas there was migration, ultimately connecting Madagascar, India and western Australia.

Aptian–Albian. North of the Walvis Ridge, the Neocomian–Barremian non-marine sediments are abruptly overlain by paralic/marine strata with marine ostracods (Fig. 1). In west-central Africa these are represented by Aptian halites (equivalent to the Ezanga Formation in Gabon) (Duarte-Morais 1996; Vernet *et al.* 1996), and the first sediments with marine ostracods and planktonic foraminifera occur in the earliest Albian age strata (lower Madiela Formation) (Dupont 1996). In Brazil, lower Aptian sabkha and salina facies with anhydrites (upper Muribeca Formation) pass upwards into upper Aptian carbonate mudstones and shales with *Hedbergella* (lower Riachuelo Formation) (Reyment & Tait, 1972; Koutsoukos *et al.* 1991*b*). On both sides of the proto-Atlantic, Albian strata consist of alternating nearshore marine/lagoonal facies that pass upwards into more open water marine sediments (lateral equivalents of Madiela Formation in Gabon and Riachuelo Formation in Sergipe).

Published data on the early marine ostracods are more complete from the African basins, and the faunas clearly reflect the onset and establishment of marine conditions, with progressive increases in specific and generic diversity from paralic to open marine environments: seven species in six genera (late Aptian); 11 species in nine genera (early Albian, including all the Aptian taxa), and a large increase in the more open marine middle–upper Albian Madiela Formation (24 species in 20 genera) (see Tambareau 1982). Mussard (1996) has applied factor analysis to the microfossils and lithofacies and showed that changes from paralic to open water habitats were characterized by the occurrence of particular taxa. Similar facies/ostracod associations were suggested by Koutsoukos & da Silva-Telles (1992) from the Muribeca and Riachuelo Formations of Brazil. A feature of the marine sequences was the presence of two paralic species which occurred on both sides of the Atlantic: *Sergipella transatlantica* Krömmelbein and *Aracajuia benderi* Krömmelbein. The former was an early South Atlantic trachyleberid-type, superficially similar to the more open marine Cenomanian–Turonian '*Rehacythereis*' species recorded by Grosdidier (1979) from Gabon, but which has not been reported elsewhere. *Aracajuia* was monospecific, but had several apparently close relatives comprising a plexus of schizocytherid forms around *Sondagella* and *Acrocythere* (see Musacchio & Simeoni 1996), which were characteristic of the Hauterivian–Valanginian of southern Africa and the Neuquén Basin. *Aracajuia* does not have a schizodont hinge, but may have been a high-salinity adaptive that evolved in the northern sector basins.

Contemporaneous Aptian–Albian strata south of the Walvis Ridge are known from southern Argentina (*Majungaella* and related taxa) (Ballent & Whatley 1997), the African margin and the upper Falkland Plateau (Dingle 1984). The latter are intermediate depth faunas [a black shale deep water facies existed in the southern sector until early Albian time (Bolli *et al.* 1978)], but contain elements that are characteristically diverse and common in the shallower marine environments of southern and eastern Africa (e.g. *Majungaella*, *Isocythereis*, *Pirileberis* and *Arculicythere*). The earliest known marine sediments off southwest Africa were deepwater, organic-rich Aptian quartz sands and shales (McMillan 1990), from which only four species have been recorded (including two species of *Arculicythere*) (Dingle 1996). These assemblages were quite distinct from their contemporaries to the north of the Walvis Ridge which, in addition to *Sergipella* and *Aracajuia*, included *Veenia*, ?*Buntonia* and *Paracypris* (Grosdidier 1979), all taxa unknown from the southern sector at this time.

Cenomanian and younger. Ostracod faunas both north and south of the Walvis Ridge underwent marked changes during early Cenomanian–late Turonian time. These are best documented in the Gabon Basin, where only 19% of the species are common to both Albian and Cenomanian sequences: 67% of the early Cenomanian Cap Lopez Formation species are new appearances. At, or just below the top of the Albian Madiela Formation, *Sergipella* and *Aracajuia* become extinct, while *Haughtonileberis*, *Brachycythere* and *Majungaella* (= *Neocythere*) appear (Grosdidier 1979). *Brachycythere sapucariensis* Krömmelbein was a remarkably successful species which rapidly migrated and speciated, with closely similar forms in both Brazil and particularly west Africa (*B. ekpo*, *angulata*, *armata*, *oguni*), sometimes in monospecific assemblages (Krömmelbein 1976) (although it did not reach Argentina). *Haughtonileberis* (not yet reported from Brazil) became a relatively diverse taxon (five species) in the Cenomanian–Turonian of Gabon, while the widely distributed South Gondwana taxon *Majungaella* gained a foothold north of the Walvis Ridge, although it did not speciate significantly (e.g. Ballent *et al.* 1997). A similar faunal change occurred at the Cenomanian–Turonian boundary, where 63% of the species in the Azile Formation were new (with a large influx of trachyleberids, such as '*Rehacythereis*', as well as *Ovocytheridea*), although there is a somewhat higher degree of communality between the late Cenomanian and early Turonian assemblages (29%) than in the corresponding Albian–Cenomanian faunal change.

A less complete picture can be obtained from areas south of the Walvis Ridge because of breaks in the record. *Brachycythere* (*B. longicaudata*) and *Haughtonileberis haughtoni* are known from the Turonian of Tanzania (Bate & Bayliss 1969) and the Coniancian of southeast Africa (Dingle 1984), while a single valve of *Haughtonilberis*? is known from the Cenomanian off southwest Africa (Dingle 1996). By early Coniacian times at the latest, a rich and diverse trachyleberid-dominated fauna was established in southern Africa (Dingle 1984).

Discussion

Regional biogeographies

Figure 2 shows the ranges and distributions of some ostracod taxa germane to the documentation of marine migrations between the northern and southern sectors of the South Atlantic.

The Neocomian–Barremian lacustrine faunas of the northern sector appear largely to have been isolated from their terrestrial contemporaries south of the Walvis Ridge, implying that a physical barrier existed in pre-Aptian times. The first ingression of the sea into the northern sector occurred in Aptian time and may have been partly from the south across the Walvis Ridge, as the presence of *Aracajuia* suggests adaptation of the widespread South Gondwanan *Sondagella* to paralic/hypersaline conditions (Figs 1 and 2). *Neocythere*/*Majungaella* could also have used this route but, given the diversity and ubiquity of this plexus in South Gondwana, a more vigorous influx might have been expected. Similarly, the previously widespread South Gondwana taxon *Rostrocytheridea* did not move northwards. Consequently, the major faunal influx at this time was probably from the Equatorial Atlantic,

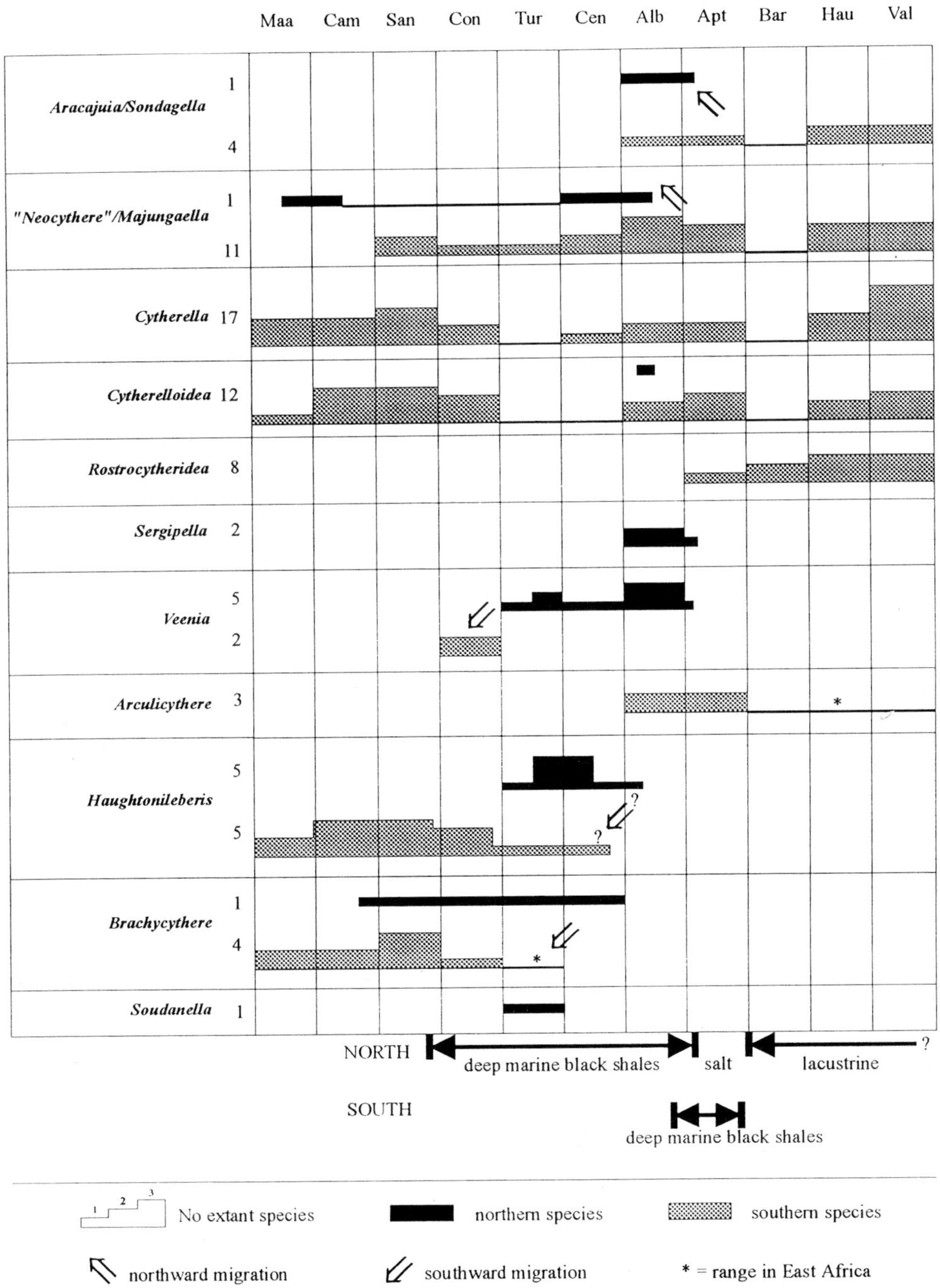

Fig. 2. Temporal and spatial distribution of important mid-Cretaceous South Atlantic ostracod taxa. Data from Bate & Bayliss (1969), Krömmelbein (1967, 1976), Brenner & Oertli (1976), Malumián & Báez (1976), Grosdidier (1979), Musacchio (1979), Tambareau (1982), Dingle, (1984, 1996), Valicenti & Stephens (1984), Koutsoukos & da Silva-Telles (1992), Musacchio & Simeoni (1996), Mussard (1996) and Ballent *et al.* (1997). Numbers in taxon boxes are total recorded number of species. Thin solid lines indicate no record, but assumed presence (except where East African data are indicated). Northward and southward migrations relate to faunal movements across the Walvis Ridge barrier. Published data on *Cytherella* and *Cytherelloidea* are incomplete. Periods of deepwater black shale deposition either side of the Walvis Ridge are based on Bolli *et al.* (1978) and Stow & Dean (1984).

where *Veenia* (which rapidly diversified in Albian times in the northern sector) already existed, along with several Aptian progonocytherids (Swain 1976). Currently published data do not allow an assessment of the distribution, and hence migrations, of the platycopids (*Cytherella* and *Cytherelloidea*) north of the Walvis Ridge (e.g. Grosdidier 1979), although it is known that they were abundant and diverse to the south of the barrier (e.g. Tambareau 1982).

A further faunal influx into the northern sector of the South Atlantic in late Albian time from a west African provenance may account for the establishment of the trachyleberid line leading to the earliest *Haughtonileberis* It is the subsequent migration of this taxon into the southeastern sector of the South Atlantic (where it later became a major element) that provides possible dates for the large-scale breaching of the barrier by surface waters and the migration of northern sector species into the South Gondwana bioprovince. This may have taken place in two stages, initially in ?early Cenomanian time and/or in late Cenomanian–early Turonian time, when influx of the aggressive colonizer *Brachycythere* occurred. Evidence suggests that these faunal migrations across the Walvis Ridge barrier were partitioned because characteristic taxa that colonized the southwestern South Atlantic regions (e.g. *Togoina*, *Ovocytheridea*, *Nigeria*) do not appear to have moved into the southeastern South Atlantic, and vice versa (e.g. Bertels 1977; Tambareau 1982). Presumably, this reflected the presence of deepwater environments over the mid-ocean ridge which had already led to an east–west faunal divide in the northern sector.

By the end of Cenomanian time, plate tectonic reconstructions suggest that the northern sector of the South Atlantic was connected to the Equatorial Atlantic (e.g. Nürnberg & Müller 1991) (Fig. 1). Faunal studies confirm this (Koutsoukos 1992), at least at intermediate water depths, and Coniacian faunas in the southern South Atlantic region have a cosmopolitan aspect (e.g. Dingle 1988). It would seem, however, that connections at intermediate–deep water depths were complete across neither the Equatorial narrows nor the Walvis Ridge until early Santonian time, when the deepwater black shale facies immediately north of the Walvis Ridge were finally displaced by oxygenated waters (Stow & Dean 1984) (Fig. 3a).

Ostracod faunas and total organic carbon-rich source rock distributions

Dark, organic-rich, argillaceous Cretaceous sediments occur in both shallow and deepwater marine environments north of the Walvis Ridge. Their geochemistry, sedimentology and stratigraphy have been described by Bolli *et al.* (1978), Dean *et al.* (1984), Stow & Dean (1984), Koutsoukos *et al.* (1991*b*) and Koutsoukos & Bengtson (1993), amongst others. Figure 3a summarizes the temporal and spatial distribution of these sediments, along with published total organic carbon (TOC) measurements. Genesis of these rocks and similar strata has been the subject of considerable debate, particularly in view of their economic potential and apparent worldwide distribution in mid-Cretaceous sediments (the so-called 'oceanic anoxic events') (Dean *et al.* 1984; Arthur *et al.* 1990), while possibly modern analogues have been investigated by Pedersen & Calvert (1990).

Studies of modern and fossil benthic and planktonic organisms associated with this lithofacies have produced a consensus that low sea floor dissolved oxygen levels are an integral factor in the preservation of the high organic carbon contents, and that variations in these levels can be monitored by observing time-series in faunal and floral changes (e.g. Kauffman & Sageman 1990; Oschmann 1994). Microfossils are ideally suited to this, but sufficiently detailed data from strata north of the Walvis Ridge are available only for foraminifera. Koutsoukos (1992) and Massala *et al.* (1996) concluded that oxygen-poor benthic environments can be characterized by the dominance of various species of *Gabonita*, while the presence of locally abundant and well-preserved planktonic assemblages indicates that water with low dissolved oxygen values periodically occurred at depths well above the deep basin floors. In addition, both authors cite evidence for the formation of such habitats in outer shelf environments considerably later (Campanian–Maastrichtian) than the time at which deepwater circulation from the Equatorial Atlantic is thought to have dispelled abyssal oxygen-poor conditions north of the Walvis Ridge (Santonian) (Fig. 3a). This suggests that the younger Cretaceous TOC-rich sediments, at least, were generated by enhanced near-surface productivity related to continental margin upwelling, in contrast to deep, oxygen-poor water masses (see Dean *et al.* 1984; Pedersen & Calvert 1990). While critical data are currently not available on the ostracod faunas from north of the Walvis Ridge, it is germane to comment on faunal changes recorded in similar-aged sediments from elsewhere. Figure 3b has been compiled from data cited in Jarvis *et al.* (1988) across the Cenomanian–Turonian oceanic anoxic event in southern England. The TOC-rich event can be identified as a strong (> 3 ppt) $\delta^{13}C$ excursion,

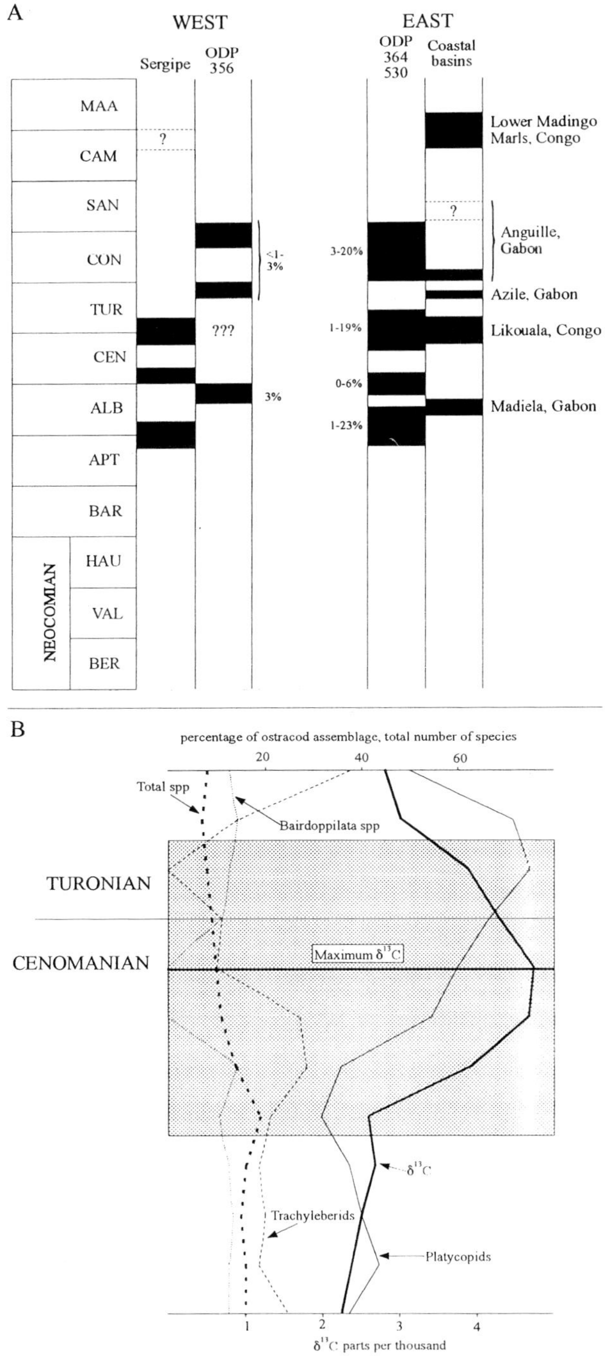

Fig. 3. Upper Cretaceous 'total organic carbon' (TOC)-rich sediments. (**a**) TOC-rich sediments (wt %) in the South Atlantic Ocean north of the Walvis Ridge. Data sources: Sergipe Basin, Koutsoukos (1992): ODP sites, Dean *et al*, (1984); coastal west-central Africa, with formational names, Dupont (1996), Massala *et al.* (1996), Vernet *et al.* (1996). ???, Hiatus. (**b**) Variations in ostracod taxa (% total species), total number of species, and $\delta^{13}C$ values across the Cenomanian–Turonian so-called 'oceanic anoxic event' in southern England (no vertical scale, total section is *c.* 11 m). Stippled zone is vertical extent of $\delta^{13}C$ anomaly. Compiled from data in Jarvis *et al.* (1988).

which reached a maximum in the uppermost Cenomanian, and to which ostracod assemblages responded by a progressive decrease in simple species diversity, although there was no specific response across the $\delta^{13}C$ peak. The diversity decline was accompanied by a simultaneous decrease in the number of extant species of trachyleberids and *Bairdoppilata*, and a concomitant strong relative increase in the number of platycopids (*Cytherella* spp. and *Cytherelloidea* spp.) as a percentage of the total species count. A marked reverse trend is evident above the $\delta^{13}C$ anomaly in the lower part of the Turonian as the fauna recovered with amelioration of the adverse benthic environment. Jarvis *et al.* (1988) considered various possibilities for the success of the platycopids in coping with the apparently harsh conditions at the height of the oceanic anoxic event and proposed that it might be connected with their appendage morphology and reproductive strategies. Whatley (1991) has developed this hypothesis and suggested that platycopid distribution can be used as a proxy for oxygen-poor environments. It should be noted, however, that in the oceanic anoxic event described by Jarvis *et al* (1988) nine taxa have a first appearance within the TOC-rich zone, and that seven of these species were confined to it, suggesting that they were specifically adapted to environments therein. Only one of these new arrivals was a platycopid. A careful analysis of particular environments may be necessary to distinguish between the different influences of low-oxygen water masses on the one hand, and sediment–water boundaries with high organic matter values caused by upper water-column enhanced productivity, on the other. Cases in point might be the faunas illustrated by Einsele & Wiedmann (1983) from the Turonian black shales of Morocco, which are not dominated by platycopids, and the apparent success of *Palmoconcha walvisbaiensis* under the modern upwelling cells off Namibia (Dingle & Giraudeau 1993). In view of the results presented by Jarvis *et al.* (1988), quantitative data on the distribution of ostracods, and platycopids in particular, are needed to monitor the oxygen-poor and/or organic-rich coastal basin sediments north of the Walvis Ridge.

M. R. A. Thomson (BAS) is thanked for critically reading the manuscript, which was further improved by the comments of two referees: A. R. Lord (UCL) and J. E. Whittaker (NHM).

References

ARTHUR, M. A., JENKYNS, H. C., BRUMSACK, H. J. & SCHLANGER, S. O. 1990. Stratigraphy, geochemistry, and palaeoceanography of organic carbon-rich Cretaceous sequences. *In*: GINSBURG, R. N. & BEAUDOIN, B. (eds) *Cretaceous Resources, Events and Rhythms*. Kluwer, 75–119.

BALLENT, S. C. & WHATLEY, R. C. 1997. Ostrácodas del Albiano de la perforáción SC-1 (Santa Cruz-1), provincia de Santa Cruz, Argentina. *Ameghiniana*, **34**, 251–253.

—— RONCHI, D. I. & WHATLEY, R. C. 1997. The ostracod genus *Majungaella* Grekoff in Argentina. *South Atlantic Mesozoic Correlations IGCP Project No. 381 News*, **7**, 11.

BATE, R. H. & BAYLISS, D. D. 1969. An outline account of the Cretaceous and Tertiary foraminifera and of the Cretaceous ostracods of Tanzania. *Proceedings of the 3rd African Micropaleontological Colloquium 1968*, Cairo, National Information and Documentation Centre, 113–164.

BERTELS, A. 1977. Cretaceous ostracodes, South Atlantic. *In*: SWAIN, F. M. (ed.). *Stratigraphic Micropaleontology of Atlantic Basin and Borderlands*. Elsevier, 271–304.

BOLLI, H. M., RYAN, W. B. F., FORESMAN, J. B. *ET AL.* 1978. Angola continental margin – Sites 364 and 365. *Initial Reports of the Deep Sea Drilling Project*, **40**, 357–455.

BRENNER, P. & OERTLI, H. J. 1976. Lower Cretaceous ostracodes (Valanginian to Hauterivian) from the Sundays River Formation, Algoa Basin, South Africa. *Bullétin Centres Recherches Exploration–Production Elf Aquitaine*, **10**, 471–533.

DEAN, W. E., ARTHUR, M. A. & STOW, D. A. 1984. Origin and geochemistry of Cretaceous black shales and multicolored claystones, with emphasis on Deep Sea Drilling Project Site 530, southern Angola Basin. *In*: HAY, W. W., SIBUET, J. C., BARRON, E. *ET AL.* 1984. *Initial Reports of the Deep Sea Drilling Project*, **75**, 819–844.

DINGLE, R. V. 1982. Some aspects of Cretaceous ostracod biostratigraphy of South Africa and relationships with other Gondwanide localities. *Cretaceous Research*, **3**, 367–389.

—— 1984. Mid-Cretaceous Ostracoda from southern Africa and the Falkland Plateau. *Annals of the South African Museum*, **93**, 97–211.

—— 1988. Marine ostracod distributions during the early breakup of southern Gondwanaland. *In*: HANAI, T., IKEYA, N. & ISHIZAKI, K. (eds) *Evolutionary Biology of Ostracoda*. Kodansha, 841–854.

—— 1996. Cretaceous Ostracoda of the SE Atlantic and SW Indian Ocean: a stratigraphic review and Atlas. *In*: JARDINÉ, S., DE KLASZ, I. & DEBENEY, J. P. (eds) *Géologie de l'Afrique et de l'Atlantique Sud*. Bullétin Centres Recherches Exploration–Production Elf Aquitaine, Mémoire, **16**, 1–17.

—— & GIRAUDEAU, J. 1993. Benthic Ostracoda in the Benguela System (SE Atlantic): a multivariate analysis. *Marine Micropaleontology*, **22**, 71–92.

DUARTE-MORAIS, M. L. 1996. Information on the geodynamic evolution of the Kwanza Basin (Angola) during the Lower Cretaceous, based on the sedimentological interpretation of new field and subsurface data. *In*: JARDINÉ, S., DE KLASZ, I. & DEBENEY, J. P. (eds) *Géologie de l'Afrique et de l'Atlantique Sud*. Bullétin Centres Recherches Exploration–Production Elf Aquitaine, Mémoire, **16**, 19–28.

DUPONT, G. 1996. Main planktic foraminifera of the Cretaceous of Gabon (Aptian to early Campanian). *In*: JARDINÉ, S., DE KLASZ, I. & DEBENEY, J. P. (eds) *Géologie de l'Afrique et de l'Atlantique Sud*. Bulletin Centre Recherches Exploration–Production Elf Aquitaine, Mémoire, **16**, 83–121.

EINSELE, G. & WIEDMANN, J. 1983. Cretaceous upwelling off northwest Africa: a summary. *In*: THIEDE, J. & SUESS, E. (eds) *Coastal Upwelling, Its Sediment Record. Part B: Sedimentary Records of Ancient Coastal Upwelling*. Plenum, 485–499.

GROSDIDIER, E. 1979. Principaux ostracodes marins de l'intervalle Aptien–Turonien du Gabon (Afrique occidentale). *Bullétin Centres Recherches Exploration–Production Elf Aquitaine, Mémoire*, **3**, 1–35.

—— BRACCINI, E., DUPONT, G. & MORON, J. M. 1996. Non-marine Lower Cretaceous biozonation of the Gabon and Congo Basins. *In*: JARDINÉ, S., DE KLASZ, I. & DEBENEY, J. P. (eds) *Géologie de l'Afrique et de l'Atlantique Sud*. Bullétin Centres Recherches Exploration–Production Elf Aquitaine, Mémoire, **16**, 67–82.

JARVIS, I., CARSON, G. A., COOPER, M. K. E. *ET AL*. 1988. Microfossil assemblages and the Cenomanian–Turonian (late Cretaceous) Oceanic Anoxic Event. *Cretaceous Research*, **9**, 3–103.

KAUFFMAN, E. G. & SAGEMAN, B. B. 1990. Biological sensing of benthic environments in dark shales and related oxygen-restricted facies. *In*: GINSBURG, R. N. & BEAUDOIN, B. (eds) *Cretaceous Resources, Events and Rhythms*. Kluwer, 121–138.

KOUTSOUKOS, E. A. M. 1992. Late Aptian to Maastrichtian foraminiferal biogeography and palaeoceanography of the Sergipe Basin, Brazil. *Palaeogeography, Palaeoclimatology, Palaeoecology*, **92**, 295–324.

—— & BENGTSON, P. 1993. Towards an integrated biostratigraphy of the Upper Aptian–Maastrichtian of the Sergipe Basin, Brazil. *Document Laboratoire Géologique Lyon*, **125**, 241–262.

—— & DA SILVA-TELLES, A. C. 1992. Stratigraphic and palaeoenvironmental distributions of mid-Cretaceous Ostracoda from the Sergipe Basin, NE Brazil, and biogeographic affinities with NW Africa. *In*: CURNELLE, R. (ed.) *Géologie Africaine, 2e Colloque African de Micropaléontologie*. Bullétin Centres Recherches Exploration–Production Elf Aquitaine, Mémoire, **13**, 410.

—— MELLO, M. R. & DE AZAMBUJA FILO, N. C. 1991*a*. Micropalaeontological and geochemical evidence of mid-Cretaceous dysoxic–anoxic palaeoenvironments in the Sergipe Basin, northeastern Brazil. *In*: TYSON., R. V. & PEARSON, T. H. (eds) *Modern and Ancient Continental Shelf Anoxia*. Geological Society, London, Special Publications, **58**, 427–447.

——, ——, ——, HART, M. B. & MAXWELL, J. R. 1991*b*. The upper Aptian–Albian succession of the Sergipe Basin, Brazil: an integrated paleoenvironmental assessment. *AAPG Bulletin*, **75**, 479–498.

KRÖMMELBEIN, K. 1966. On 'Gondwana Wealden' Ostracoda from NE Brazil and West Africa. *In*: VAN HINTE, J. E. (ed.) *Proceedings of the Second West African Micropaleontological Colloquium*. E. J. Brill, 113–118.

——, 1967. *Sergipella transatlantica* n.g., n.sp., und *Aracjuia benderi* n.g., n.sp., aus dem Ober-Aptium/Albium. *Senckenbergiana lethaea*, **48**, 525–533.

——, 1976. Remarks on marine Cretaceous ostracodes of Gondwanic distribution. *Proceedings of the 5th African Colloquium on Micropalaeontology*. Revista Española de Micropaleontologia, 539–551.

MALUMIÁN, N. & BÁEZ, A. M. 1976. Outline of Cretaceous stratigraphy of Argentina. *Annales du Museum d'Histoire Naturelle de Nice*, **4**, 1–10.

MASSALA, A., BELLIER, J. P., MAGNIEZ-JANNIN, F. & LAURIN, B. 1996. Biostratigraphy (planktonic foraminifera) and environments of the Upper Cretaceous from two wells in the offshore Congo Basin. *In*: JARDINÉ, S., DE KLASZ, I. & DEBENEY, J. P. (eds) *Géologique de l'Afrique et de l'Atlantique Sud*. Bullétin Centres Recherches Exploration–Production Elf Aquitaine, Mémoire, **16**, 29–38.

MCLACHLAN, I. R., BRENNER, P. W. & MCMILLAN, I. K. 1976. The stratigraphy and micropalaeontology of the Cretaceous Brenton Formation and the PB/A-1 well, near Knysna, Cape Province. *Transactions of the Geological Society of South Africa*, **79**, 341–370.

MCMILLAN, I. K. 1990. Foraminiferal biostratigraphy of the Barremian to Miocene rocks of the Kudu 9A-1, 9A-2 and 9A-3 boreholes. *Communications of the Geological Survey of Namibia*, **6**, 23–29.

MUSACCHIO, E. A. 1979. Datos paleobiogeográficos de algunas asociaciones de foraminíferos, ostrácodos y carofitas del Jurásico medio y el Cretácico inferior de Argentina. *Ameghiniana*, **16**, 247–271.

—— & SIMEONI, M. 1995. Ostrácodos de Argentina y el diseño de las áreas oceánicas durante el Cretácico. *Iero Workshop Sobre Paleoceánografica con enfase en Micropaleontologia e Estratigrafia. Boletim de Resumos Expandos*, Gramado RS, Brasil, 63–69.

—— & —— 1996. Biogeographic relationships of Lower Cretaceous calcareous microfossils from Patagonia: an approach to causal factors. *South Atlantic Mesozoic Correlations IGCP Project No. 381 News*, **5**, 49–50.

MUSSARD, J. M. 1996. Palynomorphs as indicators of relative sea-level variations: quantitative aŋalyses in the Upper Albian of the Congo Republic. *In*: JARDINÉ, S., DE KLASZ., I. & DEBENEY, J. P. (eds) *Géologie de l'Afrique et de l'Atlantique Sud*. Bullétin Centres Recherches Exploration–Production Elf Aquitaine, Mémoire, **16**, 57–66.

NÜRNBERG, D. & MÜLLER, R. D. 1991. The tectonic evolution of the South Atlantic from Late Jurassic to present. *Tectonophysics*, **191**, 27–53.

OSCHMANN, W. 1994. Adaptive pathways of benthic organisms in marine oxygen-controlled environments. *Neues Jahrbuch für Geologie und Paläontologie. Abhandlungen*, **191**, 393–444.

PEDERSEN, T. F. & CALVERT, S. E. 1990. Anoxia vs. productivity: what controls the formation of organic-carbon-rich sediments and sedimentary rocks? *AAPG Bulletin*, **74**, 454–466.

REYMENT, R. A. 1980. Paleo-oceanology and paleobiology of the Cretaceous South Atlantic. *Oceanologica Acta*, **3**, 127–133.

—— & TAIT, E. A. 1972. Biostratigraphical dating of the early history of the South Atlantic Ocean. *Philosophical Transactions of the Royal Society of London. B. Biological Series*, **264**, 55–95.

SIBUET, J. C., HAY, W. H., PRUNIER, A., MONTADERT, L., HINZ, K. & FRITSCH, J. 1984. Early evolution of the South Atlantic Ocean: role of the rifting episode. *Initial Reports of the Deep Sea Drilling Project*, **75**, 469–481.

STOW, D. A. V. & DEAN, W. E. 1984. Middle Cretaceous black shales at site 530 in the southeastern Angola Basin. *In*: HAY, W. W., SIBUET, J. C., BARRON, E. ET AL. 1984. *Initial Reports of the Deep Sea Drilling Project*, **75**, 809–817.

SWAIN, F. M. 1976. Lower and Middle? Cretaceous Ostracoda from the Atlantic Ocean off Guiana and off West Africa. *Journal of Paleontology*, **50**, 734–753.

TAMBAREAU, Y. 1982. Les ostracodes et l'histoire géologique de l'Atlantique sud au Crétacé. *Bullétin Centres Recherches Exploration–Production Elf Aquitaine*, **6**, 1–37.

VALICENTI, V. H. & STEPHENS, J. M. 1984. Ostracods from the Upper Valanginian and Upper Hauterivian of the Sundays River Formation, Algoa Basin, South Africa. *Revista Española de Micropaleontologia*, **16**, 171–239.

VERNET, R., ASSOUDA-WANDE, C., MASSAMBA, L. & SORRIAUX, P. 1996. Palaeogeography of the Cretaceous (Albian–Maastrichtian) in the offshore Congo Basin. *In*: JARDINÉ, S., DE KLASZ, I. & DEBENEY, J. P. (eds) *Géologie de l'Afrique et de l'Atlantique Sud.* Bullétin Centres Recherches Exploration–Production Elf Aquitaine, Mémoire, **16**, 39–55.

WHATLEY, R. C. 1991. The platycopid signal: a means of detecting kenoxic events using Ostracoda. *Journal of Micropalaeontology*, **10**, 181–185.

Detachment faulting and petroleum prospectivity in the Rio Muni Basin, Equatorial Guinea, West Africa

JONATHAN P. TURNER

University of Birmingham, School of Earth Sciences, Birmingham B15 2TT, UK

Abstract: The change from rifting to drifting is generally signified by a characteristic end-rift, or break-up, unconformity. In the Rio Muni Basin, two such unconformities bound an Aptian–Cenomanian transitional sequence of allochthonous carbonates and clastics considered unique among the South Atlantic passive margins. The transitional sequence comprises allochthonous thin-skin 'rafts', up to 2500 m thick, of shallow water carbonate and clastic sediment emplaced in a relatively deepwater setting along shallow detachments. A new model for the rift to drift development of the Rio Muni Basin interprets the raft tectonics as a response to a combination of end-rift–early post-rift uplift of the basin margins and deep offshore bathymetry *c.* 1800 m. Seismic interpretations reveal spectacular hanging wall geometry in the rafts as a consequence of their movement across a geometrically complex, ramp-flat basal detachment. In places, the hanging walls of the rafts display an 'eggbox' topography of closely spaced anticlines, often with four-way dip closure. Information from wells and analogy with adjacent basins indicates potential source rocks in the Aptian and Turonian. Several migration paths are identified, including the juxtaposition of hanging wall reservoirs with the source-bearing stratigraphy across transverse fault zones. Raft tectonics has considerable isostatic and thermal consequences, with attendant implications for predicting the timing of potential source rock maturity. Modelling maturity at variously located wells and pseudowells indicates the principal effect of the raft tectonics to have been a very rapid episode of burial in the basin centre and isostatic rebound of the unloaded margins.

Adjacent sub-basins of the South Atlantic passive margins exhibit a fascinating diversity in their tectonic development, conjugate basin architecture, geological structure, sedimentary and seismic stratigraphy, and volcanism. Deep seismic and gravity data from passive continental margins from around the world, and detailed mapping of commercial seismic datasets, demonstrate the important role of long-lived, transversely oriented crustal-scale fault zones along these margins. It is increasingly clear that transverse lineaments have exerted a fundamental influence over basin development throughout continental stretching, rifting and oceanic opening.

Gravity driven collapse of marginal platforms, during which rafts of shallow water carbonates in excess of 2000 m thick have been translated some 6–10 km basinward, was a characteristic stage in the development of several sub-basins of the South and Central Atlantic passive continental margins (e.g. Rio Muni, Equatorial Guinea: Turner 1995, 1996; Douala, Cameroon; North Gabon; South Gabon; Seranne *et al.* 1992; Kwanza, Angola; Duval *et al.* 1992; Lundin 1992). Gravitational collapse of the basin margins was an important mechanism of large-scale mass wasting and it has significant implications for isostatic uplift at continental margins affected by this mode of deformation. Gravity driven collapse of West African basin margins is often intimately related to major unconformity development – allochthonous carbonate rafts represent the first seismic sequence above the end-rift unconformity and their upper surfaces are eroded discordantly by a regionally correlatable drift unconformity. The rafts' proximity to the end-rift unconformity, and the fact that they comprise exclusively marine sediment, means that they record the earliest products of post-rift deposition on a marginal platform largely eroded during subsequent rift shoulder uplift. This contribution firstly addresses the implications of raft tectonics for basin development. Subsequent investigation of the role of raft tectonics in determining the petroleum prospectivity of the 'frontier' Rio Muni Basin draws on comparisons with the known hydrocarbon plays of the adjacent Gabon Basin system.

Rio Muni Basin in its regional context

Between North Gabon and Cameroon, the West African margin is composed of rift sub-basins separated at intervals of 50–150 km by northeast trending fracture zones (Fig. 1). The margin consists of stepped normal-faulted segments bounded by continental transcurrent fault zones that extend westward into oceanic transform structures. The Rio Muni Basin forms a small subcompartment of this segment of the West African passive margin. It is bounded by two

TURNER, J. P. 1999. Detachment faulting and petroleum prospectivity in the Rio Muni Basin, Equatorial Guinea, West Africa. *In:* CAMERON, N. R., BATE, R. H. & CLURE, V. S. (eds) *The Oil and Gas Habitats of the South Atlantic.* Geological Society, London, Special Publications, **153**, 303–320.

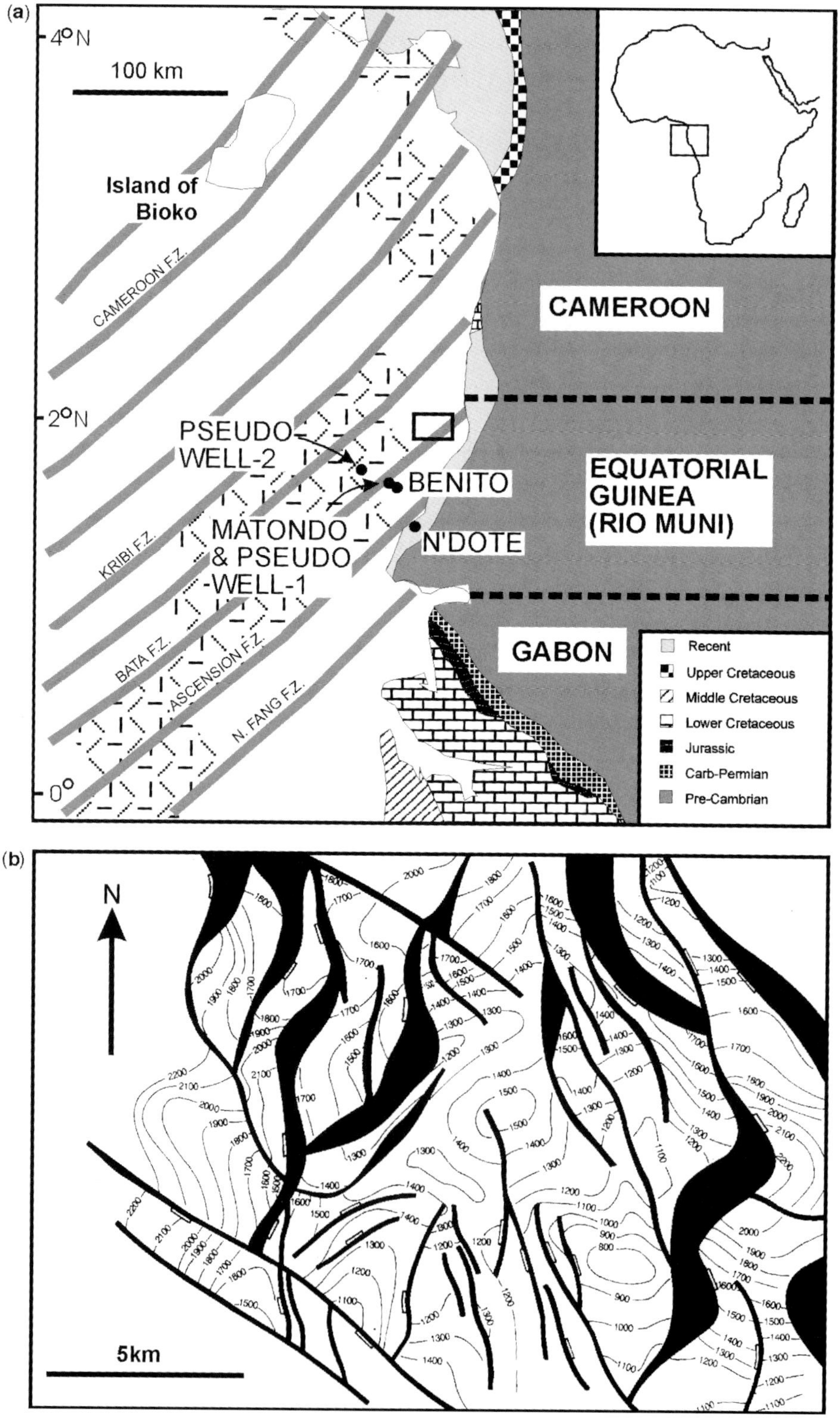
(a)
4°N
2°N
0°
100 km
Island of Bioko
CAMEROON F.Z.
KRIBI F.Z.
BATA F.Z.
ASCENSION F.Z.
N. FANG F.Z.
PSEUDO-WELL-2
MATONDO & PSEUDO-WELL-1
BENITO
N'DOTE
CAMEROON
EQUATORIAL GUINEA (RIO MUNI)
GABON
Recent
Upper Cretaceous
Middle Cretaceous
Lower Cretaceous
Jurassic
Carb-Permian
Pre-Cambrian
(b)
N
5km

transcurrent fault zones spaced some 100 km apart, intersecting the coast at roughly 30° – to the north the Bata Fracture Zone defines Rio Muni's boundary with the Douala Basin, Cameroon, and to the south the North and South Fang Fracture Zones separate Rio Muni from North Gabon. The Rio Muni Basin is intersected by one of the largest of these transforms, the Ascension Fault Zone, across which occurs an abrupt change in the free-air gravity signature (based on data in Sandwell & Smith 1995).

Information on the relation between oceanic transform fault zones and the structure of the West African continental margin between North Gabon and Cameroon has recently emerged from deep-penetrating multichannel seismic data (Rosendahl *et al.* 1991; Meyers *et al.* 1996). These data show the Ascension Fault Zone forming a 200 km long transform-faulted boundary between attenuated, rifted continental crust and proto-oceanic crust. Investigations of Rio Muni Basin development must therefore take account of the fact that post-rift thermal subsidence was probably compounded by flexural subsidence driven by the downbending of the continental plate due to its coupling, across a narrow zone of 'transitional crust', with oceanic crust at the transform fracture zones (cf. Basile *et al.* 1993; Lorenzo & Wessel 1995; Clift *et al.* 1997).

Rio Muni's conjugate passive margin is the Sergipe–Alagoas Basin in northeast Brazil. It forms one of a series of linked rift basins in northeast Brazil whose rift to drift evolution is complex and equivocal. The structure of the Sergipe–Alagoas Basin is dominated by east facing rift faults, contrasting from neighbouring basins where the faults face westward (e.g. Interior, North Gabon and Recôncavo Basins: Bradley & Fernandez 1992; Rabinowitz & La Breque 1979). This asymmetry is also expressed in the contrasting widths of the Sergipe–Alagoas and Rio Muni Basins: measured parallel to the orientation of the transform fracture zones, the width of the Cretaceous rift basin on the Brazilian margin is *c.* 190 km, compared to *c.* 125 km in Rio Muni (based on data in Sandwell & Smith 1995; Meyers *et al.* 1996; Davison 1997).

Total sediment thickness in the Sergipe–Alagoas Basin is as much as 9000 m (Koutsoukos *et al.* 1991), of which the syn-rift succession comprises up to 3500 m (Chang *et al.* 1992, pers. comm.). Although there have been no penetrations of the full thickness of the Rio Muni syn-rift succession, Meyers *et al.* (1996) report greater attenuation of rifted continental crust beneath the Sergipe–Alagoas Basin than its conjugate African margin. Furthermore, evidence from modelling and plate reconstructions (Rabinowitz & La Breque 1979) suggest that the magnitude of crustal extension accommodated by Rio Muni rifting is minor compared to contemporaneous faults on the Brazilian margin. Collectively, these observations are interpreted as evidence for an asymmetric basin system in which the locus of rifting was concentrated on the Brazilian margin whilst mantle thinning was focused beneath the African margin. Other branches of the northeast Brazilian rift system have yielded comparable interpretations of asymmetric, 'simple shear' rifting:

- a low-angle detachment geometry has been documented from the Sergipe–Alagoas Basin where syn-rift sediment and continental basement slivers overlie an east dipping detachment (Chang *et al.* 1992);
- gravity modelling by Ussami *et al.* (1986) detected no shallowing of the mantle beneath the Tucano Graben, adjacent to the Sergipe–Alagoas Basin; they interpreted the Tucano Graben to be linked to the Sergipe–Alagoas Basin and African basins by an east dipping mega-detachment;
- Castro (1987) used gravity modelling to identify 160 km offset between the Tucano Graben and its paired zone of mantle shallowing; he presented a model in which the northeast Brazilian rift basins were situated above east dipping detachments that flatten at lower crustal levels before stepping down to the Moho.

A comprehensive investigation of the Atlantic margin basins of northeast Brazil and equatorial West Africa would provide a timely contribution

Fig. 1. (**a**) Geology and crustal structure of the equatorial West African margin. The map gives the location of wells referred to in the text and models of Figs 9–12. The random stippled ornament depicts Meyers *et al.*'s (1996) interpretation of proto-oceanic crust marking the transition between rifted continental crust on the landward side and oceanic crust on the oceanward side. Offshore structure based on a compilation of Meyers *et al.* (1996); onshore geology from Aubouin *et al.* (1985). (**b**) Composite structure map of a part of the northern Rio Muni Basin (location shown by the small box in the map above). Contours give seismic TWTT to intra-Aptian markers within the transitional sequence. Fault trends at this relatively shallow depth indicate that the NW–SE deep crustal structure trends, as exemplified by the oceanic fracture zones in (a), have exerted little or no influence on structure development in the rift and post-rift successions.

to the debate surrounding the validity of the simple shear model of continental rift basin development. In particular, comparisons of the distribution of palaeo heat flow, uplift rates and amounts, and deep crustal structure between the opposing margins are much needed.

Tectono-stratigraphy of the Rio Muni Basin

Rifting in the Rio Muni Basin proceeded in response to extension oriented *c.* 30° to the rift axis. The initiation of the rift episode is unclear due to the lack of deep well penetrations. At most passive continental margins, constraining the age of the end-rift unconformity provides the best estimate of the cessation of crustal extension and the timing of the transition from rifting to drifting. Along this part of West Africa, rifting took place between the Neocomian and the Middle Aptian (Bradley & Fernandez 1992). However, several authors (e.g. Seranne *et al.* 1992; Brice *et al.* 1982) have noted the difficulty in identifying a single end-rift unconformity in this area.

In the Rio Muni Basin, the progression from rifting to post-rift thermal subsidence and marine sedimentation is marked by a transitional sequence bounded by two major unconformities. In the Matondo and Benito Wells (Fig. 1), the lower, end-rift unconformity signifies the cessation of rifting and it is overlain by: (1) *c.* 400 m thick succession of intercalated salt, sandstone and dark shale; (2) *c.* 1000 m of turbiditic clastics; and (3) an allochthonous succession of oolite-rich carbonate rocks 400 m thick. Elsewhere in the basin these allochthonous carbonates attain thicknesses > 2500 m (Turner 1995), recording deposition on a foundered peripheral shelf. The transitional sequence is bounded above by the drift unconformity, marking the onset of continental drift and oceanic opening. It is of Late Cretaceous age and defines the onset of open marine deposition in a thermally subsiding post-rift basin.

As noted above, this segment of the African continental margin is also unusual for the strongly oblique orientation of rift extension relative to the orientation of the rift axis. With the changes in palaeostress azimuth that accompany rift and passive continental margin basin development, obliquely oriented rift lineaments will be susceptible to repeated increments of reactivation. Based on seismic interpretations in the neighbouring North Gabon Basin, where clear examples of Late Cretaceous transtension are identified (D. A. Kirkland pers. comm.), it is speculated that the transitional sequence owes its origin to transpressional uplift of the Rio Muni rift margins, erosion, and redeposition in small pull-apart basins. Similar interpretations of lacustrine deposition in pull-apart basins are invoked to explain the rift to drift development of the Côte D'Ivoire–Ghana Transform Margin (Mascle *et al.* 1996; Clift *et al.* 1997).

The geology, depicted on the cross-sections in Figs 2 and 3, is largely dominated by allochthonous carbonate rafts sandwiched between the end-rift unconformity and the post-rift succession. The following features of the cross-sections have a direct bearing on understanding the mode of emplacement of the rafts.

End-rift unconformity

A distinctively undulating unconformity is interpreted truncating landward dipping syn-rift strata at the base of the carbonate rafts in Fig. 2. In the Rio Muni Basin, the high relief of the unconformity expresses directly underlying normal fault blocks in the syn-rift succession. Compared to its planar geometry elsewhere in West Africa [e.g. more or less planar offshore Namibia (Clemson 1999), Angola (Duval *et al.* 1992), Congo (McGinnis *et al.* 1993) and South Gabon (Seranne *et al.* 1992)], the Rio Muni end-rift unconformity preserves the buried depositional topography of end-rift times. It is interpreted as an expression of the restricted thickness of the Rio Muni syn-rift succession and limited erosion during footwall uplift at the close of the syn-rift episode.

Rafts

Using spectacular seismic examples from the Rio Muni Basin, offshore Equatorial Guinea, Turner (1995) has described in detail the geometry and geological setting of these rafts. They are not slumps. The rafts display clear examples of deformational and stratigraphic features, showing that they developed as cohesive blocks of shelfal sediment, subsequently emplaced in a deepwater, offshore setting at the commencement of the post-rift stage. Furthermore, they are not products of halokinesis, Despite the occurrence of the thick Upper Aptian Ezanga Formation salts in the adjacent Gabon basin system (> 800 m: Teisserence & Villemin 1989), salt in the time-equivalent succession of the Rio Muni Basin is volumetrically negligible.

Information from three wells that penetrate the rafted facies demonstrate them to be composed of the following four principal lithostratigraphic sub-divisions.

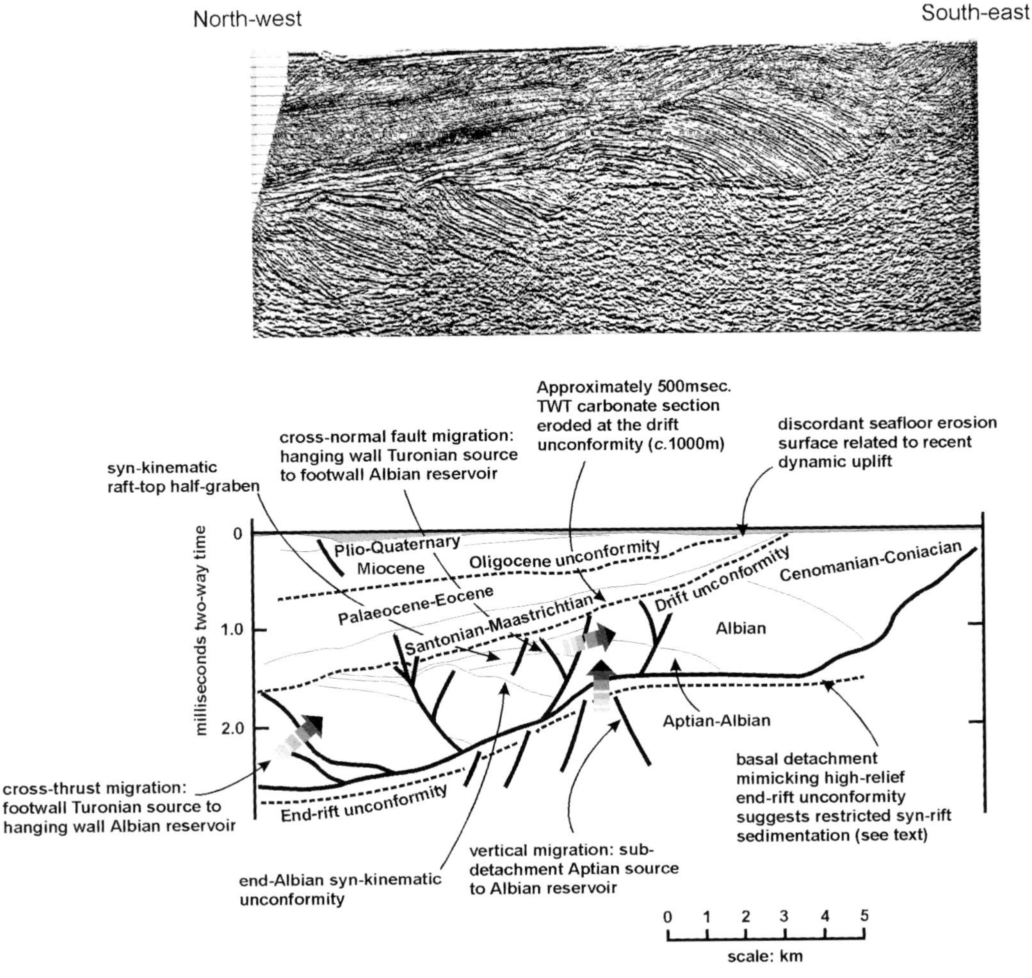

Fig. 2. Time-migrated data and interpretation of a coast-normal oriented seismic profile located offshore the southern part of the Rio Muni Basin. It depicts an allochthonous raft bounded above and below by unconformities marking the transition from rifting to drifting. Also shown are arrows indicating possible migration paths between potential source-bearing intervals and reservoirs in the hanging wall of the raft.

- lowermost Late Aptian intercalated black shale and evaporitic salt;
- sandstone-rich facies of latest Aptian age;
- Albian carbonates containing oolitic limestones (comprising *c.* 30% of the thickness of the carbonate succession);
- uppermost grey–black Cenomanian claystone, bituminous in places and with thin sandstone stringers.

The rafts are up to 2500 m thick and exclusively allochthonous, bounded along their bases by ramp-flat detachment faults. The basal detachment utilizes the intercalated salt and shale succession as a decollement horizon, which also constitutes the basin's premier source rock (Ross & Hempsted 1993). These salts and shales accumulated during the earliest stages of post-rift thermal subsidence and marine sedimentation. By analogy with equivalent salt layers in the Upper Aptian of the conjugate Sergipe–Alagoas Basin, Brazil (Koutsoukos *et al.* 1991), and in Gabon (Leyden *et al.* 1972; Teisserenc & Villemin 1989; Hardie 1990; Meyers *et al.* 1996) and Angola (Pautot *et al.* 1973), the Rio Muni salts are interpreted as products of precipitation in shallow water, transform fault-bounded sub-basins, periodically isolated during intermittent eustatic lowstands.

The rafts display all the geometric features of

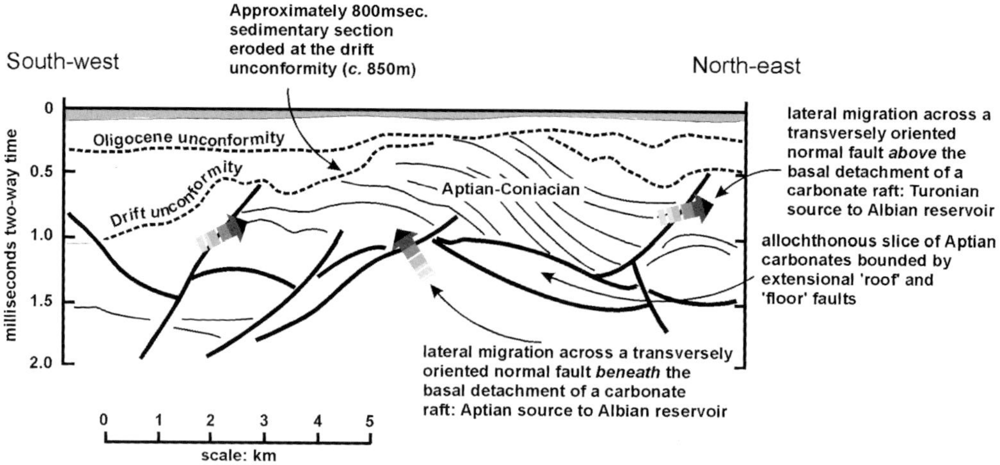

Fig. 3. Interpretation of a seismic profile oriented sub-parallel to the West African coastline, Rio Muni Basin, offshore Equatorial Guinea. Also shown are arrows indicating possible migration paths between potential source-bearing intervals and reservoirs in the hanging wall of the raft.

tectonic faulting – ramp anticline and syncline development, duplexes and imbricate zones (Fig. 4), and systematic relations between hanging wall deformation and the depth to detachment. Collectively, these features indicate that the rafts comprise blocks of cohesive, lithified sediment. Distinctive deformational geometry resulted from the movement of the rafts over the high-relief topography of the basal detachment (e.g. Fig. 3). Today, the carbonates record the only evidence of a now-foundered shelf. Onshore seismic data in the vicinity of the N'Dote Well (Fig. 1), and restoration of the raft stratigraphy to its pre-rafting configuration (Fig. 5), show that deposition and preservation of the thick carbonates was accommodated by west-dipping listric normal faulting detaching at relatively deep levels in the Lower Cretaceous pre-rift succession. These faults accommodated syn-kinematic expansion of the carbonates toward the eastern basin margin.

Mapping of the oldest sediment onlapping the rafts, and the identification of small, hanging wall basins that filled the accommodation space created by the rafting, places constraints on the timing of raft emplacement which took place during the Cenomanian, between 95 and 92 Ma. Palinspastic restorations of the rafting process give a minimum estimate of 6 km of total downslope movement. This gives a time-averaged minimum slip rate of 2 mm/year along the rafts' basal detachment. The Rio Muni rafting slip rate compares very closely with the Late Cretaceous and Tertiary rafting in the Kwanza Basin, Angola, where Duval *et al.* (1992) have shown that thin-skin extension resulting from the rafting was accommodated by an oceanic spreading rate of *c.* 20 mm/year.

Drift unconformity

The upper surfaces of the carbonate rafts are truncated by a deeply erosive surface marking the base of the autochthonous post-rift succession. The drift unconformity signifies the complete cessation of lithosphere extension, the onset of continental divergence (or 'drifting') and the development of the passive margin. The drift unconformity dips basinward at inclinations of up to 50°. It erodes discordantly across the complex structure of the sub-cropping carbonate rafts, whilst it is onlapped by Late Cretaceous and Tertiary marine facies. Consequently, the unconformity is interpreted as a surface of subaerial erosion. However, it is not clear from the well data whether the unconformity is onlapped by shallow littoral facies or deep marine turbidites. It is likely, therefore, that subaerial erosion was enhanced by subsequent submarine erosion and slumping during marine flooding and burial. Formation of the drift unconformity is bracketed between 92 and 86 Ma and reconstructions of the thickness of missing section, using seismic data, suggest that it represents a minimum 660 m erosion. The unconformity is most erosive toward the basin margin, closest to the site of deposition of the carbonate rafts, where isostatic rebound was most pronounced.

Thus, the transition from active rifting to post-rift thermal subsidence was gradual and is

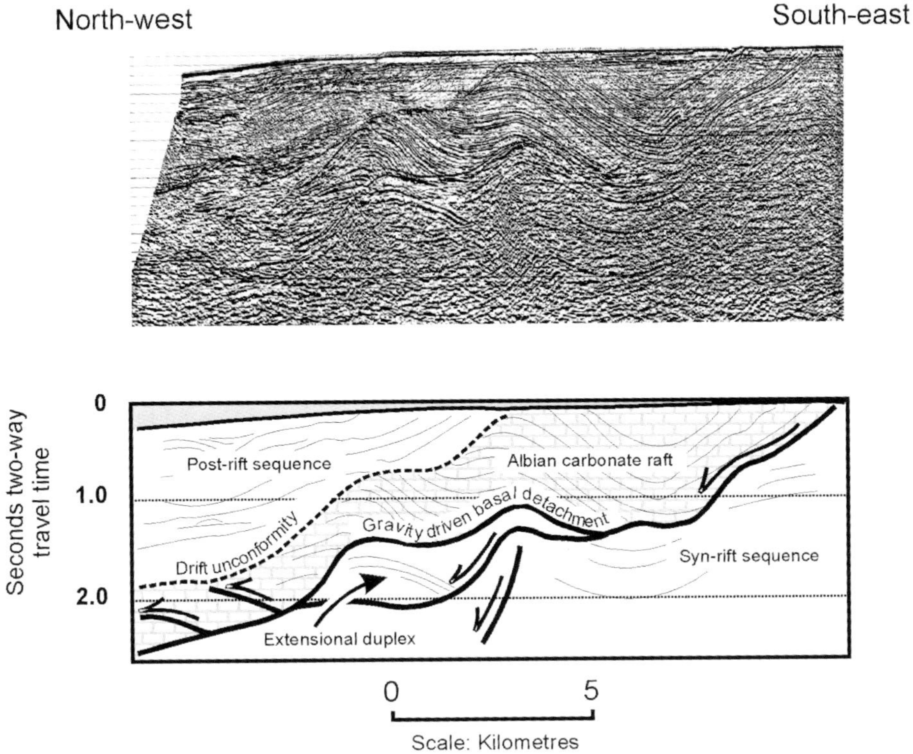

Fig. 4. Time-migrated data and interpretation of a coast-normal oriented seismic profile located offshore the central part of the Rio Muni Basin. It depicts an allochthonous carbonate raft bounded along part of its base by an extensional duplex. This interpretation suggests that the raft tectonics affected cohesive, lithified material and that these are not submarine slump deposits.

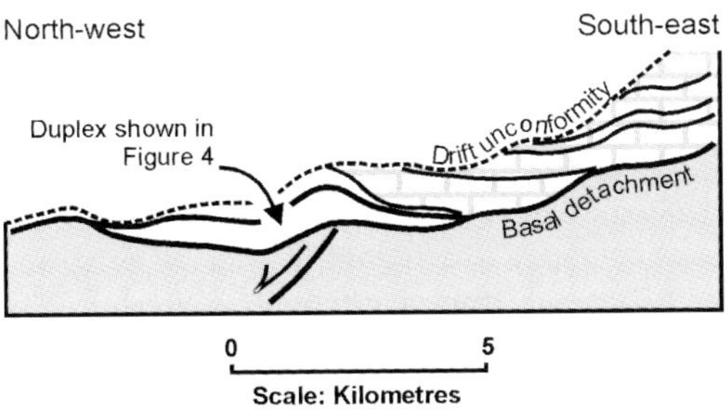

Fig. 5. Depth converted and palinspastically restored interpretation of the cross-section shown in Fig. 4. Following stripping-off of the post-rift sequence, the carbonate raft (brick ornament) has been restored to its depositional position by translating it back up the basal detachment. The divergent geometry of strata within the raft suggest that the Albian carbonates accumulated in the hanging wall of a seaward dipping, syn-depositional normal fault that defined the eastern basin margin.

marked by at least two unconformities – a lower, end-rift event in which uplift and erosion was driven by isostatic forces, and a drift event recording thermal doming as the precursor to oceanic spreading. An analogous mechanism to that by which isostatic rebound enhanced the Late Cretaceous unconformity in the Rio Muni Basin is invoked for the 'Type 1' Oligocene unconformity offshore Congo. McGinnis *et al.* (1993) modelled the Late Eocene unconformity as a product of eustatic sea-level fall and subaerial exposure, compounded by isostatic uplift in response to the erosion of material from the shelf.

Modelling the palaeobathymetric significance of the raft tectonics

Aims

Geologically rapid emplacement of the rafts suggests they glided downslope into an existing bathymetric 'deep' that owes its existence to rapid subsidence and/or restricted sedimentation. Palinspastic restorations of near-sea-level markers in the Rio Muni carbonate rafts allowed Turner (1996) to speculate that the submarine relief required to initiate and sustain their downslope movement implies deep, early post-rift bathymetry. Inverse modelling of the post-rift seismic stratigraphy of the Rio Muni Basin tests this idea by reconstructing post-rift subsidence, sedimentation and palaeobathymetry.

Modelling rationale

The modelling assumes that all post-rift subsidence is thermally driven. In view of the proximity of the study area to the Neogene–Recent Cameroon volcanics, and the presence of the related regional Oligocene unconformity, this may be an oversimplification. However, given that this study focuses on the early post-rift (Aptian Cenomanian) stage of basin development, some 70–80 Ma earlier, Neogene subsidence-driving mechanisms have been ignored. The modelling also assumes instantaneous rifting and therefore incorporates maximum post-extensional isostatic uplift.

Most extensional sedimentary basins possess a relatively small effective elastic thickness (T_e) of 5 km or less (Barton & Wood 1984; Watts 1988; Fowler & McKenzie 1989). Assuming average lithosphere, flexural backstripping is therefore fairly insensitive to estimates of T_e, crustal thickness and depth to the brittle–ductile transition. The quality of the inverse model is far more sensitive, however, to the β factor. Overestimates of β will lead to a model which elevates the post-rift stratigraphy to unrealistic heights above palaeo-sea level and vice versa.

The most common geological constraints that may be placed on the inverse modelling are palaeo-sea-level markers, such as coals, deltaics and onlapped syn-rift fault block crests. Such markers must be 'returned ' to their correct palaeoelevation with respect to sea level at the appropriate stage in the model. In the Rio Muni Basin, reliable palaeo-sea level markers in the post-rift stratigraphy are scarce. The drift unconformity is used as a zero bathymetry marker since the seismic data suggest that it records sub-aerial exposure and erosion *c.* 92 Ma (see earlier discussion). An additional palaeo-sea-level datum is provided by the Late Aptian black shales and salt. Consequently, the modelling shown in Fig. 7 uses a β profile that returns the drift unconformity and Aptian black shales and salt to palaeo-sea level at 92 Ma.

Method

Flexural backstripping sequentially strips off layers of stratigraphy in order to visualize how bathymetry and basement depth evolved during subsidence. The modelling incorporates changes in eustatic sea level through geological time, using the eustatic sea-level curve of Haq *et al.* (1987). Ultimately, a restoration is achieved that gives the bathymetry or topography along the line of cross-section at the end of the syn-rift stage. The flexural backstripping technique has been adopted by many recent studies of passive continental margins and extensional basins (e.g. Marsden *et al.* 1990; Kusznir *et al.* 1991; Roberts *et al.* 1993, 1995, 1997; Nadin *et al.* 1995), and a full account of the methodology and sensitivities of the inverse post-rift modelling procedure is given by Kusznir *et al.* (1995).

For the purposes of modelling basin development, the rafts are treated as the earliest 'deposits' of the post-rift sequence whose accumulation took place over a 3 Ma period. The modelling uses the depth-converted interpretation of a multichannel, 60-fold seismic reflection profile (acquired in 1984, reprocessed in 1990) from the Rio Muni Basin shown in Fig. 6. Interpretation is tied chiefly to wells Evoindo-1, Matondo-1 and Benito-1. The sedimentary succession of interest to this study accumulated during a prolonged period of more or less constant subsidence and burial without evidence of major basin inversion or halokinetic uplift. Consequently, an uncomplicated depth conversion employed the following interval velocities: seawater, 1500 m s^{-1}; post-rift, 2125 m s^{-1}; Albian carbonates, 3900 ms^{-1}; Aptian black shales and salt, 3050 m s^{-1}. The inverse modelling flexurally backstrips a present-day cross-section (Fig. 6) in order to derive palinspastic restorations of the section at specific times during its evolution. It incorporates the flexural strength of the lithosphere, thermal subsidence, decompaction,

Table 1. *Decompaction parameters for the modelling shown in Fig. 7*

Layer	Age of base (Ma)	Surface porosity (%)	Decompaction constant (km^{-1})	Matrix density (g cm^3)
Post-rift	92	56	0.27–0.39	2.68
Albian carbonates	95	70	0.71	2.71
Syn-rift	n.a.	49	0.27	2.65

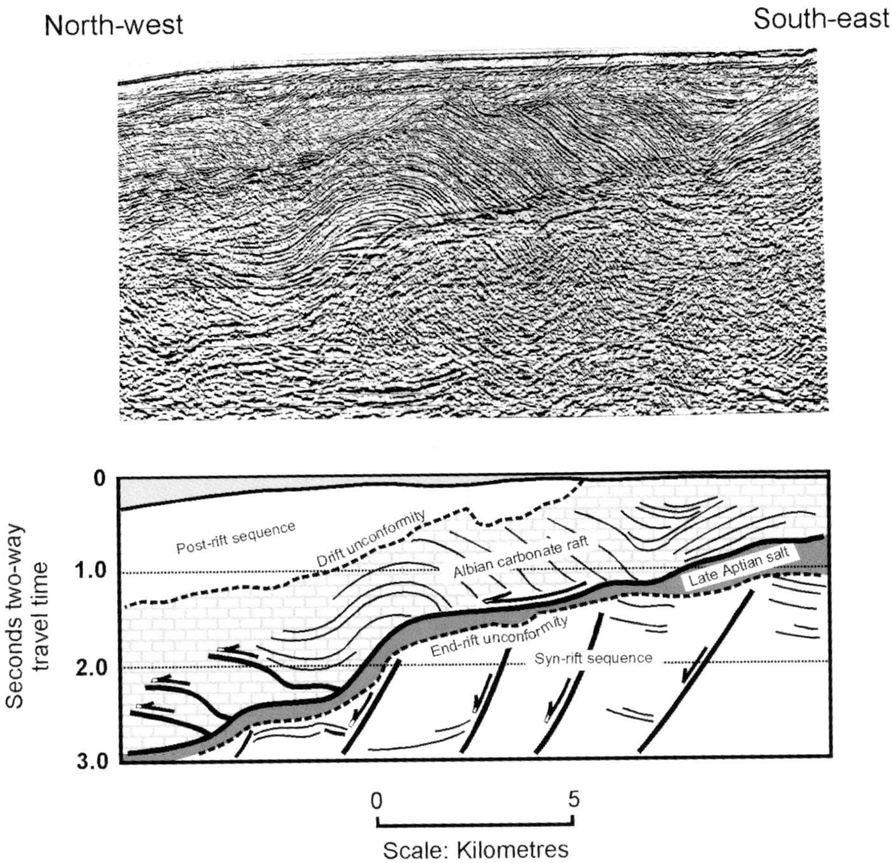

Fig. 6. Seismic profile and interpretation of a coast-normal line from the northern Rio Muni Basin. This profile is used in the modelling shown in Fig. 7. Note the substantial depositional thickness of the Albian carbonates.

sediment unloading and known palaeobathymetry using the parameters given in Table 1.

Results

Turner (1995, 1996) interprets the Rio Muni rafts as having formed in response to a significant gravitational potential that existed between the edge of the carbonate shelf and the centre of the basin at the end of the syn-rift stage. In reconstructing palaeobathymetry, a structural method was used to restore the rafts to their pre-emplacement position on the now eroded peripheral carbonate shelf. Applying this approach to a profile located only 1 km to the south of the profile modelled in this paper, Turner estimated 1600 m of immediately pre-raft bathymetry. The inverse modelling shown in Fig. 7 suggests that raft emplacement was accommodated by deep, earliest post-rift bathymetry of 1800 m. Thus, a good match between palaeobathymetry inferred from inverse modelling, and from structural restoration, provides a best estimate of the magnitude of the gravitational potential that triggered and sustained the rafts' movement.

Tectonic scheme for the rift to drift transition

The scheme shown in Fig. 8 is an attempt to reconcile the following geo-tectonic features interpreted from the Rio Muni Basin and its Brazilian conjugate, the Sergipe–Alagoas Basin: (1) high-relief end-rift unconformity sub-cropped by a syn-rift succession of restricted thickness; (2) asymmetric, east facing megatectonic basin architecture; (3) significant unfilled end-rift bathymetry, recording either a pulse of rapid subsidence and/or restricted sedimentation; (4) accumulation of a thick, gravitationally unstable pile of shallow water carbonate sediment on a now eroded marginal shelf, subsequently emplaced in a relatively deepwater setting by thin-skin detachment faulting; (5) thick (up to

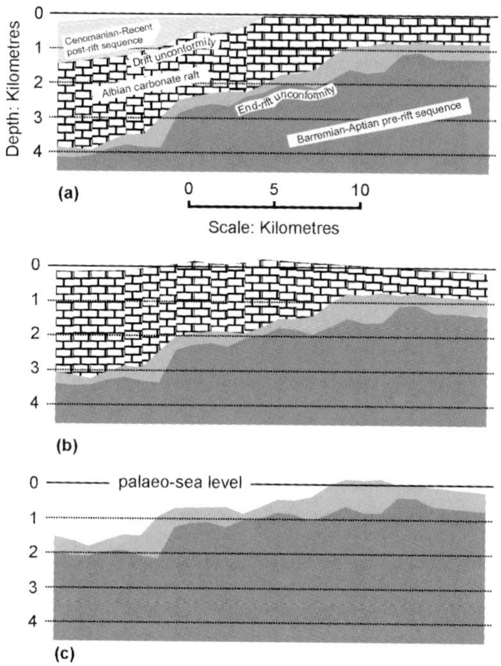

Fig. 7. Depth converted interpretation with vertical exaggeration (**a**) and flexurally backstripped models (**b**) and (**c**) of the cross-section given in Fig. 6. The models show the stratigraphic configuration of the basinfill 92 Ma (**b**), following some 3 Ma of post-rift subsidence, and 95 Ma (**c**), following the cessation of active rifting and the deposition of the salt. By flexurally backstripping successive layers of the post-rift sequence, the models effectively reverse the process of thermal subsidence. They are constrained by the need to 'return' palaeo-sea-level markers to their palaeodatum (eustatic sea level incorporated from Haq *et al.* 1987). In the Rio Muni Basin, the principal palaeo-sea-level markers are the strongly erosive drift unconformity recording sub-aerial exposure some 95 Ma, following the emplacement of the carbonate rafts.

3000 m) and strongly tapering post-rift succession (i.e. post-Cenomanian) or restricted areal extent; (6) discordant Late Cretaceous drift unconformity, immediately post-dating raft emplacement; (7) truncation of Late Cretaceous and Palaeogene post-rift stratigraphy by the Oligocene unconformity.

A new model for this sector of the West African passive margin invokes the following principal stages in the transition from rift to drift (Fig. 8).

1. **Syn-rift (Barremian–mid-Aptian)** Normal faulting and contemporaneous sedimentation, with the focus of maximum subsidence beneath the Sergipe–Alagoas and crustal thinning concentrated beneath the Rio Muni Basin.
2. **Early drift (Late Aptian–Albian)** An asymmetric pattern of basin development continued beyond the cessation of active rifting, with end-rift footwall uplift concentrated beneath the Rio Muni Basin, where syn-rift subsidence was at a minimum. Therefore, the basin preserves significant relief at the level of the end-rift unconformity because limited syn-rift extension beneath the Rio Muni Basin, and deep, unfilled bathymetry, meant that end-rift uplift was limited and only elevated the shallowest parts of the syn-rift basin above sea level. Following end-rift thermal uplift, the earliest marine incursion is recorded by intercalated salts and shales, succeeded by a 1000 m thick interval of turbiditic sandstone.
3. **Gravitational collapse (Late Cenomanian)** Rapid subsidence allowed the accumulation of a shallow marine shelf carbonate succession in excess of 2000 m thickness. Carbonate sedimentation was arrested by regional thermal uplift linked to the development of the South Atlantic spreading ridge. It is difficult to be certain of the initial trigger that induced the large-scale collapse of the marginal carbonate shelf. By analogy with the mechanics of salt-floored thrust belts, and depending on the basal decollement lithology and the presence of overpressured fluids, raft tectonics may only require a modest palaeoslope of $< 1°$ (Turner 1996). During the translation of the shelf carbonates from their autochthonous position on the basin margin, pockets of syn-kinematic Cenomanian clastic sediment accumulated in small 'raft-top' basins.
4. **Peneplanation (end-Coniacian)** The cessation of the raft tectonics was marked by the formation of a deeply erosive drift unconformity of Late Cenomanian–Santonian age. Understanding of the mechanism of formation of this unconformity remains incomplete. Isostatic rebound related to the unloading of the basin margin during raft tectonics would have enhanced the erosiveness of the drift unconformity.
5. **Post-rift (Santonian–Recent)** Absence of Oligocene strata, and truncation of the post-rift succession at the Oligocene unconformity, suggests an additional, epeirogenic process that has served to uplift the post-rift basin. Oligocene erosion is documented throughout equatorial West Africa (e.g. Morocco: Hinz *et*

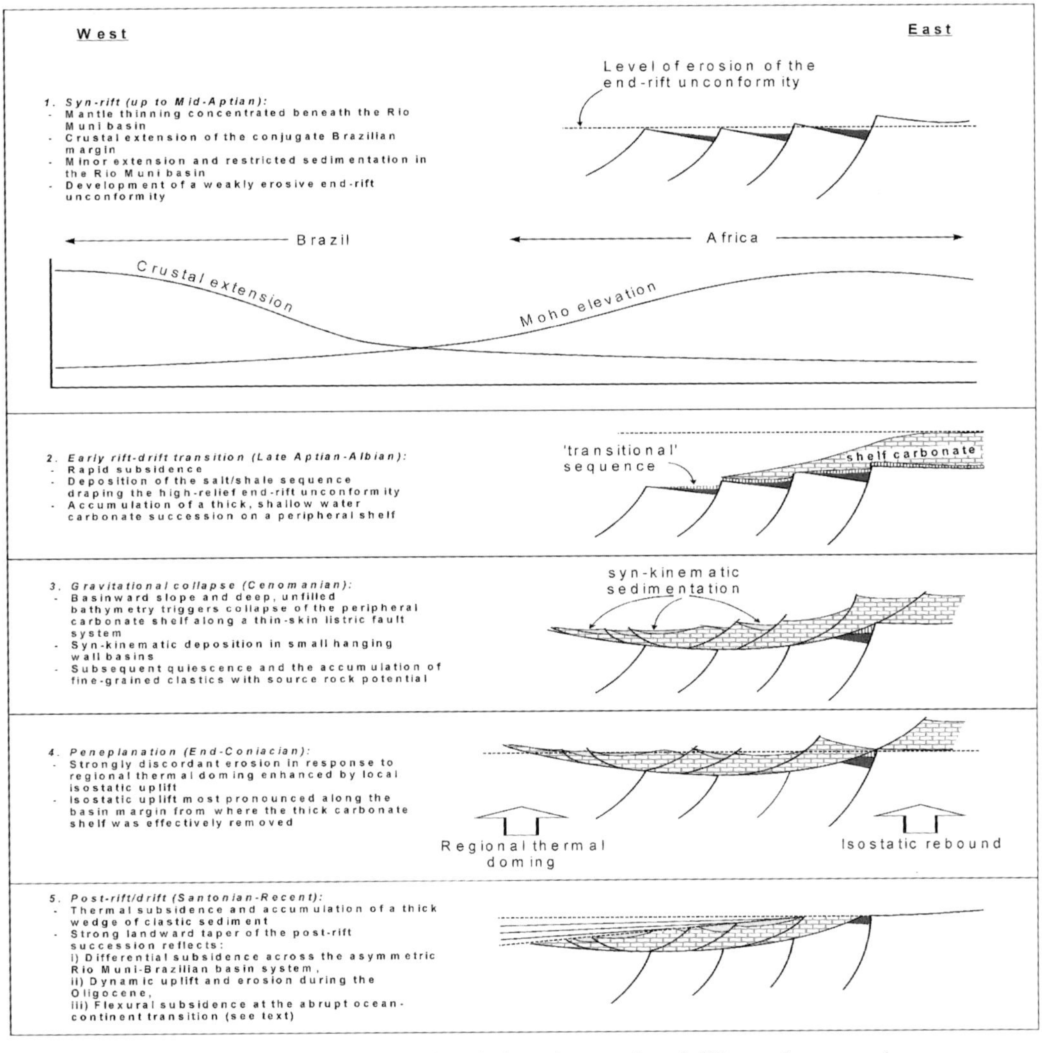

Fig. 8. Schematic development of the Rio Muni Basin from the cessation of rifting to the present day.

al. 1982; Côte D'Ivoire; Simon & Amadou 1984; Gabon: Teisserenc & Villemin 1989; Seranne *et al.* 1992; Congo: McGinnis *et al.* 1993; Cabinda, Angola: Brice *et al.* 1982; Brognon & Verrier 1966). On the basis of disparity between the global eustatic curve and the magnitude of hiatus and erosion recorded by the West African Oligocene unconformity, Seranne *et al.* (1992) conclude that it formed in response to the overprinting of eustatic effects by regional West African tectonics. In particular, igneous underplating during lithosphere thinning and volcanism, associated with the formation of the Cameroon volcanic arc in the Neogene, would have been a major factor.

Petroleum geology

Source rocks and the timing of maturity

Table 2 shows that good source rocks may be expected in several intervals within the Rio Muni succession. By analogy with adjacent Gabon and Brazil, the Late Aptian and Turonian intervals are considered to have the greatest potential. Given the geologically rapid emplacement of the rafts, one-dimensional basin modelling has been used to assess the impact of raft tectonics on the evolution of source rock maturity.

In Fig. 9, windows of source rock maturity are superimposed on a plot showing the decompacted

Table 2. *Regional source rock occurrence in West African basins and the conjugate Brazilian margin*

	Regional source rocks			
Age	Rio Muni	Gabon	Sergipe–Alagoas	Douala, Cameroon
	(Ross & Hempstead 1993; Turner 1995; unpublished reports)	(Teisserenc & Villemin 1989)	(Koutsoukos *et al.* 1991)	(Albreecht *et al.* 1976)
Turonian		Azile Formation; Type I–II TOC 3–5%		Present
Cenomanian		Cap Lopez Formation		Present
Aptian–Albian	(1) Uppermost Aptian (upper transitional sequence): TOC up to 3% (2) Upper Aptian (lower transitional sequence): Type II and I; TOC up to 8%	Albian Madiela Formation	(1) Basal Albian marine shales (2) Upper Aptian Muribeca Formation; Type II; TOC 3–5% (< 12%); S_2 up to 46 kgHC/ton	
Barremian		Melania Formation; Type I–II; TOC 4–6% (< 20%)		
Neocomian		Kissenda Formation Type III–II; TOC 1.5–2%	Present	Present

burial history of the succession encountered by the Matondo Well (Fig. 1). The plot incorporates lithological and thickness information from the well, and a realistic heat flow declining from 80 to 40 mW m^{-2} between the late Aptian and Recent (in agreement with generalized values from other margins undergoing the transition from active rifting to thermal subsidence: Allen & Allen 1990). The model has been calibrated using vitrinite reflectance data from Matondo which show a progressive increase from sub-mature in the Upper Cretaceous to early gas maturity in Lower Aptian rocks. Palaeobathymetry evolves from *c.* 1500 m immediately preceding raft emplacement to progressively shallower water during the post-rift period. This is interrupted by a relatively sudden shallowing–deepening event during the Late Cretaceous, when the drift unconformity is interpreted to have formed during an episode of subaerial emergence and erosion.

In the absence of data by which exhumation may be quantified, this study has used seismic interpretation to estimate the magnitude of erosion represented by the end-rift and drift unconformities. These estimates are well within constraints established from passive margins of the South Atlantic and elsewhere, where up to 4 km of sedimentary section has been lost from coastal regions during the transition from rift to drift (Brown *et al.* 1994; Gallagher *et al.* 1994). Thus, in Fig. 9, Late Cretaceous exhumation is modelled to be of the order of 1500 m and end-Eocene exhumation at 300 m.

The burial history plot of Matondo shows that the allochthonous Upper Aptian bituminous claystone interval (uppermost transitional sequence) first became mature in the latest Cretaceous (*c.* 70 Ma), entering the main oil generation window during the late Eocene (*c.* 40 Ma). The saliferous lower transitional sequence is presently in the gas generation window, with the main phase of oil generation having occurred between 80 and 50 Ma. Turonian source rocks are sub-mature. Burial history models from wells located in the central part of the basin, where most of the raft tectonics took place, reveal a very rapid phase of basin subsidence that accommodated the deposition and allochthonous emplacement of the transitional sequence. During this period, Upper Aptian source rocks attained their maximum burial depth. However, in spite of this rapid burial, subsequent uplift and erosion that accompanied the formation of the drift unconformity arrested the progress of source rock maturity. Consequently, oil generation in the Upper Aptian source intervals did not begin until *c.* 65 Ma, with maximum maturity attained today.

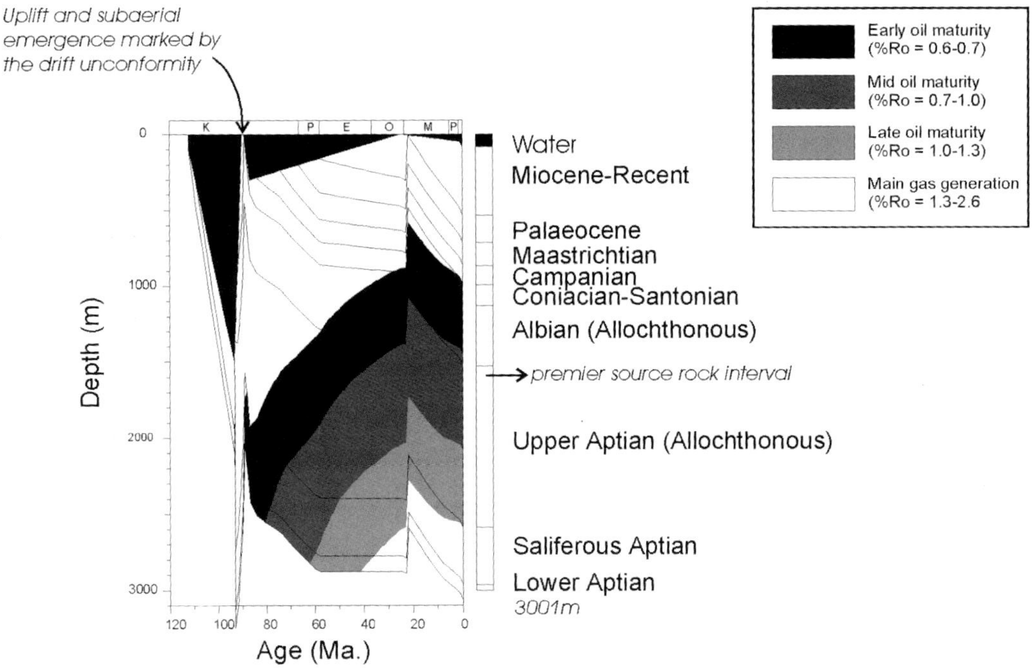

Fig. 9. Decompacted burial history of the stratigraphy encountered in the Matondo Well. Also shown are windows of source rock maturity based on modelling vitrinite reflectance data. The graph indicates that, following Late Cretaceous uplift, the premier Upper Aptian source rock interval progressed through the windows of early–mid oil maturity throughout the Cenozoic. Note that the timing of maximum burial (i.e. 92 Ma) does not in this case coincide with maximum palaeotemperature. The *BasinMod* software used in this modelling incorporates a transient heat flow option. This takes account of scenarios where rapid burial followed by rapid uplift means that the sedimentary section has insufficient time to heat up to its equilibrium temperature before subsequent uplift and cooling.

The impact of the raft tectonics on potential source rock maturity has also been assessed using a model based on the stratigraphy encountered in the Matondo Well, but replacing the allochthonous rafts with an estimate of how much sediment would have been deposited in the absence of the raft tectonics. Thus, Pseudo-Well-1 (Fig. 10) substitutes the 1460 m of allochthonous raft stratigraphy encountered by Matondo with some 500 m of Aptian–Albian shales and carbonates. Seismic data to the west of Matondo are used to establish the *c.* 500 m thickness of sedimentary succession that accumulated during raft tectonics. The chief impact of raft tectonics on the maturity of the upper transitional sequence source interval was to speed up the onset of generation by 10 Ma, whilst main generation occurs some 5 Ma earlier. However, modelling of potential sub-raft source rocks in the saliferous lower transitional sequence shows a greater difference between the rafted and non-rafted scenarios. The saliferous Upper Aptian interval in Matondo is presently mature for gas, having passed through the main oil generation window between 80 and 50 Ma. In the pseudo-well, however, the same interval has been mature for oil since *c.* 55 Ma.

Wells located at the present basin margins provide an exception to the general picture of present-day maximum maturity. Thus, the onshore N'Dote Well shows maximum maturity to have been attained during the latest Cretaceous (Fig. 11). Earlier maturity at these marginal positions within the basin are due to the fact that they were the depositional sites of up to 2500 m of transitional sequence sediment, subsequently displaced basinward during Cenomanian raft tectonics. N'Dote is modelled to have been overlain by some 4000 m of syn-rift and transitional sequence strata before their removal during the Late Cretaceous. Two kilometres of this missing section is attributed to the raft tectonics; the remaining 2 km of erosion resulted from rift flank uplift and erosion.

Using the parameters constrained by data from the Matondo Well, an investigation was also made of burial history and potential source rock maturity in the deeper basin, where the influence of raft tectonics is considerably less. Pseudo-

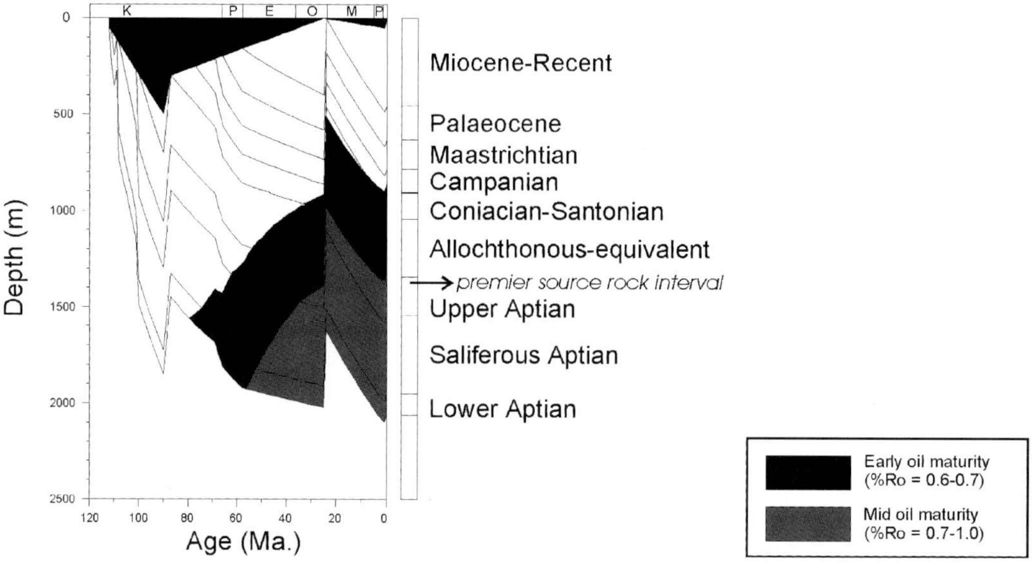

Fig. 10. Decompacted burial history of the stratigraphy encountered in the Matondo Well, with windows of source rock maturity superimposed. This model investigates the impact of the raft tectonics on potential source rock maturity by replacing the allochthonous rafted units with a 500 m autochthonous succession that would have been deposited in the absence of the rafting (cf. Fig. 9).

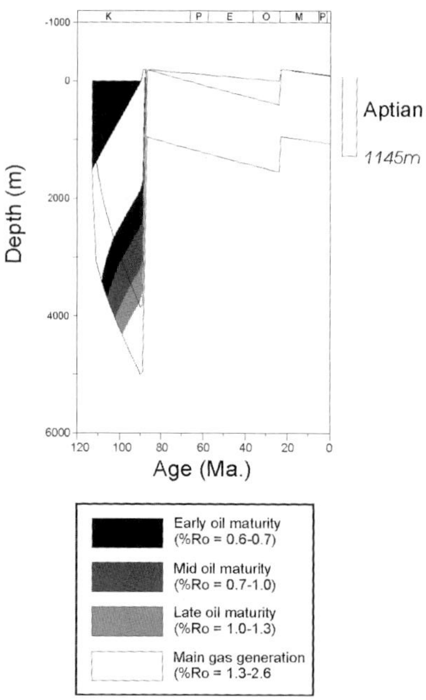

Fig. 11. Decompacted burial history of the stratigraphy encountered in the N'Dote Well, with windows of source rock maturity superimposed. This location, at the margin of the Rio Muni Basin, was severely affected by Late Cretaceous uplift and erosion related to thermal doming and isostatic rebound following the raft tectonics. The graph indicates that any Aptian source rocks would have passed through windows of oil and gas maturity during the Late Cretaceous.

Well-2 (Fig. 12) is based on the depth converted stratigraphy interpreted from a seismic profile some 25 km west of Matondo. The pseudo-well is located in 770 m of water and, in comparison to Matondo, it penetrates a thinner raft sequence (700 m) and a thicker autochthonous post-rift succession (3000 m). The onset of early maturity for the latest Aptian source interval in the pseudo-well is similar to Matondo at *c.* 70 Ma, with main oil generation occurring between 60 and 20 Ma. Gas generation is predicted at the present day. Significantly, the Turonian source interval in this more basinward setting is presently in the main oil generation window which it entered *c.* 50 Ma.

Reservoirs

By analogy with adjacent basins of the West African salt province, reservoir-quality intervals can be expected in: (1) Neocomian–Barremian pre- and early syn-rift succession (*c.*> 61.5 billion barrels reserves from equivalent lacustrine and fluvial facies sandstones in Gabon: Ross & Hempstead 1993); (2) transitional sequence Albian marine shoal carbonates and sandstones (cf. Congo, Angola); (3) Turonian–Cenomanian shallow marine sandstones (cf. Congo); and (4) Late Cretaceous–Palaeogene turbidite sandstones (*c.* 1.4 billion barrels reserves in Gabon: Ross & Hempstead 1993). Furthermore, Ross &

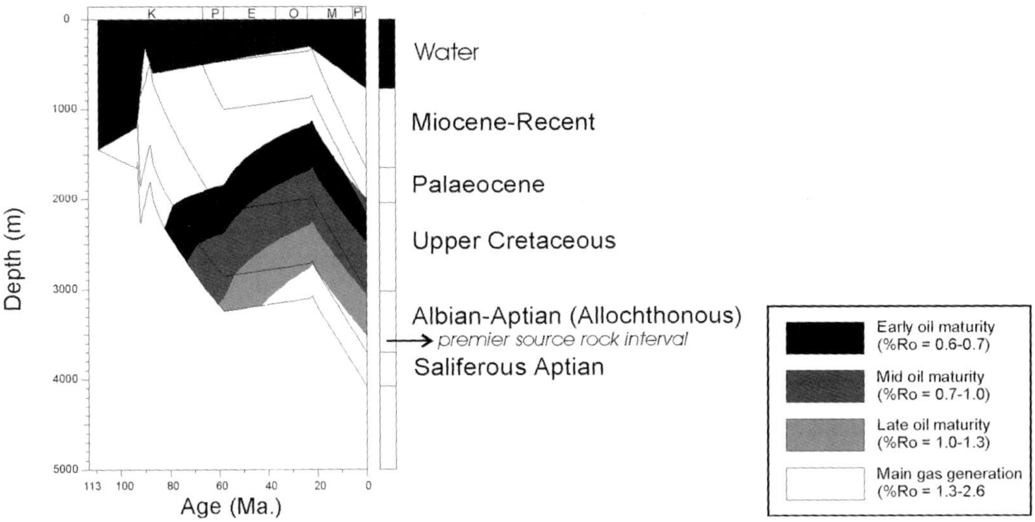

Fig. 12. Simulated decompacted burial history of the stratigraphy at the site of Pseudo-Well-2 (Fig. 1), with windows of source rock maturity superimposed. The pseudo-well is located in deep water (black), beyond the area affected by raft tectonics. Consequently, a relatively thin allochthonous sequence is modelled. The plot shows that the timing and progression of source rock maturity compares with that modelled for the Matondo Well (Fig. 9). However, unlike in the area of most intensive rafting, the potentially source-bearing Turonian interval is presently mature for oil.

Hempstead (1993) and Turner (1995) suggest that Late Aptian lacustrine sandstone of the transitional sequence constitutes a new play for the West African salt basin. These sandstones were deposited in closed lacustrine basins during a period of high clastic influx, following the relatively quiescent conditions that accompanied deposition of the intercalated salts and shales of the lower transitional sequence.

Good quality reservoir intervals within the Albian carbonate rafts were encountered in several Rio Muni wells (Ross & Hempstead 1993; Turner 1995). Predicting the basinwide distribution of reservoirs within the carbonate rafts requires an appreciation of their sedimentology and sequence stratigraphic setting, plus an understanding of the kinematics of raft emplacement – during their downslope translation, subaerial emergence and erosion of the rafts would have led to localized karstification and enhanced reservoir quality.

This paper has suggested that raft tectonics may have considerable isostatic consequences. Geologically rapid transfer of > 2000 m sediment from the peripheral shelf to the deeper water will cause the unloaded shelf to rebound, with possible subaerial emergence. Lowstand fans in the lower autochthonous post-rift stratigraphy are therefore considered to have major potential in the deeper water.

Timing, geometry and access to migration pathways of raft tectonic structures

During their downslope translation, the rafts experienced a polyphase strain history as they moved across the complex three-dimensional topography of the end-rift unconformity. Consequently, their hanging walls commonly display an 'eggbox'-like configuration of closely spaced anticlines, many of which are dip-closed in four directions (Fig. 1). However, given the importance of Late Cretaceous uplift and erosion, up to half of the areal extent of the allochthonous rafts is truncated by the present seafloor, with the attendant risk of hydrocarbon leakage to the surface.

Rafting is constrained between 95 and 92 Ma, i.e. early with respect to post-rift subsidence, burial and hydrocarbon generation. Burial history modelling has shown that, following Late Cretaceous uplift and erosion, the Aptian source rock kitchens lie beneath the area affected by raft tectonics, whilst Turonian source kitchens are located further to the west beyond the locus of raft tectonics. Figures 2 and 3 depict the following four migration pathways from potential hydrocarbon source rocks to reservoirs within the rafts' hanging walls: (1) Aptian source: vertical migration across the basal detachment; (2) Aptian source: lateral migration across transversely oriented faults juxtaposing allochthonous carbonates in the hanging walls with the sub-detachment Aptian stratigraphy; (3) Turonian source: thrusts within rafts' hanging walls placing Albian carbonate on top of Turonian–Coniacian rocks; intraraft thrust faults with 800 ms two-way travel time (TWTT) vertical displacement (*c.* 1400 m) have been mapped in Rio Muni and adjacent Gabon (D. A. Kirkland pers. comm.) sufficient to juxtapose Turonian and Albian intervals; (4) Turonian source: lateral migration across normal faults in the rafts' hanging walls bounding small syn-kinematic half-graben containing Late Cretaceous stratigraphy.

Conclusions

1. The switch from rifting to thermal subsidence in the Rio Muni Basin is marked by twin unconformities that bound a transitional sequence of rocks that do not fit easily into the bipartite rift/post-rift classification. The transitional sequence consists of up to 2500 m clastic and carbonate strata, dominated by a thick succession of allochthonous 'rafts' of shelfal carbonates emplaced in a relatively deepwater setting during the Late Cretaceous.
2. Raft tectonics took place more or less in the absence of halokinesis in response to an unusual coincidence of unstable build-up of marginal carbonate sediment, deep, unfilled bathymetry in the adjacent basin centre, and Late Cretaceous uplift. It has generated a unique basin fill consisting of an intensely deformed transitional sequence resting between autochthonous rift and post-rift successions. Deformation of the rafts occurred in response to their translation over a complex three-dimensional basal detachment.
3. A new model suggests close links between raft tectonics and the bounding end-rift and drift unconformities. Raft tectonics was triggered by regional thermal uplift in the vicinity of the 'semi-transform' boundary between the Rio Muni Basin and the adjacent spreading ridge. The drift unconformity probably formed in response to a combination of thermal and isostatic processes. Its erosiveness was compounded by isostatic uplift related to the rapid unloading of the peripheral carbonate shelf.
4. The hanging walls of the rafts are intensely structured, with deformation having occurred prior to Late Cretaceous and Cenozoic post-rift burial. However, episodes of peneplanation during both the Late Cretaceous and Oligocene means that a large part of the landward portion of the rafts outcrops at the seafloor, with attendant risk of lack of seal.
5. Raft tectonics has special implications for assessing the history of source rock maturity. From well information, and by analogy with adjacent basins, potential source-bearing intervals may occur in the lower Aptian (along and beneath the base of the rafts), above the rafts in the upper transitional sequence, and in post-raft Turonian rocks. The impact of the raft tectonics on source rock maturation is demonstrated by two 'pseudo-wells' that enable comparison between maturation histories incorporating major raft tectonics and a more 'normal' basin history. They suggest that the Turonian source becomes increasingly attractive as an oil source toward the west, into the deeper water. In the shallower water, where the raft tectonics was most severe, its chief influence was to speed up the maturation process by generating a very rapid phase of sedimentation and burial.

I am grateful to Dave Kirkland (Birmingham) for useful discussions and his assistance with the burial history modelling. Alan Roberts (Badley Earth Sciences) provided advice on the flexural backstripping and, with Nick Kusznir (Liverpool), arranged free use of the *Flex* modelling software. I acknowledge LASMO plc for granting me full access to their West African database. Thanks are also extended to Nick Cameron (Geo-Insight) for organizing an excellent meeting at the Geological Society, and for encouraging me to present this work at it. This paper has benefitted from the constructive criticism of Philip Nelson (independent) and Steve Lawrence (Quad).

References

Albrecht, P., Vanderbroucke, M. & Mandeorque, M. 1976. Geochemical studies on the organic matter from the Douala basin (Cameroon) – I. Evolution of the extractable organic matter and the formation of petroleum. *Geochimica et Cosmochimica Acta*, **40**, 791–799.

Allen, P. A. & Allen, J. R. 1990. *Basin Analysis*. Blackwell.

Aubouin, J., Choubert, G., Faure-Muret, A. *et al.* 1985. *International Geological Map of Africa*, Third Edition (1:5 000 000). International Geological Mapping Bureau.

Barton, P. & Wood, R. 1984. Tectonic evolution of the North Sea basin: crustal stretching and subsidence. *Geophysical Journal of the Royal Astronomical Society*, **79**, 987–1022.

Basile, C., Mascle, J., Popoff, F., Bouillon, J. P. & Mascle, G. 1993. The Ivory Coast–Ghana margin; a marginal ridge structure deduced from seismic data. *Tectonophysics*, **222**, 1–21.

BRADLEY, C. A. & FERNANDEZ, M. N. 1992. Early Cretaceous paleogeography of Gabon/Northeastern Brazil – A tectono-stratigraphic model based on propagating rifts. *Proceedings of the First Colloquium of Stratigraphy and Paleogeography of the Sedimentary Basins of West Africa*, Libreville, 17–30.

BRICE, S. E., COCHRAN, M. D., PARDO, G. & EDWARDS, A. D. 1982. Tectonics and sedimentation of the South Atlantic Rift Sequence; Cabinda, Angola. *In*: WATKINS, J. S. & DRAKE, C. L. (eds) *Studies in Continental Margin Geology*. AAPG Memoir, **34**, 5–21.

BROGNON, G. P. & VERRIER, G. R. 1966. Oil and gas in Cuanza basin of Angola. *AAPG Bulletin*, **34**, 5–18.

BROWN, R. W., SUMMERFIELD, M. A. & GLEADOW, A. J. W. 1994. Apatite fission track analysis: Its potential for the estimation of denudation rates and implications for models of long term landscape development. *In*: KIRKBY, M. (ed.) *Process Models and Theoretical Geomorphology*. Wiley, 23–53.

CASTRO, JR 1987. The northeastern Brazil and Gabon basins: a double rifting system associated with multiple crustal detachments. *Tectonics*, **6**, 727–738.

CHANG, H. K., KOWSMANN, R. O., FIGUEIREDO, A. M. F. & BENDER, A. A. 1992. Tectonics and stratigraphy of the East Brazil rift system: an overview. *Tectonophysics*, **213**, 97–138.

CLEMSON, J. 1997. *Structural segmentation of the Namibian passive margin*. PhD Thesis, University of London.

——, CARTWRIGHT, J. & SWART, R. 1999. The Namib Rift: a rift system of possible Karoo age, offshore Namibia. *This volume*.

CLIFT, P. D., LORENZO, J., CARTER, A., HURFORD, A. J. & ODP LEG 159 SCIENTIFIC PARTY 1997. Transform tectonics and thermal rejuvenation on the Côte D'Ivoire–Ghana margin, West Africa. *Journal of the Geological Society of London*, **154**, 483–489.

DAVISON, I. 1997. Wide and narrow margins of the Brazilian South Atlantic. *Journal of the Geological Society of London*, **154**, 471–476.

DUVAL, B., CRAMEZ, C. & JACKSON, M. P. A. 1992. Raft tectonics in the Kwanza basin, Angola. *Marine and Petroleum Geology*, **9**, 389–405.

FOWLER, S. & MCKENZIE, D. P. 1979. Gravity studies of the Rockall and Exmouth Plateaux using SEASAT altimetry. *Basin Research*, **2**, 27–34.

GALLAGHER, K., HAWKESWORTH, C. & MANTOVANI, M. 1994. The denudation history of the onshore continental margin of S.E. Brazil inferred from fission track data. *Journal of Geophysical Research*, **99**, 18 117–18 145.

HAQ, B. U., HARDENBOL, J. & VAIL, P. R. 1987. Chronology of fluctuating sea levels since the Triassic. *Science*, **235**, 1156–1167.

HARDIE, L. A. 1990. The roles of rifting and hydrothermal CACL brines in the origin of potash evaporites: An hypothesis. *American Journal of Science*, **290**, 43–106.

HINZ, K., DOSTMANN, H. & FRITSCH, J. 1982. The continental margin of Morocco: Seismic sequences, structural elements and geological development. *In*: VON RAAD, P., HINZ, K., SARNTHEIN, M. & SEIBOLD, E. (eds) *Geology of the Northwest African Continental Margin*. Springer, 34–60.

KOUTSOUKOS, E. A. M., MELLO, M. R., DE AZAMBUJHA FILHO, N. C., HART, M. B. & MAXWELL, J. R. 1991. The Upper Aptian–Albian succession of the Sergipe Basin, Brazil. *AAPG Bulletin*, **75**, 479–498.

KUSZNIR, N. J., MARSDEN, G. & EGAN, S. S. 1991. A flexural cantilever simple shear/pure shear model of continental extension: application to the Jenne d'Arc Basin and Viking Graben. *In*: ROBERTS, A. M., YIELDING, G. & FREEMAN, B. (eds) *The Geometry of Normal Faults*. Geological Society, London, Special Publications, **56**, 41–60.

——, ROBERTS, A. M. & MORLEY, C. K. 1995. Forward and inverse modelling of rift basin formation. *In*: LAMBIASE, J. J. (ed.) *Hydrocarbon Habitat in Rift Basins*. Geological Society, London, Special Publications, **80**, 33–56.

LEYDEN, R., BRYAN, G. & EWING, M. 1972. Geophysical reconnaissance on the African shelf: 2. Marginal sediments from Gulf of Guinea to Walvis Ridge. *AAPG Bulletin*, **56**, 682–693.

LORENZO, J. M. & WESSEL, P. 1995. Tectonic and stratigraphic response to flexure across a continent–ocean transform: Falkland/Malvinas Plateau, South Atlantic. *EOS*, **76**, 580.

LUNDIN, E. R. 1992. Thin-skinned extensional tectonics on a salt detachment, northern Kwanza basin, Angola. *Marine and Petroleum Geology*, **9**, 405–412.

MARSDEN, G., ROBERTS, A. M., YIELDING, G. & KUSZNIR, N. J. 1990. Application of a flexural cantilever simple-shear/pure shear model of continental lithosphere extension to the formation of the North Sea Basin. *In*: BLUNDELL, D. J. & GIBBS, A. D. (eds) *Tectonic Evolution of the North Sea Rifts*. Oxford Science Publications, 236–257.

MASCLE, J., LOHMANN, G. P. & CLIFT, P. D. 1996. *Proceedings of the Ocean Drilling Program, Initial Results*. College Station TX (Ocean Drilling Program), **159**.

MCGINNIS, J. P., DRISCOLL, N. W., KARNER, G. D., BRUMBAUGH, W. D. & CAMERON, C. 1993. Flexural response of passive margins to deep-sea erosion and slope retreat: Implications for relative sea-level change. *Geology*, **21**, 893–896.

MEYERS, J. M., ROSENDAHL, B. R., GROSCHEL-BECKER, H., AUSTIN, J. A. & RONA, P. A. 1996. Deep penetrating MCS imaging of the rift-to-drift transition, offshore Douala and North Gabon basin, West Africa. *Marine and Petroleum Geology*, **13**, 791–935.

NADIN, P. A., KUSZNIR, N. J. & TOTH, J. 1995. Transient regional uplift in the Early Tertiary of the northern North Sea and the development of the Iceland plume. *Journal of the Geological Society, London*, **152**, 953–959.

PAUTOT, G, RENARD, V., DANIEL, J. & DUPONT, J. 1973. Morphology, limits, origin and age of salt layer along South Atlantic margin. *AAPG Bulletin*, **57**, 1658–1671.

RABINOWITZ, P. D. LA BREQUE, J. 1979. The Mesozoic South Atlantic Ocean and evolution of its continental margins. *Journal of Geophysical Research*, **84**, 5973–6002.

ROBERTS, A. M., LUNDIN, E. R. & KUSZNIR, N. J. 1997. Subsidence of the Vorin Basin and the influence of the Atlantic continental margin. *Journal of the Geological Society, London*, **154**, 551–559.

——, KUSZNIR, N. J., WALKER, I. M. & DORN-LOPEZ, D. 1993. Mesozoic extension on the North Sea: constraints from flexural backstripping, forward modelling and fault populations. *In*: PARKER, J. R. (ed.) *Petroleum Geology of Northwest Europe: Proceedings of the 4th Conference.* The Geological Society, London, 1123–1136.

——, YIELDING, G., KUSZNIR, N. J., WALKER, I. M. & DORN-LOPEZ, D. 1995. Quantitative analysis of Triassic extension in the northern Viking Graben. *Journal of the Geological Society, London*, **152**, 15–26.

ROSENDAHL, B. R., GROSCHEL-BECKER, H., MEYERS, J. & KACZMARICK, K. 1991. Deep seismic reflection study of a passive margin, southeastern Gulf of Guinea. *Geology*, **19**, 291–295.

ROSS, D. & HEMPSTEAD, N. 1993. Geology, hydrocarbon potential of Rio Muni area, Equatorial Guinea. *Oil and Gas Journal*, **91**, 96–100.

SANDWELL, D. T. & SMITH, W. H. F. 1995. *Map of marine gravity anomaly from satellite altimetry.* Scripps Institute of Oceanography, La Jolla, California.

SERANNE, M., SEGURET, M. & FAUCHIER, M. 1992. Seismic super-units and post-rift evolution of the continental passive margin of southern Gabon. *Bulletin Societe Geologique de France*, **163**, 135–146.

SIMON, P. & AMADOU, B. 1984. La discordance oligocene et les dépots postérieurs à la discordance dans le bassin sédimentaire ivoirien. *Bulletin Société géologique de France*, **26**, 117–125

TEISSERENC, P. & VILLEMIN, J. 1989. Sedimentary Basin of Gabon – Geology and Oil Systems. *In*: EDWARDS, J. D. & SANTOGROSSI, P. A. (eds) *Divergent/Passive Margin Basins*. AAPG Memoir, **48**, 117–200.

TURNER, J. P. 1995. Gravity-driven structures and rift basin evolution: The Rio Muni basin, offshore Equatorial West Africa. *AAPG Bulletin*, **79**, 1138–1158.

—— 1996. Gravity-driven nappes and their relation to palaeobathymetry: examples from West Africa and Cardigan Bay, UK. *In*: BUCHANAN, P. G. & NIEUWLAND, D. A. (eds) *Modern Developments in Structural Interpretation, Validation and Modelling.* Geological Society, London, Special Publications, **99**, 345–362.

USSAMI, N., KARNER, G. D. & BOTT, M. H. P. 1986. Crustal detachment during South Atlantic rifting and formation of the Tucano–Gabon basin system. *Nature*, **322**, 629–632.

WATTS, A. B. 1988. Gravity anomalies, crustal structure and flexure of the lithosphere at the Baltimore Canyon Trough. *Earth and Planetary Science Letters*, **89**, 221–238.

The interpretation of passive margin depositional processes using seismic stratigraphy: examples from offshore Namibia

JOANNE BAGGULEY[1] & SARAH PROSSER[2]

[1]*School of Construction and Earth Sciences, Oxford Brooke University, Headington, Oxford OX3 0BP, UK*
Present address: West Australia Petroleum Ltd, 250 St George's Terrace, Perth 6000, W.A., Australia
[2]*Saga Petroleum ASA, Kjørbovn 16, PO Box 490, Sandvika 1301, Norway*

Abstract: Seismic stratigraphic analysis of the post-rift megasequence of the Namibian passive continental margin has enabled depositional features to be interpreted in terms of the likely controls on their formation. The scale of the information derived from this manner of depositional system analysis is fundamental in understanding processes of sedimentation on passive margins; processes which may not be so readily apparent at field or well-log scale studies. Inner shelf channel forms are interpreted as incised valleys, created during a fall in relative sea level. The horizontal nature of the seismic facies infilling these valleys suggests that they filled during progressive periods of rising relative sea level, implying late lowstand or transgressive sediments. A series of low mounded features can also be identified on the shelf, which may be interpreted as palaeodune fields, carbonate build-ups or longshore clastic mounds. The outer shelf and upper slope positions reveal interpreted canyon and channel forms. The infill to these features contrasts with that of the incised valleys by having a more draped form (sometimes with a chaotic unit below); a geometry consistent with submarine origins. The submarine canyons may be termed attached or detached depending on whether they can be shown to link back into fluvial systems on the shelf; attached canyons will be more likely to be able to support the development of deepwater, sand-rich submarine fans. In the slope/base of slope area both regionally extensive and more localized mounds can be interpreted as mass transport deposits, the trigger for which may be attributed to a relative sea-level fall (due to their connection to the valley and canyon systems), though a transgressive origin may also be consistent in some cases. The mounds seen in positions deepest in the basin have a markedly different seismic facies to those at the base of slope and a different depositional system may be invoked, perhaps that dominated by distal turbidity currents or waning bottom currents.

Depositional systems offshore passive margins have been described in the past from both field outcrops and subsurface information. Rarely, however, is the entire suite of systems seen on a large scale as well as on the two-dimensional seismic reflection data offshore Namibia. This paper aims to show selected parts of these seismic sections to highlight the importance of this large-scale framework within which more detailed depositional studies can be undertaken. It is not the aim to provide a comprehensive discussion of each depositional environment, but rather to show the seismic expression of certain systems and the ensuing evidence that constrains the large-scale spatial and temporal nature of passive margin sedimentation.

Regional evolution and megasequence interpretation

Namibia lies on the west coast of southern Africa and is bordered to the north (17°S, 12°E) and south (28°S, 16°E) by Angola and the Republic of South Africa, respectively. The region offshore Namibia forms the passive margin to the South Atlantic which opened as a result of the separation of the African and South American continents during Early Cretaceous times (Rabinowitz & Labrecque 1979; Austin & Uchupi 1982; Sibuet *et al.* 1984; Nürnberg & Müller 1991).

Database

The data available for this study consists mainly of two regional seismic reflection surveys which together comprise a total of 14 300 km of data (Fig. 1). These surveys form a widely spaced grid along the length of the 1200 km Namibian margin with an average line spacing between 10 and 30 km. The only released well information comes from the Kudu Gas Field which is located in the Orange Basin (Fig. 2) in the most southerly part of the margin, though four further exploration wells have recently been drilled further north in the Walvis Basin (Fig. 2).

Bagguley, J. & Prosser, S. 1999. The interpretation of passive margin depositional processes using seismic stratigraphy: examples from offshore Namibia. *In:* Cameron, N. R., Bate, R. H. & Clure, V. S. (eds) *The Oil and Gas Habitats of the South Atlantic.* Geological Society, London, Special Publications, **153**, 321–344.

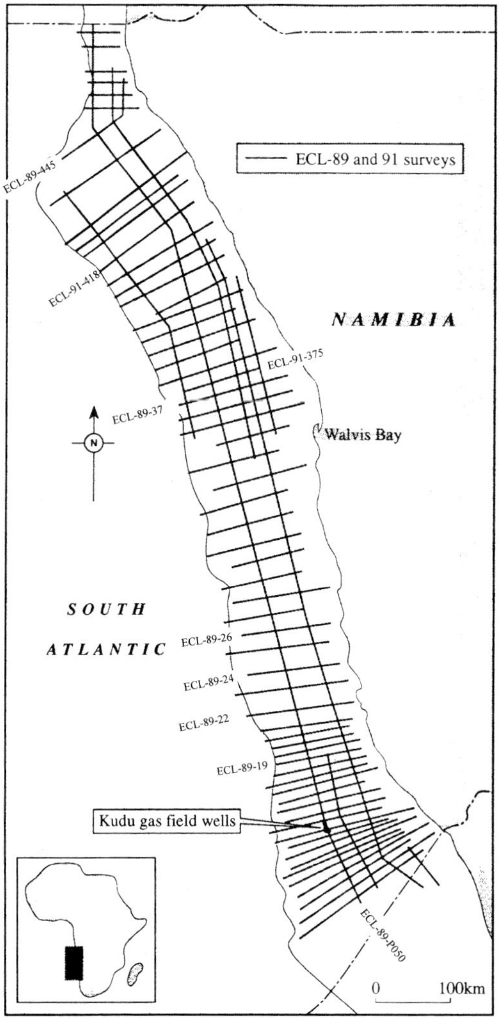

Fig. 1. Map showing the regional 2D seismic grid, offshore Namibia. The seismic surveys, which were shot in 1989 (ECL-89) and 1991 (ECL91), constitute 14 300 km of data with an average line spacing of 10–30 km.

Megasequence interpretation

Five megasequences are defined offshore Namibia on the basis of reflection terminations and geometries (Fig. 3); each of these is related to a major phase of basin evolution (as defined by Hubbard *et al.* 1985) which are numbered accordingly from MS10 to MS50. MS10–MS30 relate to early stages of rifting (MS10) and the opening of the South Atlantic (MS20 and MS30), thus broadly correlating with the events proposed by Light *et al.* (1993). These megasequences are overlain by MS40 (the transitional megasequence of Light *et al.* 1993), which is in turn overlain by the main post-rift succession (defined as MS50). This latter succession consists of a prograding clastic wedge which, on the basis of seismic reflection geometries, can be subdivided into MS50a and MS50b (Fig. 3) and dated as Turonian–Maastrichtian and Tertiary respectively, based on biostratigraphy from the Kudu wells (McMillan 1990, 1994). It is from MS50 that the examples of depositional processes are taken to form the basis of this paper.

Large-scale depositional systems

The seismic lines parallel to the main direction of depositional dip offshore Namibia are typically 150–200 km in length and cover all depositional environments from the inner shelf to the deep basin. This allows a wide spectrum of depositional systems to be interpreted from a more laterally extensive view than is usually sampled in an individual outcrop or from wells; thus, the relationship of such linked systems to one another can be observed and interpreted. The examples displayed in this paper come largely from the shelf, the slope and the deep basin environments, although components of fluvial and aeolian systems are also mentioned. Illustrations are presented from progressively proximal to distal settings along the typical passive margin depositional profile of shelf, slope and basin (though this profile becomes more ramp-like in the later stages of the Tertiary). They are by no means meant to represent a complete documentation of all possible depositional systems present on a passive margin, but rather provide a selection of the most well resolved and interesting features seen on the Namibian margin which could have implications for other margins of this type.

Depositional and erosional systems on the shelf: channels and mounds

Incised valleys

Different channel and mounded forms can be identified on the seismic reflection data in the inner shelf regime. The channel forms under discussion (Fig. 4) occur in the Orange, Lüderitz and Walvis Basins, where they are observed to develop contemporaneously with laterally extensive unconformities at the Cretaceous–Tertiary boundary and at several different stratigraphic levels within the MS50a (Turonian–Maastrichtian) unit. These channels are dominantly symmetrical in cross-section on dip and strike profiles, typically being up to

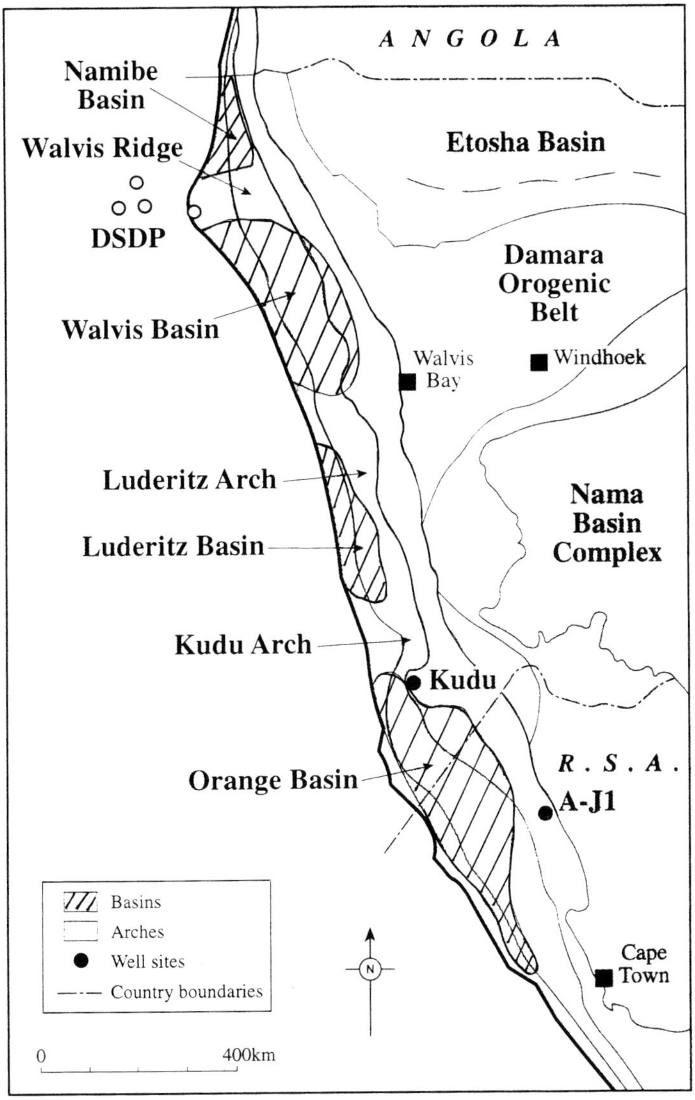

Fig. 2. Map showing the locations of the different offshore depocentres basins located along the Namibian passive continental margin (after Light *et al.* 1993).

5 km wide and up to a maximum of 200 ms two-way travel time (TWTT) (100–200 m) deep. The fill of the channel forms is dominantly transparent with occasional moderately continuous to continuous, moderate to high amplitude parallel reflections which appear to onlap the channel walls (Fig. 4). The horizontal nature of the reflections implies that the channel forms filled during progressive stages of an overall rise in relative sea level. This interpretation, when combined with the association of these features with uncomfortable surfaces, suggests that these shelfal channel forms can be interpreted as incised valleys, the fills of which are dominated by late lowstand and transgressive deposits.

Late lowstand and transgressive valley fills are thought to contain a relatively high percentage of sand, possibly resulting in the development of potentially good hydrocarbon reservoirs of estuarine, fluvial, deltaic or shallow marine origin (Van Wagoner *et al.* 1990). Potential reservoirs within similar incised valley fills have been docu-

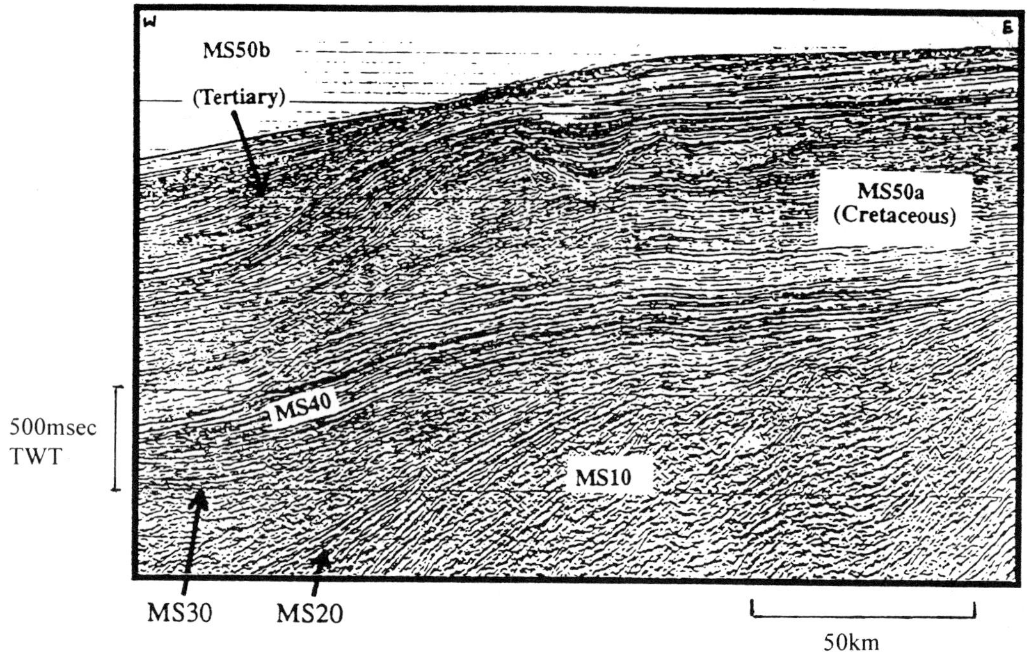

Fig. 3. Seismic profile showing megasquence development along the Namibian margin from early rift phases (MS10–MS30) to the development of the clastic post-rift wedge (MS50).

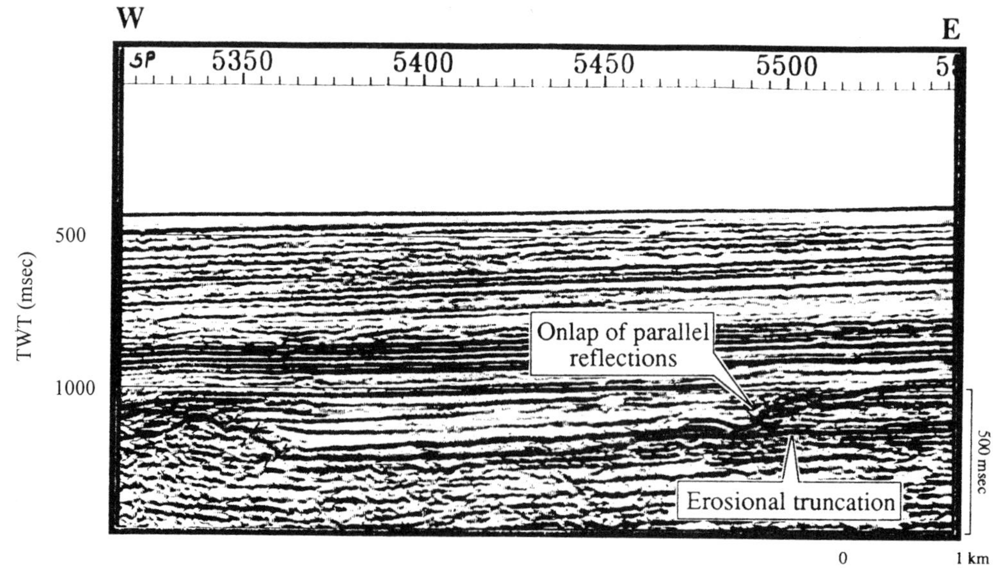

Fig. 4. Seismic expression of shelfal channels developed at the Cretaceous–Tertiary boundary. The channels, which are defined by erosional truncation, are infilled with horizontal reflections, inferring deposition during a progressive rise in relative sea level (from seismic line ECL-91-375).

mented from the South African part of the Orange Basin (Muntingh & Brown 1993; Brown *et al.* 1995). On the basis of the channel classification scheme proposed by Galloway & Hobday (1983), the Namibian incised valleys most closely resembled bed-load channels, having high width:

depth ratios and being characteristically filled with sand. Sandbodies deposited in bed-load channels stack vertically, implying that lateral continuity between adjacent sands deposited in these features may be quite high, thus improving the connectivity between possible reservoir bodies. It is recognized here that bed-load channels are of a smaller scale than the incised valleys being discussed, but the contrast in style from the slope channel forms discussed below, the comparison of the gross external forms of these features, and the documented fills of incised valleys from field- and well-based studies, suggest it is viable to predict the type and geometry of sediments in the manner described.

Shelfal mounds

The shelf is also the setting for some constructional features with broad mounded geometries generally onlapped from both sides by young reflections (Fig. 5). These features are between 12 and 54 km wide and have reliefs of up to 200 ms TWTT (100–200 m). Internally the mounds are characterized by either chaotic reflections or more progradational geometries, with progradation occurring both along (north and south) and perpendicular (east and west) to the shelf. The mounds may develop up to 120 km from the present-day coastline and are typically located just landward of the shelf break. Spatially, a mound complex can be mapped over 300 km along the northern half of the Namibian margin to the north of Walvis Bay (Fig. 6).

The shelfal mounds have been identified at three different stratigraphic levels (named A–C, oldest to youngest, respectively) all within the Tertiary (MS50b) unit of the post-rift wedge. From the map of the spatial distribution of these features (Fig. 6) it can be seen that the B mounds have the most extensive distribution, with the apparent gap in their presence being coincident with the later development of the C mounds. In cases where the A and B mounds are observed together on seismic profiles, the mounds appear to stack vertically (Fig. 7), implying that the deposition of these features occurred in approximately the same region on the shelf throughout time. The presence of irregular upper surfaces on some of the mounds suggests that some degree of erosion or reworking occurred prior to subsequent mound deposition. The A and B mounds are also observed, in some areas, to be separated by intervening reflections; further evidence that the process of mound deposition was not continuous.

The mounded geometries observed on the shelf sit directly above horizons which, on the basis of reflection termination patterns, may be interpreted as unconformities. Using the stratigraphic interpretation of the Namibian margin suggested by Bagguley (1997), these unconformities may be interpreted as being Palaeocene (in the case of the A and B mounds) and Late Miocene (C mounds) in age. The B mounds are also overlain by a downlap surface (Fig. 5).

The nature of what these seismically defined mounds represent in terms of depositional system is uncertain, but three possible interpretations may be postulated and are given below.

Palaeoaeolian dune field. A fall in relative sea level would expose the broad Namibian shelf, thus enabling a coastal palaeodune complex to shift to a new basinward position during the resulting lowstand (Fig. 8). Such a shift of dune systems in response to regressions has been documented by Corbett (1993) for the onshore palaeodune systems of the southern Namib Desert. In support of a palaeodune interpretation, the shelfal mounds observed on the seismic sections have similar dimensions (up to 200 m high, 10 km wide and several hundred kilometres long) to the coastal dunes of the present day Namib Desert. The more extensive nature of the modern desert (2000 km) compared to the spatial distribution of the shelfal mounds observed offshore Namibia (300 km) may be explained in terms of the relative aridity of the Namibian landmass at different times, with the offshore mounds representing the earliest stages of dune development in the proto-Namib Desert (Ward & Corbett 1992).

The relationship between the distribution of the B and C mounds (Fig. 6) may be explained by an aeolian environment if an analogy to the modern-day situation is considered. The constant lateral migration of the coastal dunes of the modern Namib Desert can result in the formation of a barrier to coastal drainage, thus preventing rivers from reaching the Namibian coast. During times of ephemeral stream flooding, rivers may break through this barrier and flow to the coast (Galloway & Hobday 1996). A similar situation could be envisaged during the time of deposition of the B and C mounds, whereby a flooded fluvial system breached the dune barrier to reach the continental margin, causing erosion of the previously deposited B mounds and producing a topographic low within which the preferential accumulation of the later C mounds could occur.

Longshore clastic bars. Longshore currents which transport sediment along the Namibian coast may redeposit the sediment at clastic bars, especially during times of transgression (Fig. 8). The wide geometry of the Namibian shelf will

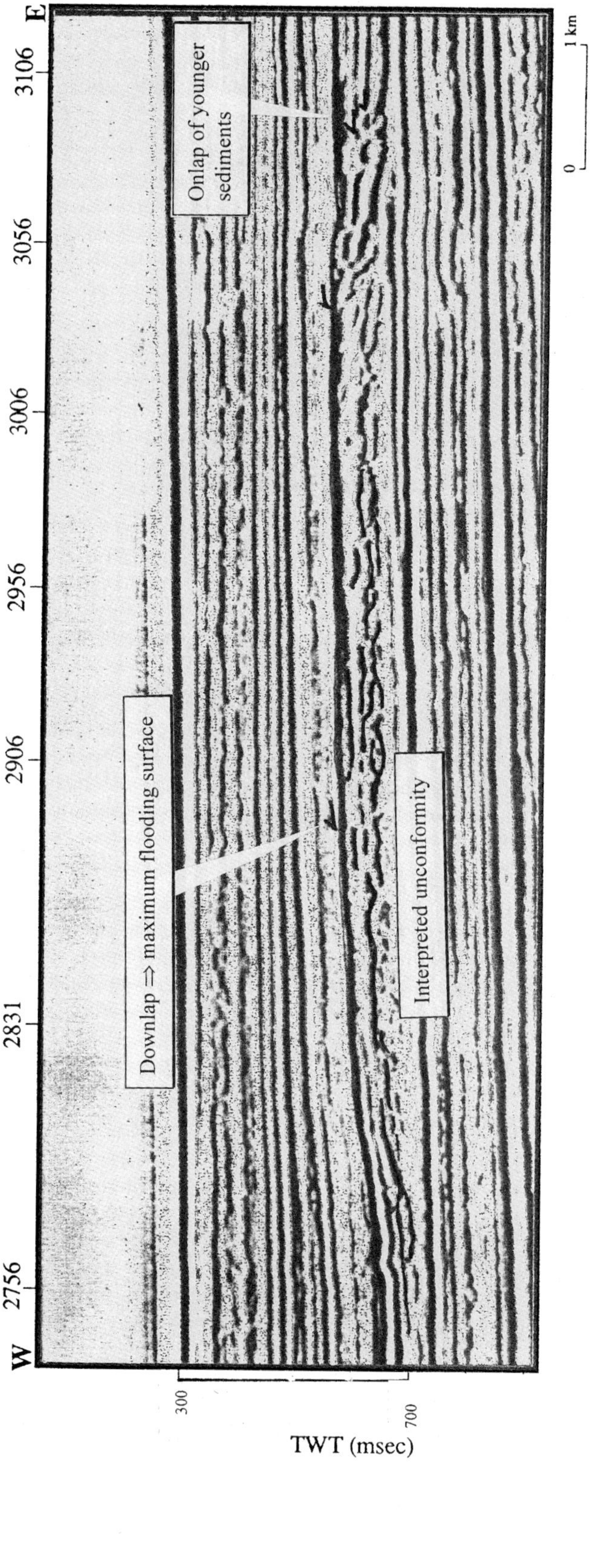

Fig. 5. Seismic expression of Tertiary shelfal mounds. The mound is underlain by an interpreted unconformity and typically overlain by an interpreted maximum flooding surface, based on the presence of downlapping reflections (from seismic line ECL-89-37).

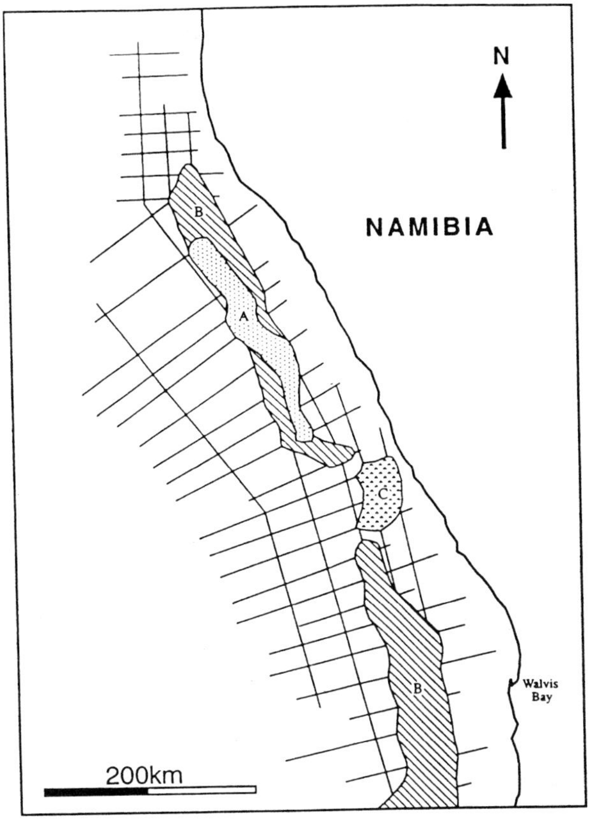

Fig. 6. Spatial distribution of Tertiary shelfal mounds. Ages of mounds are A–C from the oldest to the youngest development, respectively. The map shows that the middle-age mounds (B) have the most extensive areal distribution whilst the youngest mounds (C) only occur in a very small area, which is coincidental with the apparent gap in the distribution of the B mounds.

enhance the possibility of spatial separation of inner and outer shelf processes, the outer shelf becoming dominated by oceanic and longshore current activity. The development of outer shelf clastic bars with internal progradational geometries has been documented for the Childs Bank area offshore South Africa where bars of Miocene age can be traced for over 300 km (Dingle 1973). Similar, but smaller scale, subaqueous dune bedforms, interpreted to be deposited by the southward-flowing Agulhas Current, have been documented on the northern KwaZulu–Natal continental shelf, offshore southeast Africa (Ramsay *et al.* 1996).

The vertically stacked nature of the A and B mounds observed in some areas (Fig. 7) may provide evidence against a clastic bar origin since, after the deposition of the earliest mounds (A), subsequent mound deposition (B) would preferentially occur in areas of lower relief. An explanation of this may be that the topography produced by the earlier bar had been completely infilled prior to subsequent bar development.

Carbonate build-ups. Carbonate mounds on the shelf may develop during a relative sea-level rise. This would be consistent with the seismic stratigraphic framework if the mounds represented transgressive deposits which are overlain by a maximum flooding surface (as shown on Fig. 5). A carbonate interpretation may provide an explanation for the areal distribution of the mounds (Fig. 6) as carbonates would not be deposited at points of fluvial discharge and clastic input. Present-day rivers do drain to the coast in the region of mound distribution (although the ephemeral nature of these rivers means that clastic input to the offshore area is currently minimal); similar drainage patterns may have existed during the time of mound depression. The documented onset of aridification of the Namibian continent, and thus inferred decreased rates of clastic sedimentation, during

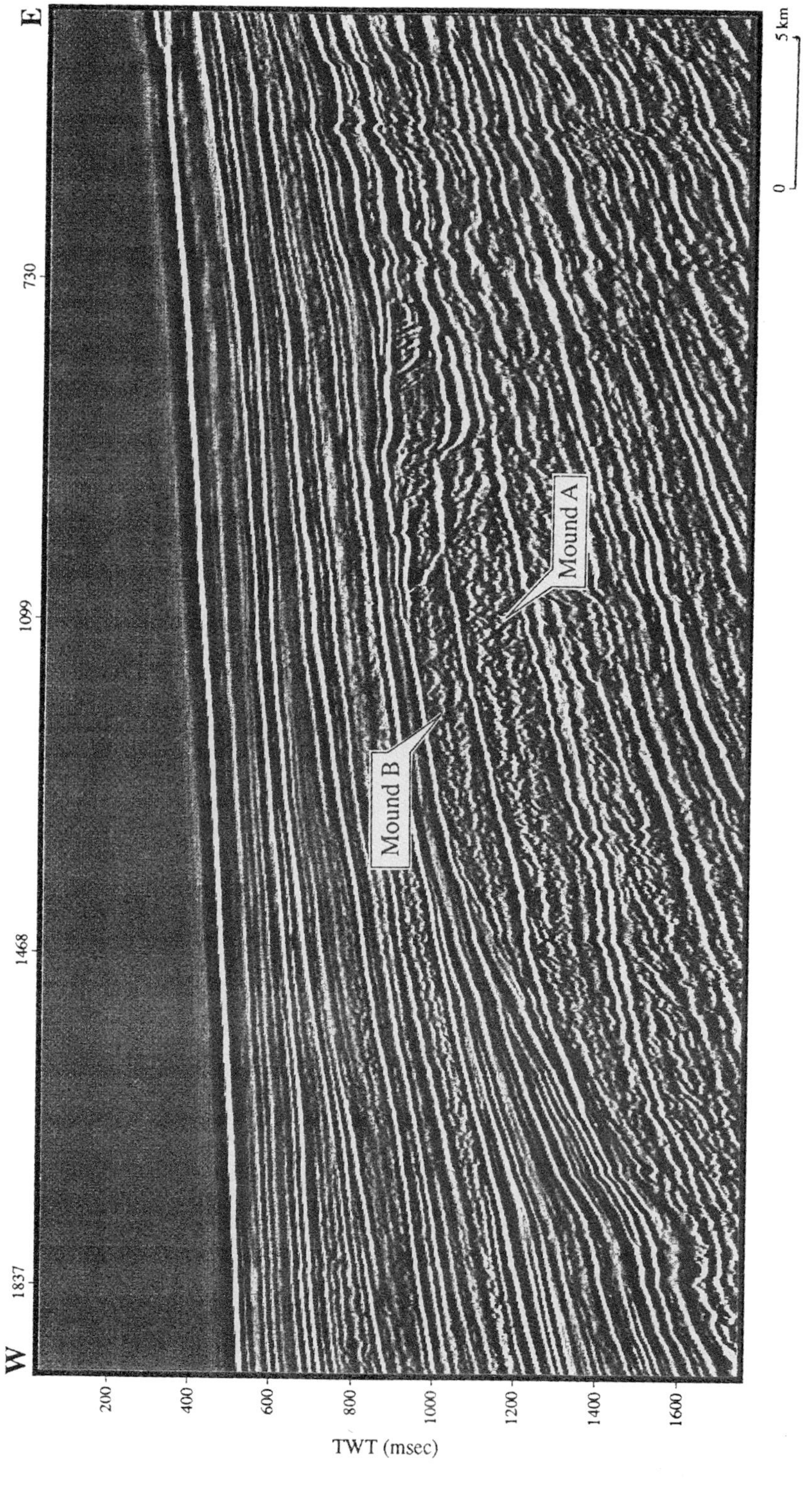

Fig. 7. Stacked nature of the A and B Tertiary mounds, suggesting similar locations of deposition on the shelf for successive mound development (from seismic line ECL-91-418).

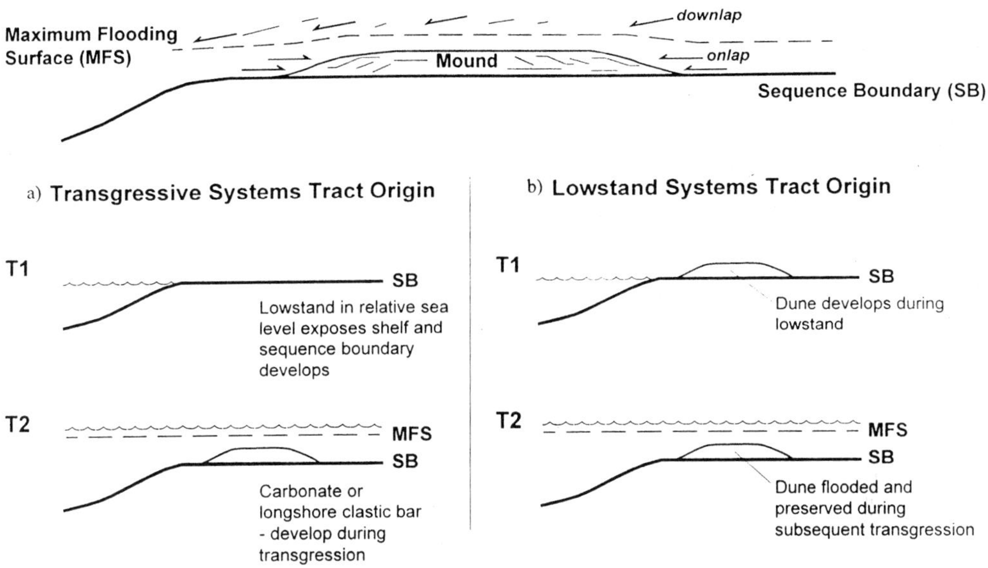

Fig. 8. Schematic diagram summarizing the transgressive and lowstand settings for shelfal mound deposition. The transgressive model encompasses the carbonate and longshore clastic bar origins of the mounds, while the dune origin requires deposition during a lowstand.

the Early–Middle Tertiary times (Ward *et al.* 1983), may also have provided favourable conditions for carbonate deposition.

Evidence against a carbonate origin includes the observation that the mounds do not appear to be associated with any kind of velocity anomaly on the seismic profiles (e.g. anomalous interval velocities or velocity pull-up effects). Strong velocity contrasts may be precluded, however, if the carbonates have a relatively low level of consolidation (e.g. oolite shoals) or if the surrounding sediments are themselves carbonate-rich.

The three alternative interpretations for the origins of the externally mounded features observed on the Namibian shelf suggest that mound deposition occurred either during lowstands in relative sea level, in the case of a palaeodune field, or during more transgressive periods, as inferred for the development of longshore clastic bars and carbonates (Fig. 8). On the basis of the data available for this study it is, however, difficult to differentiate between these possibilities and thus favour one depositional model.

Outer shelf-upper slope depositional and erosional systems: canyons and channels

Morphological features which characterize the outer shelf and upper slope include submarine canyons and channel forms. Their development can be recognized at various stratigraphic levels in the post-rift succession offshore Namibia, especially at the Cretaceous–Tertiary boundary and within the Tertiary succession. As with the shelfal mounds described above, these erosional features are restricted geographically to those parts of the margin lying to the north of Walvis Bay (Fig. 1).

Submarine canyons

The submarine canyons are up to 18 km, typically 400–500 m TWTT (400–500 m) deep and defined by the presence of erosional truncation at their lower boundaries (Fig. 9). Different canyon systems can be recognized at Turonian and Tertiary levels; the systems developed at the Base Tertiary unconformity (Canyon Systems 1–6, Fig. 10) are taken as examples for the purpose of this paper. The majority of the canyons identified are infilled by two distinct seismic facies (Fig. 9). The lowermost of these facies is characterized internally by chaotic reflections and externally by a mounded form. This basal fill is overlain by the more continuous, parallel, moderate to high amplitude reflections of the upper fill which drape the walls of the canyons. In contrast to the parallel reflections which infill the previously described incised valleys, the draped nature of

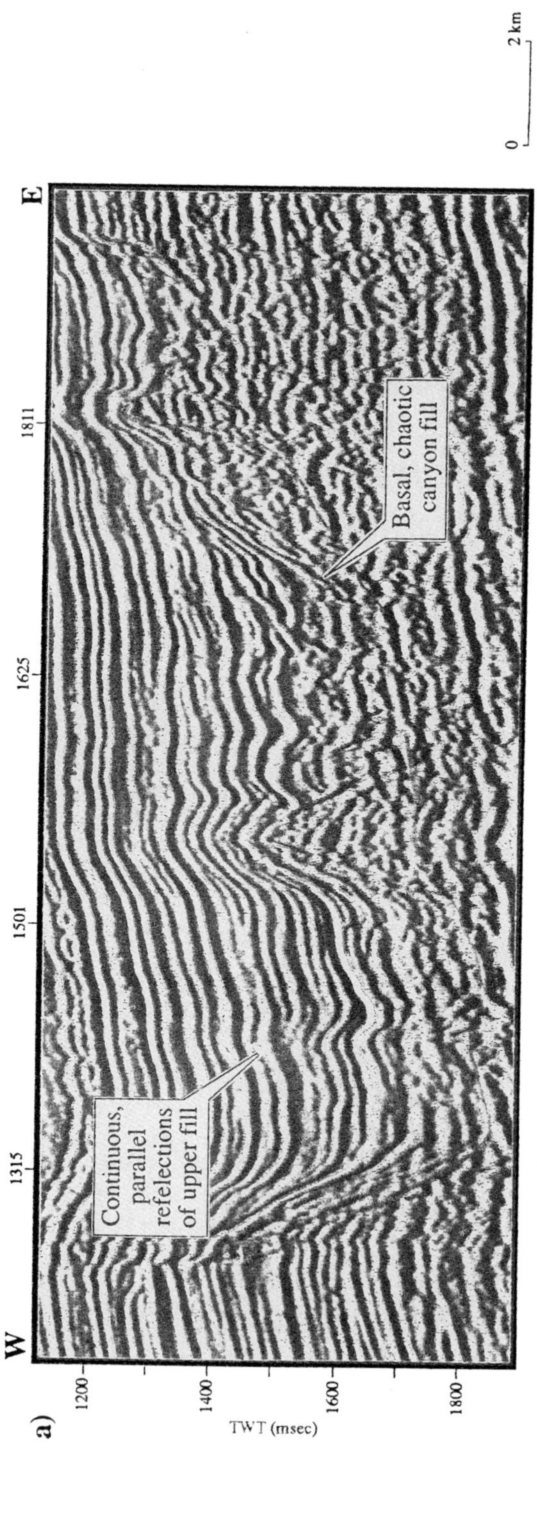

Fig. 9. Seismic expression of a submarine canyon. The basal chaotic slump fill is overlain by the more continuous, high amplitude parallel reflections of the upper fill which drape the canyon walls (from seismic line ECL-89-37).

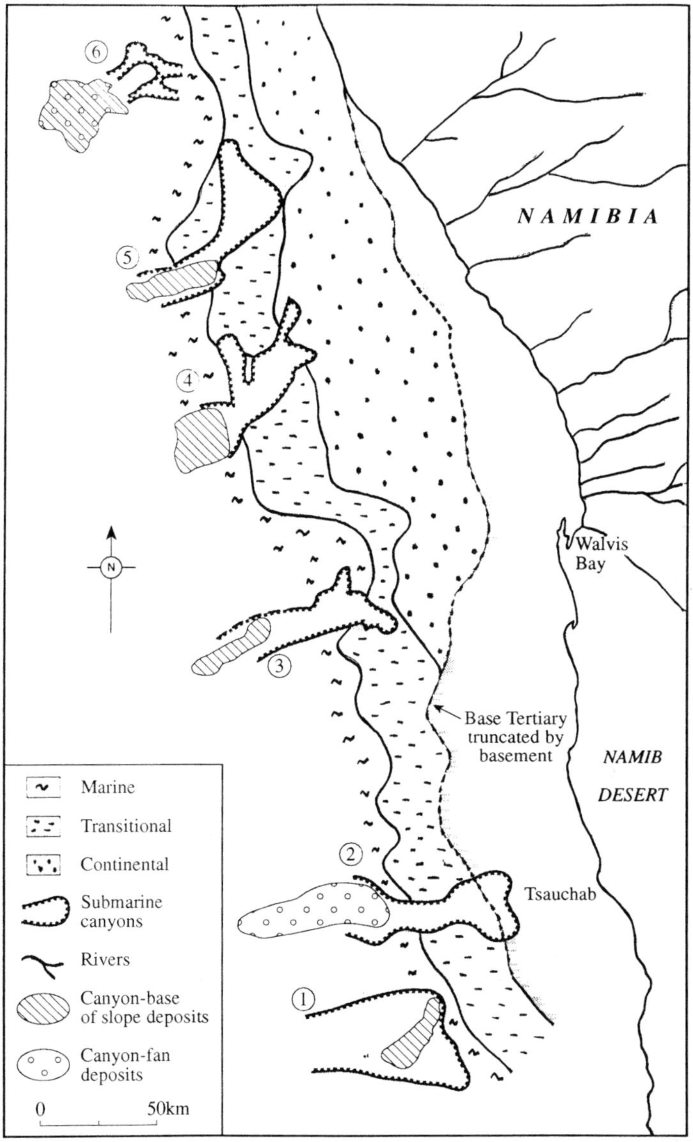

Fig. 10. Spatial distribution of submarine canyon systems (numbered 1–6) developed at the Cretaceous–Tertiary boundary. These canyon systems may be described as either attached (System 2 and 6) or detached (Systems 1 and 3–5) depending on whether or not they can be linked into fluvial systems on the shelf. Also shown are the positions of the down-dip mass transport deposits associated with canyon development.

the upper canyon fill implies that these features must have been submerged at the time of deposition, thus implying a dominantly submarine origin. Mass transport deposits are interpreted to occur at the base of the continental slope down-dip from the main canyon forms described in this section (Fig. 10); these deposits are included in the more detailed discussion of mass transport features provided below.

The main erosional surface defining the submarine canyons may have associated development of lateral terraces, the bases of which are seen in some cases to correspond in height to the position of a toplap surface developed within the canyon fill. The identification of toplap (as defined by Mitchum 1977) can provide a guide as to the location of base level at the time of deposition. The terraces and toplap surface can be interpreted as the lowest point that sea level fell to in this area, an event which may itself have caused sufficient instability in the slope area to trigger further sediment failure and slumping. This submarine type of erosion could, in turn, be extended landward through headward erosion and cannibalization processes, resulting in the creation of the majority of the accommodation space that was the canyon morphology. The development of submarine canyons may therefore be initiated by instability resulting from a fall in relative sea level (i.e. primarily during a lowstand), but they are subsequently modified and enlarged during late lowstand and transgressive times by submarine processes (e.g. retrogressive slumping and erosion by submarine currents). Similar submarine canyon development has been documented for the conjugate Brazilian margin (Bruhn & Walker 1995), whereby large-scale canyons become infilled retrogressively by coarse-grained turbidites, forming important hydrocarbon reservoirs in the Campos Basin, offshore Brazil.

The majority of the canyons observed in the post-rift megasquence offshore Namibia are interpreted to have developed to what may be termed relatively mature stages of canyon evolution (step D of Galloway & Hobday, 1983, 1996). In addition to this, early stage canyons (step B of Galloway & Hobday 1983, 1996) are also observed (Fig. 11), whereby initial sediment failure and slumping has occurred resulting in the formation of a steep slump scar and the down-dip deposition of a thick chaotic slump complex. The accommodation space created by the failed slump scar becomes rapidly infilled by later sediments, possibly with some component of longshore current deposition, and this healing process prevents any further evolution of the system into a more mature canyon.

In many documented descriptions of submarine canyons, turbidite fans are observed to develop down-dip (Cainelli 1992; Damuth *et al.* 1988) due to the fact that canyons act as conduits for sediments fed by fluvial systems across the shelf

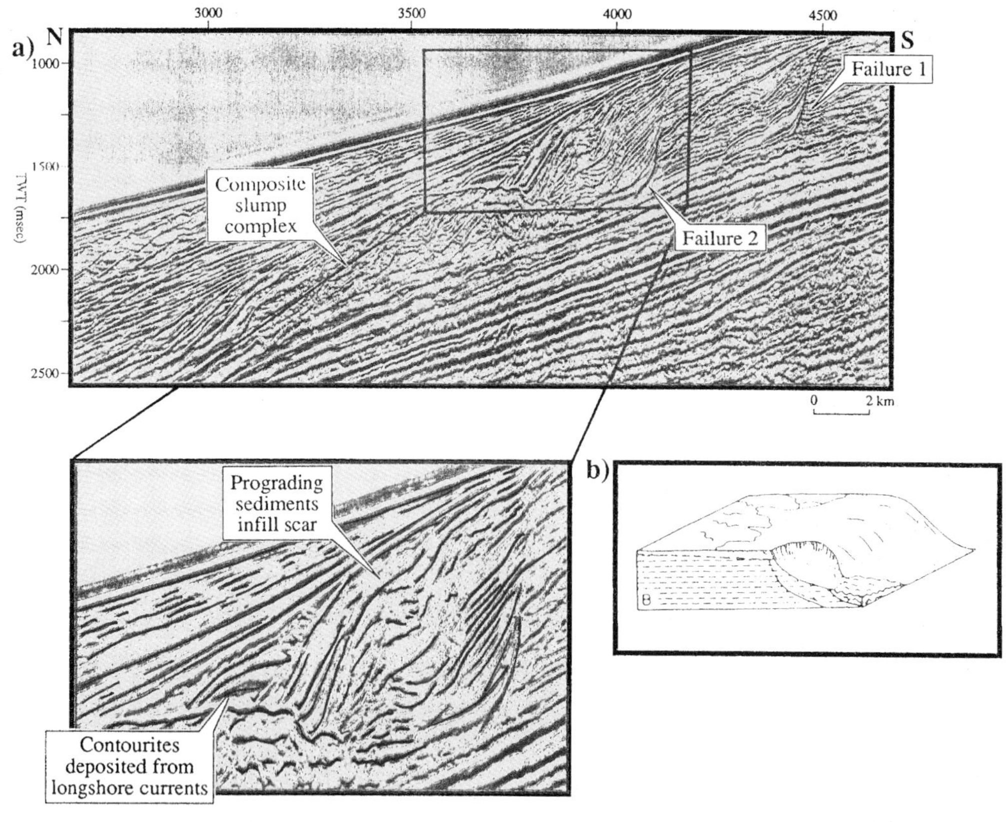

Fig. 11. Early submarine canyon development expressed as a major slump scar (400–500 m high) and the down-dip deposition of a 350 ms (TWTT 300–350 m) thick slump complex. The scar is rapidly healed by later sediments, possibly deposited by longshore currents (see enlarged part of profile) (from seismic line ECL-89-22).

and into the basin. The map distribution (Fig. 10) of the submarine canyons forming close to the Cretaceous–Tertiary boundary offshore Namibia indicates, however, that this may not always be the case. Canyon System 1 (Fig. 10), for example, does not extend onto the most landward strike lines and provides no evidence (on the seismic scale) for links into fluvial systems on the shelf via channels or incised valleys. It may be inferred therefore that canyon systems that are detached from the fluvial regime do not act as sediment conduits for fluvially derived sediments and thus do not necessarily support the development of a sand-rich submarine fan in deeper water environments (Fig. 12). Similar 'detached canyons' have been observed from the Sergipe–Alagoas Basin, offshore Brazil, e.g. the Maceio amphitheatre (Cainelli 1992), where excavation of the canyon system does not occur far enough back onto the shelf to capture fluvial systems.

In contrast to Canyon System 1, Canyon System 2 (Fig. 10) appears to be present on the most landward seismic profiles where the canyon fills are thought to have been influenced by fluvial systems due to the presence of parallel, onlapping reflections similar to those identified in the incised valleys described previously. At the present time, the equivalent onshore area corresponds with the modern Namib Desert, but at the time of canyon excavation it is possible that palaeorivers may have drained to the coast. Canyon System 2 is thus more likely to have a deepwater submarine fan associated with it and may be considered to be an attached canyon system (Fig. 12).

Submarine channels

In addition to the development of large submarine canyons both dip and strike seismic reflec-

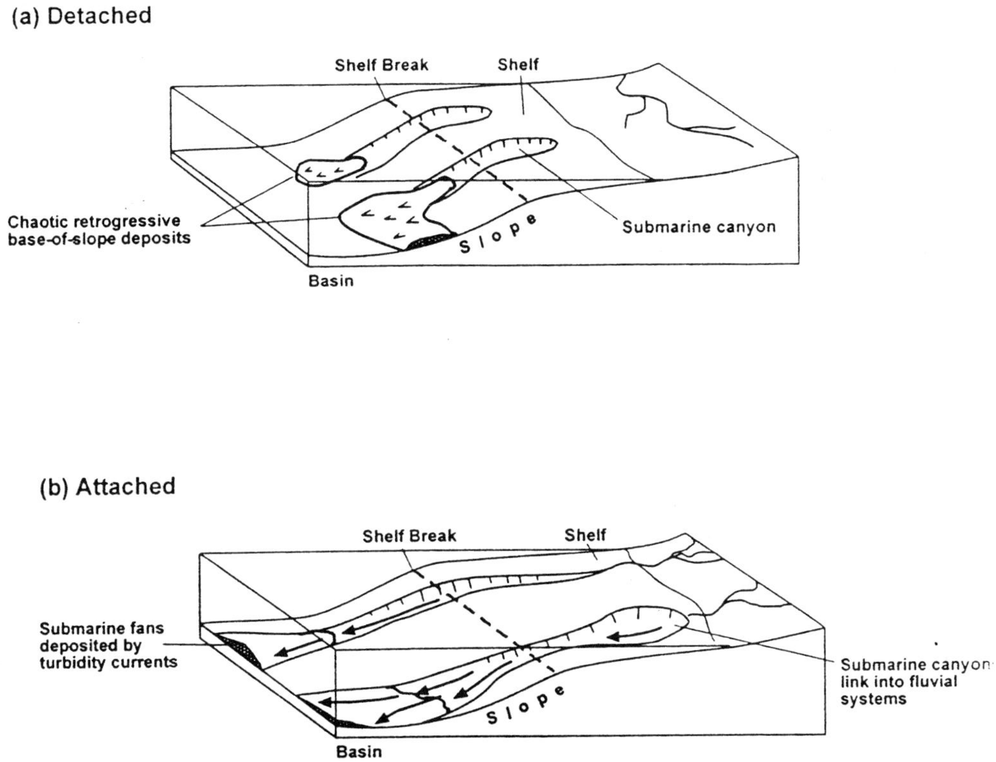

Fig. 12. Depositional settings for deepwater sediments associated with attached and detached submarine canyon systems. (**a**) Detached: breach the shelf break but do not extend far enough landward to capture fluvial systems. Deposition at the mouth of the canyon and at the base-of-slope is dominated by mass transport deposits. (**b**) Attached: extend far enough landward to link into fluvial systems, resulting in the deposition of a submarine fan in the basin. Deposition offshore Namibia, however, is dominated by mass transport processes and not by turbidity currents as suggested by current models.

tion profiles reveal smaller-scale channel forms across the upper slope environment (Fig. 13). Typically, the channels are between 1.5 and 2.5 km wide, 100–350 ms TWTT (100–300 m) deep and form exclusively within the Tertiary (MS50b) unit of the post-rift megasequence. These submarine channels occur in both the Orange and Walvis Basins, the latter of which displays a greater number of channelized events. In the Walvis Basin the most recent channelling events are observed to occur in similar locations to the older channel positions, apparently migrating laterally with time towards the northwest.

The seismic facies of the channel fills are variable. The fills in the Orange Basin, for example, are dominated by seismically transparent intervals separated by moderate amplitude, relatively low frequency reflections which drape the channel walls (Fig. 13) in a manner similar to the upper parts of the submarine canyon fills discussed previously. This again suggests that the channels, like the canyons described previously, are submarine in origin. Interpreted time-equivalent reflections in the Walvis Basin channels, however, appear to build laterally across the channel towards the northwest with some component of aggradation (Fig. 13). In contrast to the older channel events, the majority of the younger channels are filled with moderate to high amplitude, horizontal, parallel reflections which onlap the walls of the channel.

The stratigraphic framework developed for this margin implies that these channels are Middle–Upper Miocene in age (Bagguley 1997). By this time many of the rivers onshore Namibia had probably become more ephemeral in nature or shut off entirely, due to the gradual aridification of the Namibian landmass, and, although the perennial Orange River still delivered sediment to the Namibian coast, there was a marked decrease in the amount of sediment being fed to the offshore area from the continental interior (Siesser 1980; Ward *et al.* 1983). Sedimentation offshore Namibia was instead strongly influenced by both deep oceanic circulation patterns, dominated by Antarctic Bottom Water, and coast-parallel currents, such as the Benguela Current (Siesser 1980; Ward *et al*, 1983; Lavelle & Armstrong 1992).

The erosional submarine channels described from the upper continental slope of the Namibian margin are similar to those described by Galloway & Hobday (1996), which are thought to develop on the steepest parts of the continental slope due to the erosional nature of accelerating gravity flows (e.g. turbidity currents and debris flows) as they travel downslope. This would explain the clearer cross-sectional appearance of the channels on strike-oriented profiles, such as those observed in the Orange Basin (Fig. 13). The dip lines in the Walvis Basin do, however, show clear channelling (Fig. 13). This may be due to the meandering nature of the systems but could also be related to the climatic and oceanographic setting offshore Namibia at the time of channel formation, i.e. increasing aridification and the onset of strong ocean currents. Oceanic circulation patterns are thus likely to have played an important role in the initial erosion and subsequent filling of these upper slope channels. This is also implied by the increased abundance and complexity of the channel systems (on dip- and strike-oriented profiles) towards the north-east of the Walvis Basin in response to the northward flow of the currents and the overall lateral migration of subsequent channels (Fig. 13) as deposition occurs in response to waning current velocities.

Analogous channels to those observed along the Namibian continental margin have been documented from the continental margin offshore Gabon (Rasmussen 1994). The Gabon channels most closely resemble those seen in the Walvis Basin, with similar dimensions (1–3 km wide, *c.* 400 ms TWTT deep) and external forms. They are also thought to have developed during the Middle Miocene (Rasmussen 1994) and may therefore be directly analogous. The different seismic facies (lateral accretion, aggradation and parallel fill) observed within the channel fills of the Walvis Basin can also be seen in the Gabon examples. Based on these seismic facies, Rasmussen (1994) predicted the lithologies of the channel fills to be largely mud dominated, with the amount of mud increasing towards the top of the channel. Similar lithological predictions have been made by Galloway & Hobday (1983) who classified channels of this nature as depositional turbidite channels.

The largely analogous nature of the form and fills of the submarine channels observed from the Namibian and Gabon continental margins suggests that these features, particularly those observed on dip and strike profiles in the Walvis Basin, are likely to have formed under similar conditions. Rasmussen (1994) interprets the different seismic facies observed within the Gabon channels as representing the submarine response to a single cycle of relative sea-level change; i.e. the formation and filling of the channels represent one event which developed in response to an initial relative sea-level fall. From the observations made within this study, however, the submarine channels from the Namibian continental margin are interpreted to have developed in response to several periods of submarine erosion

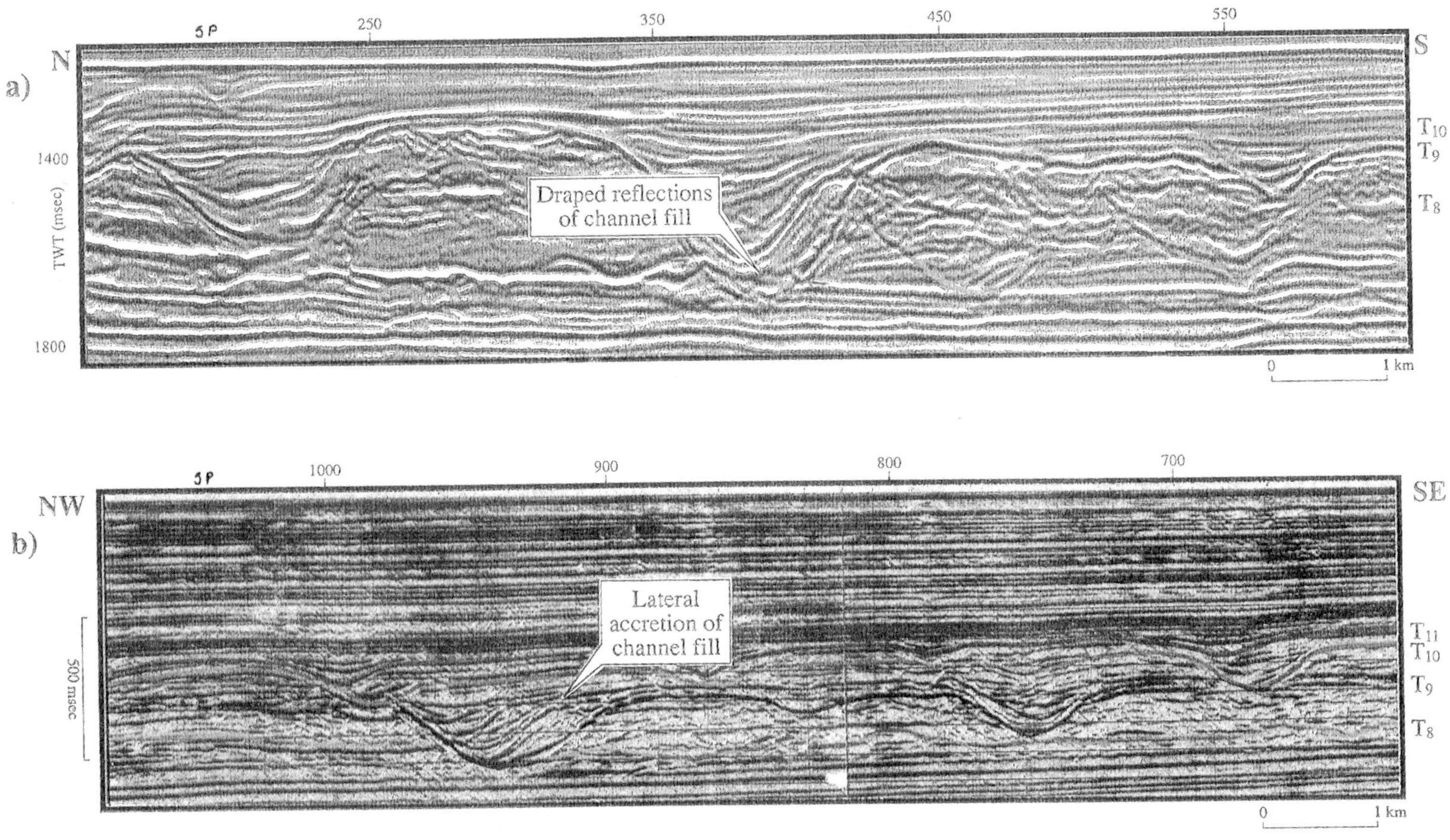

Fig. 13. Upper slope Tertiary submarine channels in the (**a**) Orange and (**b**) Walvis Basins. The draped nature of the channel fills is similar to the upper parts of the submarine canyon fills previously described, although later channel fills in the Walvis Basin appear to migrate laterally with time. [a)] is from seismic line ECL-89-P050].

and thus do not represent a single erosional–depositional event. The oceanographic conditions prevailing at the time of channel formation (Siesser 1980; Ward *et al.* 1983) suggest that the northward flowing, coast-parallel currents, such as the Benguela Current, may have been sufficient to erode and subsequently fill the channels, as suggested by Rasmussen (1994), but without the influence of a relative sea-level fall. The formation and infilling of submarine channels in the Walvis Basin is therefore interpreted as being large due to the action of submarine, longshore currents which flowed northwards along the Namibian continental margin. The submarine channels in the Orange Basin, however, have only been observed on strike-oriented profiles and may therefore have been eroded by turbidity currents in a manner similar to that proposed by Galloway & Hobday (1996), with the sediment being provided by the Orange River.

Slope and base-of-slope deposition: mass transport and bottom currents

Mass transport deposits

The seismic sections across the slope and/or base-of-slope environment can be interpreted to reveal a site where depositional systems were typically involved in transporting sediments by mass flow processes, including slumps and debris flows. Such mass transport deposits can be observed on both localized and more regional scales, the latter of which are discussed here. On a regional scale, low relief mounds are observed within the Tertiary (MS50b) post-rift unit. For the purpose of this paper, the regional mound lying above the major unconformity at the Cretaceous–Tertiary boundary is used as an example. This feature is typically up to 35 km wide, 200 ms TWTT (200 m) thick and can be mapped along-strike over an area *c.* 330 km in length within the Lüderitz and Walvis Basins (Fig. 14). Internally, the mound is characterized by discontinuous to chaotic reflections which are laterally interspersed with more transparent areas. The more proximal end of the deposits are often expressed as moderate to high amplitude, moderate frequency, sigmoidal reflections. Both the upper and lower surfaces of the Base Tertiary mounds show little evidence of erosion and are delineated by overlying continuous, high amplitude reflections, although in the distal reaches of the mound, downlap onto the underlying surface is also observed.

The regional stratigraphic framework established for this area and the association of the mounds with interpreted sequence boundaries suggests that the interpreted mass flow deposits, both regional and more localized, are initially triggered by a relative sea-level fall, although slumping during subsequent transgression [as proposed by Galloway (1989)] is also thought to be important. The general discontinuous to chaotic nature of the internal seismic facies of the mounds suggests that they are more characteristic of debris flows and slumps than turbidites (Sangree & Widmier 1977; Mitchum 1985; Posamentier & Erskine, 1991), supporting the interpretation that deposition on the continental slope is dominated by mass transport processes (e.g. slumps and debris flows).

A further classification of the Namibian slope and base-of-slope mass transport deposits can be made based on the geometry and interpreted lithologies of these deposits. The elongate nature of the regional slope mounds along depositional strike (Fig. 14) suggests that these mounds were fed by a line sediment source. The predominance of interpreted mass transport deposits within these mounds, their position on the slope and the overall absence of turbidites suggests that, using the classification scheme of Reading & Richards (1994), these deposits may be termed slope aprons. Reading & Richards (1994) propose that slope aprons can be mud rich, mud/sand rich, sand rich, or gravel rich, depending on the lithological composition of the source area. The extent of the Base Tertiary slope apron (35 km wide, 330 km long) and its linear geometry along depositional strike suggest that, on the basis of the criteria proposed by Reading & Richards (1994), this deposit may be interpreted to be a mud-rich slope apron. This implies that the sediment for the apron was transported across the broad Namibian shelf from a large source area, producing large, mass flow-dominated deposits. Other slope aprons from southwest Africa have also been interpreted as mud-rich slope aprons using the same classification scheme (Reading & Richards 1994).

In contrast to the line-sourced, mud-rich slope aprons described above, base-of-slope deposits which occur down-dip of mature submarine canyons (as briefly discussed previously) are thought to be fed by a single point source of sediment (Galloway & Hobday 1983, 1996; Mitchum 1985; Posamentier *et al.* 1991; Reading & Richards, 1994). These canyon-related deposits have been subdivided into two types, termed submarine fans and retrogressive base-of-slope mounds, each of which has a different seismic expression and origin (Galloway & Hobday 1996). Retrogressive base-of-slope mounds have been documented by Galloway & Hobday (1996) as representing shelf and slope sediments derived

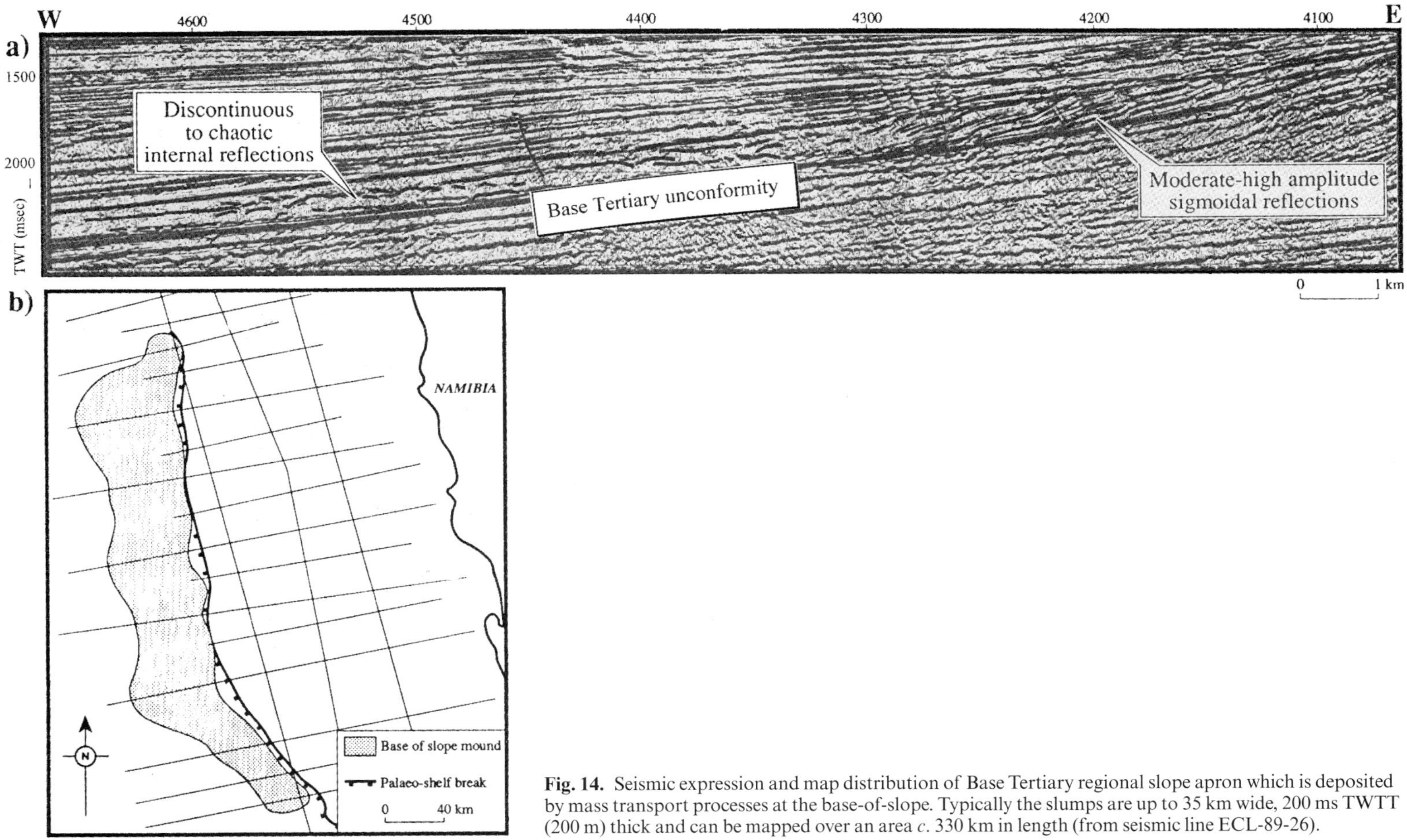

Fig. 14. Seismic expression and map distribution of Base Tertiary regional slope apron which is deposited by mass transport processes at the base-of-slope. Typically the slumps are up to 35 km wide, 200 ms TWTT (200 m) thick and can be mapped over an area *c*. 330 km in length (from seismic line ECL-89-26).

from the process of canyon excavation. These deposits, which can be observed down-dip from the detached submarine canyon systems previously described (Systems 1 and 3–5; Fig. 10), are thought to have similar seismic expressions to those described for debris flow and slumps due to the discontinuous, chaotic nature of the internal reflections. Conversely, the down-dip depositional mounds associated with attached submarine canyons (Systems 2 and 6; Fig. 10) suggests that these mounds could be termed submarine fans due to the interpreted fluvial influence (and therefore perhaps more continuous sediment flux) associated with the development of the canyons. The seismic expression of these mounds is, however, more similar to that described above for the retrogressive base-of-slope mounds; i.e. they have internally discontinuous to chaotic reflections. This suggests, therefore, that these mounds may be deposited by mass flow processes and not turbidity currents, giving credence to the arguments proposed by Shanmugan (1996) and Shanmugan *et al.* (1995) for the role of debris flows and slumps in the deposition of deepwater mounds.

Bottom current activity: contourites

Deepwater settings of the Namibian continental slope and basin are characterized, in addition to the mass transport deposits, by phases of submarine erosion and deposition which give rise to distinctive reflection packages. These are more apparent in the north of the Namibian margin, being most prevalent in the Lüderitz and Walvis Basins, and are only observed within the upper part of the Tertiary (MS50b) post-rift succession.

The distinctive depositional reflection geometries observed in the continental slope region are characterized by asymmetrical, convex-upwards, high amplitude reflections which separate areas of more transparent or moderately continuous seismic facies (Fig. 15). The reflection packages appear to stack on top of each other with some component of lateral migration being evident between successive events. In more basinal environments, depositional reflection geometries are associated with onlap onto an erosion surface at the present-day seafloor (Fig. 16). The deposit has a highly stratified expression on seismic profiles, being composed internally of high amplitude, continuous, parallel, horizontal reflections, and has a maximum observed thickness of 300 ms TWTT (*c.* 300 m).

The location of the distinctive erosional–depositional reflection geometries described and the seismic facies which comprise these features have led to the interpretation that these reflection packages represent the deposits of contour or bottom currents. The present-day water depths in which the depositional and erosional geometries occur (500–2100 m) are typical of contouritic deposits, which are thought to accumulate in water depths ranging from 500 to > 5000 m (Faugères & Stow 1993). The presence of erosional truncation and channel-like features suggests that bottom currents in this region of the margin were of a high enough intensity and velocity to cause extensive erosion of the underlying sediments. Subsequent sediment accumulation reflects a decrease in the intensity and velocity of the currents, thus enabling deposition from the waning currents to occur. Sedimentological studies suggest that contourites are typically composed of fine-grained, mud-rich sediments (Hollister 1993; Faugères & Stow 1993), though at higher current velocities, coarser sediment may be transported and subsequently deposited as sandy contourites (Faugères & Stow 1993; Shanmugan *et al.* 1993; Howe *et al.* 1994; Stoker *et al.* 1998). The deposition of sediment from bottom or contour currents is interpreted to explain the thick accumulation of Tertiary sediments observed in the north of the Namibian margin, particularly in the Walvis Basin. Strong bottom currents, such as the deep oceanic Antarctic Bottom Water current, are thought to have been in operation along the Namibian margin since the Early–Late Oligocene (Ward *et al.* 1983), although some authors argue for a later onset of longshore current activity, dating it as occurring during the Middle–Upper Miocene (Siesser 1978).

The largest contourite accumulation observed (Fig. 16) is interpreted to be either an elongate drift or a broad, contourite sheet. Both of these drift types are characterized by continuous, parallel, moderate to high amplitude reflections which exhibit downlap onto the underlying surface. Similar contourite accumulations have been documented from the Argentine Basin in the South Atlantic (Faugères *et al.* 1993) and from the Rockall Trough in the North Atlantic (Howe *et al.* 1994; Stoker *et al.* 1998). These latter elongate drifts have similar external forms and dimensions to the Namibian example, being up to 300 m thick with a maximum relief above the seafloor of 150 m. In contrast to the large-scale drift, other contourites identified on the Namibian continental slope (Fig. 15) have been interpreted as sediment waves. Similar geometries also interpreted as sediment waves have been documented from the Hatteras Outer Ridge, offshore North Carolina (Mountain & Tucholke 1988), the Argentine Basin (Faugères *et al.* 1993) and the Rockall Trough (Pickering *et al.* 1989; Damuth & Olson 1993; Stoker *et al.* 1994; Howe *et al.* 1994). Sediment

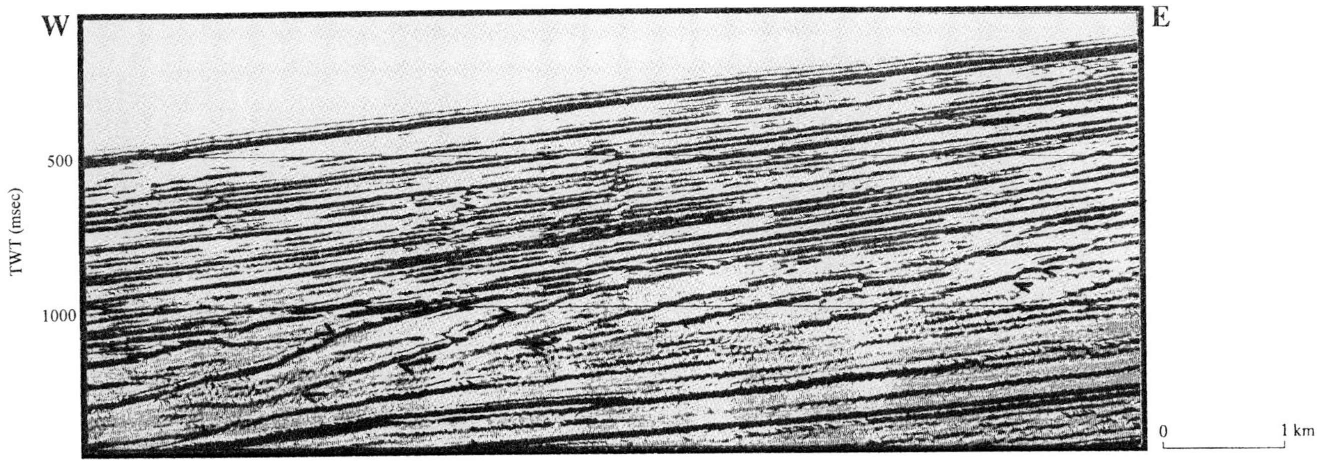

Fig. 15. Seismic expression of sediment waves. Note the asymmetric development of the sediment waves, reflecting upslope migration of the main locus of deposition. Contourite deposition offshore Namibia is largely restricted to the upper parts of the Tertiary post-rift unit when sedimentation is thought to have been influenced by strong longshore currents (from seismic line ECL-89-24).

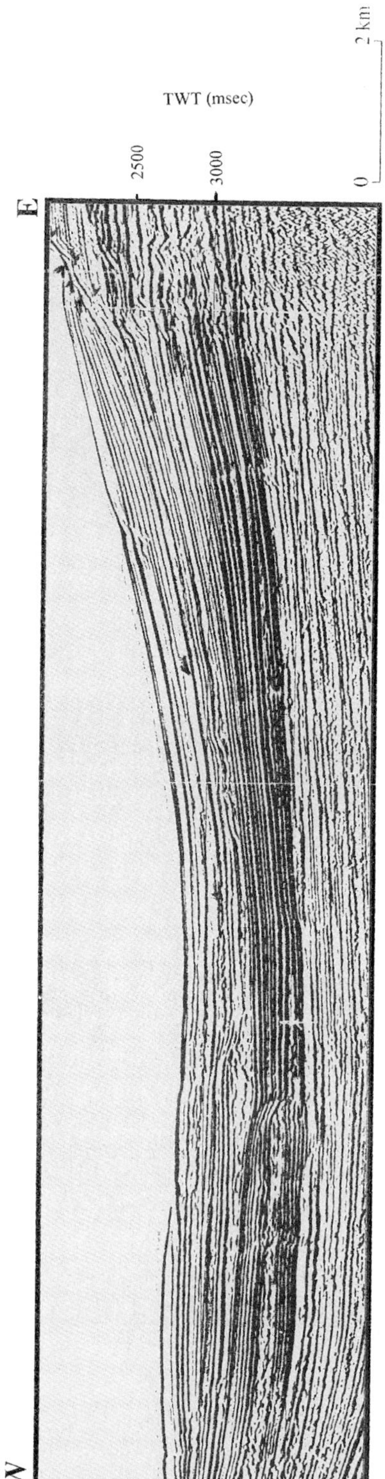

Fig. 16. Seismic expression of contourite drift. The drift onlaps the pre-existing topography of the continental slope and is characterized internally by high amplitude, continuous, parallel reflections (from seismic line ECL-89-445).

waves are commonly associated with larger contourite drifts (Stoker *et al.* 1998) and are thought to be deposited by varying current velocities in basinal to base-of-slope water depths of *c.* 1200 m (Pickering *et al.* 1989; Marani *et al.* 1993). The overall asymmetry of these features is thought to reflect progressive migration up- or downslope as deposition occurs (Stoker *et al.* 1998).

Deep basin deposition: basin mounds

The extreme length of some of the dip-parallel seismic profiles offshore Namibia means that the deep basin can occasionally be imaged, resolving deep basin mounds of a markedly different seismic facies to those on the shelf or base-of-slope discussed above. The widely spaced nature of the seismic grid means that these features are generally only observed on individual profiles and cannot be traced between adjacent lines. Seismic facies analysis has led to the identification of a lower and upper seismic facies within the mound. The lower facies is composed largely of moderate amplitude, moderately to discontinuous reflections, while the upper fill is characterized by high amplitude, parallel, continuous reflections (Fig. 17). Onlap on each side of the mound and the presence of erosional truncation suggests that the upper surface is, in part, erosional in nature.

The lowermost seismic facies of the basin mounds is similar to the facies described for both slope and base-of-slope deposits, suggesting that this unit was transported and subsequently deposited by mass flow processes. Conversely, the upper seismic facies is composed of high amplitude, continuous, parallel reflections, similar to those thought to characterize turbidite-dominated submarine fans, e.g. the Clontarf Fan, Porcupine Basin (Macurda & Nelson 1988). The absence of any submarine canyon development in the vicinity of the mound, however, suggests that any turbidity currents must have been transported through an alternative feeder system, e.g. submarine channels similar to those previously discussed. Evidence for submarine channels at this level (i.e. Latest Cretaceous-Early Tertiary) is not observed on the seismic profiles, although the development of the Base Tertiary unconformity subsequent to the deposition of the basin mound may have resulted in the erosion of any potential feeder channels. The steep flanks of the mound (*c.* 10°; Fig. 17) are also uncharacteristic of submarine fans, although the erosional nature of the upper surface suggests that the original depositional configuration of the mound has been later modified by submarine erosion. This erosion may be enhanced by the effects of differential compaction, particularly if the mound is composed of more than one lithological type, a factor which is suggested by the different seismic facies observed. An alternative interpretation for the origin of these basin mounds is that they represent the deposits of waning bottom or contour currents which are often concentrated in deep basin environments.

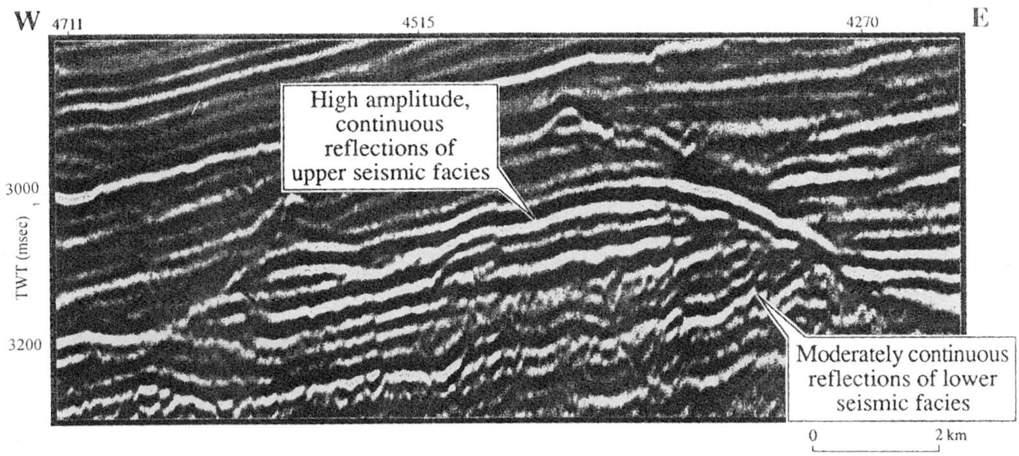

Fig. 17. Deep basin mound exhibiting both a lower chaotic and more parallel upper seismic facies. The mounds are characteristically up to 11 km wide and can be up to 300 ms TWTT (300 m) thick. The upper surface of the mound is defined by erosional truncation below and onlap above, suggesting that it is at least partly erosional in nature (from seismic line ECL-89-19).

Summary and conclusions

The high quality and long dip-parallel length of the seismic reflection profiles across the passive continental margin offshore Namibia provide an opportunity to present some examples of the seismic expression of a wide range of passive margin depositional processes. Depositional processes studied and illustrated here include the shelf, slope and deep basin, and these in turn may be influenced by processes occurring in both fluvial and possibly aeolian regimes.

Seismic facies analysis enables predictions to be made concerning the types of sediments that may characterize the fills of different channel and canyon forms. Inner shelf channels, interpreted as incised valleys, form during a relative sea-level fall and are infilled with horizontal reflections, the nature of which suggests that deposition occurred during periods of rising relative sea level. In contrast to this, the channels and canyons in the upper slope and outer shelf environment have fills dominated by draped reflections, implying that these features were largely formed and subsequently infilled in a submarine environment.

The submarine canyons observed from the Namibian passive margin may be attached or detached, depending on whether or not they can be linked back into fluvial systems on the shelf. This classification distinguishes between those canyon systems which could potentially act as feeders and support the development of a sand-rich submarine fan, and those which do not act as a conduit for sediment from more proximal areas and thus are unlikely to have a deepwater fan counterpart.

Regional mounds which develop on the slope above a major second-order sequence boundary (Cretaceous–Tertiary boundary) have been interpreted as mud-rich slope aprons, deposited by mass transport processes. The instability required to trigger such large-scale mass transport is likely to have been initiated by at least a relative sea-level fall. In contrast to these line-sourced features, mass transport deposits occurring down-dip from both attached and detached submarine canyons are interpreted to have been fed from a point sediment source.

Contourite deposits, in the form of contourite drifts and sediment waves, are frequently observed in the upper parts of the Tertiary (MS50b) post-rift unit in the region of the base-of-slope. These deposits are formed by the action of coast-parallel bottom currents (e.g. the Benguela Current). Such currents may also be responsible for the deposition of the deep basin mounds observed, although these features may have been deposited by mass transport processes and/or turbidity currents.

The authors would like to thank NAMCOR (National Oil Corporation of Namibia) and Geoquest Exploration Consultants, Henley-on Thames, for providing the data for this project. LASMO plc, London and Sasol Petroleum, South Africa, are also thanked for providing additional seismic data and well information. Thanks also go to Chevron Europe Ltd, particularly Mick Bohoslawec, and to Gordon Watt for drafting the diagrams in this paper. NAMCOR are acknowledged for permitting the publication of these observations.

References

AUSTIN, J. A. & UCHUPI, E. 1982. Continental–oceanic crustal transition off west Africa. *AAPG Bulletin*, **66**, 1328–1347.

BAGGULEY, J. G. 1997. The application of seismic and sequence stratigraphy to the post rift megasequence offshore Namibia. PhD Thesis, Oxford Brookes University.

BROWN, L. F., BENSON, J. M. *et al.* 1995. Sequence stratigraphy in offshore South African divergent basins: an atlas on exploration for Cretaceous lowstand traps by SOEKOR (Pty) Ltd. *AAPG Studies in Geology*, **41**.

BRUHN, C. H. L. & WALKER, R. G. 1995. High resolution stratigraphy and sedimentary evolution of coarse grained turbidites from the Upper Cretaceous transgressive megasequence, Campos Basin, offshore Brazil. *Journal of Sedimentary Research*, **B65**, 426–442.

CAINELLI, C. 1992. Sequence stratigraphy, canyons and gravity mass-flow deposits in the Piacabucu formation, Sergipe–Alagoas Basin, Brazil. PhD Thesis, University of Texas, Austin.

CORBETT, I. 1993. The modern and ancient patterns of sandflow through the southern Namib deflation basin. *Special Publication. International Association of Sedimentologists*, **16**, 45–60.

DAMUTH, J. E. & OLSON, H. C. 1993. Preliminary observations of Neogene–Quaternary depositional processes in the Faeroe–Shetland Channel revealed by high resolution seismic facies analysis. *In*: PARKER, J. R. (ed.) *Petroleum Geology of Northwest Europe: Proceedings of the 4th Conference*, 1035–1045.

—— *ET AL.* 1988. Anatomy and growth patterns of Amazon deep-sea fans as revealed by long-range side-scan sonar (GLORIA) and high resolution seismic studies. *AAPG Bulletin*, **72**, 885–911.

DINGLE, R. V. 1973. The geology of the continental shelf between Luderitz and Cape Town (Southwest Africa) with special reference to Tertiary strata. *Journal of the Geological Society*, **129**, 337–363.

FAUGÈRES, J-C. & STOW, D. A. V. 1993. Bottom-current-controlled sedimentation: a synthesis of the contourite problem. *Sedimentary Geology*, **82**, 287–297.

FAUGÈRES, J-C., MEZERAIS M.-L. & STOW, D. A. V. 1993. Contourite drift types and their distribution in the North and South Atlantic Ocean basins. *Sedimentary Geology*, **82**, 189–203.

GALLOWAY, W. E. 1989. Genetic stratigraphic sequences in basin analysis I: architecture and genesis of flooding surface bounded depositional units. *AAPG Bulletin*, **73**, 125–142.

—— HOBDAY, D. A. 1983. *Terrigenous Clastic Depositional Systems*. Springer.

—— & —— 1996. *Terrigenous Clastic Depositional Systems*. Springer.

HOLLISTER, C. D. 1993. The concept of deep-sea contourites. *Sedimentary Geology*, **82**, 5–11.

HOWE, J., STOKER, M. S. & STOW, D. A. V. 1994. Late Cenozoic sediment drift complex, northeast Rockall Trough. *Palaeoceanography*, **9**, 989–999.

HUBBARD, R. J., PAPE, J. & ROBERTS, D. G. 1985. Depositional sequence mapping as a technique to establish tectonic and stratigraphic framework and evaluate hydrocarbon potential on a passive continental margin. *In*: BERG, O. R. & WOOLVERTON, D. (eds) *Seismic Stratigraphy II: an integrated approach*. AAPG Memoir, **39**, 79–91.

LAVELLE, M. & ARMSTRONG, R. A. 1992. Strontium-isotope dating of marine sediments along the west coast of southern Africa. Abstract from '*South-western African continental margin: evolution and physical characteristics*', Windhoek, Namibia, 24–29.

LIGHT, M. P. R., MASLANYJ, M. P., GREENWOOD, R. J. & BANKS, N. L. 1993. Seismic sequence stratigraphy and tectonics offshore Namibia. *In*: WILLIAMS, G. D. & DOBB, A. (eds) *Tectonics and Seismic Sequence Stratigraphy*. Geological Society, London, Special Publications, **71**, 163–191.

MACURDA, D. B. & NELSON, H. R. 1988. Interactive interpretation of a submarine fan, offshore Ireland: a case history. *Geophysics: The Leading Edge of Exploration*, October, 28–34.

MARANI, M., ARGNANI, A., ROVERI, M. & TRINCARDI, F. 1993. Sediment drift and erosional surfaces in the central Mediterranean: seismic evidence of bottom-current activity. *Sedimentary Geology*, **82**, 207–220.

MCMILLAN, I. K. 1990. Foraminiferal biostratigraphy of the Barremian–Miocene rocks of the Kudu 9A-1, 9A-2 and 9A-3 boreholes. *Communications of the Geological Survey of Namibia*, **6**, 23–29.

—— 1994. Revised biostratigraphy of the Kudu wells, offshore Namibia. Unpublished data courtesy of NAMCOR.

MITCHUM, R. M. 1977. Seismic stratigraphy and global changes in sea level, part 11: glossary of terms used in seismic stratigraphy. *In*: PAYTON, C. E. (ed.) *Seismic Stratigraphy – Applications to Hydrocarbon Exploration*. AAPG Memoir **26**, 205–212.

—— 1985. Seismic stratigraphic expression of submarine fans. *In*: BERG, O. R. & WOOLVERTON, D. (eds) *Seismic Stratigraphy II*. AAPG Memoir, **39**, 117–136.

MOUNTAIN, G. S. & TUCHOLKE, B. E. 1988. Abyssal sediment waves. *In*: BALLY, A. W. (ed.) *Atlas of Seismic Stratigraphy, Volume 2*. 1.2.5.21–1.2.5.23.

MUNTINGH, A. & BROWN, L. F. 1993. Sequence stratigraphy of petroleum plays, post-rift Cretaceous rocks (Lower Aptian to Upper Maastrictian), Orange Basin, western offshore South Africa. *In*: WEIMER, P. & POSAMENTIER, H. W. (eds) *Siliciclastic Sequence Stratigraphy – Recent Developments and Applications*. AAPG Memoir, **58**, 71–98.

NÜRNBERG, D. & DIETMAR MÜLLER, R. 1991. The tectonic evolution of the South Atlantic from Late Jurassic to present. *Tectonophysics*, **191**, 27–53.

PICKERING, K. T., HISCOTT, R. N. & HEIN, F. J. 1989. *Deep Marine Environments: Clastic Sedimentation and Tectonics*.

POSAMENTIER, H. W. & ERSKIN, R. D. 1991. Seismic expression and recognition criteria of ancient submarine fans. *In*: WEIMER, P. & LINK, M. L. (eds) *Seismic Facies and Sedimentary Processes of Submarine Fans and Turbidite Systems*. 197–222.

——, —— & MITCHUM, R. M. 1991. Models for submarine fan deposition within a sequence stratigraphic framework. *In*: WEIMER, P. & LINK, M. L. (eds) *Seismic Facies and Sedimentary Processes of Submarine Fans and Turbidite Systems*. 127–136.

RABINOWITZ, P. D. & LABRECQUE, J. 1979. The Mesozoic South Atlantic Ocean and evolution of its continental margin. *Journal of Geophysical Research*, **84**, 5973–6002.

RAMSAY, P. J., SMITH, A. M. & MASON, T. R. 1996.Geostrophic sand ridge, dune fields and associated bedforms from the northern KwaZulu–Natal shelf, south-east Africa. *Sedimentology* **43**, 407–419.

RASMUSSEN, E. S. 1994. The relationship between submarine canyon fill and sea level change: an example from Middle Miocene offshore Gabon, west Africa. *Sedimentary Geology*, **90**, 61–75.

READING, H. G. & RICHARDS, M. 1994. Turbidite systems in deep water basin margins classified by grain size and feeder system. *AAPG Bulletin*, **78**, 792–822.

SANGREE, J. B. & WIDMIER, J. M. 1977. Seismic interpretation of clastic depositional facies. *In*: PAYTON, C. E. (ed.) *Seismic Stratigraphy – Applications to Hydrocarbons Exploration*. AAPG Memoir, **26**, 165–184.

SHANMUGAN, G. 1996. High-density turbidity currents: are they sandy debris flows? *Journal of Sedimentary Research*, **66**, 2–10.

——, SPALDING, T. D. & ROFHEART, D. H. 1993. Process sedimentology and reservoir quality of deep-marine bottom-current reworked sands (sandy contourites): an example from the Gulf of Mexico. *AAPG Bulletin*, **77**, 1241–1259.

——, BLOCH, R. B. *ET AL*. 1995. Basin floor fans in the North Sea: sequence stratigraphic models versus sedimentary facies. *AAPG Bulletin*, **77**, 477–512.

SIBUET, J., HAY, W. W., PRUNIER, H. *ET AL*. 1984. Early evolution of the South Atlantic: role of the rifting episode. *In*: HAY, W. W. *ET AL*. (eds) *Initial Reports of the Deep Sea Drilling Projects*. US Government Printing Office, (Washington, DC), **75**.

SIESSER, W. G. 1978. Leg 40 results in relation to continental slope and onshore geology. *In*: RYAN, W. B. F. *et al.* (eds) *Initial Results of the Deep Sea Drilling Project, Leg 40*. US Government Printing Office (Washington, DC), 965–979.

—— 1980. Late Miocene origin of the Benguela

upwelling system off northern Namibia. *Science*, **208**, 283–285.

STOKER, M. S., LESLIE, A. B. *ET AL.* 1994. A record of late Cenozoic stratigraphy, sedimentation and climate change from the Hebrides Slope, NE Atlantic Ocean. *Journal of the Geological Society of London*, **151**, 235–249.

—— AKHURST, M. C., HOWE, J. A. & STOW, D. A. V. 1998. Sediment drifts and contourites on the continental margin off north-west Britain. *Sedimentology*, in press.

VAN WAGONER, J. C., MITCHUM, R. M., CAMPION, K. M. & RAHMANIAN, V. D. 1990. Siliciclastic sequence stratigraphy in well logs, core and outcrop: concepts for high resolution correlation of time and facies. *AAPG Methods in Exploration Series* **7**.

WARD, J. D. & CORBETT, I. B. 1992. Post-breakup onshore stratigraphy and history of the south-western African margin. Abstract from *South-western African continental margin: evolution and physical characteristics*, Windhoek, Namibia, 53–57.

——, SEELY, M. K. & LANCASTER, N. 1983. On the antiquity of the Namib. *South African Journal of Science*, **79**, 175–183.

Onshore equivalents of the main Kudu gas reservoir in Namibia

I. G. STANISTREET[1] & H. STOLLHOFEN[2]

[1]*Department of Earth Sciences, University of Liverpool, Brownlow Street, PO Box 147, Liverpool L69 3BX, UK*
[2]*Institut für Geologie, Universität Würzburg, Pleicherwall 1, 97070 Würzburg, Germany*

Abstract: Radiometric dates of Jurassic and Cretaceous flood basalts of the southern South Atlantic maritimes have thrown the stratigraphic contexts of Karoo and post-Karoo sequences into considerable question. Central to the problem is the status of volcano–sedimentary units, characterized by aeolian-related sedimentary interlayers, containing evaporites and other lacustrine elements. The main Kudu gas reservoir ('Lower Gas Sand') offshore Namibia, interleaved with undated flood basalts, is a prime example. A previously proposed onshore correlate is the fluvio-aeolian Etjo Sandstone, with a central Namibian type area, and comparisons have been made, particularly with units traditionally classified as Etjo Sandstone, near the Etendeka Plateau base in northwest Namibia, now recognized to be of younger age. We time-correlate the main Kudu Sandstone reservoir, however, with evaporitic fluvio-lacustrine and aeolian interlayers beneath and interleaved with Early Jurassic Kalkrand flood basalts of southern Namibia, situated onshore almost directly opposite the Kudu Field. The main Kudu Sandstone reservoir is transitional in palaeoenvironmental setting between Early Jurassic onshore analogues. Main Kudu reservoir geometries and associated lithologies can be better constrained with knowledge derived from Kalkrand sedimentary interlayers and their tectonic and volcanic controls. Kalkrand interlayer thicknesses are controlled primarily by down-to-the-east extensional fault system half-graben geometries, antithetic to an inferred down-to-the-west master detachment. Within resulting sub-basinal accommodation space, lava topography controlled thickness variation. When Kalkrand insights are taken into account with radiometric dating and biostratigraphy, a Namibian Mesozoic framework fixes Kudu reservoirs and analogues into a southern Gondwanan context. At least in its type area in central Namibia, the Etjo Formation is Early Jurassic, discounting a previously proposed widespread <100 Ma hiatus along a sub-Etjo disconformity. Instead, sequence boundary hiatuses of Permian–Triassic, Early Jurassic and Middle Jurassic–Early Cretaceous ages represent time-stratigraphic gaps, relating to major faulting episodes, recording extensional events associated with continental margin evolution. Understanding onshore analogues and refining the Namibian Mesozoic framework should help constrain seismic analysis of the Namibian margin and hydrocarbon reservoir geometries. Consequently, exploration focus should widen to include Karoo sequences beneath the basal Early Cretaceous transgression.

From the perspective of southern African hydrocarbon exploration, understanding of the stratigraphic context and characteristics of the Namibian offshore Kudu gas reservoirs (Fig. 1) offers an important strategic advantage. This is the only hydrocarbon resource of viably economic size so far discovered in southern Africa south of the Kunene–Zambezi line. Knowledge of its age and setting would therefore feedback into continuing exploration programmes. After discovery of the Kudu reservoirs during exploration initiated in 1973, a prospective giant gas field was recognized (McLachlan 1990), and a considerable degree of subsequent offshore hydrocarbon exploration has been encouraged by these results. Nevertheless, our knowledge of the stratigraphic position and palaeoenvironmental setting of the main Kudu gas reservoir, the Lower Gas Sand, is surprisingly limited. This is, firstly, because attempts to date the interleaved flood basalts have been unsuccessful and, secondly, because the closest marine stratigraphic index fossils have only been recovered from units > 140 m above the major Lower Gas Sand reservoir. These fossils occur above an Early Cretaceous marine transgressive sequence, which contains the more minor Upper Gas Sand reservoir and all these are separated by a distinct unconformity from the underlying non-marine Lower Gas Sand. Published Kudu core data are limited and correlative lithofacies have not yet been fully established from onshore Namibia. Consequently, analogue constraints on both the palaeoenvironmental and stratigraphic context of the reservoir are also limited.

The purpose of this paper is, firstly, to identify likely onshore correlatives of the main Kudu reservoir and, with these as examples, help to constrain controls pertaining to offshore hydrocarbon reservoir geometries. Secondly, this correlation is extended further afield and the problem of distinguishing between Early Jurassic and Early Cretaceous volcano–sedimentary sequences in Namibia as a whole is addressed.

STANISTREET, I. G. & STOLLHOFEN, H. 1999. Onshore equivalents of the main Kudu gas reservoir in Namibia. *In:* CAMERON, N. R., BATE, R. H. & CLURE, V. S. (eds) *The Oil and Gas Habitats of the South Atlantic.* Geological Society, London, Special Publications, **153**, 345–365.

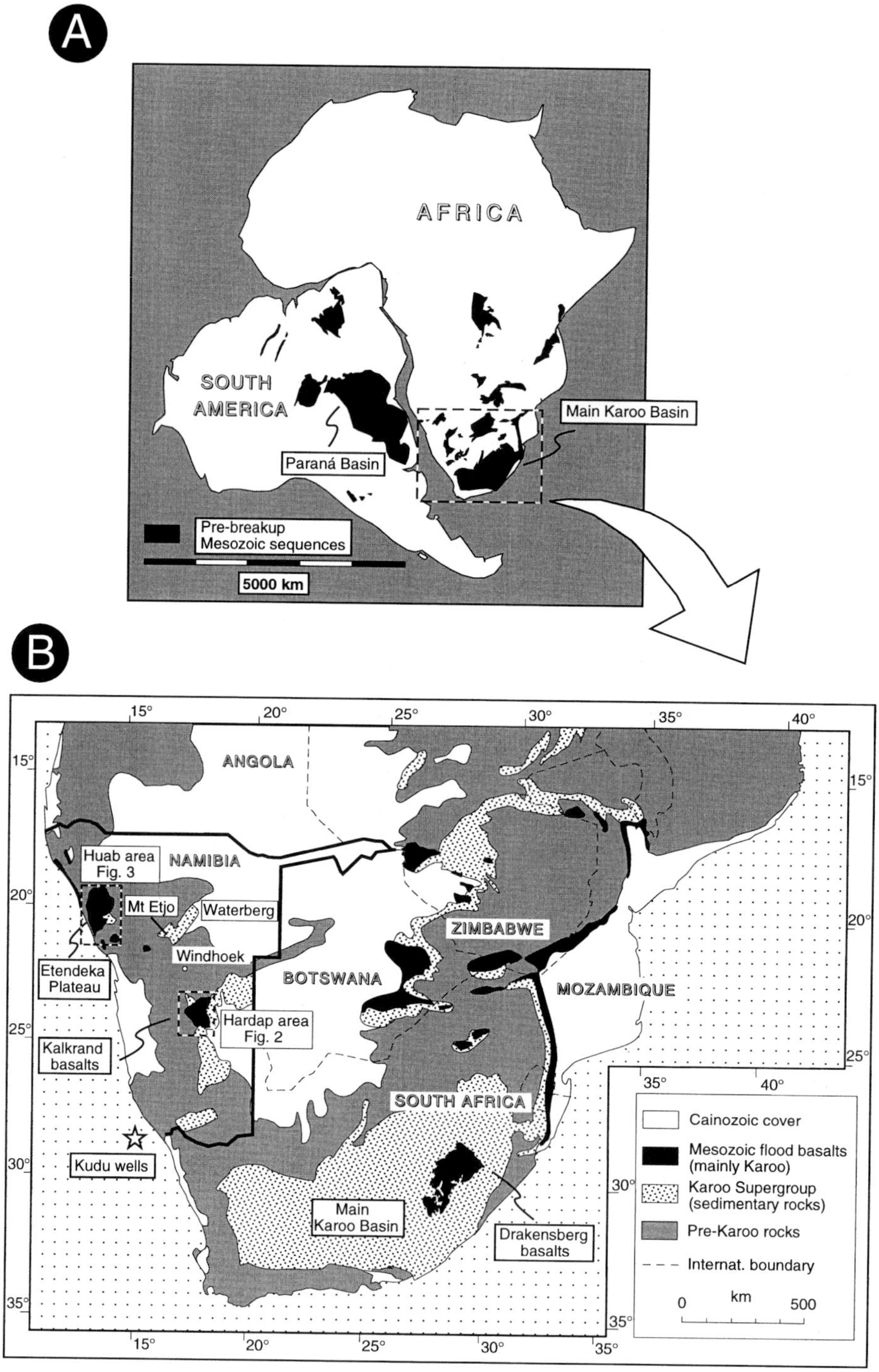

Fig. 1. (**a**) Pre-break-up reconstruction of parts of southern West Gondwana (Lambert equal area projection of De Wit *et al.* 1988) showing distribution of Karoo and post-Karoo outcrops. (**b**) Simplified geological map (modified from Eales *et al.* 1984) illustrating key Karoo volcanic and sedimentary outcrops. Sites of Kudu exploration boreholes and localities dealt with further in the text are marked. The outlined areas locate the detailed maps of Figs 2 and 3.

Characteristics and potential onshore correlatives of the main Kudu gas reservoir

Based upon limited published borehole data (Wickens & McLachlan 1990), the Lower Gas Sand reservoir comprises sedimentary interlayers, varying from 12 to 38 m in thickness, interleaved between basalt flows. The reservoir interval consists of quartzitic and volcaniclastic sandstones, the latter containing large angular and rounded basalt clasts, some exceeding 7 cm in size. The sandy units are also interbedded with thin calcareous layers, containing bivalves and algal remains, and intercalated basaltic lavas as thin as 2 m. Evaporites are common in the quartzitic sandstones, preserved as anhydritic nodules. The basalt flows have amygdaloidal flow tops, have been highly altered and are penetrated by sediment-filled cracks (Wickens & McLachlan 1990). The basalts are high-Mg (*c.* 10% MgO), high-Cr (560–810 ppm), high-Ni (140–310 ppm) tholeiites (Erlank *et al.* 1990), in which immobile minor and trace elements remain in fairly constant proportion, despite considerable alteration, indicating derivation from a similar magma source.

The well-sorted and rounded quartz grains of the sandstones, and low variance unipolar high-angle cross-bed azimuths have been considered to indicate aeolian deposition and/or sources. These characteristics and their evaporitic association were used collectively to interpret an aeolian dune complex origin (Wickens & McLachlan 1990). The volcaniclastic sandstones were discussed as possible pyroclastic material, but the conclusion was reached that they represent sedimentary reworking of basaltic extrusives. In view of the overlying marine lithofacies, and despite any direct evidence of nearby marine conditions, the units were tentatively assigned to a coastal dune complex (Wickens & McLachlan 1990). The overwhelming evidence was none the less that the basalts were exclusively terrestrial and that the sedimentary interlayers were also entirely continental.

Our recent fieldwork has identified sedimentary units in the closest onshore Mesozoic outliers to the Kudu occurrences, interleaved with Early Jurassic flood basalts that are equivalent in multiple characteristics to the main Kudu gas reservoir. Previous comparisons and correlations had been made between the Kudu main reservoir-bearing volcano–sedimentary sequence and interleaved volcanic and sedimentary units in the Etendeka outliers (Fig. 1), much further (> 800 km) to the north, where in fact only the aeolian-related characteristics of the Lower Gas Sand are in evidence. This correlation had led previously to the tentative assumption of an Early Cretaceous age for the Kudu lavas and the interleaved main reservoir sandstone.

However, the marine transgression that unconformably overlies the Kudu basalts indicates that they are older than the 133–131 Ma (Renne *et al.* 1996) Etendeka basalts. The transgression is dated biostratigraphically as pre-Barremian (McMillan 1990) but sits above transitional crust (Nürnberg & Müller 1991) varying from 100% continental in the east to 100% oceanic in the west (Martin 1984), developed south of the Walvis Ridge (Dingle *et al.* 1983; Dingle 1992/1993) during the Valanginian (M11 and M12 according to Rabinowitz & LaBrecque 1979; Nürnberg & Müller 1991). The marine incursion is here related to thermal subsidence of this transitional crust and the southern South Atlantic, during the transgression, is imagined as an oceanic gulf similar in dimensions and tectonic style to the modern-day Red Sea. The transgression also correlates with the main transgressions into the Bredasdorp and Pletmos sub-basins of the Outeniqua Basin in the South African offshore (Malan *et al.* 1990), recognized similarly as an oceanic on-set unconformity (Turner 1990). This subsidence and resulting oceanic incursion can be identified with the so-called Valanginian Event, registered in basins of the Falklands Plateau and South American margin (Richards *et al.* 1996).

Our time correlation of the Kudu sequence with Early Jurassic onshore analogues has given additional insight into the tectono-stratigraphic relationships in southern Namibia, in turn leading to a reassessment of the relationships between the Etendeka Group (Milner *et al.* 1994) and Karoo Supergroup throughout onshore Namibia.

Evolution of basins related to Gondwanan break-up

The succession of Mesozoic sedimentary basins of southern Africa is an expression of the long-lived break-up history of the Gondwanan entity (Fig. 1). The eventual break-away of South America from southern Africa at 114 Ma ended ultimately in the latter's isolation from all neighbouring continents. As in other areas of southern Africa, it is possible to distinguish between two major periods of basin formation associated with regional extensional events: (1) those sequences associated with Pangean bisection along the Tethys and interior Gondwanan extension prior to the bisection of Gondwana along the line of the Somali and Mozambique basins (> 160 Ma), developed within late Palaeozoic to early Mesozoic Karoo basins; (2) those sequences following the bisection of Gondwana and including subsequent separation of Africa from other West Gondwanan continental elements (< 160 Ma), chiefly during the Cretaceous, developed in post-Karoo Mesozoic basins.

The need to distinguish between Karoo and post-Karoo basin developments and the associated geochemical distinction between the 133–131 Ma (Renne *et al.* 1996) Etendeka flood basalts of northern Namibia and the 186–183 Ma (Duncan *et al.* 1997) Kalkrand flood basalts in southern Namibia (Figs 1–3), encouraged Milner *et al.* (1994) to promote the Early Cretaceous Etendeka Group, including interfingering and underlying aeolian sandstones as a separate tectono-stratigraphic entity related to the Paraná igneous province. As a sequence boundary between the Karoo and post-Karoo, Milner *et al.* (1994) selected the closest disconformity that could easily be identified beneath the Etendeka flood basalts in the Huab area. This was between the aeolian sandstones, traditionally classified as the Etjo Formation and underlying red beds, the latter represented by the Gai-as and Doros Formations in the Etendeka–Huab area. However, the sub-Etjo disconformity is generally difficult to justify throughout Namibia as a time-stratigraphic break on the scale of the implied 100 Ma extent. Until sufficient biostratigraphic information is considered to constrain a more refined time frame, it seems to have been premature to interpret the inter-regional significances of sequences directly above and below the sub-Etjo unconformity.

Onshore exposures of the Hardap area in southern Namibia preserve only sequences associated with Karoo-aged basin developments, capped by the 186–183 Ma Kalkrand flood basalts in the vicinity of Mariental (Fig. 2). The latter correlate in age and geochemical character with the Drakensberg basalts (Fig 1) of the main Karoo basin in South Africa (Duncan *et al.* 1984), equivalent (Batoka) flood basalts of Zimbabwe (Bond *et al.* 1970), and the Bobonong Lava Formation of the Tuli area, eastern Botswana (Aldiss *et al.* 1984). Kalkrand flood basalt volumes are severely under-represented in their outcrop area because of extensive post-Atlantic break-up paeneplanation and subsequent mantling by thin Cainozoic Kalahari Basin deposits to the east of the outcrop area (Figs 1 and 2). The flood basalt sequence is more fully developed beneath the centre of the Kalahari Basin in Botswana (Smith 1984). However, the Kalkrand outcrops exhibit conspicuous sedimentary interlayers (Gerschütz 1996; Stollhofen *et al.* 1998 in press) that provide a sequence framework (Fig. 4).

The flood basalt sequences in the context of the Namibian Mesozoic record

Prior to detailed studies of the Kalkrand flood basalt sedimentary interlayers, initial comparisons were made between the Kudu main gas reservoir and onshore exposures associated with what were then regarded as correlates of the Etjo Sandstone Formation. These are outside the Etjo-type area in central Namibia and are exposed within the Huab area in northwest Namibia at the base of the volcanic sequences of the Etendeka Plateau (Fig. 3). Aeolian sandstones both underlie and interfinger with the lowermost basalt sequences (Fig. 4). Conventionally, the Etjo Sandstone in its central Namibian type area was correlated lithostratigraphically with the Clarens Formation of the Karoo main basin, the Ntane Sandstone of Botswana and the Forest Sandstone of Zimbabwe as of Late Triassic–Early Jurassic age. It was only when Early Cretaceous $^{40}Ar/^{39}Ar$ ages were derived from lavas above the Etjo correlates in the Etendeka sequence (Fitch & Miller 1984) that doubts were voiced about the true stratigraphic standing of the Etjo Sandstone. In the 1:1 000 000 geological map of Namibia (Miller & Schalk 1980) the Etjo Sandstone and its then correlates were uniformly labelled as Jurassic in age.

Previous assessments of Mesozoic stratigraphy in northern Namibia

The term Etjo Beds was applied by Gevers (1936) to encompass the Mesozoic sediments underlying the Mesozoic flood basalts of Namibia. Only later was the Etjo Formation restricted to the uppermost aeolian dominated unit [referred to by Gevers (1936), as the Plateau Sandstone]. Keyser (1973) suggested the name Omingonde Formation for red-bed sequences beneath the aeolian sequence in central Namibia that had previously been referred to as lower Etjo Beds by Gevers (1936). These were separated by a considerable time-stratigraphic gap from the underlying Early Permian sediments, comprising Dwyka and Ecca Group equivalents (Fig. 5). This is a time-stratigraphic gap that was recognized throughout Namibia and excludes much of the stratigraphic equivalent of the Beaufort Group, which is developed so fully in the main Karoo basin in South Africa. A separate stratigraphy was proposed for the Huab area in northwest

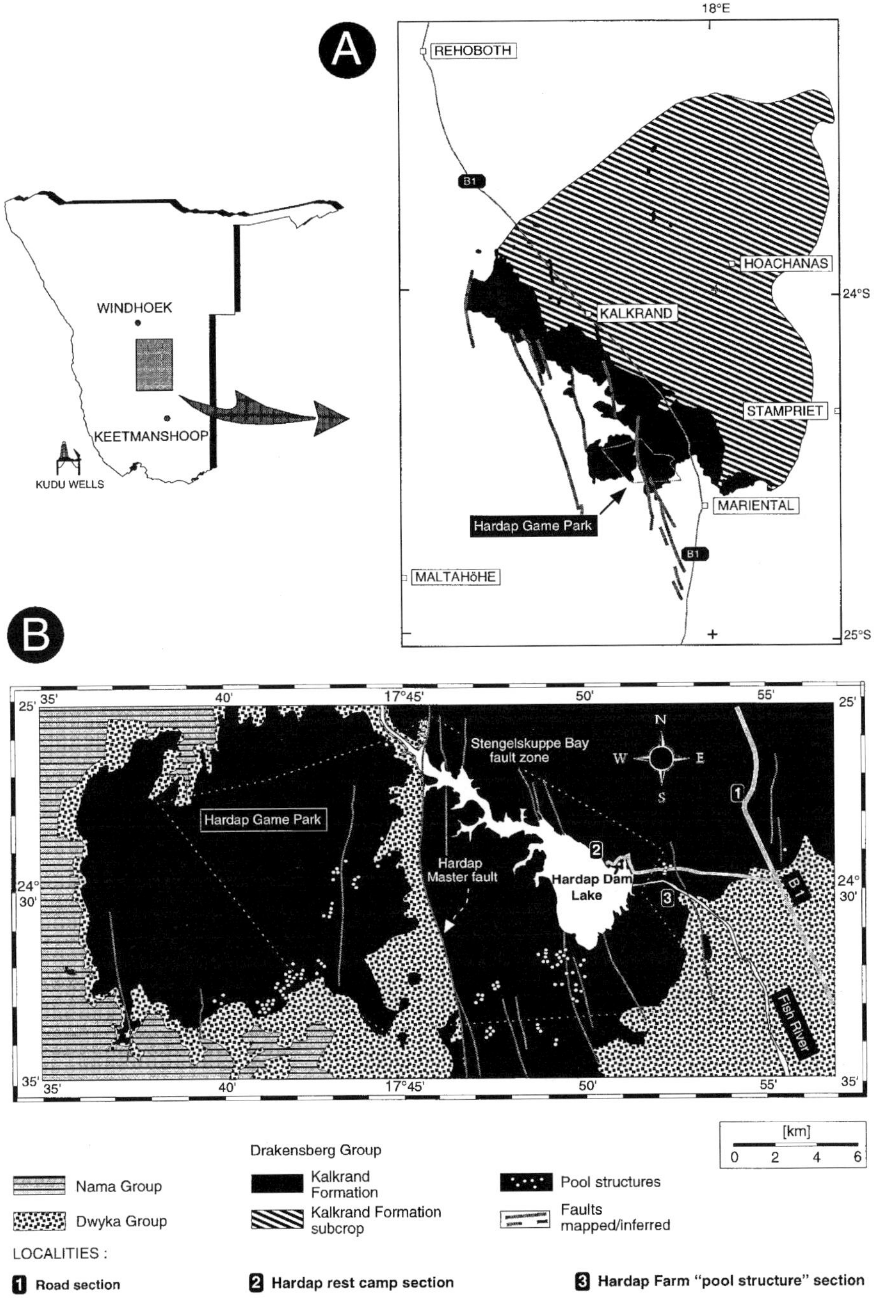

Fig. 2. (**a**) The outcrop and subcrop area of the Kalkrand Formation north of Mariental (based on Miller & Schalk 1980); (**b**) detailed geological map of the Hardap area (based on Schalk & Germs 1980; Gerschütz 1996) indicating major structural elements.

Namibia by Horsthemke *et al.* (1990), who classified all the red beds of that area as Gai-as Formation, correlating them with Permian sequences in Brazil, implying a time-stratigraphic gap of the order of 50 Ma between the Gai-as red beds and then correlates of the Etjo Formation.

Attempts by Milner *et al.* (1994) to reconcile the Early Cretaceous dates derived from

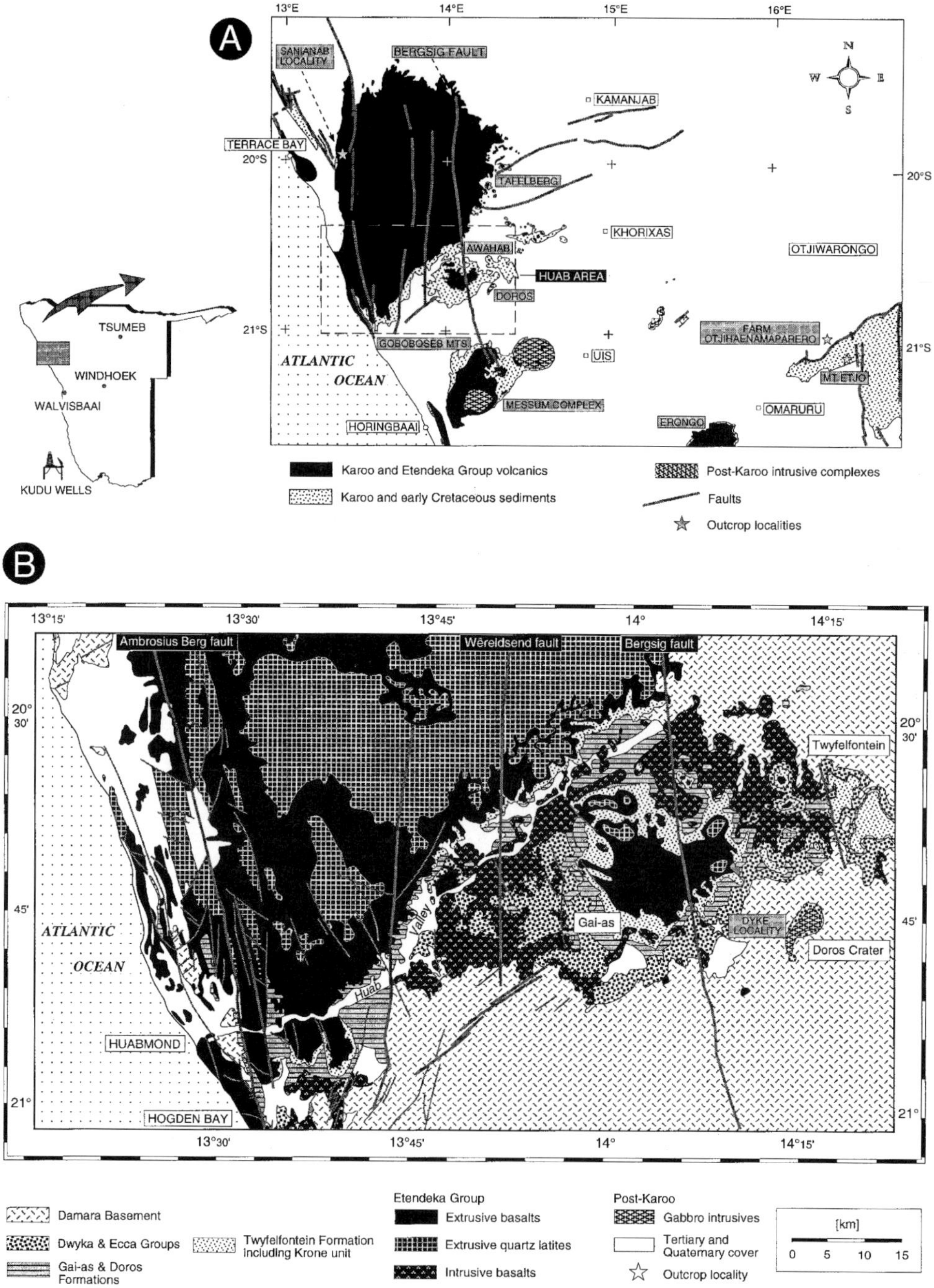

Fig. 3. (**a**) The Etendeka–Otjiwarongo outcrop area northwest of Windhoek [based on Miller & Schalk (1980)]; (**b**) detailed geological map of the Huab area [compiled using data from Miller (1988); Milner *et al.* (1990) and Horsthemke (1992)] illustrating the Huab structural framework.

Etendeka quartz latites and basalts, and the Karoo stratigraphy beneath, invoked the intercalation of the top of the underlying sandstone body with lowermost flows (Tafelkop Basalts) of the volcanic succession. These earliest basalt flows are conformable with, and even pond

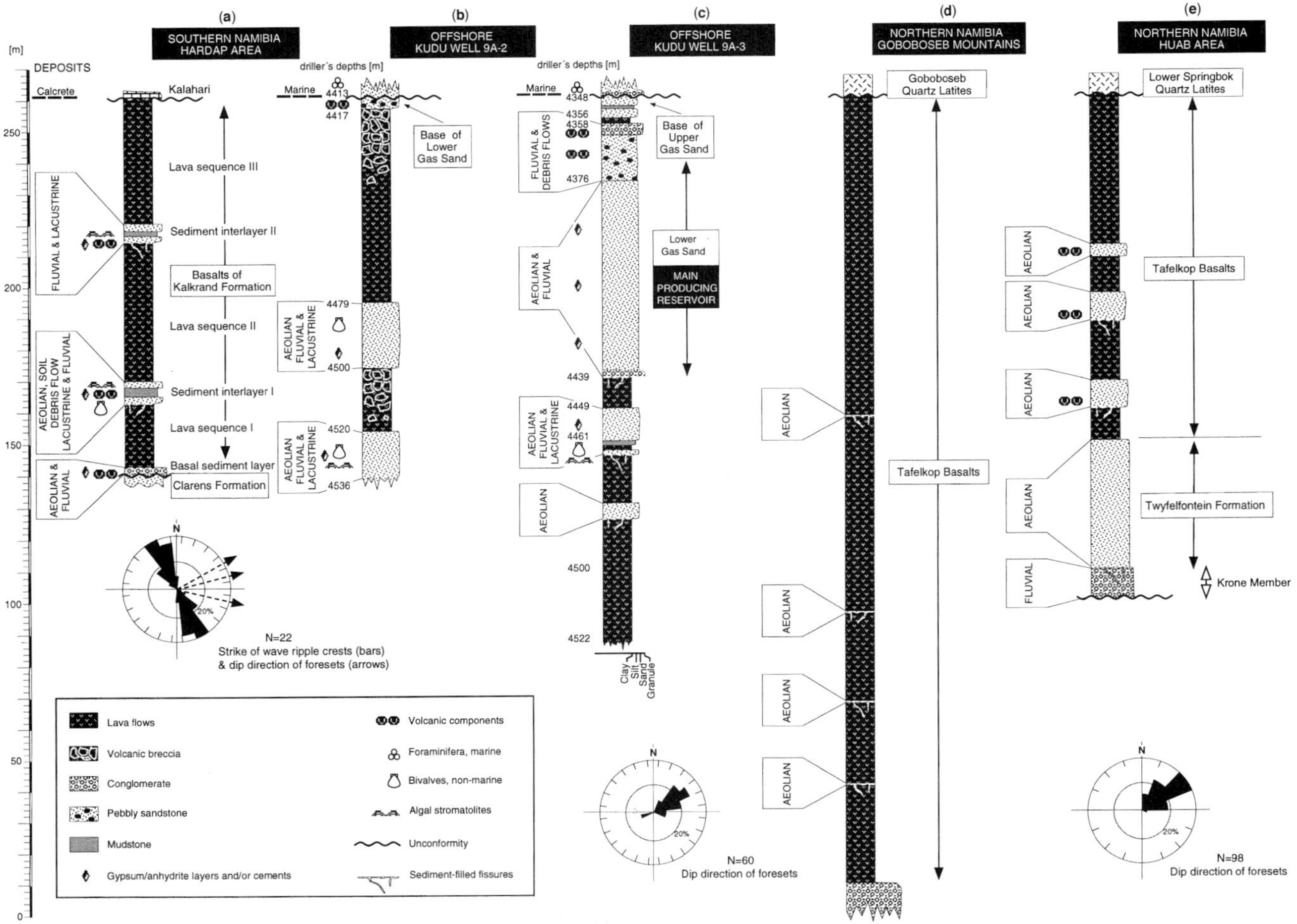

Fig. 4. Comparative logs of Jurassic and Cretaceous volcano–sedimentary successions: (**a**) southern Namibian Hardap area north of Mariental (Stollhofen *et al.* 1998); (**b**) offshore boreholes: Kudu 9A-2 and (**c**) Kudu 9A-3 (after Wickens & McLachlan 1990); (**d**) Tafelkop basalts in the northern Namibian Goboboseb Mountains; (**e**) Tafelkop equivalents of the Huab area in northern Namibia. Palaeowind indicators for the Hardap area in southern Namibia comprise strike measurements of wave ripple crests (bars) and dip directions of dune forests (arrows). Data of palaeowind directions are taken from Wickens & McLachlan (1990) for the Kudu wells and from Hodgson (1970) for the Huab area in northern Namibia. See Figs 1–3 for locations.

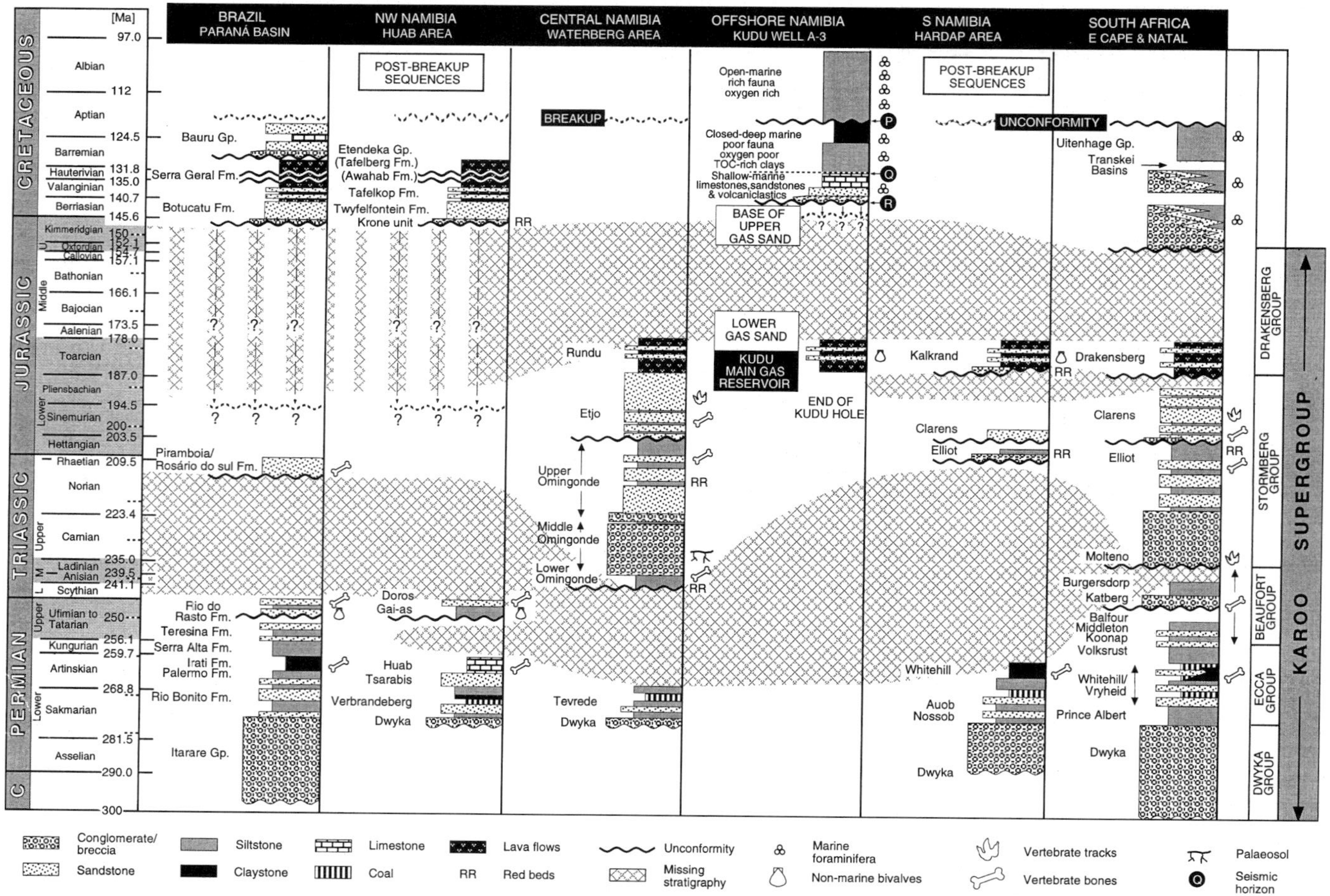

Fig. 5. Reconstructed tectono-stratigraphic framework, constraining the stratigraphic position of the main Kudu gas reservoir. The importance of pre- and post-Etjo disconformities in the early Mesozoic of Namibia, and throughout southern Gondwana, is apparent. Stratigraphic columns based on: Paraná Basin: Zalán *et al.* (1990), Ledendecker (1992), França *et al.* (1995) and Milner *et al.* (1995); Kudu boreholes: Wickens & McLachlan (1990) and McMillian (1990); South African: Kitching (1977), SACS (1980) and Kitching & Raath (1984). Ages based on Harland *et al.* (1990).

between, aeolian dune forms at the top of the sandstone unit. They therefore deduced that the aeolian sandstones were of Early Cretaceous age and, on this basis, promoted the Etendeka Group, including the assumed Etjo Sandstone correlates, as tectonostratigraphically distinct from the underlying Karoo Supergroup. This would imply that the time-stratigraphic gap represented by the sub-Etjo disconformity was now extended to between 45 Ma and 100 Ma throughout most of Namibia.

The sub-Etjo disconformities and the vertebrate record

In the case of the Etjo Sandstone in its type area and its correlates elsewhere in southern Africa, the associated basal disconformity is persistently underlain by Late Triassic red beds. Stormberg Group equivalents, including red beds, are even recognized as erosional remnants beneath the Kalkrand basalts (Stollhofen *et al.* in press), lying disconformably on Early Permian units of the Dwyka and Ecca Groups (Fig. 5) in the Hardap area of southern Namibia (Fig. 1). The concept of an all-encompassing sub-Etjo unconformity throughout Namibia therefore needs to be considered more critically.

The most precise time framework in the main Karoo basin is provided by the vertebrate biostratigraphic record (Kitching 1977; Kitching & Raath 1984; Olsen & Galton 1984). Only relatively few publications record vertebrate finds in Namibia (e.g. Keyser 1973; Keyser & Brink 1979; Pickford 1995), however, these have generally not been taken into account by most recent stratigraphic reconstructions.

Von Huene (1925) and Keyser (1973) refer to dinosaur tracks in the Etjo Sandstone from Otjihaenamaparero Farm in its type area (Fig. 3a), that we have revisited and recorded in this study. The tracks compare (Fig. 6) to late Triassic *Qemetrisauropus princeps* tracks (Ellenberger 1970) illustrated by Olsen & Galton (1984), who have used such dinosaur tracks to correlate between the main Karoo basin of South Africa and the Newark Supergroup in the USA and dated them as Late Triassic. Similar trackways have also been reported (Pickford 1995) from the Etjo at Waterberg Plateau, southeast of Otjiwarongo (Fig. 3a), and were interpreted as indicating a Late Triassic age (Dingle *et al.* 1983). Both occurrences are preserved within the aeolian-dominated upper portion of the type Etjo Formation. In both localities the trackways are imprinted on an even, laterally extensive surface, whose concentration of coarser sand grains reveals it as a major deflation surface.

Even more direct evidence of dinosaurs in Namibia has been the recent discovery of an impression of a skeleton of the prosauropod *Massospondylus* in the Etjo Formation at Waterberg (Holzförster *et al.* 1998). The skeleton compares to Early Jurassic faunal elements of the lowermost Clarens and middle–upper Elliot Formations (Kitching & Raath 1984) in the main Karoo basin, South Africa. Faunal assemblages of the lower and middle Omingonde Formation suggest upper Beaufort (*Cynognathus* zone) and lower Stormberg (Molteno; cf. Keyser 1973) equivalents of Middle–Late Triassic age. Overall, therefore, the Etjo and uppermost portions of the Omingonde Formation in their central Namibian type area correlate with the Stormberg Group of the main Karoo basin and correspond with the Clarens and Elliot Formations, respectively (Fig. 5). However even in areas of Namibia with a completely preserved Omingonde Formation, an Early Permian–Late Triassic time-stratigraphic gap of at least 25 Ma exists within the Karoo succession. In such areas the magnitute of this disconformity is emphasized by a distinct shift in palaeoclimatic and palaeoenvironmental conditions from high latitude marine cold temperate to low latitude continental aridity and semi-aridity. This gap extends to > 30 Ma in areas of enhanced stratigraphic incompleteness, such as along the Namibian continental margin and in southern Namibia. Furthermore, unless the Etjo Formation can be demonstrated to have spanned almost the entire Jurassic, *c.* 40 Ma must be accounted for by stratigraphic gaps between the true Etjo Formation and associated Jurassic Rundu basalts and the much younger Early Cretaceous sequence exposed in the Etendeka/Huab region (Fig. 5).

Red beds, lying directly beneath the previously supposed Etjo correlates of the Etendeka area, are well exposed within the Huab area 6 km north of Doros (Fig. 3b). This stratigraphic unit has recently yielded faunal elements and U-Pb zircon ages from tuff beds that would place them as Late Permian to earliest Triassic. In order to describe this sequence we restrict the term Gai-as Formation to lacustrine red mudstones forming the lower half of the sequence and introduce the term Doros Formation for sandstone and gravel dominated units that coarsen upward from those mudstones (Fig. 5). The Doros Formation preferentially preserves as wedge-shaped bodies reflecting successive rotational block movements associated with north–south trending normal faults. Although Ledendecker (1992) and Horsthemke (1992) suggested that the Gai-as Formation was exclusively Permian, on the basis of comparisons of bivalve assemblages with the

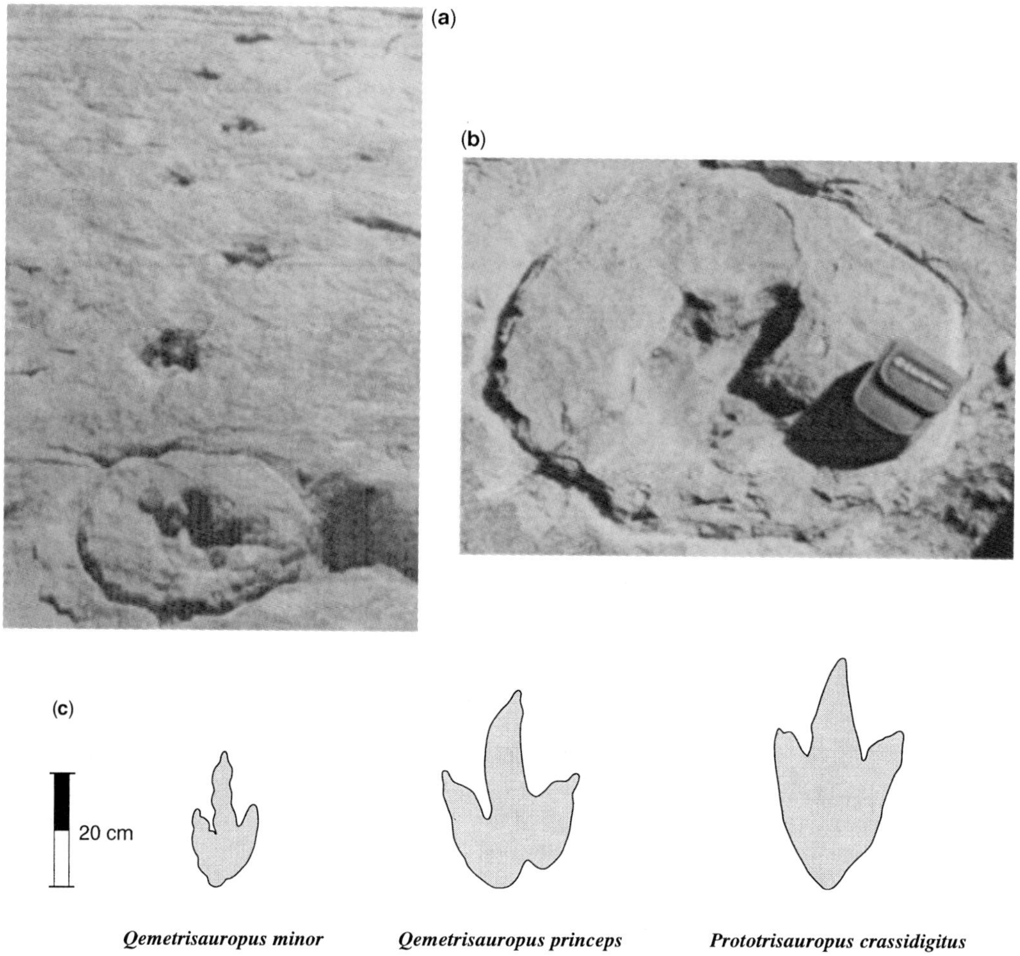

Fig. 6. (**a**) Trackway and (**b**) detail of dinosaur footprint at the farm Otjihaenamaparero, compared to (**c**) published tracings of *Qemetrisauropus minor*, *Qemetrisauropus princeps* and *Prototrisauropus crassidigitus* (Ellenberger 1970), suggesting the late Triassic–early Jurassic (Stormberg equivalent) age of the Etjo Sandstone in its type area. See Fig. 3a for location of Farm outcrop.

Serra Alta, Teresina and Rio do Rasto Formations of the Paraná Basin of South America, faunal elements discovered in Namibia suggest that the sequence extends into the Early Trassic.

Onshore Kalkrand sedimentary interlayers as Kudu main gas reservoir time-correlatives

Certain features (Wickens & McLachlan 1990) of the Kudu 9A-3 Lower Gas Sand (Fig. 4) distinguish it from previously proposed time-stratigraphic equivalents in northern onshore Namibia (Light *et al.* 1993). Evaporites, particularly anhydritic nodules, are ubiquitous; mudstone interlayers are present and shelly fossils occur at the base of the lower sediment layer beneath the reservoir (Fig. 4). The last includes bivalves and algal remains both within shales and laterally in sandstone. Volcaniclastic sandstones and breccias are developed, in which the widespread presence of non-volcanic grains indicates the mass-flow remobilization of both sedimentary and volcanogenic deposits from the flanks of topographic highs interpreted by Wickens & McLachlan (1990) as 'eruption centres'.

All these features are present in the Kalkrand flood basalt sedimentary interlayers (Figs 4 and 7). For a complete description of the lithofacies the reader is directed to Stollhofen *et al.* (1998), but the general characteristics and the more important pronounced variation in lateral lithofacies will be used for comparisons here. For example, Stollhofen *et al.* (1998) describe chicken-wire textured gypsum, layers of fibrous gypsum and euhedral crystals of gypsum, both original and as pseudomorphs, in the forms of replacive crystals and as rosettes grown at the sediment surface. These evaporites are developed in interbedded shales and algal limestones (Fig. 7), as well as in red quartzitic sandstones that exhibit either aeolian bedding and grain size characteristics or subaqueous reworking of aeolian-sourced sand. Mudstone layers are, however, far more frequent and numerous than in the Kudu reservoir, and are preferentially developed towards the middle of each of the two sedimentary interlayers. Bivalves have been discovered (Stollhofen *et al.* 1998) at the base of the lower of the two Kalkrand sedimentary interlayers, remarkably in a similar stratigraphic position as those described (Wickens & McLachlan 1990) from the bases of the sedimentary layers in the Kudu 9A-3 and 9A-2 intersections, directly beneath the Lower Gas Sand (Fig. 4).

Volcanic mass flow breccias are also present in the Kalkrand sedimentary interlayers and are particularly well exposed in a road cutting north of Mariental (Fig. 7c). Stollhofen *et al.* (1998) record and interpret these flows as having been derived from volcanic ridge topography (see their fig. 4) and syn-volcanic and syn-sedimentary fault scarps.

Kalkrand and Kudu sedimentary settings: aeolian dune-fringed rift valley lakes

The Kalkrand sedimentary interlayers were identified by Stollhofen *et al.* (1998) as deposited in rift valley-related lakes, which varied between ephemeral and permanent. Debris flow, fluvial and aeolian deposition both preceded and fringed lacustrine deposition. The aeolian quartz sand source dominated to provide sand input to the lake body and this indicates that aeolian dune fields persisted externally from the lake body throughout its history.

The Kudu palaeoenvironmental setting is analogous with such an external source terrain. In the Kudu area initially developed lakes hosted bivalves and algae, but these were soon buried by wind-blown sands. Presumably, lakes were still maintained, perhaps of a more ephemeral nature, to provide high ground water levels (cf. Marot 1990) that precipitated the evaporites; this would have contributed to dune stabilization. Southwesterly wind directions derived from Kudu 9A-3 dipmeter measurements accord with aeolian foreset directions and wave ripple trends from the Kalkrand interlayers (Fig. 4).

The similarities between Kalkrand and Kudu Lower Gas Sand palaeoenvironmental settings argue against the previous interpretation of a coastal dune setting for the latter. Wickens & McLachlan (1990) point to the existence of pronounced unconformity between the highest Kudu basalt and the overlying Cretaceous transgressive marine sequence. The unconformity is represented on top of the lava by a bored and pitted surface, resembling a marine hardground, and representing the boundary between continental and marine deposition. A pronounced stratigraphic gap at this position would additionally explain the extremely altered and oxidized state of the underlying lavas.

Kudu compared to Kalkrand lavas

The altered nature and much of the geochemical character of the Kudu basalts is shared with those of the onshore Kalkrand succession (Fig. 8). In addition, data are presented for comparison from Jurassic subvolcanic feeder dykes and sills in the Keetmanshoop area, and the basalts at Erongo that are other possible onshore correlatives of the Kalkrand Basalts. Fresh samples of Kalkrand and Kudu basalt sequences comprise olivine–normative tholeiitic basalts (Erlank *et al.* 1990; Gerschütz 1996) and differ markedly from the quartz–normative tholeiitic basalts and quartz latites (Duncan *et al.* 1990) of the higher portions of the Etendeka Group (Fig. 8a–d). In addition, Kalkrand and Kudu basalts are both less 'evolved' than the Etendeka basalt samples in terms of their Mg and SiO_2 contents, rather high concentrations of Mg, Ni and Cr, and low concentrations of incompatible trace elements. As shown by the spidergram in Fig. 8d, the geochemical features of the Kalkrand and Kudu basalts differ markedly from average MORB on first appearance in that they are relatively enriched in large-ion lithophile elements such as K, Rb and Ba. However, by considering additional isotopic and compositional characteristics, the overall affinities with enriched MORB and OIB become apparent (Gerschütz 1996). When data of both suites are plotted on tectonomagmatic discriminant diagrams only the Kudu samples classify entirely as 'within-plate' basalts (Fig. 8e, f), as might be expected from their geodynamic setting, whereas most of the Kalkrand samples plot in the calc–alkali basalt field (Fig. 8f). Compared to the Kudu basalts this could reflect either differences in source composition and/or differences in the degree and depth of partial

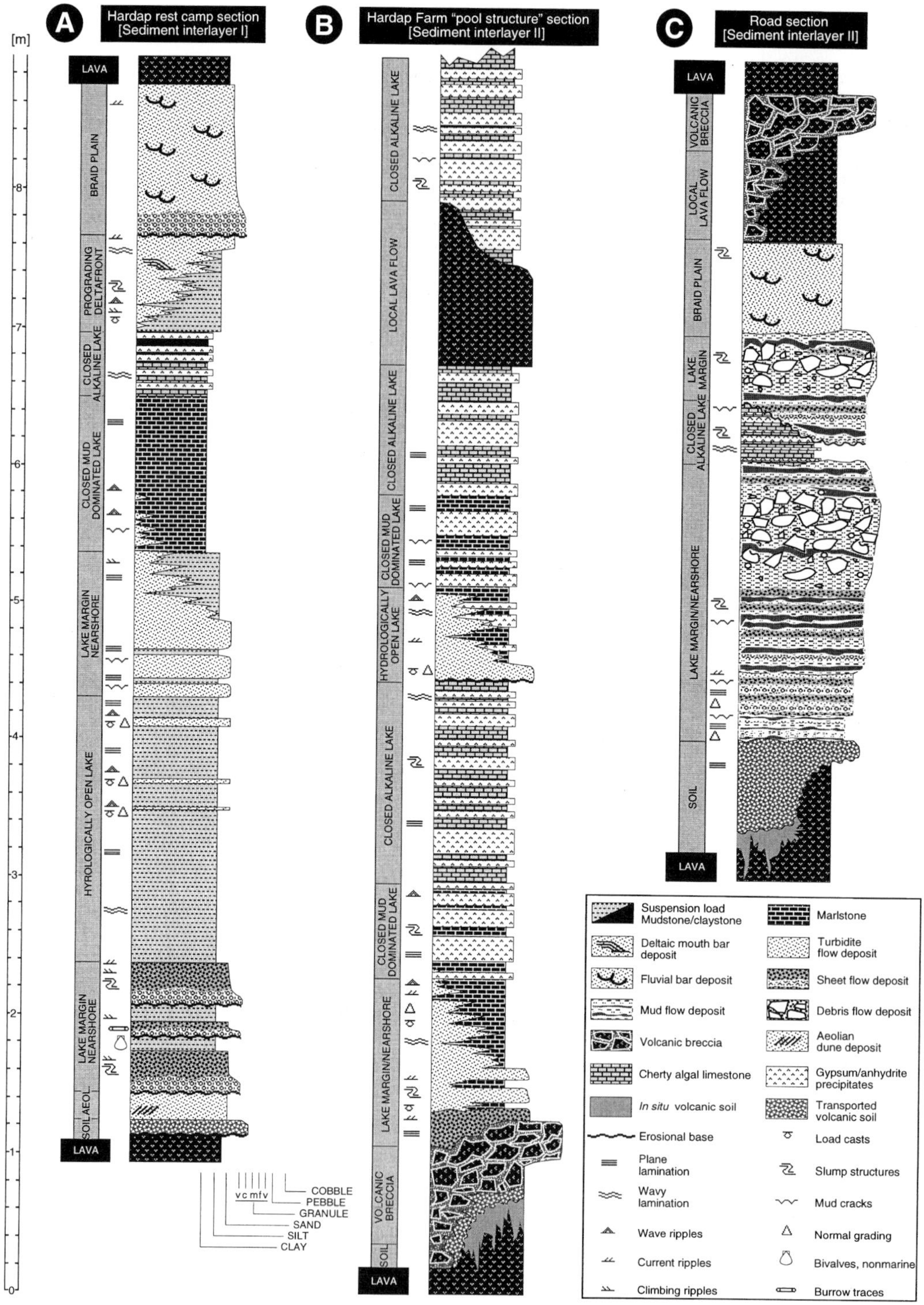

Fig. 7. Facies architecture variability of sedimentary interlayers within the Kalkrand volcano–sedimentary succession of the Hardap area, southern Namibia: **(a)** the mixed siliciclastic and (bio)chemical sediments of the Hardap rest camp section through Sedimentary Interlayer I, comprising a variety of aeolian, fluvial, deltaic and lake sediments; **(b)** the 'pool structure' section through Sedimentary Interlayer II dominated by (bio)chemical lake sediments (top of interlayer not completely preserved). Note the weathered lava surfaces, including soil development, and the occurrence of volcanic breccia. **(c)** The siliciclastic–volcaniclastic fluvial and mass-flow dominated road section through Sedimentary Interlayer II. See Fig. 2b for locations.

melting of the Kalkrand basalts as discussed with samples of similar age and composition (Duncan 1987) from the Karoo Igneous Province.

Considering volcanic architecture, Wickens & McLachlan (1990) typify the Kudu basalts as comprising a number of different lava flows, showing significant lateral thickness variations, and laterally correlating with the volcaniclastic sandstones. From the considerable variation in sediment layer thickness they deduce a pronounced lava topography. With reference to the Kalkrand sequences, the pronounced topography can be explained by the existence of elevated lava feeder ridges, sourcing minor local flows that onlap against the ridges or extrude into the adjacent lake sequences (Stollhofen *et al.* in press). This topography also sources local volcaniclastic debris flows into the lakes (Fig. 7c).

Kalkrand lava flow tops are usually debris strewn surfaces covered by large angular to rounded amygdaloidal lava fragments incorporated into residual and transported soils. Stratigraphic gaps are indicated prior to subsequent sedimentation and burial. These relationships may explain the rounded and angular vesicular volcanic clasts evident at the tops of flows in the Kudu cores. Both Kalkrand and Kudu lava flows are characteristically penetrated by aeolian sand- or soil-filled fissures up to 1.5 m wide that can be traced downward as much as 8 m in the onshore Kalkrand exposures.

Controls on Kalkrand and Kudu accommodation space

In addition to the considerable thickness variation of the main Kudu reservoir lithologies, due to lava topography (Wickens & McLachlan 1990), the geometry of the Kudu section takes the form of a westward dipping and westward thickening wedge, situated in a half-graben structure (Miller 1995). Similarly, lateral variation of the onshore Kalkrand interlayers record extensive lava topographic effects, but additionally reflect the controls of regional to subregional extensional fault systems (Fig. 2). The latter provide accommodation space in localized half-grabens associated with horst and graben geometries developed above a regional down-to-the-west detachment system. The fault systems themselves are thought to have also controlled the siting of volcanic fissure eruptions. The subsequent directions of lava flows were along the axes of the half-graben sub-basins, controlled by mapped (Fig. 2) major north-northwest trending faults and subsidiary orthogonal transfer faults.

It is suggested that the same interactions between tectonism and volcanism on both regional and sub-regional scales underpinned both the onshore sequences and the offshore reservoir geometries. On this basis, the predicted elongate strike of sandstone reservoir bodies would be controlled primarily by north-northwest fault-induced basinal accommodation and secondarily by lava ridges following the half-graben axial trends. The elongated geometries would be limited laterally by northeast trending transfer faults and, more locally, by lava damming.

Revised status of fluvio-aeolian sandstones in the Etendeka Plateau and Huab areas

A major consequence of the biostratigraphic identification of the Etjo Sandstone in its central Namibian type area (Mount Etjo and Waterberg) to be of Early Jurassic age is that it most probably does not time-correlate with the fluvio-aeolian sandstones exposed beneath the Etendeka basalts to the northwest. In the latter case, the basalts interleave with the sandstones and even pond within fossilized interdune areas and their 133–131 Ma dates indicate that they are of Early Cretaceous age. The true Etjo Sandstone belongs to the Karoo Supergroup (Fig. 5), while the fluvio-aeolian sandstones of the Etendeka and Huab areas correlate with younger, Early Cretaceous aeolian sandstones. An appropriate correlate to the letter is the Pirgua Group preserved within the Salta Rift system of northwest Argentina, initiating during the Valanginian (Comínguez & Ramos 1995). The contained lowermost La Yesera and Las Curtiembres Formations comprise fault-related marginal alluvial fan deposits that interfinger with aeolian, fluvial and distal playa deposits towards the basin centre. On this basis, it is suggested that the fluvio-aeolian sandstones of the Etendeka–Huab area be given a separate designation to identify them as a distinct entity, which is characterized below. The name Twyfelfontein Formation is used for this unit, reflecting its excellent exposure in this famous and easily accessible locale. The name Krone Member remains as a name for its basal, conglomeratic, fluvial unit.

The lowermost Etendeka (Tafelkop) basalt flows intercalate with the upper portions of the fluvio-aeolian Twyfelfontein Formation sandstones (Fig. 4) in the Etendeka and Huab regions. In most localities of the Tafelkop Member, and its lateral equivalents that have been examined, at least two intercalated sandstone layers are preserved; however, some localities develop up to four interlayers. The sedimentary interlayers consistently

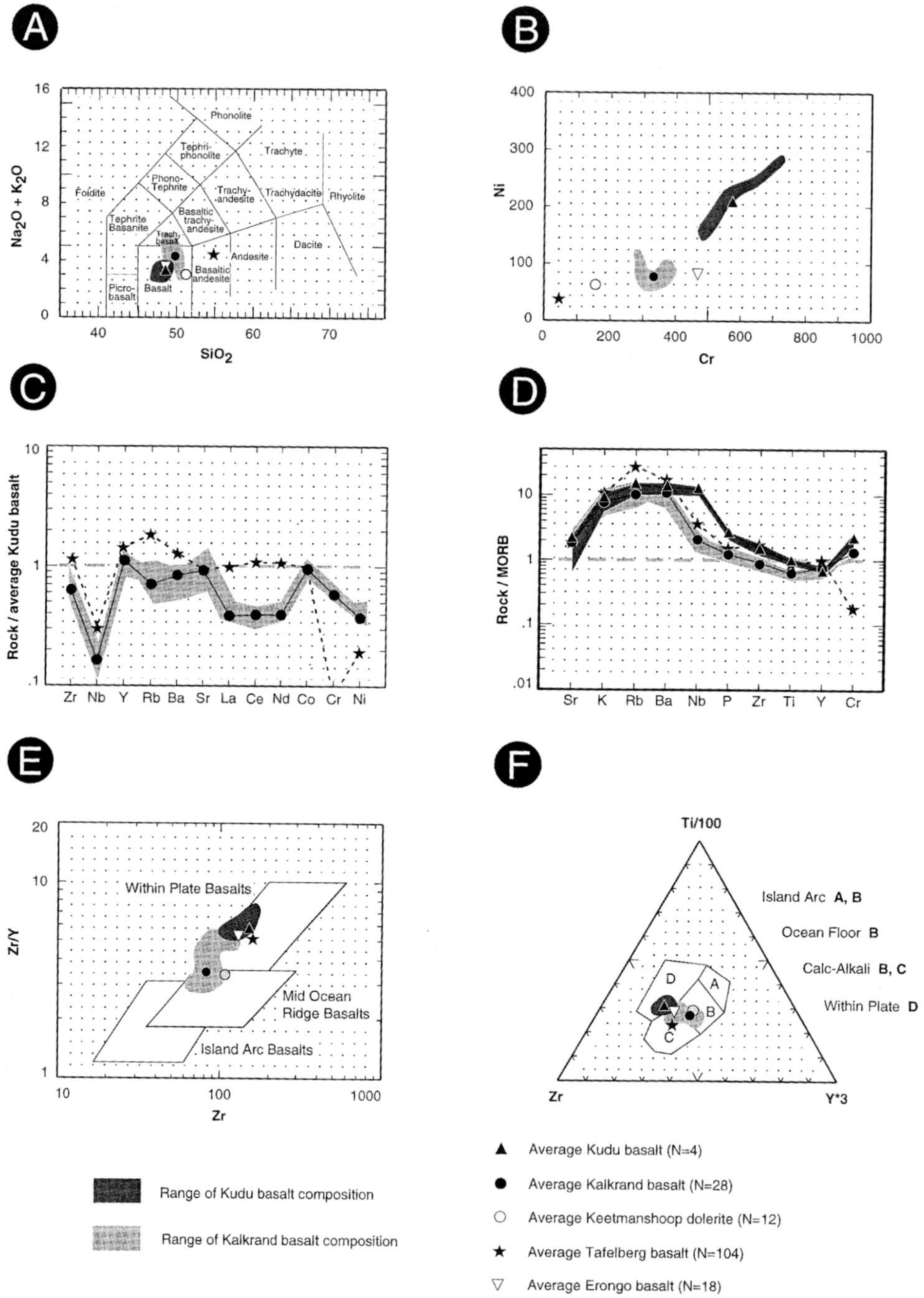

Fig. 8. Geochemistry of Kudu basalts (data from Erlank *et al.* 1990), Kalkrand basalts, Keetmanshoop intrusive dolerites (data from Gerschütz 1996), Erongo basalts (data from Blümel *et al.* 1979) and Tafelberg basalts (data from Milner *et al.* 1990). (**a**) Classification of volcanic rocks, based on the total alkali–silica diagram of Le Bas *et al.* (1986); (**b**) discrimination between Kudu, Kalkrand, Erongo and Tafelberg basalts, based on Cr and Ni contents; (**c**) comparative geochemistry relating average Kalkrand basalt to average Kudu and Tafelberg basalt compositions; (**d**) spidergrams comparing trace and minor element abundances of Kudu, Kalkrand and Tafelberg basalts to mid-ocean ridge basalts (MORB); (**e**) and (**f**) tectonomagnatic discriminant diagrams for the Kalkrand, Kudu, Tafelberg and Erongo basalts and Keetmanshoop dolerite samples after Pearce & Cann (1973).

reveal the grain and bedding characteristics of aeolian sandstones (Ward 1989; Namcor 1996), no evaporites or subaqueous deposits are evident. Thickness variation is considerable. At one extreme, thick continuous interlayers are produced when the main erg continued to grow after the first lava eruption, ultimately to cover and swamp the underlying lava body (Fig. 9a). Towards the other extreme, thin (< 1 m) discontinuous lenses may be intercalated, representing continued transport of small isolate barchans (Jerram *et al.* 1999; Mountney *et al.* 1998) after the first lavas had pooled into the interdune areas of the main Twyfelfontein aeolian sandstone and produced scoriaceous breccia over the tops of the neighbouring dunes (Fig. 9b). Taking this further, the entire erg can be buried and preserved by the volcanism, preserving dune morphologies (Fig. 9a), including the original slope of their avalanche faces (Namcor 1996; Jerram & Stollhofen 1997). Such different responses reflect not only the varying availability of sand (Jerram *et al.* 1999) but also changes in subsidence. Differential degrees of subsidence are produced by the rotational effects of fault-blocks beneath the half-grabens and the resulting variation in accommodation space made available (Fig. 9). In half-graben sub-basinal depocentres accommodation space was maximal and allowed the propagation of the erg; further up the hanging wall, block accommodation was minimal and aeolian sand accumulation was terminated rapidly. In addition, available accommodation space can be entirely occupied by enhanced lava thicknesses, generated closer to the focus of the fissure eruptions that fed the lavas and their inflated feeder ridges. The volcanic control on sedimentary accommodation space can cause variation, even on the local scale, from interlayers comprising thick aeolian sand sheets to thin drapes of sand only a few centimetres thick, ultimately to areas were sediment either bypasses, or is transported over, the volcanic sequence to leave behind a deflated rock surface. Evidence for the latter can only be found in sand deposited in topographic traps, fissures or cooling cracks (Fig. 9c).

At the base of the sediment interlayers, sand has frequently been recorded to have filled fissures which penetrate as much as a few metres into the underlying basalt flows. In fact, at the Tafelkop exposures in the Goboboseb Mountains these fissures are intrinsically important because they preserve information that sedimentation has taken place and a time-stratigraphic gap can be inferred. Fissure-bearing horizons can be traced laterally to provide important markers within the volcanic stratigraphy, even when an interlayer in itself has not been preserved. Such fissures exhibit a preferred north-northwest trend that parallels the predominant trends of regional faults (Miller & Schalk 1980; Horsthemke 1992). Although these faults cross-cut the entire Etendeka sequence of lavas (Fig. 3b), it has been established that they were operating as a syn-sedimentary fault system prior to any lava extrusion. This is revealed by examining details of the northerly trending Bergsig Fault Zone that can be traced across the Huab Valley 12 km east of Gai-as. It uplifts Early Permian units on its eastward side to form an immediate footwall directly beneath the Krone unit and overlying sandstones, thereby eliminating the entire Gai-as and Doros Formations. The same episode of syn-sedimentary faulting is evident on a local scale on the hanging wall side of the same fault zone on the south side of the Huab Valley, where major sandstone-filled fissures extend up to 12 m deep and 20 cm wide down into the Gai-as Formation, and record individual extensional fault displacements of up to 1 m. In the same area, dolerite sheets can be seen to have exploited the same fault zone (Fig. 3b; dyke locality) to feed the overlying lava sequence.

Contrasting Early Jurassic and late stage Early Cretaceous Namibian rift settings

The Early Jurassic rifting that hosted the sedimentary interaction between lacustrine and aeolian environments within the Kalkrand and Kudu volcano–sedimentary sequences of southern Namibia has analogues in the depositional settings developed presently across the southwesterly continuation of the modern African rift system in northwest Botswana (Stanistreet & McCarthy 1993). The latter represents a more recent propagation of the presently active African rift system, and reactivates a major detachment fault, the Thamalakane Fault System. The latter has a long-lived history extending back to the Mesozoic, but exploits a lithospheric weakness dating back to the late Precambrian.

The Early Jurassic rift-related Kalkrand and Kudu volcano–sedimentary sequences of southern Namibia indicate that alternately permanent and ephemeral lakes (with evaporites) developed both within the north–south active rift zone and that these lakes extended eastwards across the rift shoulder towards the cratonic interior. The lakes were fringed by areas of intermittently active sand ergs which also provided a source for sand reworked into the water bodies. The Kalkrand, Kudu and central Namibian Etjo type areas, therefore, complement one another in terms of source terrains and depocentres. In the extensional

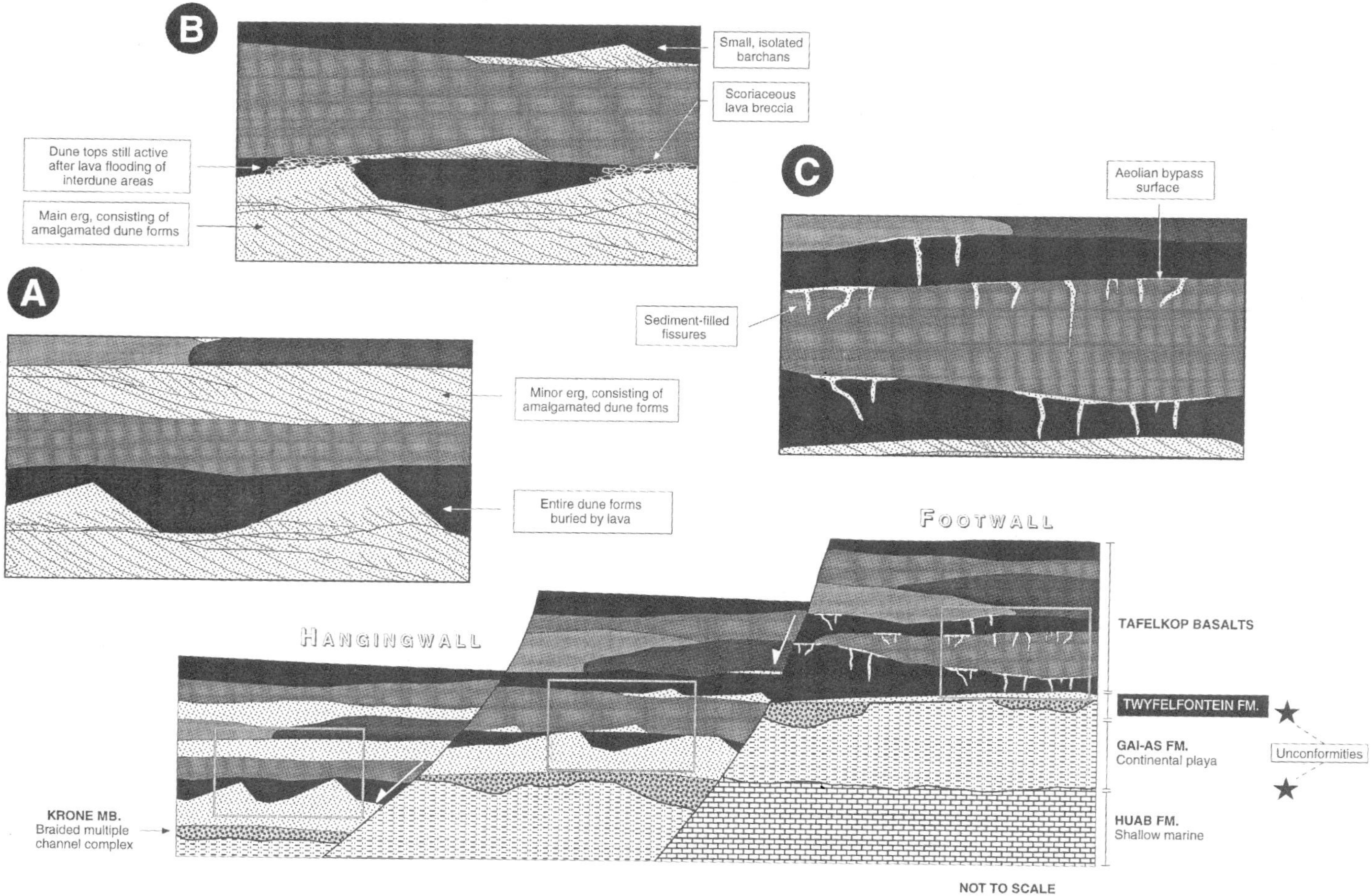

Fig. 9. How a combination of fault controlled and volcanically controlled accommodation space can produce a variety of sandstone geometries. The conceptual model is reconciled with recorded sandstone geometries of the Early Cretaceous Twyfelfontein Formation produced by volcano–sedimentary intercalation at: (**a**) northern Huab area (Huab valley); (**b**) central Huab area (east of Gai-as); (**c**) Goboboseb Mountains. See Fig. 3a and b for locations.

rift to cratonic interior topographies: (1) depocentres could be occupied by closed or open permanent lakes (Stollhofen *et al.* 1998, in press) that respond to the tectonically and volcanically controlled drainage, as recorded in the Kalkrand region; (2) areas around the lake margin could become overrun by the advance of aeolian dune forms to create sand ergs as evidenced by the Kudu occurrences; or (3) in tectonically elevated areas fluvial aggradation was totally replaced by aeolian sand erg advance as in the Waterberg–Mount Etjo region.

For comparison, in northwest Botswana during the last 40 000 years there is evidence (Shaw 1988) for the existence of an extensive permanent lake body that extended across the northeast– southwest Thamalakane fault system detachment that has controlled rifting. Lake Palaeo-Mkgadikgadi occupied both the subsident area of the hanging wall block and extended southeastward onto the uplifted footwall and towards the cratonic interior. The lake, at its maximum, covered in excess of 60 000 km^2 (Shaw 1988) and was up to 55 m deep. It occupied much of the northern half of Botswana and was of lateral dimensions similar to rift valley lakes still existent in Africa further to the northeast. Nowadays, climatic changes have reduced the Botswanan lake via a series of reducing levels (Shaw 1988) to modern remnants in the form of smaller ephemeral playa lake areas, the Mkgadikgadi Pans. These, the Ntwetwe and Nata Pans, nevertheless have impressive lateral dimensions in which active evaporite precipitation is taking place. Further change to greater aridity would reactivate the sand dune systems of the Kalahari desert which surrounds the lake system (Lancaster 1981). At present, dune sands provide a predominant sediment source, inputting into the lake depository both through local wind driven processes and through river floods. The latter vary from those of small-scale rivers to continental-scale rivers such as the Okavango and its distributary system on the Okavango Fan (Stanistreet & McCarthy 1993; Stanistreet *et al.* 1993).

By contrast, in the Etendeka area to the North, the Early Cretaceous fluvial–aeolian depositional system identified in the Twyfelfontein Formation beneath the Etendeka flood basalts shows no involvement of lacustrine bodies. This system is related here to the cut back of rivers associated with the subsequent final stages of rifting just prior to South Atlantic oceanic on-set. Rift shoulders had by then achieved their maximum thermal uplift and river system cut-back developed inboard from the incipient continental edge to produce the Krone fluvials (Fig. 5; Mountney *et al.* 1998). Subsequent onlap of aeolian sands (Swart 1992/93), without major nearby bodies of standing water or elevated ground water levels, resulted in a lack of evaporites and no outstanding evidence of tetrapod activity in what is deduced to have been an extremely arid and generally more hostile environment.

Reiterated extension and flood basalt extrusion along an entire intracontinental rift

It is deduced that a Late Triassic–Early Jurassic extensional tectonic episode affected the entire intracontinental rift system that was later to evolve into the South Atlantic ocean. Flood basalts showing affinities with enriched MORB (Duncan *et al.* 1984; Gerschütz 1996) were produced along its length, of which the Kudu and Kalkrand basalts in southern Namibia (Fig. 5) are still preserved, as are equivalents within South America. The Jurassic units there are characterized by sediment interlayers in the same way as their southern African counterparts. Preservation of the Kudu basalts was due to thermal subsidence along the entire southern South Atlantic margin, whereas the Kalkrand basalts were preserved beneath fill of the interior Cainozoic Kalahari Basin associated with subsidence during its long-lived thermal cooling history. It was the subsequent oceanic opening and the related northward rift propagation that initiated the Etendeka volcano–sedimentary sequences.

Such a delay from Early Jurassic to Early Cretaceous flood basalts displayed in Namibia is entirely in accord with equivalents in South America in areas north and south of the Paraná Basin and adjacent areas. In particular, the basalts of the Roca Blanca and Bajo Pobre Formations (Reimer *et al.* 1996) are contemporary with the Early Jurassic flood basalts of Namibia and South Africa. Additionally, basalts containing sedimentary interlayers are known from the Apoteri and Mosquito Formations of the Tacutu, North Savannas and Parnaíba rift basins in northern Brazil and Guyana, respectively (Berrangé & Dearnley 1975; Eiras *et al.* 1994; Góes & Feijó 1994). This indicates that the Early Jurassic basaltic volcanism extended even to the north of the Paraná Basin, to the verge of the then westwardly propagating Tethys Ocean itself.

Mesozoic context for the Kudu Gas Field within the Namibian stratigraphic framework

The compiled stratigraphic framework for Namibia (Fig. 5) takes account of both radiometric and biostratigraphic constraints discussed above, and places the Kudu Field in its regional stratigraphic context. The framework can now also be extended southeastward to Botswana, as far as the South African main Karoo basin, and northwestward into the Paraná Basin of South America.

The reconstructed stratigraphic framework supports the promotion (Milner *et al.* 1994) of Early Cretaceous components of the Etendeka lava units and underlying sandstones to a separate group status (Fig. 5). However, the concept of an all-encompassing time-stratigraphic gap between Etendeka–Etjo units and Permian sediments, amounting to < 100 Ma, is rejected. Rather, the stratigraphy documents three significant stratigraphic hiatuses identified variously by volcanic dates, vertebrate fossil and microfossil evidence. The Permo-Triassic hiatus, generally of *c.* 30–50 Ma, is represented more completely by the majority of the Beaufort Group in the South African main basin (Fig. 5). Basal Beaufort-equivalent units are, however, preserved in Botswana (Smith 1984) and uppermost Beaufort-equivalent units are preserved in central Namibia. On a biostratigraphic basis, only a minor Early Jurassic hiatus occurs beneath the major extrusive unit of the Kalkrand flood basalts (Gerschütz 1996; Stollhofen *et al.* 1998) and its correlatives (Fig. 5), typified by the Drakensberg basalts in South Africa. Similarly, on a biostratigraphic basis, only a minor time-stratigraphic gap occurs beneath the Etjo and its equivalent Clarens Formation, which had previously been suggested (Milner *et al.* 1994) to represent a major hiatus. The Middle Jurassic–earliest Cretaceous hiatus of up to 40 Ma in Namibia (Fig. 5) is in total accord with that recorded in South African stratigraphy as a whole, related to the progressive disintegration of Gondwana. The latter ultimately isolated Africa along the line of the South Atlantic, resulting in sequential and segmental thermal subsidence that accommodated marine transgressions across pronounced unconformities. A Namibian example of such an unconformity is that well recorded and referred to in this paper, developed beneath the Cretaceous portions of the Kudu boreholes. The thermal subsidence occurred prior to the outpouring of the Etendeka flood basalts (Fig. 5) much further to the north.

The identification of a Karoo age for the main Kudu reservoir should encourage a wider focusing of hydrocarbon exploration attention in the Namibian offshore to include underlying Karoo rift-fill sequences.

Grateful acknowledgements are made to the Geological Survey of Namibia, Namibian Nature Conservation, and the rangers at Skeleton Coast and Hardap National Parks for logistical support. We greatly appreciate input from our colleagues Volker Lorenz, Dougal Jerram, Stephan Gerschütz, Frank Holzförster, Bruce Rubidge, Mike Raath, Andy Duncan, Simon Milner and Roger Swart, and the detailed review of Ed Purdy. We also thank the Geoscience Graduiertenkolleg and the Jubiläumsstiftung of the Würzburg University for financial support of this research.

References

ALDISS, D. T., BENSON, J. M. & RUNDLE, C. C. 1984. Early Jurassic pillow lavas and palynomorphs in the Karoo of eastern Botswana. *Nature*, **310**, 302–304.

BERRANGÉ, J. P. & DEARNELY, R. 1975. The Apoteri volcanic formation – tholeiitic flows in the North Savannas Graben of Guyana and Brazil. *Geologische Rundschau*, **64**, 883–899.

BLÜMEL, W. D., EMMERMANN, R. & HÜSER, K. 1979. Der Erongo. *Wissenschaftliche Forschung in Südwestafrika*, **16**, 1–107.

BOND, G., WILSON, J. F. & RAATH, M. A. 1970. Upper Karoo pillow lava and a new sauropod horizon in Rhodesia. *Nature*, **227**, 1339.

COMINGUEZ, A. H. & RAMOS, V. A. 1995. Geometry and seismic expression of the Cretaceous Salta Rift System, Northwestern Argentina. *In*: TANKARD, A. J., SUÁREZ-SORUCO, R. & WELSINK, H. J. (eds) *Petroleum Basins of South America*. AAPG Memoir, **62**, 325–340.

DE WIT, M. J., JEFFERY, M., BERGH, H. & NICOLAYSEN, L. 1988. *Geological map of sectors of Gondwana*. AAPG map, 2 sheets, 1:10 000 000.

DINGLE, R. V. 1992/1993. Structural and sedimentary development of the continental margin off southwestern Africa. *Communications of the Geological Survey of Namibia*, **8**, 35–43.

——, SIESSER, W. G. & NEWTON, A. R. 1983. *Mesozoic and Tertiary Geology of Southern Africa*. Balkema.

DUNCAN, A. R. 1987. The Karoo igneous province – a problem area for inferring tectonic setting from basalt geochemistry. *Journal of Volcanology and Geothermal Research*, **32**, 13–34.

——, ERLANK, A. J. & MARSH, J. S. 1984. Regional geochemistry of the Karoo igneous province. *In*: ERLANK, A. J. (ed.) *Petrogenesis of the Volcanic Rocks of the Karoo Province*. Geological Society of South Africa, Johannesburg, Special Publications, **13**, 355–388.

——,——,—— & ARMSTRONG, R. A. 1990. MORB-related dolerites associated with the Etendeka volcanics, Northwestern Namibia. *Abstract Geocongress '90 Cape Town*. Geological Society of South Africa, Johannesburg, 143–146.

——, HOOPER P. R., REHACEK, J., MARSH, J. S. & DUNCAN, A. R. 1997. The timing and duration of the Karoo igneous event, southern Gondwana. *Journal of Geophysical Research*, **102**, 18 127–18 138.

EALES, H. V., MARSH, J. S. & COX, K. G. 1984. The Karoo Igneous Province. An introduction. *In*: ERLANK, A. J. (ed.) *Petrogenesis of the Volcanic Rocks of the Karoo Province*. Geological Society of South Africa, Johannesburg, Special Publications, **13**, 1–26.

EIRAS, J. F., KINOSHITA, E. M. & FEIJO, F. J. 1994. Bacia do Tacutu. *Boletim de Geociências da Petrobrás*, **8**, 83–89.

ELLENBERGER, P. 1970. Les niveaux paleontologiques de premiere apparition des Mammiferes primordiaux en Afrique du Sud et leur ichnologie: Establissement de zones stratigraphiques détaillées dans le Stormberg du Lesotho (Afrique du Sud) (Trias superieur a Jurassique). *In*: HAUGHTON, S. H. (ed.) *IUGS, 2nd Symposium on Gondwana Stratigraphy and Palaeontology. Pretoria*, Council for Scientific and Industrial Research, 343–370.

ERLANK, A. J., LE ROUEX, A. P., HARRIS, C., MILLER, R. MCG. & MCLACHLAN, I. 1990. Preliminary note on the geochemistry of basalt samples from the Kudu boreholes. *Communications of the Geological Survey of Namibia*, **6**, 59–61.

FITCH, F. J. & MILLER, J. A. 1984. Dating Karoo igneous rocks by the conventional K–Ar and $^{40}Ar/^{39}Ar$ age spectrum methods. *In*: ERLANK, A. J. (ed.) *Petrogenesis of the Volcanic Rocks of the Karoo Province*. Geological Society of South Africa, Johannesburg, Special Publications, **13**, 247–266.

FRANÇA, A. B., MILANI, E. J., SCHNEIDER, R. L. ET AL. 1995. Phanerozoic correlation in southern South America. *In*: TANKARD, A. J., SUÁREZ-SORUCO, R. & WELSINK, H. J. (eds) *Petroleum Basins of South America*. AAPG Memoir, **62**, 129–161.

GERSCHÜTZ, S. 1996. *Geology, volcanology, and petrogenesis of the Kalkrand Basalt Formation and the Keetmanshoop dolerite complex, southern Namibia*. Dr. rer. nat.-thesis, University of Würzburg.

GEVERS, T. W. 1936. The Etjo beds of northern Hereroland, South West Africa. *Transactions of the Geological Society of South Africa*, **39**, 317–329.

GÓES, Á. M. O. & FEIJÓ, F. J. 1994. Bacia do Parnaíba. *Boletim de Geociências da Petrobrás*, **8**, 57–67.

HARLAND, W. B., ARMSTRONG, R. L., COX, A. V., CRAIG, L. E., SMITH, A. G. & SMITH, D. G. 1990. *A Geologic Time Scale*. Cambridge University Press.

HODGSON, F. D. I. 1970. The geology of the Karoo System in the Southern Kaokoveld, South West Africa. *In*: HAUGHTON, S. H. (ed.) *IUGS, 2nd Symposium on Gondwana Stratigraphy and Palaeontology. Pretoria*, Council for Scientific and Industrial Research, 233–240.

HOLZFÖRSTER, F., STANISTREET, I. G., STOLLHOFEN, H., RAATH, M. & RUBIDGE, B. 1998. First dinosaur skeleton from Namibia and its biostratigraphic implications. *Abstracts of the Palaeontological Society of southern Africa meeting*, Geological Survey of Namibia, Windhoek, 13.

HORSTHEMKE, E. 1992. Fazies der Karoosedimente in der Huab-Region, Damaraland, NW-Namibia. *Göttinger Arbeiten zur Geologie and Paläontologie*, **55**, 1–102.

——, LEDENDECKER, S. & PORADA, H. 1990. Depositional environments and stratigraphic correlation of the Karoo Sequence in northwestern Damaraland. *Communications of the Geological Survey of Namibia*, **6**, 63–73.

JERRAM, D. A. & STOLLHOFEN, H. 1997. Interactions between aeolian sand and lava flows of the Etendeka Flood basalts, NW Namibia. *Journal of Conference Abstracts*, **2**, 40.

——, MOUNTNEY, N. & STOLLHOFEN, H. 1999. Facies architecture of the Etjo Sandstone Formation and its interaction with the basal Etendeka flood basalts of northwest Namibia: implications for offshore prospectivity. *This volume*.

KEYSER, A. W. 1973. A new Triassic vertebrate fauna from South West Africa. *Palaeontologia Africana*, **16**, 1–15.

—— & BRINK, A. S. 1979. A new bauriomorph (*Herpetogle marsupialis*) from the Omingonde Formation (Middle Triassic) of South West Africa. *Annals of the Geological Survey of the Republic of South Africa*, **12**, 91–105.

KITCHING, J. W. 1977. The distribution of the Karoo vertebrate fauna. *Memoir of the Bernard Price Institute for Palaeontological Research*, **1**.

—— & RAATH, M. A. 1984. Fossils from the Elliot and Clarens Formations (Karoo Sequence) of the northeastern Cape, Orange Free State and Lesotho, and a suggested biozonation based on tetrapods. *Palaeontologia Africana*, **25**, 111–125.

LANCASTER, N. 1981. Palaeoenvironmental implications of fixed dune systems in southern Africa. *Palaeogeography, Palaeoclimatology, Palaeoecology*, **33**, 327–346.

LE BAS, M. J., LE MAITRE, R. W., STRECKEISEN, A. & ZANETTIN, B. 1986. A chemical classification of volcanic rocks based on the total alkali–silica diagram. *Journal of Petrology*, **27**, 745–750.

LEDENDECKER, S. 1992. Stratigraphie der Karoosedimente der Huabregion (NW-Namibia) und deren Korrelation mit zeitäquivalenten Sedimenten des Paranàbeckens (Südamerika) und des Großen Karoobeckens (Südafrika) unter besonderer Berücksichtigung der überregionalen geodynamischen und klimatischen Entwicklung Westgondwanas. *Göttinger Arbeiten zur Geologie und Paläontologie*, **54**, 1–87.

LIGHT, M. P. R., MASLANYJ, M. P., GREENWOOD, R. J. & BANKS, N. L. 1993. Seismic sequence stratigraphy and tectonics offshore Namibia. *In*: WILLIAMS, G. D. & DOBB, A. (eds) *Tectonics and Seismic Sequence Stratigraphy*. Geological Society, London, Special Publications, **71**, 163–191.

MALAN, J. A., MARTIN, A. K. & CARTWRIGHT, J. A. 1990. The structural and stratigraphic development of the Gamtoos and Algoa Basins, offshore South Africa. *Abstract Geocongress '90 Cape Town*. Geological Society of South Africa, Johannesburg, 328–331.

MAROT, J. E. B. 1990. Petrography and porosity of the gas-bearing sandstones intersected by the Kudu 9A-2 and 9A-3 boreholes. *Communications of the Geological Survey of Namibia*, **6**, 41–48.

MARTIN, A. K. 1984. Propagating rifts: Crustal extension during continental rifting. *Tectonics*, **3**, 611–617.

MCLACHLAN, I. R. 1990. Introductory note on the history of the Kudu prospect. *Communications of the Geological Survey of Namibia*, **6**, 5.

MCMILLAN, I. K. 1990. Foraminiferal biostratigraphy of the Barremian to Miocene rocks of the Kudu 9A-1, 9A-2 and 9A-3 boreholes. *Communications of the Geological Survey of Namibia*, **6**, 23–29.

MILLER, R. MCG. 1988. *Geological map area 2013 Cape Cross: scale 1:250 000*. Geological Survey of Namibia, Windhoek.

—— 1995. Namibia's offshore petroleum potential. *Proceedings of the Centennial Geocongress 1995 Johannesburg*. Geological Society of South Africa, Johannesburg, 1049–1052.

—— & SCHALK, K. E. L. 1980. *South West Africa/Namibia geological map 1:1 000 000*. Geological Survey of the Republic of South Africa and South West Africa/Namibia, Windhoek.

MILNER, S. C. & EWART, A. 1989. The geology of the Goboboseb Mountain volcanics and their relationship to the Messum Complex, Namibia. *Communications of the Geological Survey of Namibia*, **5**, 31–40.

——, DUNCAN, A. R., EWART, A. & MARSH, J. S. 1994. Promotion of the Etendeka Formation to Group status: A new integrated stratigraphy. *Communications of the Geological Survey of Namibia*, **9**, 5–11.

——, ——, WHITTINGHAM, A. M. & EWART, A. 1995. Trans-Atlantic correlation of eruptive sequences and individual silicic volcanic units within the Paraná–Etendeka igneous province. *Journal of Volcanology and Geothermal Research*, **69**, 137–157.

——, ——, MARSH, J. S., MILLER, R. MCG. & BAILEY, M. M. 1990. Field excursion guide to the Etendeda volcanic rocks and associated intrusions, N.W. Namibia. *Field Guide Geocongress '90 Cape Town*. Geological Society of South Africa, Johannesburg.

MOUNTNEY, N., HOWELL, J., FLINT, S. & JERRAM, D. 1998. Aeolian and alluvial deposition within the mesozoic Etjo Sandstone Formation, north west Namibia. *Journal of African Earth Sciences*, **27**, 175–192.

NAMCOR, 1996. The Etjo Fm. and its relationship to Etendeka Grp. lavas in the Huab basin. *Field guide prepared for Shell Exploration and Production, Namibia*.

NÜRNBERG, D. & MÜLLER, R. D. 1991. The tectonic evolution of the South Atlantic from Late Jurassic to present. *Tectonophysics*, **191**, 27–53.

OLSEN, P. E. & GALTON, P. M. 1984. A review of the reptile and amphibian assemblages from the Stormberg of southern Africa, with special emphasis on the footprints and the age of the Stormberg. *Palaeontologia Africana*, **25**, 87–110.

PEARCE, J. A. & CANN, J. R. 1973. Tectonic setting of basic volcanic rocks determined using trace element analyses. *Earth and Planetary Science Letters*, **19**, 290–300.

PICKFORD, M. 1995. Karoo Supergroup palaeontology of Namibia and brief description of a thecodont from Omingonde. *Palaeontologia Africana*, **32**, 51–66.

RABINOWITZ, P. D. & LABREQUE, J. 1979. The Mesozoic South Atlantic Ocean and evolution of its continental margins. *Journal of Geophysical Research*, **84**, 5973–6002.

REIMER, W., MILLER, H. & MEHL, H. 1996. Mesozoic and Cenozoic palaeo-stress fields of the South Patagonian Massif deduced from structural and remote sensing data. *In*: STOREY, B. C., KING, E. C. & LIVERMORE, R. A. (eds) *Weddell Sea Tectonics and Gondwana Break-up*. Geological Society, London, Special Publications, **108**, 73–85.

RENNE, P. R., GLEN, J. M., MILNER, S. C. & DUNCAN, A. R. 1996. Age of Etendeka flood volcanism and associated intrusions in southwestern Africa. *Geology*, **24**, 659–662.

RICHARDS, P. C., GATLIFF, R. W., QUINN, M. F., WILLIAMSON, J. P. & FANNIN, N. G. T. 1996. The geological evolution of the Falkland islands continental shelf. *In*: STOREY, B. C., KING, E. C. & LIVERMORE, R. A. (eds) *Weddell Sea Tectonics and Gondwana Break-up*. Geological Society, London, Special Publications, **108**, 105–128.

SACS (South African Committee for Stratigraphy). 1980. *Stratigraphy of South Africa. Part 1: Lithostratigraphy of the Republic of South Africa, South West Africa/Namibia and the Republics of Bophuthatswana, Transkei and Venda* (L. E. Kent, Compiler). Handbook Geological Survey of South Africa, **8**.

SCHALK, K. E. L. & GERMS, G. J. B. 1980. *Geological map area 2416 Mariental: scale 1:250 000*. Geological Survey of Namibia, Windhoek.

SHAW, P. A. 1988. After the flood: the fluvio-lacustrine landform of northern Botswana. *Earth Science Reviews*, **25**, 449–456.

SMITH, R. A. 1984. The lithostratigraphy of the Karoo Supergroup in Botswana. *Geological Survey of Botswana Bulletin*, **26**.

STANISTREET, I. G. & MCCARTHY, T. S. 1993. The Okavango Fan and the classification of subaerial fan systems. *Sedimentary Geology*, **85**, 115–133.

——, CAIRNCROSS, B. & MCCARTHY, T. S. 1993. Low sinuosity and meandering bedload rivers of the Ikavango Fan: channel confinement by vegetated levées without fine sediment. *Sedimentary Geology*, **85**, 135–156.

STOLLHOFEN, H., STANISTREET, I. G. & GERSCHÜTZ, S. in press. Early Jurassic rift valley-related lake deposits interbedded with Karoo flood basalts, Southern Namibia. *In*: GIERLOWSKI-KORDESCH, E. & KELTS, K. (eds) *Global Geological Record of Lake Basins, Volume 2*. AAPG Memoir, in press.

——, GERSCHÜTZ, S., STANISTREET, I. G. & LORENZ, V. 1998*b*. Tectonic and volcanic controls on early Jurassic rift-valley lake deposition during emplacement of Karoo flood basalts, southern Namibia. *Palaeogeography, Palaeoclimatology, Palaeoecology*, **140**, 185–215.

SWART, R. 1992/93. Note: Cretaceous synvolcanic conglomerates on the coastal margin of Namibia related to the break-up of West Gondwana. *Communications of the Geological Survey of Namibia*, **8**, 137–141.

TURNER, J. R. 1990. The geological evolution of the upper Jurassic syn-rift interval within the Infanta Embayment, off-shore Republic of South Africa. *Abstract Geocongress '90 Cape Town*. Geological Society of South Africa, Johannesburg, 558–561.

VON HUENE, F. 1925. Ausgedehnte Karoo-Komplexe mit Fossilführung im nordöstlichen Südwestafrika. *Centralblatt für Mineralogie, Geologie und Paläontologie*, **5**, 151–156.

WARD, J. D. 1989. *Aeolianites of the Mesozoic Etjo Formation, Damaraland and Contemporary Dunes of the Skeleton Coast*. Dunes '89 Excursion 4B Field Guide, Geological Survey of Namibia, Windhoek.

WICKENS, H. DE V. & MCLACHLAN, I. R. 1990. The stratigraphy and sedimentology of the reservoir interval of the Kudu 9A-2 and 9A-3 boreholes. *Communications of the Geological Survey of Namibia*, **6**, 9–22.

ZALÁN, P. V., WOLFF, S., ASTOLFI, M. A. M. *ET AL.* 1990. The Paraná basin, Brazil. *In*: LEIGHTON, M. W., KOLATA, D. R., OLTZ, D. F. & EIDEL, J. J. (eds) *Interior Cratonic Basins*. AAPG Memoir, **51**, 681–708.

Facies architecture of the Etjo Sandstone Formation and its interaction with the Basal Etendeka Flood Basalts of northwest Namibia: implications for offshore prospectivity

DOUGAL A. JERRAM[1,3], NIGEL MOUNTNEY[2] & HARALD STOLLHOFEN[1]

[1]*Institut für Geologie, Universität Würzburg, Pleicherwall 1, 97070 Würzburg, Germany*
[2]*Stratigraphy Group, Department of Earth Sciences, University of Liverpool, Brownlow Street, Liverpool L69 3BX, UK*
[3]*Department of Geological Sciences, University of Durham, South Road, Durham DH1 3LE, UK*

Abstract: The Basal Etendeka Flood Basalt stratigraphy in the Huab Basin of northwest Namibia comprises a series of lava flows interleaved with aeolian sandstone bodies of the Etjo Sandstone Formation. The sandstone units are characterized by three main types: (1) the major erg – a mixed aeolian and fluvial facies up to 150 m thick; (2) minor ergs – aeolian facies which occur directly above the first volcanic units and are up to 60 m thick; (3) isolated bodies – multidune, single dune and lava topography infills. A variety of bypass surfaces identified by sand-filled cracks and sediment–lava breccias occur on lava top surfaces. Preserved ripples and pahoehoe lava imprints indicate that the aeolian sand dunes were actively migrating during basalt emplacement.

Observations recorded in the Basal Etendeka Flood Basalts which may be of relevance to offshore hydrocarbon exploration include: a major–minor erg relationship resulting in large sandbodies up to 60 m thick which occur directly after the first volcanic units; the occurrence of sand-filled fissures up to 36 m in depth which would greatly influence connectivity in an offshore setting; the identification of bypass surfaces as marker horizons which may laterally correlate with isolated sandbodies.

Lava flows erupting onto and interacting with unconsolidated sediments have been documented from a variety of geological settings (Busby-Spera & White 1987; Yamaguchi & Lawrence 1993; Riggs *et al.* 1997). Sediment interlayers that occur between successive volcanic units are important because they often provide the only record of the environment and climatic conditions into which the volcanics were erupting. The volcanic units themselves are important because they have the potential to preserve sedimentary features not normally recorded in the geological record (Stollhofen *et al.* 1998*a*). Of recent interest is the recognition of hydrocarbon plays which involve a stratigraphy composed of sediment–lava interbeds, such as the Kudu Gas Field, offshore Namibia (Wickens & McLachlan 1990; Miller 1996). Understanding the relationship between sediments interbedded within lava flows, how such facies architecture is characterized, and the various types of interaction which occur between the lava and sediment in such a system, will have important implications for exploration and development of hydrocarbon reservoirs which contain lava–sediment stratigraphy.

This study reports on a well-preserved sequence of aeolian sandstones interbedded within basaltic lava flows in the Basal Etendeka Flood Basalt stratigraphy of northwest Namibia. It is based on data collected from the Huab outliers and Huab River basin areas of Damaraland (Fig. 1), collectively termed the Huab Basin (Horsthemke *et al.* 1990), and forms part of an ongoing study on the Basal Etendeka Flood Basalt stratigraphy in northwest Namibia (Jerram & Stollhofen 1997). In the context of this study, the flood basalts are defined as the stratigraphic units between the Karoo–Etendeka unconformity marking the base of the Etendeka Group (Milner *et al.* 1994), and the occurrence of the first acidic volcanic unit, the Goboboseb quartz latite (Milner *et al.* 1992). This 300–400 m thick section encompasses the entire zone where lava sediment interaction is preserved within the region, and forms the base of the Awahab Formation (Milner *et al.* 1994). The main aims of this paper are: (1) to investigate the syndepositional interaction between basalt lava and aeolian sand; (2) to characterize the stratigraphy in terms of vertical and lateral continuity and connectivity of sandbodies; (3) to consider the implications of the Basal Etendeka Flood Basalt stratigraphy on the development and understanding of offshore analogues for sediment–lava type hydrocarbon reservoirs.

Geological setting and general stratigraphy

The study area is located in the Rocky Namib Desert in northwest Namibia (Fig. 1) and comprises an area in excess of 5000 km^2. Figure 2a

JERRAM, D. A., MOUNTNEY, N. & STOLLHOFEN, H. 1999. Facies architecture of the Etjo Sandstone Formation and its interaction with the Basal Etendeka Flood Basalts of northwest Namibia: implications for offshore prospectivity. *In:* CAMERON, N. R., BATE, R. H. & CLURE, V. S. (eds) *The Oil and Gas Habitats of the South Atlantic.* Geological Society, London, Special Publications, **153**, 367–380.

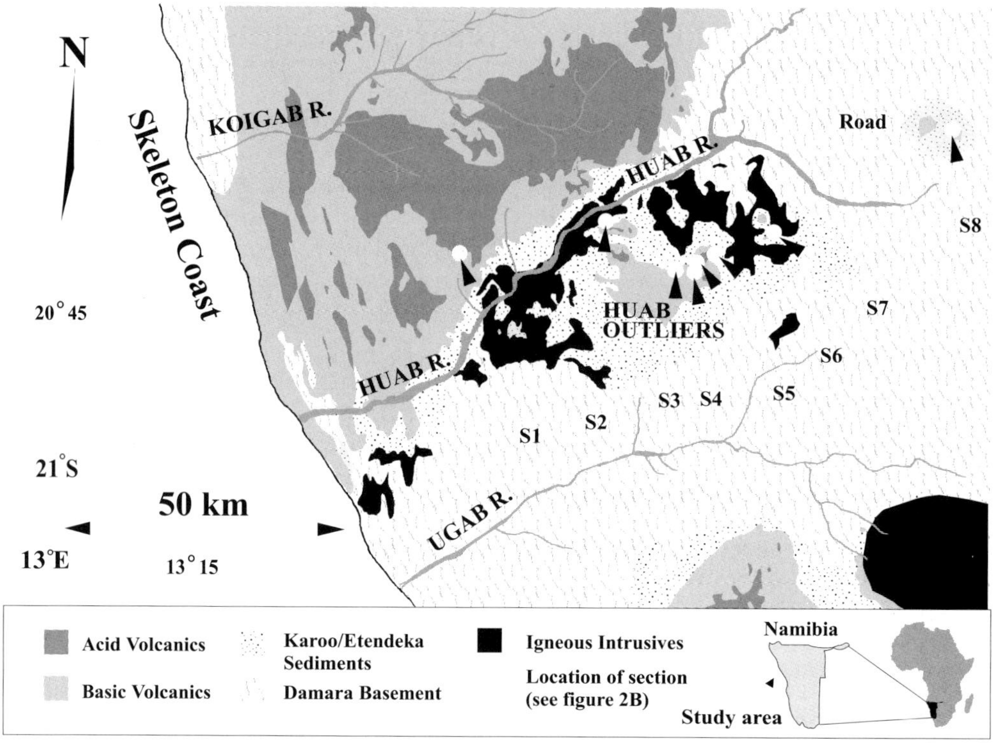

Fig. 1. Simplified geological map of the Huab Basin adapted from Milner (1992). See Fig. 2b for details of logged sections S1–S8.

shows a schematic section through the Huab River Basin, adapted from Horsthemke *et al.* (1990), which indicates the major stratigraphic units within the region. Previous stratigraphic studies within the area have concentrated on the Late Proterozoic Damara sequences (Miller 1983; Swart 1992), the Upper Palaeozoic Karoo sediments (Horsthemke *et al.* 1990) and the higher Etendeka Volcanic units (Milner *et al.* 1992). There is no detailed stratigraphic work, however, on the lowermost basalt units of the Etendeka, and studies on the Etjo Sandstone Formation have tended to be restricted to more regional projects.

Frets (1969) and Hodgson (1970, 1972) were the first to map the geology of the Huab region in any detail, and to develop a detailed stratigraphic subdivision. The South African Committee for Stratigraphy (SACS) describes the Etjo Sandstone Formation as 'a prominent sandstone unit of presumed aeolian origin which overlies any... Karoo sediments that may be present and underlies the volcanic Etendeka [lavas]' (SACS 1980, p. 543). Horsthemke *et al.* (1990) concede that the stratigraphic position of the Etjo Sandstone Formation is somewhat uncertain and that any age between Upper Triassic and Lower Cretaceous could be possible. The Etjo Sandstone Formation and overlying Etendeka lavas have traditionally been assigned to the upper part of the Karoo Supergroup (SACS 1980). However, Milner *et al.* (1994) have upgraded the Etendeka succession to group status and have assigned the Etjo Sandstone Formation as the basal formation of the Etendeka Group (Fig. 2a). The upper part of the Etjo Sandstone Formation interfingers both laterally and vertically with the flood basalts of the Etendeka–Paraná Igneous Province. Importantly, the volcanic units directly above the Basal Etendeka Flood Basalts have been dated, using $^{40}Ar/^{39}Ar$ on mineral seperates, at 131–133 Ma (Renne *et al.* 1996) which, providing there are no significant periods of non-deposition within the volcanic sequence, would date the sediment interlayers in this region as early Cretaceous in age. This is taken to be the age for the cessation of aeolian dune activity at the top of the Etjo Sandstone Formation. Although the duration, and thus the age of the base, of the Etjo Sandstone Formation is unclear, it seems likely that a considerable length of time is encapsulated within a major unconformity that divides the base of the Etjo

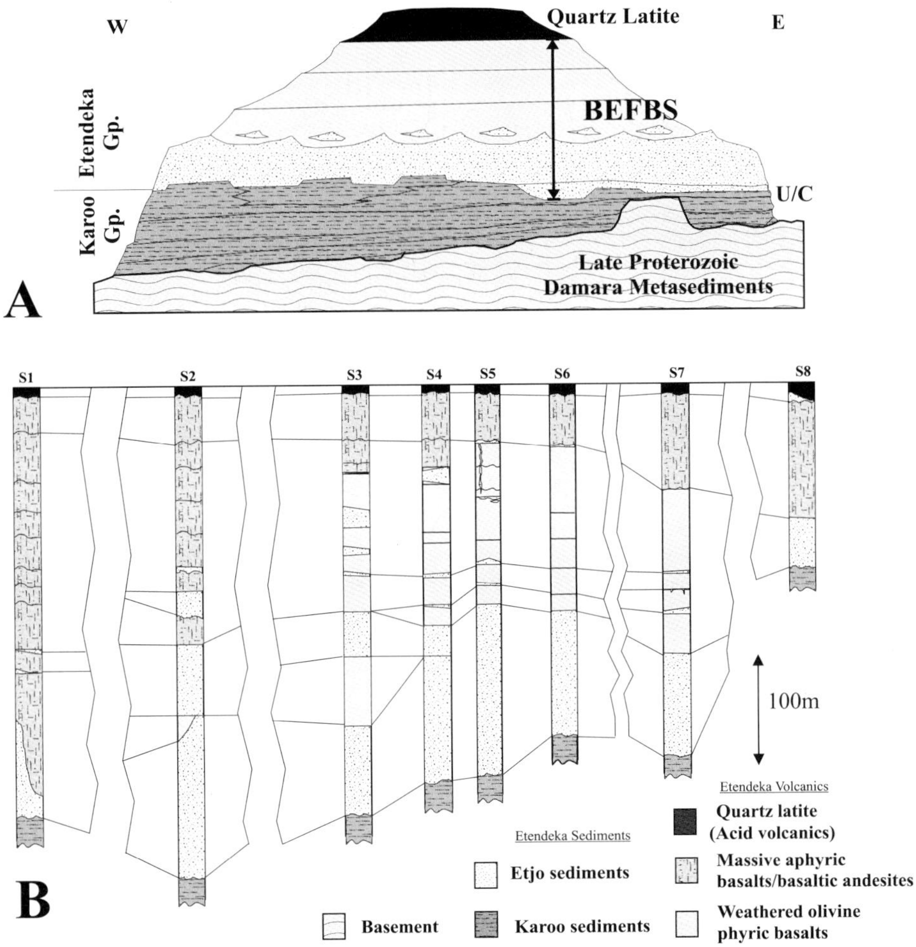

Fig. 2. Stratigraphic subdivision of the Basal Etendeka Flood Basalt stratigraphy in the Huab Basin. (**a**) Schematic east–west cross-section of the Huab area adapted from Horsthemke *et al.* (1990). (**b**) Eight lithostratigraphic sections (maximum thickness 400 m) in an approximate west–east transect *c.* 100 km across the Huab Region (see Fig. 1 for section locations).

Sandstone Formation from sediments of the underlying Karoo Group. The presence of this major hiatus together with the intimate association of the aeolian sands with the Etendeka lavas, indicates that the Etjo Formation is more closely affiliated with the Etendeka Group than with the Karoo Group (Milner *et al.* 1994).

The Etendeka volcanic units comprise a bimodal suite of basaltic lavas and quartz latite rheoignibrites (rhyolites), which have been correlated with units of the Paraná volcanic sequence of South America (Milner *et al.* 1995). Most of the work on the Etendeka volcanic units in the Huab Basin has concentrated on the acid volcanic units, a series of quartz latite flows, which occur in the middle to upper parts of the volcanic stratigraphy (Milner *et al.* 1992). The geochemistry and age of these quartz latite flows was a critical factor which enabled correlation across the African and South American continents (Milner *et al.* 1995). It should also be noted that similar stratigraphy, i.e. aeolian sands interbedded with lava, has been reported from the Paraná Basin (Milani *et al.* 1994).

The units within the Huab Basin comprise the Awahab Formation with units of the Tafelberg Formation onlapping in the area north of the Huab River (Milner *et al.* 1992, 1995). As yet there is no detailed stratigraphic study of the Basal Etendeka Flood Basalts in the Huab Basin. However, a recent study by Ewart *et al.* (1998) on similar stratigraphy from the Goboboseb

Mountains to the south indicates the occurrence of two types of basalt. Based on geochemical discrimination these basalts have been grouped into the high Ti–Zr ratio (olivine phyric, termed LTZ.H basalts) and the low Ti–Zr ratio (termed LTZ.L basalts) (Ewart *et al.* 1998). The basalts in the Huab Basin show two distinctive field relation types and may correlate with the two geochemical types observed in the Goboboseb sequence. The lowermost basalts consist of olivine phyric compound lava flows which locally display pahoehoe lava structures (possibly LTZ.H equivilent). These lavas readily weather in the field leaving poorly exposed scree covered outcrops. In contrast, the upper basalts form well-exposed outcrops and consist of fine-grained aphyric basalts, with a massive lower and middle part of the flow, and a rubbly vesicular-rich top (possibly LTZ.L equivilent). For simplicity, in the rest of this study, the lower olivine phyric basalts will be termed the 'weathered' basalts and the fine-grained upper basalts the 'massive' basalts. As is shown later, the various types of basalt lava, and consequently the style of volcanic eruption, has had an important effect on the distribution and preservation of sediment interlayers.

Facies architecture of the Etjo sediment interlayers

The Basal Etendeka Flood Basalts stratigraphy is highlighted in detail in Fig. 2b by a series of eight sections in an approximate east–west transect. The preserved thickness of stratigraphy containing sands of the Etjo Sandstone Formation varies from 100 to 400 m across the study area. Lava–sediment interaction occurred in parts of the basin throughout the flood basalts and can comprise as much as 300 m of the stratigraphy. So far, no sediment interlayers have been found in the higher volcanic units above the first quartz latite flow. In this paper the morphology of the sand units within the flood basalts, the various lava–sediment contacts which are preserved, and the characteristic vertical and lateral sandbody relationships observed in the Huab Basin are described.

Major Erg

The main dune complex occurs at the base of the sequence, immediately overlying either Karoo sediments or Damaran basement, but underlying the lowermost lavas. It is characterized by up to 150 m of quartz-rich, medium- to coarse-grained, yellow–white sandstone, that is both well sorted and well rounded, commonly exhibiting a 'millet seed' texture. The sandstones are usually only moderately cemented, though are occasionally more tightly cemented by secondary calcite growth, which markedly reduces porosity. Beds are characterized by well developed large-scale planar–tabular cross-bedding (Fig. 3a) and are typically 5–10 m in thickness, though sets of up to 50 m are observed. Throughout the region, palaeocurrent analysis reveals an average foreset dip of 24° towards 040°, findings that closely match those of Hodgson (1972). Cross-bedding is accentuated by the presence of well developed grain flow and grain fall horizons (Hunter 1977).

Sand flows comprise upper medium to lower coarse-grained, moderate to well sorted, steeply dipping, cross-laminated sands (25–34°) that are typically 5 cm in thickness. The units commonly display a slight inverse grading, although the lack of variance in the grain size of particularly well sorted examples may preclude this. The high degree of sorting, together with the relatively coarse grain size, results in porosities in excess of 30%. Sand flows from the most significant lamination type within the major erg, accounting for 80% of the deposits. Grain fall deposits comprise very fine to fine, well sorted, homogeneous, steeply dipping cross-laminae, that are typically 5 mm in thickness. Individual laminae are tabular and, where bed-scale permits may be traced both down depositional dip and along-strike for tens of metres. In rare examples, where grainfall laminae are underlain by irregularly bedded horizons, they drape the irregularity rather than lap against its margins. Grain fall accounts for 10% of the deposits of the major erg. Ripple cross-laminated sands with low ripple amplitudes (< 5 mm and high ripple indices (RI 10–35) are also observed. Deposits are commonly bimodally sorted and may display a distinctive pinstripe lamination. Ripple cross-laminated deposits account for 10% of the deposits of the major erg, though are restricted in extent to low angle sandsheet and interdune regions.

Sets of cross-strata comprising lamination of one or more of the above types make up individual beds. Cross-strata may be either simple or compound in geometry. Simple cross-strata lack internal angular truncation surfaces whilst compound cross-strata are split into a series of sets, each of which is bounded by an angular truncation surface.

Well sorted and rounded, large-scale cross-stratification, comprising grain fall and grain flow lamination, is interpreted to represent aeolian dune deposition. The narrow range of palaeomigration readings indicates a sediment transport direction to the northeast, via unidirec-

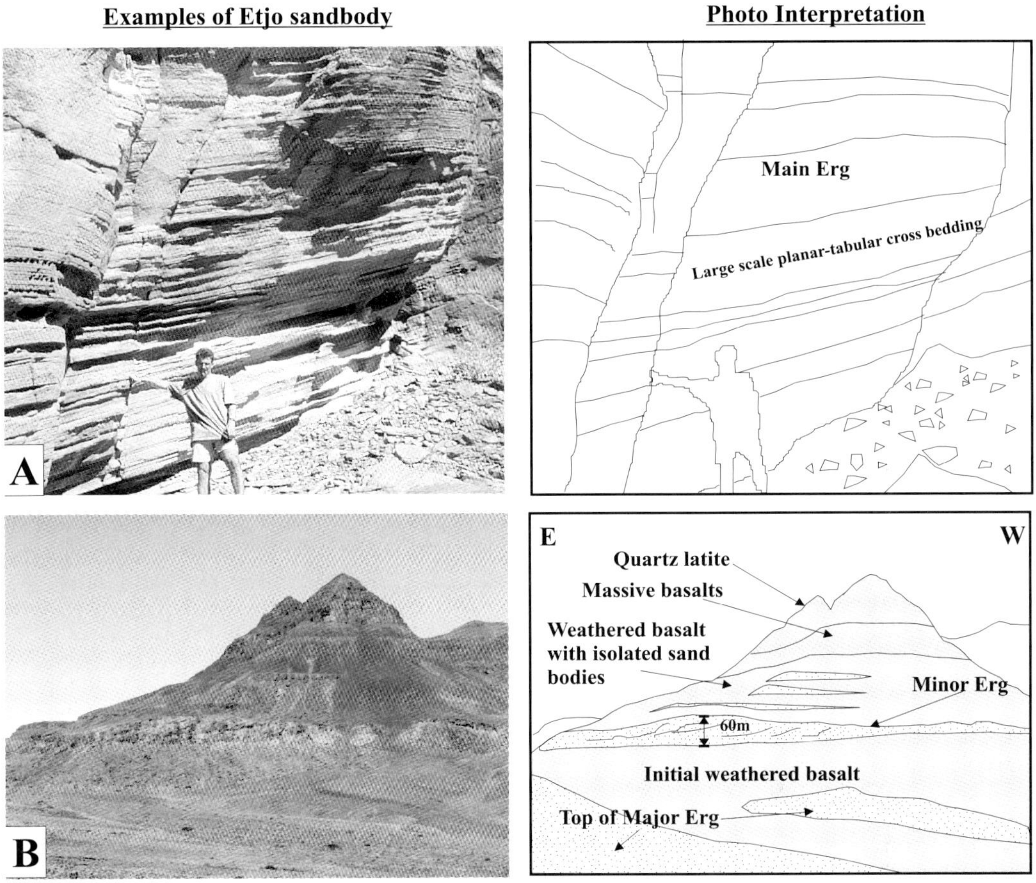

Fig. 3. (**a**) View of major erg showing shallow-dipping large-scale planar–tabular cross-bedding. (**b**) View across first volcanic unit from the top of the major erg. A minor erg can be seen in the distance reaching a thickness of 60 m.

tional winds. The lack of spread in the readings also implies the migration, across much of the region, of large transverse draa bedforms with the dip of slipfaces aligned parallel to the wind and straight-crested ridge lines aligned perpendicular to the dominant wind direction (McKee 1979). Whilst bed boundaries represent draa-scale (first-order) migration surfaces, lower order angular truncations that divide sets of cross-strata are considered to represent reactivation surfaces that result from minor changes in wind orientation (Kocurek 1996).

Minor Ergs

These are laterally continuous major units of aeolian sand dunes which occur directly above the main erg in the succession. They are distinct from the main erg because they occur after the very first period of basaltic lava activity in the region. This lava activity was laterally discontinuous and subsequently the minor ergs may lie either directly on the lowermost basaltic lavas or directly on the main erg itself. In areas where minor ergs lie directly on the main erg it is very difficult to distinguish the two units. In this case, the two successions are separated by a bounding surface. The minor ergs may be up to 60 m thick and are bound on their top surface by flood basalts. In the example shown in Fig. 3b, the lowermost lava unit thins to the east and further eastwards the lava pinches out leaving this minor erg lying directly on the major erg. The migration directions of bed foresets are predominantly toward the northeast, as in the major erg.

Isolated aeolian sandbodies

Sedimentary interlayers preserved above the minor ergs occur as a series of small isolated bodies.

These take on three distinct forms; multidune, single dune and lava topography infills (Fig. 4), and are mainly concentrated in the Huab Outliers area.

Isolated multidunes. Isolated multidunes comprise more than one sand dune and are distinguished from the minor ergs by their size, and by the fact that they are entirely enclosed within the basalts. These units reach up to 20 m in thickness and may have a lateral extent of up of 1 km. They occur mainly within the middle portions of the Basal Etendeka Flood Basalts stratigraphy. The isolated multidune bodies represent a transition from the major sand units of the lower flood basalts and more isolated single dune bodies which occur mainly in the upper flood basalts. This feature indicates a diminishing sediment supply into the basin, with continued lava flooding.

Isolated single dunes. These are solitary migrating sand dunes which have been engulfed by overlying basalt flows. These units are laterally discontinuous and commonly show characteristic barchanoid or solitary barchan dune geometries. Figure 4c shows an isolated barchan dune form, preserved *in situ*, as it migrated towards the east-northeast by overlying basalts which have subsequently weathered away to the level of the dune. Isolated dune forms have been recorded up to 10 m thick. These can occur throughout the middle–upper Basal Etendeka Flood Basalts stratigraphy.

Lava topography infills. Lava flow top surfaces are horizontal on a regional scale, though locally show small topographic variation. In some cases the topographic variation can be as much as 20 m between lava ridges, i.e. where flows have been inflated by lateral magma injection (Gerschütz 1996; Stollhofen *et al.* 1998*b*), but more commonly is < 5 m (Hon *et al.* 1994; Keszthelyi & Denlinger 1996). Aeolian sand blowing onto newly erupted lava flows will infil topography to form this type of interlayer, termed lava topography infills. These interlayers are characterized by a planar top surface and an uneven base where the sand has blanketed the irregular lava top surface collecting in the topographic lows (Fig. 4b, diii). It is also possible that topographic variation enhances the entrapment of sand leading to the onset of isolated dune bodies.

The petrology of the isolated sediment interlayers is similar to that of the major erg. The sands are predominantly over 95% quartz and show no obvious signs of inclusion of volcanic material into the sediment source, although a few > 2 cm thick beds have been observed containing basalt scoria clasts, but these are very localized occurrences. The migration direction of the aeolian sandbodies changes from a predominant northeast migration direction, similar to that of the main and minor erg systems, to a south-southeast migration direction. This inferred change in wind direction may represent a major change (i.e. not just seasonal variation) in primary wind direction. However, it may also be controlled by the restriction of sediment supply from the dominant upwind direction (the southwest) by lava increasingly flooding and covering the source region. Potential lava sources exist to the south of the region [e.g. the Messum Complex (Milner & Ewart 1989)] which would have the capacity to cover large parts of a southwest source region, resulting in a lower sediment yield carried by secondary winds from a source to the west and northwest of the Huab region.

Bypass surfaces

Bypass surfaces are lava flow top surfaces which contain evidence of aeolian sand migration across them. The majority of the sand is inferred to have bypassed the surface, with only minor amounts of sand being preserved within sand-filled cracks and in isolated pockets of brecciated lava with a sand matrix, and very thin (< 1 m) highly discontinuous drapes of sediment. An example of such a bypass surface is presented in Fig. 5, which shows a complicated arrangement of sand-filled cracks. These bypass surfaces can be found throughout the Basal Etendeka Flood Basalts stratigraphy but are mainly concentrated in the upper units. Often, such surfaces can be followed between individual isolated coeval sandbodies and therefore can be used as correlative bypass surfaces to identify sandstones of the same age.

The lava–sediment contact

Commonly, the contact between the lava and the sand is brecciated where the lava has cooled rapidly on contact with the unconsolidated sand. Such breccia zones are typically < 1 m in thickness, though can occur up to 4 m thick (Fig. 6a), and show evidence of sand baking and basalt loading into soft sand.

Locally, the aeolian sand has been indurated to form a quartzite by the hot lava. On other surfaces the lava flow has left striations in the sand, indicating local flow directions. Figure 6b shows an example from the top of the main erg where bifurcating ripples have been preserved by the

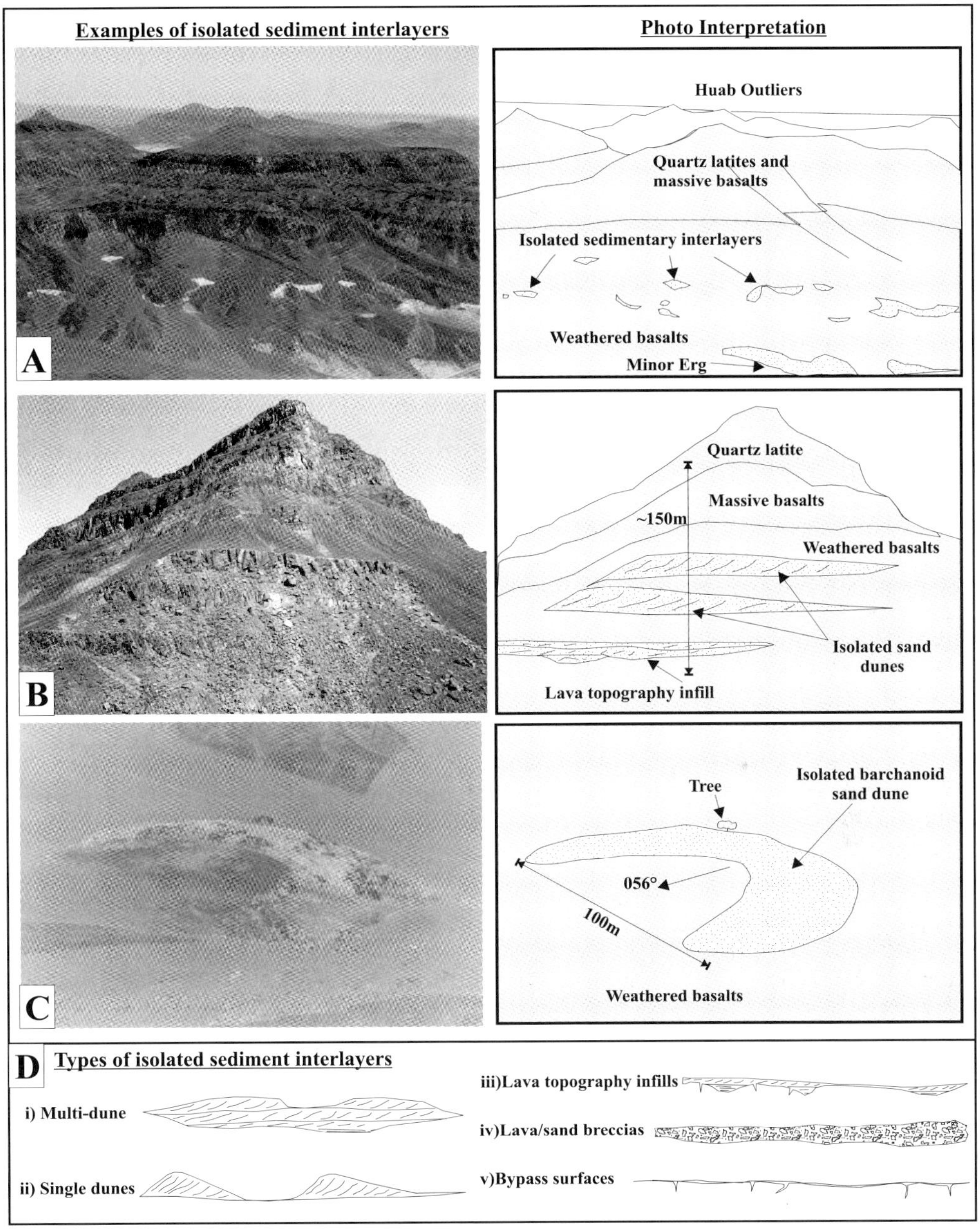

Fig. 4. Examples of the sediment interlayer stratigraphy preserved in the Huab Outliers. (**a**) View looking southwest from the top of Awahab. (**b**) Cross-sections through two isolated dunes and a lava topography infill. (**c**) An example of an isolated migrating barchanoid sand dune 'frozen' *in situ* by lava. (**d**) Schematic examples of the types of interlayer and lava surface found.

flooding of a low viscosity pahoehoe basalt flow over the surface. The surface shows imprints from the basalt flow of pahoehoe structures (ropy texture), and fine striations on the sand surface indicating the local flow direction.

The idealized stratigraphic section

It is possible to build up idealized stratigraphic sections for the Huab Outliers and Huab River areas (Fig. 7) based on detailed field observa-

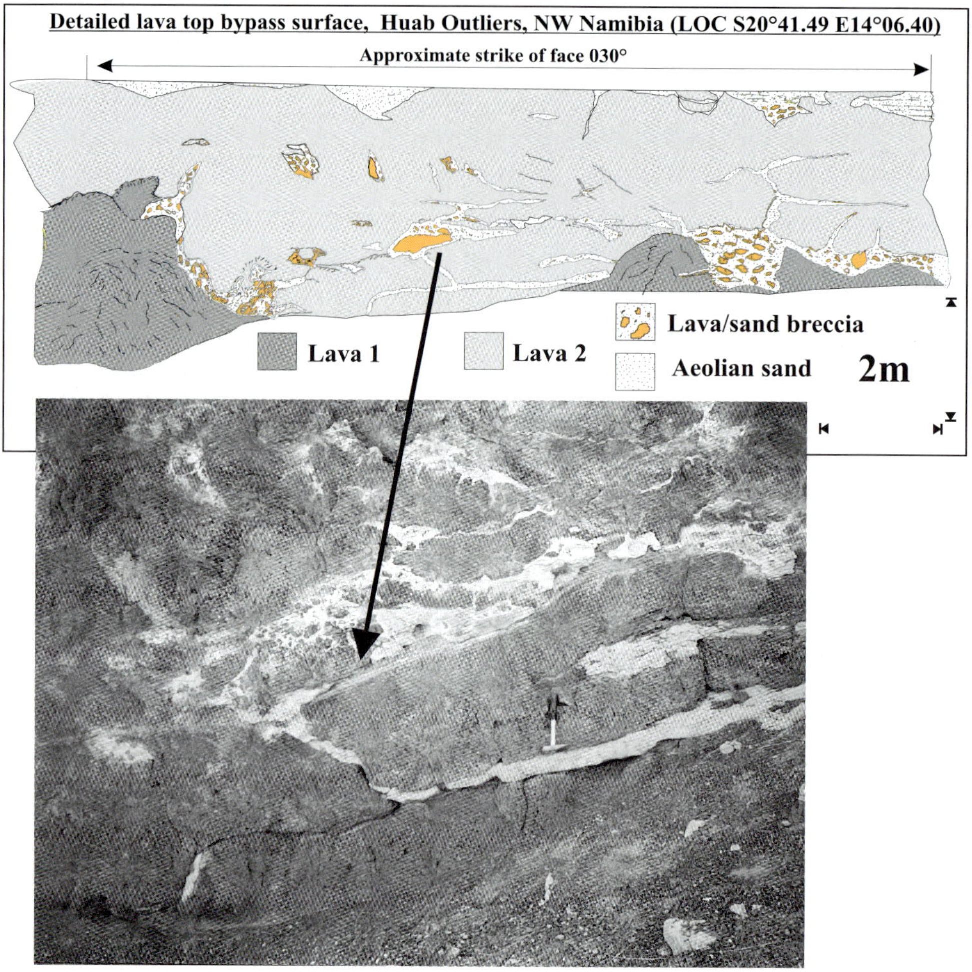

Fig. 5. Detailed section of a bypass zone showing the complex network of sand-filled cracks, sand–lava breccias and lava morphology preserved.

tions, lithostratigraphic (thickness) sections and the discrimination of volcanic flow types.

Huab Outliers section

The Huab Outliers section is presented in Fig. 7a and comprises the stratigraphy with the most intensive sediment–lava interaction preserved. The Basal Etendeka Flood Basalts stratigraphy in this part of the basin is between 250 and 300 m thick. The main erg lies uncomformably on top of Karoo sediments and is overlain by a minor erg system directly after the first volcanic incursion. Basalts in the Huab Outliers section consist mainly of the olivine–phyric weathered basalt type. It is within these compound flow basalts that the sediment–lava interaction is concentrated, with isolated multidune, single dune and lava topography infill units. Massive basalt units occur at the top of the stratigraphy immediately below the first quartz latite.

Huab River section

The Huab River section contains more massive basalt units. Again, the main erg is overlain by a minor erg relationship in the lower parts of the stratigraphy. However, there is very little preserved sediment–lava interaction in the middle and upper parts of the section. In this case, the

Fig. 6. Examples of the contact zones between lava and sand. (**a**) Complex lava breccia zone up to 4 m thick. Aeolian sand forms the matrix of the breccia, and a number of localized features (such as glassy rinds on the lava clasts) suggest that the sand and lava were intermingling as the lava flow was emplaced. (**b**) Top surface of the major erg where the lava has baked the sand and preserved bifurcating ripples. The surface also shows striations left by the lava migrating over the surface.

style of volcanic unit appears to be a controlling factor on the preservation of sediment interlayers. The more tabular geometries of the massive basalt units and their greater lateral extent restricts the amount of sand entering and being entrapped by topography.

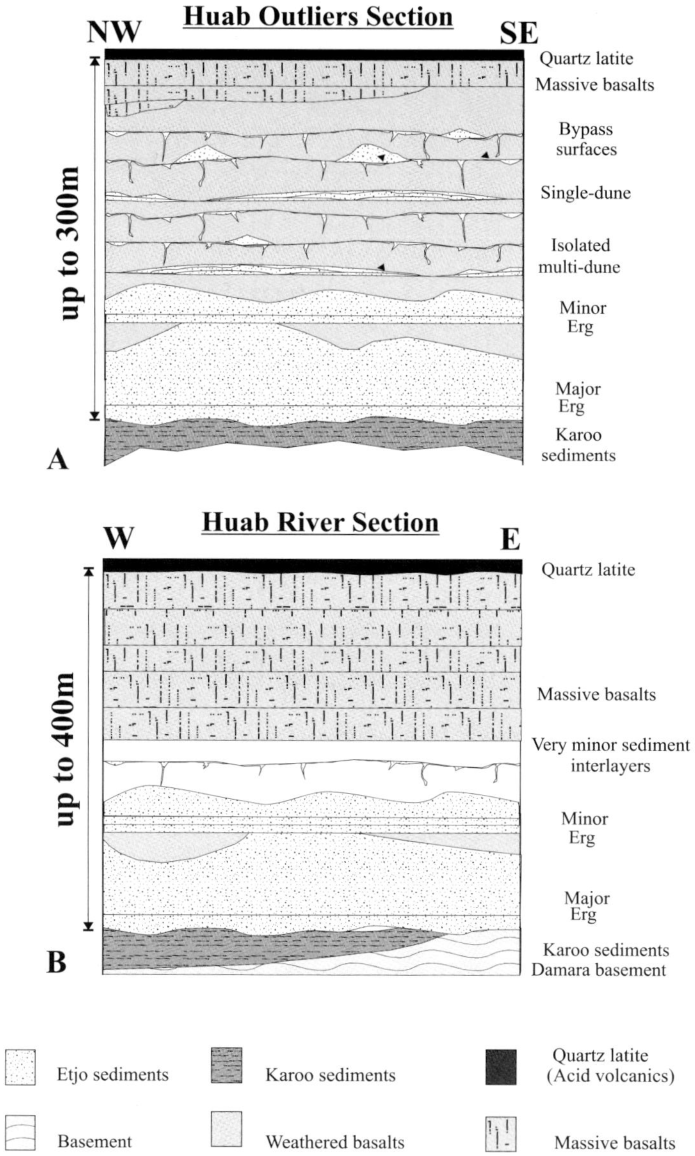

Fig. 7. Diagrammatic representations of the characteristic stratigraphy found in the Huab Basin. (**a**) The Huab Outliers section. (**b**) The Huab River section.

Lava–sediment interactions: Implications for offshore analogues

The Huab Basin provides an ideal opportunity to study volcano–sedimentary interactions between basaltic lava flows and interbedded aeolian deposits. Outlined below are the main implications of this study with regard to improving the understanding of offshore analogues. Particular reference is given to published data from the Kudu Gas Field, offshore Namibia.

The Major Erg–Minor Erg relationship: understanding the system

The occurrence of minor ergs in a volcanic–aeolian system are potentially an integral part of the stratigraphy which develops as the erg system is progressively flooded by basalts. Minor ergs are also good potential reservoirs. A schematic representation of the volcanic–aeolian system is presented in Fig. 8. The initial sand system, prior to the first lava flows, can be described according to a sand budget which contains a zone of aeolian accumulation and a zone of sediment transport where sand bypasses the system after some residence time in migrating dunes. These two zones are separated along the base level of the system. The sediment flux is the difference between sand entering the system from the source region and sand leaving the system at the downwind basin margin, and may be considered in terms of area or volume. For the purpose of this exercise an initial neutral sediment flux is assumed, i.e. no net long-term deposition prior to lava emplacement, and development of the system as it becomes progressively flooded by lava flows considered (Fig. 8).

The initial topographic variation in a major erg, between dune crests and interdune lows, can be as much as 250 m (e.g. the modern Namib desert in central-southern Namibia), and is indicated in Fig. 8a as the height of sand (Hs). The earliest lava flows entering such an erg system are unlikely to have the capacity to flood the entire topography, and would pond and flow within the interdune depressions of the erg. Sand above the lava level (Hl) infilling the erg will remain mobile and may be potentially reworked between eruptions. Providing that there is a reasonable time gap between successive eruptions, smaller 'minor erg' systems will evolve along the level of the first lavas from both mobile sand already in the system and from any sand entering the system from an outside source (Fig. 8b). Due to decreased sand availability, the minor ergs will be volumetrically smaller than the major erg system. Subsequent lava flows will therefore have a greater potential to engulf the minor erg topography, a case where lava height (Hl) approaches sand height (Hs). Once this has occurred, sand in transport will be severely restricted and may originate from the extreme highs of the minor ergs not engulfed by the lava, and from external sources not affected by lava coverage. The further volcanic units may preserve minor sandbodies of various types (e.g. isolated dunes, lava flow topography infills, etc.) to a greater or lesser extent. The identification of stratigraphic position, geometry and sandbody type within an aeolian sand–lava system is thus critical to the location of large 'economic' sandbodies. An understanding of the geometry will aid potential identification of sand bodies on two- and three-dimensional seismic profiles.

Correlative bypass surfaces

The discontinuous and isolated nature of the sediment interlayers in the middle and upper portions of the Basal Etendeka Flood Basalts stratigraphy makes it difficult to identify sandbodies of the same age, and thus the same stratigraphic position. The isolated nature of these sandbodies, which could potentially make good hydrocarbon reservoirs, means that they would be easily missed in an offshore well. It will, therefore, be of great importance in any well examination to recognize bypass surfaces between successive lava flows (e.g. Fig. 5). The recognition of a bypass surface gives a marker horizon within the stratigraphy where sand was clearly migrating over an exposed lava flow top. Such an horizon may correlate laterally with isolated sandbodies of the same age, as has been found in the Huab Outliers, and may thus form important correlative bypass surfaces leading to volumetrically important, sealed sandbodies. For example, Wickens & McLachlan (1990) report the occurrence of sandstone veins (sand-filled cracks) in the basalt interval of the Kudu 9A-3 well, which may represent similar such horizons as reported here.

Sand-filled fissures

Sand-filled fissures occur throughout the lower weathered basalts in the Huab Outliers. Examples of such sand-filled fissures are illustrated in Fig. 5, and schematically in Fig. 2b) (section S5). Although the majority of these fissures only penetrate a few metres into the top surface of lava flows, some examples (such as that in section S5, Fig. 2b) are up to 1 m in width and exceed 36 m in depth, cutting down through at least two different flow units. Large sand-filled fissures tend to be found in close proximity to each other in concentrated zones of tectonic influence. These zones run often in a north–south orientation, parallel to the rift axis along which South America and Africa finally separated (Milner *et al.* 1995), and suggest a tectonic control on fissure distribution. The uppermost volcanic units show a north–south striking flow direction, identified by Glen *et al.* (1997) using anisotropy of magnetic susceptibility (AMS), which further supports some rift influence on lava distribution. The majority of fissures appear to have been filled some time after the cooling of the lava. However, a few examples of baked sand-filled fissures can be found where sand has been in transport immediately prior to the emplacement of the lava, thus infilling the fissures as they developed and while they were still hot.

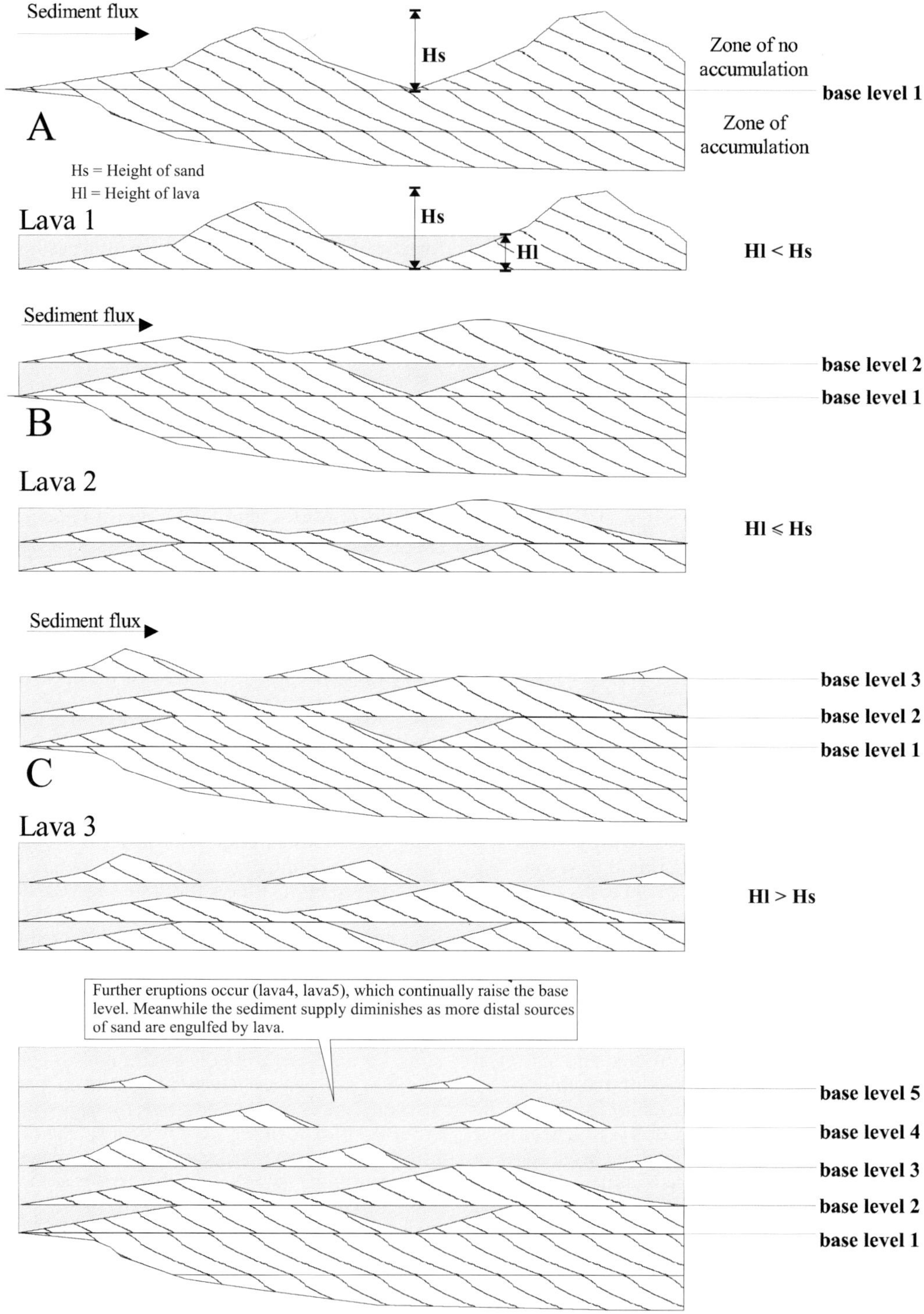

Fig. 8. Schematic representation of the development of the sediment–lava system. See text for full explanation.

Although the tectonic settings of the Huab Basin (rift shoulder and associated basins) and the Kudu reservoir (rift valley) are different, it is possible that volcano-tectonic movements during successive lava eruptions and syn-depositional extension that occurred during the formation of the Kudu stratigraphy will have resulted in similar fissure development. However, the spatial distribution and extent of fissure development may be different. The very nature of the large sand-filled fissures, 1 m thick in an approximately vertical position, means that it will be highly unlikely to intersect one while drilling a well. However, such features are considered to be potentially extremely important conduits for hydrocarbon migration, which will greatly affect vertical connectivity within a reservoir.

Conclusions

This contribution highlights the complex nature of aeolian and volcanic interaction in the Huab Basin, with a view to developing an understanding into how this relates to offshore reservoirs with similar stratigraphy. As yet, the major exploration–production reservoir containing similar stratigraphic relations, i.e. aeolian–lava interaction, is the Kudu Gas Field, offshore southern Namibia. However, sedimentary horizons interleaved with the Kudu basalts are not entirely characterized by fluvio-aeolian facies, but also show a variety of lacustrine lithologies, including elements such as algal stromatolites, bivalves and evaporites (Wickens & McLachlan, 1990). The main facies assemblage of the Kudu gas reservoir is suggested to be transitional in its palaeoenvironmental setting between the predominantly aeolian interbeds of the basal Etendeka flood basalts in northwest Namibia, and the fluvio-lacustrine interbeds of the Kalkrand Basalt Formation in southern Namibia, as discussed by Stanistreet & Stollhofen (1999).

This research has been undertaken during a Deutsche Forschungs-Gemeinschaft (DFG)-funded post-doctoral position at the Geoscience Graduiertenkolleg of the University of Würzburg, and their sponsorship is gratefully acknowledged. All of us benefited from logistical support from the Geological Survey of Namibia, the Ministry of Environment and Tourism, and Nature Conservation (particularly from the rangers D. Rudman and J. Paterson, from the Skeleton Coast Park). D. A. J. would like to thank O. Harvey, S. Knight, J. Garland and R. Jerram for field assistance. S. C. Milner, R. Swart, I. G. Stanistreet and A. R. Duncan are thanked for initial introductions to aspects of the geology in the Huab Basin. This study benefited from comments on the original manuscript by A. Duncan, J. Clempson and A. A. Dijksman. The research presented in this manuscript has benefited greatly from discussions with the above-mentioned people and also with V. Lorenz, J. Howell, F. Holzförster, A. Warne and A. Wanke. N. M. thanks British Gas plc and Conoco UK Ltd for their joint sponsorship of his postdoctoral research position at the University of Liverpool.

References

EWART, A., MILNER, S. C., ARMSTRONG, R. A. & DUNCAN, A. R. 1998. Etendeka Volcanism of the Goboboseb Mountains and Messum Igneous Complex, Namibia. Part I: Geochemical Evidence of Early Cretaceous Tristan Plume Melts and the Role of Crustal Contamination in the Paraná-Etendeka CFB. *Journal of Petrology*, **39**, 191–225.

FRETS, D. C. 1969. Geology and structure of the Huab–Welwitschia Area, South West Africa. *Bulletin of Precambrian Research*, **5**.

GERSCHÜTZ, S. 1996. *Geology, volcanology, and petrogenesis of the Kalkrand Basalt Formation and the Keetmanshoop dolerite complex, southern Namibia.* Dr.rer.nat. Thesis, University of Würzburg.

GLEN, J. M., RENNE, P. R., MILNER, S. C. & COE, R. S. 1997. Magma flow inferred from anisotropy of magnetic susceptibility in the costal Parana–Etendeka igneous province: evidence for rifting before flood volcanism. *Geology*, **25**, 1131–1134.

HODGSON, F. D. I. 1970. The Geology of the Karroo System in the southern Kaokoveld, South West Africa. *Proceedings Paper IUGS Second Gondwana Symposium*, Pretoria, South Africa, 233–240.

—— 1972. *The Geology of the Brandberg–Aba Huab Area, South West Africa.* PhD Thesis, University of Orange Free State, Bloemfontain.

HON, K., KAUAHIKAUA, J., DENLINGER, R. & MACKAY, K. 1994. Emplacement and inflation of pahoehoe sheet flows: Observations and measurements of active lava flows on Kilauea Volcano, Hawaii. *Geological Society of America Bulletin*, **106**, 351–370.

HORSTHEMKE, E., LEDENDECKER, S. & PARODA, H. 1990. Depositional environments and stratigraphic correlations of the Karoo Sequence in northwestern Damaraland. *Communications of the Geological Survey of Namibia*, **6**, 63–75.

HUNTER, R. E. 1977. Basic types of stratification in small eolian dunes. *Sedimentology*, **24**, 361–387.

JERRAM, D. A. & STOLLHOFEN, H. 1997. Interactions between aeolian sand and lava flows of the Etendeka Flood Basalts, NW Namibia. *Journal of Conference Abstracts*, **2**, 40.

KESZTHELYI, L. & DENLINGER, R. 1996. The initial cooling of pahoehoe flow lobes. *Bulletin of Volcanology*, **58**, 1–18.

KOCUREK, G. 1996. Desert aeolian systems. *In*: READING, H. G. (ed.) *Sedimentary Environments: Processes, Facies and Stratigraphy*. Blackwell, 125–153.

LEDENDECKER, S. 1992. Stratigraphie der Karoosedimente der Huabregion (NW Namibia) und deren Korrelation mit zeitäquivalenten Sedimenten des Paranabeckens (Südamerika) und des Großen Karoobeckens (Südafrika) unter besonderer Berücksichtigung der überregionalen geodynamischen und klimatischen Entwicklung Westgondwanas. *Göttinger Arbeiten zur Geologie und Paläontologie*, **54**.

McKee, E. D. 1979. An introduction to the study of global sand seas. *In*: McKee, E. D. (ed.) *Global Sand Seas*. US Geological Survey, Professional Paper **1052**, 1–19.

Milani, E. J., Franca, A. B. & Schneider, R. L. 1994. Bacia do Parana. *Boletim de Geociencias da Petrobras*, **8**, 69–82.

Miller, R. McG. 1983. The Pan-African Damara Orogen of South West Africa/Namibia. *In*: Miller, R. McG. (ed.) *Evolution of the Damara Orogen*. Geological Society, South Africa, Special Publication, **11**, 431–515.

Miller, R. M., Carstens, H. & Wolden, T. 1994. Namibia hydrocarbon potential being opened for further exploration. *Oil and Gas Journal*, **92**, 123–128.

Milner, S. C. & Ewart, 1989. The geology of the Goboboseb Mountain volcanics and their relationship to the Messum Complex. *Communications of the Geological Survey of Namibia*, **5**, 31–40.

——, Duncan, A. R. & Ewart, A. 1992. Quartz latite rheoignimbrite Flows of the Etendeka Formation, North-Western Namibia. *Bulletin of Volcanology*, **54**, 200–219.

——, ——, —— & Marsh, J. S. 1994. Promotion of the Etendeka Formation to group status: A new integrated stratigraphy. *Communications of the Geological Survey of Namibia*, **9**, 5–11.

——, ——, Whittingham, A. M. & Ewart, A. 1995. Trans-Atlantic correlation of eruptive sequences and individual silicic volcanic units within the Parana–Etendeka Igneous Province. *Journal of Volcanology and Geothermal Research*, **69**, 137–157.

Renne, P. R., Glen, J. M., Milner, S. C. & Duncan, A. R. 1996. Age of Etendeka flood volcanism and associated intrusions in Southwestern Africa. *Geology*, **24**, 659–662.

South African Committee for Stratigraphy (SACS) 1980. *Stratigraphy of South Africa. Part I. Lithostratigraphy of South Africa, South West Africa/Namibia, and the Republics of Bophuthatswana, Transkei and Venda*. Handbook of the Geological Survey of South Africa, **8**.

Stollhofen, H., Stanistreet, I. G. & Gerschütz, S. 1998*a*. Early Jurassic rift valley alkaline lake deposits interbedded with Karoo flood basalts, Southern Namibia. *In:* Gierlowski-Kordesch, E. & Kelts, K. (eds.) *Global Geological Record of Lake Basins, Vol. 2* AAPG Memoir, in press.

——, Gerschütz, S., Stanistreet, I. G. & Lorenz, V. 1998*b*. Tectonic and volcanic controls on early Jurassic rift-valley lake deposition during emplacement of Karoo flood basalts, southern Namibia. *Palaeogeography, Palaeoclimatolology and Palaeoecolology*, **140**, 185–215.

Stanistreet, I. G. & Stollhofen, H. 1999. Onshore equivalents of the main Kudu gas reservoir in Namibia. *This volume*.

Swart, R. 1992. The sedimentology of Zerrissene Turbidite System, Damara Orogen, Namibia. *Geological Survey of Namibia Memoir*, **13**.

Wickens, H. de V. & McLachlan, I. R. 1990. The stratigraphy and sedimentology of the reservoir interval of the Kudu 9A-2 and 9A-3 boreholes. *Communications of the Geological Survey of Namibia*, **6**, 9–22.

The Namib Rift: a rift system of possible Karoo age, offshore Namibia

J. CLEMSON[1], J. CARTWRIGHT[1] & R. SWART[2]

[1]*Department of Geology, Royal School of Mines, Imperial College of Science, Technology and Medicine, Prince Consort Road, London SW7 2BP, UK*
Present address: Amerada Hess Ltd, 33 Grosvenor Place, London, SW1 7HY, UK
[2]*National Petroleum Corporation of Namibia (NAMCOR), Private Bag 13196, Windhoek, Namibia*

Abstract: This paper describes a hitherto poorly documented, deep-seated rift system offshore Namibia, here named the Namib Rift. The structural history and rift geometry of this system are determined from reflection seismic data. No wells penetrate the rift-fill succession but from its stratigraphic context, regional thickness relationships and by analogy with onshore Karoo stratigraphy, the age of the rift is interpreted as broadly equivalent to the Karoo Supergroup. The rift is divisible into two major arms, with rift segmentation evident in the north. An angular unconformity separates a rift and post-rift sequence. This is interpreted as a Mid-Triassic hiatus related to a regional compressional events. The basement comprises the Pan-African Fold Belt along the coast, with the Damaran arm extending inland. In the Cape Cross–Walvis Bay area, the Autseib Lineament separates the Kaoko and Damaran arms. This lineament, and the adjacent basement fabrics, are shown to have strongly influenced rift geometry in this area. South of the Autseib Lineament, Damaran basement fabrics largely determined fault trends within the Namib Rift. To the north, Kaoko basement fabrics strongly influenced basement extensional faulting. This area was also subjected to Mid-Triassic compression. South of the Autseib Lineament, large-scale uplift was distributed across large folds whereas to the north, uplift occurred close to pre-existing faults. North-northeast directed compression was common to both areas, but the Autseib Lineament partitioned the structural response. McKenzie-type pure shear extension is favoured as the rift mechanism, with Mid-Triassic uplift modifying thermal subsidence. The final compressive event of the Cape Orogeny, coeval with the regional Mid-Triassic hiatus, terminated rifting in offshore Namibia.

The Karoo of southern Africa and South America is generally considered to be the cratonic succession of a large sag basin (Kent 1980) extending across West Gondwana from the Karoo Basin of South Africa to the Paraná Basin of South America. Smaller outcrops of this marine, lacustrine, fluvial and aeolian succession occur in Namibia. The Dwyka and Ecca Groups of the Huab area of Namibia have been correlated with similar successions in the Paraná Basin of South America (Horsthemke *et al.* 1990), i.e. across offshore areas of Namibia. Light *et al.* (1993) suggested that Karoo rift basins may be present offshore Namibia, but they are not described or mapped. These postulated rift basins would represent a precursor phase of extension to the Atlantic rift system that led to continental break-up and the formation of the South Atlantic Ocean.

This paper describes the geometry of a major rift system that pre-dates the main half-graben of the late Jurassic–early Cretaceous Atlantic rift. This system is here named the Namib Rift, to distinguish it from the much better described Atlantic rift system of offshore Namibia (Maslanyj *et al.* 1992; Light *et al.* 1993). Its syn-rift history and the subsequent basin inversion that terminated rift-controlled subsidence is described, and a case presented to suggest that this Namib Rift is most likely of Karoo age. The influence that basement fabrics and lineaments of the Pan-African Fold Belt exerted upon subsequent rift geometries are also highlighted. Inherited basement influence is a well known control on rift architecture in many extensional basins. However, relatively few examples have been documented from orogenic triple junction regions, where complex intersections of orogenic basement structure result in complementary complexity in extensional fault systems.

The regional data comprises 10 430 line-km of speculative data acquired during 1989–1991 in the offshore waters of Namibia. To the southwest of Walvis Bay, a further 430 line-km of 1995 speculative data locally allows a more comprehensive structural interpretation. In addition, there are 2860 line-km of closely spaced proprietary data extending over the formerly Ranger-operated Block 2213A northwest of Walvis Bay. Descriptions of magnetic data are from the Aeromagnetic Anomaly Map, Geological Survey of Namibia 1993. No offshore wells have reached the Karoo succession in Namibia. Onshore, the Toscanini borehole (Figs 1 and 2) penetrated

CLEMSON, J., CARTWRIGHT, J. & SWART, R. 1999. The Namib Rift: a rift system of possible Karoo age, offshore Namibia. *In:* CAMERON, N. R., BATE, R. H. & CLURE, V. S. (eds) *The Oil and Gas Habitats of the South Atlantic.* Geological Society, London, Special Publications, **153**, 381–402.

Cretaceous volcanics and Karoo clastics and terminated after drilling almost 1000 m of Pan-African basement.

Tectonic setting

Structural framework

The Pan-African Fold Belt occurs along much of the Namibian continental margin (e.g. Porada 1979, 1989) (Fig. 1). In central parts of the coastline in the Walvis Bay area, the northeast–southwest oriented Damaran arm of the Pan-African Fold Belt extends offshore and to the south, swings back onshore to form the northwest–southeast oriented Gariep Fold Belt. To the north of Walvis Bay, the north-northwest–southsoutheast oriented Kaoko arm of the Pan-African Fold Belt constitutes the basement terrane. These arms of the Pan-African Fold Belt are arranged to form an orogenic triple junction in the Walvis Bay–Cape Cross area.

It is widely considered that Pan-African basement fabrics exercised significant control on late Mesozoic rift geometries along the South Atlantic passive margin (e.g. Uchupi 1989; Uchupi & Emery 1991; Binks & Fairhead 1992; Maslanyj *et al.* 1992; Standlee *et al.* 1992). They are also demonstrated here to have influenced the fault geometry of the Namib Rift system in Namibia, particularly in the Cape Cross–Walvis Bay area. The onshore basement terranes of this area and their interpretation offshore are therefore described in some detail (see below).

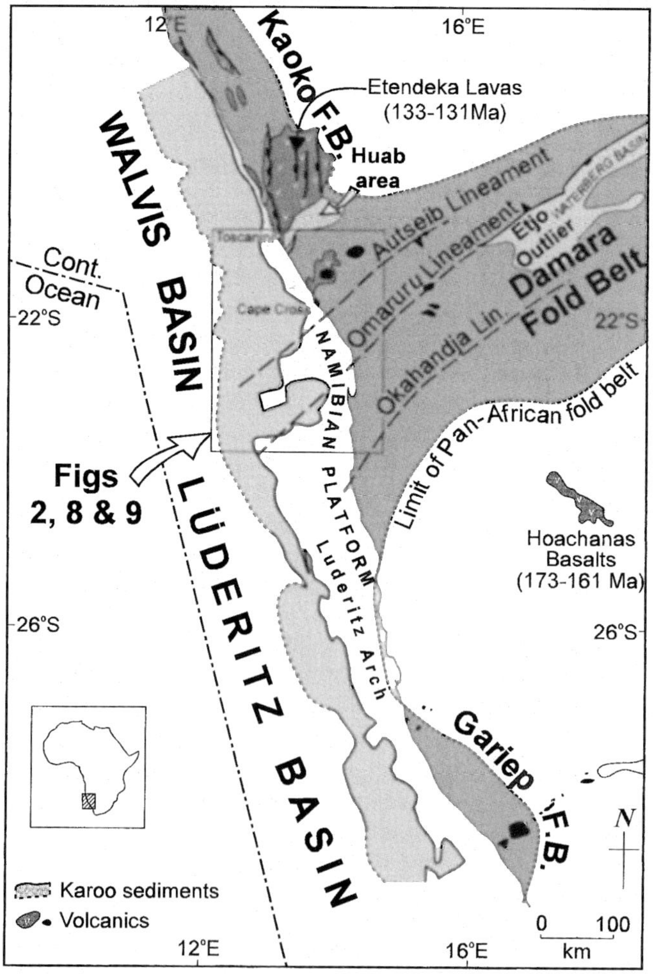

Fig. 1. Location map of the Karoo in west-central and offshore Namibia.

The Lower Cretaceous Etendeka flood basalts, that obscure southern areas of the Kaoko Fold Belt, were extruded during the early phase of rifting that led to continental break-up and formation of the South Atlantic Ocean (Fig. 3) (Clemson 1997). South of the major Etendeka flood basalt outcrops, the Karoo sediments of the Huab area also overlie the Kaoko basement terranes. Offshore, the Atlantic Lüderitz and Walvis rift basins and post-break-up drift depocentres (Fig. 1) generally coincide with the underlying Namib Rift basin, although this basin also extends landward of the Atlantic rift basin in the Walvis Bay area. The Atlantic Lüderitz and Walvis rift basins thin eastward onto the Namibian Platform, while the Lüderitz Arch part of the Namibian Platform, lies east of the Lüderitz Basin.

Basement terranes in west-central Namibia

The Purros Terrane of the Kaoko Fold Belt occurs along the Namibian coast to the north of the Etendeka Flood Basalts, while the Ogden Rocks Domain comprises the basement along a short coastal trip south of the flood basalts (Fig. 2). These two basement units may link beneath the Etendeka Flood Basalts (unpublished field guide, Geological Society of Namibia 1996). The eastern margin of the Purros Terrane is tectonically bounded by the Purros Lineament and therefore, by inference, the Purros Lineament also extends south to the coast as the tectonic boundary to the eastern margin of the Ogden Rocks Domain. Both the Purros Terrane and Ogden Rocks Domain comprise an east verging, moderately steeply dipping basement fabric, while to the east of the Ogden Rocks Domain, the basement comprises the west verging Ugab Terrane (unpublished field guide, Geological Society of Namibia 1996).

The Autseib Lineament separates the Kaoko Ugab Terrane from the Damaran Swakop Terrane while another important lineament, the Omaruru Lineament, lies within the Swakop Terrane (Fig. 2). The offshore extension of the Autseib Lineament coincides with a change in the magnetic lineations seen southwest of Cape Cross. North of this lineament, magnetic lineations have an overall north-northwest–south-southeast orientation (Clemson 1997); this trend is supported by proprietary seismic data showing a basement fabric dipping 30–70° to the west. To the south of the Autseib Lineament, magnetic lineations are oriented southwest–northeast (Clemson 1997); in this area, west-southwest–east-northeast oriented seismic lines indicate apparent west verging thrust nappes. These thrust nappes may actually verge northwest towards the Autseib Lineament and mark the northwest margin of the Damara Fold Belt. This magnetic and seismic evidence may therefore suggest that, offshore, the extension of the Autseib Lineament is a major lineament juxtaposing the Damara and Kaoko Fold Belts. This lineament and the differing basement fabrics to the north and south will be shown to have had a fundamental control on later rift geometry.

Stratigraphy of the Namib Rift and adjacent areas

Seismo-stratigraphy of the Namib Rift

The Namib Rift sequence is identified for 900 km along the entire Namibian continental margin. It is best seen in the Walvis Bay–Cape Cross area where it is not obscured by the Atlantic rift succession. In this area, the Namib Rift sequence extends landwards beyond the eastern limit of the Atlantic rift succession, where a thick pre-Atlantic sequence of generally post-rift affinities is also preserved. This appears to be absent from the north of the Walvis and Lüderitz Basins.

The Namib Rift sequence is up to 7.5 km thick in the Walvis Basin, reaching *c.* 8–10 km in the Swakop Basin, a sub-basin of the Walvis Basin (Fig. 1). It comprises moderately continuous, moderately high frequency and amplitude reflections beneath Cretaceous post-break-up transitional and drift phase sediments on the Namibian Platform (Fig. 4). In south-central parts of the Walvis Basin, large north-north-west–south-southeast oriented, west-dipping faults bound the eastern margin (Fig. 5) of intra-basinal tilted fault blocks. In the northern Walvis Basin, the Namib Rift sequence generally thins eastward before being truncated beneath Lower Cretaceous post-rift sediments. The western margin of the rift succession is not identifiable in the Walvis Basin as the sequence passes below the base of seismic data. The Namib Rift sequence is thin or absent on the northwestern flank of the Lüderitz Arch while to the south it again thickens up to 4 km in the Lüderitz Basin, but is generally thinner than the Walvis Basin. It is preserved beneath the eastern flank of the Lüderitz Basin and in downthrown outliers on the Namibian Platform. Moreover, the possible truncated western margin of the rift succession is sometimes evident beneath Atlantic rift sequences in the Lüderitz Basin. It is folded and possibly reverse-faulted in places, but there is less evidence of rifting than in the Walvis Basin. Low-angle faulting

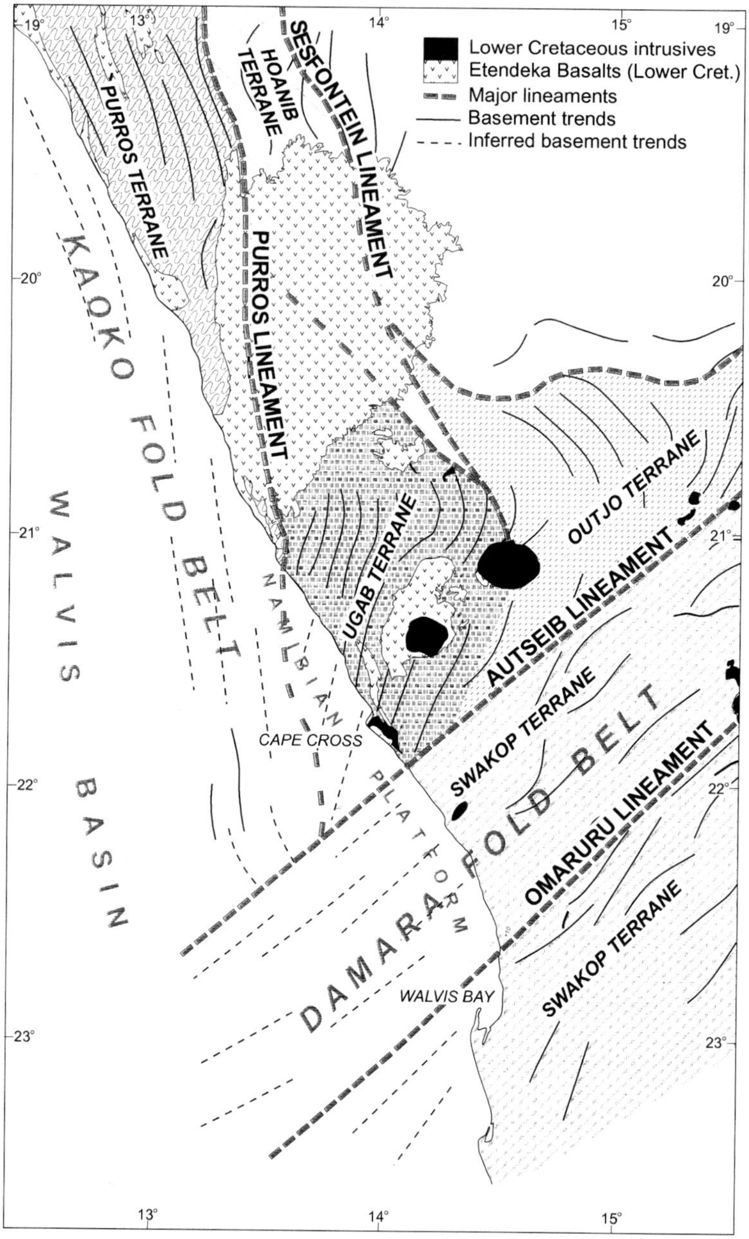

Fig. 2. Basement terranes and structural trends of the Kaoko and Damaran arms of the Pan-African Fold Belt in west-central Namibia.

in the southeast of the Lüderitz Basin preserves possible Karoo in the fault hanging wall (Light *et al.* 1993), but sub-parallel seismic reflections suggest that this faulting is associated with the Atlantic rift phase.

At the base of the Namib Rift sequence, an angular unconformity truncates the Pan-African basement which is characterized by a steep, westerly dipping, reflective fabric. A marked angular unconformity also separates the Namib Rift sequence from the overlying post-rift sequence (Fig. 4). This is thinner than the rift sequence and only reaches 3–4 km in thickness. This unconformity is less reflective in character than the Namib

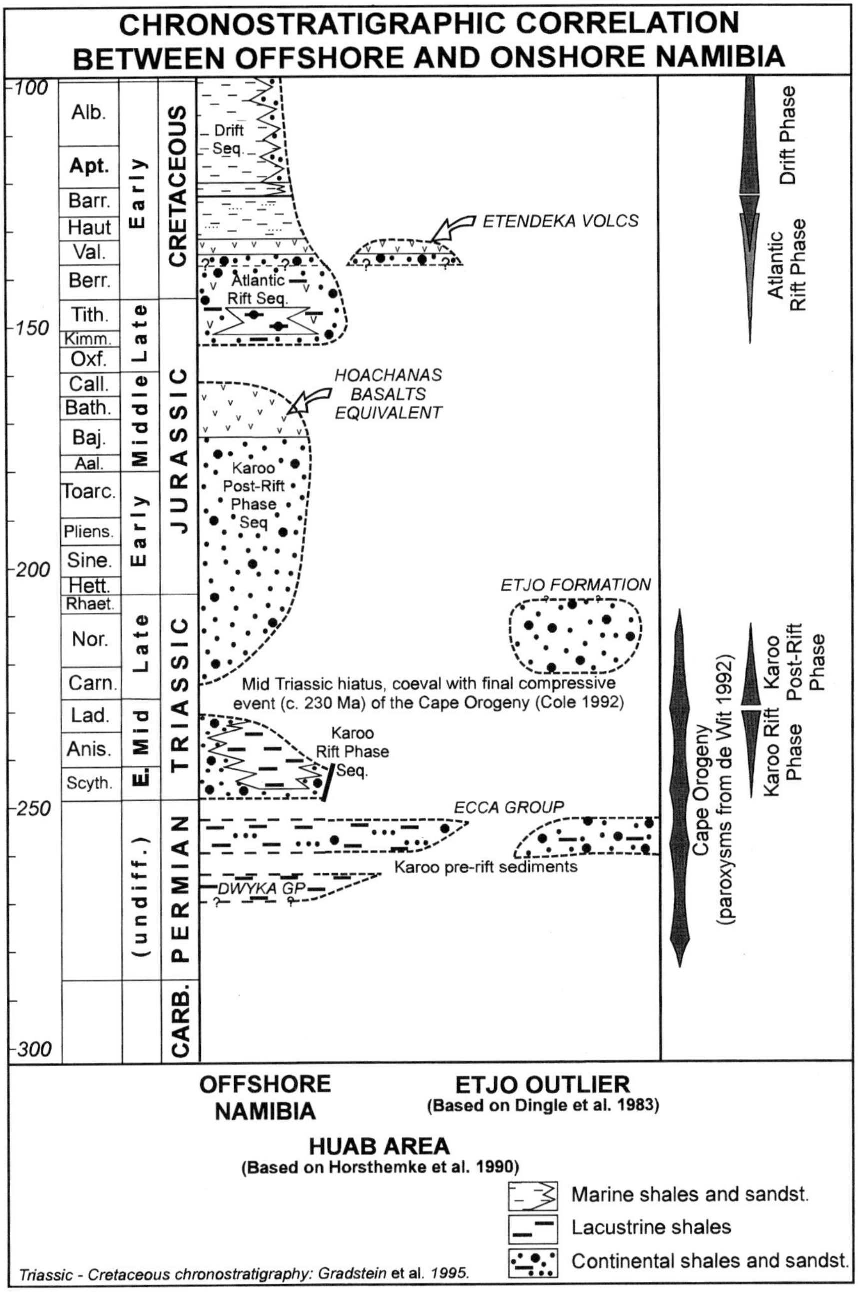

Fig. 3. Permian–Cretaceous chronostratigraphic correlation between offshore and onshore Namibia.

Rift sequence, with generally lower amplitude and less continuous reflections (Fig. 4), and is relatively undisturbed by faulting. The post-rift sequence onlaps onto the flanks of uplifted areas of the underlying rift sequence, pinching out beneath the Atlantic transitional and drift sediments. This pinch-out usually lies west of the eastern margin of the Namib Rift.

The Namib Rift sequence is faulted and folded so that the composite unconformity at the base of the drift cover may sometimes be quite angular. Uplift and erosion separate the Namib Rift and

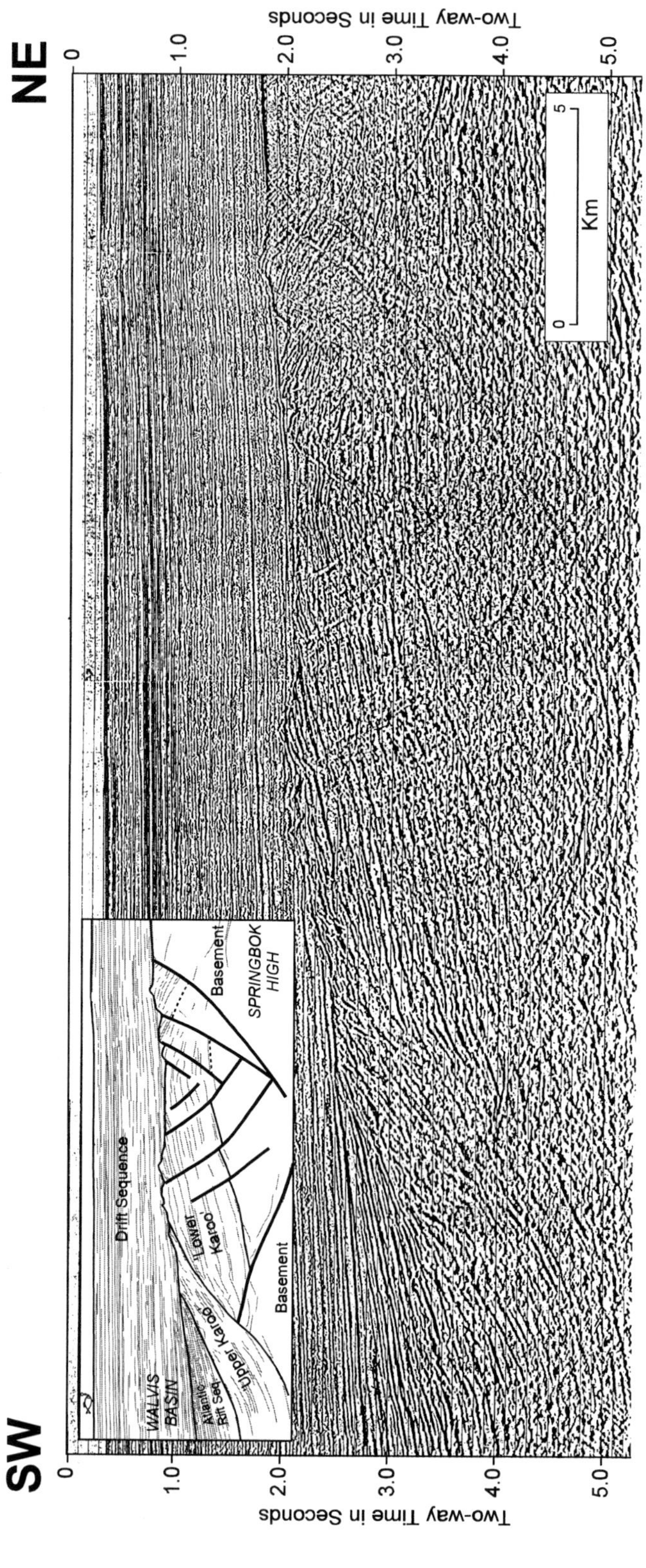

Fig. 4. Seismic line illustrating stratigraphic relationships of the 'Lower' and 'Upper Karoo' (see Fig. 8 for location).

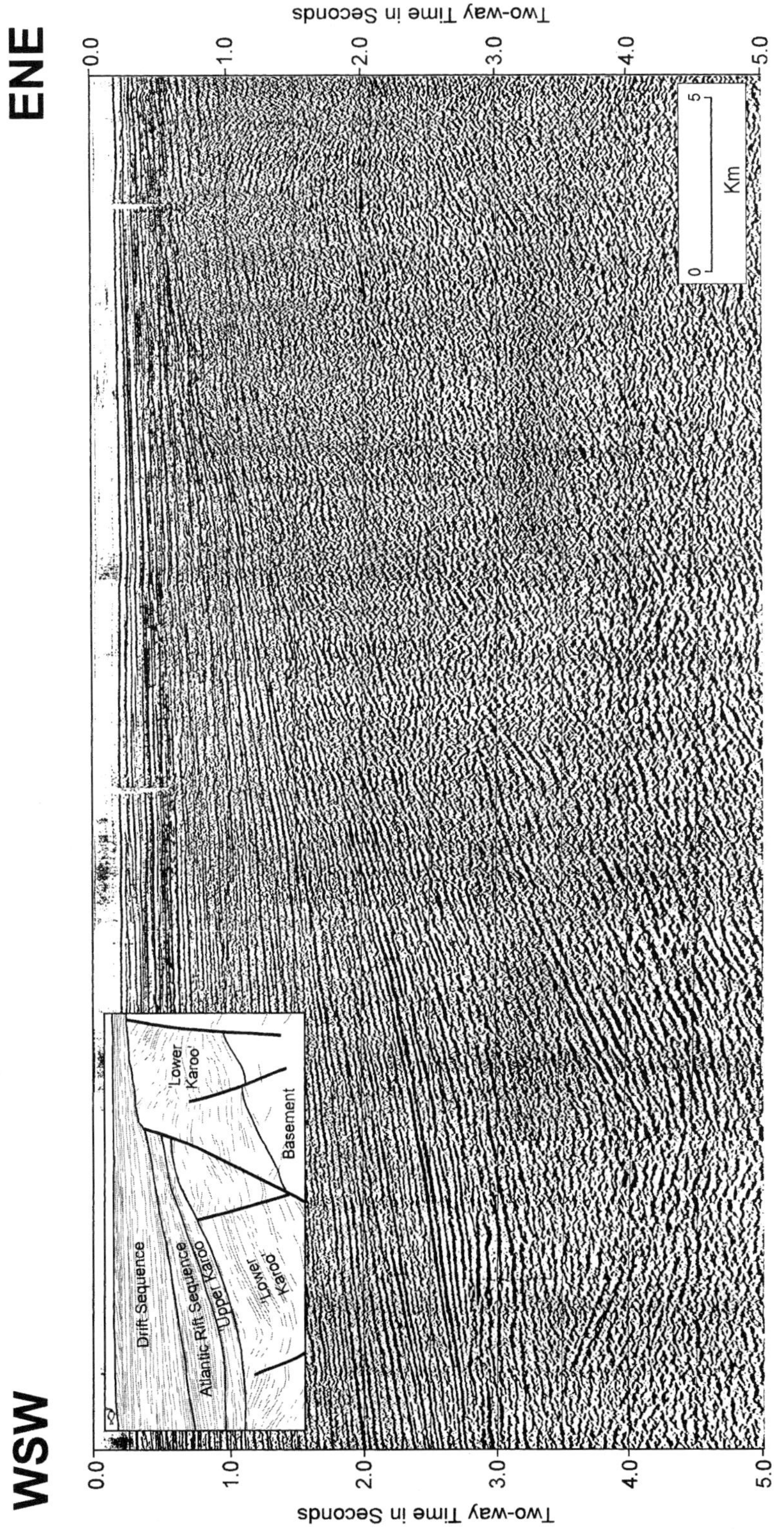

Fig. 5. Seismic line across the fault-bounded eastern margin of the Namib Rift (see Fig. 8 for location).

post-rift sequences leading to deep, semi-isolated basins and bald structural highs in the Cape Cross–Walvis Bay area.

Palaeozoic–Mesozoic cover stratigraphy beyond the Namib Rift

The late Proterozoic to early Cambrian Nama Group succession comprises 3000 m of shallow marine and fluvial sediments covering extensive areas of central and southern Namibia (Saylor *et al.* 1995). They were deposited in a foreland basin along the western margin of the Kalahari Craton, flanking the Pan-African Damara and the Gariep Fold Belts to the northwest and west, respectively (Gresse & Germs 1993).

In northern Namibia, the ?Carboniferous–Triassic Karoo succession (Fig. 3) occurs in two major areas; the Etjo Outlier in the Waterberg Basin of central Namibia (Fig. 1) and the Huab area south of the main Etendeka flood basalt outcrops (Fig. 1). A regional Mid-Triassic hiatus subdivides the succession into the Lower and Upper Karoo (Dingle *et al.* 1983), the Lower Karoo corresponding to the Omingonde Formation (Beaufort Group) of the Etjo Outlier (Dingle *et al.* 1983), and the Gai-as Formation and older Karoo (Ecca Group) of the Huab area (Horsthemke *et al.* 1990). The Upper Karoo corresponds to the Carnian–Rhaetian Etjo Formation sandstones (Dingle *et al.* 1983), while the intervening Mid-Triassic hiatus is coeval with the last phase of compression and uplift (*c.* 230 Ma) in the Cape Fold Belt (Cole 1992).

The Lower Karoo succession outcropping in the Huab area reaches a stratigraphic thickness of 250 m (Horsthemke *et al.* 1990), and comprises a minor sequence of Permian and possibly Carboniferous basal tillites, unconformably succeeded by fluvio-lacustrine sediments of the Ecca Group. Ice movement directions are interpreted to be from Namibia westwards to the Paraná Basin (Martin 1975 in Horsthemke *et al.* 1990), i.e. across the location of the Namib Rift. The succeeding fluvio-lacustrine sediments were deposited in a cratonic basin and are correlated with the Tubarão and Passa Dois Groups of the Rio Grande do Sul area of the Paraná Basin (Horsthemke *et al.* 1990). In Namibia, this succession is Permian in age but a significant part may be eroded. In the Paraná Basin, the equivalent succession extends into the early Triassic. Close to the Namibian coast near the mouth of the Huab River, the Toscanini borehole is located on the downthrown side of a fault close to the Purros Lineament (Fig. 2), where basement occurs at outcrop in the footwall (Fig. 1). It penetrated a thickness of 430 m of Karoo [well summary sheet, Du Preez Exploration (1972) – unpublished NAMCOR report], a succession much thicker than outcropping intervals in the Huab area (Horsthemke *et al.* 1990) to the east, and may indicate syn-sedimentary fault activity.

Stoakes *et al.* (1991) describe similar Karoo successions in the Colorado Basin of South America. Here, over 1500 m of sediments consist of a thick monotonous sequence of dark shales and poorly sorted sandstones of possible lacustrine origin. Palaeontological data suggests a Permian age. The sediments are typical of those from the Permo-Triassic of South Africa. Significantly, the relationship of this sequence with the overlying Atlantic rift fill is that of a coherent layered succession truncated by an angular discordance. A similar relationship is evident offshore Namibia. In the Cape Cross–Walvis Bay area, however, the Namib Rift fill is truncated by the basal unconformity of the post-rift sequence as well as by the base of the Atlantic drift sequence.

In the Etjo Outlier, the Mid-Triassic hiatus extends from the Anisian to Carnian (Dingle *et al.* 1983). This corresponds to a hiatus of Anisian–Landinian age seen elsewhere in southern Africa (Dingle *et al.* 1983). Veevers (1991) links this hiatus to epeirogenic uplift of southern Africa, related to the final coalescence of Pangaea, and also corresponding to the last phase of crustal shortening and uplift in the Cape Fold Belt (Cole 1992).

Age interpretation and discussion of the Namib Rift

Given the complete absence of well control, the age of the Namib Rift sequence and the post-rift sequence can only be provisionally dated from neighbouring sedimentary sequences and regional tectonic events. These sequences clearly post-date the Proterozoic–Cambrian Pan-African Fold Belt and pre-date the late Jurassic–early Cretaceous Atlantic rift sequences described by Maslanyj *et al.* (1992) and Light *et al.* (1993). They may, therefore, be correlated with three major possible sedimentary sequences of the Paraná Basin or Namibia:

- The late Proterozoic–early Cambrian Nama Group foreland basin sequence with sediments derived from highlands of the Pan-African Fold Belt.
- The Ordovician–Devonian and Carboniferous sequences of the Paraná Basin, like the Karoo equivalents in the Paraná Basin, were

deposited in an intracratonic basin with little apparent fault-controlled sedimentation. Some Ordovician faulting does occur in the Paraná Basin, as does some faulting of Carboniferous age (Milani 1992). Any equivalents of these sequences are not present in Namibia. It can be argued that the Namib Rift is a rift-fill sequence coveal with the mainly intracratonic sedimentation in South America.

- The ?Carboniferous–Permo-Triassic Karoo Supergroup is present in Namibia (see above) as well as in the Paraná Basin where its thicker equivalents, the Tubarão and Passa Dois Groups, occur. The faulting of the Namib Rift either occurred during sedimentation or is arguably an undocumented precursor period of Atlantic rifting that faulted the cratonic Karoo sequence before eruption of the Etendeka basalts and deposition of the Atlantic rift sequence.

The Nama Group is unlikely to represent the syn-rift fill of the Namib Rift, as the Namib Rift was was located in the axial regions of the Pan-African Fold belt during Nama Group times. It is far more probable that sediments were derived from the region and transported into the Nama Basin.

No sediments of Ordovician–Devonian and Carboniferous age are found in adjacent areas of onshore Namibia. Carboniferous–Permian ice extended west from the Namibian highlands into the Paraná Basin, and any topographical vestiges of an Ordovician–Devonian rift are unlikely to have remained during this glaciation. An Ordovician–Devonian rift-fill sequence coeval with the minor faulting in the Paraná Basin is possible but, on balance, it is considered to be less likely than a Karoo-age rift fill. The Karoo Supergroup, unlike the Ordovocian–Devonian and Carboniferous sequences, does occur in Namibia, as do its equivalents in the Paraná Basin. Furthermore, there are fault-controlled Karoo basins in Namibia (Waterberg Basin) and East Africa (Mid-Zambesi and Cabora Bassa Basins) (Johnson *et al.* 1996).

What is the evidence that faulting in the Namib Rift is Karoo in age, or was faulting associated with Atlantic rifting? Three main lines of evidence need to be considered:

- The regional seismic data quality largely precludes the identification of fault-controlled sedimentation. This is verified northwest of Walvis Bay, where unequivocal Early Cretaceous fault-controlled sedimentation seen on proprietary datasets is not identifiable on regional datasets. There is, however, some thickening towards major Karoo faults that may indicate growth faulting. The evidence of recognizable divergence over short distances does not preclude syn-sedimentary fault activity as, for example, is the case with the thick, readily identifiable fault-controlled sequences in the South Tucano Basin of South America. These show little thickening towards faults over a distance of 12 km (Milani & Davison 1998, fig. 10).
- In many cases, the faults do not cut Atlantic rift sequences or Namib post-rift sediments.
- If the major faults that juxtapose very thick Karoo intervals with crystalline basement are Atlantic rift faults, the complete removal of a similar thickness of Karoo sediments from the footwall must be invoked. Such large-scale faulting would create substantial (> 5 km) accommodation space in the hanging wall to these faults. There is no evidence for such a precursor Atlantic fill of this potential accommodation space and no indication as to where sediments shed from the footwall have gone.

These observations, taken together, suggest that most faulting in the Namib Rift preceded Atlantic rifting and is Karoo in age. In the absence of any geological data from offshore wells, this conclusion is not certain, but can be used as a working hypothesis.

What is the relationship between the Namib Rift and post-rift sequences, and the Lower and Upper Karoo of Namibia? The Mid-Triassic hiatus separating the Lower and Upper Karoo (Dingle *et al.* 1983) coincides with the last phase of compression of the Cape Fold Belt (Cole 1992). This tectonic event may be correlated with the unconformity seen offshore between the Namib Rift and post-rift sequences. This suggests that the offshore Namib Rift and post-rift sequences may correlate with the Namibian onshore Lower and Upper Karoo, respectively, with a phase of inversion uplifting and eroding the ‘Lower Karoo’ rift sequence before deposition of the ‘Upper Karoo’ post-rift sequence.

The lower age for the Namib Rift sequence, if ‘Lower Karoo’ in age, is confined by the age of Carboniferous–Permian Gondwana glaciation preceding rifting (see above). However, Karoo rifting in East Africa, although not well constrained due to imprecise palynological dating (d’Engelbronner 1996; Johnson *et al.* 1996), started in earliest Triassic times (Hiller & Buttkuss 1996). If the initiation of Namibian rifting was coincident in time, then the latest lower age limit for rift onset is the earliest Triassic. If rifting was terminated by the final compressive event of the Cape Orogeny (see above) then rifting continued until the very late Mid-Triassic. A tentative Middle Jurassic upper age limit for the post-rift sequence is confined by possible volcanics at the top of the sequence (see below). The Namib post-rift sequence may be interpreted as a lateral equivalent of the cratonic Upper Karoo Etjo Formation sandstones.

Structure of the Namib Rift

The Namib Rift is clearly divisible into two major rift arms, with a rift arms, with a right-stepping offset of 100 km in the Walvis Bay area (Fig. 6). These rift arms coincide with the Walvis and Lüderitz Basins. There is evidence for rift seg-

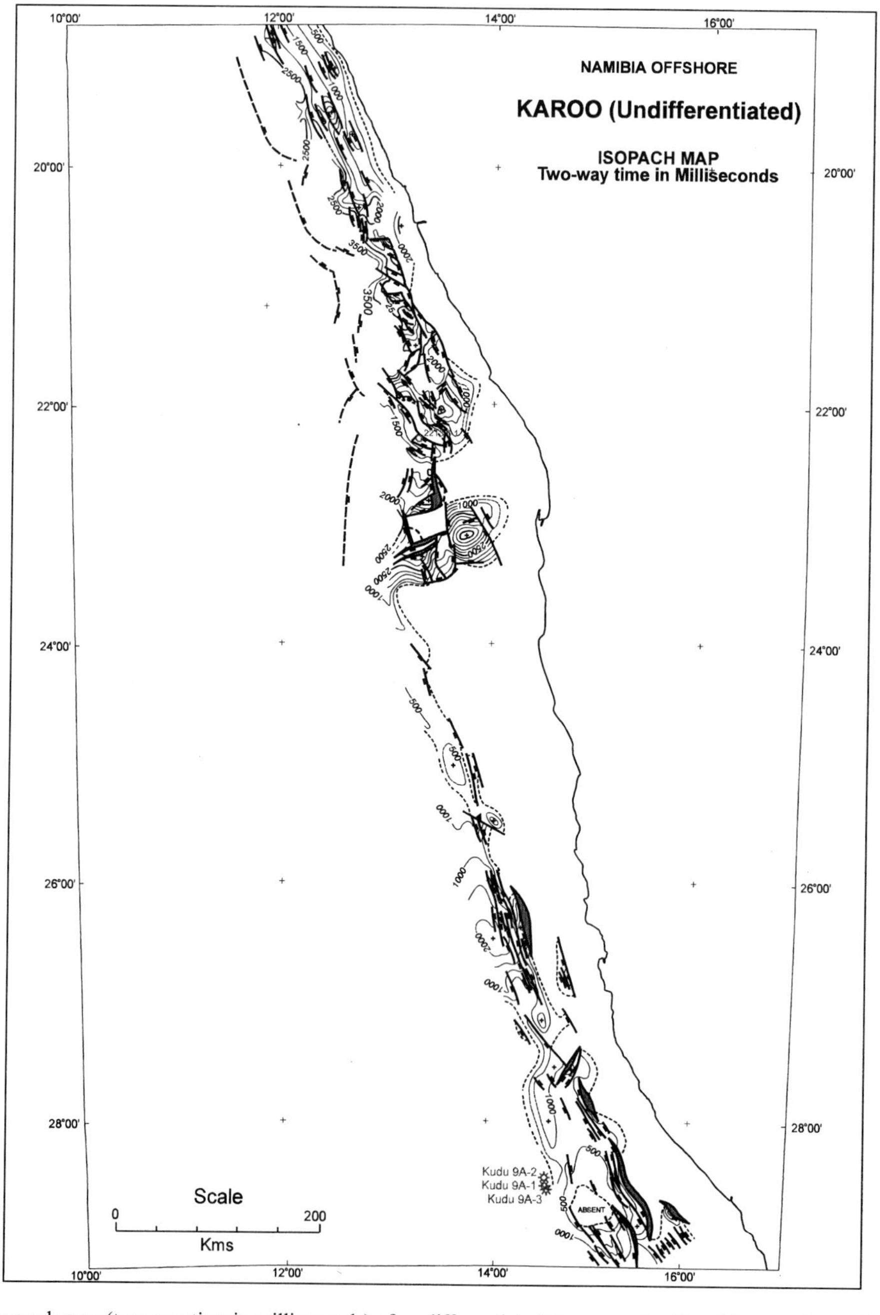

Fig. 6. Isopach map (two-way time in milliseconds) of undifferentiated Karoo on the Namibian continental margin.

mentation and changes in the polarity of faulting in the Walvis Basin (Fig. 7) that is akin to rift segmentation of the East African Rifts (Bosworth 1985; Chorowitz 1989; Morley *et al.* 1990; Nelson *et al.* 1992; Gawthorpe & Hurst 1993). Rift segmentation of the Lüderitz Basin is not evident; whether its absence is due to a different basement structural province or a looser seismic grid with a poorly visible 'Lower Karoo' sequence beneath the Atlantic rift sequences is unclear. Rift segmentation in the Walvis Basin can only be seen along the east of the rift system, as the 'Lower Karoo' dips westwards beneath the Atlantic rift sequence and cannot be identified. In the north of the Walvis Basin the 'Lower Karoo' thickens westwards away from the crest of tilted fault blocks and, in central-southern parts of the basin, major faults are west dipping while the east of the rift system is fault-bounded for 120 km by the north-northwest–south-southeast oriented Unjab Fault Zone (Fig. 8). Nearby, the northwest–southeast oriented Koigab Fault Zone is a major southwest dipping intra graben fault.

The change in fault polarity between these two northern rift segments occurs near the mouth of the Unjab River (20°20′S) (Fig. 6), but the western bounding fault is not visible. A change in trend of Atlantic rift phase faulting at 21°S occurs in the west of the Walvis Basin, producing two arcuate sets of faults (Fig. 6) (Clemson 1997). This pattern of Atlantic faulting may have been inherited from underlying Namib Rift structures and, if so, suggests the presence of a segmented, arcuate western boundary fault system to these rift segments. In the east of a third southern segment, much of the 'Lower Karoo' sequence has been eroded, but there is no evidence for a faulted eastern margin. North-north-east–south-southwest oriented faults of the Atlantic rift phase in the Walvis Basin may reflect the trend of a western boundary fault for this rift segment. These rift segments are 120–150 km long and 80–130 km across, and widen to the south of the Walvis Basin. Their size is similar to the major segments of the Suez Rift, which are comparable in width and up to 140 km long (Colletta *et al.* 1988).

The Namib Rift in the Cape Cross–Walvis Bay area

The Cape Cross–Walvis Bay area has had a long and complex history, with Namib and Atlantic rift sequences both modified to a greater or lesser extent by inversion, uplift and erosion. Consequently, although older sequences are incomplete and fault histories are unclear, enough section is preserved to reconstruct generalized palaeogeographies for the 'Lower' and 'Upper Karoo' Namib Rift and post-rift

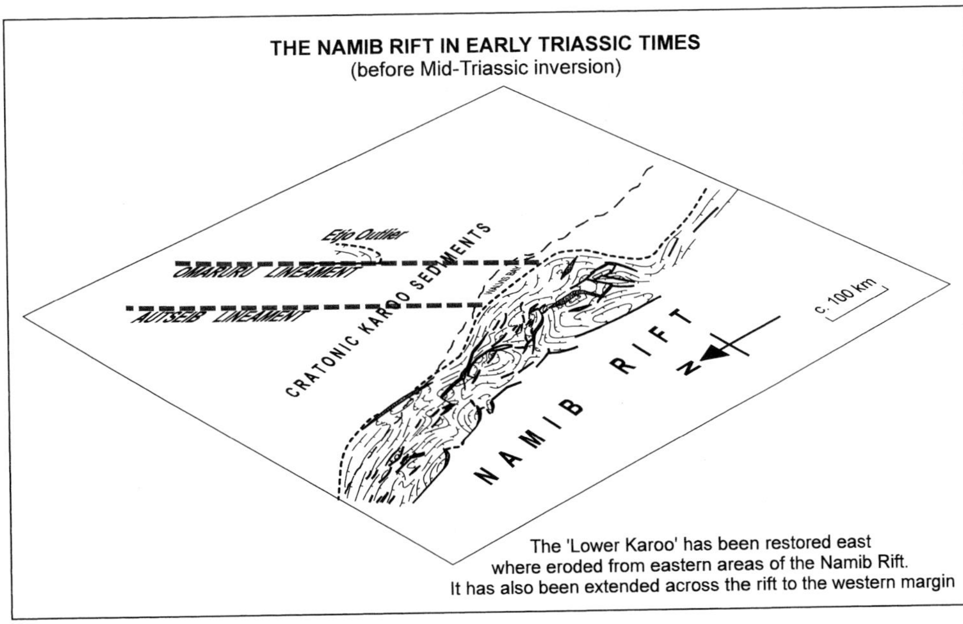

Fig. 7. Restored oblique view looking southeast of the Namib Rift in Early Triassic times, prior to Mid-Triassic inversion.

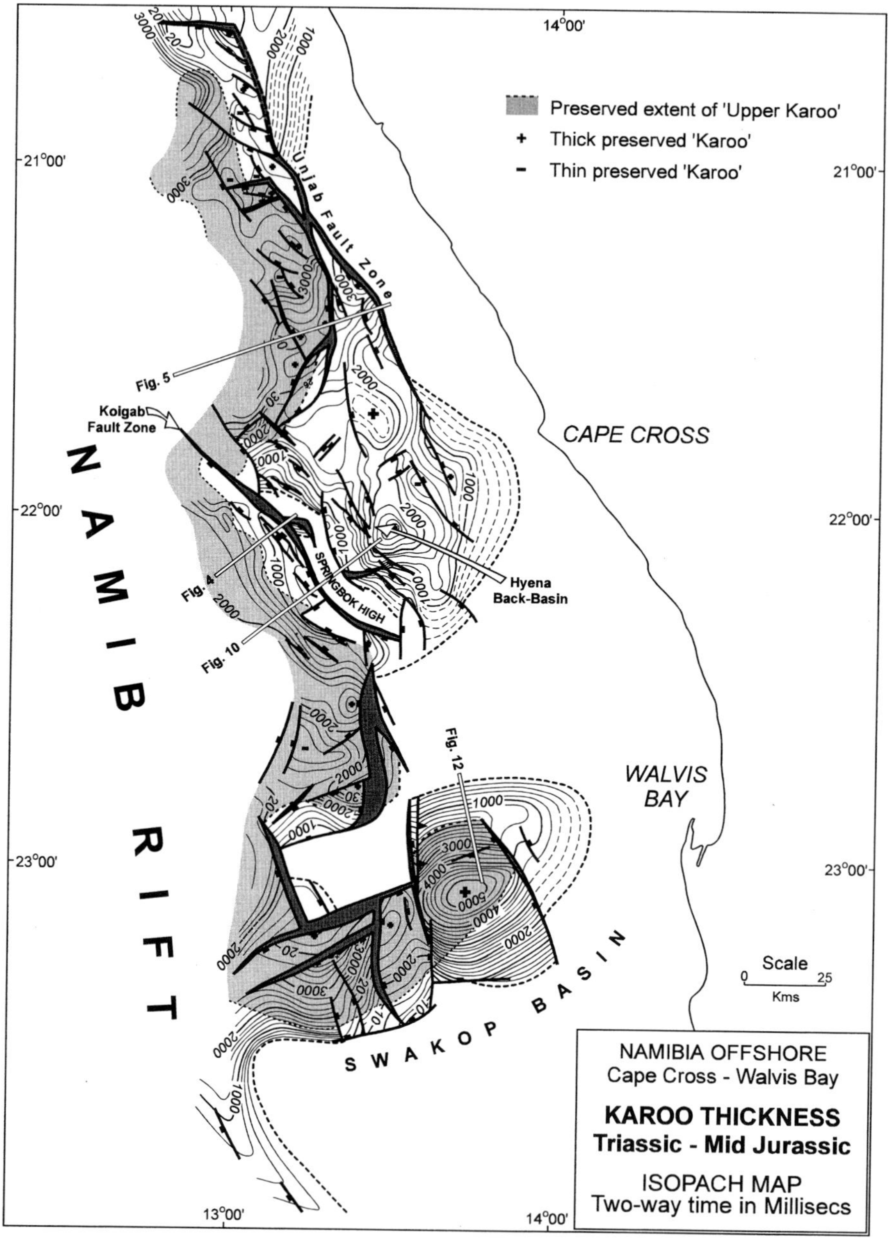

Fig. 8. Isopach map of undifferentiated Karoo in the Cape Cross–Walvis Bay area.

sequences. Due to a combination of data quality, depth and subsequent erosion, only a gross isopach of the Karoo is mapped in detail in this area (Fig. 8); however, the general structure with time thickness values is produced for both the 'Lower' and 'Upper Karoo' sequences (Fig. 9).

This area forms part of the Cape Cross Segment Boundary Zone, where Namib Rift segmentation and the partitioning of Atlantic rifting have been identified (Clemson 1997; Clemson *et al.* 1997). Several small sub-basins of varying age occur on the Namibian Platform to the east of the Walvis Basin. In the south, near the offshore extension of the Omaruru Lineament, the very deep and partly faulted synclinal Swakop Basin (Fig. 8) is mainly filled with 'Lower Karoo' sedi-

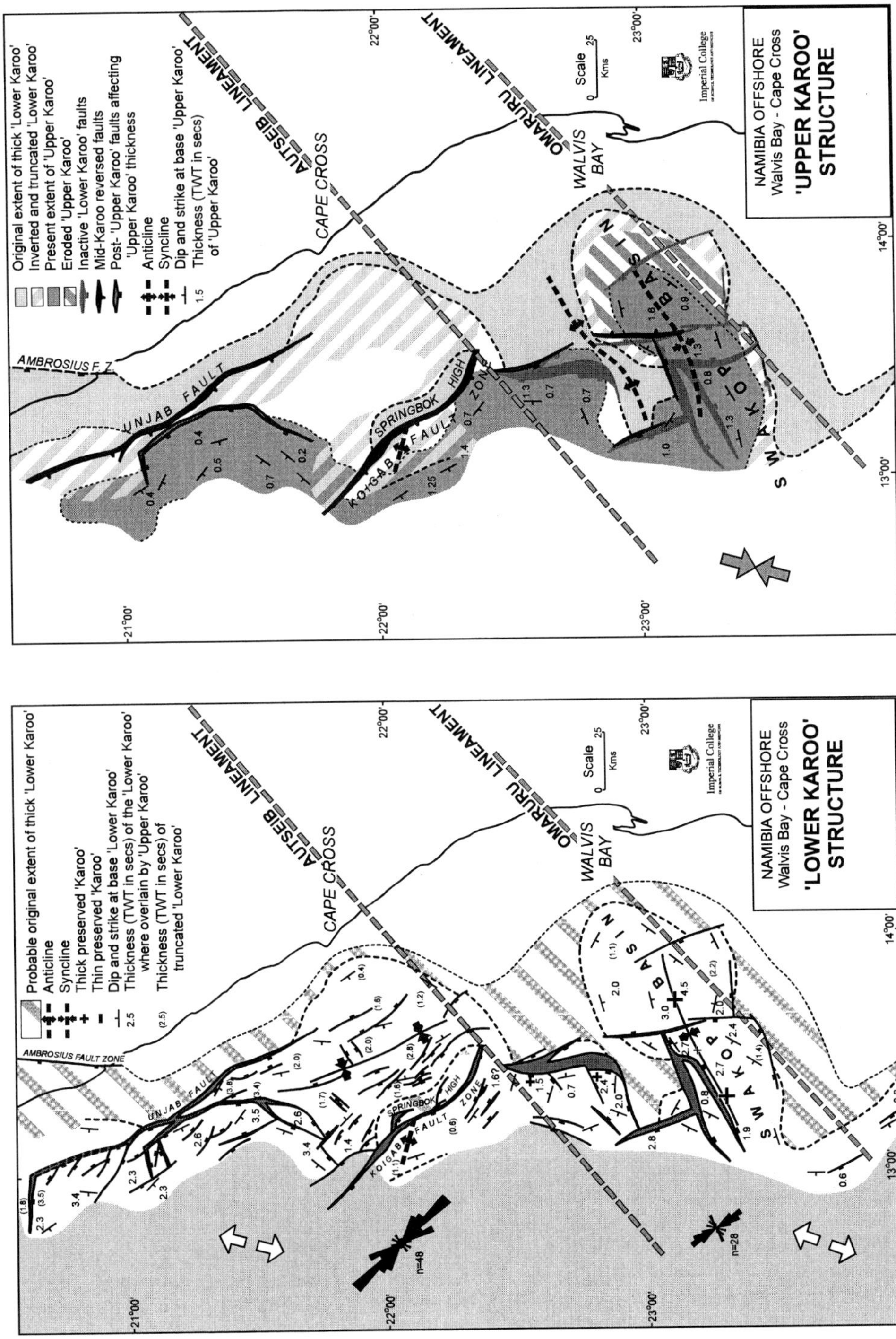

Fig. 9. Structure of the (**a**) 'Lower' and (**b**) 'Upper Karoo' in the Cape Cross–Walvis Bay area.

ments, preserved in areas less subject to Mid-Triassic uplift and erosion. To the north, the Springbok High isolates the early Cretaceous Hyena Back-basin (Fig. 8) from the Walvis Basin. This half-graben back-basin allowed the preservation of a thick 'Lower Karoo' sequence whereas surrounding areas were peneplaned in the early Cretaceous. The northwest–southeast oriented Koigab Fault Zone bounds the west of the Springbok High, while further to the north-east the Unjab Fault Zone bounds thick 'Lower Karoo' rift sediments on the eastern margin of the Walvis Basin. During development of the Namib Rift, these two major faults may have acted as a relay ramp (cf. Trudgill & Cartwright 1994) that was later uplifted and eroded during the Mid-Triassic.

The 'Lower Karoo'

In the south of this area, up to *c*. 8–10 km of the Namib Rift sequence, interpreted to be early Karoo in age, is preserved in the east-north-east–west-southwest oriented Swakop Basin. The east of this basin was originally part of a broader basin whose preserved flanks are now strongly discordant beneath the Cretaceous drift cover, while its southern boundary may have been fault-bounded in part. To the west, across a complex north–south oriented accommodation zone, the basin becomes a west-southwest–east-northeast oriented half-graben, fault-bounded to the north (Fig. 8). This sequence is truncated at the base of the 'Upper Karoo' post-rift sequence in the Swakop Basin.

For some distance north of the Swakop Basin, north–south faults bound much of the preserved margin of the Namib Rift sequence (Figs 8 and 9). This complex area has thick sections of 'Lower Karoo' faulted against bald highs. Faults bound a large horst on three sides, but some faults do not appear to be normal and may have a strike-slip component, as large but local faults with up to *c*. 6.0 km of fault-controlled sediments die out quite rapidly. In a large area north of the Swakop Basin, the seismically opaque character on regional data suggests that crystalline basement lies beneath the drift cover.

The Koigab and Unjab Fault Zones dominate the area north of the Autseib Lineament (Fig. 9), where the 'Lower Karoo' is more extensive and much better preserved, despite uplift and erosion. Its absence on the footwall to the Koigab Fault Zone may be due to early Cretaceous peneplanation as well as Mid-Triassic uplift and erosion. Significant Mid-Triassic tilting and erosion appear to be confined to major faults and the rift margin. Northeast–southwest oriented faulting between the Koigab and Unjab Fault Zones (Fig. 8) may be a failed attempt to breach the relay ramp (cf. Trudgill & Cartwright 1994).

'Early Karoo' faulting is strongly listric in style in the west of the Hyena Back-basin and across the Springbok High where, arranged as a fault fan, they sole out onto a southwest dipping low angle shear zone within the crystalline basement (Fig. 10). This low angle shear zone juxtaposes steeply dipping basement reflections above with a flatter lying reflective package beneath, and is rooted in the steeply southwest dipping Koigab Fault Zone. Fault blocks in the listric fault fan, of which the master fault is located 25 km northeast of the Koigab Fault Zone, are highly rotated with stratal dips sometimes exceeding 45° (after depth conversion). Internal reflections and the maximum preserved thickness emphasize the large scale of this fault system and the degree of subsequent erosion. Up to *c*. 6 km of 'Lower Karoo' are preserved in the east, but may been up to 10 km thick before Mid-Triassic and Early Cretaceous erosion.

The fault system, as interpreted, developed in two phases (Fig. 11) with little evidence for syn-sedimentary thickening across the listric fault fan. Internal reflection converge southwest towards the Koigab Fault Zone, with the 'Lower Karoo' interpreted as thinning to *c*. 4 km in the footwall. Initial activity was therefore probably confined to the steeply dipping Koigab Fault Zone. The subsequent listric faulting was due to collapse of the footwall to the Koigab Fault Zone to form the broader Koigab Fault System. Whether the Koigab Fault Zone itself remained active is unknown, as much of the 'Lower Karoo' is absent from the hanging wall. The low angle shear zone, upon which the listric faults sole out, is interpreted to have been 10–15 km beneath the syn-extensional surface. By line-balancing the section, it is found that the Koigab Fault System was extended by a factor of 1.4 before Mid-Triassic inversion. Fault-block rotation led to tectonic thinning exceeding 1 km. Estimates are based on Fig. 11 with the system pinned in the east at the footwall of the master fault.

Several authors have suggested that strongly listric basement extensional faults are spoon-shaped in three-dimension, with curvilinear map traces (Bosworth 1985; Rosendahl 1987). Morley (1995) stresses the importance of pre-existing basement fabrics influencing faults and how theoretical rules may not apply. For example, the Lupa Fault in the Rukwa Rift is listric in cross-section, but follows foliations along-strike to result in a linear surface trend. The Koigab Fault System consists of essentially linear segments at the scale of mapping carried out in this area. It is

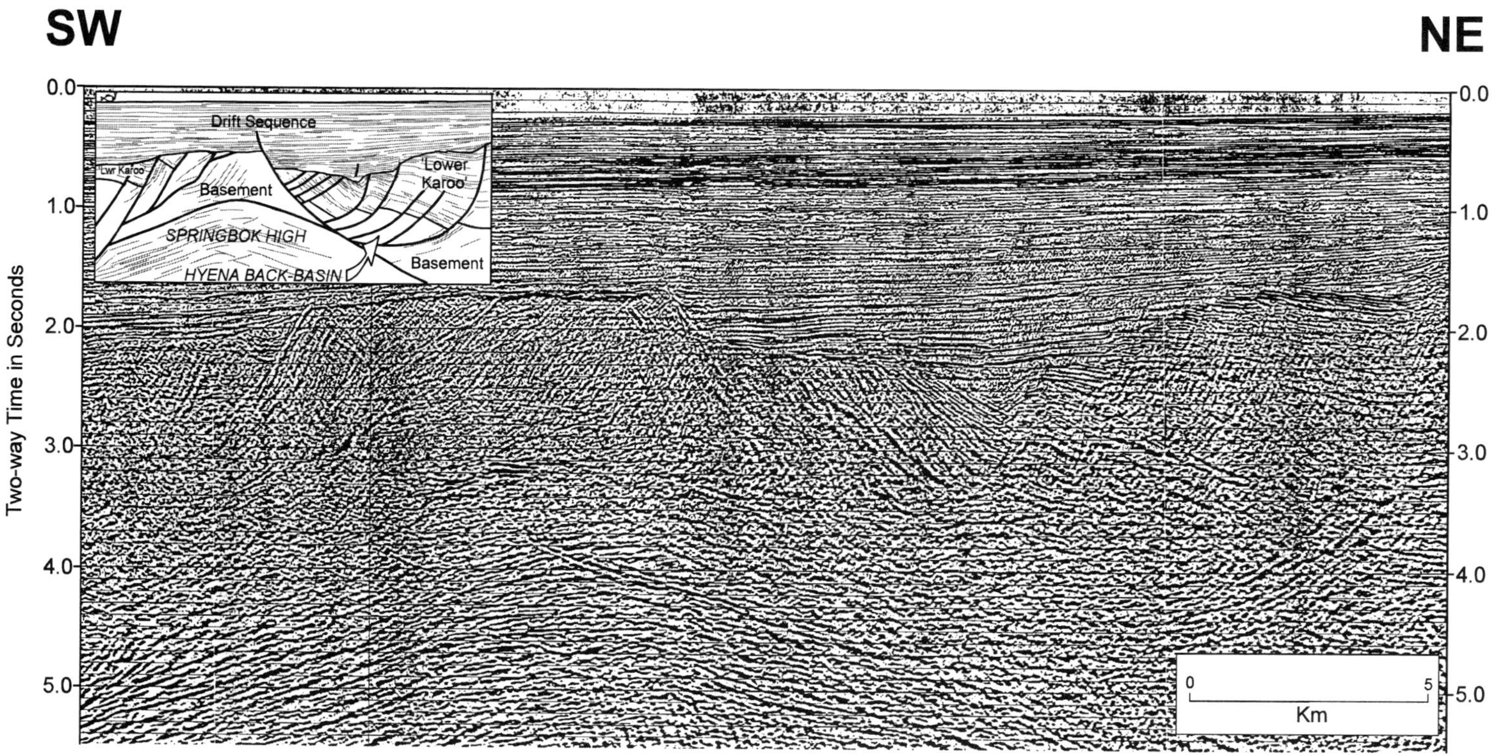

Fig. 10. Seismic line illustrating listric faulting beneath the Springbok High and Hyena Back-basin (see Fig. 8 for location).

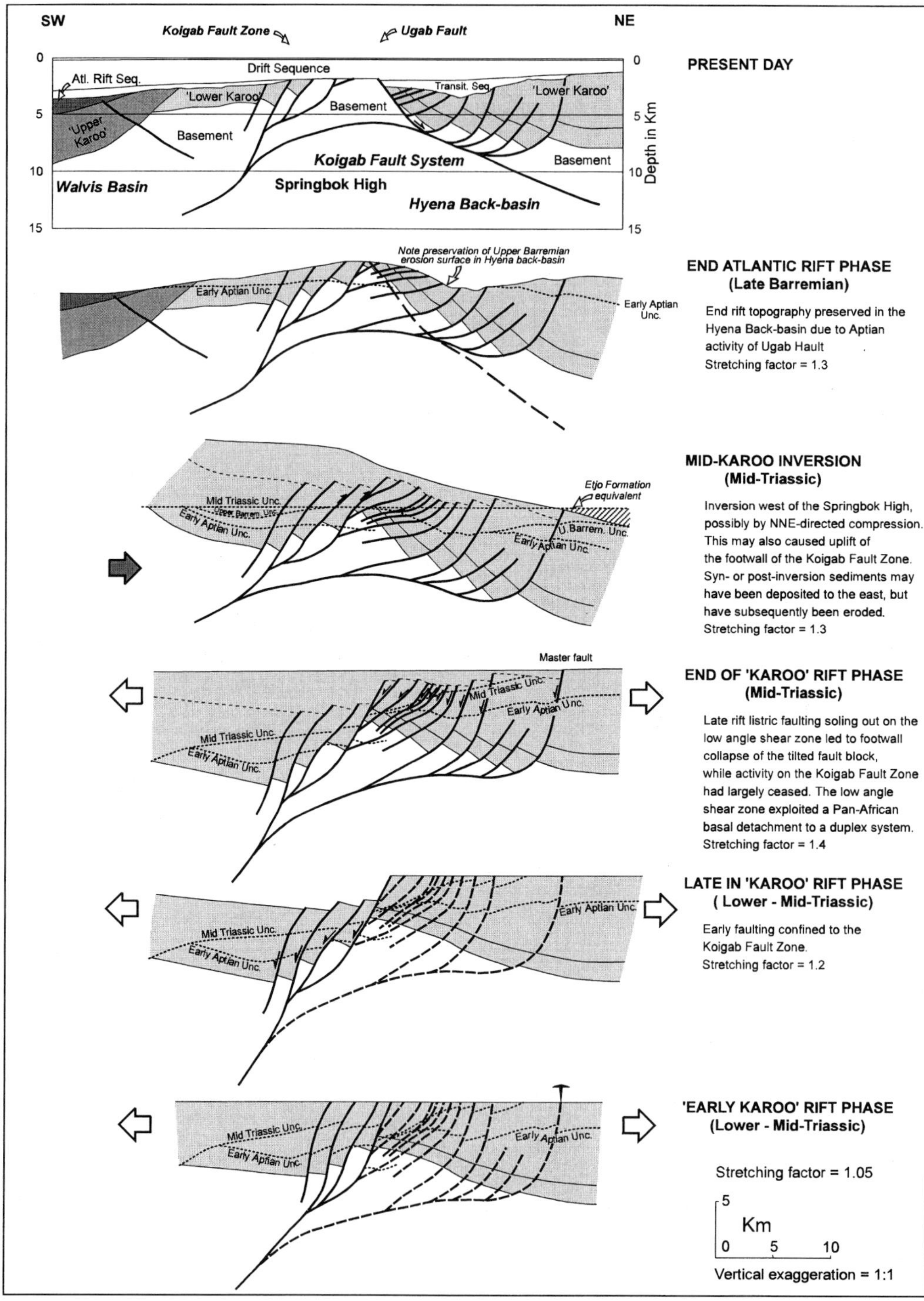

Fig. 11. Structural reconstruction of 'Lower Karoo' low angle faulting in the Cape Cross–Walvis Bay area, based on Fig. 10 (see Fig. 8 for location).

suggested, therefore, that the listric faults of the Koigab Fault System also follow foliations along-strike, with the Kaoko basement fabric influencing subsequent faulting.

Internal reflections of the 'Lower Karoo' to the southwest of the Koigab Fault Zone are sub-parallel towards the Walvis Basin, but this sequence is only present for 12–24 km to the west of the Springbok High before being truncated by erosion. Beyond this area, the regional interpretation defines major east dipping Mesozoic syn-rift faults 30–50 km to the west of the Koigab Fault Zone. If these were reactivated Namib Rift Faults, the preserved section with parallel internal reflections located west of the Springbok High may form the deepest part of a large inverted graben whose eastern boundary would be the Koigab Fault Zone.

Possible Mid-Triassic inversion and the 'Upper Karoo'

Uplift of the 'Lower Karoo' is interpreted to have occurred in the Mid-Triassic. In the south, it is distributed across large east–northeast–west-southwest oriented folds such as the Swakop Basin (Fig. 12) with local evidence in the northeast of the Swakop Basin for reversal of 'Lower Karoo' faults during the Mid-Triassic. To the north of the Autseib Lineament, uplift and erosion are more often concentrated close to discrete, pre-existing fault zones, with large-scale uplift to the west of the Springbok High giving rise to a notable angular Mid-Triassic unconformity (Fig. 4). Other evidence for inversion includes more localized folding west of the Koigab Fault Zone, where erosion of this uplifted area accounts for the thinner section of 'Lower Karoo'. A west-northwest–east-southeast oriented anticlinal fold axis is oblique to, and terminates against, the Koigab Fault at *c.* 50° (Fig. 9), with dips on the eastern limb progressively increasing upwards through the 'Lower Karoo' section. West of the Unjab Fault Zone, the 'Lower Karoo' shows evidence of folds buttressed against and juxtaposing basement to the east (Fig. 5).

North of the Autseib Lineament, short, discordant, high amplitude reflections in the uppermost part of the 'Upper Karoo' suggest that volcanics may be present. In the Swakop Basin, high seismic interval velocities and the presence of a large positive magnetic anomaly suggest that volcanics may also be present in the 'Upper Karoo', and are tentatively correlated with the Namibian Hoachanas basalts (173–161 Ma; Gidskehaug *et al.* 1975) (Figs 1 and 3). If this correlation is valid, it strongly suggests that the 'Upper Karoo' is Upper Triassic–Middle Jurassic in age.

The 'Upper Karoo' sequence is much better preserved than the 'Lower Karoo' rift sequence and, with some fault-bounded exceptions, the sequence pinches out or thins towards the Namibian Platform before a truncated pinch-out. Atlantic faulting and erosion of the 'Upper Karoo' from the footwall gave rise to local occurrences of apparently fault-bounded 'Upper Karoo'.

Structural evolution of the Namib Rift in the Cape Cross–Walvis Bay area

The Namib Rift is divided into two major arms in the Cape Cross–Walvis Bay area. Here, 'Lower Karoo' rifting extended further east than subsequent Atlantic rifting. Although Karoo sediments may have been present to the south on the Lüderitz Arch, an unquantifiable thickness was probably removed during Mid-Triassic and Early Cretaceous erosion. However, convergence of reflection towards the Lüderitz Arch within the Namib Rift sequence suggests that, if the Karoo was present, it was relatively thin. This is certainly not the case to the north in the Cape Cross–Walvis Bay area and the Walvis Basin. Therefore, the offshore projection of the Omaruru Lineament may have both offset the major southern and northern arms of the Namib Rift and acted as a hinge with increased subsidence to the north.

In the Cape Cross–Walvis Bay area two discrete zones, separated by the Autseib Lineament, exhibited contrasting behaviour during Karoo rifting (Figs 8 and 9). To the north, the Purros Terrane basement fabric (Fig. 2) may have conditioned the north-northwest–south-southeast orientation of Karoo faulting. The Koigab Fault Zone appears to be an isolated example of low-angle faulting, with little evidence for similar faulting elsewhere in this area. To the south of the Autseib Lineament, some faults are closer to a north–south orientation (Figs 8 and 9) and cross-cut the northeast–southwest oriented Swakop Terrane basement fabric. Other faults on the flanks of the Swakop Basin are oriented west-southwest–east-northeast, close to the basement trends of the Swakop Terrane (Fig. 2). There is no identifiable discrete structure that can be correlated with the offshore projection of the Omaruru Lineament. This appears to have acted as a hinge where to the south, the Lüderitz Arch remained positive.

Mapping of reflective basement fabrics suggest that dips of the basement structural fabric in the

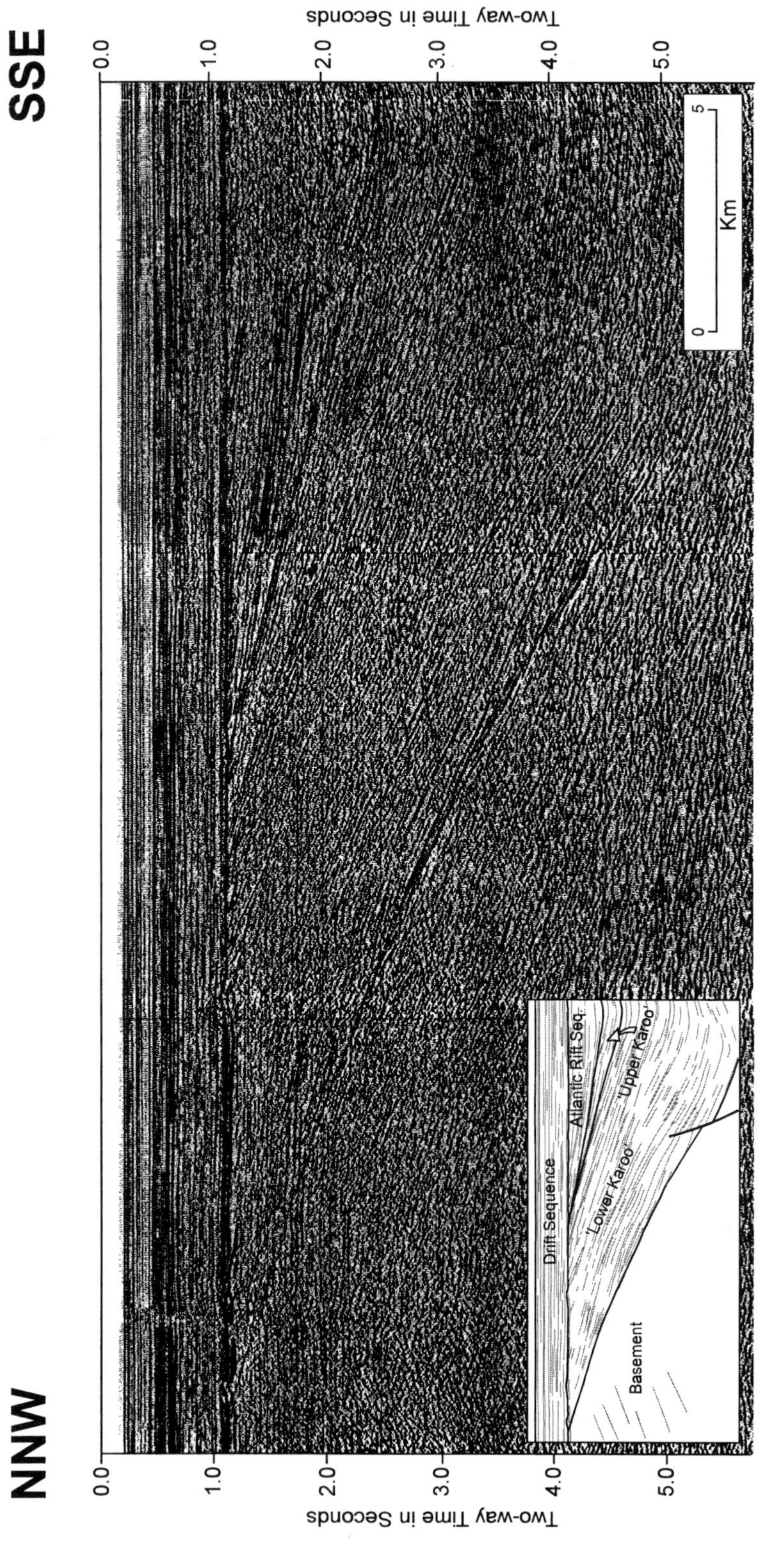

Fig. 12. Seismic line illustrating the northern flank of the Swakop Basin (see Fig. 8 for location).

northern area influenced the preferred orientation of 'Early Karoo' faulting. In the more complex southern area, the northeast–southwest basement fabric of the Swakop Terrane may have determined some 'Early Karoo' fault trends. West-southwest–east-northeast oriented faults may be related to the fault-bounded Waterberg Basin, 350 km onshore to the northwest (Fig. 1), where the reversed west-southwest–east-northeast oriented Waterberg Fault bounds the Etjo Outlier. This reversal, at least in part, may have occurred in early Cretaceous times (Clemson 1997). This fault lies on the Damaran Omaruru Lineament and was reactivated as a normal fault (Johnson *et al.* 1996) in Permian–Early Triassic times. Reactivation of the Omaruru Lineament during offshore 'Early Karoo' rifting may have created the right-stepping margin of the Namib Rift north of the Lüderitz Arch, preserved as the Swakop Basin. The Autseib Lineament was apparently not active during 'Early Karoo' rifting on the Namib rift margin or inland, and appears to have merely acted passively as a locus for a change in rift geometry within an accommodation zone.

The complexity and variety of faulting colours the kinematic setting of the Namib Rift in the Cape Cross–Walvis Bay area. Superficially, the relatively simple area north of the Autseib Lineament suggests southwest–northeast directed extension normal to the Koigab and Unjab Fault Zones. In the area south of the Autseib Lineament, the complex 'Lower Karoo' fault arrays are apparently contemporaneous, and the opposing trends and bald structural highs faulted against thick Lower Karoo suggest a strike-slip component to some faults. A north-northeast–south-southwest direction of extension reconciles the differing fault trends in this area and to the north of the Autseib Lineament. This direction of extension would have led to dominantly normal-sense faulting (with some oblique component) on the Koigab and Unjab Fault Zones in the north. Strike-dominant oblique slip between the Autseib and Omaruru Lineaments may have led to the complex structures described above. The different fault sets reflect the influence of the underlying basement trends and their response to Karoo extension.

'Mid-Karoo' inversion

It is postulated that the entire area was subjected to inversion, uplift and erosion during the Mid-Triassic, but the exact direction of crustal shortening cannot be precisely defined. However, evidence for faults subjected to inversion of the hanging wall, the relation of rollovers to these faults in the north and the orientation of broad folds in the south suggest that, taken together, the area was subjected to north-northeast–south-southwest compression. More regional considerations, such as Mid-Triassic inversion in central Africa (Daly *et al.* 1991, 1992) and coeval compression in the Cape Fold Belt (Cole 1992), furthermore suggest that this compression was north-northeast directed. The response to this postulated Mid-Triassic compression was different in the north and south of the Cape Cross–Walvis Bay area, echoing contrasts in the earlier rift phase structural evolution. The Autseib Lineament can again be seen to separate two areas of differing structural style.

Across the northern area of the Cape Cross–Walvis Bay area, uplift as a whole has occurred but, away from major faults, much of the 'Lower Karoo' is preserved, with uplift confined to the footwalls of major faults. Apparent rollover into the Koigab Fault Zone is oriented obliquely to the fault and increases upward through the 'Lower Karoo' section. It is interpreted to be due to north-northeast directed sinistral transpression exerted on the Koigab Fault Zone. In contrast, in the south, uplift occurred along broad, open west-southwest–east-northeast oriented folds with axes oblique to the southwest–northeast oriented Omaruru Lineament. North-northeast directed compression may have imposed dextral transpressive stress conditions along the Omaruru Lineament. The very long wavelength of these folds may be due to an underlying structural control. Much, if not all, the 'Lower Karoo' was removed from both north and south of the Swakop Basin as a direct consequence of this folding.

This north-northeast direct compression is thought to have been common to areas to the north and south of the Autseib Lineament, but the lineament partitions the structural response to this stress. The differing response in each area was due to a tectonic framework established in Lower Karoo times that was, in turn, conditioned by the Pan-African basement fabrics.

There is evidence for an increase in Mid-Triassic deformation and erosion towards the south of the Walvis Basin, and in the Lüderitz Basin the preserved Namib Rift fill is much thinner. This, and localized evidence for folding and possible reverse faulting, may suggest that Mid-Triassic uplift, deformation and erosion were still more severe in the Lüderitz Basin.

'Late Karoo' subsidence

The 'Upper Karoo' sequence has a much simpler geometry, and always onlaps and pinches out against 'Lower Karoo' rift structures affected by the Mid-Triassic uplift. The pinch-out is often modified by subsequent erosion, especially west of the Springbok High. This post-rift subsidence is not coincident with areas of 'Early Karoo' rifting, but always lies to the west (Figs 8 and 9). This leads to the question: what extension model can be applied to 'Early Karoo' faulting that accounts for the differing distributions of 'Lower' and 'Upper Karoo'?

If McKenzie-type pure shear (McKenzie 1978) was the underlying mechanism for 'Early Karoo' extension, post-rift sediments would be expected to onlap the rift shoulders and progressively overstep the rift margins through time, but is not the case in the Cape Cross–Walvis Bay area. Alternatively, extension via simple shear may account for the differences in distribution of the Namib Rift and post-rift sequences. The low-angle faulting identified in the Koigab Fault System (Figs 10 and 11) is a crustal-scale structure that may have dictated extension via simple shear (Wernicke 1985). The pinch-out of 'Upper Karoo' lies several kilometres west of the Koigab Fault Zone and 40 km west of the master fault, but this is much less than the *c.* 240 km in Wernicke's hypothetical model. This would suggest a steeper shear zone or, alternatively, that this was the distal basin of a proximal–distal basin system. However, no proximal fault system and no other low-angle faulting are evident in this area; most other fault planes appear to be steep. Low-angle faulting appears to be an isolated feature confined to the Koigab Fault System, therefore, simple shear is not favoured as the extension mechanism in the Cape Cross–Walvis Bay area, or elsewhere in the Namib Rift. Alternatively, tectonic and isostatic uplift may have offset any thermal subsidence arising from McKenzie-type pure shear extension. Thermal sag successions are present in many Karoo basins of southern and eastern Africa (Johnson *et al.* 1996). In the Namib Rift, the Mid-Triassic inversion has evidently modified post-rift subsidence.

Why is the 'Upper Karoo' post-rift sequence found only in the Cape Cross–Walvis Bay area, and not in the Lüderitz Basin or the northern part of the Walvis Basin? It may be absent in the Lüderitz Basin due to large-scale Mid-Triassic uplift more than compensating for any post-rift subsidence. Increasing degrees of Mid-Triassic deformation towards the Cape Fold Belt may have caused this large-scale uplift in the Lüderitz Basin, with deformation decreasing northwards into the Walvis Basin. The 'Upper Karoo' may have been deposited in the north of the Walvis Basin as well as in the Cape Cross–Walvis Bay area, but may have been removed prior to Atlantic rifting. This may have been a consequence of thermal doming associated with the Tristan hotspot.

Conclusions

The Namib Rift is a previously undocumented continental-scale rift system extending along the Namibian continental margin from the Walvis Ridge to the mouth of the Orange River. It is recognized from the partially preserved rift-fill and post-rift successions, separated by an angular unconformity. The stratigraphy of this rift fill pre-dates syn-rift sequences of the Atlantic rifting phase (Jurassic–Cretaceous). The preferred interpretation is that the earlier rift and post-rift sequences are time-equivalent to the Lower and Upper Karoo of southern Africa. The inversion, uplift and erosion separating the rift and post-rift sequences offshore is correlated with the Mid-Triassic hiatus seen over much of southern Africa, although the looseness of this correlation is acknowledged. The post-rift sequence of the Cape Cross–Walvis Bay area, if Late Karoo in age, may have extended north but eroded prior to Atlantic rifting that led to continental break-up. Its absence to the south may be due to increasing Mid-Triassic uplift towards the Cape Fold Belt.

The Damaran and Kaoko basement terranes of the Pan-African Fold Belt in west-central Namibia are correlated offshore using magnetic data and seismic reflection data. In the Cape Cross–Walvis Bay area, north-northwest–south-southeast oriented basement trends correlated with the Kaoko Fold Belt are separated from southwest–northeast oriented basement terranes correlated with the Damara Fold Belt by the Autseib Lineament. Within the Damara Fold Belt, the southwest–northeast oriented Omaruru Lineament can also be inferred to extend offshore.

The Namib Rift in the Walvis Basin is believed to be subdivided into three rift segments. Two rift segments are interpreted to have arcuate faulted western boundaries and one is fault bounded in the east. Their segmentation and size resemble rifts in East Africa and the Gulf of Suez. In the Cape Cross–Walvis Bay area, faults are divided into two distinct trends by the Autseib Lineament. This lineament acted as an accommodation zone during Karoo rifting, separating differing styles of faulting controlled by underlying Pan-African basement fabrics. Faults are generally steep dipping but some low-angle faulting is present in the north of this complex area. The basement structure may have also influenced the behaviour of the rift basin during subsequent

Mid-Triassic inversion. In the south, broad folds developed, oriented sub-parallel to the Damaran basement trends, with structural highs later stripped of probable Lower Karoo sediments to isolate the Swakop Basin. North of the Autseib Lineament, some regional uplift was compounded by strong inversion focused along large faults striking along the Namib Rift.

Regional fault trends suggest oblique north-northeast–south-southwest oriented extension, interpreted as pure shear McKenzie-type rifting despite the Karoo post-rift sequence being less extensive than the rift sequence. This departure from the McKenzie model is ascribed to modifications in post-rift subsidence by the Mid-Triassic uplift, inversion and erosion, driven by north-northeast–south-southwest oriented compression arising from the final compressive event of the Cape Orogeny.

This work was sponsored by Amerada Hess Ltd. We are also grateful to Amerada Hess Ltd and partners, Ranger UK Ltd, and Hardy Oil and Gas plc for permission to publish this paper. We are also grateful to Roy Miller, Jean Malan and Simon Milner of NAMCOR for many stimulating discussions, and to NAMCOR for permission to publish this paper.

References

BINKS, R. M. & FAIRHEAD, J. D. 1992. A plate tectonic setting for Mesozoic rifts of West and Central Africa. *Tectonophysics*, **213**, 141–151.

BOSWORTH, W. 1985. Geometry of propagating continental rifts. *Nature*, **316**, 625–627.

BROWN, R. 1992. *A fission track study of the sub-aerial continental margins of southern Africa*. PhD Thesis, La Trobe University, Bundoora, Victoria.

CHOROWICZ, J. 1989. Transfer and transform fault zones in continental rifts: examples in the African–Arabian system. Implications of crustal breaking. *Journal of African Earth Sciences*, **8**, 203–214.

CLEMSON, J. 1997. *Segmentation of the Namibian Passive Margin*. PhD Thesis, Imperial College, University of London, UK.

——, CARTWRIGHT, J. & BOOTH, J. 1997. Structural segmentation and the influence of basement structure on the Namibian passive margin. *Journal of the Geological Society, London*, **154**, 477–482.

COLE, D. I. 1992. Evolution and development of the Karoo Basin. *In*: DE WIT, M. J. & RANSOME, I. G. D. (eds) *Inversion Tectonics of the Cape Fold Belt, Karoo and Cretaceous Basins of Southern Africa*. Balkema, 87–100.

COLLETTA, B., LE QUELLEC, P., LETOUZEY, J. & MORETTI, I. 1988. Longitudinal evolution of the Suez rift structure (Egypt). *Tectonophysics*, **153**, 221–233.

DALY, M. C., LAWRENCE, S. R., DIEMU-TSHIBAND, K. & MATOUANA, B. 1992. Tectonic evolution of the Cuvette Centrale, Zaire. *Journal of the Geological Society, London*, **149**, 539–546.

——, ——, KIMUN'A, D. & BINGA, M. 1991. Late Palaeozoic deformation in central Africa: a result of distant collision? *Nature*, **350**, 605–607.

D'ENGELBRONNER, E. R. 1996. New palynological data from Karoo sediments, Mana Pools Basin, northern Zimbabwe. *Journal of Africa Earth Sciences*, **23**, 17–30.

DE WIT, M. J. 1992. The Cape Fold Belt: A challenge for an integrated approach to inversion tectonics. *In*: DE WIT, M. J. & RANSOME, I. G. D. (eds) *Inversion tectonics of the Cape Fold Belt Karoo and Cretaceous basins of southern Africa*. Balkema, Rotterdam, 3–12.

DINGLE, R. V., SIESSER, W. G. & NEWTON, A. R. 1983. *The Mesozoic and Tertiary Geology of Southern Africa*. Balkema.

GAWTHORPE, R. L. & HURST, J. M. 1993. Transfer zones in extensional basins: their structural style and influence on drainage development and stratigraphy. *Journal of the Geological Society, London*, **150**, 1137–1152.

GIDSKEHAUG, A., CREER, A. M., & MITCHELL, J. G. 1975. Palaeomagnetism and K/Ar ages of the South-west African Basalts and their Bearing on the Time of Initial Rifting of the South Atlantic Ocean. *Geophysical Journal of the Royal Astronomical Society* **42**, 1–20.

GRADSTEIN, F. M., AGTERBERG, F. P., OGG, J. G., HARDENBOL, J., VAN VEEN, P., THEIRRY, J. & HUANG, Z. 1995. A Mesozoic time scale. *Journal of Geophysical Research*, **99**, 24 051–24 074.

GRESSE, P. G. & GERMS, G. L. B. 1993. The Nama foreland basin: sedimentation, major unconformity bounded sequences and multisided active margin advance. *Precambrian Research*, **63**, 247–272.

HILLER, K. & BUTTKUS, B. 1996. Structural style and sedimentary thicknesses in the Zambezi Rift Valley, Zimbabwe – investigation of the potential for hydrocarbons. *Zeitschrift fur Angerandte Geologie*, **42**, 132–137.

HORSTHEMKE, E., LEDENDECKER, S. & PORADA, H. 1990. Depositional environments and stratigraphic correlation of the Karoo sequence in northwestern Damaraland. *Communications of the Geological Survey of Namibia*, **6**, 63–73.

JOHNSON, M. R., VAN VUUREN, W. F., HEGENBERGER, W. F., KEY, R. & SHOKO, U. 1996. Stratigraphy of the Karoo Supergroup in southern Africa: an overview. *Journal of African Earth Science*, **23**, 3–16.

KENT, L. E. 1980. *Stratigraphy of South Africa*. Geological Survey of South Africa, Handbook **8**.

LIGHT, M. P. R., MASLANYJ, M. P., GREENWOOD, R. J. & BANKS, N. L. 1993. Seismic sequence stratigraphy and tectonics offshore Namibia. *In*: WILLIAMS, G. D. & DOBB, A. (eds) *Tectonics and Seismic Sequence Stratigraphy*. Geological Society, London, Special Publications, **71**, 163–191.

MCKENZIE, D. 1978. Some remarks on the development of sedimentary basins. *Earth and Planetary Science Letters*, **40**, 25–32.

MASLANYJ, M. P., LIGHT, M. P. R., GREENWOOD, R. J. & BANKS, N. L. 1992. Extension tectonics offshore Namibia and evidence for passive rifting in the South Atlantic. *Marine and Petroleum Geology*, **9**, 590–601.

MILANI, E. J. 1992. Intraplate tectonics and the evolution of the Paraná Basin, SE Brazil. *In*: DE WIT, M. J. & RANSOME, I. G. D. (eds) *Inversion Tectonics of the Cape Fold Belt 'Karoo' and Cretaceous Basis of Southern Africa*. Balkema, 101–108.

—— & DAVISON, I. 1988. Basement control and transfer tectonics in the Reconcavo–Tucano–Jatoba rift, Northeast Brazil. *Tectonophysics*, **154**, 41–70.

MORLEY, C. K. 1995. Developments in the structural geology of rifts over the last decade and their impact on hydrocarbon exploration. *In*: LAMBIASE, J. J. (ed.) *Hydrocarbon Habitat in Rift Basins*. Geological Society, London, Special Publications, **80**, 1–31.

——, NELSON, R. A., PATTON, T. L. & MUNN, S. G. 1990. Transfer zones in the East African Rift System and their relevance to hydrocarbon exploration in rifts. *AAPG Bulletin*, **74**, 1234–1253.

NELSON, R. A., PATTON, T. L. & MORLEY, C. K. 1992. Rift–segment interaction and its relation to hydrocarbon exploration in continental rift systems. *AAPG Bulletin*, **76**, 1153–1169.

PORADA, H. 1979. The Damara–Ribeira Orogen of the Pan-African-Brasiliano cycle in Namibia (southwest Africa) as interpreted in terms of continental collision. *Tectonophysics*, **57**, 237–265.

—— 1989. Pan-African rifting and orogenesis in southern to Equatorial Africa and eastern Brazil. *Precambrian Research*, **44**, 103–136.

ROSENDAHL, B. R. 1987. Architecture of continental rifts with special reference to east Africa. *Annual Review of Earth and Planetary Sciences*, **15**, 445–503.

SAYLOR, B. Z., GROTZINGER, J. P. & GERMS, G. J. B. 1995. Sequence stratigraphy and sedimentology of the Neoproterozoic Kuibis and Schwarzrand Subgroups (Nama Group), southwestern Namibia. *Precambrian Research*, **73**, 153–171.

STANDLEE, L. A., BRUMBAUGH, W. D. & CAMERON, N. R. 1992. Controlling factors in the initiation of the South Atlantic Rift System. *Géologie Africaine: Coll. Géol. Libreville, recueil des communic.*, **6–8 May 1991**, 141–152.

STOAKES, F. A., CAMPBELL, C. V., CASS, R. & UCHA, N. 1991. Sequence stratigraphic analysis of the Punta Del Este Basin, offshore Uruguay, South America. *AAPG Bulletin*, **75**, 219–240.

TRUDGILL, B. D. & CARTWRIGHT, J. A. 1994. Relay ramp forms and normal fault linkages – Canyonlands National Park, Utah. *Bulletin of the Geological Society of America*, **106**, 1143–1157.

UCHUPI, E. 1989. The tectonic style of the Atlantic Mesozoic rift system. *Journal of African Earth Sciences*, **8**, 143–164.

—— & EMERY, K. O. 1991. Pangaean divergent margins: historical perspective. *Marine Geology*, **102**, 1–28.

VEEVERS, J. J. 1991. Mid-Triassic lacuna on the Gondwanaland Platform during the final coalescence and incipient dispersal of Pangaea. *Gondwana Seven Proceedings*, Sao Paulo, Brazil, 603–613.

WERNICKE, B. 1985. Uniform-sense normal simple shear of the continental lithosphere. *Canadian Journal of Earth Sciences*, **22**, 108–125.

Falkland Plateau evolution and a mobile southernmost South America

P. F. BARKER

British Antarctic Survey, Madingley Road, Cambridge CB3 0ET, UK

Abstract: Assessment of southwest Gondwana break-up and its implications for regional hydrocarbon prospectivity must now take into account the origin of the southeast margin of the Falkland Islands as a volcanic rifted continental margin, and of the floor of the major part of the Falkland Plateau Basin as elevated oceanic crust. A reconstruction of the Falkland Plateau against southern Africa shows a southward extensional widening of the Outeniqua Basin across the line of the Falkland–Agulhas Fracture Zone, changing from stretched continental to oceanic crustal structure. Such a model for Outeniqua Basin opening, and the independent westward and clockwise rotation of the Falkland Islands block, suggests that southernmost South America was also a collection of microplates moving independently within a generally extensional environment in the Late Triassic–Early Jurassic. This is incompatible with assumptions of a rigid southernmost South America over this time, and a dominant role for a continuous dextral strike-slip Gastre Fault.

For almost 30 years, models of the evolution of southwest Gondwanaland have had to deal with the unacceptable overlap of the Falkland Plateau and West Antarctica, pointed out by Smith & Hallam (1970). They have done so by regarding West Antarctica as a mosaic of mobile microplates at the margin of a stable East Antarctic craton during the early part of post-break-up evolution (e.g., Barker & Griffiths 1977; Storey *et al.* 1988). The stability of the Falkland Plateau itself had been questioned implicitly by Adie (1952) on the basis of comparisons of Falkland Islands and southern African geology. However, this option was only raised again when palaeomagnetic data from onshore dykes of Karoo age (and chemistry) suggested that the Falkland Islands had undergone independent post-emplacement rotation (Taylor & Shaw 1989; Mussett & Taylor 1994). Although many workers now accept such rotation for its insights into onshore geology (e.g. Cole 1992; Curtis & Hyam 1998) a minority reject it (e.g. Richards *et al.* 1996), citing uncertainties in the palaeomagnetic data. Other aspects of this problem, relevant to regional hydrocarbon prospectivity, were not pursued: the nature and origin of much of the Falkland Plateau remained enigmatic (Lorenzo & Mutter 1988), and consequences for the evolution of southernmost South America (e.g. Rapela & Pankhurst 1992) were not fully examined. However, a recent analysis of magnetic, gravity and seismic data from the western Falkland Plateau allows these aspects to be considered further.

The geophysical data have been interpreted by Barker (1999) to show that the southeastern margin of the Falkland Islands block is a volcanic rifted continental margin, and that the central basin province of the Falkland Plateau is largely oceanic, produced by 'subaerial' seafloor spreading shortly after 190 Ma. The linear gravity and magnetic anomalies at the Falkland Islands margin, the crustal structure of the basin constrained by gravity and seismic reflection data, the existing seismic refraction data set from the basin (when corrected for the effects of thermal subsidence and sediment load) and the character of deep seismic reflectors, all support this interpretation. Further, the existence of dolerite dykes onshore, with chemistry, age and magnetic remanence implying formation within the Karoo–Bouvet hotspot province, and linked offshore dykes detected by their magnetic signature, make such a structure and origin entirely plausible.

Strictly, identification of a volcanic rifted margin southeast of the Falkland Islands, and designation of an oceanic origin for most of the Falkland Plateau Basin, do not demonstrate independent rotation of the Falkland Islands block. However, if the Falkland Islands block has rotated away from southern Africa, as the balance of geological evidence now overwhelmingly suggests, then rotation most probably began with creation of the volcanic rifted southeastern block margin, and continued with formation of the largely oceanic basin province. These associations bear on several aspects of the evolution of southwest Gondwana: in particular, on the other margins of the Falkland Islands block and the environment in which it may have moved, on the relationship of the Falkland Plateau Basin to the Outeniqua Basin of South Africa, and on the role of the Gastre Fault of South America.

Margins of the Falkland Islands block

If the association of linear gravity high and strongly lineated magnetic anomalies, together with a dyke swarm offshore, serve to define the

BARKER, P. F. 1999. Falkland Plateau evolution and a mobile southernmost South America. *In:* CAMERON, N. R., BATE, R. H. & CLURE, V. S. (eds) *The Oil and Gas Habitats of the South Atlantic.* Geological Society, London, Special Publications, **153**, 403–408.

southeastern (ocean–continent) boundary of an independent Falkland Islands block, and if, as seems most likely, this block migrated westward and rotated about a near pole to the north (i.e. clockwise) during the earliest stages of Gondwana break-up, then what were its other boundaries? Near-pole rotation is not unknown, but the larger the block, the more difficult is a large, near-pole rotation to envisage.

A free-air gravity high and associated shallow-origin magnetic anomalies are seen elsewhere close to the Falkland Islands: a gravity high extends west-northwestward from the southern end of West Falkland, and a high with a similar orientation lies to the north of both islands. Platt & Philip (1995) report magnetic anomalies approximately coincident with the gravity high west of the Falkland Islands, and attribute them to igneous rocks, citing the mid-Jurassic basaltic dykes found onshore (Taylor & Shaw 1989; Mussett & Taylor 1994) and the essentially acidic late Jurassic Serie Tobifera volcanics found at depth beneath the Magallanes Basin to the west, and elsewhere in Patagonia (e.g. Winslow 1982). Neither gravity high is so well formed as that at the southeast margin, but they are similar in length and peak gravity value. The west-northwest trending dykes exposed onshore on West Falkland (Greenway 1972), dated at 189 Ma by Mussett & Taylor (1994), would run into one of these gravity highs if extended westward offshore. The gravity highs may mark block margins created at the same time as the southeast margin, and in the same way. They probably define the smallest block that is plausible, and are used in what follows: for a larger block, the consequences of block rotation become greater and more difficult to explain.

Reconstruction to 130 Ma

The position of the Falkland Plateau before South Atlantic opening, and relations to South African geology, have been considered in great detail by Dingle *et al.* (1983) and by many others. Figure 1a is a reconstruction to *c.* 130 Ma of a rigid Falkland Plateau against southern Africa that shows the Outeniqua Basin, the outline of Maurice Ewing Bank and the proposed Falkland Islands block, and the extent of the Falkland Plateau Basin. A shallow water, presumed mainly continental, area of the Falkland Plateau lies north of the Falkland Islands block (the northwest Falkland Plateau). Basement in the northwest Falkland Plateau is generally shallow, but it contains the narrow, north–south North Falkland Basin. It appears to be a southward extension of the Columbine–Agulhas Arch of southern Africa. The Falkland Plateau Basin may be seen as a southward extension of the Outeniqua Basin, and Maurice Ewing Bank as a southward extension of the Port Alfred Arch (east of the Outeniqua Basin) and the region to the northeast. Lorenzo & Mutter (1988) map a small area of thinner sediment cover and block-faulted basement close to the northern margin of the Falkland Plateau that lies adjacent to the Outeniqua Basin in the reconstruction, and is probably a southward extension of its thinned continental crust.

Outeniqua Basin, South Africa

The Outeniqua Basin (Dingle *et al.* 1983) is a collection of smaller basins, exposed onshore, that expand and merge southeastward on the east coast of South Africa between the Columbine–Agulhas Arch in the west and the Port Alfred Arch in the east. The basins are asymmetric: each has a bounding normal fault on its northern side, downthrown to the south. The faults are west–east, curving around to northwest–southeast at their eastern end, and to north–south offshore. Basement, where seen, is downfaulted continental and the style is of continental extension, increasing in extent to the southeast. The only constraint on basin age is a whole-rock K–Ar date of 162±7 Ma on lavas within a basal conglomerate (McLachlan & McMillan 1976; Dingle *et al.* 1983). The Falkland Plateau Basin may be seen as a southward extension of the Outeniqua Basin, that was therefore wider still, opening sufficiently to create ocean floor. The transition between continental stretching in the north and seafloor spreading in the south is taken here to be the southern boundary of the thinned continental crust province in Fig. 1a, on the basis of seismic character. A modern analogue is the zone of stretched continental crust in North Island, New Zealand, in the south, leading to oceanic crust in the Lau-Havre Basin to the north (Karig 1970; Lewis & Pantin 1984; Gamble & Wright 1995).

Reconstruction to 190 Ma

Figure 1b shows a reconstruction to 190 Ma, before any independent rotation of the Falkland Islands block. The reconstruction is incomplete but illustrates the consequences of what is now known and the associated uncertainties. The stretched continental crust of the Outeniqua Basin is closed (nominally, its southern end is

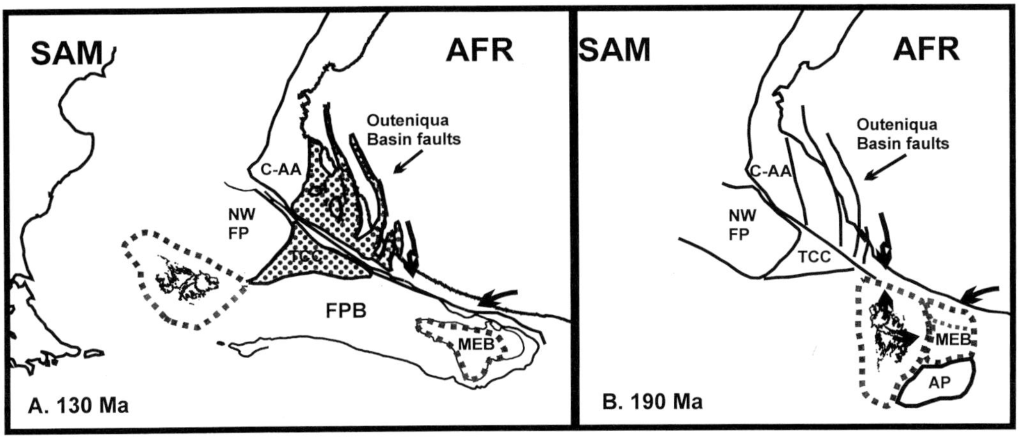

Fig. 1. Schematic reconstruction of Falkland Islands block and Maurice Ewing Bank against southern (**a**) *c.* 130 Ma before South Atlantic opening and (**b**) *c.* 190 Ma before Falkland Islands block rotation. Shaded areas in (a) are the Outeniqua Basin and corresponding thinned continental crust (TCC) of Falkland Plateau. Arrows mark strike projections of Lafonia and Dwyka tillites. Dotted line around Falkland Islands marks positions of block boundaries based on magnetic and gravity anomalies. MEB, Maurice Ewing Bank; FPB, Falkland Plateau basin province; AP, Agulhas Plateau; NWFP, northwest Falkland Plateau, either [as drawn in (b)] unmodified continent attached to Columbine–Agulhas Arch (C–AA), or part of a more mobile, microcontinent province of southernmost South America.

shortened by 80 km, representing an original extension of *c.* 150%). The northernmost basin (TCC) province is shortened also, but the northwest Falkland Plateau is unchanged. Maurice Ewing Bank is against the Africa margin (closing the St Johns Basin: Dingle *et al.* 1983), and made more compact [it was stretched by block faulting (Lorenzo & Mutter 1988), presumably during break-up]. The Falkland Islands block is restored to the position favoured on geological grounds (Adie 1952; Visser 1987; Curtis & Hyam 1998) and supported by palaeomagnetic measurement (Taylor & Shaw 1989; Mussett & Taylor 1994). The arrows on the Falkland Islands block and the African margin mark the line of outcrop of the correlated Lafonian and Dwyka tillites, which is insensitive to exposure level as the tillites have a regional dip into the Karoo Basin. The conjugate margin to the volcanic southeast margin of the Falkland Islands block is a section of the African margin off East London. The line of opening runs from the western end of the (closed) Outeniqua Basin faults, north of the Falkland Islands block, then south between there and Maurice Ewing Bank. The Agulhas Plateau, partly continental (Tucholke *et al.* 1981) and separating from the Falkland Plateau during South Atlantic break-up, is located next to Maurice Ewing Bank. The region to the south is not considered here, but the Cape Fold Belt may continue towards Antarctica with the present northwest trending chain of islands off West Falkland, with associated gravity low, forming a useful marker for West Antarctic reconstruction (e.g. Curtis & Storey 1996).

Pre-Cretaceous southernmost South America

Figure 1b illustrates the general problem of Falkland Islands block rotation. If the block rotated clockwise away from South Africa and Maurice Ewing Bank, as many lines of evidence now suggest, then one of two main conditions must have been fulfilled. Either the block converged upon a stable southernmost South America to the west, to produce underthrusting and ultimately a suture, or the motion took place within an essentially mobile and extensional southern South American environment, involving other, equally mobile, small blocks to the west of it and behind a west-migrating subducting Pacific margin. For example, in Fig. 1b, either the Falkland Island block converged upon a static northwest Falkland Plateau block as it rotated, or the plateau and its surroundings were themselves also in process of migration and possible deformation, so that their positions 190 Ma ago were not as drawn here, but farther east, and their shapes were probably different too. Figure 1b poses the question but does not solve it.

Rapela & Pankhurst (1992) did not appreciate

this dilemma. They hypothesized a northwest–southeast trending dextral Gastre Fault Zone in Patagonia, based on two features: at Gastre, some 600–700 km along-strike from the Pacific margin, a 30 km wide mylonite zone cuts plutonic rocks that yield a 208 Ma Rb–Sr age for dextral strike-slip displacement. At the Atlantic coast, pervasive normal faulting with the same strike affected acidic volcanics, Rb–Sr dated at 178 Ma. The curved line joining these two occurrences separates regions of different Mesozoic geology, characterized as a stable platform to the north, with a narrow Pacific margin component, and a more mobile environment to the south with a more widely distributed subduction signature. Rapela & Pankhurst (1992) pointed out that this line is virtually homopolar with the Falkland–Agulhas Fracture Zone, which formed the northern boundary of a by-then, rigid Falkland Plateau during South Atlantic opening (130 Ma onward). They hypothesized a precursor to the Falkland–Agulhas Fracture Zone, continuous with the Gastre Fault Zone and active during a 208–178 Ma strike-slip episode, as the means whereby the Falkland Islands block moved westward. Under this hypothesis, a rigid, more easterly southern Patagonia and Falkland Plateau lay south of the Gastre Fault Zone–Falkland-Agulhas Fracture Zone line before 208 Ma, and the Pacific margin of Gondwanaland had a pronounced corner at present-day 38°S. In view of the key role envisaged for the Gastre Fault Zone–Falkland-Agulhas Fracture Zone and the assumed rigidity of South America to the south, this may be called the 'single slice' model of Triassic–Jurassic evolution of southernmost South America.

Falkland Islands block rotation and Falkland Plateau Basin opening at the southern end of the Outeniqua Basin (a 'splay' model) are incompatible with this role for the Gastre Fault. The apparent continuity of older structures across the line of the Cretaceous strike-slip motion along it, and the rotation involved in opening the Outeniqua Basin, would have degraded an earlier coincidence of Euler poles. Both alternatives for the environment of Falkland Islands rotation, convergence with suturing or pervasive extension, are incompatible with the extent of Gastre Fault strike-slip envisaged by Rapela & Pankhurst (1992).

This is not to deny the reality of dextral strike-slip in Patagonia, or the associated ages. These observations can easily be reconciled with Falkland Plateau development: widespread westward extension at the southwest Gondwana margin, from before 200 Ma until ultimate major break-up, could have been taken up partly by dextral shear at the margins of several small rigid blocks, as at the Gastre Fault and others within southern South America farther south, as well as by the kind of oroclinal bending and near-pole extension represented by the 'splay' model for Outeniqua Basin opening. The Gastre Fault may have marked the northern boundary of a more mobile Triassic–Jurassic southernmost South America, but was not necessarily long-lived, or more extensive eastward, or the sole site of dextral strike-slip.

Others have considered the regional tectonic environment in which Falkland Island block motion might have taken place (e.g. Ben-Avraham *et al.* 1993; Curtis & Storey 1996). The initial tectonic regime could have involved a combination of Karoo hotspot-induced divergence and regional back-arc extension close to the subducting Gondwana margin. Extension was initiated after convergent deformation of the Cape Fold Belt in the Triassic and (in South America) may have lasted until the late Early Cretaceous (Albian) start of closure of the Rocas Verdes back-arc basin in southern Chile (Barker *et al.* 1991; Dalziel 1992). Within this environment, convergence may be rare, but should be detectable since it would most probably create local uplift, and thus subaerial, sutured outcrop. To date, this has not been described. Detailed examination of Late Triassic–Early Jurassic outcrop and drilled subcrop in Patagonia seems essential, but large areas will remain inaccessible.

Outeniqua Basin opening

There is significant difference between the 162 Ma whole-rock K–Ar age for a lava in basal conglomerates in an Outeniqua sub-basin onshore (McLachlan & McMillan 1976; Dingle *et al.* 1983), and the 190 Ma age of doleritic dykes on the Falkland Islands. Cox (1992) notes a difference between the North Atlantic Volcanic Province where volcanism and spreading are near-synchronous, and the Karoo province where long delays are apparent in places. In partial resolution, he postulates an early (*c.* 190 Ma) North Atlantic Volcanic Province-like opening phase more directly related to early volcanism. This involved the Lebombo Monocline and possibly the Explora Wedge off Dronning Maud Land, and presumably the Falkland Islands block. The onset of Falkland Island block rotation may have preceded Outeniqua Basin opening, which could eliminate the geometric offset apparent in Fig. 1b; however, better age control would be useful here, since a chemical affinity is noted of the dated Outeniqua Basin lavas and Lebombo Monocline volcanics (Dingle *et al.* 1983).

Otherwise, constraints on Falkland Plateau Basin opening are loose: after 190 Ma and before 130 Ma. Seismic evidence based on DSDP Site 330 on Maurice Ewing Bank (Barker 1976) suggests that the basin existed by Oxfordian times, and initial opening some 7–15 Ma after dyke injection (see White 1977) seems most likely. Falkland Plateau Basin within 15 Ma would have involved half-spreading rates < 20 mm a^{-1}, so that opening could have been complete by 165–160 Ma.

Regional hydrocarbon prospectivity

It is beyond the scope of this brief note to discuss fully the implications for regional prospectivity of a rotated Falkland Islands block and an oceanic Falkland Plateau Basin. The main implications concern possible ages of early basin development, early thermal history and provenance of basin fill, and may extend beyond the Falkland Plateau itself, over large parts of southernmost South America. In general terms, widespread Late Triassic or Early Jurassic extension would lead to an earlier and broader distribution of terrigenous reservoir sands than has been assumed. So far as is known, circulation within the intra-Gondwana basins was restricted into the Aptian, so there is no shortage of source rocks. The age and origin of a small, isolated basin, such as the North Falkland Basin, remain enigmatic: a Jurassic opening, post-rotation and perhaps effectively black-arc, is not ruled out.

Conclusions

The identification of the southeast margin of an independent Falkland Islands block and of an oceanic origin for the central basin province of the Falkland Plateau raises questions about southwest Gondwana evolution that are relevant to regional hydrocarbon prospectivity. The basin province of the Falkland Plateau may be seen as a southward extension of the Outeniqua Basin of South Africa, which opened further, creating ocean floor, after 190 Ma. This is incompatible with a view of the Gastre Fault of southern South America as a singular, long-lived strike-slip fault that accommodated all of the westward movement of the Falkland Islands block, and of a rigid southernmost South America ahead of it. Rather, Falkland Islands block rotation implies either convergence upon other parts of southernmost South America (with consequences that are not seen in outcrop) or a more widespread extensional environment within southernmost South America in the Late Triassic–Early Jurassic, involving the independent westward movement of other small blocks behind a west migrating subducting Pacific margin. The Gastre Fault could mark the boundary between this mobile southern zone and a more rigid northern South America but would have been only one of several zones of relative motion between blocks.

I thank D. I. M. Macdonald and S. R. Lawrence for helpful suggestions in review.

References

ADIE, R. J. 1952. The position of the Falkland Islands in a reconstruction of Gondwanaland. *Geological Magazine*, **89**, 401–410.

BARKER, P. F. 1976. Correlations between sites on the eastern Falkland Plateau by means of seismic reflection profiles, Leg 36, DSDP. *In*: BARKER, P. F. & DALZIEL, I. W. D. (eds) *Initial Reports of the Deep Sea Drilling Project*. US Government Printing Office (Washington, DC), **36**, 971–990.

—— 1999. Evidence for a volcanic rifted margin and oceanic crustal structure for the Falkland Plateau Basin. *Journal of the Geological Society*, submitted.

—— & GRIFFITHS, D. H. 1977. Towards a more certain reconstruction of Gondwanaland. *Philosophical Transactions of the Royal Society*, *B*, **279**, 143–159.

——, DALZIEL, I. W. D. & STOREY, B. C. 1991. Tectonic development of the Scotia Arc region. *In*: TINGEY, R. J. (ed.) *Geology of Antarctica*. Oxford University Press, 215–248.

BEN-AVRAHAM, Z., HARTNADY, C. J. H. & MALAN, J. A. 1993. Early tectonic extension between the Agulhas Bank and the Falkland Plateau due to the rotation of the Lafonia microplate. *Earth and Planetary Science Letters*, **117**, 43–58.

COLE, D. I. 1992. Evolution and development of the Karoo Basin. *In*: DE WIT, M. J. & RANSOME, I. G. D. (eds) *Inversion Tectonics of the Cape Fold Belt, Karoo and Cretaceous Basins of Southern Africa*. Balkema, 87–99.

COX, K. G. 1992. Karoo igneous activity, and the early stages of the break-up of Gondwanaland. *In*: STOREY, B. C., ALABASTER, T. & PANKHURST, R. J. (eds) *Magmatism and the Causes of Continental Break-up*. Geological Society, London, Special Publications, **68**, 209–220.

CURTIS, M. L. & HYAM, D. M. 1998. Late Palaeozoic to Mesozoic structural evolution of the Falkland islands: a displaced segment of the Cape Fold Belt. *Journal of the Geological Society*, **155**, 115–129.

—— & STOREY, B. C. 1996. A review of geological constraints on the pre-break-up position of the Ellsworth Mountains within Gondwana: implications for Weddell Sea evolution. *In*: STOREY, B. C., KING, E. C. & LIVERMORE, R. A. (eds) *Weddell Sea Tectonics and Gondwana Break-up*. Geological Society, London, Special Publications, **108**, 11–30.

DALZIEL, I. W. D. 1992. Antarctica: a tale of two supercontinents? *Annual Reviews of Earth and Planetary Science*, **20**, 501–526.

DINGLE, R. V., SEISSER, W. G. & NEWTON, A. R. 1983. *Mesozoic and Tertiary Geology of Southern Africa.* Balkema.

GAMBLE, J. A. & WRIGHT, I. C. 1995. The southern Havre Trough. *In* TAYLOR B. (ed.) *Back-arc Basins: Tectonics and Magmatism.* Plenum, 29–62.

GREENWAY, M. E. 1972. The geology of the Falkland Islands. *British Antarctic Survey Science Reports,* **76**.

KARIG, D. E. 1970. Ridges and basins of the Tonga–Kermadec island arc system. *Journal of Geophysical Research,* **75**, 239–254.

LEWIS, K. B. & PANTIN, H. M. 1984. Intersection of a marginal basin with a continent: structure and sediments of the Bay of Plenty, New Zealand. *In*: KOKELAAR, B. P. & HOWELLS, M. F. (eds) *Marginal Basin Geology*. Geological Society, London, Special Publications, **16**, 121–135.

LORENZO, J. M. & MUTTER, J. C. 1988. Seismic stratigraphy and tectonic evolution of the Malvinas/Falkland plateau. *Revista Brasileira de Geociencias*, **18**, 191–200.

MCLACHLAN, I. R. & MCMILLAN, I. K. 1976. Review and stratigraphic significance of southern Cape Mesozoic palaeontology. *Transactions of the Geological Society of South Africa*, **79**, 197–212.

MUSSETT, A. E. & TAYLOR, G. K. 1994. ^{40}Ar–^{39}Ar ages for dykes from the Falkland Islands with implications for the break-up of southern Gondwanaland. *Journal of the Geological Society*, **151**, 79–81.

PLATT, N. H. & PHILIP, R. R. 1995. Structure of the southern Falkland Islands continental shelf: initial results from new seismic data. *Marine and Petroleum Geology*, **12**, 759–771.

RAPELA, C. W. & PANKHURST, R. J. 1992. The granites of northern Patagonia and the Gastre Fault System in relation to the break-up of Gondwana. *In*: STOREY, B. C., ALABASTER, T. & PANKHURST, R. J. (eds) *Magmatism and the Causes of Continental Break-up*. Geological Society, London, Special Publications, **68**, 209–220.

RICHARDS, P. C., GATLIFF, R. W., QUINN, M. F., WILLIAMSON, J. P. & FANNIN, N. G. T. 1996. The geological evolution of the Falkland Islands continental shelf. *In*: STOREY, B. C., KING, E. C. & LIVERMORE, R. A. (eds) *Weddell Sea Tectonics and Gondwana Breakup*. Geological Society, London, Special Publications, **108**, 105–128.

SMITH, A. G. & HALLAM, A. 1970. The fit of the southern continents. *Nature*, **225**, 139–144.

STOREY, B. C., DALZIEL, I. W. D., GARRETT, S. W., GRUNOW, A. M., PANKHURST, R. J. & VENNUM, W. R. 1988. West Antarctica in Gondwana: crustal blocks, reconstruction an break-up processes. *Tectonophysics*, **155**, 381–390.

TAYLOR, G. K. & SHAW, J. 1989. The Falkland Islands: new palaeomagnetic data and their origin as a displaced terrane from southern Africa. *In*: HILLHOUSE, J. W. (ed.) *Deep Structure and Past Kinematics of Accreted Terranes*. AGU Geophysical Monograph Series, **50**, 59–72.

TUCHOLKE, B. E., HOUTZ, R. E. & BARRETT, D. M. 1981. Continental crust beneath the Agulhas Plateau, southwest Indian Ocean. *Journal of Geophysical Research*, **86**, 791–3806.

VISSER, J. N. J. 1987. The palaeogeography of part of southwestern Gondwana during the Permo-Carboniferous glaciation. *Palaeogeography, Palaeoclimatology, Palaeoecology*, **61**, 205–219.

WHITE, R. S. 1997. Mantle plume origin for the Karoo and Ventersdorp flood basalts, South Africa. *South African Journal of Geology*, **100**, 271–282.

WINSLOW, M. A. 1982. The structural evolution of the Magallanes Basin and neotectonics in the southernmost Andes. *In*: CRADDOCK, C. (ed.) *Antarctic Geoscience*. University of Madison, Wisconsin, 143–154.

Tectono-stratigraphic evolution of the North Falkland region

S. R. LAWRENCE[1], M. JOHNSON[2], S. R. TUBB[1] & S. J. MARSHALLSEA[3]

[1]*Quad Consulting Limited/Exploration Consultants Ltd, Highlands Farm, Greys Road, Henley-on-Thames, Oxon RG9 4PR, UK*
[2]*Spectrum Energy and Information Technology Ltd, Spectrum House, 56 Goldsworth Road, Woking, Surrey GU21 1LQ, UK*
[3]*Geotrack International Pty Ltd, 37 Melville Road, Brunswick West, Victoria 3055, Australia*

Abstract: An interpretation of regional seismic data acquired between 1993 and 1995 has been used to establish a regionally consistent tectono-stratigraphic framework for the North Falkland Shelf. This incorporates three rift (extensional) and two post-rift sequences separated by important tectono-stratigraphic breaks representing unconformities at near-Base Upper Jurassic, near-Base Cretaceous, Mid-Cretaceous and Early Tertiary levels. In the structural framework derived from seismic interpretation two major zones are recognized. The northeast Graben Zone constitutes three tectono-stratigraphic provinces, the north–south orientated North Falkland Graben, and the Western Rift Complex and Southern Half-graben which are dominated by northwest–southeast rift trends. The Falkland Islands form part of the southwest Uplift Zone which has a tectono-stratigraphic history, shown by the interpretation of apatite fission track data, to be compatible with that of the North Falkland Shelf and southern South America. A regional northeast–southwest seismic line shows an extensional northern edge to the Falkland Plateau and no apparent westerly extension of the Agulhas–Falkland Fracture Zone. A 'rigid' or 'non-rotational' reconfiguration of southwest Gondwanaland (i.e. with no independent rotation of a Falkland Islands microplate) shows a regional consistency of Late Triassic–Mid-Jurassic back-arc and Late Jurassic–Early Cretaceous South Atlantic rift trends. This is presented as circumstantial evidence for a 'non-rotational' model for the break-up of southwest Gondwanaland.

Between 1993 and 1995 Spectrum Energy and Information Technology Ltd acquired over 11 000 km of speculative seismic data over the continental shelf north of the Falkland Islands. These data were acquired in two phases (see Fig. 1): (1) 1993–1994 – 7426 km in a northeast–southwest/northwest–southeast reconnaissance grid processed to 90-fold stack, 3128 km of which were reprocessed in 1995; (2) 1995 – 3735 km of north–south/east–west infill grid processed to 70-fold stack, 1043 km of which were reprocessed in 1996/97.

The interpretation of these data, which include a linking northeast–southwest regional seismic line (Line 100 on Fig. 1) transversing the northern margin of the Falkland Plateau, has provided some important new information on regional geology and the break-up of southwest Gondwanaland. In particular, it has provided a contribution to the argument concerning the original position of the Falkland Islands (as part of a Falkland Islands microplate) in the 'rotational' v. 'non-rotational' debate (see *inter alia* Ben-Avraham *et al.* 1993; Marshall 1994).

The study area lies over the north-central part of the Falkland Plateau to the north of the Falkland Islands (Fig. 2). The present-day plate tectonic setting shows the Falkland Islands occupying a central position on the Falkland Plateau which forms a 1200 km long continental promontory of South America. The South Atlantic margin of South America represents a passive margin which compliments the western coast of southern Africa, having formed by perpendicular opening of the South Atlantic during Gondwanaland break-up. Part of the northern edge of the Falkland Plateau is defined by the Falkland Escarpment which marks the Agulhas–Falkland Fracture Zone, a continent–ocean transform which became active during South Atlantic opening in the Early Cretaceous (Rabinowitz & La Brecque 1979). The eastern end of the Falkland Plateau is formed by a bathymetric feature, the Maurice Ewing Bank, which is separated from the main part of the Plateau by the Falkland Plateau Basin. Formation of this basin has been attributed to Mid–Late Jurassic rifting (Lorenzo & Mutter 1988). The southern margin of the Falkland Plateau is defined by another transform-related zone characterized by the Burdwood Bank, a transpressional feature (Platt & Philip 1995) formed during the indentation of the Scotia Plate beginning in the Early Tertiary.

LAWRENCE, S. R., JOHNSON, M., TUBB, S. R. & MARSHALLSEA, S. J. 1999. Tectono-stratigraphic evolution of the North Falkland region. *In:* CAMERON, N. R., BATE, R. H. & CLURE, V. S. (eds) *The Oil and Gas Habitats of the South Atlantic.* Geological Society, London, Special Publications, **153**, 409–424.

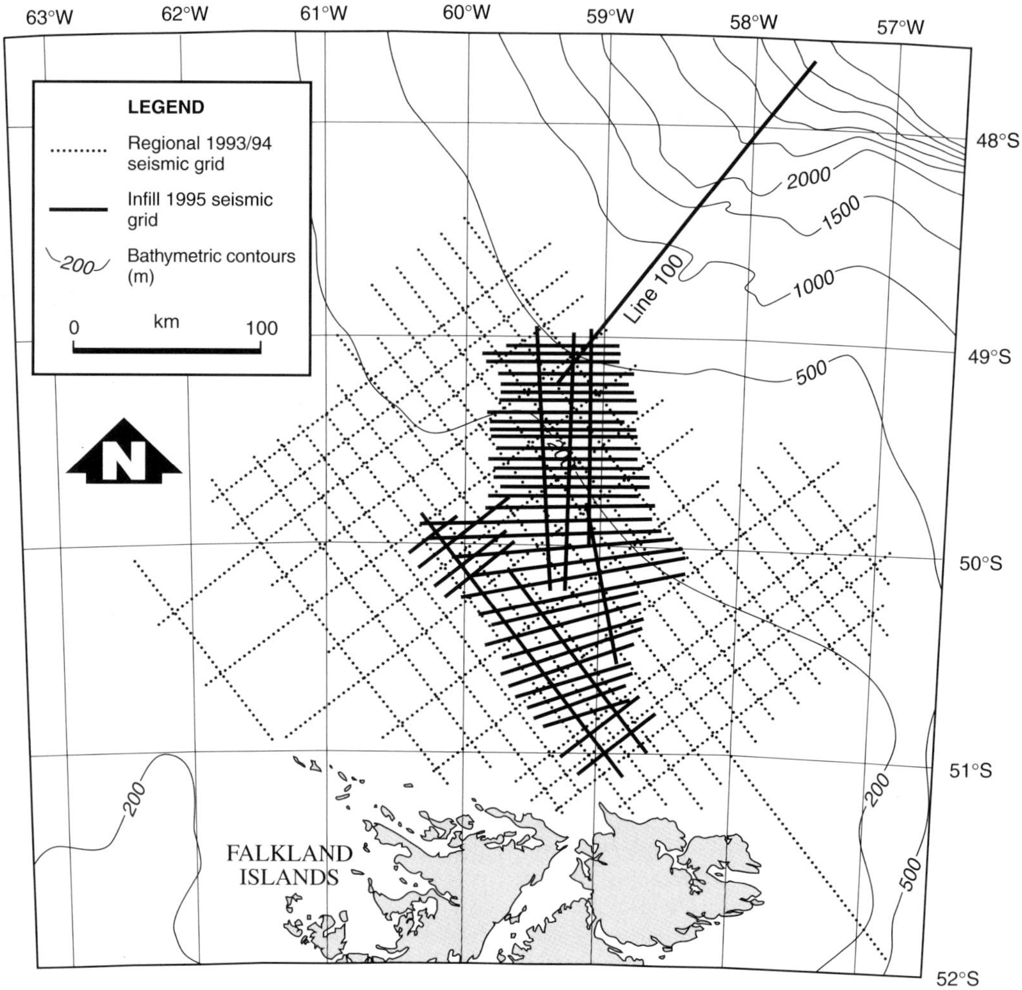

Fig. 1. Seismic data coverage over the North Falkland Shelf (data proprietary to Spectrum Energy and Information Technologies Ltd). Reconnaissance data acquired in 1993/1994 and infill data acquired in 1995.

Tectono-stratigraphy of the North Falkland Shelf

In the absence of direct stratigraphical control from well data, the stratigraphy of the North Falkland Shelf has been analysed tectono-stratigraphically, i.e. in terms of major sequences representing phases of basin subsidence and unconformities representing uplift or deformational events in the structural history. Thus, the stratigraphy has been analysed with a regional perspective and in the context of a basinal area developing as a consequence of Gondwanaland break-up and South Atlantic opening, but with Andean compressional influences. It should be noted that the authors do not view this approach as an alternative to seismic stratigraphic or sequence analysis but rather as an essential framework builder within which higher resolution studies can be conducted.

Correlation has been made with adjacent and related basins in southern South America and southern Africa, the outcrop geology of the Falkland Islands (Greenway 1972) and the DSDP boreholes drilled in Legs 36 (327, 329 and 330) and 71 (511 and 512) on the flanks of the Maurice Ewing Bank and Falkland Plateau Basin (Barker *et al.* 1977). Regional scientific seismic data acquired by the Lamont–Doherty Geological Observatory over the central and eastern Falkland Plateau (Lorenzo & Mutter 1988) have also been helpful in regional extrapolation. However, it should be recorded that there is no direct tie between the Lamont–Doherty and Spectrum datasets.

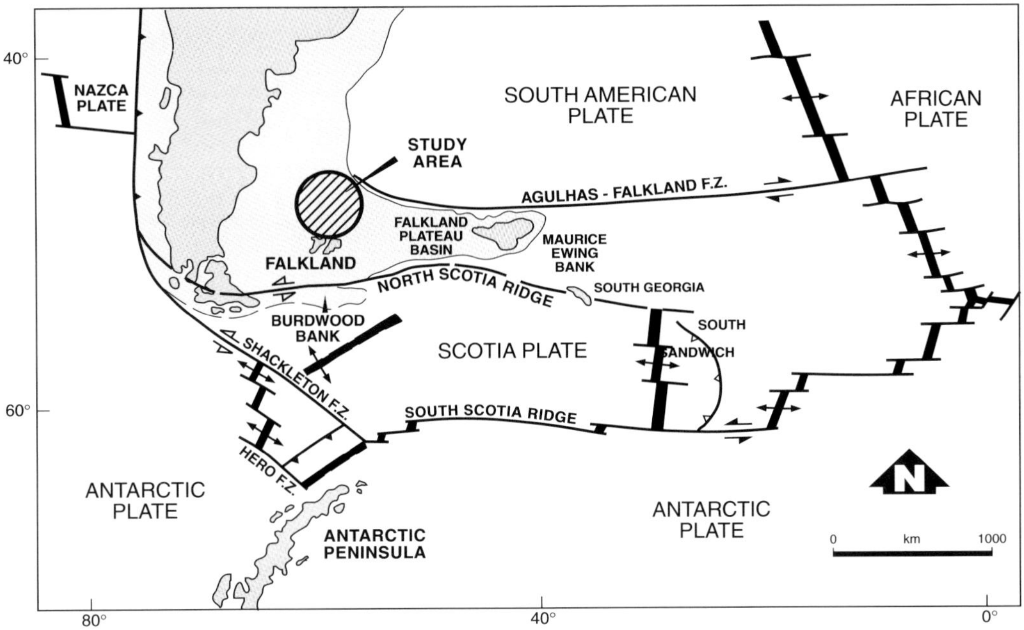

Fig. 2. The plate tectonic setting of the Falkland Plateau showing the location of the study area in the North Falkland Shelf and north-centrally within the Falkland Plateau.

Using this methodology a basic tectono-stratigraphic framework has been established (Fig. 3), which has helped to define two major zones: the southwest Uplift Zone, which includes the southern part of the North Falkland Shelf and the Falkland Islands; and the northeast Graben Zone where three major tectono-stratigraphic provinces are defined: North Falkland Graben, Western Rift Complex and Southern Half-graben (Fig. 4). The postulated stratigraphic significance of the key seismic horizons is described below.

Base Permo-Carboniferous

This horizon can only be seen over restricted parts of the database outside the main graben areas. It is thought to form the base of a pre-rift Upper Carboniferous–Permian sedimentary sequence [equivalent to the Lafonian Supergroup of the Falkland Islands (Greenway 1972)] which was folded as a result of Gondwanide (Permo-Triassic) orogenic compression. Little seismic character is observed in the underlying section, it being equated with metasediments of the Gran Malvinas Group of the Falkland Islands which are dominated by quartzites and quartzitic conglomerates (Greenway 1972).

Medial rift unconformity (near-Base Upper Jurassic)

This horizon represents a syn-rift unconformity. It is best identified in the Western Rift Complex (see Fig. 4) where it is associated with seismic character which has been attributed to volcanics. For this reason it is correlated with the regional Middle–Upper Jurassic volcanics of southern South America, the Tobifera and Chon-Aike Formations dated at *c.* 165 Ma (Gust *et al.* 1985).

In the Western Rift Complex, and the western sub-basin of the North Falkland Graben, seismic data (supported by gravity/magnetics modelling) indicate that there is a sedimentary section below this horizon. This postulated section is equated with the Late Triassic–Early Jurassic syn-rift section of southern South America (Uliana & Biddle 1988; Uliana *et al.* 1989).

Late rift unconformity (near-Base Cretaceous)

This is generally a strong reflector, forming the interface between a poorly imaged lower rift sequence and an upper rift section characterized by stronger and more continuous parallel reflec-

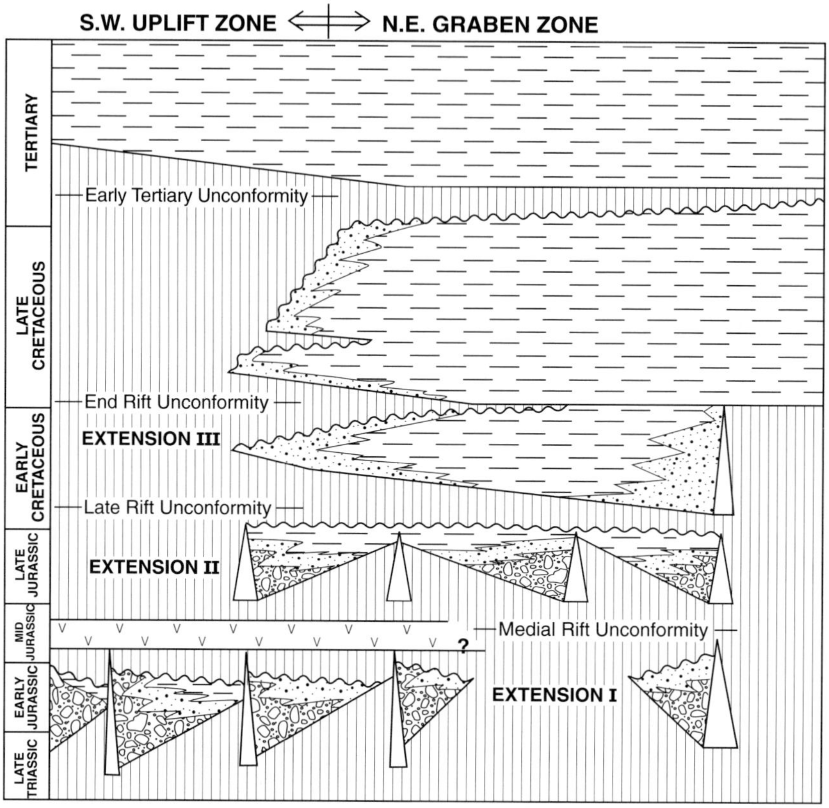

Fig. 3. Tectono-stratigraphy of North Falkland Shelf outlining major tectono-stratigraphic sequences separated by seismically defined unconformities. Three 'syn-rift' sequences are recognized separated by 'intra-rift' unconformities. The end-rift unconformity corresponds to a phase of Andean contraction in the Mid-Cretaceous.

tors. The nature of this late syn-rift unconformity is best observed in the eastern sub-basin of the North Falkland Graben (see Fig. 5a).

This horizon is equated with the so-called break-up unconformity of South Africa and southern South America. It therefore broadly corresponds to the beginning of seafloor spreading in the South Atlantic, where the earliest positively identified magnetic anomalies are M10 or M10N (Rabinowitz & La Brecque 1979). This places break-up, or the first oceanic crust formation, in the Valanginian or early Hauterivian, depending on the timescale used. In the Bredasdorp sub-basin of the Outeniqua Basin of South Africa the break-up unconformity has been designated the 1At1 sequence boundary which is dated as Valanginian (Burden 1992). In the San Jorge Basin, Fitzgerald *et al.* (1990) place the break-up unconformity at the base of the D-129 formation (121.5 Ma sequence boundary) at the Valanginian–Hauterivian boundary.

This seismic horizon is indirectly correlated with the U2 unconformity of the Falkland Plateau Basin which corresponds to a recognized Early Cretaceous (Tithonian–Barremian) lacuna in the DSDP boreholes (Barker *et al.* 1977; Lorenzo & Mutter 1988).

End rift unconformity (Mid-Cretaceous Unconformity)

This seismic horizon represents a regional unconformity which can be identified and mapped over most of the North Falkland Shelf (Fig. 4). Unconformable relationships are best seen in the Western Rift Complex where the underlying section has been compressionally deformed and eroded. In the North Falkland Graben this horizon is strongly unconformable across the basin margins, the western sub-basin and around the intrabasin structure represented by the central ridge (Fig. 5). Here it forms the base of a relatively well-layered sequence sometimes showing

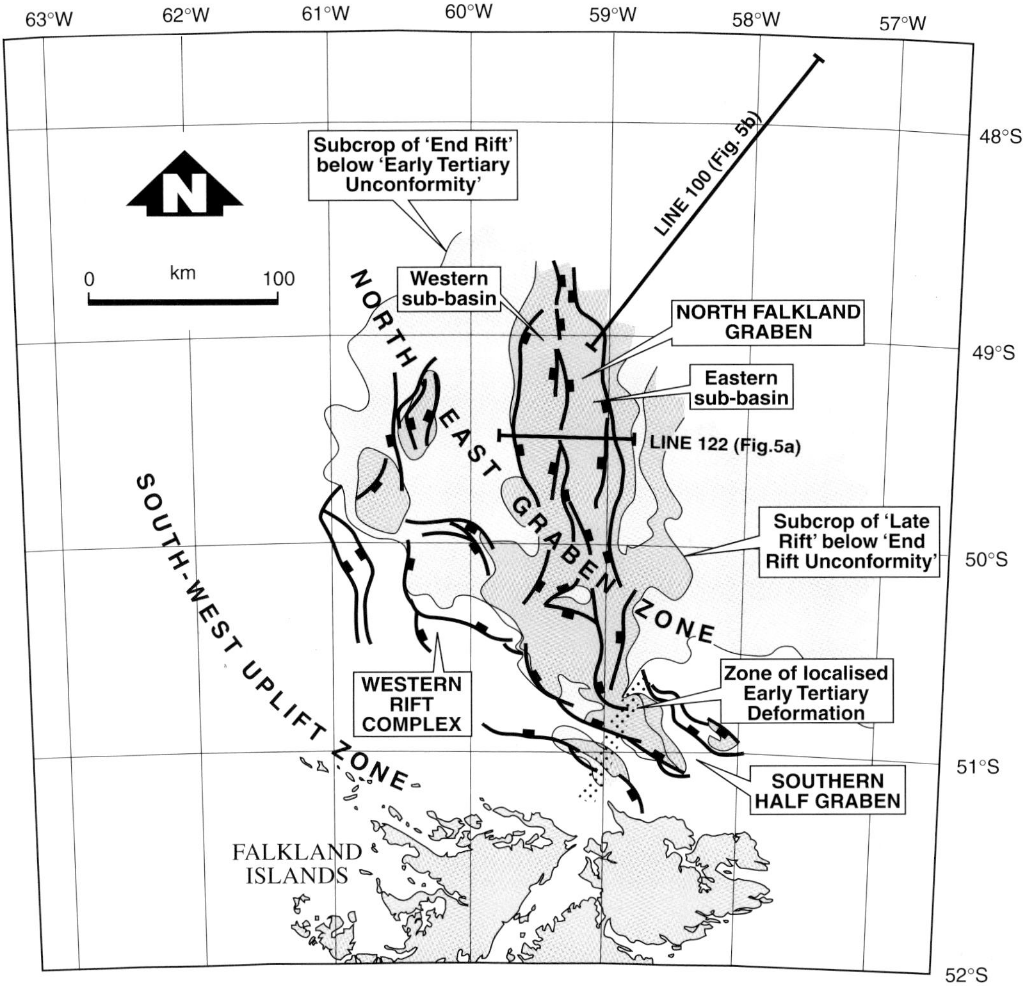

Fig. 4. Structural framework of the North Falkland Shelf highlighting two major geological terranes – the northeast Graben Zone and the southwest Uplift Zone – and three tectono-stratigraphic provinces – the North Falkland Graben, the Western Rift Complex and the Southern Half-graben.

onlap-'infill' relationships. In the Southern Half-graben this horizon is again seen to be truncational over intrabasin structures as well as basin margins.

This seismic horizon represents a major stratigraphic break marking the end of active rifting, possibly effected by a Mid-Cretaceous ('Sub-Hercynian') contractional phase recognized in the Andean province (Dalziel 1981, 1986; Dalziel & Forsythe 1985). It is correlated with the base of the Middle Inoceramus Formation in the Magallanes Basin which is placed within the Cenomanian by Biddle *et al.* (1986). By extrapolation it is correlated with the U3 unconformity of the Falkland Plateau Basin, which corresponds to a Mid-Cretaceous lacuna identified in DSDP boreholes (Barker *et al.* 1977; Lorenzo & Mutter 1988).

Early Tertiary Unconformity

This seismic horizon represents a regional unconformity which forms the base of a widespread shallow seismic sequence. It truncates deeper horizons to the southwest, south and southeast, and its surface subcrop depicts 'late' uplift of the Falkland Islands (Fig. 4). In the Southern Half-graben this horizon truncates strata affected by wrench-related deformation (see Lawrence & Johnson 1995).

This seismic horizon cannot be directly correlated with the U4 unconformity of the Falkland Plateau Basin, which represents a hiatus spanning the Cretaceous–Tertiary boundary (Maastrichtian–Palaeocene age) (Lorenzo & Mutter 1988). An Early Tertiary age is attributed to this unconformity for this study on the basis of its presumed causal relationship with the first phases of Scotia Plate indentation (see later).

Structural framework of the North Falkland Shelf

The structural framework established for the North Falkland Shelf by seismic interpretation shows a southwest Uplift Zone and a northeast Graben Zone with three tectono-stratigraphic provinces (Fig. 4).

The North Falkland Graben forms a north–south orientated graben with eastern and western sub-basins separated by a central ridge (Fig. 5a). The eastern sub-basin has the geometry of a basic half-graben with the controlling fault on its eastern side. Two thick rift sequences are positively identified corresponding to tectono-stratigraphic extensional phases II and III (see Fig. 3). The central ridge appears to be a fundamental rift structure (running most of the length of the graben), although its geometrical configuration and fault polarity varies along its strike length. The western sub-basin shows more frequent easterly and westerly throwing faults and is occupied by a shallower and thinner stratigraphic section, probably representing tectono-stratigraphic extensional phases I–III. The end-rift unconformity forms the truncational base to the post-rift sequence which oversteps the faulted margins of the basin (Fig. 5a).

The Western Rift Complex is characterized by a deformed sedimentary section below the end-rift unconformity. The deformation is interpreted as rift fault reactivation/inversion and associated folding. Although structure is complex, seismic and gravity/magnetics interpretation has determined basic northwest–southeast structural trends. Sediments deposited during extensional phases I and II are represented with possible remnants of phase III. The inversion of rift structures is here related to the Mid-Cretaceous deformation reported in the Andes (Dalziel 1981, 1986; Dalziel & Forsythe 1985).

The Southern Half-graben province is formed of a series of well defined arcuate half-graben with listric controlling faults. Graben trends are northwest–southeast with mostly, but not exclusively, northeast throwing controlling faults. Syn-rift sequences representing extensional phases II and III positively and phase I are tentatively identified. The Early Tertiary unconformity forms a distinct truncational unconformity delimiting what is interpreted to be strike-slip related deformation in the underlying section (Lawrence & Johnson 1995; Richards & Fannin 1997). This has formed a localized deformation zone (Fig. 4) which is postulated to be the product of transpressional reactivation of a northeast–southwest trending rift transfer zone.

Structural history of the North Falkland Shelf

The structural history derived from this tectono-stratigraphic interpretation indicates that the following key structural episodes affected the North Falkland Shelf:

- Folding of the 'pre-rift' sedimentary section depicted by the Base Permo-Carboniferous seismic marker and attributed to compression during the Permo-Triassic Gondwanide orogeny.
- Late Triassic–Mid-Jurassic rifting of extensional phase I, culminating in Mid–Late Jurassic volcanicity. Rifting of this phase is associated with back-arc extension in relation to the Samfrau active margin of South America (Dalziel 1981, 1986; Uliana & Biddle 1987; Uliana *et al.* 1989).
- Late Jurassic–Early Cretaceous rifting in extensional phases II and III, related to the opening of the South Atlantic. Phase II rifting (Late Jurassic–Valanginian) corresponds to regional continental separation precursor rifting. Phase III (Valanginian–Mid Cretaceous) represents a continuation of rifting into the continental separation phase associated with active movement on the Agulhas–Falkland Fracture Zone.

Fig. 5. (**a**) Regional seismic profile (Line 100) showing broad extensional nature of the edge of the Falkland Plateau north of the Falkland Islands. (**b**) Seismic section (Line 122) of the North Falkland Graben showing basic half-graben architecture with central ridge structure. Tectono-stratigraphic sequences representing extensional phases II and III are recognized. Post-rift section (above end-rift unconformity) oversteps basin margins.

- Mid-Cretaceous deformation evinced by compressional reactivation in the Western Rift Complex and movement associated with the central ridge of the North Falkland Graben. This is a well documented contractional episode in the Andes which was responsible for the closure of the Rocas Verdes Basin (Dalziel 1981, 1986; Dalziel & Forsythe 1985) and the beginnings of foreland basin development in the Magallanes Basin (Biddle *et al.* 1986). This deformational episode effectively terminated active rifting in the North Falkland Shelf. The effects are more strongly developed in the Western Rift Complex because of its closer proximity to the Andean contractional realm.
- Post-rift development was punctuated by two phases of compression related to Andean contraction and uplift. In the Early Tertiary a transpressional reactivation of rift transfer zones created a localized deformational zone in the Southern Half-graben province. This deformation has been related to the beginnings of indentation of the Scotia Plate, as recognized in Tierra del Fuego where strata as young as Eocene have been affected (Biddle et al. 1986). A late and gentle 'basin axis' inversion has affected the eastern sub-basin of the North Falkland Graben. Richards & Fannin (1997) mistakenly reported that Lawrence & Johnson (1995) had attributed this subtle deformation to Mid-Cretaceous inversion. This late episode is possibly related to a Mio-Pliocene phase of compression recognized in the Andes and responsible for the San Bernardo Foldbelt inversion zone in the western part of the San Jorge Basin (Fitzgerald et al. 1990).

Tectono-stratigraphic context of the Falkland Islands

The Falkland Islands form an emergent expression of the southwest Uplift Zone (Fig. 4). Structural history and style should be consistent with that of the North Falkland Shelf but there should be evidence for discrete episodes of uplift and erosion. The basic geology and structure of the Falkland Islands has been described by Greenway (1972) and more recently by Marshall (1994). Devonian metasediments in the western part of the West Falkland and the northern part of East Falkland are deformed in a series of folds with northwest–southeast axial trends. The eastern edge of West Falkland has a northeasterly structural grain manifest as the Hornby Mountain anticline and a coastal monocline with near-vertical dips into the Falkland Sound. Across a northwest–southeast monocline in central East Falkland, dips shallow-out in the Upper Lafonian Supergroup sediments which occupy the southern part of the island. These structural trends and relationships have been used to support a Falklands–Natal coast 'rotational' reconstruction of southwest Gondwanaland (Adie 1952; Ben-Avraham *et al.* 1993; Marshall 1994).

As described, the structural evidence from the immediate offshore is of northwest–southeast listric faults (of the Southern Half-graben province – Fig. 4) approaching the coast and conforming to the morphology of the northern coastline of both East and West Falkland, as well as Lower Lafonian Supergroup outliers mapped in northern West Falkland. On this basis, it is suggested that this listric faulting can be extrapolated across both Falkland Islands. In the model presented here, 'pre-rift' Lafonian Supergroup sediments in the southern half of East Falkland are preserved in a series of half-graben with listric northwest–southeast controlling rift faults (Fig. 6). This also explains the shape of the arcuate coastal inlets of East Falkland.

As described, the offshore half-graben have been subject to strike-slip tectonics in the Early Tertiary along the line of a possible rift transfer zone (see Fig. 4). A similar transpressional reactivation of an original rift transfer zone, it is proposed, would explain the structural effects seen along the Falkland Sound monocline.

The thermal history of the Falkland Islands, shown by the results of apatite fission track analysis (AFTA® – registered trade mark of Geotrack International) carried out by Geotrack International for this study, has been used to corroborate the tectono-stratigraphy history proposed. Apatite fission trade analysis and vitrinite reflectance data from a total of 23 outcrop samples ranging over the full stratigraphic section of East and West Falkland show clear evidence of cooling from elevated palaeotemperatures at some time in the geological past. The timing constraints from apatite fission track analysis for each sample are shown in Fig. 7.

The apatite fission track analysis data are consistent with an early palaeothermal episode during which the rocks reached their maximum palaeotemperatures. The data show that cooling from this early episode commenced during the 250–170 Ma period (Late Palaeozoic–mid-Mesozoic). Bearing in mind that the timing information in most apatite fission track analysis samples represents the time of cooling below *c.* 110°C, it is likely that in reality the thermal history was complex. From inspection of the allowed ranges in the timing of the earlier event

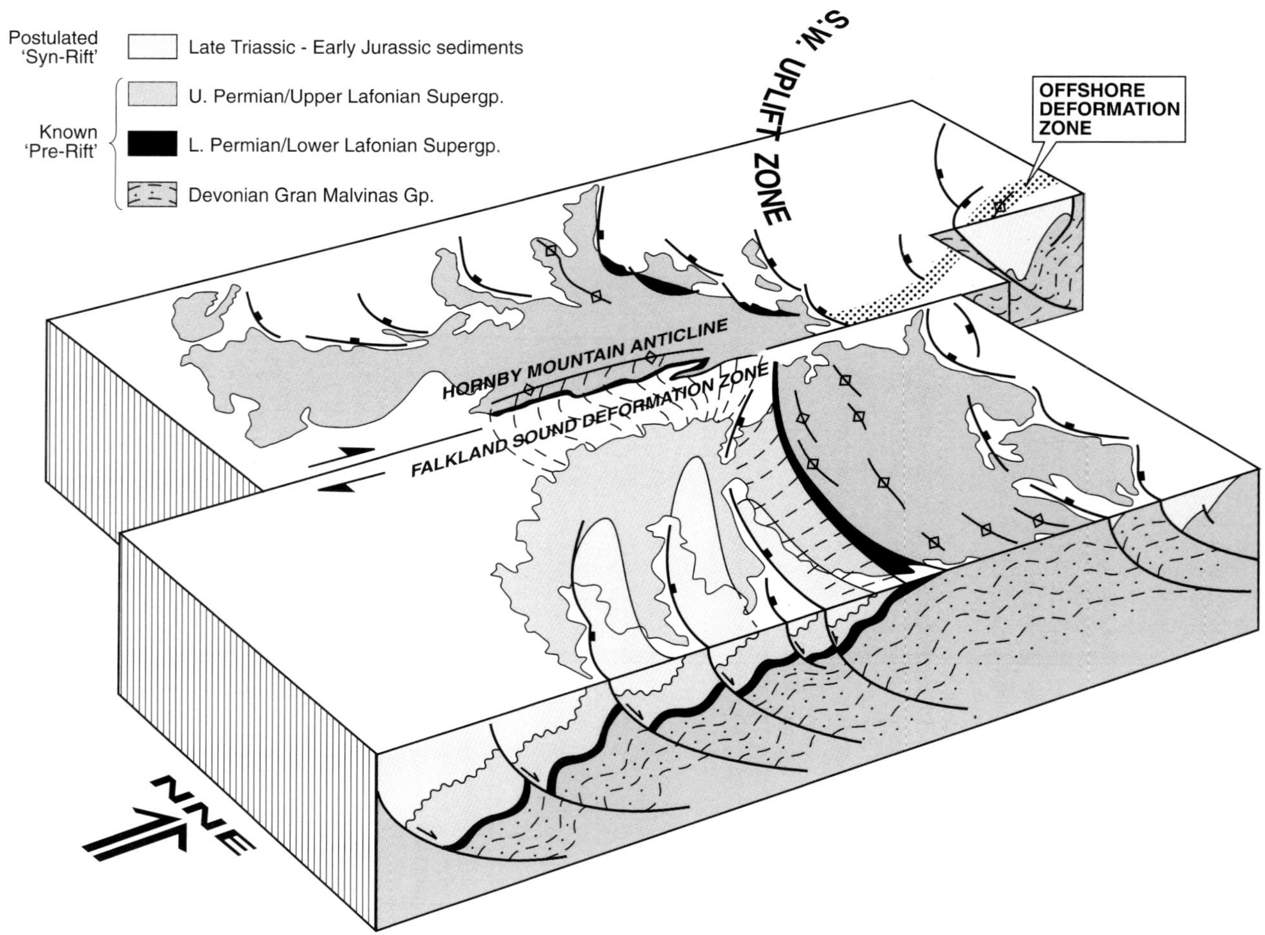

Fig. 6. Structural model proposed for the Falkland Islands showing basic half-graben morphology superimposed on Gondwanide fold structure with interfering transfer zone reactivation along Falkland Sound.

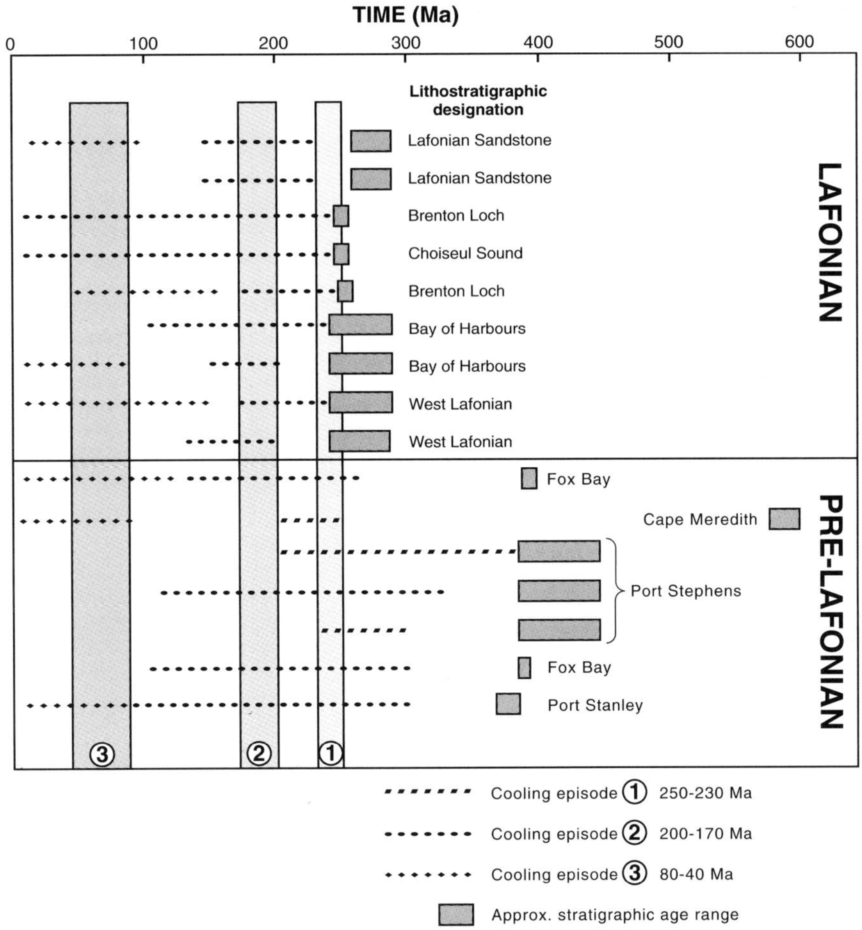

Fig. 7. Thermal history from AFTA® suggesting three episodes of cooling. 1 Permo-Triassic (Gondwanide); 2 Early–Mid Jurassic; 3 Late Cretaceous–Early Tertiary.

in each sample (Fig. 7) it can be seen that the data are compatible with two episodes. All the pre-Lafonian Supergroup samples are compatible with cooling commencing between 250 and 230 Ma. In contrast, the data from the Lafonian Supergroup suggest that the time of cooling may have been slightly later, with data from these samples being consistent with 200–170 Ma.

The apatite fission track analysis derived maximum palaeotemperatures during the early episode range from 90 to 95°C to > 110°C. Vitrinite reflectance values of 0.76–6.06% indicate maximum palaeotemperatures from 120 to > 250°C and are thus consistent with the apatite fission track analysis palaeotemperature values. Given this large range in maximum palaeotemperatures it can be envisaged that, if the early cooling episode was due to a prolonged period of emergence and erosion, the rocks from the highest palaeotemperatures would cool through the 110°C isotherm later than those which were *c.* 110°C when cooling commenced, which may explain the range in the time of cooling indicated by the data. No clear relationship between palaeotemperatures and stratigraphic age is noted, but this may be due simply to the paucity of sample coverage.

The apatite fission track analysis data from some samples clearly identify a later palaeothermal episode involving cooling from peak palaeotemperatures at some time from 80 to 40 Ma (latest Cretaceous–Early Tertiary). The nitrinite reflectance data contain no record of this later event, as reflectance levels in all samples were set by the maximum palaeotemperatures during the earlier heating episode, but the apatite fission track analysis data show peak palaeotemperatures of up to 90°C during this later episode in some samples.

Because all the data are from outcrop samples, it has not been possible to reconstruct palaeotemperature profiles or measure palaeogeothermal gradients in this study, and therefore no conclusions can be drawn directly from the thermal history data on the mechanisms of heating and cooling for each event. However, the timing of the early and late cooling episodes constrained by the apatite fission track analysis correspond generally with the age of regional unconformities. Thus, regional uplift and erosion would be a reasonable explanation for some of these events. For any geologically reasonable palaeogeothermal gradient, the amount of section removed from across the Falkland Islands would have been the order of kilometres to explain the relatively high palaeotemperatures experienced.

In the tectono-stratigraphic model presented here the cooling episodes in the Falkland Islands (and therefore the southwest Uplift Zone) can be accounted for as follows:

- 250–230 Ma (Late Permian–Early Triassic) – Gondwanide late–post orogenic uplift.
- 200–170 Ma [Early Jurassic (Pliensbachian)–Mid-Jurassic (Bathonian)] – corresponds to the end of extensional phase I (i.e. Late Triassic–Mid-Jurassic rifting) and Mid–Late Jurassic volcanicity. Cooling could therefore have been the result of the end of active rifting and the cessation of volcanicity. This suggests that this phase of development, which is well documented on mainland South America (Uliana & Biddle 1987), not only extended to the Western Rift Complex but also to the Falkland Islands. In this interpretation this phase is associated with northwest–southeast rifting as seen offshore (Southern Half-graben and Western Rift Complex), which are similar trends to those seen on mainland South America.
- 80–40 Ma [Late Cretaceous (Campanian)–Early Tertiary (Late Eocene)] – corresponding to Early Tertiary uplift of islands and transpressional reactivation of rift transfer zones, i.e. the Falkland Sound Monocline structure and the offshore deformational zone. This tectonic phase is attributed to the beginnings of indentation of the Scotia Plate.

It should be noted that the apatite fission track analysis data would allow a cooling episode in the Mid-Cretaceous (corresponding to uplift and erosion below the end-rift unconformity as seen in the Western Rift Complex) but the palaeotemperatures attained before cooling would have to have been less than those reached before the Late Cretaceous–Early Tertiary cooling episode.

These episodes of uplift and erosion, and their effect on the thermal history of not just the Falkland Islands but the Falklands Shelf as a whole, obviously have important implications for source rock maturity and therefore petroleum prospectivity in the Falklands region. However, that is beyond the scope of this paper.

Reassessment of Regional Geological Setting and South Atlantic Reconstruction

The new linking regional seismic line (Line 100) shows the nature of the northern edge of the Falkland Plateau to the north of the Falkland Islands (Fig. 5b). Seismic markers have been extrapolated across Line 100 from the North Falkland Graben on the southern end of the line. On this basis, the regional extent of the Early Tertiary and end-rift (Mid-Cretaceous) unconformities can be seen. Two small isolated half-graben can be observed in the northern Falkland Plateau, and from seismic character the rift-fill is interpreted to be dominantly representative of extensional phase II. No appreciation of structural trend can be gained but, as an expression of Late Triassic–Mid-Jurassic rifting, northwest– southeast trends are anticipated. The northern edge of the Falkland Plateau appears to be extensional and to have a similar overall geometry to the North Falkland Graben with a shallow western sub-basin separated from an eastern half-grabenal sub-basin by a central ridge. Unfortunately, no trend can be established for the extensional faults despite the existence of neighbouring Lamont–Doherty Line 138. These data are of poor quality. However, on the basis of this tectono-stratigraphic interpretation, it is proposed that this is the rifted edge of a graben paralleling the North Falkland Graben. As such it was created by precursor rifting to the opening of the South Atlantic. If this interpretation is correct Line 100 is crossing the rifted edge obliquely which would explain the low angle of the extensional faults.

Late uplift of the northern margin of the Falkland Plateau in this area is marked by a small escarpment and the subcrop to seafloor of the Tertiary and part of the Upper Cretaceous section (Fig. 5a). Uplift of the same generation characterizes the Marginal Fracture Ridge adjacent to the Falkland Escarpment to the east (Lorenzo & Mutter 1988) where transform movement has taken place along the Agulhas–Falkland Fracture Zone (Fig. 8). Despite this, there appears to be no positive evidence or requirement for the extension of the fracture

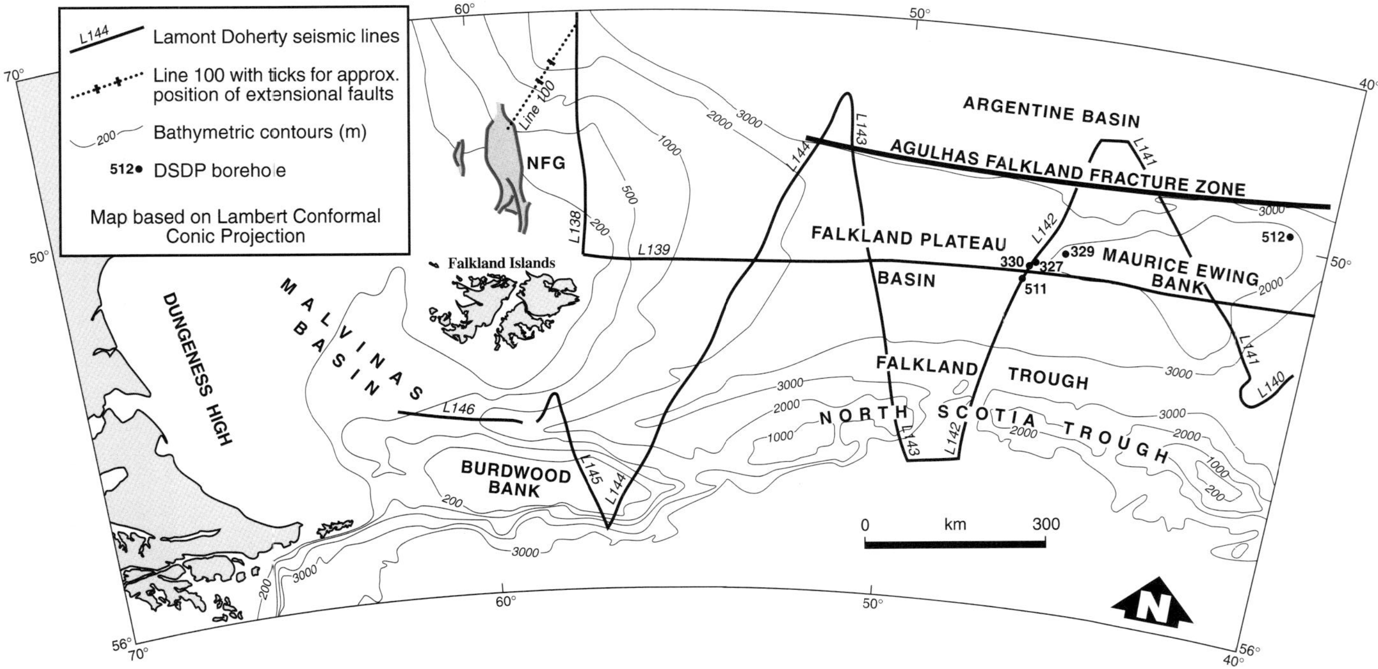

Fig. 8. Structure of the northern edge of the Falkland Plateau showing the Agulhas–Falkland Fracture Zone defined by Lamont–Doherty lines (e.g. L144), extent of North Falkland Graben (NFG) and position of Line 100 with ticks denoting approximate surface trace positions of main extensional faults.

zone in this area and the proposed link-up with the continental Gastre Fault Zone (*sensu* Rapela & Pankhurst 1992). This link forms an important element of some 'rotational' models (see MacDonald et al. 1999).

A straightforward 'rigid' or 'non-rotational' South Atlantic reconstruction, *sensu* De Wit et al. (1988), shows a possible pre-break-up structural configuration of the tectono-stratigraphic elements discussed (Fig. 9). The structural make-up of the reconstituted southwest Gondwanaland shows a regional consistency of structural trends inherited from the end of the Gondwanide orogeny.

Late Triassic–Mid-Jurassic (phase I) rifting characterizes a broad back-arc terrain stretching across South America and at least the eastern part of the Falkland Plateau. Rifting probably inverted Gondwanide compressional structures. Late Jurassic–Early Cretaceous rift trends (phases II and III) are slightly oblique to earlier rift trends resulting from an extensional stress field associated with the opening of the South Atlantic and beginnings of movement on the Agulhas–Falkland Fracture Zone. Rifts of this generation are typified by the North Falkland Graben and the Outeniqua Basin of South Africa, and have

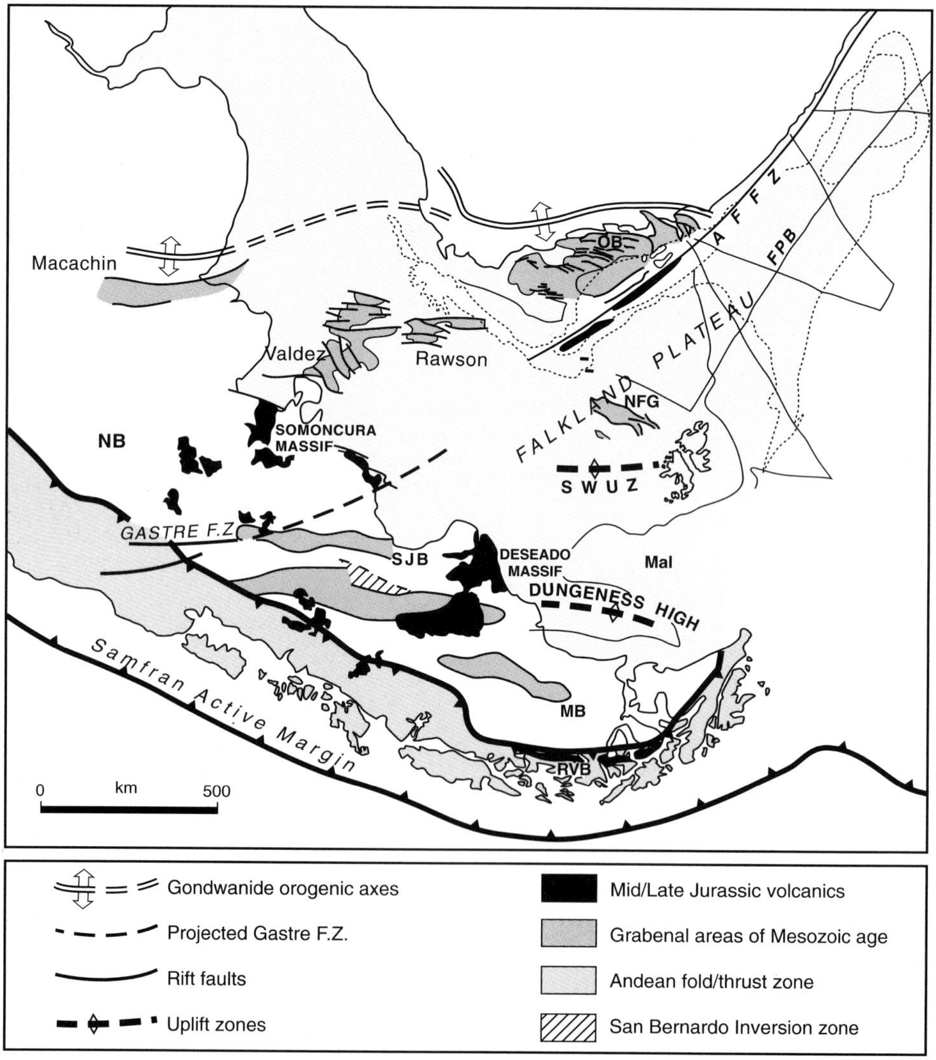

Fig. 9. South Atlantic reconstruction based on straightforward fit by De Wit *et al.* (1988), showing consistency of major regional tectono-stratigraphic elements and trends. FPB, Falkland Plateau Basin; NB, Neuquen Basin; NFG, North Falkland Graben; Mal, Malvinas Basin; MB, Magallenes Basin; OB, Outeniqua Basin; RVB, Rocas Verdes Basin; SJB, San Jorge Basin; SWUZ, southwest Uplift Zone.

been reported in the Falkland Plateau Basin (Lorenzo & Mutter 1988).

Uplift zones recognized across South America and the Falkland Plateau probably also reflect inherited Gondwanide structural trends. These include the Deseado Massif–Dungeness High and the Somoncura Massif–southwest Uplift Zone (Fig. 9). Late Triassic–Mid-Jurassic rifts were first inverted by compression in the Mid-Cretaceous and then were probably further reactivated in the Early Tertiary and Late Tertiary (Mio-Pliocene).

Break-up model for southwest Gondwanaland

On the basis of the interpretation of the North Falkland Shelf seismic data, a break-up model is proposed for southwest Gondwanaland incorporating a 'rigid' Falkland Plateau/South America configuration with no internal rotation involving a Falkland Islands microplate (Fig. 10).

The Gondwanide, or Samfrau, margin of South America developed as an Andean-type active margin during the Palaeozoic (Dalziel & Forsythe 1985). Flat-slab subduction resulted in the development of a contractional retro-arc system and the intracratonic fold belt of the Sierra de Ventana–Cape Fold Belt – Transantarctic system in the Late Permian–Early Triassic Gondwanide orogeny. Associated continental escape tectonics led to rifting/pull-apart basin development in the Triassic (Upper Karoo) in East Africa, north of the Gondwanide orogenic zone (Daly *et al.* 1991).

After the Gondwanide orogeny subduction steepening led to the formation of a back-arc rift system in South America in the Late Triassic–Mid-Jurassic (Dalziel 1986; Uliana *et al.* 1989) (Fig. 10a). Tankard *et al.* (1995) compared this phase of development to that of the basin-and-range province in the western USA. This was probably a response to patterns of mantle upwelling in a broad back-arc regime. It has been postulated that the development of a broad zone of mantle melting resulted from a reduction in plate boundary forces [due to the slowing of plate convergence following the Gondwanide orogeny (Storey *et al.* 1992)]. Middle–Upper Jurassic volcanics of mainland South America, which formed in the terminal phases of back-arc rifting, have been geochemically linked to crustal melting resulting from subduction in a back-arc regime (Bruhn *et al.* 1978; Gust *et al.* 1985). The Mozambique Basin opened as a response to continental escape and Bouvet plume development in East Africa in the Early Jurassic (Cox 1992) and then southwards propagation of Tethyan spreading in the Mid–Late Jurassic.

Continental separation in southwest Gondwanaland began in the Valanginian–Hauterivian when northward and southward propagating rift/wrenching towards the Bouvet Plume allowed a breakthrough of the geosuture represented by the Cape Fold Belt. Plume interaction and associated gravitational forces may have been responsible for the ridge jump which was effected from this point in the Natal Valley to the back-arc rift network in the South Atlantic (Fig. 10b). Dextral motion along the Agulhas–Falkland Fracture Zone commenced at drift onset at M10 times.

The geometrical constraints of southern Africa/Falkland Plateau proximity were released in the Mid-Cretaceous as the eastern extremity of the Falkland Plateau cleared the tip of the Agulhas Bank. The seafloor spreading centre driving the Falkland Plateau on its eastern side passed the Agulhas Bank shortly before magnetic anomaly C34 (*c.* 90 Ma). Atlantic opening accelerated and was promoted by propagation towards the Tristan mantle plume to the north. The knock-on change in the rate of plate convergence along the Samfrau active margin to the south (Dalziel *et al.* 1987) effected a gradual switch to flat-slab subduction and culminated in the regional compressional events of the Mid-Cretaceous. These events led to closure and compression in the Rocas Verdes back-arc basin, the initiation of a foreland stage of development in the Magallanes Basin, and the compressional effects seen in the San Jorge Basin, the Falkland Plateau Basin (Lorenzo & Mutter 1988) and (now with this study) the North Falkland Shelf.

Conclusions

Interpretation of seismic data over the North Falkland Shelf has shown a tectono-stratigraphic history consistent with rifting related to southwest Gondwanaland break-up. This incorporates Late Triassic–Mid Jurassic back-arc rifting and Late Jurassic–Early Cretaceous extension related to the initiation and opening of the South Atlantic. Episodes of Andean contraction in the Mid-Cretaceous, Early and Late Tertiary are also recognized in the tectono-stratigraphic evolution of the North Falkland Shelf. The Falkland Islands has a tectono-stratigraphic history shown by interpretation of apatite fission track analysis data to be compatible with that of the North Falkland Shelf and mainland South America.

A regional seismic line shows an extensional northern edge to the Falkland Plateau and no apparent westerly extension of the Agulhas–

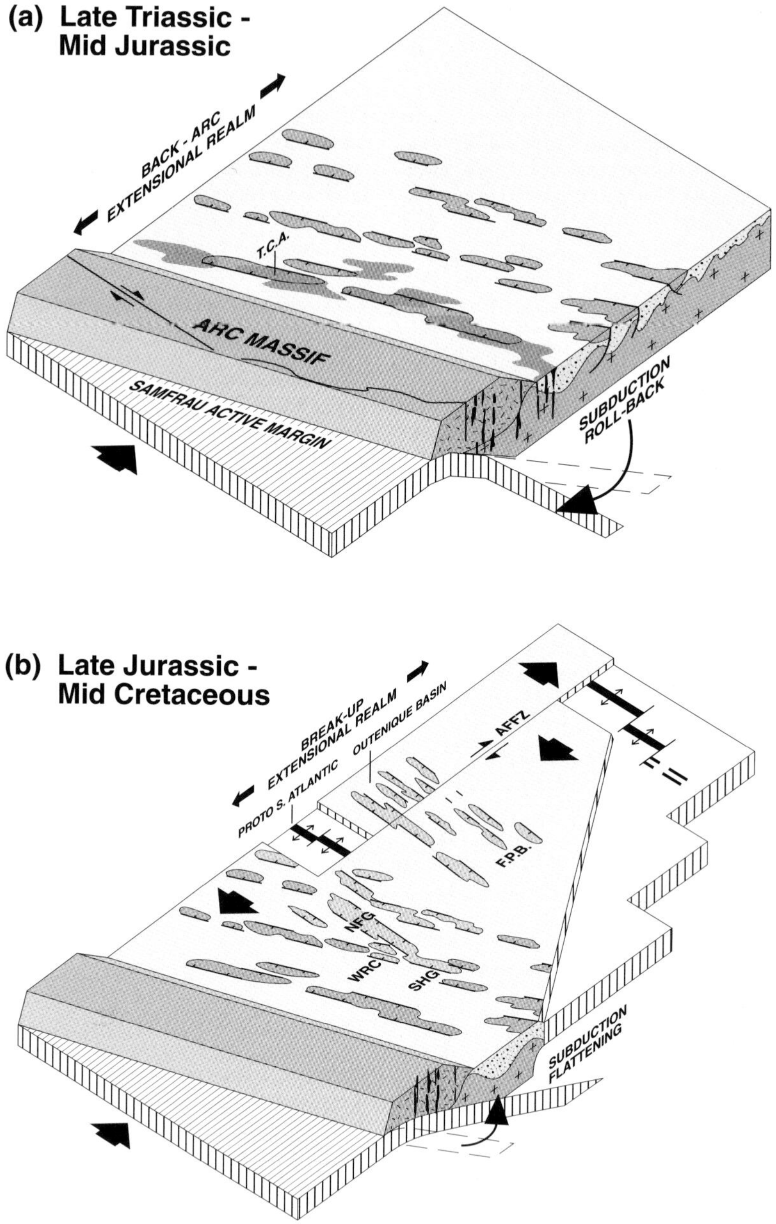

Fig. 10. 'Break-up' model for southwest Gondwanaland showing (**a**) early back-arc rifting followed by (**b**) South Atlantic rift propagation and Atlantic opening. F.P.B., Falkland Plateau Basin; NFG, North Falkland Graben; SHG, Southern half-graben; TCA, Mid–Upper Jurassic volcanics; WRC, Western Rift Complex.

Falkland Fracture Zone. No trend can be established in the observed extensional faults but, as a product of South Atlantic rifting, they can be expected to have north–south trends consistent with the North Falkland Graben.

A 'rigid' or 'non-rotational' reconfiguration of southwest Gondwanaland shows a regional consistency of Late Triassic–Mid Jurassic back-arc, and Late Jurassic–Early Cretaceous South Atlantic rift trends. This is presented as circumstantial evidence in favour of a 'non-rotational' model for southwest Gonwanaland break-up.

The work which forms the basis for this paper was carried out under the aegis of Spectrum Energy and Information Technology Ltd who acquired the seismic data across the North Falkland Shelf. We are grateful to Spectrum for permission to publish this paper. Spectrum would like to thank the Falkland Islands government for allowing the conduct of speculative seismic surveys. We are grateful to ARK Geophysics who carried out gravity/magnetics data processing and two-dimensional modelling.

References

ADIE, R. J. 1952. The position of the Falkland Islands in a reconstruction of Gondwanaland. *Geological Magazine*, **89**, 401–410.

BARKER, P. F. & DALZIEL, I. W. D. *et al.* 1977. *Initial Reports of the Deep Sea Drilling Project 36*. US Government Printing Office, (Washington, DC).

BEN-AVRAHAM, Z., HARTNADY, C. J. H. & MALAN, J. A. 1993. Early tectonic extension between the Agulhas Bank and the Falkland Plateau due to the rotation of the Lafonia microplate. *Earth and Planetary Science Letters*, **117**, 43–58.

BIDDLE, K. T., ULIANA, M. A., MITCHUM, R. M. JR., FITZGERALD, M. G. & WRIGHT, R. C. 1986. The stratigraphic and structural evolution of the central and eastern Magallanes Basin, southern South America. *In*: ALLEN, P. A. & HOMEWOOD, P. (eds) *Foreland Basins*. International Association of Sedimentologists, Special Publication, **8**, 41–61.

BRUHN, R. L., STERN, C. R. & DE WIT, M. J. 1978. Field and geochemical data bearing on the development of a Mesozoic volcano-tectonic rift zone and back-arc basin in southernmost South America. *Earth and Planetary Science Letters*, **41**, 32–46.

BURDEN, P. L. A. 1992. Soekor, partners explore possibilities in Bredasdorp basin off South Africa. *Oil and Gas Journal*, **Dec. 21**, 109–112.

COX, K. G. 1992. Karoo igneous activity, and the early stages of the break-up of Gondwanaland. *In*: STOREY, B. C., ALABASTER, A. & PANKHURST, R. J. (eds) *Magmatism and the Causes of Continental Break-up*. Geological Society, London, Special Publications, **68**, 137–148.

DALY, M. C., LAWRENCE, S. R., KIMUN'A, D. & BINGA, M. 1991. Late Palaeozoic deformation in central Africa: a result of distant collision. *Nature*, **350**, 605–607.

DALZIEL, I. W. D. 1981. Back-arc extension in the southern Andes: a review and critical reappraisal. *Philosophical Transactions of the Royal Society, London, A*, **300**, 319–335.

— 1986. Collision and Cordilleran orogenesis: an Andean perspective. *In*: COWARD, M. P. & RIES, A. C. (eds) *Collision Tectonics*. Geological Society, London, Special Publications, **19**, 389–404.

— & FORSYTHE, R. D. 1985. Andean evolution and the Terrane concept. *In*: HOWELL, D. G. (ed.) *Tectonostratigraphic Terranes of the Circum-Pacific Region*. Circum-Pacific Council for Energy & Mineral Resources, Earth Science Series **1**, 555–564.

—, STOREY, B. C., GARRETT, S. W., GRUNOW, A. M., HERROD, L. D. B. & PANKHURST, R. J. 1987. Extensional tectonics and the fragmentation of Gondwanaland. *In*: COWARD, M. P., DEWEY, J. F. & HANCOCK, P. L. (eds) *Continental Extensional Tectonics*. Geological Society, London, Special Publications, **28**, 433–441.

DE WIT, M. J., JEFFERY, M., BERGH, H. & NICOLAYSEN, L. 1988. *Geological Map of Sectors of Gondwanaland*. AAPG.

FITZGERALD, M. G., MITCHUM, R. M., ULIANA, M. A. & BIDDLE, K. T. 1990. Evolution of the San Jorge Basin, Argentina. *AAPG Bulletin*, **74**, 879–920.

GREENWAY, M. E. 1972. *Geology of the Falkland Islands*. British Antarctic Survey Science Report No. 7, Natural Environment Research Council.

GUST, D. A., BIDDLE, K. T., PHELPS, D. W. & ULIANA, M. A. 1985. Associated Middle to Late Jurassic volcanism and extension in Southern South America. *Tectonophysics*, **116**, 223–253.

LAWRENCE, S. R. & JOHNSON, M. 1995. Shelf North of Falkland Maybe New S. Atlantic Petroleum Province. *Oil and Gas Journal*, **Mar. 6**, 52–55.

LORENZO, J. M. & MUTTER, J. C. 1988. Seismic stratigraphy and tectonic evolution of the Falkland/Malvinas Plateau. *Revista Brasileira de Geociências*, **18**, 191–200.

MARSHALL, J. E. A. 1994. The Falkland Islands: a key element in Gondwanaland paleogeography. *Tectonics*, **13**, 499–514.

PLATT, N. H. & PHILIP, P. R. 1995. Structure of the southern Falkland Islands continental shelf: initial results from new seismic data. *Marine and Petroleum Geology*, **12**, 759–771.

RABINOWITZ, P. D. & LA BRECQUE, J. 1979. Geophysical study of the continental margin of Southern Africa. *Journal of Geophysical Research*, **84**, 5973–6002.

RAPELA, C. W. & PANKHURST, R. J. 1992. The granites of northern Patagonia and the Gastre Fault System in relation to the break-up of Gondwanaland. *In*: STOREY, B. C., ALABASTER, T. & PANKHURST, R. J. (eds) *Magmatism and the Causes of Continental Break-up*. Geological Society, London, Special Publications, **68**, 209–220.

RICHARDS, P. C. & FANNIN, N. G. T. 1997. Geology of the North Falkland Basin. *Journal of Petroleum Geology*, **20**, 165–183.

STOREY, B. C., ALABASTER, T., HOLE, M. J., PANKHURST, R. J. & WEVER, H. E. 1992. Role of subduction-plate boundary forces during the initial stages of Gondwanaland break-up: evidence from the proto-Pacific margin of Antarctic. *In*: STOREY, B. C., ALABASTER, T. & PANKHURST, R. J. (eds) *Magmatism and the Causes of Continental Break-up*. Geological Society, London, Special Publications, **68**, 149–163.

TANKARD, A. J., ULIANA, M. A., WELSINK, H. J. *ET AL.* 1995. Petroleum basins of South America. *AAPG Memoir*, **62**, 5–52.

ULIANA, M. A. & BIDDLE, K. T. 1987. Permian to late Cenozoic evolution of Northern Patagonia: Main tectonic events, magmatic activity and depositional trends. *In*: McKenzie, G. D. (ed.) *Gondwanaland Six – Structure, Tectonics and Geophysics*. Geophysical Monograph, **40**, 271–286.

—— & —— 1988. Mesozoic–Cenozoic paleogeographic and geodynamic evolution of southern South America. *Revista Brasileira de Geociências*, **18**, 172–190.

——, —— & Cerdan, J. 1989. Mesozoic extension and the formation of Argentine sedimentary basins. *In*: Tankard, A. J. & Balkhill, H. R. *Extensional Tectonics and Stratigraphy of North Atlantic Margins*. AAPG Memoir, **46**, 599–614.

Evidence for multi-phase rifting in the North Falklands Basin

P. J. E. BRANSDEN[2], P. BURGES[1], M. J. DURHAM[1], & J. G. HALL[2]

[1]*LASMO PLC 101 Bishopsgate, London EC2M 3XH, UK*
[2]*LASMO Indonesia, Ratu Plaza Tower, Jl Jend. Sudirman 9, Jakarta 10270*

Abstract: The post-Gondwanide basin development of the North Falklands Shelf has been evaluated using seismic sequence stratigraphy to determine the structural and stratigraphic evolution of the region.

Two areally distinct post-Gondwanide basin systems (megasequences) are identified on the North Falklands Shelf, each with an independent structural and stratigraphic history. This contrasts with published models which interpret all basin areas to have been initiated in the early–mid Jurassic.

Megasequence I is developed in a series of northwest and subsidiary north and east trending half-graben, and typically shows sedimentary growth into faults indicating an active rifting environment. Pervasive small-scale synthetic faulting is largely non-rotational, suggesting that it represents later deformation. No post-rift sediments associated with this phase of extension are identified. Comparisons with seismic geometries of other basins in the region, based on a restored plate model, indicate that Megasequence I represents part of a widespread mid–late Jurassic rifting event, ultimately leading to the opening of the Rocas Verdes (oceanic) Basin.

Megasequence I is cross-cut by a north trending half-graben system which represents a failed arm of the South Atlantic Rift. The sedimentary fill of this basin, Megasequence II, can be subdivided into two units. The lower unit locally demonstrates sedimentary growth into the main eastern bounding fault of the half-graben and is thought to represent the syn-rift phase of Atlantic extension. Within the upper unit of Megasequence II there is no evidence of fault-controlled sediment growth into the main easterly bounding fault and the overall geometry is indicative of passive fill of pre-existing topography, supporting a post-rift origin for this section.

Similar seismic geometries are observed in parts of the North Atlantic Rift System and the northern North Sea. In particular, the North Shetland Basin has a Permo-Triassic to late Jurassic rift and post-rift section. Renewed extension in the early Cretaceous resulted in break-up and domino faulting, with the generation of high relief fault blocks. The intervening lows were sites of deposition for onlapping deepwater sediments.

Initial publications established a structural and stratigraphic framework for the North Falklands Basin based on the Spectrum Energy regional speculative seismic survey (Lawrence & Johnson 1995; Richards *et al.* 1996*a*, *b*). Subsequent to this phase of analysis, Spectrum undertook acquisition of infill seismic data and reprocessing of the entire seismic dataset, completed in late 1995. This enhanced imaging, particularly of deeper stratigraphic sections, where complex geometries and velocity fields were recognized.

Numerous studies were undertaken both by industry and academic groups in 1996 as a precursor to licensing of much of the North Falklands Graben area and, in late 1996, LASMO and its partners were awarded Tranches C and D covering the central part of the graben. Subsequent to the First Licence Round a comprehensive review of the North Falklands area was published by Richards & Fannin (1997*a*).

The aim of this paper is an analysis of the megasequence distribution and sequence stratigraphy of the North Falklands area, with the primary objective of evaluating structural controls on basin development. The megasequence stratigraphy is linked to the causative plate tectonic controls, allowing dating of the principal seismic events, and projection of known regional marker beds and source facies into the basin. Seismic sequence stratigraphy (sequence geometries and seismic facies) allows prediction of facies (and hence reservoir/seal and source distribution) within the context of predicted sediment supply and base-level fluctuations.

Extensive analysis of the geology, geomorphology and plate tectonic development of the Falklands continental shelf and its pre-drift adjacent terranes is contained in summary publications (De Wit & Ransome 1992; Story *et al.* 1996; Lorenzo & Mutter 1988). The North Falklands area described here (Fig. 1) is constrained by the limits of the Spectrum reconnaissance seismic data. This does not necessarily reflect the limits of the basin areas, which extend beyond the seismic control to the west, the north and, possibly, the east.

Following the terminology of Lawrence & Johnson (1995), the North Falklands Shelf has

BRANSDEN, P. J. E., BURGES, P., DURHAM, M. J. & HALL, J. G. 1999. Evidence for multi-phase rifting in the North Falklands Basin. *In:* CAMERON, N. R., BATE, R. H. & CLURE, V. S. (eds) *The Oil and Gas Habitats of the South Atlantic.* Geological Society, London, Special Publications, **153**, 425–442.

three basin areas with distinctive structural styles; the North Falklands Graben, the Western Rift Complex and the Southern Half-graben. A further set of potential sedimentary basins occurs to the east of the North Falklands Graben. Currently, insufficient seismic data is available to tie these areas to other basins outside the Falklands territorial waters. The Western Rift Complex can be interpreted to show a connection with the San Julian Basin to the west. The structural relationship between the North Falklands basins and the main South Atlantic basin system to the north is not yet established.

Megasequence identification

Seismic line Falk 18 (Fig. 2) demonstrates the overall seismic character and geometries of the North Falklands area. The North Falklands offshore is covered by an undeformed, although stratigraphically complex, sedimentary blanket of between 0.5 to *c.* 1.5 s two-way travel time (TWTT) thickness, which onlaps a prominent regional unconformity surface, identified as Horizon 30, provisionally identified as Late Cretaceous–Early Tertiary in age. Beneath this, a number of sedimentary basins are recognized, separated by areas where Horizon 30 has high amplitude and no coherent seismic data beneath, which are considered to be acoustic basement.

Southward projection of this surface on seismic lines near the Falklands coast suggest that economic basement comprises folded and locally cleaved Gondwanide (Ordovician–Upper Permian) rocks. The relatively low angle dip-slip faults controlling the Southern Half-graben have previously been interpreted as relaxed Gondwanide thrust planes, due to the continuity of strike and dip with onshore fault traces (Lawrence & Johnson 1995). This supports the interpretation that the half-graben fill is post-Gondwanide and that acoustic basement is an eroded Gondwanide fold–thrust terrane.

The sedimentary basins beneath Horizon 30 are differentiated by their internal sequence stratigraphy and their relationship with the sediment package above. The Western Rift Complex and the Southern Half-graben, for the most part, show a shallow major angular relationship with Horizon 30 beneath which, in places at least, 500 ms TWTT of section has been removed. The sediments above Horizon 30 show minimal compaction drape over the western and southern areas. In the North Falklands Graben, however, the sediments above and below Horizon 30 are generally conformable, apart from locally at the basin margins where a slight angularity is observed. Marked drape of post-Horizon 30 sediments over the Falklands Graben suggests the pre- and post-Horizon 30 sequences have a common structural history in this basin.

Western and Southern Areas

Line Falk 12 (Fig. 3) is typical of the seismic character of both the western and southern areas. Beneath the major angular unconformity of Horizon 30 the preserved sections generally show marked sediment expansion into the controlling faults, indicating an active rift control. There are no palaeobathymetric indicators (e.g. shelf breaks) and the rift rill, on the basis of seismic geometries alone, is predicted to be non-marine to marginally marine. Internally, these syn-rift sequences are broken by closely spaced fault sets which are non-rotational and show little or no sediment expansion. These are interpreted to post-date the main rift phase. A third distinctive feature is the presence of one or more discrete bands of high amplitude, discontinuous 'rubbly' reflectors, which tend to mask reflectivity of the section beneath. These reflector packages show similarities to the Tobifera Formation in the Malvinas and Austral Basins of Argentina, and are therefore interpreted as extrusive igneous sequences.

Compared to the denser grids over the North Falklands Graben, the relatively wide spacing of seismic lines over the western and southern areas hampers the determination of fault trends (refer to database map in Richards 1997, fig. 1). However, it is likely that east–west, northwest–southeast and north–south trends are present (Fig. 1), of which the northwest–southeast trend of the Southern Half-graben is the best defined. The controlling faults are relatively closed spaced (10–15 km) and the half-graben are typically elongate and narrow. Within the western and southern areas it has not been possible to subdivide the pre-Horizon 30 syn-rift section, unlike the North Falklands Graben where, using a similar grid spacing, consistent genetic seismic packages and bounding events can be widely mapped.

The western and southern areas are interpreted as a complex of short wavelength rift basins, affected by later tensional faulting. No passive post-rift sections have been identified in these areas.

North Falklands Graben

Within the North Falklands Graben, up to 3 s TWTT of sedimentary section can be identified beneath Horizon 30 (Fig. 4). This section is best

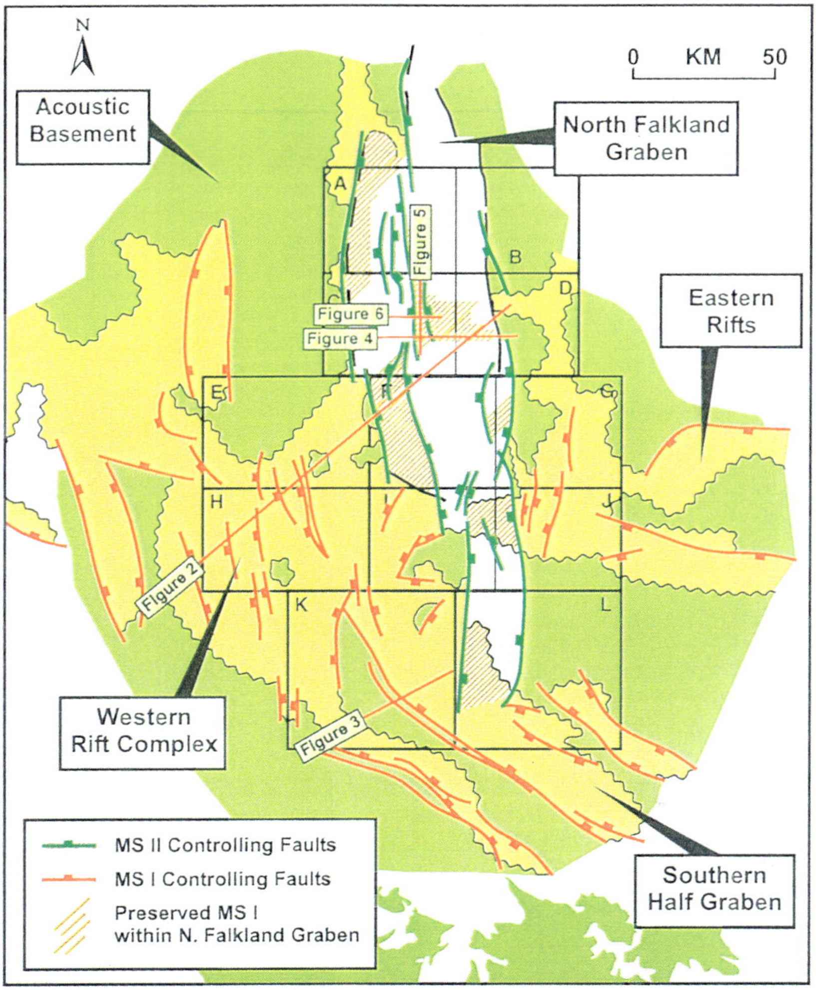

Fig. 1. Distribution of post-Gondwanide sedimentary basins on the North Falklands Shelf. The Western Rift Complex, Southern Half-graben and Eastern Rifts are interpreted as being Megasequence I (mid–late Jurassic). Hatched areas indicate where Megasequence I is preserved beneath the Megasequence II (early Cretaceous) North Falkland Graben. Green areas are where acoustic basement (Gondwanide) underlies Horizon 30 (refer to Fig. 2).

developed in the eastern depocentre of the basin and constitutes Megasequence II. In contrast to Megasequence I, both syn- and post-rift sequences are recognized within this megasequence. Compared with the western and southern rift complexes, the basin fill of the eastern depocentre is relatively undeformed. An antithetic fault set cuts the western flank; this fault set tends to die out upsection, although some die out downward suggesting soling out within the section; the faults are non-rotational. This pre-Horizon 30 section can be divided into three discrete units described below.

Unit 70–60. This is a wedge-shaped unit characterized by marked sediment expansion into the main eastern controlling fault of the North Falklands Graben, and depositional thinning toward the central ridge. The base of the unit is well imaged on only a few lines (e.g. Fig. 4). The section demonstrating sediment expansion ceases immediately above Horizon 60, which is a high amplitude seismic doublet, mappable throughout the deeper eastern depocentre. Therefore, Unit 70–60 is interpreted as a syn-rift section and is assigned a Ryazanian–Valanginian age. So far, this unit has only been identified in the eastern

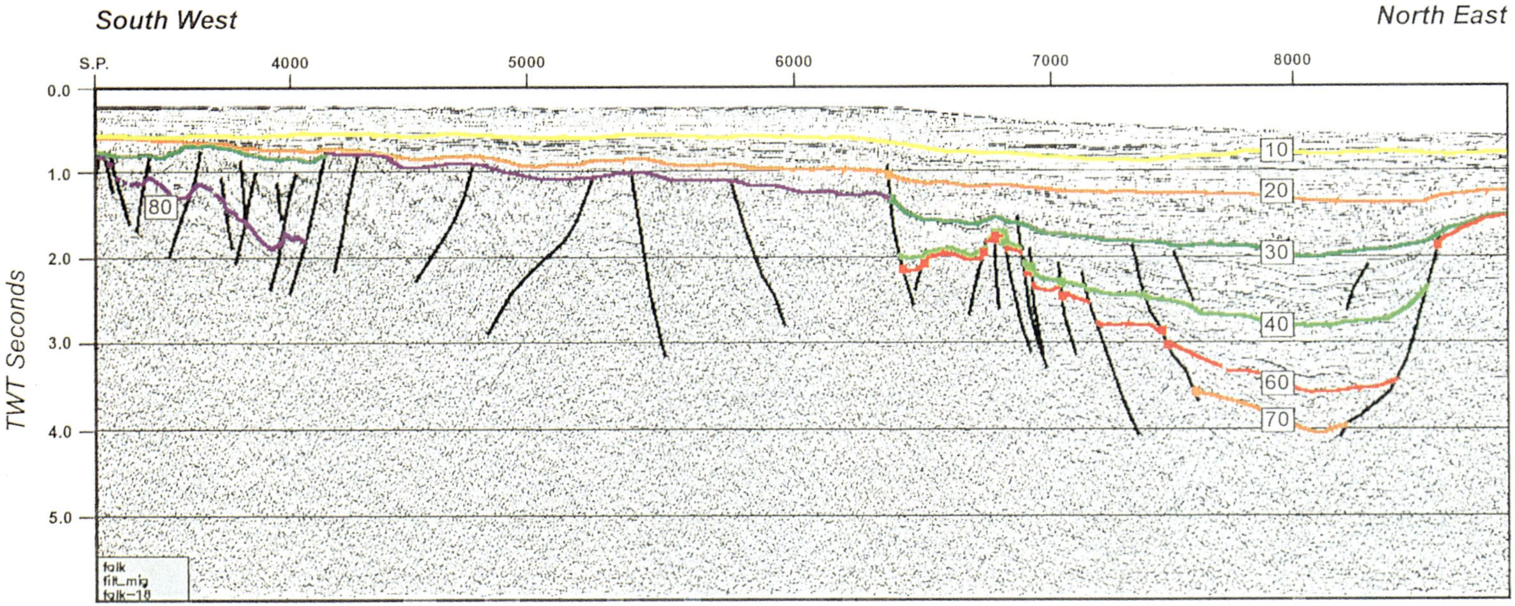

Fig. 2. Seismic line Falk 18 (by courtesy of Spectrum Energy). Pre-Horizon 30 section in the Western Rift Complex (S.P. 3200 to 4200) displays marked subcrop and post-Horizon 30 shows no compaction. In contrast, in the North Falklands Graben (S.P. 6400 to 8600) pre- and post-Horizon 30 are largely conformable, with significant compaction drape.

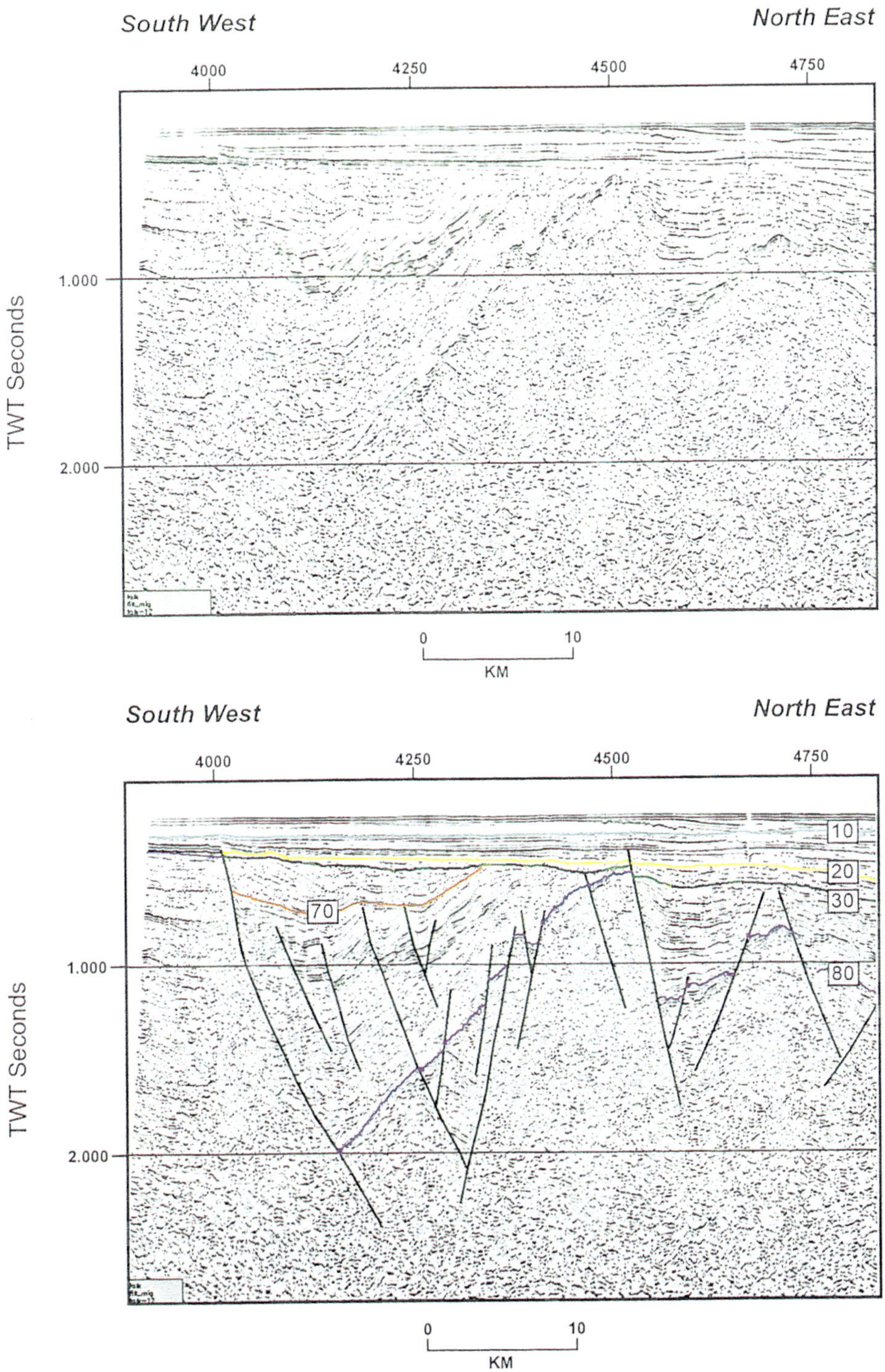

Fig. 3. Seismic line Falk 12 (by courtesy of Spectrum Energy) showing typical seismic character of Megasequence I.

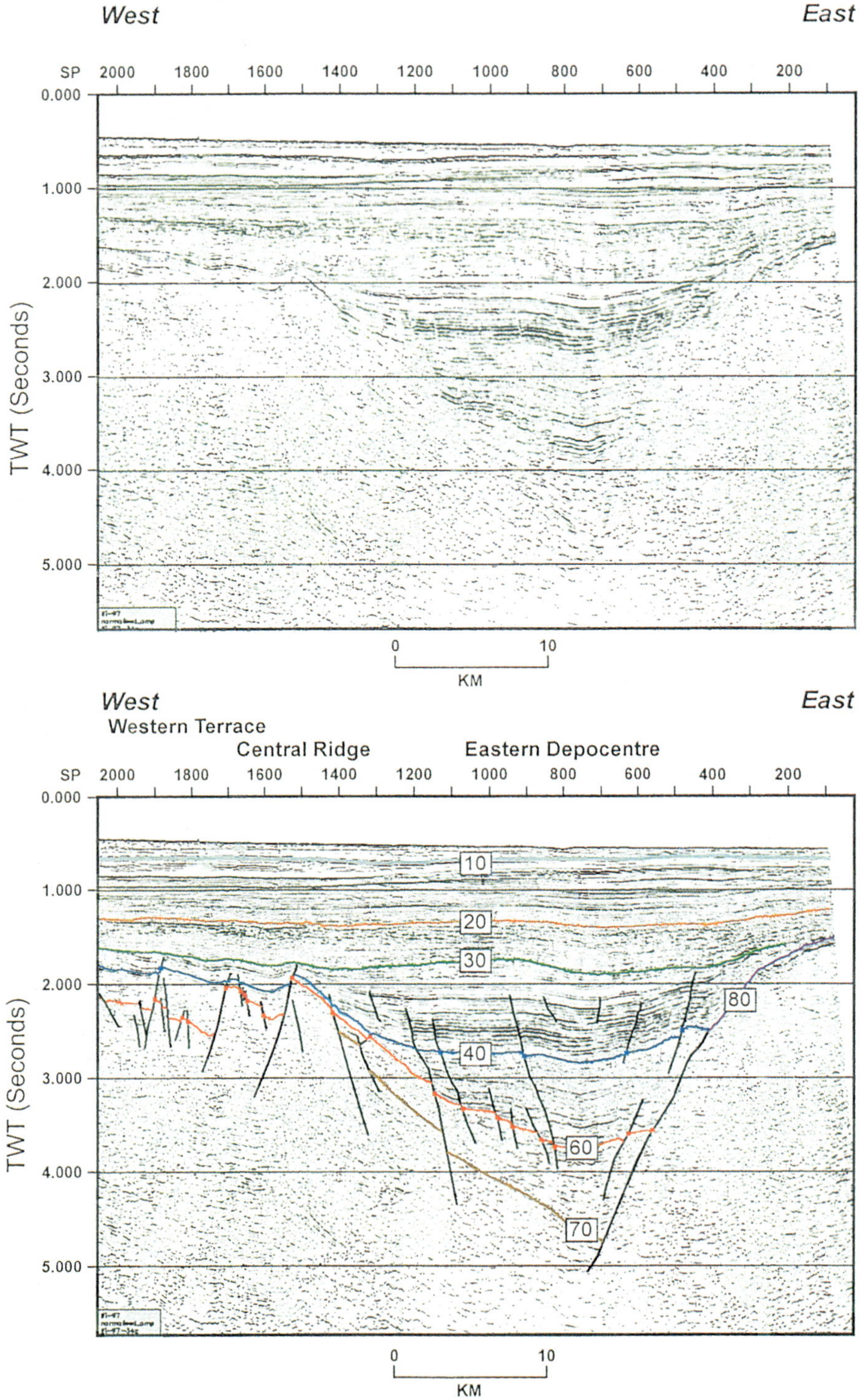

Fig. 4. Seismic line LF1 97 34z across the North Falkland Graben. Unit 70–60 syn-rift displays marked sediment expansion into the eastern depocentre controlling fault. Note most of Unit 60–40 post-rift shows marked onlap onto the central ridge tilted fault block.

depocentre. Thinly developed equivalents may be developed in the western terrace and southern extension (Tranches J and L), although these areas are likely to contain only post-rift sections.

Unit 60–40. This unit oversteps Unit 70–60 and is widespread in the eastern depocentre, and mapped into the southern extension of the North Falklands Graben (Tranches J and L). The upper surface (Horizon 40) is a widespread downlap surface for the progradational Unit 40–30. Unit 60–40 is characterized by parallel to sub-parallel continuous reflectors and, where data quality is good, displays marked onlap onto the eastern flank of the central ridge, thinning toward the eastern controlling fault. This relationship is only observed on a few lines where the section is unbroken by younger antithetic faults. There is no evidence of active fault control and structural modelling demonstrates that no rotation of the central tilted fault block is required to restore the section; therefore, the onlap is attributed to post-rift passive infill of probable Hauterivian–Maastrichtian age. Generally, the entire thickness of Unit 60–40, and much of Unit 40–30, onlap the eastern flank of the central ridge above a composite unconformity which equates to the base post-rift. Locally, however, the upper section of Unit 40–60 and Unit 40–30 overstep the central ridge.

Unit 40–30. This sequence marks a dramatic change from passive infill sedimentation of Unit 60–40 to major progradation of sigmoidal foresets, from the west and north into the western terrace area, and from the north into the eastern depocentre (see Richards *et al.* 1997*a*, figs 9 and 11). This unit is assigned a Maastrichtian–Palaeocene age. In the northwest and north of the North Falklands Graben, topset, foreset and basinal seismic facies are identified, whereas over the eastern depocentre of the graben, parallel continuous, high amplitude basinal seismic facies are dominant. The latter areas have anomalously low seismic stacking velocity and shallow faults sole out into this unit. Coupled with the predominance of progradation v. aggradation (sediment supply exceeded subsidence), it is likely that the unit was rapidly deposited, and is undercompacted and possibly overpressured. The low clinoform angle, low interval velocity and the inferred high palaeolatitude indicate that Unit 40–30 is comprised of siliciclastics rather than carbonates.

Estimates of decompacted prograde heights show a regular decrease from 350 to 450 m during initial downlap to Horizon 40, to < 100 m immediately beneath Horizon 30 in the basin centre. This demonstrates a continuous reduction in accommodation space. On the basin margins the uppermost parts of both the topset facies, and locally the basinal facies, are bevelled beneath Horizon 30.

The overall geometry of the North Falklands Graben, as delineated by the distribution of Units 70–30 (and the associated drape above Horizon 30), is an asymmetric rift of maximum dimensions 240 km north–south and 60 km east–west. The eastern depocentre and central ridge have the geometry of a domino fault block with dimensions of 30–40 km, significantly wider than those observed in the western and southern areas.

Post-Horizon 30

This section has complex sequence geometries but shows little or no tectonic control. The only exception is the presence of broad low-relief inversion features in the axis of the eastern depocentre. Sequential flattening demonstrates that they began to elevate after Horizon 30 and growth continued until around Horizon 10, indicating that basin inversion spanned Palaeocene–Late Miocene.

Initial post-Horizon 30 sedimentation shows progressive onlap of the basin margins within the North Falklands Graben and into the surrounding footwall areas. Near the top of this package, Horizon 20 (possible Oligocene?) (Fig. 2) is a regional marker immediately beneath a downlap surface which marks renewed progradation into the North Falklands area, in this case from the southwest. Cessation of progradation was followed by transgressive onlap of the shelf–slope system; a high amplitude, probable flood, surface within this system is mapped as Horizon 10. This is referred to as a base Tertiary tectonic event (Lawrence & Johnson 1995; Richards *et al.*, 1997*a*). Above this, major mounded sediment packages were deposited beneath the palaeoshelf edge. The upper surface of these is a high amplitude, locally incised event. The final phase of basin fill is further progradation from the southwest, resulting in an abandoned shelf edge within Tranches F and G.

Chronology of rifting – evidence for multiple rift phases

Megasequence analysis indicates that syn-rift packages are present within the Western Rift Complex, Southern Half-graben and North Falklands Graben, of which only the latter has an associated post-rift section. Figure 1 displays the distribution of faults which control the distribution of the western, southern and northern graben complexes, distinguished by the characteristic sequence geometries and seismic character detailed above. The interference of the two trends has been noted in previous studies (e.g. Richards *et al.* 1996*a*). This is best demonstrated in the western part of Tranche L where north–south fault trends, with compaction above Horizon 30, obliquely cut a northwest–southeast narrow half-graben which has no compaction effects.

Further evidence of the two trends is observed on line LFI 97 21z (Fig. 5). This is a strike line along the axis of a major north–south trending titled fault block with associated draper closure in the overlying post rift sequences. On LFI 97 21z a major unconformity at 1.6–1.9 ms TWTT is identified as Horizon 60. A relatively undeformed package is underlain by a faulted and more steeply dipping unit which subcrops both to the north and south. This sequence geometry is incompatible with a possible Megasequence II syn-rift origin in that it is apparently unrelated to the north–south controlling faults of the southern basin extension. Therefore, the preferred interpretation is that the subcropping sequence beneath Horizon 60 is a preserved extension of the truncated east–west basin in the footwall area, which also demonstrates erosional truncation to the north and south. This sequence geometry is direct evidence for the presence of two megasequences.

Many seismic lines crossing the central ridge show reflector packages beneath the onlap surface of the post-rift basin fill (Horizon 60) which appear to be structurally unconnected with the main syn-rift package to the east (Fig. 4). This section also shows evidence for high amplitude, discontinuous 'rubbly' seismic facies, as observed in the western and southern graben areas. On the west flank of the central ridge the pre-Horizon 60 sections are in fault juxtaposition with Unit 30–40 distal facies. General dip within the pre-Horizon 60 sections is easterly, although other sections show marked counter-regional dips to the west (Fig. 6). This example shows high frequency banded seismic facies and some evidence for internal faulting characteristic of the Western Rift Complex fill. The structural position and variable dip of the pre-Horizon 60 sections in the central ridge are incompatible with a syn-rift origin related to the North Falklands Graben trend. The major fault displacement on the western flank (and juxtaposition against Unit 30–40) indicates a pre-Atlantic rift origin (i.e. Megasequence I), since the equivalent section in the downthrown block must have been displaced by a later rift event.

These seismic geometries suggest that the faults controlling the North Falklands Graben offset older stratigraphic sections and that two megasequences are present. Furthermore, the North Falklands Graben (Megasequence II) has a very different structural history from the surrounding basins in that it has post-Horizon 30 compactional drape, whereas the western and southern/eastern areas do not, suggesting that significant compaction occurred in the latter prior to denudation. It is proposed that basins of Megasequence I age, including portions of the western and southern graben systems, are preserved beneath the North Falklands Graben. The distribution of preserved Megasequence I is shown on Fig. 1. The orientation of these preserved sections is probably oblique or perpendicular to the later north–south trend of Megasequence II, although any north–south trends in Megasequence I rifts are likely to have been reactivated during Megasequence II rifting.

Plate tectonic controls

Initial post-Gondwanide break-up occurred by separation of Antarctica from South America–Africa in a dextral shear regime [refer to Storey *et al.* (1996) for a compilation of relevant papers]. This occurred under a high heat flow associated with mantle plume activity, with very widespread basaltic (southern Africa) and acidic (Argentina) extrusive and intrusive activity.

Figure 7a demonstrates the plate configuration resulting from late Triassic–Liassic break-up. The model accepted here is that developed by the British Antarctic Survey group (Storey *et al.* 1995), which assumes that all sedimentary basins offshore the Falklands post-date this phase. Any microplate rotation that may have occurred will have been restricted to this period of high heat flow. Rotational theories will not be discussed further here, save to say that the authors concur with Richards *et al.* (1996*a*) in seeing no evidence on offshore seismic datasets for collisional tectonics (apart from the Tertiary North Scotia arc system).

Further extensional stress affected the South America–Africa plate in mid-Jurassic times leading to widespread rifting (Fig. 7a). The syn-rift fill in Argentine coastal basins to the west is extrusive volcanic rocks, demonstrated by thickness changes across faults in the San Jorge Basin (Gust *et al.* 1985), and seismic observations of expansion of the Tobifera Formation volcanics into faults in the Magallanes (Biddle *et al.* 1986) and Malvinas Basins. Within the Outeniqua Basins (southern Africa) and the Falklands Plateau Basin the basin fill is largely clastic; non-

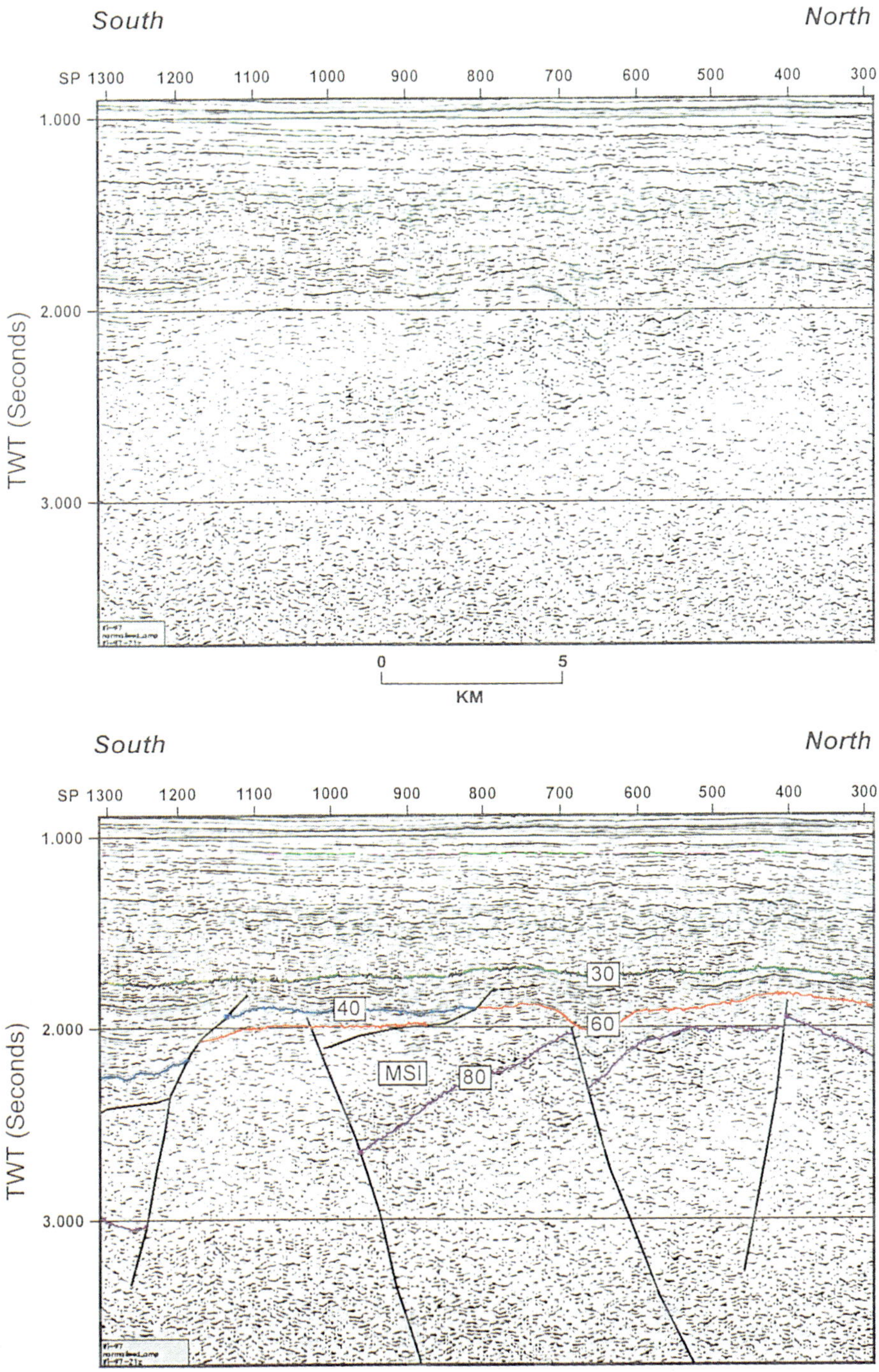

Fig. 5. Seismic line LFI 97 21z. Line section is oriented nearly north–south along the major north–south trending tilted fault block in the North Falklands Graben (Fig. 1). Beneath Horizon 60 a subcropping unit (Megasequence I) pinches out to the north and south. In the immediate eastern footwall of the North Falklands Graben, a truncated Megasequence I graben is aligned east–west and also subcrops to the north and south.

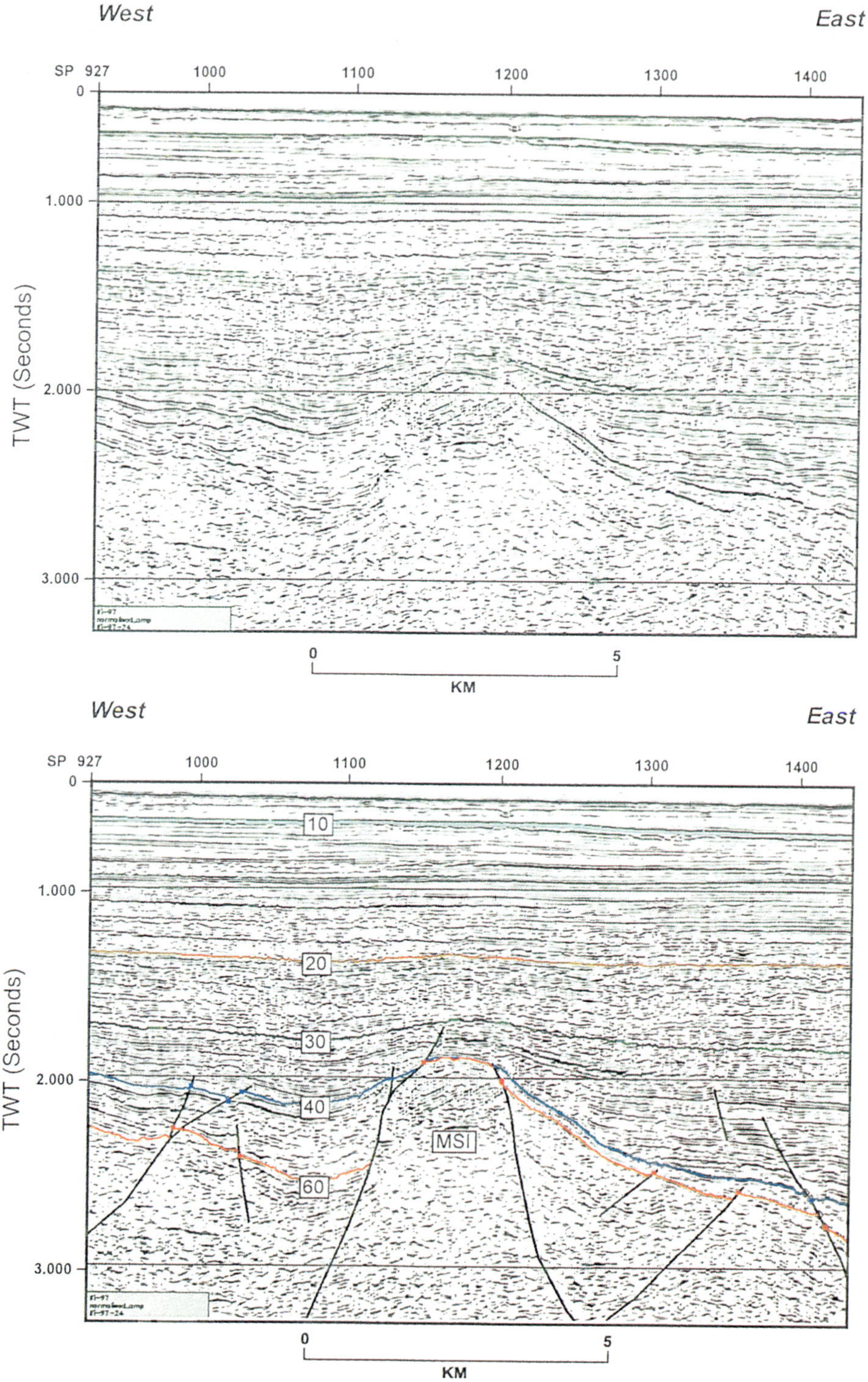

Fig. 6. Seismic line LFI 97 24 across the central ridge tilted faulted block. Unit 60–40 overlaps the steep eastern flank of the central ridge onto the base post-rift unconformity surface. Beneath this surface both east and west dipping (counter-regional) sediment packages are observed. Structural position and seismic character are most consistent with a pre-rift interpretation, i.e. Megasequence I.

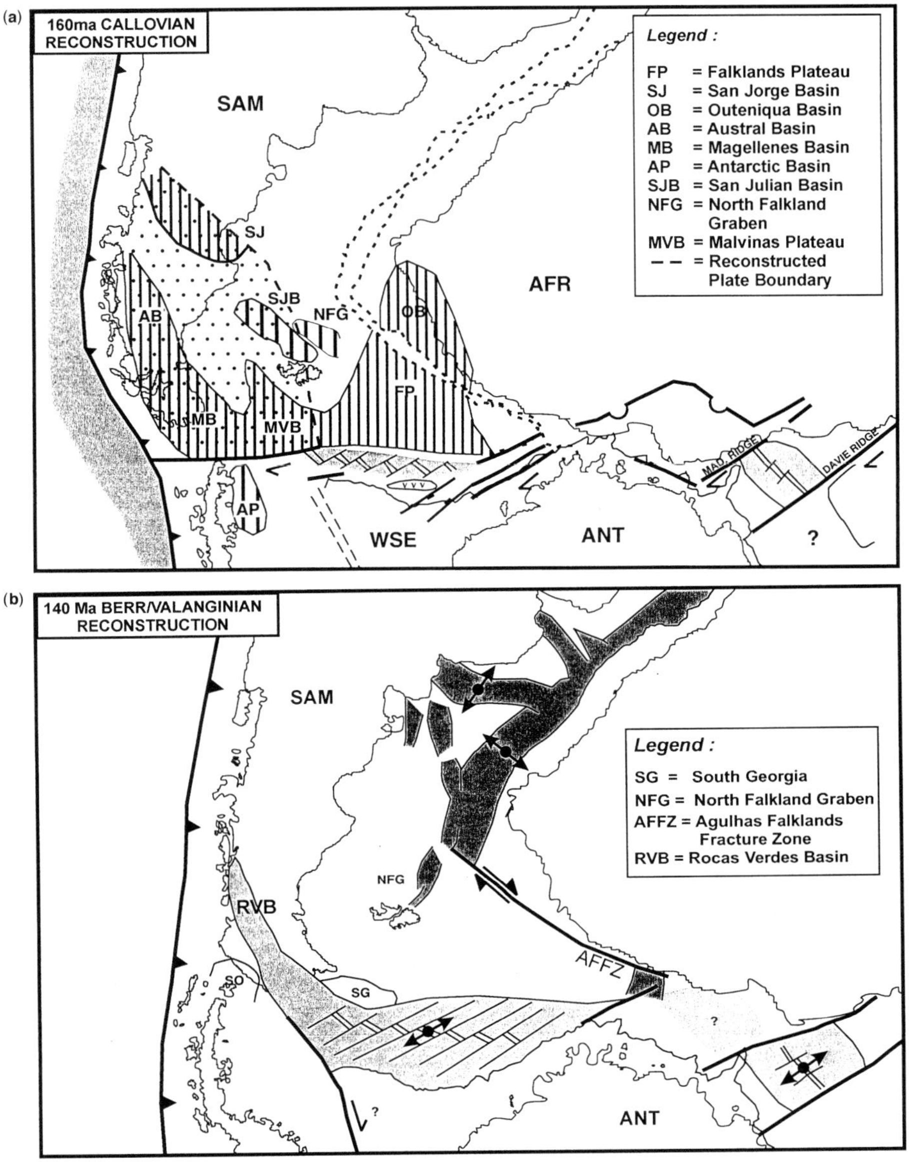

Fig. 7. (**a**) South Atlantic and Weddell Sea reconstruction at 160 Ma [modified from Livermore & Hunter (1996)] showing distribution of Megasequence I (mid–late Jurassic) syn-rift basins (vertical stripes). Bold dots indicate distribution of partly, or wholly, volcanic syn-rift fill. Stipple represents oceanic crust. (**b**) Reconstruction at 140 Ma showing distribution of Megasequence II (early Cretaceous) basins. The North Falklands Graben is developed west of the termination of the Agulhas–Falkland Fracture Zone. To the east, areas adjacent to the fracture zone display marked uplift.

marine in proximal basins and shallow to deep marine in distal areas. Oldest penetrated sediments are Callovian in the Falklands Plateau Basin and Oxfordian in the offshore Outeniqua Basins. These are younger than most of the Tobifera volcanic syn-rift section, although the deepest parts of some southern African half-grabens with possible older sections have not been penetrated.

This period of extension led to early Cretaceous seafloor spreading in the Rocas Verdes back-arc basin (Fig. 7b) and thermal subsidence with post-rift sedimentation in the Magallanes–Austral and Malvinas Basins, marked by the diachronous basal transgressive Springhill Formation.

Relative dextral movement between Antarctica and South America–Africa continued and in the earliest Cretaceous a second major phase of plate break-up was initiated. Widespread rifting of Neocomian age occurred in the South Atlantic area, connected to a triple junction in the Natal Valley (offshore southern Africa) by the Agulhas Transform Fault. This led to three-plate spreading from the Valanginian onwards.

Dating of Rift Events

Three principal rift events could have affected the North Falklands area but there is seismic evidence for only two separate rift phases. The oldest break-up event (late Trias–Liassic) is recorded in parts of the San Jorge and Austral Basins. Sediment-filled rifts of this age are not widespread but their presence offshore the Falklands cannot be ruled out. The only direct evidence of the late Trias–Liassic events in the Falklands plateau are early Jurassic dykes onshore the Falklands.

The mid- and early–late Jurassic rifts commonly have controlling fault orientations strongly associated with existing Gondwanide trends, suggesting a commonality of regional extensional stress vector. The Megasequence I basins in the North Falklands area lie within the distribution of rifts of this age, and are consistent with the structural grain of the reconstructed Magallanes, Malvinas and Outeniqua Basins, and are also demonstrated to reactivate Gondwanide contractional faults (Lawrence & Johnson 1995). The most likely age of Megasequence I syn-rift is therefore mid–late Jurassic (e.g. Lawrence & Johnson 1995), in contrast with Richards *et al.* (1996*a*, *b*, 1997*a*) who suggest a late Trias–early Jurassic age.

Along-strike to the northwest of the Western Rift Complex, the San Julian Basin (Richards 1997*b*) contains a sedimentary section which is highly faulted and subcrops with major angularity beneath a shallow, unstructured section. Although probably not directly connected with the Western Rift Complex, it has very similar seismic geometries. In the single-well penetration, the truncated and faulted section contains very tuffaceous clastics dated as mid–late Jurassic (and possibly earliest Cretaceous); the unstructured shallow section is Tertiary in age.

In contrast with the Megasequence I basins, the North Falklands Graben (Megasequence II) has principal controlling fault trends which clearly relate to reconstructed South Atlantic rift orientations (Fig. 7b). The linkage with the main South Atlantic rift will remain unclear without further seismic acquisition. The North Falklands Graben represents a failed attempt of the South Atlantic to propagate southward through the Falklands area. To the east of this point, the divergence between South America and Africa was taken up solely on the Agulhas Fracture Zone. Most areas adjacent to the fracture zone in the Outeniqua and Falklands Plateau Basins underwent uplift and erosion rather than extension, with the exception of minor areas of the Outeniqua Basin where transtensional reactivation occurred in the Barremian–Aptian.

Basin Model for the North Falklands area

Rift history

Two rift phases are identified (Fig. 8). Megasequences I comprises a northwest–southeast rift complex with lesser north–south and east–west rifts, which are contemporaneous with rifting in the Malvinas and Outeniqua Basins (Fig. 9). This was terminated by the onset of rifts with a north–south (Megasequence II) which cut across, or locally reactivate, the earlier rift system. This differs from other models (Lawrence & Johnson 1995; Richards *et al.* 1996*a*) which recognize two rift phases, although often controlled by the same fault sets. Lawrence & Johnson (1995) interpret rifting in the Western Rift Complex as mid–late Jurassic, and in the North Falklands Graben and Southern Half-graben as late Jurassic–early Cretaceous. They interpret continuous deposition in all areas, until terminated in the western and southern areas by a mid-Cretaceous transpressional deformation phase. Richards *et al.* (1997*a*) identify early syn-rift of Triassic–early Jurassic age in the North Falklands Graben, and in the southern and western areas. A second rift of mid–late Jurassic age terminated deposition in the Western Rift Complex, and deposited widespread syn-rift sequences in the North Falklands Graben and southern areas. This is a modification of their earlier model (Richards *et al.* 1996*a*, *b*) which suggests a single unbroken rifting period from Trias to early–mid Jurassic (and possibly as young as early Cretaceous).

It is proposed here that the western and southern rift areas have a common tectonic history, comprising mid-Jurassic extension, termination of sedimentation in the late Jurassic (with tensional faulting and possible associated uplift) and regional uplift in the latest Cretaceous. There is no evidence in these areas of thrust or inversion tectonics and all geometries can be explained in

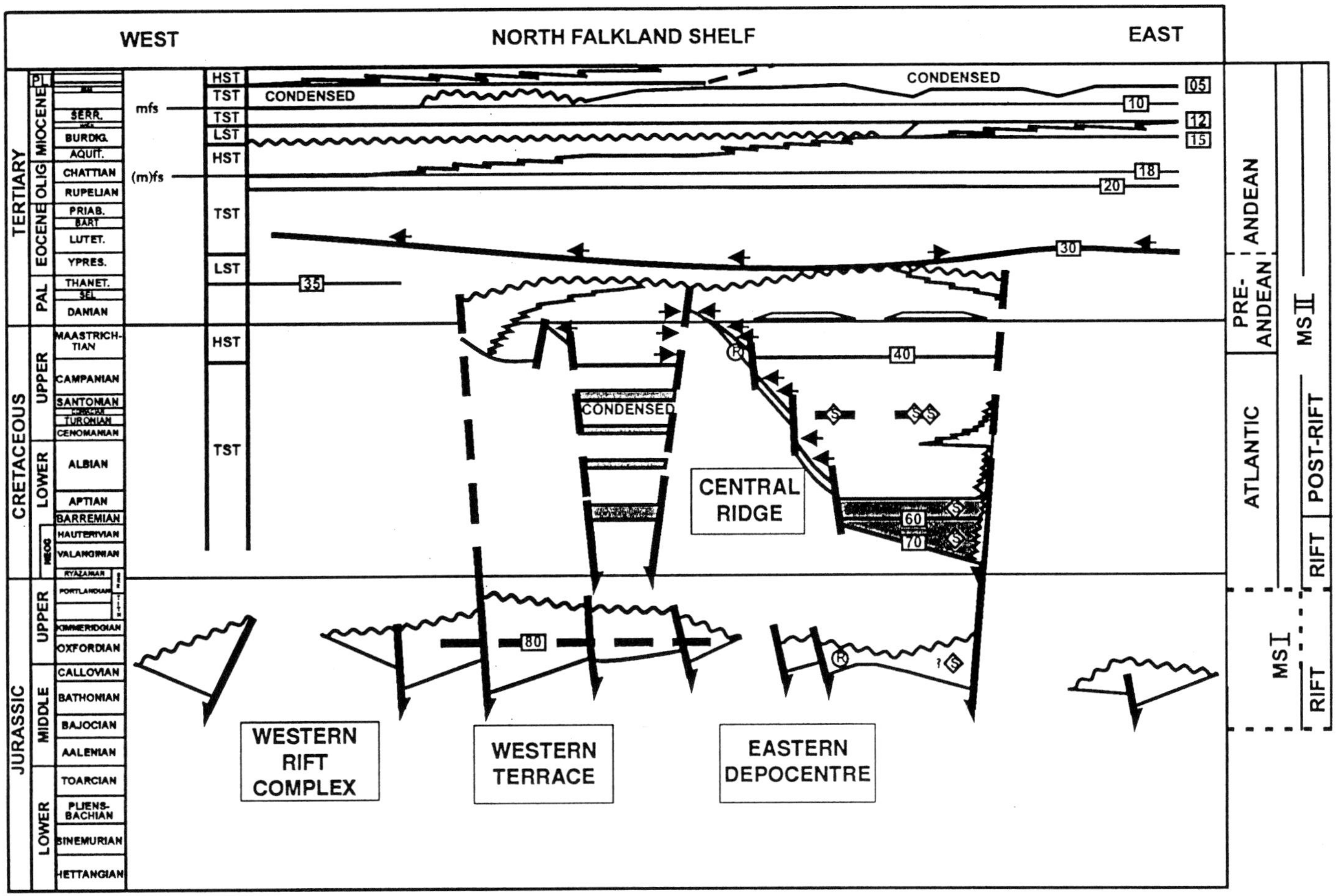

Fig. 8. Chronostratigraphic chart over the North Falklands Shelf. Within Megasequence II, Horizon 40 marks the onset of major clastic influx into the North Falklands area, attributed to regional uplift. Continued uplift led to exposure and bevelling of all areas except the depocentres of the North Falklands Graben.

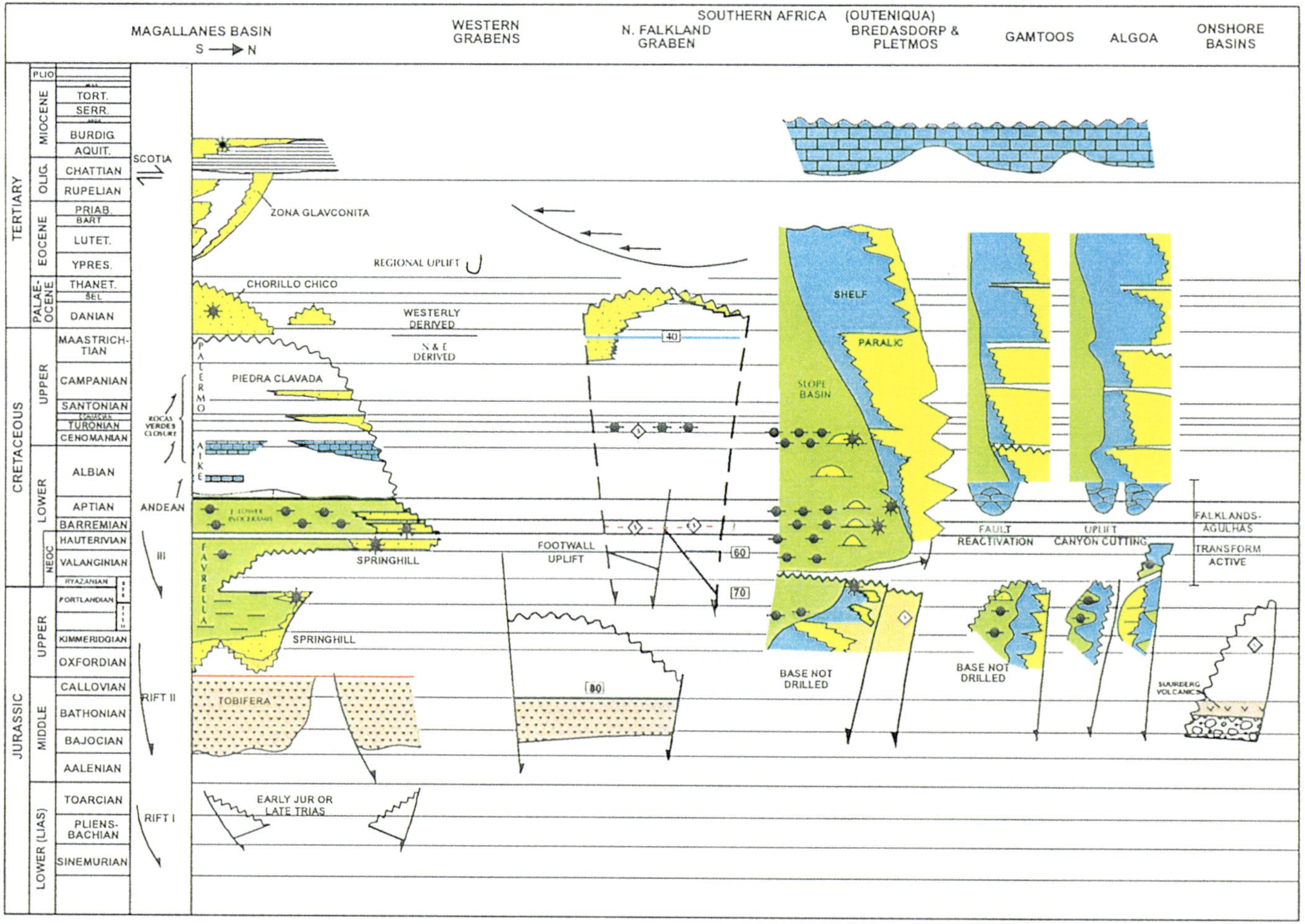

Fig. 9. Chronostratigraphic chart linking the Magallanes Basin (Argentina), the North Falklands Shelf and the Outeniqua Basins (South Africa). The North Falklands area has extensional and transpressional structural controls in common with the southern South America plate (Magallanes Basin).

the context of deeply truncated extensional fault blocks, with associated hanging wall rollover and footwall drape features. Apparent folding and flower structures are probably a function of line orientation, e.g. Richards *et al.* (1996, fig. 12) which is near to the strike of a major subsurface fault in the Southern Half-graben. Termination of deposition is interpreted here to be due to the change in extensional stress at the initiation of Atlantic rifting. If sedimentation had continued in all areas until uplift at Horizon 30 [as suggested by Lawrence & Johnson (1995)] then the western and southern areas should show post-Horizon 30 compaction effects comparable with the North Falklands Graben.

There is no unequivocal seismic evidence for footwall uplift adjacent to the Megasequence II North Falklands Graben, and if it did occur the effects have probably been masked by later erosion at Horizon 30. Major basin margin uplift associated with development of the South Atlantic is recorded from the Namibia and southern Africa margin (Brown 1990) in apatite fission track analysis data. The uplift is dated as early Cretaceous and estimated at 1.5–3 km depending on thermal gradient assumptions. However, on the Falkland Islands, apatite fission track analysis data show cooling from Permo-Trias and marked cooling events in mid–late Cretaceous and early Tertiary (Lawrence & Johnson 1997), but no evidence for accelerated early Cretaceous cooling. This is consistent with the observed low stretching factor over the North Falklands Graben, which would result in the magnitude and lateral extent of any footwall uplift being very modest. For example, Young (1992) modelled a maximum of 500 m uplift over the Magnus Fault Block as a result of 112% extension over a 100 km section of the Northern North Sea Viking Graben.

Atlantic megasequence

Sequence stratigraphic analysis of Megasequence II points to the development of a starved basin in the North Falklands Graben. A syn-rift sequence is only positively identified in the eastern depocentre adjacent to the main controlling fault. This approximately equates with the 'lower syn-rift' of Richards *et al.* (1997*a*, fig. 5). Elsewhere in the North Falklands Graben seismic packages observed beneath the composite base post-rift event are considered to be Megasequence I. Early post-rift Unit 60–40 is also restricted to the axis of the eastern depocentre, although may be present in western terrace. At Horizon 40 significant palaeobathymetry is indicated by prograde heights within the overlying Unit 40–30. Decompaction and structural restoration has confirmed that the internal geometries of Units 60–40 and 40–30 are incompatible with any movement on the eastern boundary fault, and rotation of the central ridge titled fault block. Significant palaeorelief on the eastern boundary fault is demonstrated by modelling.

Basin starvation probably resulted from the isolation of North Falklands Graben from areas of major sediment supply. Widespread extension during Megasequence I (Fig. 7a) would have diminished large areas of the Falklands Plateau to near sea level. As discussed above, footwall uplift associated with the subsequent Megasequence II rifting was probably limited. This situation has also been reported in the North Magnus Trough (Young 1992). Several rift phases are identified in this area of the North Sea; a widespread Permian–mid-Jurassic section was affected by further extension in the Callovian–Oxfordian with a north–south orientation. The final Portlandian–Hauterivian rift phase was oriented northwest–southeast and produced major fault offsets. However, no syn-rift fill is identified with this final extension phase and initial deposits in the North Magnus Trough are starved early Cretaceous passive infill post-rift. Reconstruction and decompaction indicates up to 3–4 km of palaeobathymetry on the Magnus Fault. Restricted deposition continued through the Cretaceous and the submerged tilted footwall blocks were not overstepped until within the late Cretaceous.

Andean events

The most significant seismic event over the North Falklands shelf is Horizon 30, which is strongly truncational over all areas apart from the eastern depocentre of the North Falklands Graben, where a correlative conformity exists. A similar distinctive event is identified in the Magallanes Basin within the foreland sequence (Biddle *et al.* 1986) and in the offshore Malvinas Basin (Yrigoyen 1989). This is a composite unconformity, which in areas proximal to the Andean and Scotia Arc contractional systems demonstrates a rapid deepening of the basin and passive onlap. Deepwater erosion is also recorded beneath the unconformity in the Malvinas Basin (Galeazzi 1994). The time gap is smallest close to the deformation front and is dated as Thanetian–Ypresian (late Palaeocene–early Eocene) (Fig. 9). To the north the time gap increases and in the northern Magallanes Basin Upper Oligocene rocks rest on Barremian beneath the unconformity as a result of deep subaerial erosion.

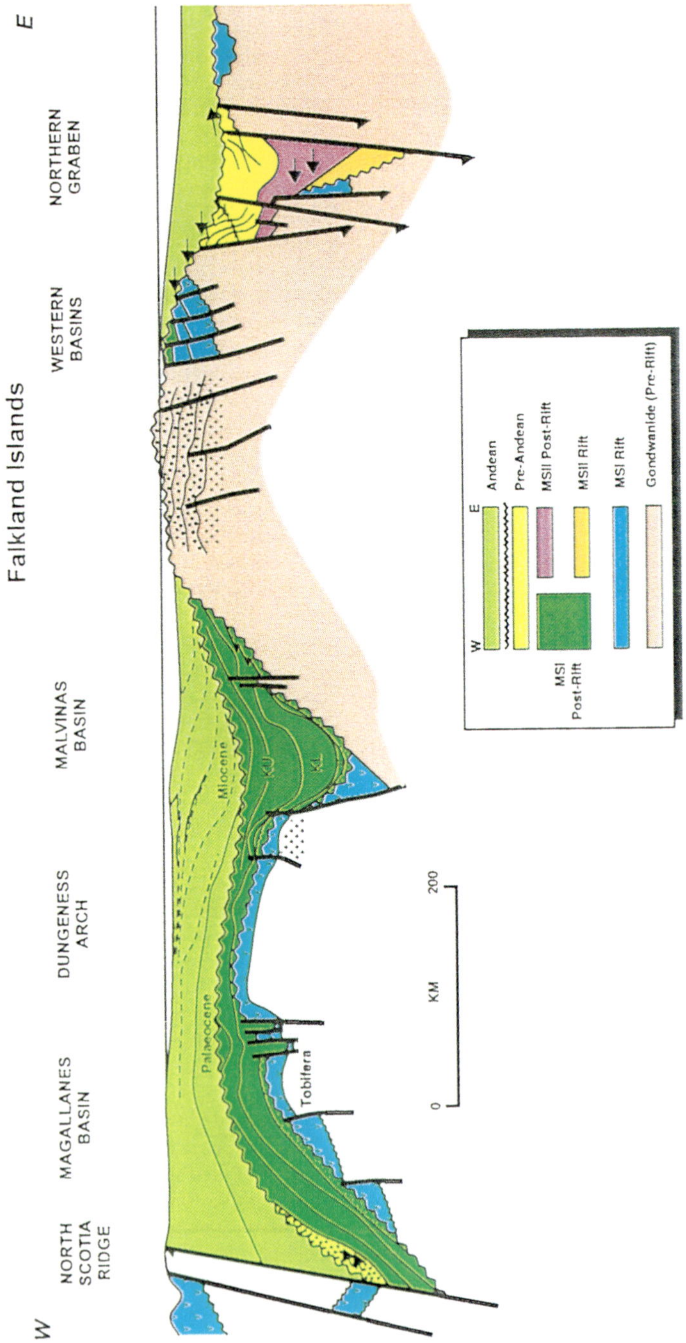

Fig. 10. Megasequence identification in Magallanes, Malvinas and North Falklands Basins.

This event is interpreted here as resulting from final closure of the Rocas Verdes Basin and initiation of the Scotia Arc plate boundary system. Closure of the Rocas Verdes Basin was initiated in late Albian, probably contemporaneous with plate reorganization in the Atlantic, which resulted from the final clearance of southern Africa by the Falklands Plateau. Subsequently, relative plate motion between South America and Antarctica changed from dextral to sinistral. Foreland sedimentation is observed in the westernmost Magallanes during Albian–Santonian (Biddle *et al.* 1996), contemporaneous with deformation in the Cordillera Darwin. This may have been subduction related, and the subsequent extremely rapid uplift of the Cordillera Darwin in latest Cretaceous was the result of final Rocas Verdes closure and arc–continent collision. Continued relative eastward movement of the Antarctica Peninsula with respect to South America resulted in transfer of stress to the southern Falklands margin with the development North Scotia transpressional plate boundary.

Significant unconformities related to Andean tectonics have been discussed by Lawrence & Johnson (1995, 1997). Both mid-Cretaceous and early Tertiary events are predicted, supported by evidence of accelerated cooling (and hence uplift) interpreted from apatite fission track analysis data onshore the Falklands.

The major unconformity recognized in the Palaeogene in the Magallanes and Malvinas Basins are correlated here with Horizon 30 over the North Falklands Shelf (Fig. 10), and are interpreted to be the result of flexure of the South American plate (including the Falklands Plateau) as a result final closure of the Rocas Verdes Basin. In the late Palaeogene a deep transpressional foreland basin developed on the southern margin of the Falklands Plateau and the northern margin (northern Magallanes Basin, San Julian Basin, North Falklands Shelf) was subaerially exposed due to regional tilting. Prior to this, Andean effects were largely limited to the western Magallanes Basin; in the Malvinas Basin well data does not demonstrate a time gap associated with the widespread Aptian reflector or at any other level in the Upper Cretaceous. It is suggested that the middle Cretaceous age predicted for Horizon 30 by Lawrence & Johnson (1995) is less likely. Richards *et al.* (1997*a*) ascribe an Aptian age to Horizon 30, suggesting the major overstep of the basin margins by post-Horizon 30 sediments was due to foundering of the Falklands Plateau after final separation of southern Africa and South America.

The predicted flexural uplift and exposure of the northern Magallanes, Malvinas and Falklands areas led to widespread truncation. Only pre-existing deepwater areas are likely to have escaped the effects and will show continuous sedimentation. Sequence geometries and restoration of the North Falklands Graben has demonstrated that significant palaeobathymetry had developed in the early post-rift (Unit 60–40). This sediment starvation prior to Horizon 40 was abruptly terminated by a major sediment influx and all pre-existing accommodation space ultimately became filled by Unit 40–30 progradational clastics. The reduction in accommodation space was probably partly due to contemporaneous uplift, since parts of the topset facies themselves become truncated beneath Horizon 30.

Unit 40–30 is interpreted as a forced regression complex with primary control of sediment supply due to regional uplift and no active fault control. Progradation from the north and west (rather than the southwest as seen post-Horizon 20) supports the regional uplift model, since no other sediment source exists between this area and the Argentine mainland present day. Other examples of forced regressions include the Palaeogene of the northern North Sea (Anderton 1993), where thermal uplift of the adjacent Scottish mainland was responsible for the change from a deep sediment-starved basin to a high energy depositional environment, accumulating thick deltaic and deep marine clastic sediments.

The age of Horizon 40 is predicted to be late Cretaceous (?Maastrichtian), which is the age assigned by Biddle *et al.* (1986) to the first southerly derived sediments in the southern Magallanes Basin of Chile. Older Andean foreland sequences are present in the western Magallanes but not represented in the southern Magallanes or Malvinas Basins. The Chorillo Chico Formation of southern Chile lies stratigraphically beneath the regional late Palaeocene unconformity and is considered to be contemporaneous with sequence to Unit 40–30, although depositional environments and facies probably differ.

Discussion

Three independent analyses of essentially similar datasets have generated marked differences in interpretation, particularly in the hierarchy of seismic sequences, and their rift history. This, in turn, directly impacts on perceptions of prospectivity.

There is general consensus that the larger part of the North Falklands Graben fill is 'Atlantic' in age, and that equivalents of Barremian–Aptian and Cenomanian source horizons will be present. However, the top syn-rift pick in this basin differs widely between the interpretations, hence the position of early post-rift potential source rock varies. In the Richards *et al.* (1996) model, top syn-rift is picked near the base of Unit 40–30, which does not exceed 2.3 s TWTT. Lawrence & Johnson (1995, 1997) pick a horizon within upper Unit 60–40 which probably goes no deeper than 2.8 s TWTT. In this model the source horizons could be present as deep as 3.8 s TWTT in the interpreted early post-rift.

The stratigraphy of the central ridge tilted fault blocks depends on the megasequence model adopted. The interpretation of a starved depositional system requires that major palaeobathymetry was developed and the basin was isolated from sediment supply. This palaeorelief can only be developed during active faulting, hence, any rock units offset by these faults must be pre-rift by definition. For this reason the sediment packages beneath the onlapping post-rift of the central fault blocks cannot be syn-rift units deposited by North Falkland Graben controlling faults. In this two-phase rift model there is no Megasequence II syn-rift predicted in the central ridge tilted fault block but instead the opportunity for subaerial exposure of Megasequence I and possible enhancement of reservoir quality. In the Lawrence & Johnson (1995, 1997) model a single Atlantic syn-rift is mapped over the entire basin including the central ridge. In the Richards *et al.* (1996) model 'early' and 'late' rift phases in the North Falkland Graben are superimposed and sequential, with no predicted uplift and/or exposure, except at the basin margins.

We are grateful to Spectrum Energy and Information Technology for the permission to publish the seismic data used in Figs 2, 3 and 5, and to our partners Desire Petroleum and Clyde Petroleum.

References

ANDERTON, R 1993. Sedimentation and basin evolution in the Paleogene of the Northern North Sea and Faeroe-Shetland basins *In:* PARKER, J. R. (ed.) *Petroleum Geology of Northwest Europe: Proceedings of the 4th conference.* Geological Society, London.

BIDDLE, K. T., ULIANA, M. A., MITCHUM JR, R. M., FITZGERALD, M. G. & WRIGHT, R. C. 1986. The stratigraphic and structural evolution of the central and eastern Magallanes Basin, southern South America, International Association of Sedimentology, Special Publications, **8**, 41–61.

BROWN, R. W., RUST, D. J., SUMMERFIELD, M.A., GLEADOW, A. J. W. & DE WIT, M. C. J. 1990. An early Cretaceous phase of accelerated erosion on the south-western margin of Africa: evidence from apatite fission track analysis and the offshore sedimentary record. *Nucl. Tracks Radiat. meas.*, **17/3**, 339–350.

DE WIT, M. J. & RANSOME, L. G. D. 1992. *Inversion Tectonics of the Cape Fold Belt, Karoo and Cretaceous Basins of Southern Africa.* Balkema, Rotterdam.

GALEAZZI, J. S. 1994. *Stratigraphic and Structural Evolution of the Western Malvinas and Southeastern Magallanes Basins, Argentina.* MA dissertation, Rice University.

GUST, D. A., BIDDLE, K. T., PHELPS, D. W. & ULIANA, M. A. 1985. Associated Middle to Late Jurassic vulcanism and extension in southern South America. *Tectonophysics*, **116**, 223–253.

LAWRENCE, S. R. & JOHNSON, M. 1995 Shelf north of the Falklands may be new S. Atlantic petroleum province. *Oil and Gas Journal*, March 6, 52–55.

——, JOHNSON, M. & TUBB, S. R. 1997. The structural evolution of the North Falkland region and its bearing on the formation of the southern South Atlantic. *Oil and Gas Habitats of the South Atlantic.* (*abstract*)

LIVERMORE, R. A. & HUNTER, R. J. 1996. Mesozoic seafloor spreading in the southern Weddell Sea. *In*: STOREY, C. C., KING, E. C., LIVERMORE, R. A. (eds) *Weddell Sea Tectonics and Gondwana Break-up*. Geological Society, London, Special Publications, **108**, 105–128.

LORENZO, J. M. & MUTTER, J. C. 1988 Seismic stratigraphy and tectonic evolution of the Falkland/Malvinas plateau. *Revista Brasileria de Geosciencias*, **18**, 191–200.

RICHARDS, P. C., GATLIFF, R. W., QUINN, M. F., WILLIAMSON, J. P. & FANNIN, N. G. T. 1996*a* Petroleum potential of the Falkland Islands offshore area. *Journal of Petroleum Geology*, **19** 161–182.

—— & FANNIN, N. G. T. 1997*a*. Geology of the North Falkland Basin. *Journal of Petroleum Geology,* **20 (2)**, 165–183.

—— 1997*b*. The North Falkland Basin. *Oil and Gas Habitats of the South Atlantic.* (*abstract*)

——, GATLIFF, R. W., QUINN, M. F., WILLIAMSON, J. P. & FANNIN, N. G. T. 1996*b* The geological evolution of the Falkland Islands continental shelf. *In*: STOREY, C. C., KING, E. C. & LIVERMORE, R. A. (eds) *Weddell Sea Tectonics and Gondwana Break-up*. Geological Society, London, Special Publications, **108**, 105–128.

STOREY, B. C., CURTIS, M. & THISTLEWOOD, L. 1995. Structural development of the Falkland Islands region. Unpublished report. British Antarctic Survey.

——, KING, E. C. & LIVERMORE, R. A. (eds.) 1996. *Weddell Sea Tectonics and Gondwana Break-up*. Geological Society, London Special Publications, **108**

YOUNG, R. 1992. Restoration of a regional profile across the Magnus Field in the northern North Sea. *In*: LARSEN, R. M., BREKKE, H., LARSEN, B. T. & TALLERAAS, E. (eds) *Structural and Tectonic*

Modelling and its Application to Petroleum Geology. NPF Spec. Pub. 1, pp. 221–229. Elsevier, Amsterdam Norwegian Pet. Soc., (NPF).

YRIGOYEN, M. R. 1989. Cuenca de Malvinas. *In*: CHEBLI, G. A. & SPALETTI, L. A. (eds) Cuencas Sedimentarias Argentinas. Serie Correlacion Geologia No. 6, Universidad Nacional del Tucuman, 481–491.

Basin evolution and sedimentary fill in the Palaeozoic sequences of the Falkland Islands

NEIL S. MEADOWS

Geochem Group Ltd, Chester Street, Chester CH1 8RD, UK
Present address: Redrock Associates Ltd, 38 Queens Drive, Prenton, Wirral L43 0RP, UK

Abstract: The sedimentary sequences exposed on the Falkland Islands can be referred to two major episodes of basin evolution. On the basis of fieldwork and recent palynological determinations carried out as part of the study from which this publication is taken, the older of these is largely Middle Devonian in age and comprises a marine transgressive–regressive system culminating in extensive fluvial braidplain deposits which are referred to the Gran Malvinas Group. The younger is largely Permian in age, commences with a range of glacial deposits and includes lacustrine, shallow marine and deeper marine deposits before also culminating in fluvial channel deposits and is referred to the Lafonian Supergroup. The facies identified at outcrop and their relationships are described and presented herein as depositional models for the two major systems, while inferences derived from these models are used both to suggest the structural and tectonic history of the islands, and to erect a revised lithostratigraphy which relates to these episodes of basin evolution. It is recognized that this is an interim measure pending much more intensive work on the geology of the Falkland Islands which is currently taking place.

The Falkland Islands have never been geologically mapped in any detail and the most recent and only available geological map, published by the British Antarctic Survey (Greenway 1972), was compiled entirely by examination of aerial photographs and with reference to previous publications, many of which date from the late nineteenth and early twentieth centuries. These include the work of Darwin (1846), Thomson (1877), Halle (1912) and Baker (1924). The first significant contribution to an understanding of the geology of the Falkland Islands was provided by the work of Halle (1912) who identified the major lithostratigraphical divisions present on the islands. This author determined, from macrofossil evidence, that the two major sedimentary divisions were Devonian and Permo-Carboniferous in age, respectively, although both of these were still very generalized. Halle (1912) was also the first to recognize the similarity between the Falkland stratigraphical succession with its distinct fossil species and that of southern Africa. The first official geological survey was carried out under the auspices of the Falkland Island government between 1920 and 1922. The results were published two years later and provided the first comprehensive stratigraphy and structure (Baker 1924). In particular, Baker (1924) defined subdivisions within successions of both Devonian and presumed Permo-Carboniferous–Triassic age.

The Devonian sedimentary sequence was divided into four units comprising, from the base, the Port Stephens, Fox Bay, Port Philomel and Port Stanley Formations, while the sediments of supposed Permo-Carboniferous age comprise a basal tillite referred to the Lafonian Diamictite, overlain by the Port Sussex Formation comprising carbonaceous shales of the Black Rock Member, mudstones and sandstones of the Shepherds Brook Member and sandstones of the Terra Motas Sandstone. An overlying, presumed Permo-Triassic, succession of sandstones and mudstones was not stratigraphically differentiated but referred, on grounds of local variations in lithology, to the Bay of Harbours, Choiseul Sound, Brenton Loch and West Lafonian Beds. The whole of this presumed Carboniferous–Triassic succession was referred to the Lafonian Supergroup by Borrello (1963, 1972). Included within this was a heterolithic succession identified around Bluff Cove underlying the Lafonian Diamictite and referred to the Bluff Cove Beds (Dawson 1967), although recent proprietary palynological dating by the Geochem Group identified Devonian palynomorphs in this sequence.

The terminology erected by Baker (1924) is essentially that in use today and has been repeated in subsequent publications with some minor modification (Frakes & Crowell 1967; Greenway 1972). Further examination of outcrops on the islands were undertaken, most notably by Adie (1952) who described a range of depositional features within the presumed Permo-Carboniferous tillites and Marshall (1994*a*) who identified marginal marine to offshore shelf transitions within the Devonian Fox Bay Formation and regressive sandstones in the base of the Port Stanley Formation. These provided some information on the sedimentology of the two successions but did

MEADOWS, N. S. 1999. Basin evolution and sedimentary fill in the Palaeozoic sequences of the Falkland Islands. *In:* CAMERON, N. R., BATE, R. H. & CLURE, V. S. (eds) *The Oil and Gas Habitats of the South Atlantic.* Geological Society, London, Special Publications, **153**, 445–464.

not attempt to refine the existing lithostratigraphy or to integrate these data with concepts of basin evolution.

Regional structure

Three major structural provinces can be identified on the islands (Fig. 1). A northern belt running east to west across both islands is characterized by east–west trending, mainly upright, folds in largely Devonian sequences. Southern West Falkland comprises either flat-lying or shallow open folds in Devonian sequences and southern East Falkland, together with adjacent parts of West Falkland and northern East Falkland, comprises a largely Permian basin (Lafonia Basin) that has been partially shortened by thrusting. Post-dating this deformation and intruding both Devonian and Permian sedimentary sequences are at least two sets of dykes, dated as latest Early Jurassic (Mussett & Taylor 1994), the orientation of these tending to be roughly parallel to the major east–west and northeast–southwest fold axes. Offshore structures follow similar trends, particularly to the north of the islands (Lawrence & Johnson 1995; Richards *et al.* 1996*b*; Richards & Fannin 1997), suggesting that all of these structures may be related to basement lineaments which have undergone repeated reactivation.

The northern areas of both East and West Falkland comprise relative uplands by comparison with the south and are typified by large upright open to tight folds that trend broadly east–west to east-southeast–west-northwest. According to the mapping of Greenway (1972) these form a broad anticlinal structure that can be traced across both islands (Fig. 2) with the Port Stanley Formation exposed on the limbs and the Fox Bay Formation exposed in the core. Recent work is revealing considerably more complexity and suggests some exposures of the Port Stephens Formation along the north coast of East Falkland (D. Aldiss, pers. comm.), although whether this formation is present to the extent suggested by Curtis & Hyams (1998) is unclear. In a few localities, e.g. near Elephant Beach (Fig. 1), there are small-scale thrusts associated with the folding and, although the amount of displacement is uncertain, the direction of shortening is consistent with broadly north–south directed compression (Fig. 2). Throughout the area, folds exhibit near-vertical axial planes, although there are also commonly developed fractures, usually parallel to the axial planes, which suggest that there has been subsequent extensional reactivation.

At the southern limits of the fold belt in East Falkland sequences younger than the Port Stanley Formation are also affected by the deformation. The Lafonian Diamictite and Port Sussex Formations exhibit similar mainly upright east–west trending folds, although the extent of this folding is difficult to determine in the extremely poorly sorted and largely unbedded tillites of the Lafonian Diamictite Formation. By comparison, the Black Rock Member of the Port Sussex Formation exhibits intense deformation in response to the dominantly claystone lithology, commonly revealing cleavage development and, in places, the overturning and refolding of both primary folds and cleavage. Dips decrease to the south and the structures pass out into the relatively flat-lying sequences of Lafonia.

On West Falkland the main east–west fold trend is intersected by northeast–southwest trending structures related to a major fold axis forming the boundary between the relatively flat-lying Devonian sequences of southern West Falkland and Permian exposures associated with Lafonia. The interference of the two fold trends is most apparent at Bold Cove (Fig. 1) where a series of asymmetrical folds and associated wrench faults occur as parasitic structures around a major southerly plunging syncline, as discussed in detail by Curtis & Hyam (1998). The northeast–southwest trending structure is cored at the surface by Port Stephens Formation sandstones which form the Hornby Mountains, while a synformal monocline, which exposes successively the Fox Bay and Port Stanley Formations followed by a succession of the overlying Permian sequences, occurs on the foreshore of Falkland Sound. The western limb of this synformal structure is extremely steep with bedding commonly recorded at angles $>70°$ and, in places, near vertical. The outcrop of the resistant Port Stanley Formation sandstones in the limb of this fold form a prominent ridge trending parallel to the coastline. The relationship of bedding across Falkland Sound suggests an asymmetry to the monocline, as discussed by Frakes & Crowell (1967), with a steep western limb and shallow eastern limb (Fig. 3). The axis of the structure is parallel to Falkland Sound and, by implication, the youngest sequences on the Falkland Islands probably occur within this axis.

The Permian sequences on Lafonia are either subhorizontal or inclined at generally low angles to the west (Fig. 3) and, in northern areas, to the south (Fig. 2). Only in the extreme northwest and to the north of Brenton Loch and Choiseul Sound is bedding inclined at angles $>15–20°$, and in these areas it is locally up to 50° to the south and southwest. Deformation of these

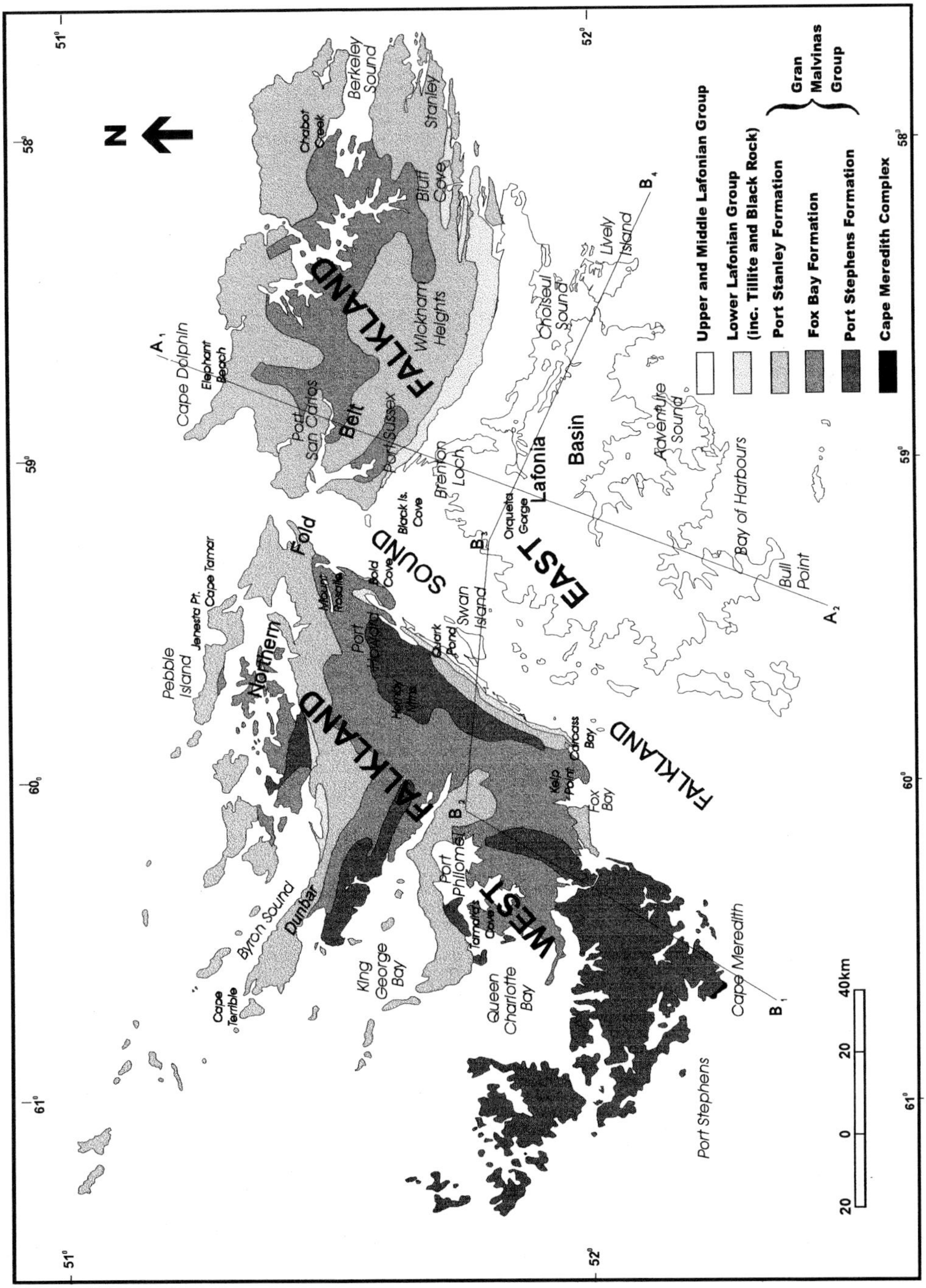

Fig. 1. General geology of the Falkland Islands (from Greenway 1972) with localities discussed in the text and the position of the schematic cross-sections in Figs 2 and 3.

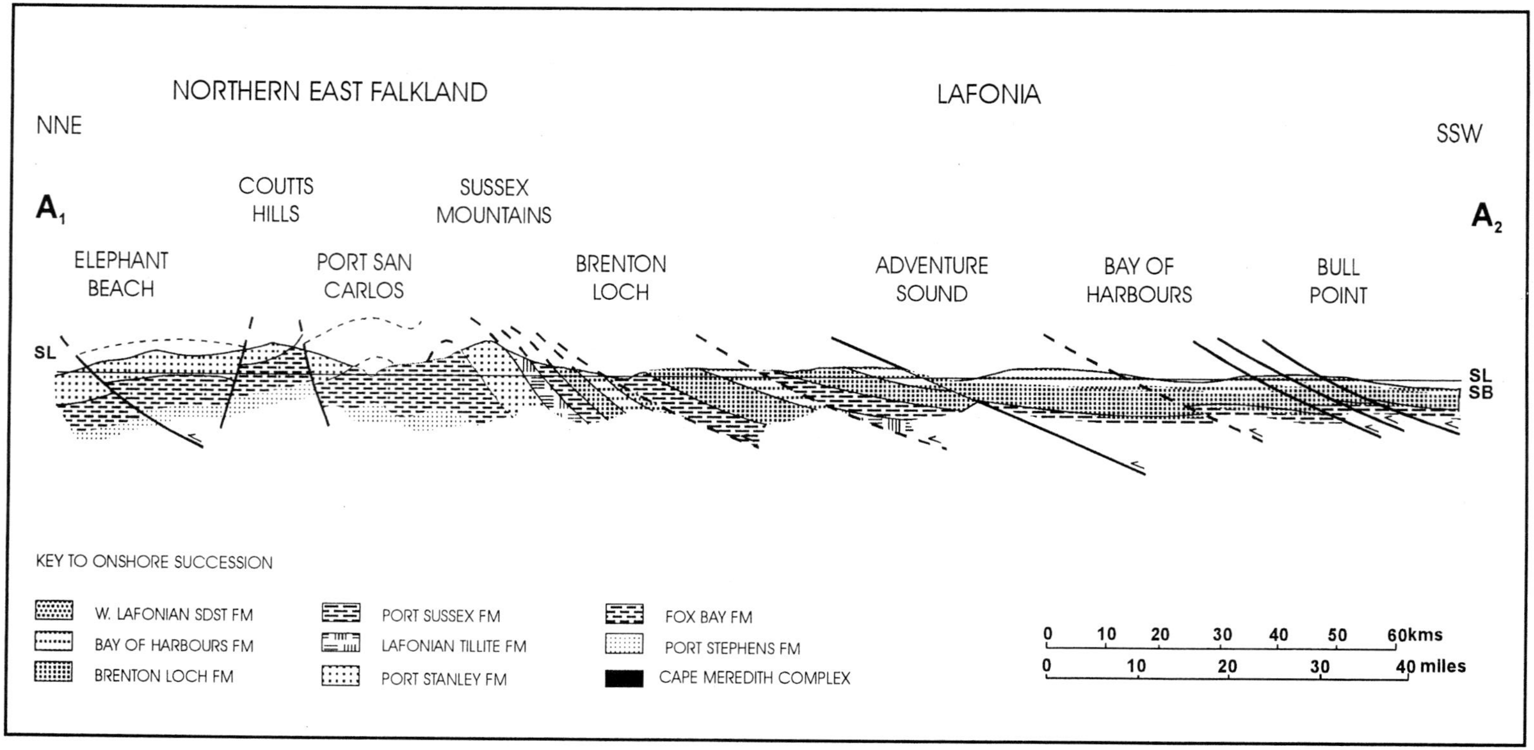

Fig. 2. Schematic cross-section based on outcrop data through East Falkland showing structure in the northern fold belt and the inferred repetition of sequences in the Permian Lafonian Supergroup due to northerly directed thrusting.

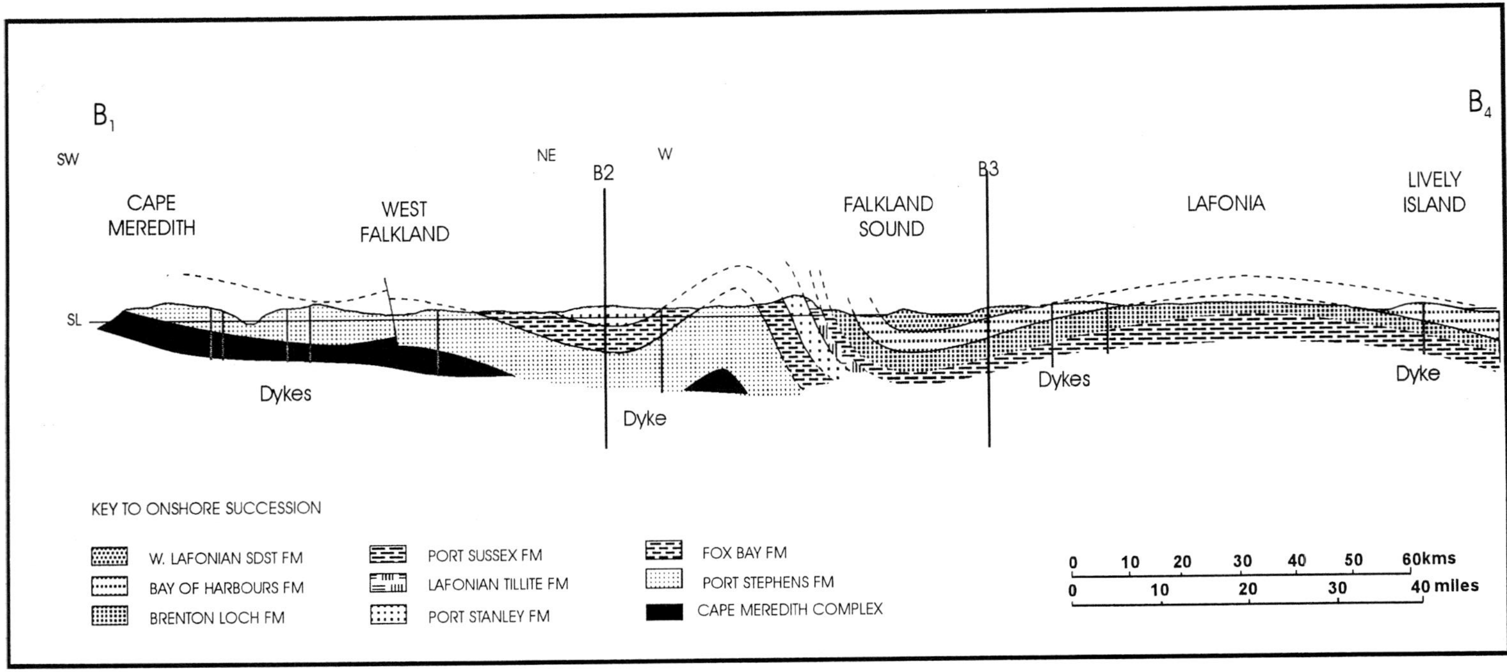

Fig. 3. Schematic cross-section from West Falkland through Falkland Sound and east Falkland showing the position of the inverted monocline forming the western boundary to the Lafonia Basin.

Permian sequences occurs mainly in the form of thrust planes which dip generally to the south (Fig. 2) and are best exposed at Bluff Cove and Bull Point. A series of these thrusts at Bull Point has resulted in repetition of the exposed sequence along the cliffline, although in each case the displacement is no more than a few tens of metres. The direction of shortening associated with the thrusting is broadly from south to north wherever they are observed and is therefore consistent with the apparent shortening recorded by fold axes in the Devonian sequences to the north. It can be suggested, therefore, that the thrusting in the Permian sequences records the same episode of deformation as that which generated the northern fold belt. The difference in the expression of the deformation can be related to differences in lithology. In particular, the capacity for decollement of the Permian sequences on the highly organic-rich claystones of the Black Rock Member of the Port Sussex Formation, although the possibility that the thrusts sole out at greater depth cannot be discounted.

Devonian sedimentology

Devonian sedimentary and metasedimentary sequences of the Gran Malvinas Group (Fig. 4) dominate exposures across much of the Falkland Islands. They comprise the present day erosion surface throughout virtually the whole of West Falkland, the majority of northern East Falkland and are inferred to form a major component of offshore sequences, locally comprising highs between Mesozoic basins (Richards & Fannin 1997). The four identified formations are described and interpreted from the base where sandstones of the Port Stephens Formation lie directly on gneissic Precambrian basement (Rex & Tanner 1982) of the Cape Meredith Complex.

Port Stephens Formation

Sandstones and pebbly sandstones with rare, thin claystone interbeds outcrop throughout the southern areas of West Falkland where they form a prominent cliffline around the coast and hills rising to around 1000 feet in height. They also crop out as inliers exposed within the cores of anticlines in central and northern West Falkland and as coastal exposures along northern East Falkland (D. Aldiss, pers. comm.). The sandstones are generally medium to coarse grained exhibiting locally moderate to good porosities and form decimetre to metre scale sets that are either massive or exhibit flat lamination to low angle cross stratification. The majority of the beds exposed at Cape Meredith are laterally continuous on a scale of hundreds of metres although those at the base of the sequence can be seen to terminate against irregularities in the unconformity, the latter having a relief along the visible section of several metres. These lowermost beds of the Port Stephens Formation contain common pebbles and cobbles, many being identifiable as derived from the Cape Meredith Complex, that form basal lags to the beds and in places are scattered along foreset surfaces. Inland exposures higher in the sequence exhibit channelling in the form of wide shallow scours up to tens of metres wide by approximately 1 m in depth with fills comprising well developed low to moderate angle cross stratified sandstones. Larger scale sedimentary structures are also observed as cross-bedded sets ranging up to some 3.5 m in preserved thickness. Many of the sandstones exhibit a range of ichnofabrics of which the burrow forms *Skolithos* and *Diplocraterion* are by far the most common and can be sufficiently intensely developed to almost completely obscure primary sedimentary features.

The Port Stephens Formation sandstones are interpreted as representing the depositional products of a marine transgression across the eroded Cape Meredith Complex palaeosurface. The sedimentological characteristics are consistent with an upper shoreface setting with the decimetre-scale bedding suggested to represent deposition in response to oscillatory currents induced within the zone of influence of fair weather wave motions. The incidence of *Skolithos* burrows indicates episodes of relative stability that allowed colonization of the substrate. The shallow channelling observed within these sequences can be related to the action of offshore rip currents eroding into the unconsolidated shoreface sands. The presence of larger, metre scale, cross-stratified units suggests the possibility of offshore bar development, although such an interdigitation of shoreface and offshore bar depositional units would imply episodic basinward migration of the shoreline within the overall transgressive setting.

Fox Bay Formation

The highly variable claystone, siltstone and sandstone lithologies of the Fox Bay Formation form a major tract of central West Falkland. They also occur as isolated outcrops throughout northern West Falkland exposed mainly as inliers in the cores of anticlinal structures in the Port Stanley Formation and as more areally extensive tracts within similar, larger structures in northern East Falkland.

Areas in the Fox Bay Formation dominated by fine-grained lithologies occur mainly in central West Falkland and the northern areas of the Fox Bay outcrop in East Falkland. They are typified by finely flat laminated dark grey to brown claystones and siltstones with low to locally moderate organic content. These are variably disrupted by bioturbation which ranges from sparse. *Chondrites* and *Planolites* at some localities to abundant bioturbation, comprising these two ichnogenera and *Thalassinoides*, resulting in near-complete sediment disruption at others. All of these features are consistent with deposition on a marine offshore shelf, probably below storm wave base, where the low rates of deposition would have allowed intensive infaunal colonization. The occurrence of disarticulated Brachiopod valves, isolated Crinoid ossicles and a rare Trilobite pygidium suggest episodic transport of shoreface and near-shore shelf detritus into this deeper water environment, although the possibility of local reworking cannot be discounted.

By comparison with the above, sandstone-rich facies of the Fox Bay Formation are typified by outcrops around Berkeley Sound on East Falkland and Tamara's Cove on West Falkland. The sandstones examined at these localities range from fine to coarse grained, although porosities are generally poor, and commonly form decimetre scale sets with either low angle cross-stratification or flat lamination. In many instances these are disrupted by a range of burrow forms including, most commonly, the ichnogenera *Chondrites, Planolites* and *Zoophycus* with sparse *Diplocraterion* burrows. In most instances these sandstones are interbedded with claystones and siltstones similar to those described above, or are closely associated with the finer-grained lithologies as adjacent outcrops where direct relationships cannot be discerned.

The abundance of sandstone at these localities suggests a more proximal setting, with respect to energy levels, than that described above and indicates the presence of shoreface, and, possibly, offshore shoal environments within the Fox Bay Formation sequences. At several localities, but principally those around Chabot Creek, some of the sandstone beds exhibit well developed hummocky cross-stratification on a scale of 30–40 cm preserved thickness and a wavelength of 1–2 m. These structures support interpretations regarding a marine shelf setting and also indicate the presence of nearshore environments from which storm surge currents could have derived the sand grade material, although no bioclastic material was observed in this instance. At Tamara's Cove the Fox Bay Formation is dominated by sandstones and, rarely for this formation, provides evidence of channelling with decimetre scale relief on set boundaries and channel geometries up to several metres across.

One of the most complete sequences of the Fox Bay Formation is that exposed at Kelp Point within Fox Bay itself where continuous coastal outcrops and cliffs with shallow southwesterly dips expose *c.* 50 m of vertical section. A series of upward-coarsening depositional units varying from 7 to 15 m in preserved thickness are exposed, each of which is interpreted as a prograding shoreface within a highstand systems tract. Each sequence commences at the base with dark grey finely laminated claystones and argillaceous siltstones, passes up through siltstones and very fine sandstones with common siderite concretions and bioturbated fine sandstones. In the more complete sequences, these terminate at the top in decimetre scale beds of flat laminated, low angle cross-stratified and sparsely bioturbated fine and fine to medium sandstones. The marine affinities of these sequences are clearly indicated by the ichnogenera recorded, including *Chondrites, Planolites* mainly within the siltstones, and well developed *Diplocraterion* mainly in the sandstones. In addition, several of the sequences contain discrete sandstone beds exhibiting hummocky cross-stratification interbedded within the siltstone intervals. There are numerous examples of both articulated and disarticulated Brachiopod valves tentatively ascribed to *Spirifer hawkinsii* (E. J. Edwards, pers. comm.), a common Gondwanoid form.

The transition back to claystones at the base of each upward-coarsening interval at Kelp Point is marked by a sharp, near-planar surface preserved in response to flooding of the underlying shoreface and, hence, forming a parasequence boundary. One of these surfaces has a considerable number of near-circular low relief mounds up to 2 m in diameter developed both on, and partially in, the surface itself. These contain an internal, extremely fine, irregular lamination and are interpreted as relict isolated algal mats, suggesting an extended episode of relative stability with little or no subsequent erosion prior to flooding at this particular boundary. In cases where comparable boundaries occur above apparently incomplete sequences it is assumed that these represent either partial aggradation of the shoreface or erosion in response to a lowering of base level prior to the subsequent flooding that was responsible for the overlying claystones.

The range of sedimentary characteristics exhibited within the Fox Bay Formation are consistent with the development of offshore marine shelf to shoreface environments and represent a continuing marine transgression when compared with the underlying Port Stephens Formation. The similarity, however, of the shallowest water facies within the Fox Bay Formation and those representing the majority of the Port Stephens Formation, together with the evidence for episodic, or possibly periodic, changes in base level suggests that these two formations are, at least in part, laterally equivalent.

Several localities on the north coast of the islands, both East and West Falkland, and at Bluff Cove, include bioturbated sandstones interbedded with the dominant fluvial channel facies of the Port Stanley Formation (see below). The most notable of these occurs in the cliff section at Jenesta Point (Fig. 1) where laterally continuous sandstone beds developed on a decimetre to metre scale are so intensely bioturbated that, in places, little trace of any primary sedimentological structures are preserved. The burrows are mainly *Skolithos*, which are 0.5–2 cm in diameter and range up to 1 m in length, although *Diplocraterion* burrows have also been recorded in similar beds at Elephant Beach. In the majority of cases the bioturbation is recorded in sequences of three to four stacked beds with the top of the uppermost bed, where this could be observed, exhibiting an intensely pock-marked texture in response to the abundance of subvertical burrows that terminated upwards at this surface. Overlying fluvial channel sandstones of the Port Stanley Formation are unaffected by bioturbation although burrows can be observed to penetrate down through the bioturbated beds into underlying cross-stratified fluvial channel sandstone beds. The bioturbated sequences are interpreted as representing shallow marine, probably upper shoreface, environments consistent with the sandy facies of the Fox Bay Formation elsewhere. The interdigitation of this facies with the Port Stanley Formation fluvial channel sequences, and their close relationship as indicated by the downward penetration of burrows into channel sandstones, suggests a coastal setting with juxtaposition of fluvial channel and shoreface environments. The alternation of facies indicates subtle, and repeated, transgressive and regressive events resulting in a close spatial relationship between the marine Fox Bay Formation and the fluvial Port Stanley Formation.

Port Philomel Formation

Greenway (1972) describes the Port Philomel Beds as forming the upper part of the 'fossiliferous series' (cf. Fox Bay Formation) and comments that they may be merely a local development and, hence not mappable. From observations and the evidence of other authors (e.g. Marshall 1994*a*) it is clear that the Port Philomel Formation only occurs sporadically as an identifiable sequence intervening between the Fox Bay and Port Stanley Formations. Elsewhere, the contact between these formations, albeit transitional, exhibits no evidence of the sequence described below.

The east side of Carcass Bay on West Falkland (Fig. 1) exposes mainly massive sandstone beds up to 1.5 m thick, suggestive of mass flow deposits, interbedded with thin claystones. Porosities within the sandstones are poor due to extensive recrystallization in response to faulting and folding at this locality. The sandstones contain substantial concentrations of finely comminuted plant material, potentially indicating proximity to a terrestrial hinterland. This is usually scattered throughout the beds but locally defines a low angle cross-stratification and, in some instances, soft sediment deformation in the form of slumping and water escape structures. The Port Philomel sequences are interpreted as delta front mass flow deposits in the light of their sedimentological characteristics and stratigraphic location. Stratigraphically they overlie the shallow marine deposits of the Fox Bay Formation and underlie the fluvial channel sandstones of the Port Stanley Formation. They therefore represent a component in the major regression which culminates in the establishment of Port Stanley Formation fluvial channel environments (see below) across much of the present-day Falkland Islands, a factor which may, in part, explain their limited preservation.

Port Stanley Formation

The dominantly sandstone and metaquartzite lithologies of the Port Stanley Formation occur throughout the northern areas of both East and West Falkland and as outliers in central West Falkland. The metaquartzites which form the main upland areas across the north of the two islands are the folded, partially tectonized and recrystallized, derivatives of the Port Stanley Formation sandstones elsewhere.

In untectonized areas the Port Stanley Formation comprises moderate to poorly sorted medium and coarse sandstones with sparse granules to, rarely, cobbles and moderately preserved porosities. These sandstones exhibit usually well defined planar and trough cross-stratification within decimetre to metre scale beds. Channel geometries, commonly developed on a scale of metres to, rarely, tens of metres, are orientated north–south to northwest–southeast with cross-stratification indicating transport directions universally to the north and northwest. The fill within the channels is generally compound, comprising stacked cross-stratified beds that commonly fine and decrease in preserved set thickness upwards. In a few recorded instances there are examples of current rippling and finely developed flat lamination associated with these uppermost channel-fill sequences, although claystones or siltstones, representing abandonment, are extremely rare and were only seen along the coastline of eastern Pebble Island south of Cape Tamar.

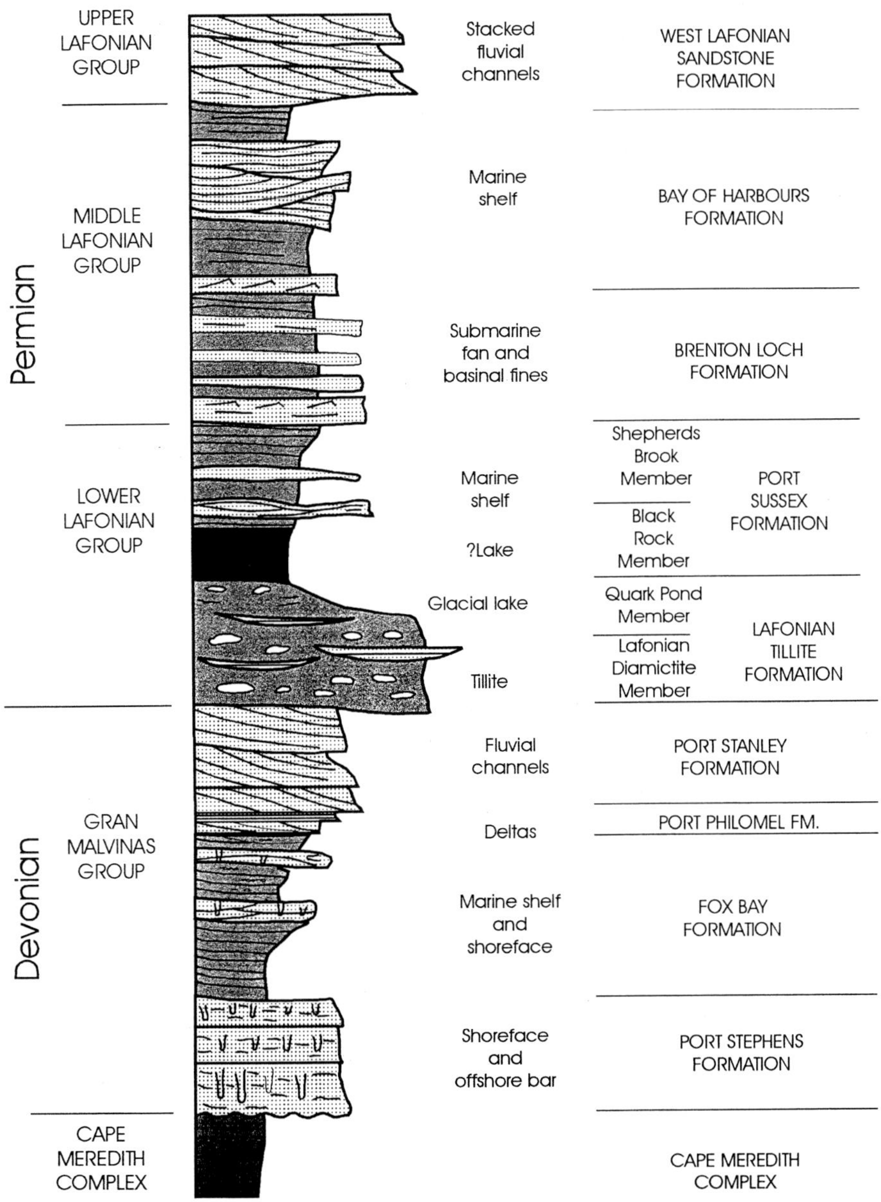

Fig. 4. Revised lithostratigraphy for the Gran Malvinas Group (Devonian) and Lafonian Supergroup (Permian) with generalized environments of deposition for each of the formations.

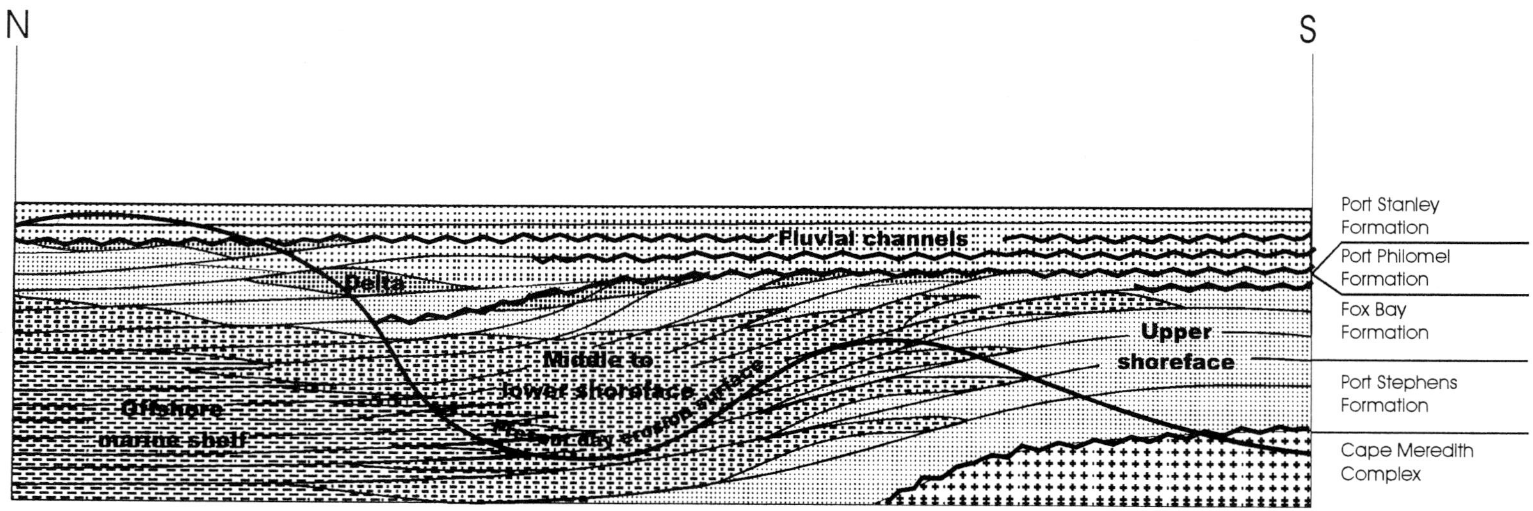

Fig. 5. Schematic facies distribution in the Devonian Gran Malvinas Group showing the inferred lateral equivalence of facies belts and their relationship to the present-day erosion surface with shallow marine sands of the Port Stephens Formation mainly in the south, marine shelf to offshore marine deposits of the Fox Bay Formation in the centre, and fluvio-deltaic to fluvial sandstones of the Port Philomel and Port Stanley Formation largely in the north.

The Port Stanley Formation sandstones are interpreted as being the deposits of a sandy, potentially braided, fluvial channel system which prograded across the entire Falkland Island area. The upward fining and thinning of channel fills indicates preservation of bedforms developed during waning flow conditions following peak discharge. Claystones and siltstones, where preserved, represent temporary abandonment of channel anabranches, although the paucity of fines preservation testifies to the activity of the fluvial system which is suggested to have swept and reworked its braidplain. Similarly, the presence of pebbles and rare cobbles gives an indication of flow competences at peak discharge levels. Examples of bar form collapse and slumping of bar fronts suggest falling stage conditions and inherent instability of braid bars as these became emergent. The rare instances of flat laminated sandstones preserved as interbeds between channel fills may represent episodes of low flow stage deposition when major bars were emergent and only channel anabranches remained active. At such times, flow acceleration between bars provides the necessary hydraulic conditions for the generation of plain bed lamination.

Devonian depositional model

Facies analysis of the sediments forming the four formations identified within the Devonian Gran Malvinas Group has revealed depositional environments ranging from marine shelf to fluviatile. The association of these facies suggests that the Gran Malvinas Group succession records an overall marine transgressive–regressive system. This first flooded the eroded surface of the Cape Meredith Complex basement and established shoreface to shallow marine conditions, progressively deepened to form a marine shelf and then regressed with local development and preservation of deltaic complexes to allow the establishment of a fluvial braidplain across the area with only marginal marine influences.

Inferences regarding the interdigitation of facies suggest that the formations, as previously established, are, in part, laterally equivalent. This is best explained in terms of a schematic facies distribution in which the present-day erosion surface has revealed varying sections through a series of compound transgressive, highstand and regressive systems tracts (Fig. 5). A combination of sedimentary features and regional data indicate that the marine transgression probably occurred broadly from the north or northwest relative to the present orientation of the islands, with clastic input to a marine shoreface from a hinterland to the south or southeast. To the north of this shoreface (basinwards), and stratigraphically higher in the section, these upper shoreface sands gave way to lower and middle shoreface silts and fine sands. These latter comprise the Fox Bay Formation which records numerous prograding highstand parasequences within the overall transgressive system. The interdigitation of the Fox Bay Formation with the overlying Port Philomel and Port Stanley Formations suggests that a similar relationship exists to that between the underlying Fox Bay and Port Stephens Formations. This is therefore suggested to represent a basinwards shift in facies (north or northwest) in response to regression with a consequent preservation of fluvial deposits across a large area of the islands.

Permian sedimentology

Sedimentary sequences identified as being entirely Permian (from palynological analyses carried out by Geochem Group, although the lowermost sequence, the Lafonian Diamictite, remains undated) form the whole of southern East Falkland (Lafonia), southerly parts of northern East Falkland and the eastern coastal strip of West Falkland on the shores of Falkland Sound. Additional outliers have been recorded in western and northern areas of West Falkland (Frakes & Crowell 1967; Greenway 1972; Marshall 1994*a*). The Permian succession is referred to the Lafonian Supergroup (Borrello 1963, 1972) which is divided informally into Upper, Middle and Lower Lafonian Groups. The Lower and Middle Lafonian have each been subdivided herein into two formations with the formations in the Lower Lafonian being divided again into two members each. These divisions of the Lafonian Supergroup represent a refinement of the scheme presented in Baker (1924) and subsequent publications, for the first time placing the identifiable formations and members of the group into a complete succession. The five formations, with their constituent members as appropriate, are described below in ascending order.

Lafonian Tillite Formation

Previously referred to as the Lafonian Diamictite (Frakes & Crowell 1967), these largely glacial deposits directly overlie the Devonian Port Stanley Formation and outcrop in a broad tract across north central East Falkland (Fig. 1). They also occur as a narrow strip of outcrop running the length of the Permian exposures on the west shore of Falkland Sound and as isolated outliers in the northwest of West Falkland. The formation is extended and divided herein into a lower Lafonian Diamictite Member and an upper Quark Pond Member (Fig. 4) on the basis of sedimentary characteristics identified in the field.

The Lafonian Diamictite Member comprises largely unstructured grey–brown highly argillaceous siltstones and fine sandstones containing a highly varied clast suite. Clasts range from small pebbles and cobbles through to boulders exceeding 1 m in diameter and comprise quartzites, variably feldspathic granites, granodiorites, dolerites, gneisses, basalts, schistose rock fragments, jasper and a range of sandstone, siltstone and claystone fragments, the latter possibly representing locally reworked Devonian rocks. For the most part, bedding is extremely difficult to determine although a crude parting, generally subhorizontal, does appear like bedding in a number of places, most notably at Bold Cove. This represents either the effects of shearing within the tillite or, possibly, pressure release effects at the present-day surface rather than any sedimentary structure. The absence of discernible structures and the random nature of the clasts, which are dispersed throughout the tillite matrix without any apparent grading or sorting, indicates wholesale dumping or slumping of these deposits. In accordance with Frakes & Crowell (1967), the Lafonian Diamictite Member is interpreted as the depositional product of glacial processes including material released directly from static or melting glaciers but also as gravity induced slumps and mass-flow deposits associated with moraines. The slumping processes in particular are suggested by the presence of boulder-sized mudballs comprising the same material as the tillite which occur within specific areas, again most notably at Bold Cove.

Bedded sequences do, however, occur within the Diamictite Member, generally as discontinuous lenses or pods within which sediment similar to the matrix of the tillite forms discrete centimetre to decimetre scale beds that can be traced for up to tens of metres. Examples of these bedded sequences exhibit flat lamination, current rippling and the presence of associated claystones representing suspension settling of fines. These intervals are suggested to represent temporary lakes formed in response to glacial outwash on the surface of the moraine and subject to episodic influxes of fluvially reworked tillite.

The overlying Quark Pond Member (introduced here) of the Lafonian Tillite Formation (Fig. 4) is identified on the basis of exposures along the eastern coast of West Falkland and mainly around Quark Pond (Fig. 1). These exposures comprise well bedded but poorly sorted siltstones and sandstones similar to the matrix of the underlying tillites that are interbedded with claystones. The former exhibit flat lamination, while the latter are very finely laminated to the point of being fissile. Although there is no definitive sedimentary or palynological evidence, it is suggested here that this upper part of the Lafonian Tillite Formation represents the development of a more substantial and potentially more areally extensive lake than that represented by the restricted sequences recorded within the Diamictite Member. This may have formed during the latter stages of glacial retreat in this area, possibly in response to irregularities in the moraine surface and/or local damming by outwash deposits. The suggestion by Frakes & Crowell (1967) that these sequences were deposited in a marine environment cannot be either proved, as these authors admit, or categorically refuted. There are no definitive marine indicators recovered in either the Quark Pond Member or the overlying Black Rock Member of the Port Sussex Formation. Hence, a lacustrine interpretation is favoured here on the basis of both the close association with the tillites and sedimentological evidence higher in the stratigraphy (see discussions on the Port Sussex Formation below).

Port Sussex Formation

The Port Sussex Formation is subdivided into lower and upper members comprising, respectively, the Black Rock Member and the Shepherds Brook Member, terms which accord with those used by Greenway (1972) and Frakes & Crowell (1967). Representatives of both members are best exposed in the area around Port Sussex on the west coast of East Falkland but also occur as discontinuous outcrops in a narrow tract overlying the Lafonian Tillite Formation across East Falkland.

The Black Rock Member is exposed on the north shore of Port Sussex as intensely deformed black shales that are folded into tight, often isoclinal, folds with an axial planar cleavage which, in places, also appears to be folded. The shales are extremely carbonaceous and fissile with common sulphur staining. Although clearly representing a subaqueous, extremely low energy, environment, the palynological and sedimentological evidence is sparse as to whether the Black Rock Member is marine or lacustrine. Presumed equivalents in South Africa, the Whitehill Formation and South America, the Irati Formation, are known to be largely marine, containing a range of marine macrofossils and shallow marine carbonates (Visser 1992*a*; Franca *et al.* 1995) that are apparently absent from the Black Rock Member.

The overlying Shepherds Brook Member is distinguished from the Black Rock Member on the basis of a transitional change in both colour and lithology exposed around the head of Port Sussex. The base of the sequence comprises extremely finely laminated dark grey claystones and thin siltstones that are interpreted as varve deposits and, hence, inferred to be of lacustrine origin. Outcrop is not continuous to stratigraphically higher sequences but similar claystones are present south of Port Sussex, although with packages of centimetre to decimetre scale fine sandstone beds that are massive or exhibit flat lamination. Similar characteristics are recorded south of Brenton Loch, although these sequences are disrupted by bioturbation, the most common forms being *Pelecypodichnus*, a bivalve escape burrow, and *Planolites*, an infaunal feeding trace, although neither are abundant. Also present are several fine sandstone beds up to 30–40 cm in preserved thickness which exhibit well defined hummocky cross-stratification. These latter structures, together with the onset of bioturbation, would appear to indicate that marine conditions were established during Shepherds Brook Member time and, in this area at least, record the development of a marine shelf below fair weather wave base.

Further round the coast from Brenton Loch at Black Island Cove a series of upward-coarsening siltstone to sandstone packages on a scale of 10–20 m thickness are exposed. These comprise finely laminated siltstones at the base of each sequence, passing up through interlaminated siltstones and flat or current rippled fine sandstones to stacked cross-stratified fine and medium sandstones exhibiting uniformly low porosities and forming decimetre scale beds at the top of complete sequences. Bioturbation is recorded by a variety of burrow forms including *Planolites, Pelecypodichnus* and *Chondrites*, all forms consistent with a variably middle to lower shoreface environment. Each of these sequences is suggested to represent a highstand deposit in which upward shoaling is reflected by the changes in lithology and sedimentary structure. Traced further to the west along the coast (stratigraphically up-section as dips are uniformly to the west) the siltstones become the dominant lithology. These suggest the establishment of lower shoreface to offshore shelf environments and, hence, the upward shoaling sequences to represent minor fluctuations in relative sea level and accommodation space within an overall transgressive system. These sequences provide further evidence and corroboration of marine conditions within the upper parts of the Port Sussex Formation.

Brenton Loch Formation

Sequences ascribed herein to the Brenton Loch Formation are exposed in northern Lafonia, south of Port Sussex and at a few places on the eastern shore of West Falkland where they are the youngest rocks recorded on this island. The sandstones and claystones of the Brenton Loch Formation are typified by the exposures north of Quark Pond and at Orqueta Gorge where the outcrops reveal laterally continuous decimetre to metre scale bedding of fine and medium sandstones with generally low porosities. These are commonly massive but locally exhibit decimetre scale and ripple cross-stratification, a range of dewatering structures, gutter and groove casts on the bases of beds and, more rarely, the incorporation of irregular claystone/siltstone intraclasts. The lower parts of the sequence include a significant proportion of claystone as thin interbeds between the sandstones and as thicker, locally metre scale, intervals separating packages of sandstone beds. Some of these claystone interbeds contain moderate to high concentrations of organic material.

The sedimentary characteristics of the sandstones are consistent with deposition from high density turbidity currents in which the massive nature of the sandstones and common dewatering features result from rapid suspension sedimentation. The interbedded claystones represent background processes involving the more gradual suspension settling of fines between turbidity current events and suggest an outer to mid submarine fan depositional environment. The thicker claystone intervals represent apparently longer episodes of quiescence on a palaeofan surface. Claystone proportions decrease up the exposed section until, close to the shoreline on West Falkland and, hence, as stratigraphically high as exposure in this area allows, the sandstone beds are stacked without intervening claystone breaks. Evidence of channelling is also recorded within these upper parts of the section in the form of wide, up to 200 m, relatively shallow, 2–3 m relief, scours that cut out underlying beds. Most of these channels exhibit form concordant drapes comprising thin fine sandstones suggesting suspension settling following channel abandonment. The upward transition to sandstone dominated fan deposits and the advent of channelling indicate the presence of a prograding submarine fan system, although whether the sequences examined represent single fans or the stacking and coalescing of adjacent fans is unclear from the available exposure. Where palaeocurrent orientations can be discerned in these sequences they are uniformly to the east and south, as indicated by current rippling, low angle cross-stratification and the orientation of channels.

Bay of Harbours Formation

Sandstone dominated sequences ascribed to the Bay of Harbours Formation are only known to outcrop in the extreme south and east of Lafonia and are best exposed at Bull Point and on Lively Island. Regional dips, where these can be observed as the majority of Lafonia is obscured by vegetation (grassland), suggest that these outcrops should be higher in the stratigraphy than the turbidite sequences of the Brenton Loch Formation. The palynological evidence, however, is highly equivocal and the possibility that these exposures equate with lower formations, the Brenton Loch Formation or, particularly, the Port Sussex Formation cannot be entirely discounted.

The section exposed at Bull Point comprises two upward-coarsening sequences that are repeated due to a series of low angle thrusts. They are dominated by planar-bedded fine sandstones forming decimetre scale beds that are commonly separated by thin silty partings and pass gradationally down into very fine silty sandstones and finely laminated argillaceous siltstones. Porosities are generally low in response to the fine grain size of the sandstones. The burrow form *Chondrites* is recorded but tends to be rare and found only near the transition zone from sandstones to siltstones. Conversely, current rippling occurs mainly within the upper parts of each sequence and, in one instance, a 1 m thick sandstone unit is formed almost entirely of climbing ripples. The upper parts of the lower sequence also contain several erosion surfaces on a scale of tens of centimetres to 1.5 m relief which tend to be draped by laminae and thin beds of fine sandstones, suggesting episodes of high energy incision followed by quiescence and suspension sedimentation. These sequences are interpreted as shoreface to offshore bar deposits in which the observed channel structures may represent either rip current or storm surge incision of the fair weather deposits. The orientation of these channels appears to be broadly east–west to southeast–northwest, although no inferences can be made with regard to shorelines or sediment transport directions.

The inferred stratigraphical position of these Bay of Harbours Formation sediments above the submarine fan deposits of the Brenton Loch Formation implies a substantial change in depositional environments with the re-establishment of shallow marine conditions following a major episode of basin infilling.

West Lafonian Sandstone Formation

Exposures of the West Lafonian Sandstone Formation are only recorded along the west coast of Lafonia bordering Falkland Sound. The inland extent of this formation is largely unknown due to the paucity of outcrop and even along this coast exposure is limited. Dips are uniformly westward and the West Lafonian Sandstone Formation, lying in the axis of the Falkland Sound Monocline, forms the youngest sequence identified to date on the Falkland Islands.

The exposures comprise mainly trough cross-stratified fine and medium grained sandstones forming channel-fill sequences on a scale of decimetres in preserved thickness and 3–8 m in width. Preserved porosities in these sandstones are variable but can be moderate to high. The channel geometries exhibit an alternate stacking pattern where they can be seen in plan view on the wave cut platform. Thin siltstones and claystones occur interbedded between some of these sandstones but these are rare and are locally cut out by incision of overlying channel bases. Flat laminated fine sandstones also occur as more laterally continuous, but rare, interbeds between the dominant channel fills. The West Lafonian Sandstone Formation sequences are interpreted as fluvial channel deposits which, from the limited outcrop available, can be related to a low sinuosity, possibly braided, fluvial system. Palaeocurrent directions derived from the orientation of cross-stratification are ubiquitously towards the west, a sediment transport direction diametrically opposed to those recorded in underlying formations.

Permian depositional model

Facies analysis of the Lafonian Supergroup sedimentary sequences has identified five formations which comprise a revised lithostratigraphy (Fig. 4) retaining the terminology of Baker (1924), Frakes & Crowell (1967) and Greenway (1972). These facies indicate a major glacial episode followed by lake development and a subsequent transition to marine conditions with the establishment of shelf environments. Progressive subsidence resulted in increasing water depths and the development of submarine fan systems which provided a major element of the basin fill. Depositional infill, or the initiation of inversion, resulted in the re-establishment of a marine shelf prior to emergence and a major change to subaerial, fluviatile deposition.

The variations in facies between the formations forming the Lafonian Supergroup have been utilized to erect a model of Permian basin evolution. Sedimentological evidence presented above indicates that the present-day onshore northern and western limits of the Lafonian Supergroup outcrop may approximate to Permian post-tillite basin margins, particularly in the Falkland Sound area. By inference, this margin may have been controlled by a basement lineament lying beneath, or close to, the present-day sound (Curtis & Hyam 1998), which has been reactivated at various times in the geological history of the islands (Fig. 6a). The position of post-tillite Lafonian Supergroup sequences further west, particularly around Byron Sound, is unclear but it is noteworthy that Richards & Fannin (1997) record the presence of another Permian basin offshore to the west and north of the islands.

The Diamictite Member of the Lafonian Tillite Formation has been correlated with the Itarare Formation of Argentina and the Dwyka Formation of South Africa (e.g. Tankard *et al.* 1995). Notwithstanding the areal extent of the glaciation, petrographical evidence suggests that a component, at least, of the tillite is derived from basement similar to the exposed Cape Meredith Complex. The presence within the tillite of bedded intervals, suggested to represent isolated lacustrine sequences developed on the irregular surface of the moraines, indicates, in part at least, that the tillite was deposited subaerially. The overlying Quark Pond and Black Rock Members of the Lafonian Tillite and Port Sussex Formations, respectively, are inferred to be more areally restricted and related to the evolving Lafonia Basin. While it is recognized that the Black Rock Member in particular has facies equivalents in the Irati Formation of Argentina and the Whitehill Formation of South Africa (e.g. Tankard *et al.* 1995), it is suggested here that both of these two members of the Lafonian Supergroup were deposited in a relatively isolated, potentially lacustrine, basin.

The transition to marine conditions is inferred to occur within either the top of the Black Rock Member or the base of the overlying Shepherds Brook Member of the Port Sussex Formation (Fig. 6b). The onset of submarine fan related sedimentation occurs within the overlying Brenton Loch Formation. Evidence from sedimentological structures and channel orientations suggests that sediment transport directions were from northwest to southeast on the western margin basin and from north to south on the northern margin in East Falkland. It is therefore suggested that the fan systems prograded south and east across the evolving basin and potentially provided a significant component of the basin fill (Fig. 6c).

An apparent return to shallow marine conditions in the Bay of Harbours Formation suggests either a lowering of base level or a reduction in subsidence rates. The lateral extent of marine shelf environments at this time is not known due to the paucity of relevant outcrop but the apparent filling of the basin is consistent with a return to subaerial conditions in the overlying West Lafonian Sandstone Formation. These sandstones outcrop only on the western side of Lafonia along the Falkland Sound coastline but westward directed palaeocurrents in channel sandstones indicate a reversal of sediment transport compared to the underlying succession. This drainage reversal reflects the onset of compressional deformation and the earliest stages in the formation of the major synformal monocline underlying Falkland Sound (Fig. 6d). Petrographical evidence shows that the detrital mineralogy of the West Lafonian Sandstone is similar to underlying sequences despite the change in transport direction which, although not conclusive, accords with this supposition. It is, therefore, considered likely that the inversion associated with this event resulted in contemporaneous erosion of underlying sequences by the channels of the West Lafonian Sandstone Formation. The inference that folding commenced contemporaneous with, and controlled, late Lafonian Supergroup sedimentation is consistent with suggestions regarding the correspondence of Falkland Island deformation and that of the Cape Fold Belt in South Africa. According to Halbich (1983), deformation in the Cape Fold Belt initiated in the Permian and continued until the Middle Triassic.

Basin evolution

Deposition of Devonian clastics has been recorded throughout much of southern South America, across the Cape Basin of South Africa and in the Falkland Islands. Tankard *et al.* (1995) regard the Devonian basins as being linked across a large area of southwestern Gondwanaland on the basis of the wide ranging Malvinokaffric invertebrate fauna (Boucot 1971) and similarity of facies. From the sedimentological interpretations presented here it is suggested that the Falkland Island area occupied a position marginal to a

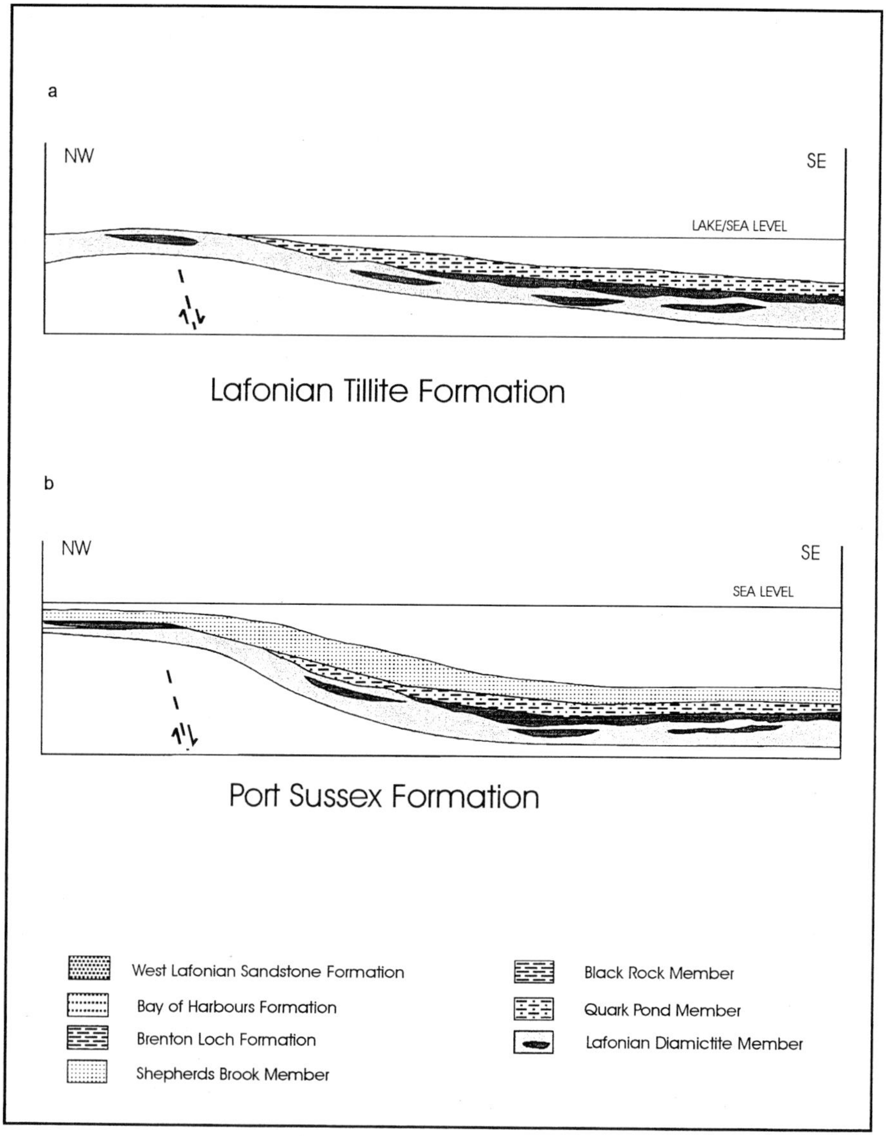

Fig. 6. Schematic facies distributions in the largely Permian Lafonian Supergroup with (**a**) initiation of the Lafonia Basin with a bounding fault underlying the present-day Falkland Sound and deposition of tillites and glacial lake deposits relating to a retreating ice-front, the Lafonian Tillite Formation, followed by the establishment of lacustrine conditions, the Black Rock Member of the Port Sussex Formation; (**b**) the establishment of a marine shelf, Shepherds Brook Member of the Port Sussex Formation, and continued basinal subsidence; (**c**) infilling of the deepened basin by submarine fans, Brenton Loch Formation, and re-establishment of shallow marine conditions, Bay of Harbours Formation; (**d**) onset of inversion, reactivation of the Falkland Sound Fault and establishment of fluvial channels, West Lafonian Sandstone Formation.

major downwarp on the southwest coast of Gondwanaland where it was subject to marine transgression and regression. There is little evidence within the sedimentary sequence for significant contemporaneous fault activity and it is inferred that the islands formed a part of a northwards sloping shelf, relative to present-day orientation. Visser (1992*b*), in discussing the sediments of the Cape Basin of South Africa, attributes an analogous situation to the thinning and weaken-

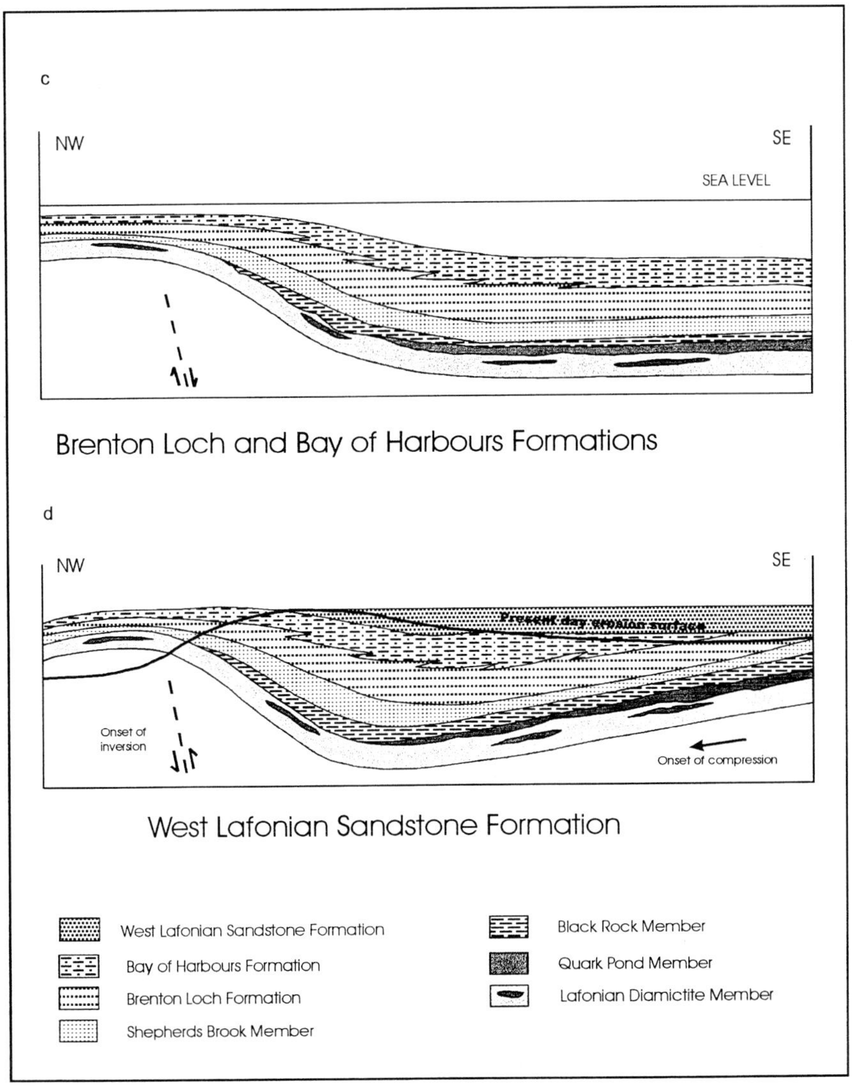

Fig. 6. (*continued*)

ing of the lithosphere along the passive cratonic margin during extension.

Following deposition of the Gran Malvinas Group sediments, there appears to have been some minor northward rotation of the islands. The extent of any erosion associated with this is difficult to determine due to later events but, from the near-conformable relationship between the Gran Malvinas Group and the overlying Lafonian Supergroup, this must have been minimal. Elsewhere in South America this period is marked by a major inversion, referred to as the Chanic Orogeny (Tankard *et al.* 1995), which resulted in a widespread unconformity. A similar event created the unconformity in South Africa between the Cape Group and the overlying Karoo Supergroup sequences (Visser 1992*b*) but the Falkland Islands seem to have been relatively protected at this time.

The initiation of subsidence leading to deposition of the Lafonian Supergroup sequences is inferred to have started either very late in the Carboniferous or, more likely, early in the Permian. The oldest Permian sediments relate to the Gondwanaland glaciation and, as such, have correlatives across a wide area of south, central and western Gondwana (Williams 1995). Tankard *et al.* (1995) relate this episode of basin formation, which is recorded throughout southern South America and Africa, to compressional, transtensional and oblique-slip processes associated with an early stage of Gondwanaland break-up. In the Falkland Islands deposition is inferred to have been controlled to some extent by movements on the Falkland Sound Fault. Although the sedimentary evidence indicates this may have been close to the western basin margin, the extent of the basin in the other directions is largely unknown. The fold belt to the north has removed any direct evidence regarding a potential basin margin in the northern area of East Falkland, although sparse palaeocurrent data suggests that it may not have extended a significant distance beyond the present outcrop. To the south and east the Lafonian Basin passes out into the offshore areas but has been tentatively mapped from seismic data by Richards *et al.* (1996*b*) as a discrete basin, or preserved remnant, some 120 km across northwest–southeast and *c.* 180 km southwest–northeast. The possibility of further preserved remnants, particularly to the north and west of the islands, has also been indicated (Richards & Fannin 1997). Contiguity between these basins, and those of similar age in South Africa and South America, has been suggested on the basis of regional tectonic modelling by Tankard *et al.* (1995), while Marshall (1994*a*) suggests a much closer relationship with the Karoo Basin. It is suggested here that the Lafonia Basin formed as a discrete, initially isolated, basin, albeit in response to regional patterns of strain and extension that were also responsible for basin formation throughout southwest Gondwanaland and may be related to the earliest stages of Gondwana break-up.

Subsidence in the Lafonia Basin through the Permian appears to have occurred progressively with the potentially fault-bounded western margin locally controlling sedimentation and the distribution of facies. Only towards the latter stages of Lafonian Supergroup deposition is there a change in basin architecture which is indicated by the reversal of the drainage system in the West Lafonian Sandstone Formation. This is suggested to reflect the onset of compression, broadly north–south to northwest–southeast directed, which subsequently caused the folding and thrusting now evident in the Falkland Islands. The initial effect appears to have been minor inversion of the basin, giving rise to the drainage reversal and the onset of folding along the western basin margin resulting from reactivation. of the basement fault underlying Falkland Sound. The timing of this compression, inferred from the available stratigraphical evidence to be latest Permian to earliest Triassic, is effectively coeval with that of the Cape Fold Belt (Visser 1992*b*) in South Africa, and the Sierra de la Ventana and Sierra Grande fold belts (Tankard *et al.* 1995) of Argentina. In a plate reconstruction of southwest Gondwana, Williams (1995) extends this correlation to include a broad fold belt running from South America through South Africa and Antarctica into what is now southeastern Australia and effectively encompasses the whole southwest margin of Pangea.

On the Falkland Islands the later and more significant effects of this orogeny are recorded in the northern fold belt and the thrusting which has decoupled sequences within the Lafonia Basin. The variation in deformation style between the fold belt and the thrust sequences of Lafonia can be related to lithological differences between the areas. The predominantly sandstone lithologies of the Devonian Port Stanley and Port Stephens Formations, being more competent than the largely interbedded sandstones and claystones of the Permian succession, are deformed by folding. Conversely, the latter were able to decouple along claystone beds and, particularly, on the organic-rich shales of the Black Rock Member. The intense deformation locally recorded in the Black Rock Member is suggested to be, in part, a response to this phase of decoupling. The majority of the structures indicate broadly north–south directed shortening, the exception being the Falkland Sound synformal monocline which trends north-northeast. Although this may have formed in response to the same episode of deformation by inversion of the basin margin fault beneath Falkland Sound, Curtis & Hyam (1998) record evidence for overprinting east–west structures by northeast–southwest orientated folds and localized thrusts.

Regional implications

Devonian

Comparison of the Devonian sequences on the Falkland Islands with those of South America and South Africa suggests a broad similarity of facies. The Bokkeveld Group of the Cape Supergroup forms a thick series of shallow epeiric sea deposits, while the Ponta Grossa Formation in the Parana Basin of northern Argentina and southern Brazil records a thinner, but similar, succession (Tankard *et al.* 1995). The apparently closer Chaco-Parana Basin of Argentina contains the deeper water and organic-rich deposits of the Los Monos Formation. No sequences of a corresponding age are recorded in southern Patagonia and the source of the clastics for the South American basins has been suggested to lie to the west and north in a precursor Andean uplift and on the Brazilian Shield (Williams 1995). Similarities between the Falklands Devonian succession and that in southern Africa appear to be more closely matched, and both Baker (1924) and Marshall (1994*a*) have suggested that the islands correspond in detail with the Bokkeveld and overlying Witteberg Groups of the Cape Supergroup. It is possible, therefore, that the Falkland Islands have a greater affinity with southern Africa than South America at this time.

Gondwanaland reconstructions for the Devonian place Patagonia, southern Africa and the Falkland Plateau on the southwestern margin of the continent close to the palaeo south pole but unaffected directly by glaciation (Williams 1995). The implications of the Devonian sedimentological model proposed above for regional palaeogeographies depend on the details of the reconstruction and the orientation of the Falkland Plateau with respect to the other continental fragments. The reconstructions of Marshall (1994*a*, *b*), which are generally supported by palaeomagnetic data, suggest that the islands, rotated through > 180°, are an integral part of the Cape Basin sequence. Those of Tankard *et al.* (1995), Williams (1995) and Richards *et al.* (1996*a*), supported by seismic data, suggest a related basin in a position south of southern Africa but receiving a similar range of facies. In the former case the proposed palaeoshoreline would correspond with gross Gondwanaland reconstructions assuming a radial drainage, whereas in the latter the proposed shoreline would indicate an embayment in the generally northwest–southeast trending coastline of Gondwanaland.

Permian

The similarity of Lafonian Supergroup facies with those of the South African Karoo Basin sequences has been used as part of the justification for proposing that the Falkland Plateau comprises a rotated microplate which formed a pre-drift extension to the Cape and Karoo Supergroups (Marshall 1994*a*, *b*). Other authors, notably Lawrence & Johnson (1995) and Richards *et al.* (1996*a*, *b*), have refuted this contention on the basis of seismic data which fails to show evidence for the rotation. Evidence presented here shows that there is a general correspondence between the Permian sequences on the Falkland Islands and those forming the fill in the Karoo Basin. There are, however, also significant differences in detail between the two sequences which suggest that the relationship is regional rather than local. In particular, the presence of phosphorites and dolomitic limestones in the lower parts of the Ecca Group, the common marine microfossils associated with the Whitehill Formation, volcaniclastic sediments which comprise up to 40% of the overlying Collingham Formation (Visser 1992*a*) and the abundance of coals higher in the Karoo sequence all indicate a divergence of basin evolution and facies. None of the above are recorded in the Lafonian sequences, while the latter does indicate a rapid transition to marine, relatively deepwater conditions with associated submarine fan processes at a time when the eastern parts of the Karoo Basin were undergoing uplift and development of coal-forming peats. It is therefore suggested here that, while the Falkland Plateau formed an integral part of southwest Gondwanaland together with southern Africa and adjacent parts of Antarctica during the Permian, the Lafonian Supergroup sediments were deposited in a separate and potentially discrete basin.

The author wishes to acknowledge consent to publish this paper by the directors of Geochem Group Ltd. The text was improved by the useful comments of Rob Crossley and David Macdonald. Additionally, a personal thanks is extended to the people of the Falkland Islands, especially the Edwards family, who provided accommodation, transport and assistance during the fieldwork, to the Falkland Islands Government who allowed me to remove a significant proportion of their islands and to Spectrum Energy who transported it back to the UK. Finally, thanks to Don Aldiss, currently carrying on the good work, for continuing discussions, and to my wife and family who tolerated my absence, both for the fieldwork and the generation of this paper.

References

ADIE, R. J. 1952. The position of the Falkland Islands in a reconstruction of Gondwanaland. *Geological Magazine*, **89**, 401–410.

BAKER, H. A. 1924. *Final Report on Geological Investigations in the Falkland Islands*. Government Press, Port Stanley.

BORRELLO, A. V. 1963. *Sobre la Geologia de las Islas Malvinas*. Ediciones Cultura Argentina, Ministerio de Educione y Justicia, Buenos Aires.

—— 1972. Islas Malvinas. *In*: LEANZA, A. F. (ed.) *Geologia Regional Argentina*. Academia Nacional de Ciencias, Cordoba, 755–770.

BOUCOT, A. J. 1971. Malvinokaffric Devonian marine community distribution and implications for Gondwana. *Anais da Academia Brasileira de Ciencias*, **43**, (suplemento), 23–49.

CURTIS, M. L. & HYAM, D. M. 1998. Late Palaeozoic to

Mesozoic structural evolution of the Falkland Islands: a displaced segment of the Cape Fold Belt. *Journal of the Geological Society, London*, **155**, 115–129.

DAWSON, J. C. 1967. *The geology of the Bluff Cove area, Falkland Islands*. MSc Thesis, University of California.

DARWIN, C. R. 1846. On the geology of the Falkland Islands. *Quarterly Journal of the Geological Society of London*, **2**, 267–274.

FRAKES, L. A. & CROWELL, J. C. 1967. Facies and paleogeography of Late Paleozoic Diamictite, Falkland Islands. *Geological Society of America Bulletin*, **78**, 37–58.

FRANCA, A. B., MILANI, E. J., SCHNEIDER, O. *ET AL.* 1995. Phanerozoic correlation in southern South America. *In*: TANKARD, A. J., SUAREZ SORUCO, R. & WELSINK, H. J. (eds) *Petroleum Basins of South America*. AAPG Memoir, **62**, 129–161.

GREENWAY, M. E. 1972. *The Geology of the Falkland Islands*. British Antarctic Survey, Scientific Report No. 76.

HALBICH, I. W. 1983. A tectogenesis of the Cape Fold Belt (CFB). *In*: SOHNGE, P. G. & HALBICH, I. W. (eds) *Geodynamics of the Cape Fold Belt*. Geological Society of South Africa, Special Publication, **12**, 165–175.

HALLE, T. G. 1912. On the geological structure and history of the Falkland Islands. *Bulletin of the Geological Institution, University of Upsala*, **11**, 115–229.

LAWRENCE, S. & JOHNSON, M. 1995. Future petroleum province in the South Atlantic. *Oil and Gas Journal*, **March 6**, 52–55.

MARSHALL, J. E. A. 1994*a*. The Falkland Islands: A key element in Gondwana paleogeography. *Tectonics*, **13**, 499–514.

—— 1994*b*. The Falkland Islands and the early fragmentation of Gondwana: implications for hydrocarbon exploration in the Falkland Plateau. *Marine and Petroleum Geology*, **11**, 631–636.

MUSSETT, A. E. & TAYLOR, G. K. 1994. ^{40}Ar–^{39}Ar ages for dykes from the Falkland Islands with implications for the break-up of southern Gondwanaland. *Journal of the Geological Society, London*, **151**, 79–81.

REX, D. C. & TANNER, P. W. G. 1982. Precambrian age for gneisses at Cape Meredith in the Falkland Islands. *In*: CRADDOCK, C. (ed.) *Antarctica Geoscience Symposium on Antarctic Geology and Geophysics*. University of Wisconsin Press, 107–110.

RICHARDS, P. C. & FANNIN, N. G. T. 1997. Geology of the North Falkland Basin. *Journal of Petroleum Geology*, **20**, 165–183.

——, GATLIFF, R. W., QUINN, M. F. & FANNIN, N. G. T. 1996*a*. Petroleum potential of the Falkland Islands offshore area. *Journal of Petroleum Geology*, **19**, 161–182.

——, ——, ——, WILLIAMSON, J. P. & FANNIN, N. G. T. 1996*b*. The geological evolution of the Falkland Islands continental shelf. *In*: STOREY, B. C., KING, E. C. & LIVERMORE, R. A. (eds) *Weddell Sea Tectonics and Gondwana Break-up*. Geological Society, London, Special Publications, **108**, 105–128.

TANKARD, A. J., ULIANA, M. A., WELSINK, H. J. *ET AL.* 1995. Structural and tectonic controls of basin evolution in southwestern Gondwana during the Phanerozoic. *In*: TANKARD, A. J., SUAREZ SORUCO, R. & WELSINK, H. J. (eds) *Petroleum Basins of South America*. AAPG Memoir, **62**, 5–52.

THOMSON, C. W. 1877. *The Atlantic. A preliminary account of the general results of the exploring voyage of H.M.S. Challenger during the year 1873 and the early part of the year 1876*. Macmillan.

VISSER, J. N. J. 1992*a*. Deposition of the Early to Late Permian Whitehill Formation during a sea-level highstand in a juvenile foreland basin. *South African Journal of Geology*, **95**, 181–193.

—— 1992*b*. Basin tectonics in southwestern Gondwana during the Carboniferous and Permian. *In*: DE WIT, M. J. & RANSOME, I. G. D. (eds) *Inversion Tectonics of the Cape Fold Belt, Karoo and Cretaceous Basins of Southern Africa*. Balkema, 109–115.

WILLIAMS, K. E. 1995. Tectonic subsidence analysis and Paleozoic paleogeography of Gondwana. *In*: TANKARD, A. J., SUAREZ SORUCO, R. & WELSINK, H. J. (eds) *Petroleum Basins of South America*. AAPG Memoir, **62**, 79–100.

Index

Page numbers in **bold** refer to tables; those in *italic* refer to figures